Embryos, Genes and Birth Defects

Second Edition

Embryos, Genes and Birth Defects

Second Edition

EDITORS

Patrizia Ferretti
UCL Institute of Child Health, London, UK

Andrew Copp
UCL Institute of Child Health, London, UK

Cheryll Tickle
University of Dundee, UK

Gudrun Moore
UCL Institute of Child Health, London, UK

John Wiley & Sons, Ltd

Telephone (+44) 1243 779777

Email (for orders and customer service enquiries): cs-books@wiley.co.uk
Visit our Home Page on www.wileyeurope.com or www.wiley.com

Other Wiley Editorial Offices

John Wiley & Sons Inc., 111 River Street, Hoboken, NJ 07030, USA

Jossey-Bass, 989 Market Street, San Francisco, CA 94103-1741, USA

Wiley-VCH Verlag GmbH, Boschstr. 12, D-69469 Weinheim, Germany

John Wiley & Sons Australia Ltd, 42 McDougall Street, Milton, Queensland 4064, Australia

John Wiley & Sons (Asia) Pte Ltd, 2 Clementi Loop #02-01, Jin Xing Distripark, Singapore 129809

John Wiley & Sons Canada Ltd, 22 Worcester Road, Etobicoke, Ontario, Canada M9W 1L1

Wiley also publishes its books in a variety of electronic formats. Some content that appears in print may not be available in electronic books.

Library of Congress Cataloging-in-Publication Data

Embryos, genes, and birth defects / editors, Patrizia Ferretti . . . [et al.].
– 2nd ed.
p. cm.
Includes index.
ISBN-13: 978-0-470-09010-7 (alk. paper)
ISBN-10: 0-470-09010-3 (alk. paper)
1. Abnormalities, Human. 2. Teratogeneis. 3. Embryology, Human.
I. Ferretti, Patrizia.

[DNLM: 1. Embryo–abnormalities. 2. Gene Expression Regulation,
Developmental. QS 675 E535 2006]
QM691.E44 2006
616′ .043–dc22 2005036439

British Library Cataloguing in Publication Data

A catalogue record for this book is available from the British Library

ISBN-13 978-0-470-09010-7 ISBN-10 0-470-09010-3

Typeset in 10.5/12.5 pt Minion by Thomson Press (India) Limited, New Delhi, India
Printed and bound in Great Britain by Antony Rowe Ltd., Chippenham, Wilts
This book is printed on acid-free paper responsibly manufactured from sustainable forestry
in which at least two trees are planted for each one used for paper production.

Image of mouse paw kindly provided by Tom Glaser and Ed Oliver of the University of Michigan.
The Bst/+ mouse has preaxial polydactyly - one extra digit on the anterior side (preceding the first digit.)

Contents

Preface to the First Edition

This book has a single purpose. It is to provide, in an intellectually accessible and concise form, an overview of contemporary understanding of the mechanisms of embryonic development, as they pertain to dysmorphogenesis or the generation of birth defects. In order to do so we will explore a variety of systems and strategic approaches to analysis, and the layout of the book is designed to facilitate this. The first six chapters cover selected modern strategies of analysis and introduce some of the major themes. The subsequent nine chapters, all of which are structured according to a common pattern, review current knowledge of developmental mechanisms in those organ systems for which there has been particular progress in our understanding. Each of these 'systems' chapters presents an agenda for future research directions. It is perhaps necessary to point out that we do not attempt to cover the topics of inherited metabolic disease or those syndromes where the phenotype is exclusively behavioural; the emphasis in this volume is largely on physical birth defects.

Recognition of the need for a book of this type has had a gradual gestation. Vague thoughts on the form that such a book might take have been brought sharply into focus through discussion with my immediate colleagues at the Institute of Child Health: Andrew Copp, Patrizia Ferretti and Adrian Woolf. It is my pleasure to be able to acknowledge with gratitude their contributions not only as chapter authors but also through our various research interactions and the general support provided as we went about our everyday tasks of running busy reserch teams. The image of the human embryo on the front cover was provided by Rachel Moore and Simon Brown. Finally, my editor at John Wiley & Sons Ltd, Dr Sally Betteridge, and her assistant, Lisa Tickner, have guided the project to completion with wisdom, common sense, but most of all with patience! Thank you.

Peter Thorogood
Institute of Child Health

Peter Thorogood
Photograph by Nicholas Geddes, Medical Illustration Unit, ICH

Preface to the Second Edition

Until the first edition of this book, *Embryos, Genes and Birth Defects*, most works published on birth defects concentrated on developmental pathology, clinical genetics, syndromology or the consequences for health care of the affected newborn, but neglected to discuss in depth the mechanisms that might have led to a particular abnormality. The late Peter Thorogood, who edited the first edition, had the vision to fill this gap and produced a very successful resource for both clinicians and basic scientists. The need for such a book is even more pressing today because considerable progress in the understanding of normal and abnormal developmental mechanisms has been made since the first edition was published, opening avenues to the development of novel therapeutic approaches for birth defects, such as gene and stem cell therapy. As colleagues of Peter and contributors to the first edition, we therefore felt that it was crucial to bring this book up to date and we dedicate this new edition to him.

The overall purpose and structure of the book have not changed in this new edition. However, additional chapters focusing on human cytogenetics, identification of genes involved in congenital malformations and specific reviews of sensory organs have been included to illustrate further strategic approaches to the study of birth defects and how basic developmental biology is providing new paradigms for understanding them.

We are grateful to all our colleagues who have managed to find the time to contribute to this book despite their busy schedules. We also wish to thank our editors at John Wiley & Sons, Joan Marsh and Andrea Baier, for their encouragement and professionalism that have made publication of this new edition possible.

Patrizia Ferretti
Andrew Copp
Gudrun Moore
Cheryll Tickle

Contributors

Michael Baraitser Formerly Consultant Clinical Geneticist, Department of Clinical Genetics, UCL Institute of Child Health, 30 Guilford Street, London WC1N 1EH, UK

Nigel A. Brown Division of Basic Medical Sciences, St. George's Hospital, University of London, London, UK

Alan J. Burns Neural Development Unit/Developmental Biology Unit, UCL Institute of Child Health, 30 Guilford Street, London WC1N 1EH, UK

Andrew J. Copp Neural Development Unit/Developmental Biology Unit, UCL Institute of Child Health, 30 Guilford Street, London WC1N 1EH, UK

J. D. A. Delhanty UCL Centre for Preimplantation Genetic Diagnosis, Department of Obstetrics and Gynaecology, University College London, 86–96 Chenies Mews, London WC1E 6HX, UK

Patrizia Ferretti Neural Development Unit/Developmental Biology Unit, UCL Institute of Child Health, 30 Guilford Street, London WC1N 1EH, UK

Deborah Henderson Institute of Human Genetics, International Centre for Life, University of Newcastle upon Tyne, UK

Mary R. Hutson Neonatal–Perinatal Research Institute, Department of Pediatrics (Neonatology), Room 157, Bell Building, Duke University Medical Center, Durham, NC 27712, USA

Margaret L. Kirby Neonatal–Perinatal Research Institute, Department of Pediatrics (Neonatology), Box 3179, Room 157, Bell Building, Duke University Medical Center, Durham, NC 27712, USA

Irene M. Leigh Centre for Cutaneous Research, St. Bartholomew's Hospital, and the Royal London Queen Mary's School of Medicine and Dentistry, London, UK

Gudrun Moore Clinical and Molecular Genetics, UCL Institute of Child Health, 30 Guilford Street, London WC1N 1EH, UK

Gillian M. Morriss-Kay Department of Human Anatomy and Genetics, University of Oxford, UK

Donald F. Newgreen The Murdoch Children's Research Institute, Royal Children's Hospital, Parkville, 3052 Victoria, Australia

Peter J. Scambler Molecular Medicine Unit, UCL Institute of Child Health, 30 Guilford Street, London WC1N 1EH, UK

Andreas Schedl INSERM U636, Centre de Biochimie, Parc Valrose, 06108 Nice, France

Jane C. Sowden Developmental Biology Unit, UCL Institute of Child Health, 30 Guilford Street, London WC1N 1EH, UK

Sarah Spiden Wellcome Trust Sanger Institute, Wellcome Trust Genome Campus, Hinxton, Cambridge CB10 1SA, UK

Philip Stanier Neural Development Unit/Developmental Biology Unit, UCL Institute of Child Health, 30 Guilford Street, London WC1N 1EH, UK

Karen P. Steel Wellcome Trust Sanger Institute, Wellcome Trust Genome Campus, Hinxton, Cambridge CB10 1SA, UK

Irma Thesleff Institute of Biotechnology, University of Helsinki, Finland

Cheryll Tickle Division of Cell and Developmental Biology, School of Life Sciences, MSI/WTB Complex, University of Dundee, Dow Street, Dundee, UK

Valerie Vidal INSERM U636, Centre de Biochimie, Parc Valrose, 06108 Nice, France

Ahmad Waseem Oral Diseases Research Centre, Department of Clinical and Diagnostic Oral Sciences, St. Bartholomew's Hospital, and the Royal London Queen Mary's School of Medicine and Dentistry, London, UK

Paul J. D. Winyard Nephro-urology Unit, UCL Institute of Child Health, 30 Guilford Street, London WC1N 1EH, UK

Heather M. Young Department of Anatomy and Cell Biology, University of Melbourne, 3010 Victoria, Australia

1

The Relationship Between Genotype and Phenotype: Some Basic Concepts

Philip Stanier and **Gudrun Moore**

Introduction

Without even considering early fetal loss, it is reported that as many as 3.5% of all live-born babies have some kind of major abnormality, referred to as a birth defect. Actual incidences may vary according to locality, culture, ethnicity and the efficiency of recognition and reporting. If minor abnormalities such as cleft lip are included, then the incidence is nearer to 5%. In the Western world, birth defects constitute the greatest single cause of infant mortality and have a major impact on national health care budgets (http://www.modimes.org/).

In this introductory chapter some basic precepts and concepts are presented and explained. For a comprehensive introduction to embryonic development *per se*, the reader is referred to any one of several excellent publications that already exist (e.g. Alberts *et al.*, 2002; Gilbert, 2003; Wolpert, 2002). What this chapter attempts to provide is the information that might be necessary for a clinician or advanced student specializing in paediatric medicine to understand and appreciate in context what follows. In that sense, an element of unorthodoxy might be discerned by some readers. However, we hope that this rationale will be justified as the reader progresses through the book.

Embryos, Genes and Birth Defects, Second Edition Edited by Patrizia Ferretti, Andrew Copp, Cheryll Tickle and Gudrun Moore

The relationship between genotype and phenotype

The term 'genotype' is generally used to refer to the genetic make-up or constitution of an individual organism, be it virus, fruit fly or human. In contrast, we use the word 'phenotype' to cover the form and functioning of an individual, to the extent that it may encompass metabolism and behaviour (and thus we can refer to 'behavioural phenotypes'). The word 'genotype' is subtly but distinctly different from the term 'genome', which refers not to the totality of genes in an individual cell but to the array of genes in a complete haploid set of genes characteristic for that species. In this sense, a genome is a species-specific concept, whereas genotype is a concept applying to an individual of the species in question.

The complexity of the phenotype reflects largely but not entirely the complexity of the genotype. However, there is not necessarily a simple and direct relationship, since genome size and genome complexity are rather different entities. Overall genome size, in terms of DNA, is to some extent determined by the relative proportion of non-protein coding sequences contained within it. Thus, some plant, insect and amphibian species contain far more total DNA in their genomes than does *Homo sapiens*, even though they are phenotypically simpler and contain fewer genes (indeed, some amphibian species contain up to 9×10^{11} nucleotide bases per haploid genome, as opposed to the 2.85×10^9 nucleotides recently sequenced in humans; International Human Genome Sequencing Consortium, 2004). Much of this increase in DNA content is thought to represent a greater than normal proportion of non-coding, repetitive sequences. If we consider genome complexity in terms of the number of genes present, then a more systematic relationship emerges. In simple organisms, such as viruses, the limited number of genes in the genome can be accurately determined. However, for more complex multicellular organisms, total gene number is an estimate based on confirmed genes and potential coding regions identified by predictive methods. Therefore, the size of these estimates has changed as our ability to visualize the DNA sequence and our understanding of genomic organization has evolved. Currently, *Drosophila melanogaster*, the fruit fly, is estimated to contain some 14 000 genes in its genome, whereas the genome of *Homo sapiens* is thought to comprise between 20 000 and 25 000. However, this latter set of figures is still subject to revision and does not take into account the considerable protein variation that can accrue from alternate usage and splicing of exons or the existence of functional non-coding RNAs.

Whereas gene mapping refers to identification of the chromosomal location of an individual gene, genome mapping is a programme of research designed to identify the chromosomal location of all genes in the genome of a particular species. Although it is the international Human Genome Project that has received wide media attention, it should be noted that genome mapping projects for other species are also under way or recently completed. These include a number of model organisms, such as the mouse, fruitfly, toad and nematode worm, as well as those of economically important food species, such as cow, pig and chicken (http://www.ncbi.nlm.nih.gov/Genomes/index.html). The mapping of individual genes, or of candidate gene loci, means that

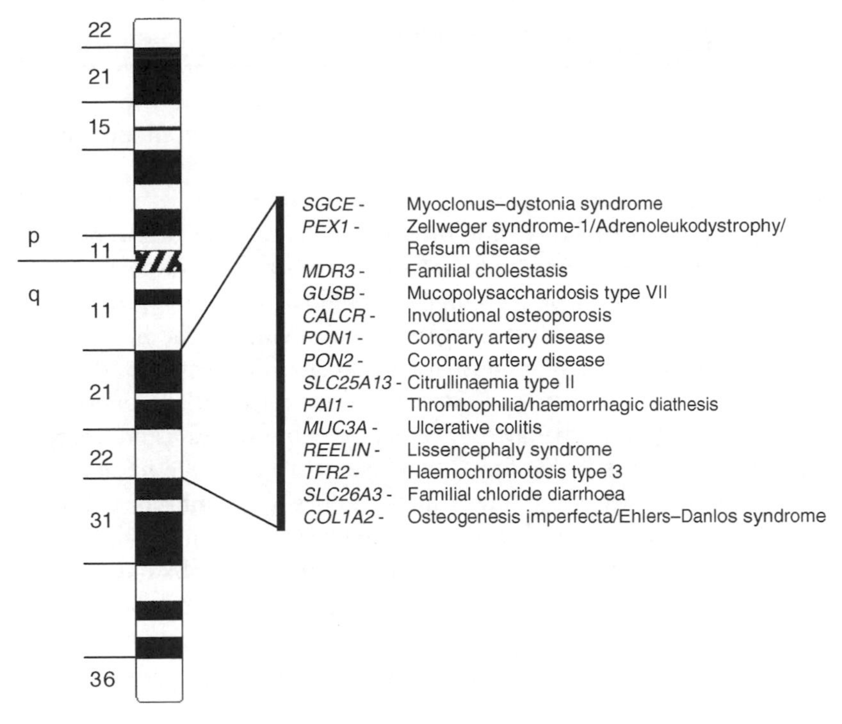

Figure 1.1 Congenital malformation with gene mutations mapping to 7q21-q22. More than 1700 have been identified throughout the genome, including >80 on chromosome 7 (see http://www.ncbi.nlm.nih.gov/LocusLink/ and Chapter 2)

chromosomal 'maps' of congenital abnormality can be drawn up (see Chapters 2, 3 and 4; also OMIM: http://www.ncbi.nlm.nih.gov/entrez/query.fcgi?db=OMIM), whereby the location of genes, in which mutation produces a particular dysmorphology or inherited metabolic disease, can be displayed (Figure 1.1).

At this point we should ask ourselves what kind of information is encoded within the genes. Are the genes really the 'blueprint' to which they are often analogized? A blueprint implies some kind of descriptive specification. Is that indeed how the genome is organized? In fact, the information content of genes is one-dimensionally complex, since it is specified by the nature of the linear sequence of nucleotide bases along the DNA molecule. In dramatic contrast, the phenotype is three-dimensionally complex (and four-dimensionally complex if we include dynamic phenomena, such as metabolism and homeostasis, rather than just morphology); yet the linear nucleotide sequence itself conveys no sense of what the phenotype might look like. To appreciate just how phenotypic complexity might be generated we have to move away from the rather dated analogy of a descriptive specification and think of the genome and its implementation as a generative programme. The more appropriate and meaningful analogy of origami has been proposed to illustrate the characteristics of a generative programme (Wolpert, 1991). Here, the instructions for creating a topologically complex shape from a sheet of paper contain within them no description of the final outcome. The complexity is generated progressively by implementing those instructions, which may in themselves be very simple, even though the outcome is complex. In this way, the genome, or at least the developmentally significant parts of it, can be seen as assembly rules for building an embryo.

In one sense, genes 'simply' encode proteins. Transcription of a gene produces a message that is translated from the four-letter alphabet (nucleotides) of the nucleic acids to the approximately 20-letter alphabet (amino acids) of the proteins, by virtue of the genetic code. The primary structure of a protein, i.e. the linear sequence of amino acids, together with any post-translational modifications, determines its secondary and tertiary structure. Proteins endow cells with properties such as characteristic metabolisms, behaviour, polarity, adhesiveness and receptivity to signals (Figure 1.2) and it is this functional level that marks the implementation of those assembly rules. Within the increasingly multicellular embryo, cell interactions and inductions are initiated, cell lineages are established, and morphogenesis, growth and histogenesis proceed. Thus, interactions of proteins, cells and tissues during

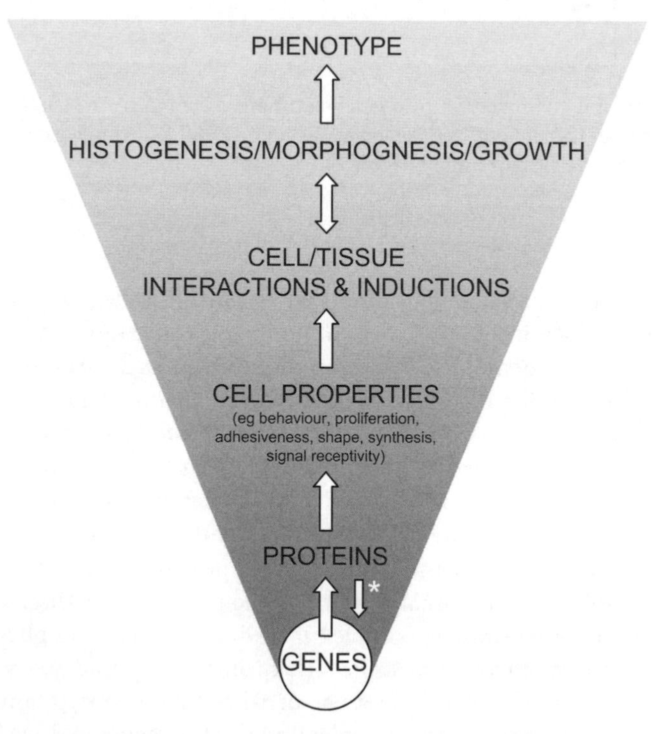

Figure 1.2 Causal relationship between genotype and phenotype. Higher-order complexity is generated progressively by the interaction of proteins, of cells and of tissues during development. The asterisked 'return' arrow between genes and proteins represents the controlling role on gene expression of transcription factors encoded by regulatory genes

development generate progressively higher-order complexity (Figure 1.2), from the one-dimensional complexity of the genotype and primary protein structure to the three-dimensionally complex phenotype. Embryonic development is therefore a typical generative programme. From a limited range of fundamental cell properties, an almost infinite range of complex phenotypes can be built, simply by deploying these cell properties in varying ways. The diverse range of phenotypic from across extant and extinct species bears witness to the morphogenetic power of these basic cell properties over an evolutionary time-scale.

Thus, it is the morphogenetic potential of cell properties and the mechanisms of embryonic development that causally link genotype and phenotype. And from this brief and perhaps simplistic rationalization, one can see that during development there will be significant, higher-order events taking place in the absence of direct genetic control but which are themselves the inevitable consequences of genetic specification (Figure 1.2, from the level of 'cell properties' upwards). Thus, the phenotypic expression of an individual's genotype is influenced by a variety of non-genetic factors that might involve variables such as diet, infection and ageing. These factors can have direct effects on gene expression or may influence more subtle control mechanisms, such as DNA or histone methylation. This class of phenomena is sometimes described as epigenetic and, clearly, much morphological complexity is generated within this so-called epigenetic domain (Alberch, 1982; McLachlan, 1986; Gottesman and Hanson, 2005).

Developmental biologists are interested in defining assembly rules and elucidating their operation at tissue, cellular and molecular/genetic levels. To understand dysmorphogenesis it is necessary to clarify what happens when certain assembly rules are either mis-specified or wrongly interpreted and a birth defect results. Clearly, understanding a particular birth defect involves much more than simply identifying a mutated gene or an environmental teratogen. It requires knowledge of the consequences of these on the mechanisms operating within the embryo, an understanding of how the generative programme has been perturbed and how that produces an abnormal phenotype. Furthermore, just as an understanding of normal development can help clarify abnormal development, so analysis of abnormal development can sometimes throw light on hitherto unknown aspects of normal mechanisms.

Before leaving this topic, it should be noted that in Figure 1.2 there is feedback indicated from proteins to genes (see reverse arrow). This reflects the fact that the role of some proteins is to bind to DNA, typically in a highly sequence-specific manner. Genes that encode such proteins are referred to as 'regulatory genes' and the proteins themselves known as 'transcription factors', since they control (either upregulate or downregulate) transcriptional activity of the gene to which they have bound. In essence, genes work in hierarchies, with regulatory genes controlling the expression of 'downstream' genes and with elements of 'cross-talk' between regulatory genes themselves. The definition of such genetic cascades and signalling pathways is a very topical issue in contemporary developmental biology and this is reflected by the prominence given to it by many of the contributors to this volume.

Such genes are, of course, pivotally important in the normal life of the cell, in its synthetic and metabolic activity, homeostasis and proliferation, but during embryonic development they have multiple and crucial roles in determining cell fate. Although many of the genes identified to date as being involved in birth defects encode enzymes or structural proteins, it is not surprising that numerous families of regulatory genes have been established as playing important roles in dysmorphogenesis (see later).

Having discussed some aspects of the genotype–phenotype relationship, it is now appropriate to point out that it can be simplistic to always interpret dysmorphogenesis on the basis of a 'one gene:one (dysmorphic) phenotype' model. It is clear that, in some cases, a diversity of phenotypes can emerge from mutations in a single gene, each disease or dysmorphic phenotype reflecting a different mutation within that gene. Thus, different mutations in the receptor tyrosine kinase gene, *RET*, can result in familial medullary thyroid carcinoma, multiple endocrine neoplasia types 2A and 2B (all of which accords with its original recognition as an oncogene) and in Hirschsprung's disease, a developmental anomaly of the gut (reviewed by Manié *et al.*, 2001; and see Chapter 11). This last disorder appears to be the consequence of a failure of RET-expressing neural crest cells to migrate normally and establish a parasympathetic innervation to the gut. The thyroid cancer-associated syndromes all result from mutations causing specific amino acid substitutions that apparently alter the functionality of the receptor tyrosine kinase encoded by *RET* (i.e. gain-of-function mutations that may lead to hyperplasia of the RET-expressing tissues). In contrast, the Hirschsprung mutations comprise deletion, insertion, frameshift, nonsense and missense mutations that lead to a loss of function. The phenotype can be explained as due to haploinsufficiency, whereby a threshold sensitivity to absence of 50% of the gene product (due to a mutated allele) is sufficient to perturb the development of the cells normally expressing that particular gene. In this case, it is the neural crest progenitors of the gut parasympathetic neurones that are affected, leaving other RET-expressing cell populations in the embryo apparently unscathed, due to tissue-specific differences in the threshold sensitivity (Manié *et al.*, 2001). Interestingly, *RET* mutations that affect one of four extracytoplasmic cysteine residues have been found in Hirschsprung's patients, as well as patients with MEN2A and familial medullary thyroid carcinoma. These findings have raised the idea that a single mutation has opposing effects, depending on the tissue in which RET is expressed, and results in uncontrolled proliferation in endocrine cell types and apoptosis in enteric neurons (reviewed in Manié *et al.*, 2001). Furthermore, mutations in *RET* are found only in about half of the familial cases of Hirschsprung's disease and then frequently with variable penetrance. This suggests a higher level of complexity, involving the interaction of other genes or non-coding variants, often referred to as modifiers. Co-inheritance of mutations in distinct loci but with additive effect gives rise to a multi- or polygenic inheritance model. In this case, each of the individual mutations alone may be considered risk factors, as they are insufficient to cause the phenotype alone but do so when inherited together.

The causality of birth defects is not necessarily genetic in origin and various aetiological categories can be recognized:

- Chromosomal anomalies (e.g. trisomies, translocations)
- Polygenic disorders
- Single gene mutations
- Environmental/teratogenic factors
- Multifactorial aetiology
- Unknown aetiology

Each of these six categories presents its own set of problems in determining how a particular birth defect is generated (see Chapters 2, 3, 4 and 6). It might be argued that events occurring within the epigenetic domain referred to earlier can be extended to environmental influences on development. The embryo does not occupy a completely protected and privileged environment and, in some respects, is as open to effects from its environment as the neonate, juvenile or adult. Indeed, the recognition that the intrauterine experience of the fetus is strongly influenced by maternal nutritional or hormonal status is pivotal in determining later susceptibility to a number of adult diseases, such as diabetes and coronary heart disease (reviewed by Barker, 1995).

Clearly, the phenotype, be it adult or embryonic, is always the product of the combined effects of genetic and environmental influences (Sykes, 1993), but the relative contributions of each can differ for each aspect of the phenotype (Figure 1.3). Thus, Down's syndrome, as a trisomy disorder, reflects a condition that is 100% genetic, whereas a neural tube defect such as spina bifida (see Chapter 8) may have a

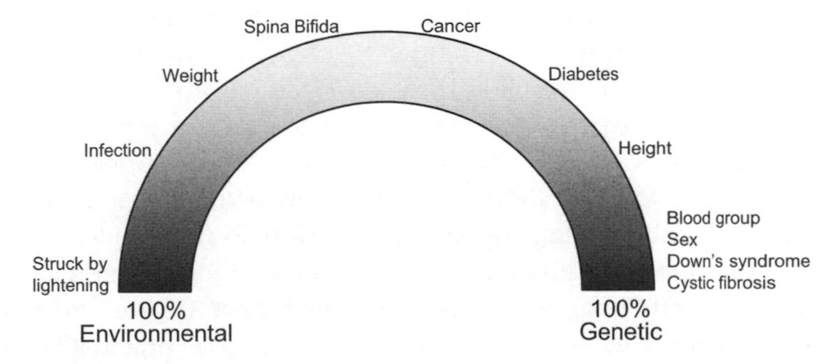

Figure 1.3 Interplay between environmental and genetic factors in the determination of phenotype. The relative importance of each will vary according to the particular phenotype, or aspect of phenotype, under consideration (after Sykes, 1993)

strong environmental component in its aetiology, coupled with a possible genetic predisposition in some cases (reviewed by Marsh, 1994).

Even though the majority of birth defects have a genetic component, the extent of interaction between genotype and environment is still poorly understood and, in research studies, often neglected. Thus, the majority of animal studies assessing the teratogenicity and reproductive toxicity of environmental factors have frequently failed to take into account the different genotypes of the various strains of animal species used (discussed by Copp, 1994). Yet we are increasingly aware of human genes, which increase susceptibility to environmental teratogens. Examples of this include common polymorphisms that affect either protein levels or the activity of enzymes that metabolize the teratogens resulting from cigarette smoking, alcohol or drug intake (Polifka and Friedman, 2002). We may conclude that, in elucidating the complex relationship between genotype and embryonic phenotype, whether it be in the context of normal development or dysmorphogenesis, environmental factors may sometimes be critical (see Chapter 6).

The role of 'model' systems

To understand the mechanisms of development inevitably means dismantling and/or perturbing the embryo in some way. Very little has ever been learnt of mechanisms by simply observing embryonic development. Traditionally, developmental biologists dismantle and reassemble embryos, or parts of embryos, at the level of gene, cell, tissue or organ. In this way we learn how the system responds to perturbation, and through that we can elucidate the functional role of the component parts, sometimes down to the level of an individual nucleotide base within a DNA codon. For example, a change in a single nucleotide in the *bicoid* gene of *Drosophila* will actually reverse the anteroposterior axis of the embryo (Fronhöfer and Nusslein-Volhard, 1986; Struhl *et al.*, 1989).

The ability to manipulate DNA in the laboratory has brought an unparalleled precision and finesse to developmental analysis, bringing exquisite control to gene-rescue, knock-out, overexpression, ectopic expression and regulatory element studies. Thus, transgenic technology (see Chapter 5) can be considered as the most sophisticated of strategies, following in the great tradition of experimental perturbation started in the nineteenth century with the emergence of experimental embryology, epitomized by the German *Entwicklungsmechanik* ('developmental mechanics') school established by Wilhelm Roux and colleagues. However, it is important to comment that molecular biology as it exists now has not rendered traditional experimental embryology redundant. The molecular biology monoculture that some feared 15–20 years ago has not prevailed and what we see emerging today, and which is well reflected in the following chapters, is a pragmatism in which molecular approaches are creatively integrated with cellular and tissue approaches. For instance, a well-designed 'cut 'n' paste' tissue grafting experiment can generate results with profound implications at the molecular level (see e.g. some of the grafting experiments described in Chapter 7), and can itself direct further analysis at the molecular level.

This theme of perturbational analysis to reveal mechanisms means that the human embryo is not a system of choice, at least not after the 14-day limit set by the regulating authorities in Britain, and manifest in the Human Fertilization and Embryology Act, 1990 (and see Burn and Strachan, 1995; Table 1.1). Most dysmorphogenesis is likely to have its inception during the major stages of morphogenesis and organogenesis, starting with neurulation during the fourth week in the human embryo (Larsen, 2001). Disruptions earlier than that are likely to result in spontaneous abortion and be lost; indeed, it has been estimated that at least 15% of all human pregnancies end as spontaneous abortions after implantation (Warburton and Fraser, 1964).

Nevertheless, human fetal tissues are being used in biomedical research, particularly in the context of somatic gene therapy, fetal cell transplantation, haematopoietic stem cell transplantation and fetal organ transplantation (reviewed by Reed *et al.*, 1995). However, many research programmes almost always use developmentally late fetal material, which is of little if any use in studying embryonic expression of genes implicated in birth defects. For this specific purpose, human embryo banks have been established (reviewed by Burn and Strachan, 1995), using material obtained from terminations and collected with full ethical approval. Reports on the expression of genes causally involved in dysmorphogenesis are now commonplace (see Chapter 2) and the use of such data in the long-term development of preventive and therapeutic clinical strategies is likely to escalate over the next few years.

However, in order to study developmental mechanisms during the crucial stages of morphogenesis and organogenesis we are, of necessity, obliged to use animal model systems. In Chapters 7–17 you will find reference to work using embryonic systems as diverse as zebrafish, *Xenopus*, chick and mouse. Are we to view these simply as models or research surrogates for the human embryo (see discussion by Monk, 1994)? In fact, developmental research is often driven by reasons of scholarship, and animal models are more typically studied for their own intrinsic interest, in the context of comparative biology and evolution (Bard, 1993). Nevertheless, several essential concepts that have significantly enhanced our understanding of human dysmorphology have emerged from analysis of animal model systems; the developmental field concept, as applied to dysmorphogenesis (Opitz, 1985), and chromosomal imprinting (Harwell imprinting site plus Otago site for human imprinted genes; Monk, 1994; see Chapter 2) are two obvious examples. However, when animal model systems are seen as 'research surrogates' for the human embryo and fetus in the sense that extrapolations are made, we must ask ourselves: 'to what extent is this justified?'.

The changing concept of homology

For many years, developmental biologists, if challenged, have sought to justify the use of animal model systems by virtue of homology of form. This is likely to have been based loosely upon the a priori argument that, at least in early development, phenotypic similarities between human and non-human vertebrate species must reflect equivalence of the underlying generative mechanism. However, only slowly has evidence for such assumptions about homology of mechanism begun to accumulate. Perhaps one of the clearest early demonstrations of this relates to the zone of polarizing activity (ZPA) – the region of a limb bud which, by release of a diffusible morphogen, polarizes the distal part of the growing limb and controls the anteroposterior pattern of digits (see Chapter 7). It was found that the ZPA taken from a human limb bud will, when assayed by grafting ectopically into a chick embryo wing bud, display the same activity as the equivalent region of a chick bud. Extra digits are formed in a predictable and organized fashion by the host, demonstrating that human and chicken ZPAs produce the same morphogen, but chicken host cells respond to it by forming additional chick digits (Fallon and Crosby, 1977). In other words, there is an equivalence of mechanism in the building of this particular bit of anatomy. Although this example deals with just a small part of the body plan (digit specification), it can be seen as exemplifying a widely held belief that similar equivalences exist at the mechanistic level in the building of much of the anatomy or, at least, that portion of it which is characteristically and uniquely 'vertebrate' in character.

This type of assumption has been cautiously held for a number of years and, in a rather piecemeal and limited fashion, evidence gradually accumulated to give it some justification. However, it has become clear in the last few years that the concept of homology is underpinned by an amazing degree of conservation of both gene sequence and function (reviewed by Scott, 2000). So fundamental is this to our understanding of the genotype–phenotype relationship and to our interpretation of data from model systems, that it is necessary to deal with the topic at some length.

The existence of *Drosophila* mutants in which body parts are transformed into recognizable structures but develop at an inappropriate site, the so-called homeotic mutants, has been known since the nineteenth century, when the phenomenon of homeosis was first discovered. Certain unidentified genes were thought to be involved in the specification of the segmented body plan of *Drosophila*, with mutation resulting in mis-specification of particular body parts. Cloning and sequencing revealed that the homeotic genes are in fact regulatory genes and contain a highly conserved motif, the homeobox (McGinnis *et al.*, 1984), encoding a DNA-binding domain that subsequently became known as the homeodomain. Further analysis of homeobox-containing genes confirmed their role in morphogenetic specification and revealed a complex and hierarchical genetic control of the body plan in this arthropod (reviewed by Akam *et al.*, 1994). The cloning of these genes provided probes with which to screen the genomes of other species, and screening revealed a surprising degree of conservation, with orthologous genes being found in a very wide and diverse range of species examined. The largest and best known of these homeobox-containing gene families are the *Hox* genes, of which there are 39, organized in four clusters on different chromosomes in all vertebrates including humans.

Sequence homology and position within each cluster is such that derivation of each gene can be traced from a single ancestral cluster similar to the *HOM-C* complex in *Drosophila.* Excellent reviews of the organization, evolution and functional roles of *Hox* genes have been published elsewhere (McGinnis and Krumlauf, 1992; Burke, 2000; Garcia-Fernàndez, 2005) and accounts of their role in specification of major features of the vertebrate body plan are given here in Chapters 7, 11, 12 and 15.

Although these genes and others like them have only been identified in vertebrate genomes by virtue of their sequence homology with their *Drosophila* counterparts (remember that *Drosophila* probes were used in the screening), conservation of gene sequence is only one aspect of this remarkable evolutionary story. If there is truly homology of function, then we might expect conservation of expression domains of the gene(s) in question across a range of species, and this is indeed often found. The most rigorous test, however, has to be an operational one in which genes are moved into the genome of another and distant species, preferably into individuals in which the orthologue has been inactivated. Will the introduced 'foreign' gene be switched on in the correct spatiotemporal pattern and will it function to produce a normal embryo?

Homeobox-containing genes provide a number of examples in which these three criteria of sequence homology, equivalence of expression domain and functional homology are satisfied. Thus, a regulatory sequence of the *Drosophila* homeotic gene, *Deformed*, which supports expression in subregions of posterior head segments, can be replaced by the equivalent mouse sequence and still result in normal embryonic development (Awgulewitsch and Jacobs, 1992). The mouse gene, *Hoxb-6*, can be moved into the *Drosophila* embryo and specify normal thoracic segments (Malicki *et al.*, 1990) and even the regulatory element of a human *Hox* gene, *HOXB4*, is expressed rostrally and supports head development when introduced into *Drosophila* (Malicki *et al.*, 1992). Finally, we should not assume that such exchanges only operate between species with segmented body plans, no matter how divergent they may be, since it has also been shown that equivalent functional homology exists between the *Hox* genes of *Drosophila* and those of the unsegmented nematode worm, *Caenorhabditis elegans* (Hunter and Kenyon, 1995).

The existence of such amazing functional homology might suggest that there has been some conservation of downstream target genes for the homeoproteins. But how would this degree of conservation of homeobox gene function across a wide range of species correlate with the diverse range of phenotypic form displayed by these species? In other words, how do we reconcile functional homology, and all that that entails, with the evolution of the disparate body plans displayed by mammals and insects, for example? Such questions are currently unresolved but various possibilities, such as homeoproteins acquiring new targets, homeobox genes changing expression domains, changes in the function of downstream target genes and the emergence of new modes of regulation, are all under consideration (Kenyon, 1994; Manak and Scott, 1994; Hughes and Kaufman, 2002). Meanwhile, similar levels of conservation for genes involved in major morphogenetic events are being discovered, with functional homology apparently being retained by other key regulatory genes and pathways, such as *goosecoid*, *Brachyury* and non-canonical *Wnt* signalling controlling the very different modes of gastrulation across species as diverse as zebrafish, *Xenopus*, chick and mouse (Beddington and Smith, 1993; De Robertis *et al.*, 1994; Tada *et al.*, 2002).

However, it is not just regulatory genes that display such conservation of sequence, expression domain and function. It is rapidly emerging that genes encoding a number of secreted molecules involved in signalling between cells have been similarly conserved. Genes homologous to the *Drosophila hedgehog* gene family (so named because of the 'spiny' appearance of the mutant larvae) encode secreted proteins that appear to have a pivotal role in patterning a number of structures in vertebrates (reviewed by Hammerschmidt *et al.*, 1997; Nybakken and Perrimon, 2002). The product of *sonic hedgehog (shh)* has a major role in notochord induction of the ventral floor plate of the neural tube (e.g. Roelink *et al.*, 1994; and see Chapter 8). A parallel signalling role for this secreted protein is seen in limb development. Thus, *shh* is expressed in the posterior region of both fin (zebrafish) and limb buds (chick and mouse) where it is thought to be active in establishing pattern across the anteroposterior axis of the bud and is a component of the ZPA (see earlier). Ectopic expression of this gene in the anterior part of the chick limb bud produces duplication of anterior structures, paralleling the mirror-image duplication of the anterior wing compartment in *Drosophila* resulting from ectopic *hedgehog* expression (Fietz *et al.*, 1994). Functional homology is even maintained amongst some of the other signalling molecules thought to be downstream from the hedgehog proteins, such as *decapentaplegic* (*dpp*) in *Drosophila*, and the related *transforming growth factor-β* (*TGF-β*) gene family in vertebrates (reviewed by Hogan *et al.*, 1994), and the proteins with which they interact during specification of dorsoventral pattern in the neural primordium (Holley *et al.*, 1995).

Another example of evolutionary conservation of function has recently been demonstrated with the signalling pathway for planar (epithelial) cell polarity. Epithelial cell orientation and cross-talk to the surrounding cells is critical for correct assembly of the *Drosophila* compound eye and uniform positioning of hairs on the wing and thorax (Strutt, 2003). This is dictated by a secreted Wnt ligand that binds to a frizzled receptor protein complex, which then signals to the nucleus via an intracellular protein called dishevelled. A similar signalling pathway with essentially the same protein components has now been found to organize cellular convergence on the dorsal midline of the mammalian embryo in order to regulate formation of the neural tube. Disruption to any of the core protein components leads to a failure of neural tube closure, as demonstrated in mice, *Xenopus* and zebrafish (reviewed in Copp *et al.*, 2003; and see Chapter 8). This pathway is also required for correct orientation of the stereociliary bundles found in the mammalian inner ear (Montcouquiol *et al.*, 2003; Curtin *et al.*, 2003). In contrast to neural tube development, this represents a vertebrate phenotype more closely resembling the invertebrate wing hairs.

Similarly conserved function through evolution is elegantly illustrated by the study of mutations in different orthologues of the *PAX6* gene (see Chapter 9). Mutations in

PAX6 give rise to the eye defect aniridia, while the mouse orthologue turned out to be a gene formerly known as *Small eye*, since a loss-of-function mutation produced a microphthalmic phenotype. Both of these genes are the functional orthologues of the *Drosophila Eyeless (ey)* gene (Quiring *et al.*, 1994); *ey* is also involved in eye development, and a loss-of-function mutation eliminates the compound eye. As a result, *Aniridia, Small eye* and *Eyeless* are collectively regarded as *Pax-6* homologues with pivotal roles in eye development, whether it be the compound eye of an arthropod or the vertebrate eye (Quiring *et al.*, 1994). This has been assessed by ectopic expression of the *ey* gene, which results in ectopic compound eyes with relatively normal facet organization and arrays of photoreceptor cells (Figure 1.4a). More relevant to this discussion is the finding that ectopic expression of the mouse *Pax-6/Small eye* gene introduced into *Drosophila* will also generate ectopic compound eyes that are morphologically equivalent to the normal compound eye (Figure 1.4b; Halder *et al.*, 1995). In other words, the generative programme for assembling an arthropod compound eye can be activated and controlled by a mouse *Pax-6* gene. It is concluded that these various *Pax-6* homologues constitute master genes, arising from a common ancestral gene and with conserved function in controlling eye morphogenesis.

With the advent of more efficient positional cloning strategies, mouse knockout technology and the development of large scale ENU mutagenesis programmes, more and more genes are being assigned to function and phenotype. As a consequence, the extent of regulatory gene involvement in birth defects is becoming better defined. There are now numerous examples of regulatory gene families that are grouped

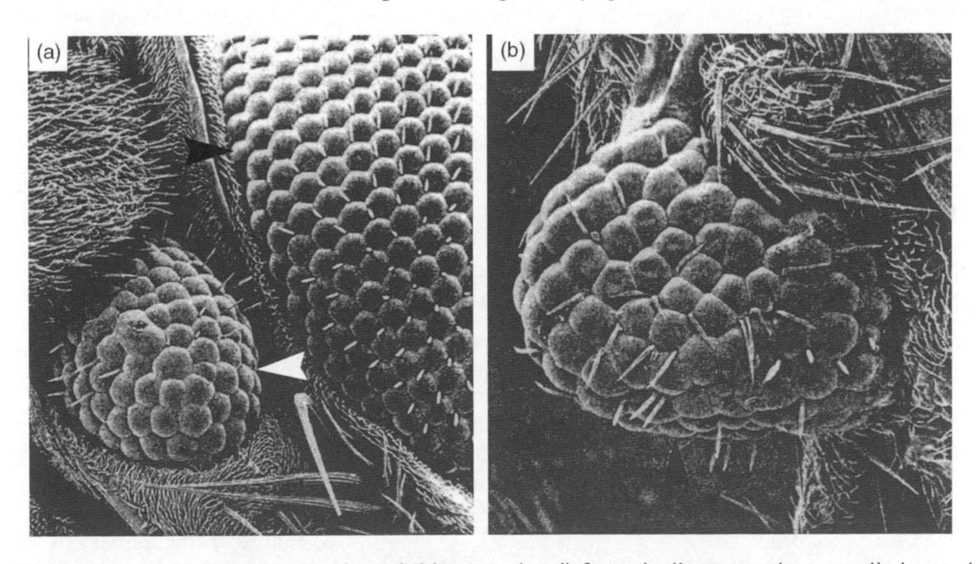

Figure 1.4 (a) Ectopic compound eye (white arrowhead) formed adjacent to the normally located compound eye (on the right, black arrowhead) in the head of a *Drosophila* fly; this is the result of the ectopic expression of the *ey* gene. (b) Ectopic compound eye formed, in this case, on the leg of a fly, under the control of an ectopically expressed mouse *Pax-6* gene introduced experimentally. In both (a) and (b), note the similarity of the ommatidial organization and interommatidial bristles, in the ectopic eyes and in their normal counterpart in (a). Photographs supplied by Professor Walter Gehring

through both sequence and functional homology that are also directly implicated in dysmorphogenesis and neoplasia. These include the *PAX*, *HOX*, *ZIC*, *ZNF*, *SOX*, *FOX* and *TBX* families. When the first T-box gene, *T* (*Brachyury*), was identified (Herrmann *et al.*, 1990) it was thought to be unique. However, most species studied have multiple family members and mammals contain a total of 17 different functional *T*-like genes (see Table 1.1). Family members are based on their

Table 1.1 The mammalian T-box gene family, with mouse and human mutant phenotypes

Gene	Location (human)	Human phenotype	Mouse phenotype
T	6q27	Possible risk factor for NTD	Lack notocord and posterior somites
Tbx1	22q11	DiGeorge syndrome	DiGeorge syndrome-like
Tbx2	17q23	Potential oncogene implicated in breast cancer	Atrioventricular canal and septation of the outflow tract; hindlimb digits
Tbx3	12q24	Ulnar–mammary syndrome	Mammary gland, limb and yolk sac defects
Tbx4	17q23	Small patella syndrome	Hindlimb bud outgrowth failure
Tbx5	12q24	Holt–Oram syndrome	Cardiac defects and forelimb malformations
Tbx6	16p11		Abnormal patterning and specification of the cervical somites posterior paraxial mesoderm
Tbx10	11q13		Cleft lip and palate*
Tbx15	1p12		Coat pigmentation anomalies and skeletal development
Tbx18	6q14		Somite compartment boundary formation
Tbx19	1q24	ACTH deficiency	ACTH deficiency
Tbx20	7p14		Defects in cardiac chamber differentiation
Tbx21	17q21	Physiological and inflammatory features of ashma	
Tbx22	Xq21	X-linked cleft palate	
Tbr1	2q24		Defects in neuronal migrations and axonal projection
Eomes	3p24		Failure of trophoblast differentiation and mesoderm formation
MGA	15q15		Unknown

In addition, an intronless pseudogene (*TBX23*) is present on chromosome 1 and a truncated T-box gene similar to *TBX20* is present on chromosome 12.
*Gain of function.

conservation of a 180–200 amino acid sequence that encodes a DNA-binding domain, the so-called 'T domain' or 'T-box'. These genes function as transcription factors and are found throughout metazoans, including primitive species such as *Caenorhabditis elegans* and hydra. Most of the T-box genes are expressed early in development, regulating cell fate and behaviour, thereby specifying regional tissue characteristics. For example, *Tbx4* and *Tbx5* play major roles in limb identity (see Chapter 7). While the genes are specifically expressed in hindlimb and forelimb, respectively, ectopic expression of either has the ability to at least partially reprogramme cell fate decisions into the opposite limb type (Rodriguez-Esteban *et al.* 1999; Takeuchi *et al.* 1999). Most of the human phenotypes resulting from various *TBX* gene mutations occur as a consequence of haploinsuficiency, indicating that the tissues involved are especially sensitive to expression levels. It is also interesting to note that significant overlap is found in the expression domains of a number of the T-box genes (see Chapter 9). This may simply be a legacy of their ancestral origin, where expanding genome size through rounds of duplication has retained the same regulatory elements that have not yet re-specified. Nevertheless, this co-expression may allow for additional protein–protein interactions, such as heterodimerization, which in turn lead to increased complexity of function.

As yet, the biochemical pathways that are regulated by T-box genes remain poorly understood. Nevertheless, there is an increasing list of developmental defects and phenotypes being associated with the various family members (see Table 1.1). Of course, from a birth defects point of view, this is only going to be the tip of the iceberg because of the numerous upstream and downstream genes in each pathway, amplifying the number of potential targets for mutation-induced phenotypes. However, with the use of animal models and modern expression profiling techniques, dissection of these transcription factor networks is already under way and will provide the means for further disease gene discovery.

It would seem that that long-held views on the evolution of phenotypic form are being fundamentally revised. It is not simply homology of function that has driven the development of anatomical analogues through convergent evolution. There would seem to be a basic, shared genetic 'tool-kit' for development that has been retained over many millions of years (Akam *et al.*, 1994). The different generative programmes of development have deployed this in a multitude of ways to build different phenotypes. Not surprisingly, those phenotypes are sometimes dysmorphic, as is often the case with mutations in the genes that specify developmental processes. Earlier assumptions about the extent of homology of developmental mechanisms between human and various animal model systems have been vindicated more powerfully than could have been anticipated even a few years ago.

Acknowledgement

We are most grateful to Professor Walter Gehring, who generously provided the photographs used in Figure 1.4.

References

Akam, M., Holland, P., Ingham, P. and Wray, G. (eds) (1994) Preface. *Development* (suppl: *Evolution of Developmental Mechanisms*).

Alberch, P. (1982) Developmental constraints in evolutionary processes. In *Evolution and Development*, Bonner, J.T. (ed.). Springer-Verlag: Berlin; 313–332.

Alberts, B., Johnson, A., Lewis, J. *et al.* (2002) *Molecular Biology of the Cell*, 4th edn. Garland: New York.

Awgulewitsch, A. and Jacobs, D. (1992) *Deformed* autoregulatory element from *Drosophila* functions in a conserved manner in transgenic mice. *Nature* **358**, 341–344.

Bard, J. (1993) *Embryos: Colour Atlas of Development.* Wolfe: London.

Barker, D.J.P. (1995) Intrauterine programming of adult disease. *Mol. Med. Today* **9**: 418–423.

Beddington, R.S.P. and Smith, J.C. (1993) Control of vertebrate gastrulation: inducing signals and responding genes. *Curr. Opin. Genet. Dev.* **3**: 655–661.

Burke, A.C. (2000) Hox genes and the global patterning of the somitic mesoderm. *Curr. Top. Dev. Biol.* **47**: 155–181.

Burn, J. and Strachan, T. (1995) Human embryo research in developmental research. *Nat. Genet.* **11**: 3–6.

Copp, A.J. (1994) Birth defects: from molecules to mechanisms. *J. R. Coll. Phys. Lond.* **28**: 294–300.

Copp, A.J, Greene, N.D. and Murdoch, J.N. (2003) The genetic basis of mammalian neurulation. *Nat. Rev. Genet.* **4**: 784–793.

Curtin JA, Quint E, Tsipouri V, Arkell RM *et al.* (2003) Mutation of *Celsr1* disrupts planar polarity of inner ear hair cells and causes severe neural tube defects in the mouse. *Curr. Biol.* **13**: 1129–1133.

De Robertis, E.M., Fainsod, A, Gont, L.K. and Steinbesser, H. (1994) The evolution of gastrulation. *Development* (suppl: *Evolution of Developmental Mechanisms*): 117–124.

Fallon, J.F. and Crosby, G.M. (1977) Polarizing zone activity in limb buds of amniotes. In *Vertebrate Limb and Somite Morphogenesis*, Ede, D.A., Hinchliffe, J.R. and Balls, M. (eds). Cambridge University Press: Cambridge; 55–69.

Fietz, MJ, Concordet, J.-P., Barbosa, R. *et al.* (1994) The *hedgehog* gene family in *Drosophila* and vertebrate development. *Development* (suppl: *Evolution of Developmental Mechanisms*): 43–51.

Fronhofer, H.G. and Nüsslein-Volhard, C. (1986) Organisation of anterior pattern in the *Drosophila* embryo by the maternal gene *bicoid. Nature* **324**: 120–1125.

Garcia-Fernàndez J. (2005) Hox, ParaHox, ProtoHox: facts and guesses. *Heredity* **94**: 145–152.

Gilbert, S.F. (2003) *Developmental Biology*, 7th edn. Sinauer: Sunderland, MA.

Gottesman, I.I. and Hanson, D.R. (2005) Human development: biological and genetic processes. *Ann. Rev. Psych.* **56**: 263–286.

Halder, G., Callaerts, P. and Gehring, W.J. (1995) Induction of ectopic eyes by targeted expression of the *eyeless* gene in *Drosophila. Science* **267**: 1788–1792.

Hammerschmidt, M., Brook, A. and McMahon, A.P. (1997) The world according to hedgehog. *Trends Genet.* **13**: 14–21.

Herrmann, B.G., Labeit, S., Poustka, A., King, T.R. and Lehrach, H. (1990) Cloning of the T gene required in mesoderm formation in the mouse. *Nature* **343**: 617–622.

Hogan, B.L.M., Blessing, M., Winnier, G.E., Suzuki, N. and Jones, C.M. (1994) Growth factors in development: the role of TGF-β related polypeptide signalling molecules in embryogenesis. *Development* (suppl: *Evolution of Developmental Mechanisms*): 53–61.

Holley, S.A., Jackson, P.D., Sasai, Y. *et al.* (1995) A conserved system for dorsal–ventral patterning in insects and vertebrates involving *sog* and *chordin*. *Nature* **376**: 249–253.

Hughes, C.L. and Kaufman, T.C. (2002) *Hox* genes and the evolution of the arthropod body plan. *Evol. Dev.* **4**: 459–499.

Hunter, C.P. and Kenyon, C. (1995) Specification of anterioposterior cell fates in *Caenorhabditis elegans* by *Drosophila* Hox proteins. *Nature* **377**: 229–232.

International Human Genome Sequencing Consortium. (2004) Finishing the euchromatic sequence of the human genome. *Nature* **431**: 931–945.

Kenyon, C. (1994) If birds can fly, why can't we? Homeotic genes and evolution. *Cell* **78**: 175–180.

Larsen, W.J. (2001) *Human Embryology*, 3rd edn. Churchill Livingstone: Edinburgh.

Malicki, J., Schughart, K. and McGinnis, W. (1990) Mouse *Hox-2.2* specifies thoracic segmental identity in *Drosophila* embryos and larvae. *Cell* **63**: 961–967.

Malicki, J.LC., Cianetti, C., Peschle, C. and McGinnis, W. (1992) A human HOX4B regulatory element provides head-specific expression in *Drosophila* embryos. *Nature* **358**: 345–347.

Manak, J.R. and Scott, M.P. (1994) A class act: conservation of homeodomain protein functions. *Development* (suppl: *Evolution of Developmental Mechanisms*): 61–71.

Manié, S., Santoro, M., Fusco, A. and Billaud, M. (2001) The RET receptor: function in development and dysfunction in congenital malformation *Trends Genet.* **17**: 580–589.

Marsh, J. (ed.) (1994) *Neural Tube Defects.* Ciba Foundation Symposium, vol 181. Wiley: Chichester.

McGinnis, W. and Krumlauf, R. (1992) Homeobox genes and axial patterning. *Cell* **68**: 283–302.

McGinnis, W., Garber, R.L., Wirz, J., Kuriowa, A. and Gehring, W. (1984) A homologous protein-coding sequence in *Drosophila* and its conservation in other metazoans. *Cell* **37**: 403–409.

McLachlan, J.C. (1986) Self-assembly of structures resembling functional organs by pure populations of cells. *Tissue Cell* **18**: 313–320.

Monk, M. (1994) The value of man–mouse homology in human embryology. In *Early Fetal Growth and Development*, Ward, R.H.T. Smith, S.K. and Donnai, D. (eds). Royal College of Obstetricians and Gynaecologists: London; 63–74.

Montcouquiol M, Rachel RA, Lanford PJ, Copeland NG *et al.* (2003) Identification of Vangl2 and Scrb1 as planar polarity genes in mammals. *Nature* **423**: 173–177.

Nybakken, K. and Perrimon, N. (2002) Hedgehog signal transduction: recent findings. *Curr. Opin. Genet. Dev.* **12**: 503–511.

Opitz, J. (1985) The developmental field concept. *Am. J. Med. Genet.* **21**: 1–11.

Polifka, J.E. and Friedman, J.M. (2002) Medical genetics: 1. Clinical teratology in the age of genomics. *Canad. Med. Assoc. J.* **167**: 265–273.

Quiring, R., Walldorf, U., Kloter, U. and Gehring, W.J. (1994) Homology of the *eyeless* gene of *Drosophila* to the *Small eye* gene in mice and *Aniridia* in humans. *Science* **265**: 785–789.

Reed, G.B., Rajan, K.T., Ballard, P.L., Shephard, T.H. and Wong, L. (1995) The uses *of* human embryonic and fetal tissues in treatment and research. In *Diseases of the Fetus and Newborn*, 2nd edn, Reed, G.B., Claireaux, A.E. and Cockburn, F. (eds). Chapman & Hall: London; 389–397.

Rodriguez-Esteban, C., Tsukui, T., Yonei, S., Magallon, J. *et al.* (1999) The T-box genes Tbx4 and Tbx5 regulate limb outgrowth and identity. *Nature* **398**: 814–818.

Roelink, H., Augsburger, A., Heemskerk, J. *et al.* (1994) Floor plate and motor neuron induction by vhh-1, a vertebrate homologue of hedgehog expressed by the notochord. *Cell* **76**: 7611–7775.

Scott, M.P. (2000) The natural history of genes. *Cell* **100**: 27–40.

Struhl, G., Struhl, K. and Macdonald, P.M. (1989) The gradient morphogen bicoid is a concentration dependent transcriptional activator. *Cell* **57**: 1259–1273.

Strutt, D. (2003) Frizzled signalling and cell polarization in *Drosophila* eye. *Development* **130**: 4501–4513.

Sykes, B. (1993) Introduction to medical genetics. In *Connective Tissue and Its Heritable Disorders*. Wiley-Liss: New York; 7–50.

Tada, M., Concha, M.L., and Heisenberg, C.P. (2002). Non-canonical Wnt signalling and regulation of gastrulation movements. *Semin. Cell Dev. Biol.* **13**: 251–260.

Takeuchi, J.K., Koshiba-Takeuchi, K., Matsumoto, K., Vogel-Hopker, A. *et al.* (1999) Tbx5 and Tbx4 genes determine the wing/leg identity of limb buds. *Nature* **398**: 810–814.

Warburton, D. and Fraser, C. (1964) Spontaneous abortion risks in man: data from reproductive histories collected in a medical genetics unit. *Am. J. Hum. Genet.* **16**: 1–15.

Wolpert, L. (1991) *The Triumph of the Embryo.* Oxford University Press: Oxford.

Wolpert, L. (2002) *Principles of Development.* Oxford University Press: Oxford.

2

Uses of Databases in Dysmorphology

Michael Baraitser

What is a syndrome?

A syndrome is defined as a group of malformations that tend to occur together. One of the main tasks of a clinical geneticist, and especially a dysmorphologist, is the recognition of these syndromes. There are a number of reasons why making a precise diagnosis is essential:

1. It is necessary to establish whether the combination of malformations is genetic and, if so, to determine the mechanism of inheritance, enabling assessment of the subsequent recurrence risks.

2. A precise diagnosis is needed in order to establish a prognosis and to direct the clinician to investigate other organs that might be involved.

3. A syndrome diagnosis may lead to prenatal diagnosis, with the possibility of increasing parental choice in a subsequent pregnancy.

4. Perhaps most importantly, people simply want to know the diagnosis. In many instances this opens up the possibility of meeting others and their families with the same condition. There are many lay societies dedicated to specific syndromes and parents can join them in order to discuss mutual problems. Many parents are much happier when they have a name for their child's problem, so that they can answer enquiring questions from family and friends.

Embryos, Genes and Birth Defects, Second Edition Edited by Patrizia Ferretti, Andrew Copp, Cheryll Tickle and Gudrun Moore

A clinical geneticist faces a number of problems in making a diagnosis, and this is mainly due to the large number of different and rare syndromes. There are now 2000–3000 known dysmorphic syndromes – 'known' in that they have been described previously in the literature, although often only as a single case report. Diagnosis is also difficult because syndromes are clinicaly variable. They are easier to define if they occur in siblings or in previous generations, especially if the combination of features is unusual, making the likelihood of separate conditions in the same family small. Definition is also easier if a specific diagnostic test allows the clinician to define the condition. However, in the case of some gene tests the same mutations in a given gene can give rise to what was previously thought to be two or three separate conditions. Do we define syndromes on clinical criteria, or do we lump conditions with the same mutation together under one name? Even more difficult is the recognition of a new syndrome that always occurs as a sporadic condition, as this necessitates the recollection of previous single case reports, some of which might have been written in the distant past. It is not unusual for the clinician to vaguely remember a previous patient with the same condition, either reported somewhere in the literature or seen previously at the clinic, but finding documentation can be a daunting task.

Reports of syndromes are widely spread among many different specialist journals. As familial occurrence of malformations has always intrigued clinicians, the literature stretches back for many years and there is difficulty in remembering, or finding, older reports. The majority of these conditions can only be recognized by using clinical and radiological criteria, so that there is a constant need to look at the details of previous cases of children thought to have the same problem.

Some of these problems are addressed by dysmorphology databases

Names

There are many names for the same condition. Names in the literature are often meaningless (the first author's name on the paper might be used) and it is not uncommon for syndromes to be reported in different journals under separate titles or for clinicians on different sides of the Atlantic to use different names for the same condition. A classic example is the Beckwith–Wiedemann or Wiedemann–Beckwith syndrome. It is also known as the 'EMG syndrome', denoting its main features as exomphalos, macroglossia and gigantism. This is not always useful, as these features might not always be present. The problem is resolved on the database by including all the different names as synonyms, so that the user will arrive at the same information irrespective of the name used. The syndrome list is indexed alphabetically but it is also possible to search using only part of the name in a 'keyword' search, for example 'Sheldon' of the Freeman–Sheldon syndrome will still access information about the condition.

Variability

Almost without exception, all syndromes are variable. The extent of the variability can usually be determined by looking at the same condition within a single family. Siblings who have, for instance, Laurence–Moon–Biedl (LMB) syndrome, which is characterized by post-axial polydactyly, mental retardation, obesity, renal abnormalities and a pigmentary retinopathy, might differ in that one sibling has the complete picture whereas the other might not, e.g the second sibling may have normal intelligence. It is likely that both siblings have the same genetic defect, as there is too much overlap for them to have different conditions; the boundaries of the syndrome can then begin to be delineated. Databases have no problem with this, provided that all described features are included in the feature list. If, using the above example, you search on polydactyly and mental retardation, the computer will at least remind you that Laurence–Moon–Biedl is still a possibility.

Expansion and contraction

New features are constantly being added. This might be because the syndrome is rare and the extent of the manifestations is not yet known. Alternatively, the underlying mechanism of causation is a chromosomal deletion, in which case a bigger or smaller piece missing might slightly alter the phenotype whilst retaining the facial features and hence the recognizable Gestalt. The underlying aetiology might also depend on the timing of an insult and result in similar, but not necessarily identical, phenotypes. Most databases can be readily modified and additions to the list of features can easily be made. Unlike textbooks, there is plenty of space and the whole procedure is flexible. The only problem is that any change might necessitate the reclassification of an enormous amount of data. For instance, if it is shown that children with cleft lip and palate can be usefully separated according to whether the cleft is on the left or the right, then recoding by rechecking all syndromes in the literature might be a necessary, but formidable, task. It is much easier if a new diagnostic tool, such as magnetic resonance imaging, becomes frequently used, since extra categories can just be added and only new data need to be entered.

Different severity leads to a different phenotype

In a given syndrome, even within a single family, a feature might differ in severity. For instance, a child presents with a radial aplasia. Many syndromes in which there are problems down the radial side of the limb are variable to the extent that in the same syndrome there might be a classic radial club hand, or a normal radius but an absent thumb, just a small thumb or, surprisingly, a duplicated thumb. All of these should be in the list of features of a condition called VATER syndrome, where the R stands for radial problems. The VATER syndrome is relatively common and the literature is large, which helps to determine the extent of the malformations. If the syndrome is less well delineated, the variability of the radial lesions might not be known (those updating the database can only put in what has been described) and diagnoses might be missed.

Effects of age

One of the other major problems is that syndromes can develop and change with the age of the proband. Some features are present at birth but are not recognized because they are not looked for, whereas others evolve with time. Consider the following scenario:

1. At birth a baby is found to have a post-axial polydactyly (an extra digit on the small toe side of the foot). At this stage, no conclusions can be drawn and, indeed, there might be nothing more to the situation than that. If the baby is black, then the experienced clinician might decide at this stage that this is a common, unimportant, autosomal dominant finding in those of African origin, and be cautiously optimistic. If the database were to be used at this stage, a total of 149 conditions would be displayed and each one would be found to have additional features.

2. Nine months later it might be noticed that the infant's development is slightly delayed and causing concern. If a search were done at this stage, there would be 53 possibilities. Certainly Laurence–Moon–Biedl syndrome should be considered but the evidence from the patient would be so far inconclusive.

3. At 14 months the child appears obese. The main problem at this stage is clinical judgement. The clinician might decide with some justification that the child is a little 'chubby' and not regard this new feature as significant. However, if the computer has prompted the clinician to consider Laurence–Moon–Biedl syndrome, then the weight gain might be considered in another light. If suspicion has now been raised, then it would be appropriate to ask for a renal scan to look for cystic dysplasia and to ask the ophthalmologists to investigate retinal function. If both series of tests are normal, the clinician might still want to wait.

4. At 2 years the retinal test would be repeated. If the retina is found to be abnormal, then the diagnosis is certain. If not, a further period of cautious waiting is appropriate.

During this process the computer has made other suggestions that need to be followed. In the list of possibilities, given an entry of mental retardation, obesity and a retinal dystrophy, there is Cohen syndrome. To date, no-one has described polydactyly in that condition, but all dysmorphic syndromes are rare and therefore the possibility has at least to be considered.

Time may alter the phenotype, and dysmorphologists are accustomed to recognizing facies at a particular age. For instance, the faces of patients with Williams syndrome change with age (the face becomes coarser) but provided that the feature list includes all the features, whatever the age, this should not be a problem. In addition, with the advent of pictures on the database, a sequential series of images showing these changes with age is very valuable. The database would also prompt the clinician to look (with DNA probes) for a deletion in Williams syndrome chromosome and a negative result would be a significant, finding against the diagnosis.

Importance of an individual feature

If it were possible to rank features (i.e. decide what are the essential features), according to how frequent they are found in reported cases, then this could be easily incorporated into the database and indeed some have tried to do this. If 100% of cases of TAR syndrome (AR standing for aplasia of the radius and the T for thrombocytopenia), despite having radial aplasia, always had the thumb present, then clearly a diagnosis could not be made if it were absent and the computer could be programmed to insist on this.

The classic way of approaching the problem of variability in recessive disorders has been elegantly demonstrated in a study of Meckel–Gruber syndrome by Fraser and Lytwyn (1981). The main features are polydactyly, polycystic renal disease and a posterior encephalocele. If a series of cases were to be reported by neurosurgeons, all patients would have a posterior encephalocele, as that would be the reason why they were seen by the neurosurgeons. To overcome this bias of ascertainment, Fraser and Lytwyn (1981) ignored the index case and looked only at the clinical features of the subsequently born siblings (sibs). In this way it was shown that 100% had polycystic renal disease, which was essential for the diagnosis. The other features could likewise be ranked. The problem is that many of the recessive syndromes are so rare that a sib–sib study has not been performed and data are not available. The other major problem concerning ranking is that many syndromes seem to be on the whole sporadic events, as seen with both De Lange syndrome and the CHARGE association. In the latter syndrome, patients might present to: a cardiologist (H is for heart defects); an ear, nose and throat specialist (A is for atresia choanae and E is for ear); or to a paediatrician for growth failure (R stands for retardation of growth or development). Thus, depending on who collects and reports the data, bias will enter and the ranking of individual features becomes a problem.

Where databases do not help

Familial resemblance

It is sometimes not possible to know whether or not one of the dysmorphic features is part of the clinical picture. For example, a patient with learning difficulties also has a bulbous nasal tip. The latter feature, however, is not part of the condition but is inherited from a parent who is perfectly normal but just happens to have a broad, bulbous nasal tip. Such an assessment needs clinical judgement, which is part of the art of dysmorphology. Computers will not help!

Unusual features in a patient

Good clinical judgement is essential for the appropriate use of databases. Consider a further scenario, using Laurence–Moon–Biedl (LMB) as an example. A patient has the following features: mental retardation, post-axial polydactyly, obesity, retinal dystrophy, scoliosis and renal cysts. The patient appears to have the LMB syndrome but has, in addition, a severe scoliosis. If the user includes this feature in the search, as well as all the more usual features, the correct diagnosis cannot be made by the computer, because the feature list attached to the LMB syndrome does not contain scoliosis. Is this therefore LMB? If most of the other cardinal features are present, yes; but this is a matter for clinical judgement. When a patient has, for instance, two extra malformations not previously recorded, it becomes difficult to know whether or not one is dealing with a new syndrome.

There is, however, a mechanism allowing the user to select the mandatory features from the features entered and then to search on a selection of the rest. Instead of selectively choosing 'good handles' to enter, the user can enter all of the features a patient has, but mark, in this case, mental retardation, post-axial polydactyly and retinal dystrophy as mandatory, asking the computer to search on these three with one or two of the other three non-mandatory features. If the user searches on any four out of the six features above, the correct diagnosis will also be made. If he/she searches on any three, then the correct diagnosis will still be made, but the list of possibilities becomes so long that the correct diagnosis might be hidden.

In summary, the problems faced by the clinician, and which have to be addressed by database design, are as follows:

1. Not only are syndromes rare, they are variable.

2. As there is often no test to confirm the diagnosis, precise criteria cannot be formulated.

3. Syndromes change with age.

4. Some dysmorphic signs are familial and not relevant.

All children with multiple disabilities need a chromosomal analysis and diagnosis may be confirmed in this way, without needing to use a database in order to find a match. Specialized blood tests, be they biochemical or molecular, are now starting to make a significant contribution to syndrome diagnosis, although mostly they concern 'common' rare syndromes, such as Smith–Lemli–Opitz or Williams syndromes.

Dysmorphology databases

Despite the problems outlined above, dysmorphology databases have a major clinical function in aiding doctors seeking a differential diagnosis for a patient presenting with a given array of signs. They can provide a list of all conditions with a combination, for example, of mental handicap, deafness and retinal dystrophy. The emphasis tends to be on a differential diagnosis, 'a manageable list of possible diagnoses', and although this might not seem as desirable as an exact diagnosis, searching a database for an exact diagnosis may lead to the correct condition being missed. This is because many unique features, or 'handles', will need to be entered in order to retrieve only one possibility, and for this to happen other possibilities will be excluded. In clinical dysmorphology too many syndromes overlap and it is better to view a short list of the possible diagnoses and then reject those that seem on inspection not to fit.

There are a small number of dysmorphology and related databases that have been developed over the past 20 years to help solve some of these clinical problems. Databases currently available in this field include:

- The Online Mendelian Inheritance in Man (OMIM) database (McKusick, 2004; www.ncbi.nih.gov/Omim).
- The London Dysmorphology Database (LDDB) (Winter and Baraitser, 2003) with its partners, the London Neurogenetic Database (Baraitser and Winter, 2003) and Geneeye, an ophthalmo-genetics database (www.lmdatabases.com).
- POSSUM, an Australian-based dysmorphology database (Bankier, 2003; www.possum.net.au).
- REAMS, a radiological database developed by Christine Hall and John Washbrook (2000).
- The Human Cytogenetics Database (Schinzel, 2004).

Except for OMIM, which is web-based, all these databases are commercially available. OMIM is based on the McKusick catalogue and includes all conditions showing Mendelian inheritance. It therefore excludes many of the sporadic syndromes. It can only be searched by means of keywords, and there is no search strategy for diagnosing a syndrome by using features that are not included in the abstract or reference title. It is an excellent source of reference if the diagnosis is already known and the volume of data therein is enormous.

The LDDB and POSSUM are widely used and largely cover the same ground. Both contain a comprehensive, alphabetically indexed syndrome list, with each syndrome entry attached to an abstract, a list of features found in the condition, the related references from the literature, a mode of inheritance, gene localization (if known) and a McKusick number. The abstract describes the main clinical features, so that every case suggested in the differential diagnosis need not be consulted in detail. It also includes a discussion of other similar conditions and whether the inheritance pattern is uniform or the condition is heterogeneous. Most will have the latest molecular data, with gene localization, if known, and have links to other databases, such as OMIM. Databases can now provide, on CD-ROM or videodisc, an accompanying set of photographs of patients with any condition. In instances where these have been published, this is especially useful, since visual clues are most important in dysmorphology.

Most databases have inclusion and exclusion criteria. For the LDDB it was decided to include all clinical reports of patients with multiple malformations, be they clearly genetic or simply sporadic. It was thought unnecessary to include the dysmorphology of all the chromosomal deletions and duplications, as these conditions are usually diagnosed on cytogenetic analysis. For instance, a computer was not thought to be necessary to diagnose cri-du-chat syndrome (*5p* deletion). However, this is changing and clinicians have become aware that the cytogenetic laboratory might only detect certain deletions if the clinician gives guidance on where to look. The classic example of this is the Wolf–Hirschhorn syndrome (*4p* deletion), in which the deletion at the tip of the short arm is so small that it could be missed by routine cytogenetics and only detected after that region is intensively studied. An even finer degree of resolution is needed for some other deletions; for example, the deletion now known to be involved in Williams syndrome can only be detected by fluorescence *in situ* hydridization (FISH; see Chapter 3). Subtelomeric probes have also changed clinical practice, and LDDB now includes clinical information on syndromes that can only be diagnosed with subtelomeric probes. There is, therefore, a need either to include cytogenetic microdeletions on a dysmorphology database or to establish a separate database for these. The problem has largely been solved by the creation in Zurich of the Human Cytogenetics Database (Schinzel, 2004), using similar programs to those used in the LLDB.

How databases work

Features or 'handles'

For a dysmorphology database to be useful, a comprehensive list of dysmorphic features needs to be constructed covering every possible malformation. The list of malformations can then be ordered in such a way that they can be accessed system by system or by entering a keyword. A thesaurus can be incorporated to list similar features if the one being looked for is not found. If, for example, the user enters the feature 'ante-mongoloid eye-slant', which is synonymous with 'down-slanting palpebral fissures', the thesaurus will link the two terms together.

In the LDDB, the feature list is biased towards signs rather than symptoms, as the former are more important to the dysmorphologist. As a result, certain features will not be found. Some, such as diarrhoea, might occasionally be an important feature of a condition and are included in the feature list, as is 'headache' in the neurogenetics database. It is still possible to search on a symptom if it appears in the syndrome title or abstract. Similarly, 'vomiting' and 'abdominal pain' are excluded, as these are of more importance to the paediatrician than to the dysmorphologist. The handles used in making a diagnosis are changing all the time. For instance, the behavioural phenotype is an important part of Williams syndrome, in that the children are friendly and tend to want to engage with strangers. Angelman syndrome was previously called the 'happy puppet' syndrome in order to emphasize the happy disposition that is an integral part of the diagnosis. It might be that, in the future, some mentally handicapped, non-dysmorphic children will only be characterized by patterns of behaviour.

Features are accessed in two main ways. If the patient has cataracts, then it is quickest to simply type in the word in order to perform the search. However, it is often safer and easier to browse through the feature list. The LDDB has a three-tier system. The initial subdivisions are 'build', 'stature', 'head', 'neck', 'ears', 'eyes', 'thorax' and so on. Each of the above is then broken down into the next level of complexity. For example, the first level might be 'eye'. A search on this will search for anything abnormal with the eye.

The next level divides the eye into:

- Anterior chamber.
- Conjunctiva.
- Cornea.
- Iris.
- Retina, etc.

A search on this level will search on anything wrong with the anterior chamber, the cornea, or whichever subdivision is chosen. If the user is sure that the ophthalmic problem is a coloboma of the iris, then it is better to search at the third level, which looks like this:

- Iris.
- Aniridia.
- Brushfield spots.

- Coloboma of the iris.
- Heterochromia of the iris.
- Pigmentary abnormalities of the iris.
- Iris atrophy/dysplasia.

The search

By far the most powerful attribute that databases have is the ability to search and this function should be sufficiently versatile to allow searching in a number of different ways:

- A combination of features.
- Features combined with an inheritance pattern, e.g. any X-linked disorder with a cleft palate and deafness.
- Keywords.
- An author and a keyword; it is possible that the user has a diagnosis in mind but can only remember the first author's name, or only the journal in which it was published and possibly a rough idea of the year. The programs allow a search on any of these variables.

A search in LDDB can be made at any of the three levels of codes, or a on a combination of all three. Each feature is put in a separate box and the computer will view this as the clinician asking for all syndromes with this, AND that, AND that, if three features are entered. Looking at the breakdown of codes in the previous section, the user might simply choose 'coloboma of the iris' and perform a search using this single criterion, but if 'cataract' is added into another box, the computer will search for all conditions with a coloboma plus a cataract.

There can be difficulties with the definition of features (or 'handles'). As an example, take the case of a child with Coffin–Lowry syndrome. The main clinical features are mental retardation (a reasonable handle, but there are approximately 1000 syndromes with this feature), downslanting palpebral fissures and a prominent lower lip. But there are also features of 'full lips', 'everted lips' and 'prominent lips', and this can be confusing. If you search on 'full lips', those conditions with thick lips or everted lower lips will be missed and one can never be certain whether the lip in the original case reports was correctly described. This problem can be overcome by using a search function that allows you to search on 'either/or' and, in the example given above, the computer will locate all syndromes with prominent, full or everted lips, in combination with downslanting palpebral fissures and mental retardation. This problem would not, of course, have arisen if these features had not been presented as separate categories, but some would argue that there is a difference between full lips and prominent lips. The 'either/or' technique is used in a similar way to the three-tier system, that is, by exploiting the ability to search on a general category such as 'lip-general', the second-level tier incorporating anything to do with the lip. Similarly, a patient is short but the clinician might be uncertain which limb segment is affected. The user could then search using 'either rhizomelic or mesomelic or acromelic' to cover all possibilities or, alternatively, merely search on 'short stature-general', which will pick up everything to do with short stature. Both strategies will give the same end result.

The search strategy must be focused

Database searches are not useful if the user loads the search with non-essential trivia. It is necessary to pick out the essential dysmorphic features, that is, the gross and unusual features. Absent fingers are strikingly unusual features. Extra fingers and toes are gross and unusual (for gross and unusual, the words 'good handles' can be substituted), provided that family background is taken into account. Syndactyly between toes 2 and 3 is an important feature in a condition called Smith–Lemli–Opitz syndrome, but in this condition there are also severe mental retardation and genital problems. In fact, syndactyly between toes 2 and 3 is a common familial trait of no particular significance and is therefore not a particularly good handle in the vast majority of situations in which it is encountered. It is therefore of little use to detail all the abnormal features, starting as some do at the top of the body and working in an obsessively thorough way to the bottom, and then present this list to the computer–far better to look carefully at everything and then select out the best handles before using the database.

The order of entry

No order is prescribed, but in general the most unusual feature should be entered first. If this is, for example, arrhinia (an absent nose), then there is little point in following this with three or four more features, as there are only two or three syndromes known in which the nose is totally absent and it would be worthwhile looking at all three. When one browses through the feature list, the computer displays the number of syndromes to which each feature is attached, so a user entering 'arrhinia' will realize that only a short differential diagnosis list is going to be generated. Furthermore, whatever the variability of the condition, if another condition matches on four other features but does not have an absent nose as a feature, then it probably is not worth looking at, as one would guess that arrhinia is such a cardinal feature that it should be present. However, if 'absent nose' is only providing a very small list of syndromes, and none of these gives a good match for the patient under consideration, the next step is to try 'hypoplastic nose'. The user can then go to the relevant references and look at the pictures in the published papers, since a very small nose might just have the same significance as an absent nose if all other features match. Thus, in order to overcome the variability problem, it is best not to be too precise.

Essential criteria

Having said that a good handle is an unusual feature not common to many conditions, there are exceptional features which, although common, are essential in dividing children into broad categories, and these should always be entered. Mental retardation is one of these, and severe short stature is another. Mild short stature (someone just under the 3rd centile for height) might not be an essential handle but, if someone with a syndrome is very short indeed, then this is important and is obligatory to the diagnosis.

In general, those conditions most likely to be diagnosed by the computer are those that are strikingly dysmorphic. In this situation the differential diagnosis list will be short. A child with microphthalmia (small eyes) and a smooth brain (lissencephaly) will produce a list of eight possible diagnoses, whereas microphthalmia combined with mental retardation is much less unusual and the list is long. The experienced clinician will manipulate the feature list until satisfied that all hope of making a diagnosis has gone. If, for instance, a child is born with no eyes (rather than small eyes) and a smooth brain, and the search for syndromes with a combination of anophthalmia and lissencephaly reveals none, then the clinician must think of the possibility that 'small eyes' are in the spectrum of 'no eyes' and should change the search to use microphthalmia in place of anophthalmia.

The role of pictures

There are many disabled individuals, especially those with mental retardation, in whom the handles or dysmorphic features seem very mild or subtle. Simply entering on to the database a combination of a 'big head' and 'mental retardation' is not a useful search strategy, as there are over 100 syndromes with this combination. However, by viewing the visual records, the eye can detect subtle similarities and differences and this phenomenon of 'Gestalt' recognition, together with textual information, can allow a diagnosis to be made. Pictures are therefore important and, as dysmorphology is essentially a visual subject, most databases will have a method of displaying pictures that can be accessed by syndrome. The system is especially useful when the original pictures cannot be viewed because the local library does not carry the relevant journals. Literally thousands of images can be entered and, with hard disk capacity and storage on DVD so much greater these days, the modern dysmorphology databases will contain 10 000–50 000 pictures. Clearly, electronic means of image archiving, together with the advent of digital cameras for the creation of clinical records, facilitate the ready incorporation of visual records into databases. Furthermore, electronic means of communication allow images to be transmitted between clinical centres and the use of the Internet could permit (regulated) access to databases from a distance. In recent editions of LDDB, LNDB and POSSUM there is a facility for storing one's personal collection of pictures and moving pictures around, so that pictures attached to different syndromes can be compared. Changes in the field are rapid; the nature of dysmorphology databases, and the ease with which we use them, are likely to evolve dramatically as new modes of information technology are developed.

References

Bankier, A. (2003) *POSSUM*. The Murdoch Institute, Royal Children's Hospital: Victoria, Australia.

Baraitser, M. and Winter, R.M. (2003) *London Neurogenetic Database*. London Medical Databases: London.

Fraser, F.C. and Lytwyn. A. (1981) Spectrum of anomalies in the Meckel syndrome, or: 'maybe there is a malformation syndrome with at least one constant anomaly'. *Am. J. Med. Genet.* **9**: 67–73.

Hall, C. and Washbrook, J. (2000) *Radiological Electronic Atlas of Malformation Syndromes (REAMS)*. Oxford University Press: Oxford.

McKusick, V.A. (2004) *Online Mendelian Inheritance in Man*. Johns Hopkins University Press: Baltimore, MD.

Schinzel, A. (2004) *Human Cytogenetics Database* (in press).

Winter, R. M. and Baraitser, M. (2003) *London Dysmorphology Database*. London Medical Databases: London.

3

Human Cytogenetics

J. D. A. Delhanty

Introduction

At birth, at least 1% of humans have a clinically significant chromosomal abnormality. Important though these surviving cases are, in terms of clinical, economic and social effects, they represent a small fraction of those present in early developmental stages. By the time of birth, natural selection has eliminated the vast majority of abnormal embryos. At conception, aneuploidy (extra or missing chromosomes) may affect any chromosome but only trisomies of the sex chromosomes or of autosomes 13, 18 or 21, or monosomy of the X, are to some extent compatible with survival to the end of pregnancy. Some indication of the high levels of fertilization failure, gametic abnormalities or errors in embryogenesis that result in inviability prior to implantation is given by the observation that in humans the fecundity rate (probability of achieving a clinically recognized pregnancy within a monthly cycle) is about 25% (Wilcox *et al.*, 1988). This figure was derived from studying a group of 220 women, 95% of whom were under 35 years of age and fertile, who were attempting to conceive. In this group of relatively young women the rate of clinically recognized miscarriages was only 9%, but pregnancy loss before this stage was more than double this figure. More recent studies support these findings, suggesting that in young, unselected couples who are trying to conceive, 20–25% should be successful each monthly cycle (Bonde *et al.*, 1998; Edwards and Brody, 1995). This compares with an average of 70% in captive baboons, for example (Stevens, 1997). Interestingly, the implantation rate after *in vitro* fertilization (IVF) at best averages around 20% per embryo transferred (Edwards and Beard, 1999). Evidence is steadily accumulating to prove that the major cause of implantation failure in humans after both *in vivo* and *in vitro* fertilization is the high incidence of chromosomal abnormality.

Embryos, Genes and Birth Defects, Second Edition Edited by Patrizia Ferretti, Andrew Copp, Cheryll Tickle and Gudrun Moore

For all age groups, clinically recognized pregnancy loss is usually quoted as 15–20%. It is this fraction of failed pregnancies that has been extensively studied cytogenetically and in which a chromosome anomaly rate of at least 50% has been found (Hassold, 1986). This contrasts with a figure of 5% in stillbirths, illustrating clearly the *in utero* selection process that eliminates 95% of chromosomally unbalanced conceptions. Clinical prenatal diagnosis can thus be seen as an extension of this natural process.

Combining data from cytogenetic studies of spontaneous abortions with those obtained from pre-implantation embryos suggests that chromosomal anomalies are present in 25% of conceptions, an order of magnitude higher than is found in other well-studied species, such as the mouse (Hassold and Jacobs, 1984; Jamieson *et al.*, 1994). Additionally, interphase fluorescence *in situ* hybridization (FISH) analysis of 3 day-old human embryos has shown that up to 50% are chromosomally mosaic, due to post-zygotic errors (Delhanty *et al.*, 1997; Munné *et al.*, 1998a), further increasing the chance of implantation failure.

Population cytogenetics

It is interesting to compare the known incidence of the various types of anomalies at different stages, comparing data on spontaneous abortions, stillbirths and live births (Table 3.1). These data are based upon large numbers of observations, over 56 000 in the case of live-born infants. Triploidy, the presence of a whole extra set of haploid chromosomes, occurs in 5–10% of early miscarriages and is almost totally lethal,

Table 3.1 Incidence of different trisomies at various stages of development

Trisomy (chromosome no.)	Spontaneous abortions (%) ($n = 4088$)	Stillbirths (%) ($n = 624$)	Live births (%) ($n = 56952$)	Live-born (%)
1–12	5.8	0.2	0	0
13	1.1	0.3	0.005	2.8
14	1.0	0	0	0
15	1.7	0	0	0
16	7.5	0	0	0
17	0.1	0	0	0
18	1.1	1.1	0.01	5.4
19	0	0	0	0
20	0.6	0	0	0
21	2.3	1.3	0.13	23.8
22	2.7	0.2	0	0
XXY	0.1	0.2	0.05	53.0
XXX	0.1	0.2	0.05	94.4
XYY	0	0	0.05	100
Mosaics	1.1	0.5	0.02	9.0

Data from Hassold and Jacobs (1984).

being very rare at birth. Absence of an autosome (monosomy) is clearly lethal very early on in life since, with the rare exception of an occasional monosomy for chromosome 21, none are found in the miscarriage data. X monosomy is thought to occur in 1% of conceptions but the incidence at birth is reduced to around 1 in 5000. Half of all chromosomally abnormal miscarriages are due to trisomy – the presence of an extra chromosome. There are clear chromosome-specific variations in incidence (Table 3.1). The larger autosomes (numbers 1–12) are under-represented; the one that stands out as most frequently involved is chromosome 16, followed by chromosomes 22, 21 and 15. Sex chromosome trisomies do not appear frequently in spontaneous abortion data, although almost half of conceptions with a 47,XXY karyotype do in fact miscarry, for reasons that are not well understood. This compares with X chromosome trisomy, with a survival rate of 94%, and 47,XXY, with 100% survival. For the autosomes, conceptions with trisomies of chromosomes 13, 18 and 21 are the only ones to survive to birth, to varying degrees. At birth, trisomy 21, leading to Down's syndrome, has an incidence of 1.3/1000, trisomy 13 (Patau syndrome) occurs in 0.05/1000, and trisomy 18 (Edward syndrome) in 0.1/1000. Even for Down's syndrome, the survivors represent less than one-quarter of those conceived, and for Patau and Edward's cases, a mere 3% and 6%, respectively, are survivors. Mosaic trisomies (conceptions with more than one chromosomally distinct cell line) are detected quite infrequently (1.1% of abortions, 0.02% of live-borns). This probably reflects that fact that analyses are carried out on limited tissue samples in the case of miscarried products and very few cells in the case of live-born infants; they are certainly underestimates.

Structural anomalies

Structural anomalies of the chromosomes are also common in the human population. These are caused by chromosome breakage and abnormal reunion, either following exchange of segments between non-homologous chromosomes (reciprocal translocations) or after two or more breaks within one chromosome that can lead to a shift in the position or reversal of the order (inversions) of the freed segment of chromatin. Robertsonian translocations are a particular type that involve chromosomes 13–15 and 21–22, the so-called 'acrocentrics', where the centromere is close to the end of the chromosome. The very short segments above the centromere carry little genetic information, except for ribosomal RNA sequences that are present on each of these chromosomes. Breakage at the centromeres of any of these chromosomes and reunion of the long arms with loss of the short arms is thus possible without deletion of unique genetic material. The net outcome is reduction of the chromosome number by one, but with no phenotypic effect.

Reciprocal translocations are carried by about 1/500 people; Robertsonian translocations as a group are slightly less common at about 1/1000, mostly affecting chromosomes 13 and 14 or 14 and 21. Chromosomal inversions are more rare; exact incidences are difficult to determine, as many remain undetected. The genetic effect

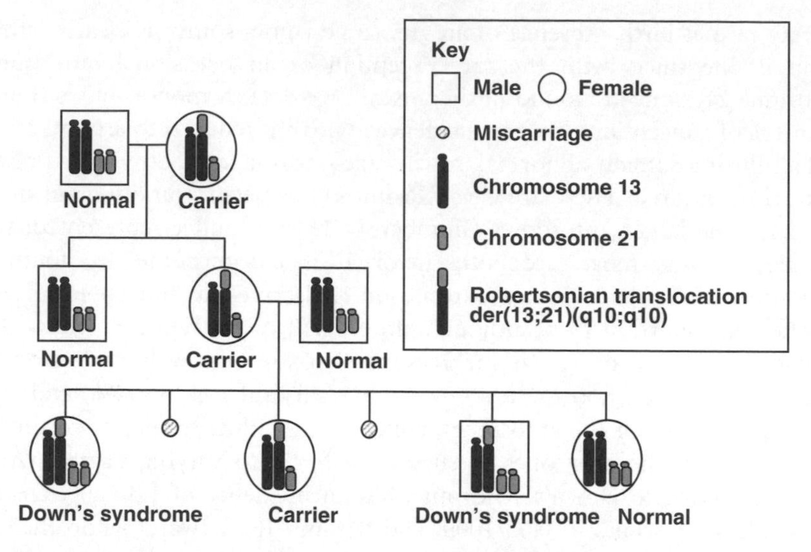

Figure 3.1 Robertsonian translocation between chromosomes 13 and 21, leading to a derivative chromosome, der(13;21), with loss of the short arms from both chromosomes. The derivative chromosome is present in three generations but the birth of infants with Down's syndrome is seen only in the third generation

of structural rearrangements is caused by the increased risk of the production of chromosomally unbalanced gametes after segregation of homologous segments at anaphase of the first meiotic division. The risk is difficult to quantify, as it is frequently unique to the family, but as a rule of thumb at least half the gametes of a carrier of a structural rearrangement are likely to be abnormal. The products of conception derived from such gametes will have a variable chance of survival, depending upon the amount of genetic material that is lost or gained. Carriers are often ascertained after the birth of an abnormal child or the occurrence of several miscarriages, but equally, many rearrangements may pass through several generations without apparent effect. Parents that carry Robertsonian translocations involving chromosome 21 are at increased risk of a conception with trisomy 21, leading to Down's syndrome (Figure 3.1). Risks of an abnormal birth are higher for female cariers (around 10%) than for males (1–3%). However, the presence in a parent of a Robertsonian translocation between the long arms of both chromosomes 21 precludes the formation of normal gametes, since each one will have either two copies of the chromosome or none at all. All live-born infants will therefore have Down's syndrome.

The genesis of chromosome abnormalities

There are essentially three developmental stages at which chromosome abnormalities may arise; gametogenesis, fertilization and embryogenesis. The process of gametogenesis in humans varies considerably between the two sexes. In males, each cell that enters meiosis produces four spermatozoa; the process is continuous, taking 64 days in all. Once past puberty, the male remains fertile into old age. In contrast, the human female is born with a complete set of oogonia – no more develop after birth. The initial stages of the first meiotic division take place early in fetal life but, after synapsis and recombination, each cell enters a period of arrest until after puberty. One egg then matures in each monthly cycle. Ovulation occurs when the oocyte is at metaphase II of meiosis and completion of the second division occurs after fertilization. Although there are several million oogonia at the outset, most are lost before birth and only a few hundred ever mature. Once the egg store is depleted, the menopause begins and the woman becomes infertile.

Errors arising during meiosis

The complexities of chromosome behaviour during the two meiotic divisions provide ample opportunity for errors to arise. Recombination between non-sister chromatids during prophase I has two functions – to recombine the genetic material and to ensure that synapsis persists long enough to allow proper alignment of the bivalent (paired chromosomes) on the metaphase spindle. In addition, cohesion needs to be maintained at the centromere of each homologous chromosome until the second anaphase, to prevent precocious separation of the two chromatids.

Molecular studies of the origin of trisomy using DNA markers are now available for over 1000 conceptions (Koehler *et al.*, 1996). Generally, errors at meiosis I of oogenesis predominate but there are notable exceptions. Among males with 47,XXY chromosomes (Klinefelter syndrome), the origin is almost equally divided between parental sexes, whereas over 80% of 45,X females lack a paternal sex chromosome (Hassold *et al.*, 1992). For the autosomes, a paternal origin is evident for a significant number of trisomies affecting the larger chromosomes, while for trisomy 18, maternal meiosis II errors predominate (Hassold *et al.*, 1996; Hassold and Hunt, 2001; see Table 3.2). The molecular studies also provide data on genetic recombination

Table 3.2 The parental origin of human trisomies determined by molecular analysis

		Paternal meiosis (%)		Maternal meiosis (%)		
Trisomy	Cases (*n*)	I	II	I	II	Mitotic (%)
2	18	28	–	54	13	6
7	14	–	–	17	26	57
15	34	–	15	76	9	–
16	104	–	–	100	–	–
18	143	–	–	33	56	11
21	642	3	5	65	23	3
22	38	3	–	94	3	–
XXY	142	46	–	38	14	3
XXX	50	–	6	60	16	18

Data from Hassold and Hunt (2001).

for the different trisomies. It is clear that aberrant recombination patterns are pertinent to the origins of human trisomy, but only a minority of cases are associated with complete absence of recombined chromatids. Reduced recombination is associated with all autosomal trisomies of maternal origin, as is advanced maternal age. There is most data for trisomy 21; particularly notable is a specific reduction in the number of proximal chiasmata (nearest to the centromere) in association with meiosis I errors, but an excess of recombination is reported for meiosis II errors. In general, the accumulating data suggest that the factors associated with non-disjunction of different chromosomes are very heterogeneous (Hassold and Hunt, 2001). Analysis of anaphase I in other species shows that short chromosomes with a single chiasma usually manage to separate, but long chromosomes with many cross-overs may have difficulty and may only succeed in the latter part of anaphase, providing a mechanism for chromosome loss by anaphase lag (White, 1954).

Studies on human gametes

The male gamete Over the past decade, FISH studies on human sperm have taken over from the far more labour-intensive method of fusing individual sperm with hamster eggs to allow visualization of the chromosome set. The use of multi-colour FISH to assess the copy number of two or three chromosomes at once has enabled chromosome-specific aneuploidy frequencies of 0.1–0.2% to be obtained (Hassold, 1998). Assuming that these rates apply to the entire complement, 1–2% of spermatozoa would be expected to have missing or additional chromosomes.

The female gamete Access to human oocytes is mainly limited to those that fail to develop following exposure to spermatozoa during IVF after ovarian hyperstimulation. These are from a selected population group, those with fertility problems, although not necessarily affecting the female. One advantage of oocytes is that they are at metaphase of meiosis II when obtained, allowing direct study of the chromosomal complement; this has allowed the accumulation of data from routine cytogenetic analysis over several years. The early data set, based on over 1000 oocytes from IVF clinics, showed aneuploidy rates as high as 20–25% (Jacobs, 1992). Due to the risk of artefactual loss of chromosomes when spreading a single metaphase, these overall rates were usually based upon doubling the hyperhaploidy rate (the presence of extra chromosome material). The assumption is that there will be an equal frequency of chromosome loss from the mature oocyte, an assumption that is not necessarily justified, given the current state of knowledge. More recent data, combining classical cytogenetic analysis with chromosome-specific analysis using FISH, suggest an overall aneuploidy frequency in oocytes of around 10% (Dailey *et al.*, 1996; Mahmood *et al.*, 2000; Pellestor *et al.*, 2002; Cupisti *et al.*, 2003). The apparent 10-fold increase in abnormality rate that is found for IVF oocytes compared with that for male gametes may also be true of ooctyes obtained from natural cycles. Ninety oocytes from unstimulated ovaries were studied by FISH analysis of four chromosomes, 16, 18, 21 and X, with 10 (unspecified) abnormalities detected (Volarcik *et al.*, 1998). However, results obtained for these particular chromosomes cannot be extrapolated to the entire set. There is evidence from FISH studies of a wider range of chromosomes (1, 9, 12, 13, 16, 18, 21 and X) that there is differential involvement of the larger and smaller pairs, with a significant excess of errors affecting the latter group (Mahmood *et al.*, 2000; Cupisti *et al.*, 2003).

Mechanisms of maternal aneuploidy Classically, aneuploidy of meiosis I origin was assumed to arise from the failure of (paired) homologous chromosomes to disjoin at anaphase I (non-disjunction). An alternative hypothesis was proposed by Angell (1991, 1997), based upon cytogenetic analysis of oocytes from an IVF programme. From her observation that oocytes contained additional or missing chromatids rather than whole chromosomes, Angell proposed that precocious separation of chromatids prior to anaphase I, with subsequent random assortment to the oocyte and first polar body, is the main mechanism of aneuploidy induction in the human female (Figure 3.2). Subsequent molecular cytogenetic analysis of IVF oocytes has shown that non-disjunction of whole chromosomes, as well as that of chromatids, does also occur (Mahmood *et al.*, 2000). The two modes of origin are genetically indistinguishable in their effects. The presence of unpaired, univalent

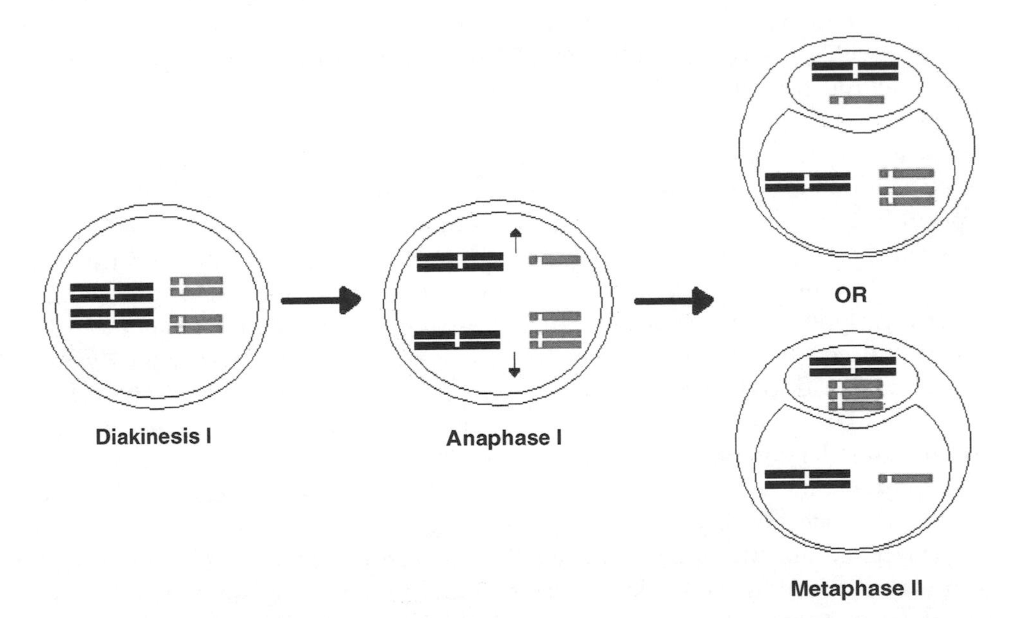

Figure 3.2 Diagram of female meiosis to illustrate premature separation of chromatids. Two pairs of homologous chromosomes are shown but one pair is not closely paired during prophase I of meiosis; this predisposes to precocious separation of the constituent chromatids of one of the unpaired chromosomes before the first anaphase. The separated chromatids can then migrate at random to the primary oocyte or first polar body, causing aneuploidy in the mature gamete

chromosomes has been shown to be a factor predisposing to aneuploidy in the mouse (Hunt *et al.*, 1995). Such chromosomes can either segregate intact, and randomly, to the spindle poles or can divide precociously into the component sister chromatids. Univalent chromosomes may exist at metaphase I because of pairing or recombination anomalies in a normal (disomic) oogonium, but they may also occur with greater frequency if the cell is originally trisomic. Fully trisomic individuals that reproduce are very rare, but gonadal mosaicism for a trisomic cell line in an otherwise normal individual may be more frequent than has been realized. The ability to use molecular cytogenetic techniques for specific chromosomal analysis of the oocyte and the corresponding first polar body has, for the first time, provided cytological evidence of gonadal mosaicism (Cozzi *et al.*, 1999; Mahmood *et al.*, 2000). In one case, a couple requested pre-implantation genetic diagnosis of trisomy 21, following three conceptions out of four with this aneuploidy. FISH analysis of cleavage-stage embryos again found three of four to be trisomic for 21. Analysis of four unfertilized oocytes showed that one had the normal, single copy of chromosome 21 present, one had an additional 21 chromosome, and two had additional chromatid 21s. The first polar body was available for one of these; this also showed an additional chromatid, proving that the precursor cell was trisomic and hence proving gonadal mosaicism (Cozzi *et al.*, 1999). In the second report, evidence for unsuspected gonadal or germinal mosaicism involving chromosomes 13 and 21 was found in two IVF patients (Mahmood *et al.*, 2000). Mosaicism for a trisomic cell line that affects only the gonads may in fact be relatively common. Analysis of many thousands of chorionic villus samples (CVS) has found discrepancies between the karyotype of the fetus and the placental tissue in 1–2% of cases; many of these involve trisomy confined to the placenta. Embryologically, the primordial germ cells are related to the chorionic stroma, suggesting that conceptuses diagnosed with confined placental mosaicism may be at increased risk of gonadal mosaicism. This suggstion was given support by the report of a case in which trisomy 16 was found in 100% of cells from cultured CV stroma, but all fetal tissues examined were found to be disomic, with the sole exception of the oocytes, 25% of which were trisomic (Stavropoulos *et al.*, 1998). Apart from pre-existing gonadal mosaicism, germinal trisomic mosaicism may arise during the early mitotic divisions of the female germ cells. Such anomalies would sporadically lead to the production of oocytes with extra or missing chromosomes. Direct evidence for this suggestion has been obtained by FISH studies of metaphase II oocytes and the corresponding first polar body (Cupisti *et al.*, 2003; Pujol *et al.*, 2003). Germinal or gonadal mosaicism would lead to an increased risk of an aneuploid conception irrespective of maternal age.

Metaphase II oocytes are also frequently observed to contain two or more well-separated chromatids. This is known as 'balanced pre-division', since there is as yet no imbalance, but clearly there is the potential for unbalanced segregation at anaphase II, after fertilization. Overall, meiosis in the female is obviously more error-prone than in the male; this could result from the lack of checkpoint control at the metaphase–anaphase transition in female mammals, a suggestion for which some experimental evidence exists in the mouse (LeMaire-Adkins *et al.*, 1997).

Fertilization Triploidy is frequent in humans, estimated to occur in 1% of conceptions (Hassold, 1986). Almost all triploid conceptions end in miscarriage during the first trimester of pregnancy, but a proportion of those where the additional haploid set is of maternal origin have a longer survival. About two-thirds of triploids are due to dispermy; the remainder are caused by failure to extrude the first or, more usually, the second polar body (Zaragoza *et al.*, 2000). The 45, X anomaly that causes Turner syndrome in live-born survivors is also present in about 1% of conceptions; again almost all miscarry. As stated previously, 80% lack the paternally contributed sex chromosome. However, the exact cause of the anomaly is not fully understood and may well frequently involve a fault at the time of, or soon after, fertilization, since the sperm data show that neither XY non-disjunction nor sex chromosome loss at meiosis appear to contribute to any significant extent.

Embryogenesis The advent of IVF over 20 years ago allowed access for the first time to the early human embryo. After fertilization *in vitro* on day 0, the embryo undergoes successive cleavage divisions, to consist of 6–10 cells by day 3 and maybe over 100 by day 5, when blastocyst formation occurs, with separation of the inner cell mass and the trophectoderm. The embryo proper is derived from the inner cell mass. Unlike mouse embryos, most human fertilized eggs in culture do not become blastocysts, but arrest in development at an earlier stage. Many studies employing routine karyotype analysis have been attempted on human pre-implantation embryos. At the cleavage stage, the embryo is analysed as a whole, after treatment to induce metaphase arrest. Usually one or two cells only are undergoing mitosis; the chromosomes are difficult to spread and not suitable for G-banding to enable precise identification. Nevertheless, it is readily apparent that there is a high frequency of chromosomal anomalies at this stage of development (Jamieson *et al.*, 1994). A technical advance was achieved by Clouston, who developed a way of obtaining good quality metaphases from blastocysts (Clouston *et al.*, 1997, 2002). G-banding was possible, proving that specific, widespread abnormalities of the chromosomes were quite compatible with development as far as the blastocyst stage, when implantation would be expected to occur *in vivo*.

Further advance was driven by the need to develop pre-implantation genetic diagnosis (PGD). By day 3 of development, it proved possible to remove one or two cells from the embryo and use these for molecular diagnosis (Handyside, 1991). Metaphase preparation was not technically possible, but the application of FISH analysis to interphase nuclei rapidly became the method of choice when sexing the embryo to avoid X-linked disease. Fluorescently labelled chromosome-specific DNA probes allowed the copy number of individual chromosomes to be determined for each cell. Very quickly it became apparent that chromosomal mosaicism, as well as aneuploidy, was rife in the day 3 embryo (Delhanty *et al.*, 1993). Spare embryos that were not transferred to the mother after diagnosis were spread whole and used for interphase FISH analysis with the same set of probes as had been used for diagnosis of sex, namely those specific for chromosomes X and Y. Later, a probe for chromosome 1

Table 3.3 Results of FISH analysis using probes for chromosomes X, Y and 1 of 93 normally developing spare pre-implantation embryos after PGD

	Normal	Abnormal	Diploid mosaic	Abnormal mosaic	Chaotic
Three patients	24	1	13	1	0
Four patients	21	1	5	3	24

Total mosaic embryos, 50%; total abnormal embryos, 52%.
Data from Delhanty *et al.* (1997).

was added as an extra indicator of ploidy, and this set was used also for the analysis of spare embryos from routine IVF cycles where there had been no embryo biopsy (Harper *et al.*, 1995). Embryos derived from both sets of patients showed the same types of abnormalities. They could be divided into different classes: completely diploid for the chromosomes examined; uniformly aneuploid; and mosaic. The mosaics could be further subdivided into those which were originally diploid but developed an aneuploid line by mitotic non-disjunction or chromosome loss, those that were originally aneuploid and similarly became mosaic, and a third group that were designated 'chaotic' because the chromosome content varied randomly from cell to cell with no discernible mechanism (Harper *et al.*, 1995). Other groups of researchers were simultaneously obtaining comparable results (Munné *et al.*, 1994). It was evident that, whichever set of three chromosome-specific probes was used, about 30% of embryos proved to be mosaic. In unselected IVF patients, about 5% came into the 'chaotic' classification but, with the greater numbers of spare embryos that are available from PGD patients, analysis showed that certain couples had a much greater tendency to produce embryos with these extreme abnormalities (Delhanty *et al.*, 1997; Table 3.3).

With the specific aim of screening for several aneuploidies simultaneously in older women undergoing routine IVF, interphase FISH employing up to eight chromosome-specific probes was developed (Munné *et al.*, 1998a). Mosaicism levels greater than 50% were then detected, raising the question of whether there were any human embryos created by IVF that had completely normal chromosomes by day 3 of development. The answer could only be obtained by finding a way to determine the chromosome constitution of every cell from a series of good quality human embryos at the cleavage stage.

Comparative genomic hybridization analysis of single blastomeres Interphase FISH analysis is severely limited by the number of chromosome-specific probes that can be used simultaneously to give reliable results. An altogether different molecular approach was needed, namely comparative genomic hybridization (CGH). Originally developed for use in cancer cytogenetics, when the tissue obtained cannot be readily induced to produce analysable metaphases, this is a DNA-based method employing FISH technology (Kallionemi *et al.*, 1992). Extracted DNA from the test sample is labelled with one fluorochrome, say green, while DNA from a normal male source is labelled with a second (red) fluorochrome. The DNAs are then competitively co-hybridized to prepared chromosome spreads from a normal male. If the test sample contains excess chromosome material, the corresponding chromosomes in the metaphases on the slide will show more green fluorescence, but if the test sample is lacking a particular chromosome, then that chromosome pair in the metaphases will show excess red fluorescence.

That is the basis of CGH in general but, for the purpose of obtaining information from single embryonic cells (blastomeres), a suitable way of first amplifying the whole genome to obtain sufficient DNA for analysis was required. A variety of methods were investigated in detail and the most appropriate was determined to be degenerate oligonucleotide-primed polymerase chain reaction (DOP-PCR). This gave reliable CGH results when tested in a blind study with trisomic fibroblasts and also provided sufficient DNA for 90 separate PCR analyses (Wells *et al.*, 1999). Twelve good-quality day 3 human embryos were then dissaggregated into single cells and the combination of DOP-PCR and CGH was then applied to obtain a comprehensive picture of the chromosome constitution of each individual cell. The results were remarkable (Wells and Delhanty, 2000). Most notable was that three of the embryos were completely euploid and had no chromosome imbalance. One was uniformly double aneuploid (trisomy 21 and X monosomy), one had three of four cells with chromosome 1 monosomy. Overall, eight were mosaic, of which two showed a 'chaotic' pattern. It seemed likely that, of the seven containing all or a majority of cells with abnormal chromosomes, four had a meiotic origin. All these types of abnormalities had been detected by interphase FISH analysis, but an unexpected finding was evidence for chromosome breakage in two embryos, with reciprocal products in sister cells in one case. In the same year, a comparable study was carried out in Australia, producing remarkably similar results (Voullaire *et al.*, 2000; see Table 3.4).

Table 3.4 Results of chromosome analysis in two series of human cleavage stage embryos by single cell CGH

	London, UK[1]	Melbourne, Australia[2]
Normal	3	3
Aneuploid throughout	2[a]	3[a]
Mosaic, at least 50% abnormal	3[a]	2[a]
Mosaic, less than 50% abnormal	2	3
Chaotic	2[a]	1[a]
Meiotic anomaly	4	3
Total mosaic	8	8
Total embryos	12	12

[a]Likely to be lethal.
Data from Wells and Delhanty (2000)[1]; Voullaire *et al.* (2000)[2].

Embryo survival

How important are these abnormalities, particularly mosaicism, for embryo survival? It is frequently stated that, since the embryo proper is derived from very few cells, the presence of a minority of chromosomally abnormal cells at the cleavage stage may not be important. While this may be true if the abnormal cells affect a very small proportion of the embryo, if greater than 50% of cells at the cleavage stage are abnormal, then it seems likely that the placental karyotype, and hence its function, will be affected. It is known that the presence of a normal cell line in the placenta greatly enhances the intrauterine survival of fetuses trisomic for chromosomes 13 and 18 (Kalousek *et al.*, 1989). In the two series of embryos analysed by single-cell CGH, it can be predicted with some confidence that in each case, five of the twelve embryos would have no chance of survival – those with complete or extensive monosomy, and those with chaotic chromosome complements or greater than 50% of cells with lethal abnormalities (Table 3.4). In each series, the three euploid embryos and those with less than 50% of cells abnormal would be predicted to have at least some chance of survival.

Some information on the survival capabilities of different types of abnormalities is provided by allowing embryos diagnosed (on the basis of a single cell analysed) as chromosomally abnormal to grow on in culture. In one study, about 20% of 247 embryos diagnosed as abnormal on day 3 by interphase analysis of chromosomes X, Y, 1, 13, 15, 16, 18, 21 and 22 (or otherwise designated unsuitable for transfer) survived to the blastocyst stage (Sandalinas *et al.*, 2001). Among the 50 blastocysts, 17 were aneuploid, 14 of which were trisomies. The three surviving monosomies were for chromosome 21 or the X. Mosaics with more than 60% abnormal cells and chaotic mosaics were not found among the best blastocysts, which consisted of more than 60 cells. Most of the embryos with a basically normal karyotype had a proportion of tetraploid cells; however, this is a common observation that may be a response to *in vitro* culture conditions but is frequently considered to reflect the process of trophectoderm development. A reasonable conclusion could be that monosomies other than for chromosomes 21 or X were likely to be lethal prior to the blastocyst stage, and that extensive mosaicism slowed development considerably, making successful implantation unlikely.

The cause of high levels of chromosome abnormality in human embryos

The incidence and type of post-zygotic errors leading to mosaicism that have been consistently observed in human cleavage stage embryos is totally unlike any observed in cultured somatic cells, suggesting that the mechanisms operating are peculiar to this stage of development. The observation that these embryonic cells in culture resemble tumour cells in terms of chromosome instability led to the suggestion that the normal cell cycle checkpoints are not operating during early cleavage (Delhanty and Handyside, 1995). Cell cycle checkpoints, first identified in yeast, would normally be expected to protect cells from genetic damage by ensuring that successive phases of the cell cycle and of mitosis are completed before the next is initiated (Hartwell and Weinert, 1989; Murray, 1992). In cancer cells or those transformed in culture, these checkpoints are often defective, allowing the sporadic accumulation of secondary chromosomal and other genetic defects. The human embryo is largely reliant on maternal transcripts until global activation of the embryonic genome at the 6–8 cell stage on day 3 (Braude *et al.*, 1988), whereas in the mouse, this takes place earlier, at the two-cell stage, possibly explaining the lack of such widespread mosaicism in that species.

Unfortunately, it is not possible to carry out similar studies on embryos from natural conceptions, but it is of great interest that the classical observations of Hertig *et al.* (1954) on *in vivo* fertilized embryos included a high proportion of cells with 'nuclear abnormalities' (binucleate cells, for example), a type of abnormality also frequently seen in embryos from IVF cycles. If the frequent occurrence of chromosomal mosaicism and chaotically dividing embryos also applies to *in vivo* conceptions, this may explain the relatively poor rate of embryonic implantation in the human species.

Relative parental risks – age, translocations, inversions, gonadal and germinal mosaics

In the population as a whole, the most important risk factor for a chromosomally abnormal conception is advanced maternal age. Among recognized pregnancies, the main association is with trisomy; there is no increased risk with age for triploidy or monosomy X, the risk for which is in fact increased in young women. When estimates of maternal age-specific rates of trisomy were calculated, the outcome suggested that in women aged 40 or more, the majority of oocytes may be aneuploid (Hassold and Chiu, 1985). The causes of age-related aneuploidy have been debated for many years and numerous hypotheses have been proposed but, although some experimental evidence has been obtained, a clear understanding of the problem remains elusive. Evidence obtained from studying recombination patterns of chromosome 21 in trisomic fetuses from younger and older women suggest that the types of susceptible configurations are similar in both age groups. This observation has led to the proposal of a 'two-hit' hypothesis, relevant at least for certain of the common trisomies (Hassold and Hunt, 2001). The first 'hit' is a recombination pattern of the type that is associated with an increased risk of non-disjunction, while the second 'hit' involves failure to resolve the difficulties created by the susceptible recombination, in some way related to the increased age of the meiotic cell. This could involve growth of the immature follicles, defective assembly of the spindle, or failure of the paired chromosomes to align correctly upon the equator of the spindle.

Couples at specifically increased risks of a chromosomal anomaly are those in which one partner carries a chromosomal rearrangement such as a translocation or an inversion, as described earlier. These couples require genetic counselling and in most cases appropriate prenatal diagnosis can be offered, ensuring that an ongoing pregnancy is chromosomally balanced. However, a minority of such couples experience repeated early miscarriages or primary infertility; for this sub-group pre-implantation diagnosis with selective transfer of embryos is appropriate (Conn *et al.*, 1998, 1999; Munné *et al.*, 1998b). It is of considerable interest that follow-up studies on embryos from such cases that are not transferred due to chromosomal imbalance have shown exceptionally high levels of mosaicism in these couples with fertility problems (Conn *et al.* 1998; Iwarson *et al*, 2000; Simopoulou *et al.*, 2003). These findings suggest that in this sub-group, couples are the victims of two pathologies, abnormal chromosome segregation at meiosis, due to the rearrangement, and an increased susceptibility to the factors that are responsible for high levels of mosaicism in human pre-implantation embryos.

A second group of couples at high risk of conceiving a chromosomally abnormal child are those in which one partner is a gonadal or germinal mosaic for a trisomic cell line. If 30% of the primary oocytes (or spermatocytes) are trisomic, then 15% of gametes formed would be expected to have an extra copy of the chromosome, since there is inevitable 'non-disjunction'. However, the evidence gained from pre-implantation diagnosis in one such case suggests that the risks are in fact higher than would be expected from classical considerations (Cozzi *et al.*, 1999). This is because the three copies of the chromosome may not associate in a trivalent in prophase of meioisis I, but as a bivalent and an unpaired monovalent (single chromosome). The monovalent is then more likely to undergo premature separation into its constituent chromatids and these may segregate at random, producing additional unbalanced gametes.

As explained earlier, germinal mosaicism may arise during mitosis in the pre-meiotic divisions of the germ cells and may affect one or several germ cells, again leading to a high-risk situation. Trisomic syndromes such as Down's that are due to chromosomal anomalies in a parent, such as a translocation or gonadal or germinal mosaicism, will occur independently of maternal age. Since most cases of this syndrome are in fact born to women who are not of advanced age, it is clearly important to understand the possible causes leading to a high risk of abnormality, so that counselling and prenatal diagnosis can be offered where appropriate.

References

Angell, R.R. (1991) Predivision in human oocytes at meiosis I: a mechanism for trisomy formation in man. *Hum. Genet.* **86**: 383–387.

Angell, R. (1997) First-meiotic-division nondisjunction in human oocytes. *Am. J. Hum. Genet.* **61**: 23–32.

Bonde, J.P.E, Ernst, E, Jenson, T.K., Hjollund, N.H.I. *et al.* (1998) Relation between semen quality and fertility: a population based study of 430 first-pregnancy planners. *Lancet* **352**: 1172–1177.

Braude, P., Bolton, V. and Moore, S. (1988) Human gene expression first occurs between the four- and eight-cell stages of pre-implantation development. *Nature* **333**: 459–461.

Clouston, H. J., Fenwick, J., Webb, A. L., Herbert, M. *et al.* (1997) Detection of mosaic and non-mosaic chromosome abnormalities in 6–8 day-old human blastocysts. *Hum. Genet.* **101**: 30–36.

Clouston, H.J., Herbert, M., Fenwick, J., Murdoch, A. P. and Wolstenholme, J. (2002) Cytogenetic analysis of human blastocysts. *Prenat. Diagn.* **12**: 1143–1152.

Conn C.M., Harper J.C., Winston R.M.L. and Delhanty J.D.A. (1998) Infertile couples with Robertsonian translocations: pre-implantation genetic analysis of embryos reveals chaotic cleavage divisions. *Hum. Genet.* **102**: 117–123.

Conn C.M., Cozzi J., Harper J.C., Winston R.M.L. and Delhanty J.D.A. (1999) Pre-implantation genetic diagnosis for couples at high risk of Down's syndrome pregnancy owing to parental translocation or mosaicism. *J. Med. Genet.* **36**: 45–50.

Cozzi J., Conn C.M., Harper J.C., Winston R.M.L. and Delhanty J.D.A. (1999) A trisomic germ cell line and precocious chromatid separation leads to recurrent trisomy 21 conception. *Hum. Genet.* **104**: 23–28.

Cupisti, S., Conn, C.M., Fragouli, E., Whalley, K. *et al.* (2003) Sequential FISH analysis of oocytes and polar bodies reveals aneuploidy mechanisms. *Prenat. Diagn.* **23**: 663–668.

Dailey, T., Dale, B., Cohen, J. and Munné. S. (1996) Association between nondisjunction and maternal age in meiosis-II human oocytes. *Am. J. Hum. Genet.* **59**: 176–184.

Delhanty, J.D.A., Griffin, D.K., Handyside, A.H., Harper, J. *et al.* (1993) Detection of aneuploidy and chromosomal mosaicism in human embryos during pre-implantation sex determination by fluorescent *in situ* hybridization (FISH). *Hum. Mol. Genet.* **2**: 1183–1185.

Delhanty, J.D.A. and Handyside, A.H. (1995) The origin of genetic defects in man and their detection in the human pre-implantation embryo. *Hum. Reprod. Update* **1**: 201–215.

Delhanty, J.D.A., Harper, J.C., Ao, A., Handyside, A.H. and Winston, R.M.L. (1997) Multi-colour FISH detects frequent chromosomal mosaicism and chaotic division in normal pre-implantation embryos from fertile patients. *Hum. Genet.* **99**: 755–760.

Edwards, R.G. and Beard, H.K. (1999) Blastocyst stage transfer: pitfalls and benefits. *Hum. Reprod.* **14**: 1–6.

Edwards, R.G. and Brody, S.A. (1995) *Principles and Practice of Assisted Human Reproduction.* W. B. Saunders: Philadelphia, PA.

Handyside, A.H. (1991) Biopsy of human cleavage stage embryos and sexing by DNA amplification. In *Proceedings of the First Symposium on Pre-implantation Genetics*, Verlinsky, Y. and Strom, B. (eds). Plenum: New York; 75–83.

Harper, J.C., Coonen, E., Handyside, A.H, Winston, R.M.L. *et al.* (1995) Mosaicism of autosomes and sex chromosomes in morphologically normal, monospermic pre-implantation human embryos. *Prenat. Diagn.* **15**: 41–49.

Hartwell, L.H. and Weinert, E.A. (1989) Checkpoints: controls that ensure the order of cell cycle events. *Science* **246**: 629–634.

Hassold, T.J. and Jacobs, P.A. (1984) Trisomy in man. *Annu. Rev. Genet.* **18**: 69–97.

Hassold, T. and Chiu, D. (1985) Maternal age-specific rates of numerical chromosome abnormalities with special reference to trisomy. *Hum. Genet.* **70**: 11–17.

Hassold, T.J. (1986) Chromosome abnormalities in human reproductive wastage. *Trends Genet.* **2**: 105–110.

Hassold, T., Pettay, D., Robinson, A. and Uchida, I. (1992) Molecular studies of parental origin and mosaicism in 45,X conceptuses. *Hum. Genet.* **89**: 647–652.

Hassold, T., Abruzzo, M., Adkins, K., Griffin, D. *et al.* (1996) Human aneuploidy: incidence, origin, and etiology. *Environ. Mol. Mutagen.* **28**: 167–175.

Hassold, T.J. (1998) Nondisjunction in the human male. *Curr. Top. Dev. Biol.* **37**: 383–406.

Hassold, T.J. and Hunt, P. (2001) To err (meiotically) is human: studies of the genesis of human aneuploidy. *Nat. Rev. Genet.* **2**: 280–291.

Hertig, A.T., Rock, J., Adams, E.C. and Mulligan, W. J. (1954) On the pre-implantation stages of the human ovum: a description of four normal and four abnormal specimens ranging from the second to the fifth day of development. *Contrib. Embryol.* **35**: 201–220.

Hunt, P., LeMaire, R., Embury, P., Sheean, L. and Mroz, K. (1995) Analysis of chromosome behaviour in intact mammalian oocytes: monitoring the segregation of a univalent chromosome during female meiosis. *Hum. Mol. Genet.* **4**: 2007–2012.

Iwarsson, E., Malmgren, H., Inzunza, J., Ährlund-Richter, L. *et al.* (2000) Highly abnormal cleavage divisions in pre-implantation embryos from translocation carriers. *Prenat. Diagn.* **20**: 1038–1047.

Jacobs, P.A. (1992) The chromosome complement of human gametes. *Oxf. Rev. Reprod. Biol.* **14**: 47–72.

Jamieson, M.E., Coutts, J.R.T. and Conner, J.M. (1994) The chromosome constitution of human pre-implantation embryos fertilized *in vitro*. *Hum. Reprod.* **9**: 709–715.

Kallioniemi, A., Kallioniemi, O.P., Sudar, D., Rutovitz D. *et al.* (1992) Comparative genomic hybridization for molecular cytogenetic analysis of solid tumors. *Science* **258**: 818–821.

Kalousek, D.K., Barrett, I. and McGillivray, B.C. (1989) Placental mosaicism and intrauterine survival of trisomies 13 and 18. *Am. J. Hum. Genet.* **44**: 338–343.

Koehler, K., Hawley, R., Sherman, S. and Hassold, T. (1996) Recombination and nondisjunction in humans and flies. *Hum. Mol. Genet.* **5**: 1495–1504.

LeMaire-Adkins, R., Radke, K. and Hunt, P.A. (1997) Lack of checkpoint control at the metaphase/anaphase transition: a mechanism of meiotic nondisjunction in mammalian females. *J. Cell. Biol.* **139**: 1611–1619.

Mahmood, R., Brierley, C.H., Faed, M.J., Mills, J.A. and Delhanty, J.D. (2000) Mechanisms of maternal aneuploidy: FISH analysis of oocytes and polar bodies in patients undergoing assisted conception. *Hum. Genet.* **106**: 620–626.

Munné, S., Grifo, J. and Cohen, J. (1994) Chromosome mosaicism in human embryos. *Biol. Reprod.* **51**: 373–379.

Munné, S., Magli, C., Bahce, M. *et al.* (1998a) Pre-implantation diagnosis of the aneuploidies most commonly found in spontaneous abortions and live births: XY, 13, 14, 15, 16, 18, 21, 22. *Prenat. Diagn.* **18**: 1459–1466.

Munné, S., Fung, J., Marquez, C. and Weier, H.U.G. (1998b) Pre-implantation genetic analysis of translocations: case-specific probes for interphase cell analysis. *Hum. Genet.* **102**: 663–674.

Murray, A.W. (1992) Creative blocks: cell cycle checkpoints and feedback controls. *Nature* **359**: 599–604.

Pellestor, F., Andréo, B., Arnal, F., Humeau, C. and Demaille, J. (2002) Mechanisms of non-disjunction in human female meiosis: the co-existence of two modes of malsegregation evidenced by the karyotyping of 1397 *in vitro* unfertilized oocytes. *Hum. Reprod.* **17**: 2134–2145.

Pujol, A., Boiso, I., Benet, J., Veiga, A. *et al.* (2003) Analysis of nine chromosome probes in first polar bodies and metaphase II oocytes for the detection of aneuploidies. *Eur. J. Hum. Genet.* **11**: 325–336.

Sandalinas, M., Sadowy, S., Alikani, M., Calderon, G. *et al.* (2001) Developmental ability of chromosomally abnormal human embryos to develop to the blastocyst stage. *Hum. Reprod.* **16**: 1954–1958.

Simopoulou, M., Harper, C.J., Fragouli, E., Mantzouratou, A. *et al.* (2003) Pre-implantation genetic diagnosis of chromosomal abnormalities: implications from the outcome for couples with chromosomal rearrangements. *Prenat. Diagn.* **23**: 652–662.

Stavropoulos, D.J., Bick, D. and Kalousek, D. (1998) Molecular cytogenetic detection of confined gonadal mosaicism in a conceptus with trisomy 16 placental mosaicism. *Am. J. Hum. Genet.* **63**: 1912–1914.

Stevens, V.C. (1997) Some reproductive studies in the baboon. *Hum. Reprod. Update* **3**: 533–540.

Volarcik, K. *et al.* (1998) The meiotic competence of *in vitro* matured human oocytes is influenced by donor age: evidence that folliculogenesis is compromised in the reproductively aged ovary. *Hum. Reprod.* **13**: 154–160.

Voullaire, L., Slater, H., Williamson, R. and Wilton, L. (2000) Chromosome analysis of blastomeres from human embryos by using comparative genomic hybridization. *Hum. Genet.* **106**: 210–217.

Wells, D., Sherlock, J.K., Handyside, A.H., Delhanty, J.D.A. (1999) Detailed chromosomal and molecular genetic analysis of single cells by whole genome amplification and comparative genomic hybridization. *Nucleic Acids Res.* **27**: 1214–1218.

Wells, D. and Delhanty, J.D.A. (2000) Comprehensive chromosomal analysis of human preimplantation embryos using whole genome amplification and single cell comparative genomic hybridization. *Mol. Hum. Reprod.* **6**: 1055–1062.

White, M.J.D. (1954) *Animal Cytology and Evolution*, 2nd edn. Cambridge University Press: Cambridge, UK.

Wilcox, A.J., Weinberg, C.R., O'Connor, J.F., Baird, D.D. *et al.* (1988) Incidence of early loss of pregnancy. *N. Engl. J. Med.* **319**: 189–194.

Zaragoza, M.V., Suri, U., Redline, R.W., Millies, E. *et al.* (2000) Parental origin and phenotype of triploidy in spontaneous abortions: prevalence of diandry and association with the partial hydatidiform mole. *Am. J. Hum. Genet.* **66**: 1807–1820.

4

Identification and Analysis of Genes Involved in Congenital Malformation Syndromes

Peter J. Scambler

Gene identification

Mapping of disease loci

The major steps involved in the identification of birth defect loci have not changed substantially since the first edition of this book in 1997, but several stages have been substantially accelerated by the advances accompanying the human (and other organism) genome projects. Candidate gene approaches to developmental disorders are becoming more common as the numbers of phenotypes obtained from gene targetings increases and developmental expression profiles become known. In these instances, the investigator would move straight to mutation screening. For instance, the Edinburgh Mouse ATLAS (EMAP) provides a series of three-dimensional models of mouse embryos at successive stages of development, linked to a standard anatomical nomenclature (http://genex.hgu.mrc.ac.uk/). The Jackson laboratories provide several search tools for retrieving expression data from GXD (the gene expression database: http://www.informatics.jax.org/menus/expression_menu.shtml). TBASE, also curated by the Jackson laboratories, allows searches of mouse mutations created predominantly by gene targeting (http://tbase.jax.org/; Anagnostopoulos *et al.*, 2001). However, it still remains usual for approximate localizations to be obtained using cytogenetic methods or genetic linkage analyses.

Embryos, Genes and Birth Defects, Second Edition Edited by Patrizia Ferretti, Andrew Copp, Cheryll Tickle and Gudrun Moore

Linkage analysis

Mendelian disorders are susceptible to linkage analysis if an appropriate pedigree or series of pedigrees is available. The loci underlying most of the more frequent Mendelian disorders have now been identified by positional cloning. Identification of the genes mutated in rare dominant disorders can be complicated by intergenetic heterogeneity and, in order to refine the disease interval, assumptions about penetrance are required to take account of informative meioses in unaffected individuals. Mapping of rare recessive gene loci is usually limited by the lack of families with multiple affected children. This difficulty can be circumvented by using consanguineous families coupled with homozygosity mapping. In practice, this involves a genome-wide linkage scan in order to identify a region of the genome inherited identically by descent (IBD) in affected individuals. Assuming no intergenetic heterogeneity, this is extremely powerful and, even in the presence of heterogeneity, a single large family of appropriate structure can establish linkage. It is often useful to have access to families with a range of inter-relatedness, using first-cousin matings to establish linkage and additional families with more distantly related parents to refine the region IBD. A discussion of optimization strategies in homozygosity mapping has been presented (Genin *et al.*, 1998). To date, the vast majority of homozygosity scans have been accomplished using short tandem repeat polymorphisms (STRPs). Increasingly, this strategy is being replaced by the use of single nucleotide polymorphism (SNP) microarrays, which are now available commercially. While the individual SNPs are less likely to be informative, the high density of the arrays more than compensates for this shortcoming, as has been demonstrated by the identification of a locus for neonatal diabetes (Sellick *et al.*, 2003). A two-step strategy to maximize cost-effectiveness might involve analysing affecteds using SNP arrays, then using STRP analysis of parents and unaffected sibs to confirm which candidate regions are IBD. Once confidence in the use of SNP microarrays increases, it is likely that the use of pooling strategies will enhance cost-effectiveness still further. An alternative but technically more demanding approach is to combine genomic mismatch scanning and comparative genome hybridization (CGH) microarray analysis to identify regions IBD without genotyping multiple individual loci (Smirnov *et al.*, 2004).

Smaller groups of SNPs can be used to refine a disease interval, and appropriate polymorphisms are easily found using websites such as the SNP database (dbSNP) at the NCBI (http://www.ncbi.nlm.nih.gov/SNP). They can also be displayed on human genome browsers, such as Ensemble (http://www.ensembl.org/). SNPs are likely to prove instrumental in the analysis of birth defects with complex genetic aetiology, e.g. predisposition to congenital heart defect, neural tube defect or cleft lip and palate (Blanton *et al.*, 2004). Here, genome-wide screens are likely to produce wide (10–30 cM) minimum genetic intervals. Fine mapping would then proceed with linkage disequilibrium mapping or case-control association analyses, using a high density of markers. The HapMap project aims to record patterns of sequence variation within the human genome and determine how this variation differs between racial groups (The International HapMap Consortium, 2003). Allelic association maps will facilitate the choice of markers (haplotype tags) that allow linkage studies to have the maximum power. While there are no examples to date, it is possible that SNP arrays representing these haplotype tags will be used to further the genetic analysis of birth defects. Combined linkage and cDNA microarray work has been used to identify candidate genes within specific genetic intervals, for instance in the detection of *Cd36*, a gene implicated in spontaneous hypertension in a rat model (Aitman *et al.*, 1999; Pravenec *et al.*, 2001). Similar approaches using mouse models for developmental disorders are easy to envisage (Dobrin and Stephan, 2003).

An illustrative example of a birth defect with complex inheritance is aganglionic megacolon, or Hirschprung's disease, where mutations in eight genes have been implicated in the disorder. It is worth noting that these genes, *EDNRB, EDN3, ECE1, SOX10, RET, GDNF, NRTN* and *ZFHX1B*, each have role in the development of the enteric nervous system and highlight the fact that genes operating with distinct biochemical or developmental pathways can produce a similar phenotype. Inbred populations should have lower genetic heterogeneity than outbred populations, and it was with this in mind that one team conducted a genome-wide association analysis in Old Order Mennonite families (Carrasquillo *et al.*, 2002). A multipoint linkage disequilibrium method was used to analyse data from over 2000 microsatellite and SNP loci identifying three susceptibility loci. An epistatic interaction between genes at two of these loci, *EDNRB* and *RET*, was postulated and mouse mutant crosses established to support this hypothesis. However, severe *RET* mutations seem sufficient to cause long segment disease, but milder mutations do so in conjunction with additional susceptibility encoded at a locus at 9q31 (Bolk *et al.*, 2000). Short segment HSCR involves three major susceptibility loci at 3p, 19q and 10q (*RET*) (Gabriel *et al.*, 2002). Thus, long segment disease is inherited in a predominantly autosomal dominant fashion, with reduced penetrance, whereas the short segment form is oligogenic. As an example of a SNP association study in a rare genetic disease, Emison and co-workers were able to identify a common, low-penetrance variant within an intron 1 enhancer or *RET* which makes a 20-fold greater contribution to risk than rare coding region alleles (Emison *et al.*, 2005).

In a similar vein, there are at least eleven loci causing the Bardet–Biedl syndrome (BBS) and in a small proportion of families the disorder is found is association with homozygous mutations at one locus and heterozygous mutations (or rare predisposition alleles) at another (Katsanis *et al.*, 2001). While some BBS genes share motifs, others do not, although a common theme underlying pathogenesis might be a role in basal body or primary cilial function (Ansley *et al.*, 2003). Thus, these congenital defects provide important paradigms for the study of complex genetic disorders and blur the distinction between Mendelian monogenic disorders and multifactorial conditions.

Chromosome analysis

Analysis of karyotype is part of the standard work-up of patients with congenital malformation, especially where the disorder is an 'unknown' syndrome or accompanied by learning difficulty and multiple dysmorphisms. The location of a disrupted gene or genes would be suggested by the position of a chromosome deletion or the breakpoints of any balanced translocation detected (see Chapter 3). Fluorescence *in situ* hybridization (FISH) is still a useful tool for the analysis of chromosome structure (Min and Swansbury, 2003). BAC clones are available for the entire human genome and specific sets of clones have been produced to detect rearrangements of certain regions of the genome, such as recurrent deletions or duplications, telomeres and centromeres. FISH can be used at different resolutions, depending upon the size and nature of the rearrangement suspected. Duplications can be particularly difficult to detect with metaphase chromosome spreads, for example, and interphase, extended chromatin or fibre FISH can be used to map rearrangements down to the level of kilobase pairs. Confirmation of rearrangement is often accomplished using southern analysis of DNA fractionated through standard or pulsed-field gel electrophoresis.

Classically, translocations and deletions help identify genes mutated in dominant disorders, but it is important to remember that such rearrangements may uncover a recessive mutation at the other allele. For instance, the gene mutated in Alstrom syndrome was identified following the observation that a t(2;11) translocation disrupted a gene carrying a loss of function mutation on the other allele (Hearn *et al.*, 2002). A t(5;11) translocation was identified in a patient with the Klippel–Trenaunay syndrome (KTS), a disorder involving diverse blood vessel malformations that may, for example, be associated with limb overgrowth. The translocation disrupted the *VG5Q* gene. *VG5Q* mutations appear to predispose to KTS, but not to be sufficient for its development. Thus, translocations can also help identify susceptibility alleles. It is also noteworthy that evidence that VG5Q is involved in angiogenesis came from a yeast two-hybrid screen that demonstrated an interaction with the angiogenic factor TWEAK, expression analysis showing transcripts in the vascular endothelium, as well as a bioassay (Tian *et al.*, 2004), providing an applied example of some of the functional approaches described in more detail below.

Translocations and deletions do not necessarily directly disrupt the gene whose function is affected by the rearrangement. In the aniridia and the campomelic dysplasia autosomal sex reversal syndromes, for instance, balanced translocations lie several hundred kb from the gene known to be haplo-insufficient in the disorder (*PAX6* and *SOX9*, respectively; Fantes *et al.*, 1995; Pfeifer *et al.*, 1999). Elegant work using YAC complementation of the *se* (*small eye*) mouse, a model for aniridia, demonstrated the presence of regulatory elements separated from the target gene by the translocation (Kleinjan *et al.*, 2001).

Rarely, a chromosomal or subchromosomal isodisomy can result in homozygosity for a recessive mutation. Detection of hetero- or homo-isodisomy is made following analysis of the inheritance of polymorphic markers. In an interesting example of this approach, the location of the Bloom's syndrome gene was refined following detection of maternal uniparental isodisomy for chromosome 15 (Woodage *et al.*, 1994) in a patient who also had features of Prader–Willi syndrome.

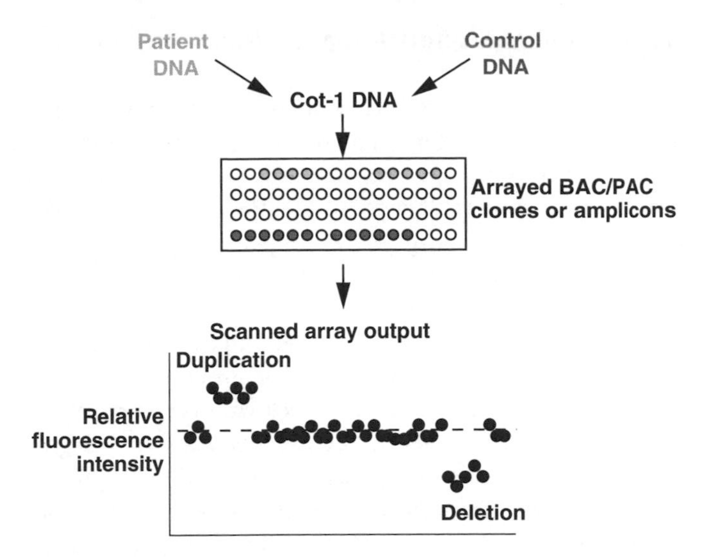

Figure 4.1 Array comparative genome hybridization. DNA from a control individual and patients are differentially labelled with fluorescent probes. After addition of C_0t1 DNA to prevent hybridization to repeat sequences, the mixture is hybridized to an array of genomic clones (or PCR-derived amplimers from those clones) on a microscope slide. Following washes, the slide is scanned and the relative fluorescence plotted across chromosomes to identify regions of the genome potentially deleted or duplicated

A major advance in the field of molecular cytogenetics is the development of array-based comparative genome hybridization (aCGH). In standard CGH, differentially labelled probes from patients and control genomic DNA are hybridized to metaphase spreads to detect deletion and duplication where there is a relatively higher or lower signal strength in one region of the genome (Figure 4.1). While useful in tumour cytogenetics, where the deletions and duplications are often large and/or involve more than a 50% dosage alteration, resolution for detecting rearrangements in birth defect syndromes was insufficient. The advent of a high-resolution physical map comprising overlapping tiles of BACs has allowed the array of clones representing the human genome at 1 Mb intervals, or giving coverage to specific chromosomes at higher resolution. Improvements to the resolution and removal of clones giving artefactual results will improve these arrays further, but already progress has been made in identifying novel deletions in patients with facial dysmorphism and learning difficulty (Shaw-Smith *et al.*, 2004; Vissers *et al.*, 2003) or non-deletion 22q11 DiGeorge/velocardiofacial syndrome (our unpublished data). Resolution may be improved by using flow-sorted chromosomes as template material for labelling (Gribble *et al.*, 2004). A tiling resolution array of overlapping BACs and mouse 1 Mb resolution array have recently been described (Ishkanian *et al.*, 2004; Chung *et al.*, 2004) and use of genomic clone microarrays is likely to have a major impact on the analysis of birth defect syndromes over the next 5 years.

Gene identification and characterization

The availability of the human genome sequence and, increasingly, the genome sequence of model organisms, means that the laboratory techniques aimed at the identification of gene-encoding sequences within genomic DNA have become largely redundant. That is not to say that all genes in the human genome have been correctly identified and, while annotation is improving daily, much progress is still required. In addition, EST sequences are biased to representation of 3′-coding sequences. Experimental verification of gene predictions is often lacking, and investigators would be well advised to conduct exon connection and/or 5-RACE (rapid amplification of cDNA ends) in order to deduce the full-length coding sequence and splicing variation of candidate genes. Comparison of cDNA with genomic sequence is a simple and effective means of determining gene structure, necessary for designing primers for mutation screening. If the tissue(s) affected in the disorder studied are not routinely available, the work could be conducted in mouse, extrapolating genomic organization in the mouse to man as a first approximation of the likely gene structure.

Once a genetic or physical interval of interest has been identified, various tools can be used to examine the genes within this region. A step-by-step guide to bioinformatic approaches is presented in 'Users Guides' produced by *Nature Genetics* (Packer, 2003). In essence, the investigator uses genome browsers such as Ensemble (http://www.ensembl.org/), UCSC (http://genome.ucsc.edu) or the NCBI (http://www.ncbi.nlm.nih.gov/mapview/) map viewer to provide a graphical interface to the region, with the investigator selecting from menus the features he/she wishes to display within the interval. Most workers will want to know which annotated and unannotated genes map to the area, and which BAC or PAC clones provide the tile (perhaps for FISH studies). As mentioned above, SNPs can be mined, STRPs identified and sequence files downloaded to search for unannotated di-/tri-/tetra-nucleotide repeats that might be polymorphic. These browsers can also display the genome organization of a region, highlight sequences conserved in other species, indicate repeat sequences and output from gene prediction programs. In particular, intron–exon boundaries can be indicated for the design of primers for mutation screening, and alternative splicing highlighted. Putative alternative splice forms can be detected simply by examining the different ESTs and gene predictions vs. the genome sequence. Of course, such predicted variants should be validated at the laboratory bench.

Comparative analysis of genomic sequences from different species can be useful in identifying genes not so far annotated or linked to ESTs, and in the identification of regulatory regions. One useful tool for mining potential regulatory motifs is Theatre, which provides output from a suite of programs (http://www.hgmp.mrc.ac.uk/Registered/Webapp/theatre/). In the absence of any biochemical or physiological clue, investigators could make use of the PROSPECTR program, which was trained on the set of genes known to be involved in human genetic disease and prioritizes genes within any defined region for generic similarities to the disease gene set (http://www.genetics.med.ed.ac.uk/prospectr/).

Mutation screening

Identification of mutations within a candidate gene is required to prove that a disease-causing or disease-predisposing allele has been identified. A reasonable first step is southern analysis of patient vs. control DNA to rapidly scan for rearrangements or, in the most serendipitous cases, point mutations. Usually, however, DNA sequencing with or without a priori sequence screening procedure is required, and the approach taken will depend upon a combination of the material available, the number and complexity of the genes in the disease interval, the likely expression profile of the disease gene and the resources available to the investigator. PCR is employed almost ubiquitously in these procedures. Computer programs are available to assist in the design of primers appropriate for the task in hand and, where relevant, predict assay conditions, e.g. for denaturing high-performance liquid chromatography (dHPLC).

In some instances it may be possible to sequence the candidate cDNAs from affected individuals, especially where the gene is expressed in circulating lymphocytes. Buccal cells, skin biopsies, gut biopsies and post-mortem material may be useful in this regard. An advantage of this approach is that mutations leading to splicing defects may be detected, and promoter mutations may give reduced levels of expression quantifiable by real-time PCR. The inability to amplify from patient but not control DNA may indicate that a premature termination codon has led to nonsense-mediated decay (NMD) of the corresponding mRNA. Thus, a rapid pre-screen for NMD is often useful where the disease interval is gene-rich, the patients are IBD for the mutation (both alleles reduced) or where a SNP can be used to determine the relative expression of each allele. Urbach–Wiethe disease, or lipoid proteinosis, provides a good example. Standard genetic linkage identified a candidate region on chromosome 1q21, following which cultured fibroblasts from patients and controls were tested for expression of genes within the recombination interval. This revealed downregulation of extracellular matrix protein 1 gene (*ECM1*), with subsequent detection of mutations in genomic DNA (Hamada *et al*., 2002). cDNA sequencing may also be considered where a large gene is composed of many small exons and a source of appropriate mRNA is available. It may be possible, using nested primers, to amplify overlapping cDNA fragments for sequencing from tissues that do not express the protein. This procedure capitalizes upon the presence of 'illegitimate transcripts' which, depending upon the gene, are present at low levels in cells such as Epstein–Barr virus-transformed lymphoblastoid lines (Cooper *et al*., 1994).

Many mutation screening methods have been developed and usually involve electrophoresis of single-stranded or double-stranded PCR-amplified genomic DNA in order to detect abnormally migrating species indicative of sequence variation (Kristensen *et al*., 2001). Gel electrophoresis is being replaced by higher throughput techniques making use of automated capillary array electrophoresis or dHPLC. Sensitivities of single-stranded sequence polymorphism (SSCP) and heteroduplex analysis (HA) have been improved by the use of specific matrices, and have the advantage of being relatively inexpensive and making use of widely available sequencers (Hoskins *et al.*, 2003; Kourkine *et al.*, 2002). dHPLC is particularly useful where a large number of patients have to be screened for mutations in a relatively small number of exons, as it takes some effort to establish the optimal parameters for each exon. However, once this is achieved the technique is very sensitive and inexpensive (Xiao and Oefner, 2001). The recent development of capillary array electrophoresis chips offers the potential to improve throughput by an order of magnitude, and if costs are reduced resequencing chips may find wide application in the future (Andersen *et al.*, 2003). The protein truncation test (PTT) is designed to detect mutations that introduce frameshifts, splice site or premature termination mutations. Coding sequence is amplified and translated *in vitro*, the resulting proteins checked for size by SDS polyacrylamide gel electrophoresis. Mutations give rise to truncated or elongated products (Wallis, 2004).

Once a potential disease-causing variant has been discovered, steps need to be taken to validate the sequence change as a mutation. This should involve a screen for the variant in a large number of control chromosomes (at least 200), matched for ethnic background where necessary. Functional evidence may be harder to produce, unless the encoded protein has a known activity that can be assayed straightforwardly. Missense mutations in transcription factors, for instance, may reduce the transcriptional effect in reporter gene assays, or alter the DNA binding in electrophoretic mobility shift assays. Growth factors may be tested in bioassays. Where the function of the gene is entirely unknown, a first characterization often involves tagging the gene in order assess whether proteins carrying the mutation are localized identically to wild-type. This approach proved particularly informative in assessing the effect of mutations in the Treacher–Collins syndrome protein (Marsh *et al.*, 1998). In contradistinction to the wild-type GFP fusion protein, proteins containing disease-causing mutations failed to localize to the nucleolus. Quite how a defective nucleolar protein causes a neural crest defect and subsequently the birth defect syndrome remains to be elucidated. In some cases, where there is a phenotype detectable in cell lines carrying disease-causing mutations, genetic complementation may provide compelling data that the correct gene has been identified. Indeed, for certain disorders, it may be possible to identify directly the gene mutated in a disorder by complementation, for instance in DNA repair defects. Corroborative evidence may also be obtained from animal models (see below).

Splice site mutations may be validated by RT-PCR or western analysis if appropriate patient tissue samples are available. If not, then the genomic DNA flanking the putative mutation can be used in an exon amplification assay, although it is difficult to be sure that the *in vivo* situation is recapitulated in a tissue culture cell line. Even more difficult to validate are mutations that might act via a long-range effect on transcription. In facio-scapular-humeral muscular dystrophy (FSHD), amplification of the D4Z4 repeat element is associated with the disease. In muscle, the expansion has been shown to associated with a depression of genes at 4q35 (Gabellini *et al.*, 2002), although which gene or genes is/are important is unknown.

Recent years have shown that a wide variety of diseases may be caused by mutations that affect RNA metabolism, as opposed to mutations that alter the protein-encoding sequence of the gene. Some missense and synonymous mutations may occur in *cis*-acting elements that regulate splicing, resulting in exon skipping, inefficient splicing of introns or usage of cryptic splice sites. If such mutations are suspected, a bioinformatic approach can be used to assess the likelihood of any variant affecting an intronic or exonic splicing enhancer (ISE or ESE), e.g. http://exon.cshl.edu/ESE/. mRNA stability may also be affected by mutations in the 3′-UTR, e.g. Fukuyama congenital muscular dystrophy can be caused by an insertion of a retrotransposon which reduces steady-state transcript levels (Kobayashi *et al.*, 1998). 5′-UTR mutations may affect translational efficiency, such as those seen in hyperferritinaemia/cataract syndrome, hereditary thrombocythaemia (Cazzola and Skoda, 2000) and Charcot–Marie–Tooth disease (Hudder and Werner, 2000). In a rare case of α-thalassaemia, transcription of an antisense RNA has been shown to result in silencing and methylation of the *HBA2*-associated CpG island, the effect of which was biologically determined in ES cells and a transgenic model (Tufarelli *et al.*, 2003).

Biological analysis of genes implicated in birth defect syndromes

Having identified a gene defective in a particular syndrome, questions arise as to when and where the gene is expressed and what the encoded protein does. Of course, there have been instances, e.g. *FGFR*s when discovered to be involved in craniosynostosis syndromes, where the relevant genes have already been subject to a great deal of investigation. Even in this situation, a consideration of the mutational mechanisms in man can provide a useful insight into how to approach biological problems in cellular and animal models but, not infrequently, disease gene loci are found to encode proteins of unknown function. In these cases interrogation of sequence and motif databases is an essential step in formulating testable hypotheses.

Structural considerations

As a first step towards an analysis of a conceptually translated sequence, LocusLink (http://www.ncbi.nlm.nih.gov/LocusLink/) provides a single query interface to curated sequence and descriptive information about genetic loci. It presents information on official nomenclature, aliases, sequence accessions, phenotypes, EC numbers, MIM numbers, UniGene clusters, homology, map locations and related websites. From a query, users can use the BL (Blast link) to obtain a graphical representation of related proteins, together with a link (CDD-Search), which can be accessed to retrieve conserved domains. Various databases contain sequence motifs typical of certain structural domains, e.g. InterPro, and matches to one or more domains may suggest protein functions or interactions (http://www.ebi.ac.uk/interpro; Mulder *et al.*, 2003). InterPro provides information from a number of sources, such as PROSITE,

PRINTS, Pfam and ProDom, and allows text searches of domain names. One problem with this kind of analysis is that many of the functional motifs involve few amino acids with a high degree of degeneracy and a high frequency of false positive assignments can be made (the same problem arises when searching for transcription factor binding sites in promoter regions). One can also be misled by rarer events, for instance some genes encode more than one protein by frameshifts, e.g. α-enolase and τ-crystallin, or other mechanisms, to create so-called 'moonlighting proteins' (Jeffery, 2003). Divergent evolution has resulted in similar structures adopting different roles in different proteins, e.g. WD domains, TIM barrels and zinc fingers.

Despite the fact that there is a huge number of possible ways of stringing together amino acids to produce proteins with variable secondary and tertiary structures, experimental methods have delineated approximately 500 three-dimensional (3-D) configurations called structural folds. It is still difficult to predict 3-D structure from primary sequence, but a combination of structural biology, itself becoming the subject of high-throughput approaches, and computation is likely to introduce structure–function correlations of increased sensitivity and specificity in the future. One problem with the interpretation of protein similarities is knowing whether extrapolating functional equivalence is valid. For instance, cytokines of the same family can be selective or promiscuous in binding their receptor partners, and some FGSs are intracellular rather than secreted proteins. In the computational prediction of whether protein interactions are conserved across members of protein families, it has been shown that 3-D modelling of the interaction is a useful filter for ranking such interactions and thus prioritizing laboratory experiments (Aloy *et al.*, 2003; Aloy and Russell, 2002).

Gene expression

Standard techniques such as northern blotting may provide valuable information concerning the relative abundance of transcripts in different tissues or cell lines and evidence of alternative splicing that might be of biological significance. However, most investigators will wish to have a higher resolution analysis of expression in space and time. Most commonly, whole-mount and/or tissue section hybridizations will be employed. If a suitable antibody is available, immunological staining can be employed and can demonstrate the persistence of protein expression once mRNA levels have decayed.

As indicated below, it is becoming more common to couple introduction of a reporter with gene targeting approaches. Together with modern imaging techniques, this promises novel insights into the parallel analysis of gene expression and phenotype. A recent advance has been the development of techniques that can assemble gene expression into 3-D maps, and layer these onto images of embryos at various stages (see URLs for Emage, above). Microscopy based upon optical projection tomography (OPT) can produce high-resolution 3-D images of both fluorescent and non-fluorescent biological specimens with a thickness of up to 15 mm (Sharpe *et al.*, 2002). Thus, the tissue distribution of RNA and protein expression can be determined in intact embryo explants and related to mutant phenotype or gene function. Images can also be obtained using magnetic resonance imaging (MRI) using an MRI contrast agent activated by reporter gene expression in living animals, e.g. β-galactosidase (Louie *et al.*, 2000).

Implicit in the previous section is that vertebrate, probably mouse, embryos provide a good model for the human condition under study. However, it is well documented that this is not always the case (Fougerousse *et al.*, 2000). Of course, we are limited by ethical considerations and tissue availability in any investigation of gene expression in humans. Therefore, certain centres have established banks of carefully selected human embryos, from which sections can be requested for specific projects. (e.g. http://www.ncl.ac.uk/ihg/research/developmental/vertebrate/project/653; http://www.mrc.ac.uk/index/current-research/current-resources/current-hdbr.htm). Several human disease genes have been studied in this way (Clement-Jones *et al.*, 2000; Crosnier *et al.*, 2000; Lai *et al.*, 2003).

Microarray technologies have allowed more global analyses of gene expression changes in models of birth defect syndromes. For instance, the Ts65n mouse provides a model for Down's syndrome by virtue of a segmental trisomy for part of MMU16. This study demonstrated that the trisomy resulted in small but widespread changes in the cerebellar transcriptome, rather than large changes in a small number of genes (Saran *et al.*, 2003).

Analysis of proteins encoded at birth defect loci

An important aspect of understanding protein function is to understand where within the cell the protein can be found. Structural considerations often offer good clues, e.g. a signal peptide in the absence of a transmembrane domain may indicate that the protein is secreted. Investigators should also be aware of dynamic changes in protein localization, for instance cytoplasmic to nuclear shuttling, relocalization or turnover during different phases of the cell cycle or upon certain stimuli, and proteolytic cleavage of subfragments with biological function. In order to follow such events it is advantageous to have an antibody, or panel of antibodies, raised against the native protein. These can be raised *in vivo* or selected from libraries using techniques such as phage display. In the absence of antibody, useful information can often be gained by using constructs that will express a tagged version of the protein within the cell. Such tags can be short oligopeptides, such as FLAG, c-myc or HA, or they can be biologically active, such as green fluorescent protein (GFP). The advantage of GFP and its relatives is that fusion protein localization can be followed in live cells in order to track stimulus responses. The disadvantages of tags include artefacts due to overexpression and interference with protein function.

Many investigators have examined protein interactions as a way of identifying pathways involved in human genetic disease, especially where the function of the protein under study is unknown. The rationale is based on the idea that interaction with a protein of known function would implicate the interactor in a similar pathway. Perhaps the most widely used assay is the yeast two-hybrid (Y2H). Here, various reporters, such as auxotrophic markers or β-galactosidase, are used to select yeast clones harbouring clones encoding interacting proteins. The reporters are activated upon interaction of the bait protein fusion, which contains sequences from the protein of interest, and the target protein fusion, which contains the putative interactor (Figure 4.2). The first Y2H systems involved variations on the theme of transcription factor reconstitution, the bait being fused to a DNA-binding domain (specific for the promoters upstream of the reporters) and the prey library being fused to a transcriptional activation domain (Brent and Finley, 1997; Gietz and Woods, 2002). Haploid yeast strains of different mating types and containing baits and prospective preys, respectively, can be mated to produce diploids. This procedure can be used to increase the throughput of the procedure and, with other technologies, begin to establish protein interaction networks (Stagljar, 2003).

If the bait protein acts as a transcriptional activator in a standard Y2H setting, then other systems not based on transcriptional read-out can be used. The ras/sos recruitment method reconstitutes the activity of a guanine nucleotide exchange factor (GEF). The bait is fused to a sub-membrane localization domain and the target to the GEF catalytic domain, and bait–target interaction enables complementation of a temperature-sensitive mutation in the Cdc25ts protein (Broder *et al.*, 1998; Huang *et al.*, 2001). A number of other systems with split enzymes have been described (Mendelsohn and Brent, 1999).

Reverse two-hybrid systems can be used to identify mutations that disrupt protein interaction and thus facilitate mapping of interaction domains. Potentially, such techniques can be used to identify small molecules that interfere with protein–protein interactions. 'Bridge' hybrid systems can be used to identify interacting proteins where the binding requires a third protein (Gordon and Buchwald, 2003) or RNA (Jaeger *et al.*, 2004) to establish the complex, or post-translationally modify one of the proteins.

All protein interaction screens produce false positives and a number of techniques are available to enrich for true positives and corroborate the interaction. Additionally, different techniques are required to explore the molecular interaction in more detail. In the Y2H screen itself, multiple reporter systems can be used (James *et al.*, 1996) and positives compared with databases of known Y2H false interactors (for a useful summary, see http://www.fccc.edu/research/labs/golemis/InteractionTrapInWork.html). Putative interactors should be expressed at the same time in the same cellular compartment. Corroboration of interaction is often achieved by biochemical techniques such as affinity capture, or by co-immunoprecipitation using antibodies directed against native protein, or tags if a cell transfection overexpression system is used. More recently, fluorescence resonance energy transfer (FRET) has been employed. This relies on the use of fusion proteins carrying fluorescent moieties such as GFP, BFP and YFP. If the two fusion proteins are brought into apposition via protein interaction, the fluorophores will likely be within 100 Å of each other.

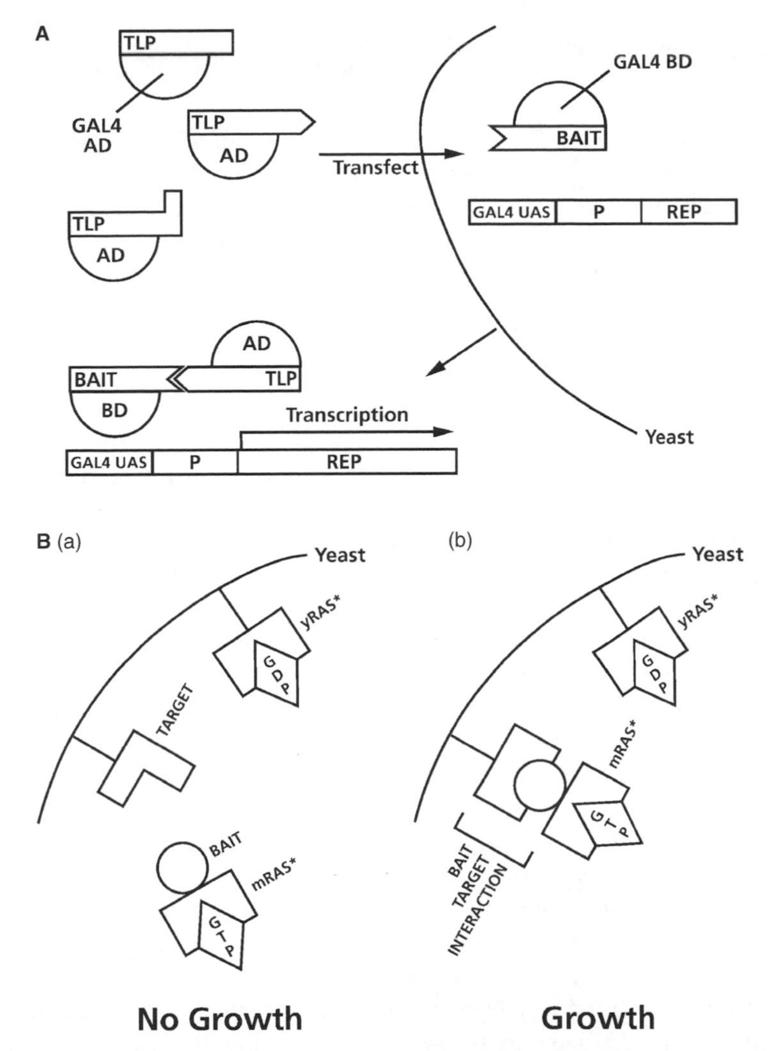

Figure 4.2 Yeast two-hybrid screening. (A) Standard GAL4 reconstitution. The bait protein represents the protein for which interacting proteins will be isolated. It is expressed in a yeast strain fused to the GAL4 (or other) DNA-binding domain (GAL4 BD). A cDNA library is created which, when transfected into yeast, will direct synthesis of the GAL4 DNA transcriptional activation domain (GAL4 AD) fused to a library of target proteins (TLP). If any target library protein interacts with the bait, then the GAL4 AD and BD are brought into apposition. The complex then binds to the GAL4 upstream activation sequence (GAL4 UAS) of the GAL4 promoter (P), activating transcription of the reporter gene (REP, bottom left), which is then detected by growth and/or colony filter assay. Clones encoding the TLP can then be isolated and sequenced. (B) Ras recruitment. The yeast ras (yRas) is maintained in its inactive GDP-bound form due to a mutation in the Cdc25 exchange factor, which is penetrant at the restrictive temperature. In (a), as the bait fails to interact with the protein expressed from the cDNA library, the bait-fused activated murine Ras (mRAS) is unable to promote growth (the mRAS is marked with an asterisk to denote the fact that it lacks its CAAT box). In (b), the presence of an interacting protein at the membrane recruits activated mRas via the bait, which allows the yeast to grow

Below this distance stimulation of the higher-energy donor fluorphore allows energy transfer to the lower-energy fluorphore, which then emits photons (Sekar and Periasamy, 2003).

Surface plasmon resonance is a method which can quantify interactions between known proteins, and can be used to compare wild-type with mutant proteins. Bait protein is bound to a thin metallic surface at the interface of glass and aqueous medium and a solution of the target protein allowed to flow over the surface. As the bait is bound by the target the refractive index alters, which in turn affects the resonance angle at which energy in a light wave is dissipated in the metal. For instance, surface plasmon resonance has been used to demonstrate that two Apert syndrome mutations of *FGFR2* result in a selective decrease in the dissociation kinetics of FGF2, but not of other FGF ligands, when compared with the wild-type protein (Anderson *et al.*, 1998). Biosensors such as those based around BIAcore technology can be used to evaluate many kinds of molecular interactions relevant to biology (Malmqvist, 1999).

High-throughput proteomic approaches are likely to have an impact on our understanding of birth defects by uncovering interactions with more complex groups of proteins than the pairwise screens afforded by techniques like Y2H (Zhu *et al.*, 2003). Proteomic approaches have the advantage that the bait–protein interactions are explored in their native cellular environment and multiprotein complexes can be isolated in a single experiment (Aebersold and Mann, 2003). Following affinity purification, often using native or anti-tag antibodies, the resulting protein mix is subjected to electrophoresis and individual spots subjected to mass spectroscopy. In order to increase signal:noise ratios, tandem affinity tags have been used (TAP-tags). Commonly, the calmodulin-binding domain is placed in series with a tobacco etch virus (TEV) protease recognition site and the immunoglobulin-binding domain of protein A. The first affinity step is followed by proteolytic cleavage of the first tag, and then the second round of affinity purification takes place (Knuesel *et al.*, 2003). As an example, a recent mass spectrometric analysis of purified pre-ribosomal ribonucleoprotein complexes yielded sequence for over 100 proteins, including the TCOF1 protein mutated in Treacher–Collins syndrome (Hayano *et al.*, 2003).

Animal models

Dominant disorders associated with gain-of-function mutations can be modelled by relatively straightforward transgenic approaches. These disorders include trinucleotide repeat expansions, which are not naturally occurring in mice, and diseases such as the skin disorder Vohwinkel syndrome, modelled by expression of the *connexin 26* D66H mutation from the keratin 10 promoter (Bakirtzis *et al.*, 2003). Gene targeting of the murine homologues of genes mutated in human congenital malformations, to produce loss of function alleles, has proved very informative in elucidating the role of these genes during development. Besides the creation of straightforward loss-of-function alleles, targeting can be used to knock in specific point mutations. More recently, it has become common practice to create conditional alleles using site-specific recombinase systems such as Cre-*lox*P, Flp-*FRT* and ϕC31-*att*. These systems can be used to circumvent early embryonic lethality, analyse the role of a gene in a particular cell lineage or subset of tissues, allow temporal and spatial induction of a lineage tracing marker, allow temporal and spatial control over the mutation via the use of ligand-regulated recombinases, produce an allelic series, and produce mosaics to enable the assessment of the requirement of a specific gene for cellular contributions to developing structures (Branda and Dymecki, 2004). Recombinases can also be used in *E. coli* to facilitate high-throughput engineering of targeting vectors (Valenzuela *et al.*, 2003). One of the most exciting prospects is the convergence of advances in gene targeting and tagging with high-resolution imaging in living embryos and animals (Marx, 2003). This promises an unprecedented ability to analyse cellular processes and interactions and protein function *in vivo*, in both wild-type and mutant animals.

Knock-in approaches have been used to establish temporal control over gene expression. For instance, in order to analyse the temporal requirement for *EDNRB*, the tetracycline transactivator tTA (and in other animals rtTA) was knocked into one allele of *EDNRB*. The responder allele contained tet-OP sites upstream of an *EDNRB* minigene within the second *EDNRB* allele. In the presence of tetracycline or doxycycline the tTA is active, binds to the tet-OP and induces transcription of the minigene. In the case of the rtTA, the antibiotics abrogate transactivator activity downregulating minigene expression. Thus, in transheterozygote animals the expression of *EDNRB* from the minigene was under the control of doxycycline within a $EDNRB^{-/-}$ background (Shin *et al.*, 1999).

While gene targeting offers a standard route to the exploration of gene function during development, and the various enhancements outlined above allow specific types of analysis within the model created, a useful alternative is offered by the increasing number of gene trap libraries created in academic and commercial laboratories. Gene trapping screens involve an essentially random insertion of constructs into the genome of ES cells, with some selection procedure employed to identify instances where the exogenous sequence has inserted into a gene (Stanford *et al.*, 2001). Commonly this involves a splice acceptor site upstream of a marker such as β-geo, which allows both selection of genes expressed in ES cells and expression analysis using LacZ staining. 5′-RACE can be used to identify the sequence of the native transcript upstream of the trap vector. Thus, in contradistinction to chemically induced mutants, the gene mutated is known prior to phenotype generation. Newer vectors have been designed to specifically identify secreted or transmembrane protein-encoding genes. Vectors have also been constructed to allow future recombinase manipulations and knock-in strategies, increasing flexibility still further. Various databases are now available to easily identify previously trapped genes (Table 4.1) and some centres and commercial concerns offer an ES cell injection service; a gene trap consortium has been established to access information from a single site (http://www.igtc.ca; To *et al.*, 2004). The Ensemble genome database can be requested to highlight trapped genes from the Skarnes laboratory as part of the

Table 4.1 Online gene trap databases

Gene trap resources	Notes
http://bsw3.aist-nara.ac.jp/kawaichi/naistrap.html	Japanese centre; BLAST is by email request
http://genetrap.gsf.de/index.html	German consortium
http://www.lexgen.com/omnibank/overview.php	Lexicon, a commercial organization; cost and IP issues, but large array of clones available
http://www.escells.ca/	Over 1000 tagged genes
http://socrates.berkeley.edu/~skarnes/resource.html	Secreted and membrane proteins
http://baygenomics.ucsf.edu/	Provides blastocyst injection service
http://www.sanger.ac.uk/PostGenomics/genetrap/	Baygenomics mirror with Ensemble interface
http://www.cmhd.ca/sub/database.htm	Registration required

display, or the sequences of trapped genes BLAST-searched at: http://www.sanger.ac.uk/cgi-bin/blast/submitblast/genetrap

The disadvantages of using gene traps are that the induced mutation may not be a true null and the reporter gene may not accurately reflect the expression pattern of the trapped gene. Even so, as the availability of ES lines expands, gene traps are likely to be a first source of material regarding gene function with subsequent bespoke targetings designed to probe specific aspects of the developmental biology of the processes affected by the mutation.

As opposed to genetically driven screens such as targeting and trapping, phenotypically driven screens allow a higher throughput of mutants, but without a molecular tag of the mutated genes (Cox and Brown, 2003). Mutations are usually chemically induced, e.g. by *N*-ethyl-*N*-nitrosourea (ENU), and banks of mutants are available for fly, zebrafish and mouse. The largest and most recent murine screens employ structured prenatal and postnatal programmes to assess the phenotypes. For instance, external appearance is scored, biochemical, radiological, haematological, immunological parameters assessed, and baseline cognitive and behavioural screens undertaken. Novel imaging techniques, such as MRI, offer the opportunity of a reasonably quick means of detecting internal malformations, such as cardiac defects (Schneider *et al.*, 2003). The main centres conducting this work maintain search engines allowing identification of mutants that potentially model human disorders (e.g. http://www.mgu.har.mrc.ac.uk/mut.html; http://www.gsf.de/ieg/groups/enu-mouse.html; http://www.gsc.riken.go.jp/e/group/thememouseE.html. The Jackson laboratories have an extensive mouse phenotype database with mutants from various sources: http://www.jax.org/resources/search_databases.html).

Engineered chromosomes (see below) can be used in conjunction with ENU-induced mutations to screen for recessive mutations in F_1 or F_2 animals, as an induced deletion reduces one region of the genome to hemizygosity. Nested deletions can subsequently be used to refine locus position. Similarly, chromosome engineering can be used to create inversions that act as balancer chromosomes. Here, one end of the inversion disrupts a gene with a recessive lethal phenotype, and the dominantly marked (e.g. by coat colour) inversion used to map recessive mutations. The inversion suppresses recombination with the homologous chromosome, and up to one-third of a single chromosome can be inverted without the likely complication of haplo-insufficiency that hemizygosity for such a large region would produce (Nishijima *et al.*, 2003). Genome-wide coverage of balancers is likely to prove a useful resource.

Gene targeting and chromosome abnormalities

Gene targeting technology has been adopted into the task of chromosome engineering in the mouse. The general principle involves the knock-in of a sequence that acts as the substrate for a subsequent chromosome recombination event; commonly, this sequence is *lox*P. If a pair of *lox*P sequences are in the same relative orientation, then a deletion or duplication can be created; *lox*P sites in opposite orientations recombine to produce an inversion. Balanced translocations can be created by targeting *lox*P to the two chromosomes of interest Alternatively, selection for *trans* recombination can be employed, for instance where recombination brings together the two halves of an active dominant selectable marker, such as *Hprt* (Ramirez-Solis *et al.*, 1995). Libraries of gene targeting vectors have been created to provide both 5′ and 3′ *Hprt* constructs with a view to creating engineered chromosomes. In one such application the two halves of the mini-gene are flanked by *Agouti* (*Ag*) and *Tyrosinase* (*Ty*) coat colour markers in order to facilitate visual identification of mice carrying rearranged chromosomes (Zheng *et al.*, 1999). Depending upon the nature of the rearrangement induced, the reconstituted minigene is found in association with either *neo* or *puro* markers, allowing sib selection of ES cells to pick the required clone (Yu and Bradley, 2001). *In vivo* recombination is preferred in some instances, for instance where a conditional allele is desired (Kochilas *et al.*, 2002) or where chromosome translocations are to be produced (germ line transmission of such rearrangements is often compromised).

The creation of engineered deletions and duplications was instrumental in identification of a gene critical for the development of Dr George/velocardiofacial syndrome (DGS/VCFS). This syndrome is usually caused by a deletion in chromosome 22q11 (22q11DS), resulting in hemizygosity for up to 50 genes (Scambler, 2003). Human genetics, in the form of the shortest region of deletion overlap mapping, had not revealed whether 22q11DS was the result of haplo-insufficiency of one gene, or a combination of genes. It was thought that some deletions might affect the transcription of genes lying outwith the deletion interval via a long-range effect on expression. Because the genes deleted in 22q11DS were all, with one exception, clustered on proximal mouse chromosome 16, chromosome engineering experiments in the mouse were able to mimic the human situation of multigene hemizygosity, even though gene order is not perfectly conserved (Sutherland *et al.*, 1998). The first targeted deletion of the region, *Df1*, encompassed 18 genes within a 1.2 Mb interval (Lindsay *et al.*, 1999). Hemizygous mice were viable and fertile, and approximately 20% died in the perinatal period. Examination of late gestation embryos revealed a series of congenital heart defects reminiscent of 22q11DS, with full penetrance for hypo/aplasia of at least one of the fourth branchial arch arteries. These defects represent the class of heart defects most specific to 22q11DS. Importantly, breeding experiments producing mice transheterozygous for the deleted chromosome 16 and the reciprocal duplication on the corresponding homologue had no heart defects. This demonstrated that a gene or genes within the *Df1* interval was responsible for the observed phenotype, and that long-range effects on transcription, such as those proposed to act in the human situation, did not have a role. Independent experiments created a larger, 1.5 Mb, deletion termed *Lgdel* (Merscher *et al.*, 2001) with a similar phenotype.

Subsequently engineered deletions and duplications allowed a shortest region of deletion overlap map to be established in the mouse, and this in turn suggested that one or more of six genes in the *Arvcf–Ufd1l* interval had a haplo-insufficient phenotype (Lindsay *et al.*, 2001; Puech *et al.*, 2000). Transgenic rescue experiments with either human or murine genes narrowed the region further, to just four genes. Based on its embryonic expression pattern in the mesodermal core of the pharyngeal arches, the transcription factor *Tbx1* appeared the best candidate, and three teams independently created single-gene targeted mutants at this locus (Jerome and Papaioannou, 2001; Lindsay *et al.*, 2001; Merscher *et al.*, 2001). In each case, the heterozygous mice had the same cardiovascular malformations that had been observed in the *Df1* strain. The $Tbx1^{-/-}$ (*Tbx1* null) phenotype comprises defects of all the main structures affected in DGS/VCFS, and these mice can perhaps be viewed as a having a severe form of the syndrome. Thus, *TBX1* was an excellent candidate for being mutated in non-deletion cases of DGS/VCFS and subsequently rare mutations of *TBX1* in such patients were identified (Yagi *et al.*, 2003).

Engineered deletions need not have defined end points. Nested deletions can be created by specifically targeting one end of the deletion as an anchor point, and then using retroviral insertions to introduce the other engineering vector to the genome. Selection for deletions using reconstitution of half-genes is likely to produce deletions of a range of sizes, which can be characterized by inverse PCR or FISH (Su *et al.*, 2000).

One of the most common human malformation syndromes is Down's syndrome, trisomy 21. Little is known about the underlying developmental biology of the condition, and transgenic analysis of single genes have, perhaps unsurprisingly, not produced particularly informative models. Many investigators have turned to mouse models involving complete or partial trisomy of MMU16 (Galdzicki and Siarey, 2003), but this produces trisomy for genes not mapping to HSS21. Future dissection of trisomy 21 is likely to make use of engineered duplications, as described above, or the creation of freely segregating supernumerary chromosomes (Tomizuka *et al.*, 1997). Chimeric mice derived from cells carrying a human chromosome 21 have been created, although chromosome segregation was unstable (Kazuki *et al.*, 2001; Shinohara *et al.*, 2001). Presence of the human chromosome was associated with the presence of heart defect and learning deficits. Germ line transmission of a human 5 Mb subchromosomal fragment was also achieved. Most recently, stable transmission of a freely segregating, stably transmitted human chromosome 21 has been observed (Professor E. M. C. Fisher, O'Doherty *et al.*, 2005) and detailed phenotyping is under way.

Cloning methodologies have been developed that allow the construction of human artificial chromosomes (HACs) with defined chromosomal region inserts. A panel of HACs harbouring inserts ranging in size from 1.5 to 10 Mb from three human chromosomes (2, 7 and 22) has been constructed, and developments such as this should permit the manipulation of genes within the cloned sequences prior to the creation of mouse models (Kuroiwa et al., 2002).

Other disease-modelling approaches

The creation of targeted mutants is still a relatively costly and time-consuming approach, and other organisms can offer certain advantages in probing developmental pathways. Steady progress has been made in refining antisense strategies for the knock-down of specific mRNAs (Scherer and Rossi, 2003). Phosphorodiamidate morpholino oligonucleotides bind to RNA and efficiently interfere with gene expression in a sequence-specific manner. Compared with previous versions of antisense oligonucleotides, they offer greater specificity and less toxicity and are relatively resistant to degradation. If no antibody is available to assess the efficacy of knock-down, the target gene sequence can be designed to span an intron–exon boundary in order that RT-PCR can be used as an alterative. *D. rerio* (zebrafish) and *Xenopus* spp. are commonly used as target organisms for this approach. A zebrafish approach has the advantage that there is a large array of mutants available, where increasingly the gene affected is known, allowing the investigation of potential epistatic relationships using a combination of knock-down and mutant.

Double-stranded RNA 21-23mers complementary to a target RNA can be used to silence gene expression via site specific cleavage (RNAi or siRNA strategies). Such siRNAs do not activate the interferon pathway (or at least not fully) and can be introduced into cells either as dsRNAs or using a polIII promoter-driven construct with a polyT stop signal. Such constructs can be used in a transgenic approach, where dominant knock-down embryonic phenotypes are directly assessed in mice entirely derived from ES cells by tetraploid rescue (Kunath *et al.*, 2003). In addition, morpholino or dsRNAi knock-downs can be used in the chick system, where *in ovo* electroporation is relatively straightforward (Pekarik *et al.*, 2003). Organ culture systems also lend themselves to dsRNAi knock-down approaches and offer one method of analysing loss-of-function effects at several different stages of organ development. One application of this strategy was the analysis of Wt1, Pax2 and Wnt4 function during renal organogenesis (Davies *et al.*, 2004).

Dominant negative approaches have been used, particularly in the frog and chick, to study molecules such as transcriptional regulators, growth factors and their receptors. For transcription factors the technique may involve replacing an activation domain with a strong repressor, e.g. that from the engrailed protein, or a repressor domain with a strong activator, e.g. VP16 (Suzuki *et al.*, 2004). Transcription factor activity can be induced, for instance using a glucocorticoid receptor ligand binding domain fusion (Horb and Thomsen, 1999).

The chick lends itself to embryological analysis because of the ease by which embryos can be manipulated and observed *in ovo*. Placement of beads containing growth factors or teratogens can be used to explore gene–environment interactions. Genetic manipulation can be achieved using electroporation of morpholinos, plasmids, siRNA or retroviruses (Krull, 2004). The gene expressed from such DNAs can be used to analyse overexpression, knock-down or dominant negative effects. Often, co-electroporation of a marker such as GFP is used to track transformed cells, and manipulations with test sequences compared with contralateral controls. A chick retroviral approach was used to complement biochemical studies which suggested a selective loss of function caused by an I47L substitution in the HOXD13 homeodomain, associated with a human brachydactyly and central polydactyly syndrome (Caronia *et al.*, 2003).

Why study rare human birth defect syndromes?

Human genetic approaches have been instrumental in the identification of novel genetic mechanisms and roles for proteins during development that were entirely unsuspected from gene targeting experiments or cell biology. For instance, work on the craniosynostosis syndromes identified FGFR gain-of-function mutations that were instrumental in the identification of critical residues within the protein, and mechanisms of paternal age effects, as well as helping to elucidate a pathway involved in craniofacial bone and suture formation (Goriely *et al.*, 2003; Wilkie *et al.*, 1993). Similarly, human genetics identified a role for the polyalanine tracts within 5′-homeodomain transcription factors and identified gain-of-function and selective loss-of-function alleles of *HOXA13* and *HOXD13* (Goodman and Scambler, 2001). Investigation of the causes of holoprosencephaly uncovered a slew of interacting gene products, one of which is *Shh* (Ming and Muenke, 2002; Roessler and Muenke, 2003). As has been seen, work on Bardet–Biedl syndrome and Hirschsprung's disease (amongst others) has helped provide models of gene interaction that should provide a bridge to the understanding of more complex gene interaction underlying common disorders, such as heart disease, stroke and cancer. Despite the millions of dollars expended on schizophrenia research, so far the genetic variation with the greatest relative risk for the disorder is deletion of chromosome 22q11, a predisposition identified by meticulous dissection of the phenotype of such patients followed by studies in cohorts of schizophrenic patients (Murphy, 2002). Some classes of mutation appear to be peculiarly human (e.g. trinucleotide repeat expansion), although the mouse may provide a useful model for analysing the effect of such mutations. Other mutations have an effect in man, but not in mouse. A case in point is sacral agenesis, part of Currarino triad. This human syndrome is caused by haplo-insufficiency of the homeodomain transcription factor *MNX1* (*HLXB9*) (Ross *et al.*, 1998). However, mice heterozygous for null mutations of *Mnx1* are apparently entirely normal and, although homozygotes have severe deficiencies in the development of motorneurones and the pancreas (Arber *et al.*, 1999; Li *et al.*, 1999), the caudal axial skeleton is normal. A few genes are unique to humans and thus mouse models cannot be easily produced. The gene *SHOX1* maps to the pseudoautosomal region of the X and Y chromosomes and is haplo-insufficient in Leri–Weill syndrome and nullizygous in Langer syndrome (Belin *et al.*, 1998; Shears *et al.*, 1998). The mouse has no *Shox1* gene, but both humans and mice have an orthologous autosomal gene *Shox2*. Perhaps *SHOX1* function can be analysed by transgenesis, or a targeted replacement of *Shox2* by *Shox1*.

Thus, human genetics should be considered as part of the armamentarium of the developmental biologist, providing novel entry points to developmental pathways. At the same time, the value of being able to provide a molecular diagnosis should not be underestimated from the patient's and family's point of view. In some circumstances it will allow antenatal diagnosis if desired, and it may provide a better idea of prognosis. Finally, access to welfare and social service support is sometimes improved by the provision of a firm diagnosis.

References

Aebersold, R. and Mann, M. (2003) Mass spectrometry-based proteomics. *Nature* **422**: 198–207.

Aitman, T.J., Glazier, A.M., Wallace, C.A., Cooper, L.D. *et al.* (1999) Identification of Cd36 (Fat) as an insulin-resistance gene causing defective fatty acid and glucose metabolism in hypertensive rats. *Nat. Genet.* **21**: 76–83.

Aloy, P., Ceulemans, H., Stark, A. and Russell, R.B. (2003) The relationship between sequence and interaction divergence in proteins. *J. Mol. Biol.* **332**: 989–998.

Aloy, P. and Russell, R.B. (2002) Interrogating protein interaction networks through structural biology. *Proc. Natl Acad. Sci. USA* **99**: 5896–5901.

Anagnostopoulos, A.V., Mobraaten, L.E., Sharp, J.J. and Davisson, M.T. (2001) Transgenic and knockout databases: behavioral profiles of mouse mutants. *Physiol. Behav.* **73**: 675–689.

Andersen, P.S., Jespersgaard, C., Vuust, J., Christiansen, M. and Larsen, L.A. (2003) High-throughput single strand conformation polymorphism mutation detection by automated capillary array electrophoresis: validation of the method. *Hum. Mutat.* **21**: 116–122.

Anderson, J., Burns, H.D., Enriquez-Harris, P., Wilkie, A.O. and Heath, J.K. (1998) Apert syndrome mutations in fibroblast growth factor receptor 2 exhibit increased affinity for FGF ligand. *Hum. Mol. Genet.* **7**: 1475–1483.

Ansley, S.J., Badano, J.L., Blacque, O.E., Hill, J. *et al.* (2003) Basal body dysfunction is a likely cause of pleiotropic Bardet–Biedl syndrome. *Nature* **425**: 628–633.

Arber, S., Han, B., Mendelsohn, M., Smith, M. *et al.* (1999) Requirement for the homeobox gene Hb9 in the consolidation of motor neuron identity. *Neuron* **23**: 659–674.

Bakirtzis, G., Choudhry, R., Aasen, T., Shore, L. *et al.* (2003) Targeted epidermal expression of mutant Connexin 26(D66H) mimics true Vohwinkel syndrome and provides a model for the pathogenesis of dominant connexin disorders. *Hum. Mol. Genet.* **12**: 1737–1744.

Belin, V., Cusin, V., Viot, G., Girlich, D. *et al.* (1998) SHOX mutations in dyschondrosteosis (Leri–Weill syndrome). *Nat. Genet.* **19**: 67–69.

Blanton, S.H., Bertin, T., Patel, S., Stal, S. *et al.* (2004) Nonsyndromic cleft lip and palate: four chromosomal regions of interest. *Am. J. Med. Genet.* **125A**: 28–37.

Bolk, S., Pelet, A., Hofstra, R.M., Angrist, M. *et al.* (2000) A human model for multigenic inheritance: phenotypic expression in Hirschsprung's's disease requires both the RET gene and a new 9q31 locus. *Proc. Natl Acad. Sci. USA* **97**: 268–273.

Branda, C.S. and Dymecki, S.M. (2004) Talking about a revolution: the impact of site-specific recombinases on genetic analyses in mice. *Dev. Cell* **6**: 7–28.

Brent, R. and Finley, R.L. (1997) Understanding gene and allele function with two-hybrid methods. *Ann. Rev. Genet.* **31**: 663–704.

Broder, Y.C., Katz, S. and Aronheim, A. (1998) The ras recruitment system, a novel approach to the study of protein-protein interactions. *Curr. Biol.* **8**: 1121–1124.

Caronia, G., Goodman, F.R., McKeown, C.M., Scambler, P.J. and Zappavigna, V. (2003) An I47L substitution in the HOXD13 homeodomain causes a novel human limb malformation by producing a selective loss of function. *Development* **130**: 1701–1712.

Carrasquillo, M.M., McCallion, A.S., Puffenberger, E.G., Kashuk, C.S. *et al.* (2002) Genome-wide association study and mouse model identify interaction between RET and EDNRB pathways in Hirschsprung's's disease. *Nat. Genet.* **32**: 237–244.

Cazzola, M. and Skoda, R.C. (2000) Translational pathophysiology: a novel molecular mechanism of human disease. *Blood* **95**: 3280–3288.

Chung, Y.J., Jonkers, J., Kitson, H., Fiegler, H. *et al.* (2004) A whole-genome mouse BAC microarray with 1-Mb resolution for analysis of DNA copy number changes by array comparative genomic hybridization. *Genome Res.* **14**: 188–196.

Clement-Jones, M., Schiller, S., Rao, E., Blaschke, R.J. *et al.* (2000) The short stature homeobox gene SHOX is involved in skeletal abnormalities in Turner syndrome. *Hum. Mol. Genet.* **9**: 695–702.

Cooper, D.N., Berg, L.P., Kakkar, V.V. and Reiss, J. (1994) Ectopic (illegitimate) transcription: new possibilities for the analysis and diagnosis of human genetic disease. *Ann. Med.* **26**: 9–14.

Cox, R.D. and Brown, S.D. (2003) Rodent models of genetic disease. *Curr. Opin. Genet. Dev.* **13**: 278–283.

Crosnier, C., Attie-Bitach, T., Encha-Razavi, F., Audollent, S. *et al.* (2000) JAGGED1 gene expression during human embryogenesis elucidates the wide phenotypic spectrum of Alagille syndrome. *Hepatology* **32**: 574–581.

Davies, J.A., Ladomery, M., Hohenstein, P., Michael, L. *et al.* (2004) Development of an siRNA-based method for repressing specific genes in renal organ culture and its use to show that the Wt1 tumour suppressor is required for nephron differentiation. *Hum. Mol. Genet.* **13**: 235–246.

Dobrin, S.E. and Stephan, D.A. (2003) Integrating microarrays into disease–gene identification strategies. *Expert. Rev. Mol. Diagn.* **3**: 375–385.

Emison, E.S., McCallion, A.S., Kashuk, C.S., Bush, R.T. *et al.* (2005) A common sex-dependent mutation in a RET enhancer underlies Hirschsprung's's disease risk. *Nature* **434**: 857–863.

Fantes, J., Redeker, B., Breen, M., Boyle, S. *et al.* (1995) Aniridia-associated cytogenetic rearrangements suggest that a position effect may cause the mutant phenotype. *Hum. Mol. Genet.* **4**: 415–422.

Fougerousse, F., Bullen, P., Herasse, M., Lindsay, S. *et al.* (2000) Human–mouse differences in the embryonic expression patterns of developmental control genes and disease genes. *Hum. Mol. Genet.* **9**: 165–173.

Gabellini, D., Green, M.R. and Tupler, R. (2002) Inappropriate gene activation in FSHD: a repressor complex binds a chromosomal repeat deleted in dystrophic muscle. *Cell* **110**: 339–348.

Gabriel, S.B., Salomon, R., Pelet, A., Angrist, M. *et al.* (2002) Segregation at three loci explains familial and population risk in Hirschsprung's's disease. *Nat. Genet.* **31**: 89–93.

Galdzicki, Z. and Siarey, R.J. (2003) Understanding mental retardation in Down's syndrome using trisomy 16 mouse models. *Genes Brain Behav.* **2**: 167–178.

Genin, E., Todorov, A.A. and Clerget-Darpoux, F. (1998) Optimization of genome search strategies for homozygosity mapping: influence of marker spacing on power and threshold criteria for identification of candidate regions. *Ann. Hum. Genet.* **62**(5): 419–429.

Gietz, R.D. and Woods, R.A. (2002) Screening for protein–protein interactions in the yeast two-hybrid system. *Methods Mol. Biol.* **185**: 471–486.

Goodman, F.R. and Scambler, P.J. (2001) Human *HOX* gene mutations. *Clin. Genet.* **59**: 1–11.

Gordon, S.M. and Buchwald, M. (2003) Fanconi anemia protein complex: mapping protein interactions in the yeast 2- and 3-hybrid systems. *Blood* **102**: 136–141.

Goriely, A., McVean, G.A., Rojmyr, M., Ingemarsson, B. and Wilkie, A.O. (2003) Evidence for selective advantage of pathogenic FGFR2 mutations in the male germ line. *Science* **301**: 643–646.

Gribble, S.M., Fiegler, H., Burford, D.C., Prigmore, E. *et al.* (2004) Applications of combined DNA microarray and chromosome sorting technologies. *Chromosome Res.* **12**: 35–43.

Hamada, T., McLean, W.H., Ramsay, M., Ashton, G.H. *et al.* (2002) Lipoid proteinosis maps to 1q21 and is caused by mutations in the extracellular matrix protein 1 gene (ECM1). *Hum. Mol. Genet.* **11**: 833–840.

Hayano, T., Yanagida, M., Yamauchi, Y., Shinkawa, T. *et al.* (2003) Proteomic analysis of human Nop56p-associated pre-ribosomal ribonucleoprotein complexes. Possible link between Nop56p and the nucleolar protein treacle responsible for Treacher–Collins syndrome. *J. Biol. Chem.* **278**: 34309–34319.

Hearn, T., Renforth, G.L., Spalluto, C., Hanley, N.A. *et al.* (2002) Mutation of ALMS1, a large gene with a tandem repeat encoding 47 amino acids, causes Alstrom syndrome. *Nat. Genet.* **31**: 79–83.

Horb, M.E. and Thomsen, G.H. (1999) Tbx5 is essential for heart development. *Development* **126**: 1739–1751.

Hoskins, B.E., Thorn, A., Scambler, P.J. and Beales, P.L. (2003) Evaluation of multiplex capillary heteroduplex analysis: a rapid and sensitive mutation screening technique *Hum. Mutat.* **22**: 151–157.

Huang, W., Wang, S.L., Lozano, G. and de Crombrugghe, B. (2001) cDNA library screening using the SOS recruitment system. *Biotechniques* **30**: 94–98, 100.

Hudder, A. and Werner, R. (2000) Analysis of a Charcot–Marie–Tooth disease mutation reveals an essential internal ribosome entry site element in the connexin-32 gene. *J. Biol. Chem.* **275**: 34586–34591.

Ishkanian, A.S., Malloff, C.A., Watson, S.K., DeLeeuw, R.J. *et al.* (2004) A tiling resolution DNA microarray with complete coverage of the human genome. *Nat. Genet.* **36**: 299–303.

Jaeger, S., Eriani, G. and Martin, F. (2004) Results and prospects of the yeast three-hybrid system. *FEBS Lett.* **556**: 7–12.

James, P., Halladay, J. and Craig, E.A. (1996) Genomic libraries and a host strain designed for efficient two-hybrid selection in yeast. *Genetics* **144**: 1425–1436.

Jeffery, C.J. (2003) Moonlighting proteins: old proteins learning new tricks. *Trends Genet.* **19**: 415–417.

Jerome, L.A. and Papaioannou, V.E. (2001) DiGeorge syndrome phenotype in mice mutant for the T-box gene, *Tbx1*. *Nat. Genet.* **27**: 286–291.

Katsanis, N., Ansley, S.J., Badano, J.L., Eichers, E.R. *et al.* (2001) Evidence for triallelic inheritance in Bardet–Biedl syndrome, a recessive, genetically heterogeneous and pleiotropic disorder. *Science* **293**: 2256–2259.

Kazuki, Y., Shinohara, T., Tomizuka, K., Katoh, M. *et al.* (2001) Germline transmission of a transferred human chromosome 21 fragment in transchromosomal mice. *J. Hum. Genet.* **46**: 600–603.

Kleinjan, D.A., Seawright, A., Schedl, A., Quinlan, R.A. *et al.* (2001) Aniridia-associated translocations, DNase hypersensitivity, sequence comparison and transgenic analysis redefine the functional domain of PAX6. *Hum. Mol. Genet.* **10**: 2049–2059.

Knuesel, M., Wan, Y., Xiao, Z., Holinger, E. *et al.* (2003) Identification of novel protein–protein interactions using a versatile mammalian tandem affinity purification expression system. *Mol. Cell Proteom.* **2**: 1225–1233.

Kobayashi, K., Nakahori, Y., Miyake, M., Matsumura, K. *et al.* (1998) An ancient retrotransposal insertion causes Fukuyama-type congenital muscular dystrophy. *Nature* **394**: 388–392.

Kochilas, L.K., Merscher-Gomez, S., Lu, M.M., Potluri, V. *et al.* (2002) The role of neural crest during cardiac development in a mouse model of DiGeorge syndrome. *Dev. Biol.* **251**: 157–166.

Kourkine, I.V., Hestekin, C.N., Magnusdottir, S.O. and Barron, A.E. (2002) Optimized sample preparation for tandem capillary electrophoresis single-stranded conformational polymorphism/heteroduplex analysis. *Biotechniques* **33**: 318–315.

Kristensen, V.N., Kelefiotis, D., Kristensen, T. and Borresen-Dale, A.L. (2001) High-throughput methods for detection of genetic variation. *Biotechniques* **30**: 318–322, 324, 326.

Krull, C.E. (2004) A primer on using *in ovo* electroporation to analyze gene function. *Dev. Dynam.* **229**: 433–439.

Kunath, T., Gish, G., Lickert, H., Jones, N. *et al.* (2003) Transgenic RNA interference in ES cell-derived embryos recapitulates a genetic null phenotype. *Nat. Biotechnol.* **21**: 559–561.

Kuroiwa, Y., Yoshida, H., Ohshima, T., Shinohara, T. *et al.* (2002) The use of chromosome-based vectors for animal transgenesis. *Gene Ther.* **9**: 708–712.

Lai, C.S., Gerrelli, D., Monaco, A.P., Fisher, S.E. and Copp, A.J. (2003) FOXP2 expression during brain development coincides with adult sites of pathology in a severe speech and language disorder. *Brain* **126**: 2455–2462.

Li, H., Arber, S., Jessell, T.M. and Edlund, H. (1999) Selective agenesis of the dorsal pancreas in mice lacking homeobox gene *Hlxb9*. *Nat. Genet.* **23**: 67–70.

Lindsay, E.A., Botta, A., Jurecic, V., Carattini-Rivera, S. *et al.* (1999) Congenital heart disease in mice deficient for the DiGeorge syndrome region. *Nature* **401**: 379–383.

Lindsay, E.A., Vitelli, F., Su, H., Morishima, M. *et al.* (2001) *Tbx1* haplo-insufficiency identified by functional scanning of the DiGeorge syndrome region is the cause of aortic arch defects in mice. *Nature* **401**: 97–101.

Louie, A.Y., Huber, M.M., Ahrens, E.T., Rothbacher, U. *et al.* (2000) *In vivo* visualization of gene expression using magnetic resonance imaging. *Nat. Biotechnol.* **18**: 321–325.

Malmqvist, M. (1999) BIACORE: an affinity biosensor system for characterization of biomolecular interactions. *Biochem. Soc. Trans.* **27**: 335–340.

Marsh, K.L., Dixon, J. and Dixon, M.J. (1998) Mutations in the Treacher–Collins syndrome gene lead to mislocalization of the nucleolar protein treacle. *Hum. Mol. Genet.* **7**: 1795–1800.

Marx, J. (2003) Imaging. Animal models: live and in color. *Science* **302**: 1880–1882.

Mendelsohn, A.R. and Brent, R. (1999) Protein interaction methods – toward an endgame. *Science* **284**: 1948–1950.

Merscher, S., Funke, B., Epstein, J.A., Heyer, J. *et al.* (2001) *TBX1* is responsible for the cardiovascular defects in velo-cardio-facial/DiGeorge syndrome. *Cell* **104**: 619–629.

Min, T. and Swansbury, J. (2003) Cytogenetic studies using FISH: background. *Methods Mol. Biol.* **220**: 173–191.

Ming, J.E. and Muenke, M. (2002) Multiple hits during early embryonic development: digenic diseases and holoprosencephaly. *Am. J. Hum. Genet.* **71**: 1017–1032.

Mulder, N.J., Apweiler, R., Attwood, T.K., Bairoch, A. *et al.* (2003) The InterPro Database, 2003, brings increased coverage and new features. *Nucleic Acids Res.* **31**: 315–318.

Murphy, K.C. (2002) Schizophrenia and velo-cardio-facial syndrome. *Lancet* **359**: 426–430.

Nishijima, I., Mills, A., Qi, Y., Mills, M. and Bradley, A. (2003) Two new balancer chromosomes on mouse chromosome 4 to facilitate functional annotation of human chromosome 1p. *Genesis* **36**: 142–148.

O'Doherty, A., Ruf, S., Mulligan, C., Hildreth, V., Errington, M.L., Cooke, S., Sesay, A., Modino, S., Vanes, L., Hernandez, D., Linehan, J.M., Sharpe, P.T., Brandner, S., Bliss, T.V., Henderson, D.J., Nizetic, D., Tybulewicz, V.L., and Fisher, E.M. (2005) An aneuploid mouse strain carrying human chromosome 21 with Down syndrome phenotypes. *Science* **309**: 2033–2037.

Packer, A. (2003) Spreading the word. *Nat. Genet.* **35**(suppl 1): 1.

Pekarik, V., Bourikas, D., Miglino, N., Joset, P. *et al.* (2003) Screening for gene function in chicken embryo using RNAi and electroporation. *Nat. Biotechnol.* **21**: 93–96.

Pfeifer, D., Kist, R., Dewar, K., Devon, K. *et al.* (1999) Campomelic dysplasia translocation breakpoints are scattered over 1 Mb proximal to *SOX9*: evidence for an extended control region. *Am. J. Hum. Genet.* **65**: 111–124.

Pravenec, M., Landa, V., Zidek, V., Musilova, A. *et al.* (2001) Transgenic rescue of defective *Cd36* ameliorates insulin resistance in spontaneously hypertensive rats. *Nat. Genet.* **27**: 156–158.

Puech, A., Saint-Jore, B., Merscher, S., Russell, R.G. *et al.* (2000) Normal cardiovascular development in mice deficient for 16 genes in 550 kb of the velocardiofacial/DiGeorge syndrome region. *Proc. Natl Acad. Sci. USA* **97**: 10090–10095.

Ramirez-Solis, R., Liu, P. and Bradley, A. (1995) Chromosome engineering in mice. *Nature* **378**: 720–724.

Roessler, E. and Muenke, M. (2003) How a hedgehog might see holoprosencephaly. *Hum. Mol. Genet.* **12**(1): R15–R25.

Ross, A.J., Ruiz-Perez, V., Wang, Y., Hagan, D.-M. *et al.* (1998) A homeobox gene, *HLXB9*, is the major locus for dominantly inherited sacral agenesis. *Nat. Genet.* **20**: 358–361.

Saran, N.G., Pletcher, M.T., Natale, J.E., Cheng, Y. and Reeves, R.H. (2003) Global disruption of the cerebellar transcriptome in a Down's syndrome mouse model. *Hum. Mol. Genet.* **12**: 2013–2019.

Scambler, P.J. (2003) Velocardiofacial/DiGeorge syndrome. In *Nature Encyclopedia of the Human Genome*, Cooper, D. (ed.). Macmillan: London.

Scherer, L.J. and Rossi, J.J. (2003) Approaches for the sequence-specific knock-down of mRNA. *Nat. Biotechnol.* **21**: 1457–1465.

Schneider, J.E., Bamforth, S.D., Farthing, C.R., Clarke, K. *et al.* (2003) Rapid identification and 3-D reconstruction of complex cardiac malformations in transgenic mouse embryos using fast gradient echo sequence magnetic resonance imaging. *J. Mol. Cell Cardiol.* **35**: 217–222.

Sekar, R.B. and Periasamy, A. (2003) Fluorescence resonance energy transfer (FRET) microscopy imaging of live cell protein localizations. *J. Cell Biol.* **160**: 629–633.

Sellick, G.S., Garrett, C. and Houlston, R.S. (2003) A novel gene for neonatal diabetes maps to chromosome 10p12.1-p13. *Diabetes* **52**: 2636–2638.

Sharpe, J., Ahlgren, U., Perry, P., Hill, B. *et al.* (2002) Optical projection tomography as a tool for 3-D microscopy and gene expression studies. *Science* **19**(296): 541–545.

Shaw-Smith, C., Redon, R., Rickman, L., Rio, M. *et al.* (2004) Microarray-based comparative genomic hybridization (array-CGH) detects submicroscopic chromosomal deletions and duplications in patients with learning disability/mental retardation and dysmorphic features. *J. Med. Genet.* **41**: 241–248.

Shears, D.J., Vassal, H.J., Goodman, F.R., Palmer, R.W. *et al.* (1998) Mutation and deletion of the pseudoautosomal gene *SHOX* cause Leri–Weill dyschondrosteosis. *Nat. Genet.* **19**: 70–73.

Shin, M.K., Levorse, J.M., Ingram, R.S. and Tilghman, S.M. (1999) The temporal requirement for endothelin receptor-B signalling during neural crest development. *Nature* **402**: 496–501.

Shinohara, T., Tomizuka, K., Miyabara, S., Takehara, S. *et al.* (2001) Mice containing a human chromosome 21 model behavioral impairment and cardiac anomalies of Down's syndrome. *Hum. Mol. Genet.* **10**: 1163–1175.

Smirnov, D., Bruzel, A., Morley, M. and Cheung, V.G. (2004) Direct IBD mapping: identical-by-descent mapping without genotyping. *Genomics* **83**: 335–345.

Stagljar, I. (2003) Finding partners: emerging protein interaction technologies applied to signaling networks. *Sci. Sign. Transduct. Knowl. Environ*: http://stke.sciencemag.org/cgi/content/full/OC_sigtrans; 2003/213/pe56.

Stanford, W.L., Cohn, J.B. and Cordes, S.P. (2001) Gene-trap mutagenesis: past, present and beyond. *Nat. Rev. Genet.* **2**: 756–768.

Su, H., Wang, X. and Bradley, A. (2000) Nested chromosomal deletions induced with retroviral vectors in mice. *Nat. Genet.* **24**: 92–95.

Sutherland, H.F., Kim, U.-J. and Scambler, P.J. (1998) Cloning and comparative mapping of the DiGeorge syndrome critical region in the mouse. *Genomics* **52**: 37–43.

Suzuki, T., Takeuchi, J., Koshiba-Takeuchi, K. and Ogura, T. (2004) Tbx genes specify posterior digit identity through Shh and BMP signaling. *Dev. Cell* **6**: 43–53.

The International HapMap Consortium. (2003) The International HapMap Project. *Nature* **426**: 789–796.

Tian, X.L., Kadaba, R., You, S.A., Liu, M. *et al.* (2004) Identification of an angiogenic factor that when mutated causes susceptibility to Klippel–Trenaunay syndrome. *Nature* **427**: 640–645.

To, C., Epp, T., Reid, T., Lan, Q. *et al.* (2004) The Centre for Modeling Human Disease gene trap resource. *Nucleic Acids Res.* **32**(database issue): D557–D559.

Tomizuka, K., Yoshida, H., Uejima, H., Kugoh, H. *et al.* (1997) Functional expression and germline transmission of a human chromosome fragment in chimaeric mice. *Nat. Genet.* **16**: 133–143.

Tufarelli, C., Stanley, J.A., Garrick, D., Sharpe, J.A. *et al.* (2003) Transcription of antisense RNA leading to gene silencing and methylation as a novel cause of human genetic disease. *Nat. Genet.* **34**: 157–165.

Valenzuela, D.M., Murphy, A.J., Frendewey, D., Gale, N.W. *et al.* (2003) High-throughput engineering of the mouse genome coupled with high-resolution expression analysis. *Nat. Biotechnol.* **21**: 652–659.

Vissers, L.E., de Vries, B.B., Osoegawa, K., Janssen, I.M. *et al.* (2003) Array-based comparative genomic hybridization for the genomewide detection of submicroscopic chromosomal abnormalities. *Am. J. Hum. Genet.* **73**: 1261–1270.

Wallis, Y. (2004) Mutation scanning for the clinical laboratory protein truncation test. *Methods Mol. Med.* **92**: 67–79.

Wilkie, A.O.M., Taylor, D., Scambler, P.J. and Baraitser, M. (1993) Congenital cataract, microphthalmia and septal heart defects in two generations: a new syndrome? *Clin. Dysmorphol.* **2**: 114–119.

Woodage, T., Prasad, M., Dixon, J.W., Selby, R.E. *et al.* (1994) Bloom syndrome and maternal uniparental disomy for chromosome 15. *Am. J. Hum. Genet.* **55**: 74–80.

Xiao, W. and Oefner, P.J. (2001) Denaturing high-performance liquid chromatography: a review. *Hum. Mutat.* **17**: 439–474.

Yagi, H., Furutani, Y., Hamada, H., Sasaki, T. *et al.* (2003) Role of *TBX1* in human del22q11.2 syndrome. *Lancet* **362**: 1366–1373.

Yu, Y. and Bradley, A. (2001) Engineering chromosomal rearrangements in mice. *Nat. Rev. Genet.* **2**: 780–790.

Zheng, B., Mills, A.A. and Bradley, A. (1999) A system for rapid generation of coat color-tagged knockouts and defined chromosomal rearrangements in mice. *Nucleic Acids Res.* **27**: 2354–2360.

Zhu, H., Bilgin, M. and Snyder, M. (2003) Proteomics. *Annu. Rev. Biochem.* **72**: 783–812.

5

Transgenic Technology and Its Role in Understanding Normal and Abnormal Mammalian Development

Valerie Vidal and **Andreas Schedl**

Introduction

The last 15 years have been exceptionally successful for developmental biologists. The enormous advances in understanding mammalian development and organ formation would have been unthinkable without the improvements in transgenic technologies. With some exceptions studies are restricted to the mouse, which, due to its small size, rapid cycle of generation and the ease with which its genome can be modified, is now the model of choice. Consequently, mouse development has become a paradigm for the study of developmental processes in mammals and has been used to generate a large variety of models for human disorders. The genetic bases of diseases are diverse and consequently requires a similarly complex technology to study them. Fortunately, the last few years have seen the development of sophisticated tools that allow the re-creation of almost any genetic alteration found in patients in the mouse genome. Moreover, transgenic strategies have been developed to confirm the function of a gene in molecular pathways.

Transgenic technologies can be essentially divided into two separate approaches: classical transgenic mice, in which genetic information (a transgene) integrates randomly into the host genome, and mice that are genetically modified using homologous recombination, thus targeting foreign DNA to a specific locus in the

Embryos, Genes and Birth Defects, Second Edition Edited by Patrizia Ferretti, Andrew Copp, Cheryll Tickle and Gudrun Moore

genome. Both methods are equally important and serve distinct purposes. In the following sections we will outline the basic approaches and applications of each technique, and describe the various improvements that have been developed over the last few years.

Transgenic mice

Principles

Transgenic mice are traditionally generated by micro-injecting linearized DNA into the male pronucleus of fertilized oocytes (Figure 5.1). Characteristically, 1–2 pl of a 2 ng/μl DNA solution is injected, which, with a standard construct, will represent up to 1000 copies of DNA. Integration of transgenes occurs usually at a single site, with as many as 1–100 copies fused in a head-to-tail fashion. When generating transgenic animals, one should be aware that insertion of the DNA occurs randomly into the host genome. This has several important implications. The integration event can occur close to or within a gene and can thus disrupt its function, creating either a loss of function or a dominant phenotype. Loss-of-function mutations are mostly

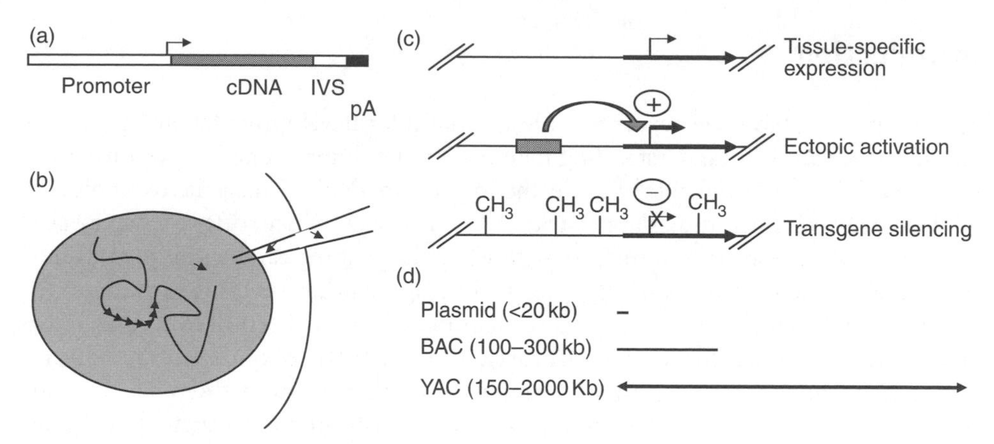

Figure 5.1 Overview of the generation of transgenic mice by microinjection. (a) Typical design of a plasmid-based transgene. A tissue-specific promoter fragment is cloned in front of a cDNA. The addition of an intronic sequence (IVS) and a polyadenylation site (polyA) increases the stability of mRNA produced. (b) Linear DNA is injected into the male pronucleus of fertilized mouse oocytes. Multiple copies of a transgene insert randomly into the host genome, usually as head-to-tail fusions. (c) Possible effects of the integration site on transgene expression. The promoter of a transgene can be ectopically activated by enhancers situated near the site of integration. Alternatively, methylation or integration into heterochromatin can inactivate expression of a transgene. (d) Comparison of transgene sizes. Traditional plasmid-based vectors allow cloning of inserts up to 20 Kb. BAC and YAC constructs are much longer, shield the transgene from position effect and allow cloning of the entire locus, including exons, introns and regulatory elements. Consequently, expression from these transgenes is usually copy-number-dependent and position-independent

recessive and a mutant phenotype can only be seen when transgenic lines are crossed to homozygosity. While this is usually undesirable, it should be noted that gene inactivation via transgenesis has been very instructive, since it leads to the identification of genes with an important function. For example, the integration of a tyrosinase minigene has interrupted the inversin gene, resulting in the *situs inversus* phenotype. Homozygous *inv* mice show a constant reversal of left/right polarity (*situs inversus*) and cyst formation in the kidneys (Mochizuki *et al.*, 1998). Compared to random mutagenesis, e.g. ENU-based, cloning of the underlying gene in transgene-induced mutants is usually easier, since the transgene can be used as a bait to fish out flanking sequences. However, it should be noted that transgenic insertions are occasionally accompanied by rearrangements or deletions at the site of integration, which may complicate the identification of the underlying gene defect.

Besides the inactivation of a gene, the expression of a transgene frequently depends on the site of integration, a feature that is referred to as a 'position effect' and is well known from *Drosophila melanogaster* genetics. Position effects come in many different flavours and can range from methylation-mediated inactivation of the transgene to ectopic activation due to the presence of an enhancer element close to the site of integration (Figure 5.1). As a consequence, transgenic lines generated with the same constructs can result in different phenotypes. Transgenic approaches have therefore to be performed and interpreted with care and should include the analysis of several independent transgenic lines to confirm the specificity of an observed phenotype.

Whereas standard (plasmid) transgenic constructs are constrained by the capacity of plasmid-based cloning vectors (insert size up to 20 kb), the use of yeast artifical chromosome (YAC; Schedl *et al.*, 1992; Jakobovits *et al.*, 1993; Strauss and Jaenisch, 1992) and bacterial artificial chromosome (BAC; Yang *et al.*, 1997) vectors significantly extended the versatility of transgenic approaches. The large size of these constructs (BAC up to 300 kb, YAC up to 2 Mb) allows the cloning of an entire locus rather than a cDNA (Figure 5.1). This has several significant implications. First, the use of a genomic locus allows the inclusion of introns into a transgenic construct. This is an important issue, since many genes are post-transcriptionally modified by alternative splicing or RNA editing. BAC and YAC transgenesis therefore allows the production of all products of a genomic locus in ratios reflecting the endogenous situation. Second, the large size of BAC and YAC constructs usually ensures the presence of regulatory elements, which in some cases can be located several hundred kb upstream or downstream of a gene. Importantly, the use of these constructs does not necessarily require a detailed knowledge of the position of regulatory elements and BAC constructs can be used to drive expression of a transgene in a specific compartment. Moreover, due to their large size and the inclusion of all regulatory elements, expression of transgenes is usually copy number-dependent and position-independent. Thus, this allows the analysis of gene dosage effects in a controlled manner. Improvements in cloning methods and the recent development of homologous recombination techniques in bacteria (Yang *et al.*, 1997; Muyrers *et al.*, 1999; Lee *et al.*, 2001) make these approaches the methods of choice for advanced and reliable analysis in transgenic mice.

Applications

There are several applications of transgenic technologies which include rescue experiments, overexpression and ectopic expression studies, and analysis of *cis*-regulatory sequences. Since it is impossible to provide a complete coverage of all approaches, we have decided to concentrate on those which we believe are of most importance in the study of developmental defects.

Rescue experiments Mouse mutants have been identified and studied for almost a century, initially as spontaneously arising mutants and later on by a more systematic approach, using large-scale mutagenesis. A large number of these mutants showed developmental defects, often mimicking human syndromes. While these models were initially used to characterize the developmental and physiological defects at a descriptive level, it was important to identify the mutation underlying the defect. Transgenic rescue or complementation of a phenotype is a crucial approach to prove that a mutation associated with a disorder is causative for a phenotype. This has been particularly important for positional cloning approaches, in which several genes are removed due to large deletions, e.g. mice that have been generated by X-ray-induced mutagenesis (Antoch *et al.*, 1997). Given the random nature of the large-scale N-ethyl-N-nitrosourea (ENU) mutagenesis screens that are currently performed at several institutes, it is likely that complementation analysis will become more and more important to confirm gene function.

To test whether a gene can rescue a phenotype, the gene in question is expressed in the same spatial and temporal pattern as the endogenous locus, ideally using its own promoter. In that respect, BAC or YAC transgenic approaches are the method of choice, since these large constructs are likely to contain all regulatory elements to confer tissue specific expression of the transgene and do not have to be modified before microinjection.

Overexpression studies

Clearly most disorders are caused by loss-of-function or dominant/dominant-negative mutations affecting gene function. A separate class of disorders is caused by 'overexpression' of a gene/genes. In nature there are different reasons for such an 'overexpression' to occur. These can include mutations in *cis*-regulatory control regions, such as enhancers or promoters, which can lead to an increase of transcriptional activity, mutations that positively affect mRNA or protein stability or those that interfere with degradation pathways. The most common reasons for overexpression-caused disorders, however, are duplications of genomic regions. Duplication of genetic material is often due to translocations, chromosomal duplications or trisomies, most notably trisomy 21 (Down's syndrome). It is clear that the various phenotypic aspects of these syndromes are caused by the overexpression of several genes. Transgenic analysis allows us to dissect the contribution of each gene mapping to a duplicated chromosomal region to a particular phenotype, and has demonstrated that the chromosome 21-encoded *Ets2* gene plays an important role in the cardiac phenotype associated with Down's syndrome (Sumarsono *et al.*, 1996). Clearly, not all genes are sensitive to gene dosage and in many ways it is surprising that a simple increase from two to three copies of a gene can induce developmental defects. How can we explain this observation? From an evolutionary point of view it makes sense to express sufficient amounts of a protein from one allele, so that mutations in the second allele do not interfere with survival. This also allows the second allele to acquire a new function and thus contribute to the evolutionary process. However, there are genes from which a precise amount of protein has to be produced. These genes are often signalling molecules or factors involved in setting up or interpreting a protein gradient. Changes in expression levels will distort this gradient and result in changes of the timing of development.

Ectopic expression studies

In contrast to overexpression studies, in which a gene is expressed according to its endogenous pattern, ectopic expression directs a transgene to tissues or at times where it is not normally found. This approach is very important, as it allows us to test whether expression of a gene is sufficient to activate a particular downstream target and whether it can direct differentiation of cells along a specific developmental pathway. As an example, we have ectopically expressed the male-specific gene *Sox9* in XX gonads using a gonad-specific promoter. Transgenic XX mice developed testes instead of ovaries (Vidal *et al.*, 2001). This study thus confirmed that *Sox9* is sufficient to induce the male-specific pathway and placed the gene at the top of a molecular cascade of male development.

Ectopic expression studies can be performed either using a well-defined promoter fragment of a heterologous promoter or by replacing the coding region of a gene (knock-in approach). Although the latter strategy is often performed using gene targeting in embryonic stem cells (see below), the advances in YAC/BAC transgenic technologies and the development of BAC cloning techniques using homologous recombination now also allow the efficient insertion of a cDNA into a locus encoded on a BAC. The advantages of this approach include the much more rapid generation of transgenic lines when compared to ES-cell gene targeting and the fact that BAC transgenesis does not usually lead to the disruption of an endogenous gene.

Analysis of *cis*-regulatory elements

During development expression of genes has to be tightly regulated. This is achieved through an intricate network of factors that bind to promoter and enhancer sequences and control the transcription of downstream target genes. Molecular pathways are not necessarily linear and there is a significant amount of cross-regulation between individual factors. While most enhancers are located in close proximity to the 5′end of a gene, some genes have very complex regulatory regions, which can extend over as much as 1 Mb. This is demonstrated in patients suffering from syndromes such as campomelic dysplasia or aniridia, in which rearrangements (translocation or inversions) up to 1 Mb upstream of *SOX9* (Pfeifer *et al.*, 1999) or 300 kb downstream of *PAX6* (Kleinjan *et al.*, 2001), respectively, result in severe developmental disorders. Besides the classical enhancer elements there are also sequences that may have a more basic role in organizing chromatin structure (chromatin organizer, matrix attachment region, locus control regions or insulator sequences) but which can be equally important for proper expression of a gene.

Understanding the regulation of genes is crucial, as it allows us to decipher molecular networks and thus to comprehend developmental processes at the molecular level. Moreover, sequences that regulate expression are essential for proper functioning of a gene and are therefore potential sites for mutations in human diseases. Although mutations are mostly identified in the coding region of a genes, this may simply represent the bias of researchers, since regulatory sequences are less obvious than open reading frames and thus less accessible for mutagenesis screens.

Traditionally, *in vitro* studies, such as co-transfection into cultured cell lines, were employed to show that a specific sequence is required for the activation of a gene. Since these approaches do not take into account chromatin structure, data obtained by *in vitro* analysis have to be interpreted with care. In transgenic mice, and particularly in YAC/BAC transgenic mice, we can analyse the importance of regulatory elements in an almost natural context. To facilitate the analysis of enhancers, a reporter gene (*lacZ* or *GFP*) is linked to a minimal promoter, which gains activity only in the context of additional expression-promoting sequences (Kothary *et al.*, 1989). This type of analysis has been used to demonstrate the modularity of activating sequences with individual tissue-specific enhancers attached to one gene.

Inducible systems

So far we have discussed the use of endogenous genomic sequences to direct transgene expression to a specific tissue. While this is adequate for a large number of assays, it is clear that many applications require a more amenable way of controlling the expression of a gene. This is particularly desirable for the analysis of gene function in the adult situation, e.g. physiological problems. Recent developments of inducible systems are beginning to address this problem and there are now several options to control activation of a gene through the administration of drugs. Probably the most frequently used approach is based on the bacterial tetracycline sensing system, using the tetracycline-controlled transactivator tTA (effector) in combination with a responsive promoter (responder construct), usually encoded on two distinct transgenes and brought together by genetic mating (Figure 5.2). tTA represents a fusion protein between the DNA-binding domain of the tetracycline repressor (TetR) coupled to the transcriptional activation domain of the viral protein

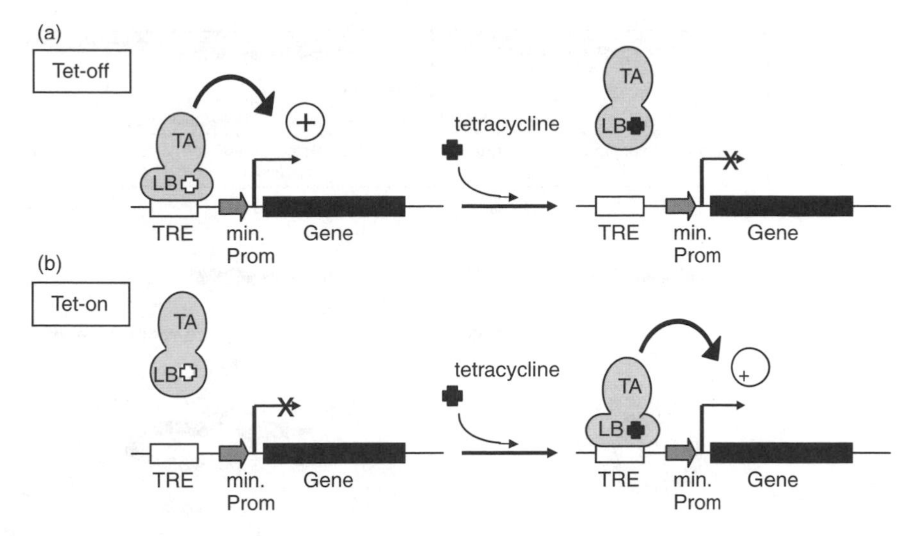

Figure 5.2 Schematic representation of the tetracycline-inducible system. Two alternative systems are available. (a) Tet-off: in the absence of tetracycline, the DNA-binding domain of the tetracycline-controlled transactivator (tTA) interacts with the recognition sequence and activates expression of the transgene. Binding of tetracycline to the ligand-binding domain of the tTA protein induces conformational changes, leading to the release of tTA from the DNA recognition sequence. The expression of the transgene is then shut down. (b) Tet-on system. The rtTA protein is initially inactive. Upon binding of tetracycline to the ligand-binding domain, rtTA gains DNA-binding activity and the transgene becomes activated

VP16. In its original version, the DNA-binding domain of the tTA protein binds to its recognition sequence in the absence of its ligand tetracycline and the target promoter is thus active. Upon binding of tetracycline, the repressor changes its conformation, loses its affinity to the DNA recognition sequence and, as a result, transcription stops (Tet-off system). The tetracycline system also exists in the opposite configuration, which allows activation of a target gene upon tetracycline administration (Tet-on system). Nowadays tetracycline is usually substituted by the commercially available compound Doxycycline (Dox). In addition to the tetracycline-inducible system, there exist some alternative hormone receptor-based strategies, including progesterone (Gardner *et al.*, 1996; Kellendonk *et al.*, 1999; Wang *et al.*, 1999) and ecdysone (No *et al.*, 1996). Since they have not been used very much in transgenic approaches, we will not discuss them in detail.

Induction using the Cre–loxP system

In addition to the above substrate-induced transgenic system, we can also induce transgene expression using a site-specific recombinase (e.g. Cre, Flp) in combination with a stop cassette. Site-specific recombinases are extremely versatile for the manipulation of the mouse genome and have been used for a wide variety of

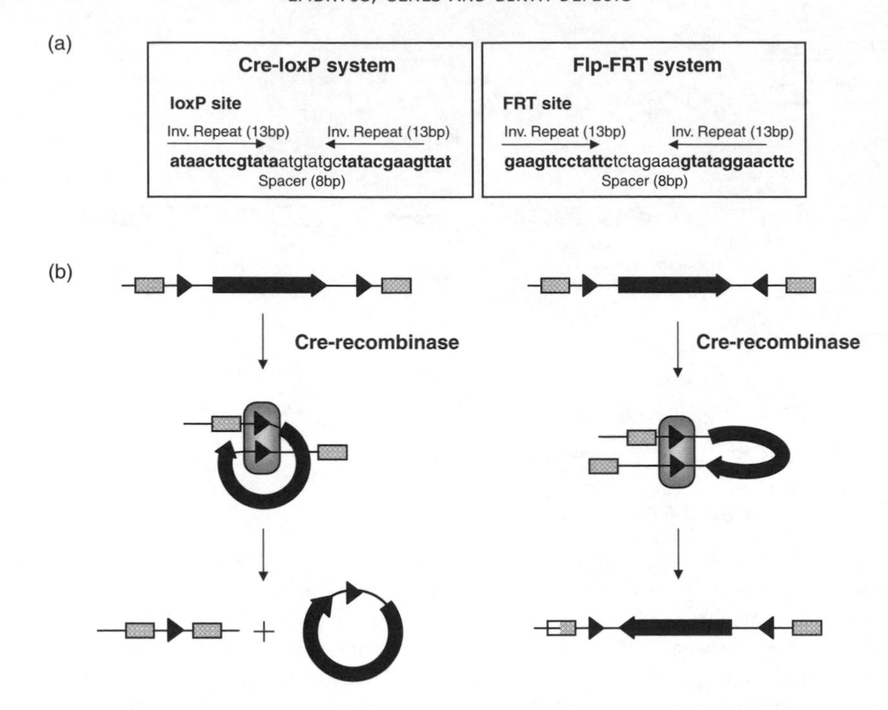

Figure 5.3 Site-specific recombinases used for conditional gene targeting. (a) LoxP and FRT are palindromic sequences interrupted by an 8 bp spacer, which are recognized by the Cre and Flp recombinases, respectively. (b) Excision vs. inversion. Excision of intervening DNA occurs when the loxP or frt sites are inserted in the same orientation. When the loxP or frt sequences are in the opposite orientation, recombination between these sites will lead to the inversion of the intervening sequence. LoxP or frt sites are represented by black triangles

approaches, including conditional gene targeting and chromosome engineering (see below). Hence it is worthwhile discussing some of their features in more detail.

Site-specific recombinases recognize short palindromic sequences (34 bp) and induce homologous recombination between two of these sites (Figure 5.3). Depending on the orientation of the pair of recognition sites, Cre-recombination results in the inversion (opposite orientation) or excision (same orientation) of the intervening sequence (Figure 5.3). Currently there are two site-specific recombination systems in use to manipulate the mouse genome: the Cre–loxP system, which was isolated from the P1 bacteriophage, and the yeast Flp–FRT system (Figure 5.3; for review, see Branda and Dymecki, 2004).

The Cre–loxP system has been used to activate genes in a tissue-specific manner. The gene of interest is cloned under the control of a ubiquitously expressing promoter. To avoid expression of the gene in all tissues, a stop cassette (inactivation cassette) flanked by loxP sites is inserted between the transcription start site and the coding region of the gene, thus rendering this transgene silent. Activation is achieved via a second transgene, which expresses the Cre-recombinase in a tissue-specific manner. In tissues where no Cre is expressed the transgene remains silent. In contrast,

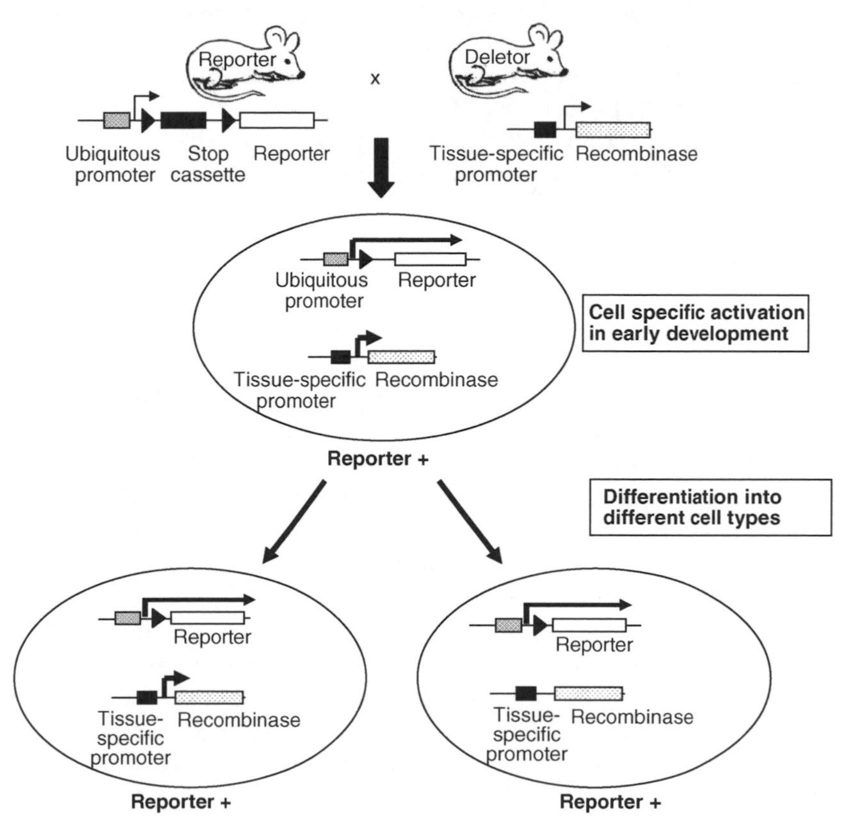

Figure 5.4 Cell lineage analysis. To determine the origin of a cell type, mice carrying an inactivated reporter gene placed under the control of an ubiquitous promoter are crossed with mice carrying the Cre recombinase under the control of a tissue-specific promoter. In double transgenic animals, the Cre recombinase will be expressed in a tissue-specific manner, leading to the excision of the inactivation cassette and expression of the reporter gene in this subset of cells. During embryogenesis these cells may differentiate along different pathways. However, all cells will maintain expression of the reporter gene, which was activated at an early time point. Thus, this analysis can be used to trace back the developmental origin of a cell type

Cre-recombinase expression leads to excision of the stop cassette and hence activation of the transgene (Figure 5.4). It should be noted that, in contrast to the tetracycline or hormone-induced systems, this activation is irreversible, since it is based on a permanent genetic modification of the transgene. Although this may be of disadvantage for some applications, developmental biologists have made use of this system by employing it for cell lineage analysis.

To achieve this, a reporter gene (*lacZ*) driven by a ubiquitous promoter was generated, which in its original form is inactive due to a *loxP*-flanked polyadenylation signal after the transcription start of the *lacZ* reporter gene. To analyse cell lineages, these reporter mice are crossed with animals transgenic for a

Cre-recombinase expressed in a tissue-specific pattern. Excision of the polyadenylation signal, and thus activation of the reporter gene, occurs only in those tissues where the Cre-recombinase is active. However, since the reporter gene is driven by a ubiquitously expressing promoter, all tissues derived from this cell will maintain expression of *lacZ*, independent of the transcription state of the *Cre* transgene. This type of analysis has provided important clues about the inter-relations of cells and tissues.

siRNA approaches

While homologous recombination provides a very direct way of identifying the function of a gene, the establishment of mice carrying a targeted allele is still a time-consuming process. siRNA (short interfering RNA) approaches may offer an attractive alternative. siRNA technology is based on the ability of short double-stranded RNA oligos, or hairpin structures, to align with cellular mRNA and induce degradation of the transcript (for review, see McManus and Sharp, 2002). Since siRNA only rarely causes degradation of 100% of transcripts, we generally refer to a 'knock-down' rather than a 'knock-out' approach. Consequently, phenotypes achieved vary with the siRNA used and often represent hypomorphs. This methodology has now been tested extensively in cell culture, organ culture (Sakai *et al.*, 2003; Davies *et al.*, 2004) and in some cases even in transgenic mice (Kunath *et al.*, 2003; Carmell *et al.*, 2003). Despite the enormous potential of siRNA approaches, we should be cautious with this new technology. This was demonstrated by a recent report that showed that siRNA can induce unexpected changes in the expression levels of untargeted proteins (Scacheri *et al.*, 2004).

Genetic manipulation using gene targeting in ES cells

Of equal importance to transgenic techniques has been the development of gene-targeting technology for the analysis of the function of genes in developmental processes. A crucial step was the establishment of totipotent embryonic stem (ES) cell lines which, when reintroduced into mouse blastocysts, can give rise to all cell types. ES cells can be easily manipulated *in vitro*, allowing the introduction of a variety of genetic alterations at an endogenous gene locus.

Principles

Site-directed mutagenesis of the mouse genome is based on homologous recombination between a targeting construct introduced by electroporation into ES cells and the corresponding endogenous sequence brought about by the cellular recombination system. Clones are selected on the basis of a selectable marker (e.g. neomycin) co-introduced with the targeting construct and can include a counter-selection (e.g. thymidine kinase) against non-homologously recombined clones (Figure 5.5). An

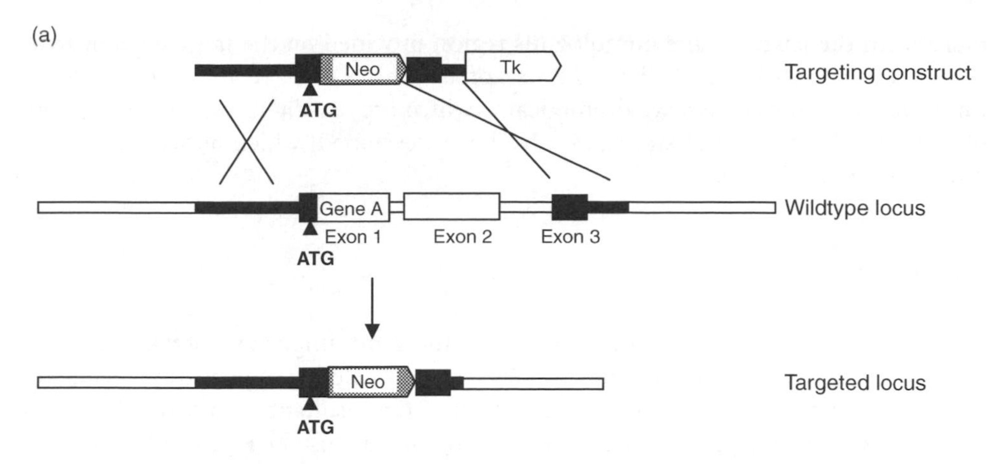

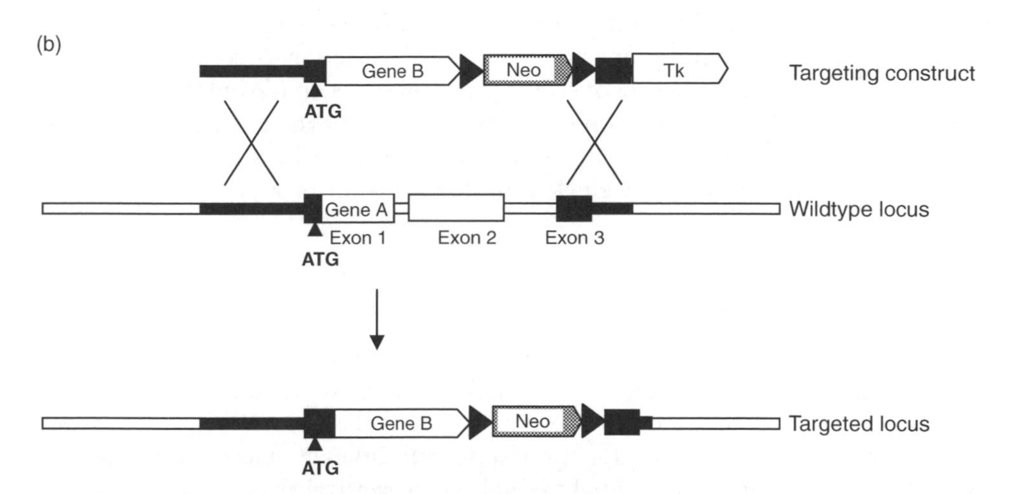

Figure 5.5 Site-directed mutagenesis of the mouse genome. (a) Design of a knock-out strategy. The targeting construct possesses a selection marker (*Neo*) flanked by sequences homologous to the target gene (Gene A), and in addition it contains a counter-selection marker (*Tk*) at its 3′-end. Upon electroporation of this construct in ES cells, recombination will take place between homologous sequences, leading to the replacement of the endogenous wild-type gene by its knock-out allele. (b) Standard design of a knock-in (gene-replacement) experiment. The targeting construct harbours the gene to be inserted (Gene B), selection (*Neo*) and counter-selection markers (*Tk*), and it possesses homology to an endogenous locus (Gene A). Upon electroporation of the targeting construct into ES cells, recombination occurs between sequences of homology, leading to the insertion of the construct into the genomic locus. Expression of the transgene will be driven by regulatory elements provided by the endogenous ('host') locus. *Neo*, neomycin resistance gene; *Tk*, thymidine kinase gene

increase in the length of the homologous region provided on the targeting construct (Hasty *et al.*, 1991b), as well as the use of isogenic DNA (te Riele *et al.*, 1992), improves targeting efficiencies dramatically, which are usually in the range 0.1–30% of selected clones. Several strategies have been developed, which allow a variety of different manipulations.

Knock-out

The standard knock-out approach has been the most important breakthrough in analysing gene function. Function is usually tested by deleting sequences that code for an important part of the protein and observing the resultant phenotype. To allow easy monitoring of gene expression, part of the gene can also be replaced by a marker gene such as *lacZ.* This insertion has two positive effects. First, it allows determination of the expression of the endogenous gene in much greater detail than conventional methods allow and has often helped in discovering new expression domains, due to its higher sensitivity compared to *in situ* hybridization. Second, the *lacZ* gene can be used to determine the fate of cells in a homozygous knock-out. For example, if a gene is required for the migration or survival of cells, a complete knock-out of this gene would result in misplacement or absence of these cells. Such an analysis can therefore be very instructive in elucidating the function of a gene during normal development.

Conditional knock-outs

Although gene targeting represents a remarkable tool to analyse gene function, it quickly became clear that there are several questions that cannot be addressed with this methodology. Genes often function at several different stages during development. However, if a gene is essential for embryonic survival at an early development stage, a complete knock-out would result in embryonic lethality and thus preclude the analysis of gene function at later stages of development. Conditional mutagenesis using site-specific recombinases provided the answer to this problem. To direct deletion of a gene in a specific tissue, transgenic mice are generated with the site-specific recombinase (Cre) cloned under the control of a tissue-specific promoter. Genetic mating of transgenic mice with animals carrying a conditional allele leads to excision (inactivation) of a gene only in those tissues were the Cre-recombinase is expressed (Figure 5.6). For example, this tissue-specific approach allowed the analysis of β-catenin function in a variety of tissues, including the hair (Huelsken *et al.*, 2001), nervous system (Hari *et al.*, 2002; Brault *et al.*, 2001) and endoderm (Lickert *et al.*, 2002), despite the fact that complete knock-out mice die at E7.5 due to gastrulation defects (Haegel *et al.*, 1995). At present the major limitation of this technique is the specificity and efficiency of *Cre* lines available, a problem which will surely be addressed in future research.

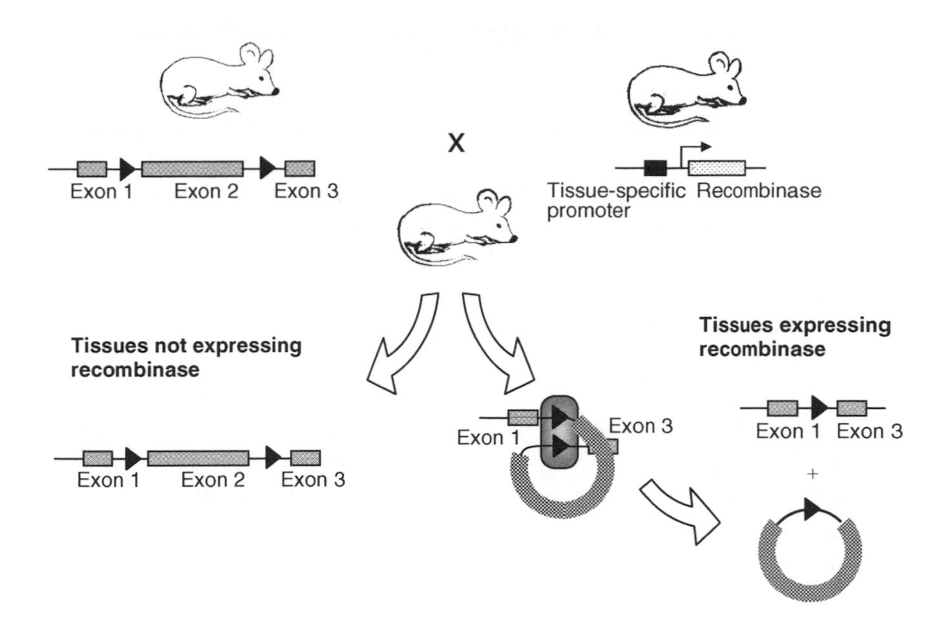

Figure 5.6 Design of a conditional knock-out experiment. Transgenic mice are generated from modified ES cells, which contain two *LoxP* sites inserted into the locus to be depleted. These mice are crossed with mice expressing the Cre-recombinase in a tissue-specific manner. In double transgenic mice, in cells expressing the Cre-recombinase, the genomic DNA flanked by the *LoxP* sites will be excised, generating a null allele

In addition to straightforward tissue-specific approaches, inducible *Cre* lines have been developed that allow stimulation of *Cre* activity upon drug administration. The most commonly used is a fusion between the oestrogen receptor ligand-binding domain and the Cre-recombinase (Cre–ER) (Brocard *et al.*, 1997; Feil *et al.*, 1996). This fusion renders the Cre-recombinase inactive. However, upon binding of its ligand the three-dimensional structure of the protein changes and the Cre-recombinase becomes active. Thus, administration of tamoxifen can be used to induce recombination, an approach that has been employed to delete the retinoid X receptor-α (RXR-α in adult mouse keratinocytes; Li *et al.*, 2000).

Introduction of subtle mutations

A high proportion of human disorders are due to single base changes rather than complete deletions. While some of these mutations result in a complete loss of gene function (frame shift, stop-codon, etc.), there exist also many other possible consequences. Point mutations can result in reduced activity of the protein and thus create 'hypomorphic alleles' or they can interfere with only one particular function in the case of multifunctional proteins. In addition, they can provoke additional functions or constitutive activation of a protein. While the molecular consequence of such a mutation can often be tested *in vitro*, the developmental aspect has to be analysed *in vivo*. Consequently, there have been an increasing number of studies using subtle gene targeting to recreate mutations found in human patients.

Introduction of point mutations is more demanding than a simple knock-out and usually requires two consecutive steps of genetic manipulation. There are several strategies available: 'hit-and-run', 'double-replacement' and 'Cre–lox-based' strategies (see Figure 5.7). The hit-and-run strategy was first described by Hasty *et al.* (1991a) and is based on an integration-type targeting vector carrying the mutation. Integration occurs through a single cross-over event and leads to a duplication of part of the gene. The second step involves the spontaneous excision of the integrated allele. Excised clones are selected for using a counter-selectable marker, such as *HPRT* or the thymidine kinase gene (*TK*). The excision event can result in either the wild-type or the mutant locus. Since this second event is rather inefficient, this methodology has not been used very often.

In contrast to the hit-and-run strategy, double replacement is achieved through two consecutive rounds of targeting events, using replacement-type vectors (Reid *et al.*, 1990). In the first targeting, a counter-selectable marker (e.g. *TK*) is inserted at a position close to the site where the subtle mutation should be introduced. The second targeting event is designed to replace the counter-selectable marker with a piece of genomic DNA carrying the desired mutation. Similarly to the hit-and-run strategy, the efficiency of this technique can be rather low, since gene conversion of the targeted to the wild-type allele significantly contributes to false positives in the second round of targeting.

By now the most frequently used technology to create subtle mutations is based on the Cre–lox system. The mutation is introduced in one step using a standard gene replacement approach. In a second step the selectable marker, which is flanked by recombinase recognition sites, is removed by transient transfection with a site-specific recombinase. Since this second step does not require homologous recombination, but just transient expression of the Cre-recombinase, deletion of the selectable marker is highly efficient. A drawback of this strategy is that one recognition site is left behind during the excision event. Hence, to confirm that this sequence does not cause any unwanted effects, it is essential to generate a control allele carrying the recombination recognition site but lacking the point mutation (see e.g. Arango *et al.*, 1999). Mice carrying this control allele should develop normally.

Knock-in strategies

We have seen above that ectopic expression of genes is often desirable to gain information about the function of a gene. Transgenic constructs are often hampered by the lack of regulatory elements driving expression of a gene in the appropriate compartment. As an alternative we can introduce a cDNA coding for a gene of interest

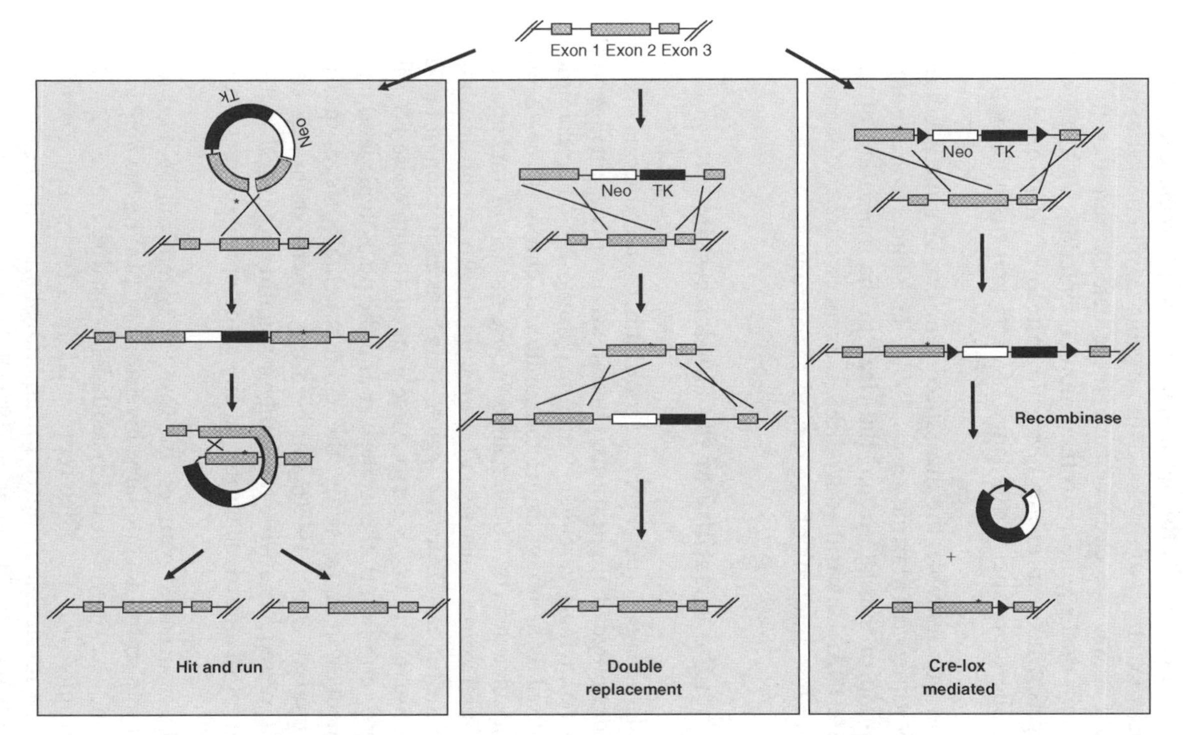

Figure 5.7 Insertion of a point mutation into the mouse genome. Hit-and-run system: the targeting vector possesses selection and counter-selection markers, as well as regions of homology harbouring a point mutation (star) to be inserted into the locus of interest. This vector is linearized within the region of homology and introduced by electroporation into ES cells. This plasmid integrates ('hit' step) into the locus by homologous recombination, leading to a duplication of the genomic sequences. Spontaneous reversion occurs ('run' step) by intra-chromosomal crossing-over. Depending on the position of the crossing over, the locus either reverts to wild-type or is replaced by the mutated version. Double replacement: in the first step, the locus of interest is tagged by insertion via homologous recombination of a selection/counter-selection cassette (*Neo/Tk*). In the second step, the tagged ES cells are electroporated with a linear fragment homologous to the tagged genomic sequence, which contains the desired point mutation (asterisk). Homologous recombination results in replacement of the selection/counter-selection cassette with the gene carrying the point mutation. Cre–lox-based system: the targeting construct consists of a sequence homologous to the endogenous locus but carrying the desired point mutation (asterisk). In addition, a selection/counter-selection marker flanked by *loxP* sites is inserted close to the mutation. Homologous recombination in ES cells replaces the endogenous locus with the targeting construct. Finally, the selection/counter-selection cassette is removed by transient expression of the Cre-recombinase in ES cells. Note that one *loxP* site remains in the mutated genomic locus

into an existing gene, using homologous recombination. This gene-replacement or 'knock-in' strategy usually results in the inactivation of the underlying gene (Figure 5.5(b)). Although this may be undesirable in some cases, this approach has been successfully used to test the functional redundancy of genes and to help to clarify molecular pathways. For example, this technique was used to replace the *Myf5* locus with myogenin cDNA (Wang *et al.*, 1996). Homozygous *Myf5* knock-out mice show rib cage defects, but it was not clear whether these defects were due to the failure of the early activation of the gene or to the unique interactions of *Myf5* with specific downstream targets. Mice expressing *myogenin* instead of *Myf5* developed normally, demonstrating the functional redundancy of *Myf5* and *myogenin* for rib formation.

If replacement of a gene is undesirable, the introduced gene can be joined with the existing locus using IRES sequences (Martinez-Salas, 1999). These internal ribosomal entry sites allow re-initiation of translation and thus lead to the translation of two proteins from the same cDNA. Unfortunately, IRES sequences often work inefficiently and produce rather low amounts of the second gene product.

Chromosome engineering: the versatility of the Cre–lox system

Although the Cre–lox system plays a crucial role for conditional gene targeting, it can be applied to a much wider range of applications. Human syndromes are frequently caused by large deletions removing several genes on one chromosome. For example, DiGeorge syndrome is caused by hemizygous deletion of human chromosome 22 and characterized by cardiovascular defects. To recreate a mouse model for this disease, two *loxP* sequences were inserted on either side of the region on mouse chromosome 16, which corresponds to the human DiGeorge region. Transient expression of *Cre* in such ES resulted in a precise deletion of 1.2 Mb. Mice generated with these ES cells mimicked the human syndrome at both the molecular and the phenotypic level, and thus provided an excellent system to study the developmental defects of this developmental disease (Lindsay *et al.*, 1999). Furthermore, when combining site-specific integration of a *loxP* sequence with randomly integrating retroviral vectors carrying a *loxP* site, a series of nested deletions on the same chromosome can be generated (Su *et al.*, 2000).

Since *loxP* sites work in an oriented manner, the Cre–lox system can be used not only to delete sequences but also to generate large inversions by placing the *loxP* sites in the opposite orientation. This approach has recently been used to create a 24 cM inversion on mouse chromosome 11. This represents the first mouse balancer chromosome and will be an invaluable resource for functional analysis in combination with ENU mutagenesis screens (Zheng *et al.*, 1999; Kile *et al.*, 2003).

Finally, integration of *loxP* sites on distinct chromosomes and subsequent recombination between these *lox* sites has been used to generate inter-chromosomal translocations similar to those found in human patients (Smith *et al.*, 1995; Van Deursen *et al.*, 1995).

Outlook and future developments

Transgenic and ES cell technology have drastically changed the way developmental problems are addressed. Despite the already impressive toolbox available, there are still new techniques evolving. Surely, chromosome engineering in combination with ENU mutagenesis will become extremely important to generate mutants for specific parts of the mouse genome. On the other hand, improvements in siRNA technology may allow rapid functional screens for genes in the transgenic system and may change the way developmental problems are addressed. Finally, transgenic technology may evolve towards mammalian artificial chromosomes (for a recent review, see Larin and Mejia, 2002), which would not only increase the efficiency of transgenesis but, more importantly, would also avoid the undesirable integration of the transgene into the host genome. The importance of such vectors for somatic gene therapy is self-evident.

References

Antoch, M.P., Song, E.J., Chang, A.M., Vitaterna, M.H. *et al.* (1997) Functional identification of the mouse circadian *Clock* gene by transgenic BAC rescue. *Cell* **89**(4): 655–667.

Arango, N.A., Lovell-Badge, R. and Behringer, R.R. (1999) Targeted mutagenesis of the endogenous mouse *Mis* gene promoter: *in vivo* definition of genetic pathways of vertebrate sexual development. *Cell* **99**(4): 409–419.

Branda, C.S. and Dymecki, S.M. (2004) Talking about a revolution: the impact of site-specific recombinases on genetic analyses in mice. *Dev. Cell* **6**(1): 7–28.

Brault, V., Moore, R., Kutsch, S., Ishibashi, M. *et al.* (2001) Inactivation of the β-catenin gene by Wnt1-Cre-mediated deletion results in dramatic brain malformation and failure of craniofacial development. *Development* **128**(8): 1253–1264.

Brocard, J., Warot, X., Wendling, O., Messaddeq, N. *et al.* (1997) Spatio-temporally controlled site-specific somatic mutagenesis in the mouse. *Proc. Natl Acad. Sci. USA* **94**(26): 14559–14563.

Carmell, M.A., Zhang, L., Conklin, D.S., Hannon, G.J. and Rosenquist, T.A. (2003) Germline transmission of RNAi in mice. *Nat. Struct. Biol.* **10**(2): 91–92.

Davies, J.A., Ladomery, M., Hohenstein, P., Michael, L. *et al.* (2004) Development of an siRNA-based method for repressing specific genes in renal organ culture and its use to show that the Wt1 tumour suppressor is required for nephron differentiation. *Hum. Mol. Genet.* **13**(2): 235–246.

Feil, R., Brocard, J., Mascrez, B., LeMeur, M. *et al.* (1996) Ligand-activated site-specific recombination in mice. *Proc. Natl Acad. Sci USA* **93**(20): 10887–10890.

Gardner, D.P., Byrne, G.W., Ruddle, F.H. and Kappen, C. (1996) Spatial and temporal regulation of a *lacZ* reporter transgene in a binary transgenic mouse system. *Transgen. Res.* **5**(1): 37–48.

Haegel, H., Larue, L., Ohsugi, M., Fedorov, L. *et al.* (1995) Lack of β-catenin affects mouse development at gastrulation. *Development* **121**(11): 3529–3537.

Hari, L., Brault, V., Kleber, M., Lee, H.Y. *et al.* (2002) Lineage-specific requirements of β-catenin in neural crest development. *J. Cell Biol.* **159**(5): 867–880.

Hasty, P., Ramirez-Solis, R., Krumlauf, R. and Bradley, A. (1991a) Introduction of a subtle mutation into the *Hox-2.6* locus in embryonic stem cells. *Nature* **350**(6315): 243–246.

Hasty, P., Rivera-Perez, J. and Bradley, A. (1991b) The length of homology required for gene targeting in embryonic stem cells. *Mol Cell Biol* **11**(11): 5586–5591.

Huelsken, J., Vogel, R., Erdmann, B., Cotsarelis, G. and Birchmeier, W. (2001). β-Catenin controls hair follicle morphogenesis and stem cell differentiation in the skin. *Cell* **105**(4): 533–545.

Jakobovits, A., Moore, A.L., Green, L.L., Vergara, G.J. *et al.* (1993) Germ-line transmission and expression of a human-derived yeast artificial chromosome. *Nature* **362**(6417): 255–258.

Kellendonk, C., Tronche, F., Casanova, E., Anlag, K. *et al.* (1999) Inducible site-specific recombination in the brain. *J Mol Biol* **285**(1): 175–182.

Kile, B.T., Hentges, K.E., Clark, A.T., Nakamura, H. *et al.* (2003) Functional genetic analysis of mouse chromosome 11. *Nature* **425**(6953): 81–86.

Kleinjan, D.A., Seawright, A., Schedl, A., Quinlan, R.A. *et al.* (2001) Aniridia-associated translocations, DNase hypersensitivity, sequence comparison and transgenic analysis redefine the functional domain of PAX6. *Hum. Mol. Genet.* **10**(19): 2049–2059.

Kothary, R., Clapoff, S., Darling, S., Perry, M.D. *et al.* (1989) Inducible expression of an *hsp68-lacZ* hybrid gene in transgenic mice. *Development* **105**(4): 707–714.

Kunath, T., Gish, G., Lickert, H., Jones, N., Pawson, T. and Rossant, J. (2003) Transgenic RNA interference in ES cell-derived embryos recapitulates a genetic null phenotype. *Nat. Biotechnol.* **21**(5): 559–561.

Larin, Z. and Mejia, J.E. (2002) Advances in human artificial chromosome technology. *Trends Genet.* **18**(6): 313–319.

Lee, E.C., Yu, D., Martinez de Velasco, J., Tessarollo, L. *et al.* (2001) A highly efficient *Escherichia coli*-based chromosome engineering system adapted for recombinogenic targeting and subcloning of *BAC* DNA. *Genomics* **73**(1): 56–65.

Li, M., Indra, A.K., Warot, X., Brocard, J. *et al.* (2000) Skin abnormalities generated by temporally controlled RXRα mutations in mouse epidermis. *Nature* **407**(6804): 633–636.

Lickert, H., Kutsch, S., Kanzler, B., Tamai, Y. *et al.* (2002) Formation of multiple hearts in mice following deletion of β-catenin in the embryonic endoderm. *Dev. Cell* **3**(2): 171–181.

Lindsay, E.A., Botta, A., Jurecic, V., Carattini-Rivera, S. *et al.* (1999) Congenital heart disease in mice deficient for the DiGeorge syndrome region. *Nature* **401**(6751): 379–383.

Martinez-Salas, E. (1999). Internal ribosome entry site biology and its use in expression vectors. *Curr. Opin. Biotechnol.* **10**(5): 458–464.

McManus, M.T. and Sharp, P.A. (2002) Gene silencing in mammals by small interfering RNAs. *Nat. Rev. Genet.* **3**(10): 737–747.

Mochizuki, T., Saijoh, Y., Tsuchiya, K., Shirayoshi, Y. *et al.* (1998) Cloning of *inv*, a gene that controls left/right asymmetry and kidney development. *Nature* **395**(6698): 177–181.

Muyrers, J.P., Zhang, Y., Testa, G. and Stewart, A.F. (1999) Rapid modification of bacterial artificial chromosomes by ET-recombination. *Nucleic Acids Res.* **27**(6): 1555–1557.

No, D., Yao, T.P. and Evans, R.M. (1996) Ecdysone-inducible gene expression in mammalian cells and transgenic mice. *Proc. Natl Acad. Sci. USA* **93**(8): 3346–3351.

Pfeifer, D., Kist, R., Dewar, K., Devon, K. *et al.* (1999). Campomelic dysplasia translocation breakpoints are scattered over 1 Mb proximal to SOX9: evidence for an extended control region. *Am. J. Hum. Genet.* **65**(1): 111–124.

Reid, L.H., Gregg, R.G., Smithies, O. and Koller, B.H. (1990) Regulatory elements in the introns of the human HPRT gene are necessary for its expression in embryonic stem cells. *Proc. Natl Acad. Sci. USA* **87**(11): 4299–4303.

Sakai, T., Larsen, M. and Yamada, K.M. (2003) Fibronectin requirement in branching morphogenesis. *Nature* **423**(6942): 876–881.

Scacheri, P.C., Rozenblatt-Rosen, O., Caplen, N.J., Wolfsberg, T.G. *et al.* (2004) Short interfering RNAs can induce unexpected and divergent changes in the levels of untargeted proteins in mammalian cells. *Proc. Natl Acad. Sci. USA* **101**(7): 1892–1897.

Schedl, A., Beermann, F., Thies, E., Montoliu, L. *et al.* (1992) Transgenic mice generated by pronuclear injection of a yeast artificial chromosome. *Nucleic Acids Res.* **20**(12): 3073–3077.

Smith, A.J., De Sousa, M.A., Kwabi-Addo, B., Heppell-Parton, A. *et al.* (1995) A site-directed chromosomal translocation induced in embryonic stem cells by Cre–loxP recombination. *Nat. Genet.* **9**(4): 376–385.

Strauss, W.M. and Jaenisch, R. (1992) Molecular complementation of a collagen mutation in mammalian cells using yeast artificial chromosomes. *EMBO J.* **11**(2): 417–422.

Su, H., Wang, X. and Bradley, A. (2000) Nested chromosomal deletions induced with retroviral vectors in mice. *Nat. Genet.* **24**(1): 92–95.

Sumarsono, S.H., Wilson, T.J., Tymms, M.J., Venter, D.J. *et al.* (1996) Down's syndrome-like skeletal abnormalities in Ets2 transgenic mice. *Nature* **379**(6565): 534–537.

te Riele, H., Maandag, E.R. and Berns, A. (1992) Highly efficient gene targeting in embryonic stem cells through homologous recombination with isogenic DNA constructs. *Proc. Natl Acad. Sci. USA* **89**(11): 5128–5132.

Van Deursen, J., Fornerod, M., Van Rees, B. and Grosveld, G. (1995) Cre-mediated site-specific translocation between nonhomologous mouse chromosomes. *Proc. Natl Acad. Sci. USA* **92**(16): 7376–7380.

Vidal, V.P., Chaboissier, M.C., de Rooij, D.G. and Schedl, A. (2001) Sox9 induces testis development in XX transgenic mice. *Nat. Genet.* **28**(3): 216–217.

Wang, X.J., Liefer, K.M., Tsai, S., O'Malley, B.W. and Roop, D.R. (1999) Development of gene-switch transgenic mice that inducibly express transforming growth factor beta1 in the epidermis. *Proc. Natl Acad. Sci. USA* **96**(15): 8483–8488.

Wang, Y., Schnegelsberg, P.N., Dausman, J. and Jaenisch, R. (1996) Functional redundancy of the muscle-specific transcription factors Myf5 and myogenin. *Nature* **379**(6568): 823–825.

Yang, X.W., Model, P. and Heintz, N. (1997) Homologous recombination based modification in *Escherichia coli* and germline transmission in transgenic mice of a bacterial artificial chromosome. *Nat. Biotechnol.* **15**(9): 859–865.

Zheng, B., Sage, M., Cai, W.W., Thompson, D.M. *et al.* (1999) Engineering a mouse balancer chromosome. *Nat. Genet.* **22**(4): 375–378.

6

Chemical Teratogens: Hazards, Tools and Clues

Nigel A. Brown (with revisions by **Cheryll Tickle**)

Introduction

Individual chemicals, a single drug or food contaminant for example, have been recognized to cause only a small proportion of birth defects, perhaps less than 5%. Nevertheless, since this represents a significant health burden, there is no excuse for not making every effort to avoid such exposures, and this chapter will review some of these human teratogens. Furthermore, the contribution of chemicals to human birth defects may be much larger. In addition, studies of chemical teratogens can certainly contribute more to understanding both normal and abnormal development. This chapter will consider how chemical teratogens can: (a) phenocopy birth defects for which there may be no convenient genetic models; (b) be used as tools to manipulate development; (c) reveal unknown components of normal development; and (d) have generated general principles applicable to human malformation.

When Etienne Geoffroy Saint-Hilaire concocted the term 'teratology' in the early nineteenth century, he meant the study of birth defects, in a broad and all-encompassing sense. However, in the 1960s, the thalidomide tragedy generated a new field of investigation dedicated to ensuring that we are not exposed to environmental influences that cause birth defects. This study of chemical and physical agents, largely an aspect of toxicology, assumed the name 'teratology'. That this field is now termed '*developmental toxicology*' (and 'teratology' has reverted to its original meaning) provides the first general principle. It is clear from experimental studies that chemically induced effects on prenatal development are manifested in many more ways than the 'monstrous' defects of Hilaire. Pre- and perinatal death, growth retardation, behavioural and functional impairment, germ cell mutation

Embryos, Genes and Birth Defects, Second Edition Edited by Patrizia Ferretti, Andrew Copp, Cheryll Tickle and Gudrun Moore

and adult-onset disease are all parts of the spectrum. There was no scientific reason to consider structural defects separately when studying the consequences of embryonic chemical insults, and there is equally no reason to be blinkered about wider effects when considering human birth defects.

Prevention of environmental (non-genetic) causes of birth defects requires methods to detect teratogens (hazard identification) and to predict their human effect (risk characterization). Testing is a thorny issue, particularly as society increasingly questions the use of animals. Eventually, we will understand mechanisms of teratogenesis and the conservation of developmental processes between species. Until then, there is no choice but to do the best we can to devise tests that balance the conflicting needs for sensitive detection and for humanity to animals. To characterize risk, we need to know about the exposure, absorption, disposition, metabolism and elimination of a chemical, both in the test system and in humans. All these aspects are essential for the active prevention of birth defects, but are outside the scope of this book.

Teratogens and human malformations

Lists of human teratogens are dangerous, and have undoubtedly resulted in the deaths of many normal fetuses through needless therapeutic abortion. They are also notoriously contentious, and for good reasons. Should a list give what we know has happened, or what might happen? Many chemicals probably would cause human malformation, given sufficient exposure. Should a list include only agents that cause structural defects? What about miscarriage, functional effects, and so on? Should a list include pharmacological effects, like the congenitally heroin-addicted baby, or neonatal meconium ileus after anticholinergics? What evidence is required to place chemicals on the list? Most importantly, what about dose? Ionizing radiation and ethanol are undoubtedly human teratogens but all embryos are exposed to both from natural sources. Lest there be any doubt: just because a pregnant woman has been exposed to a chemical listed as a human teratogen does not necessarily justify a termination of pregnancy, and just because a chemical is not on the list does not guarantee its safety.

So, Table 6.1 is offered with caution. These are chemicals that certainly have disrupted human prenatal development, meeting objective criteria for identification (Shepard, 2002). The list is not comprehensive, but selected to illustrate the range of effects and chemical classes. Perhaps the most important aspect of this list is that we have no plausible molecular mechanism for over half of these teratogens. However, successful investigation of those with known mechanisms has revealed previously unknown processes in development, as discussed below.

It is sometimes said that all human teratogens have been identified by astute clinical observation. This is both a truism and misleading: a truism in that, of course, the only conclusive proof of human teratogenicity is an affected baby; misleading in that several human teratogens were known animal teratogens before any human

Table 6.1 Teratogens that have caused human birth defects

Chemical	Use	Effects	Molecular site of action
ACE inhibitors: captopril, enalapril, etc.	Antihypertensive	Patent ductus arteriosus, oligohydramnios, renal abnormalities and dysfunction, skull hypoplasia	ACE (kininase II)
Androgens, including synthetic progestens	Antimiscarriage	Masculinization of female external genitals and urogenital sinus	Androgen receptor
Cytotoxic agents: cyclophosphamide, busulphan, etc.	Cancer chemotherapy	Multiple malformations: most organ systems	DNA integrity?
Diethylstilboestrol	Anti-miscarriage	Multiple defects of female (and male less often) reproductive tract, vaginal adenocarcinoma	Oestrogen receptor
Diphenylhydantoin	Anticonvulsant	Nail and digit hypoplasia, fetal hydantoin syndrome	?
Ethanol	Recreational drug	Growth and mental retardation, craniofacial and CNS defects, fetal alcohol syndrome	Alcohol dehydrogenase
Folic acid antagonists: aminopterin, methotrexate, etc.	Abortifacient, cancer chemotherapy	Multiple malformations: most organ systems	DNA synthesis?
Lithium	Antidepressant	Cardiac defects: Ebstein's anomaly	?
Mercury, organic	Food contaminant	Cerebral palsy, microcephaly	?
Polychlorinated biphenyls	Food contaminant	Intrauterine growth retardation, skin discoloration	?
Retinoids: isotretinoin, etretinate, etc.	Anti-acne	Multiple malformations: craniofacial, CNS, cardiac, thymic aplasia	Retinoid receptors
Streptomycin	Antituberculous	Deafness	?
Tetracyclines	Antibiotic	Tooth and bone discoloration	?
Thalidomide	Sedative	Phocomelia, external ear defects, oesophageal and duodenal atresia, tetralogy of Fallot, renal agenesis	?
Trimethadione	Anticonvulsant	Cleft palate and other craniofacial defects, cardiac defects	?

(continued)

Table 6.1 (*continued*)

Chemical	Use	Effects	Molecular site of action
Valproic acid	Anticonvulsant	Spina bifida, cardiac defects, fetal valproate syndrome	Histone deacetylase
Warfarin	Anticoagulant	Nasal hypoplasia, bone stippling	Vitamin K-dependent bone matrix protein?

ACE, angiotensin-converting enzyme.
See Shepard (2002) and Schardein (2000) for comprehensive listings, and Jones (2005) for full descriptions of syndromes.

exposure (valproate, retinoids, lithium and angiotensin-converting enzyme inhibitors in Table 6.1). Indeed, there have been no human teratogens identified over the past 30 years that were not already under suspicion from experimental studies. Who knows how many more human teratogens might be in use were it not for current testing? The pharmaceutical industry has shelves full of prospective drugs abandoned because of teratogenicity in animals.

General strategy in chemical teratogenesis

The approach to discovering a mechanism of teratogenesis obviously depends upon the properties of the chemical and the nature of the birth defects, but a general strategy is shown in Figure 6.1. This has been a well-worn path for some 40 years, but there are still few complete journeys, perhaps only for TCDD (2,3,7,8-tetrachlorodibenzo-*p*-dioxin) and the glucocorticoids (see below) in addition to those in Table 6.1. The major impact of the advances in developmental biology that fill this book has been to introduce a much more solid background in which to take the final two steps towards cellular and molecular mechanisms. The teratogenicity of the antiepileptic valproic acid (VPA; Depakene®, Epilim®) illustrates the general strategy.

Valproic acid

VPA is a short-chain carboxylic acid, 2-propyl pentanoic acid (Figure 6.2). It was first identified as a teratogen by orthodox animal testing. The site of action is embryonic and the unchanged drug is the proximate teratogen, as shown by direct effects on mammalian embryos in culture, which also showed effective concentrations close to clinical plasma levels (Kao *et al.*, 1981). This, and teratogenic doses well below those toxic to the maternal animal, characterized VPA as a likely human teratogen

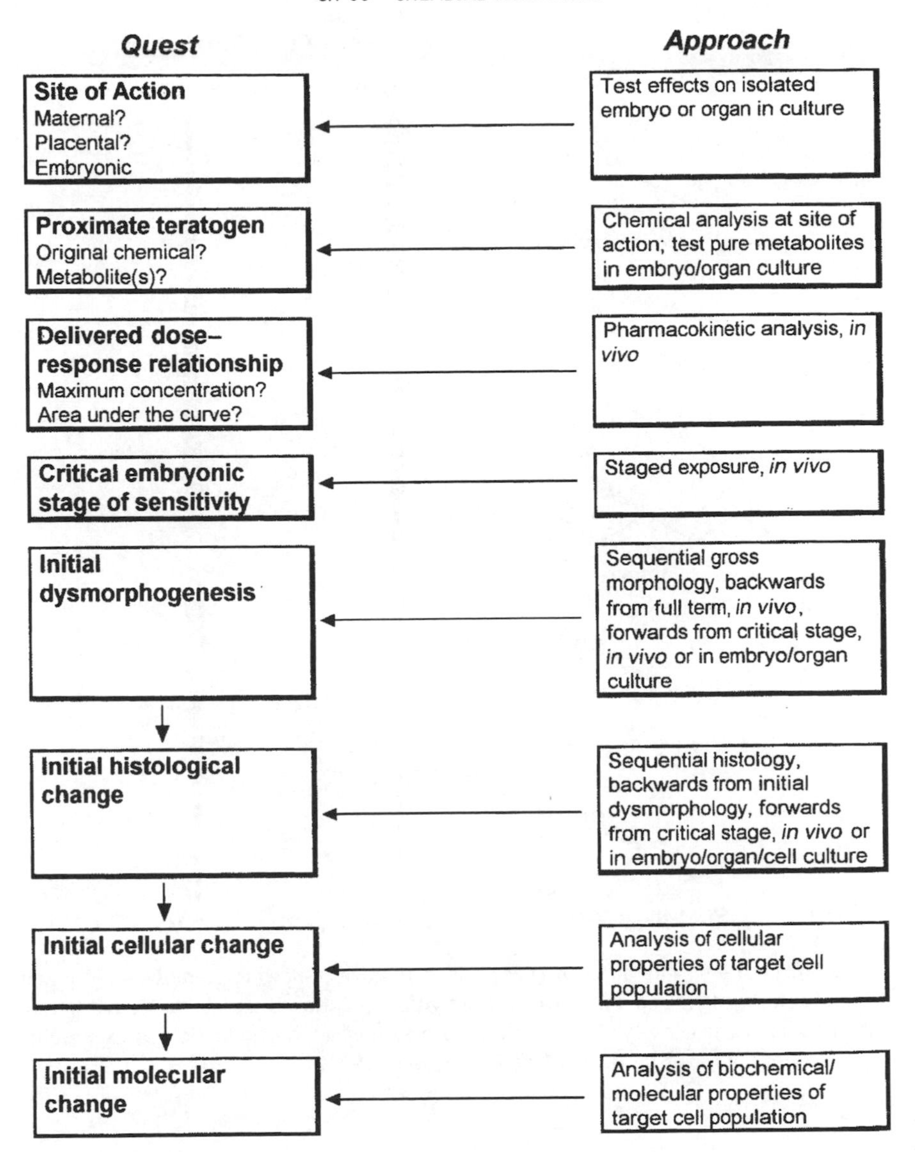

Figure 6.1 General strategy for the investigation of mechanisms of chemical teratogenesis

(Brown *et al.*, 1980). Clinically, the drug was first shown to cause spina bifida but it can also induce malformations of the heart, craniofacies, axial skeleton and limb (Thomas *et al.*, 2004). A fetal valproate syndrome has been described, with facial features in common with the fetal hydantoin syndrome (Jones, 2005).

Several antiepileptic drugs appear to cause birth defects: carbamazepine and trimethadione, and perhaps some barbiturates, as well as valproate and the hydantoins (Thomas *et al.*, 2004). This suggests a relationship between the mechanisms of pharmacological and teratological effects, but may simply reflect the fact that

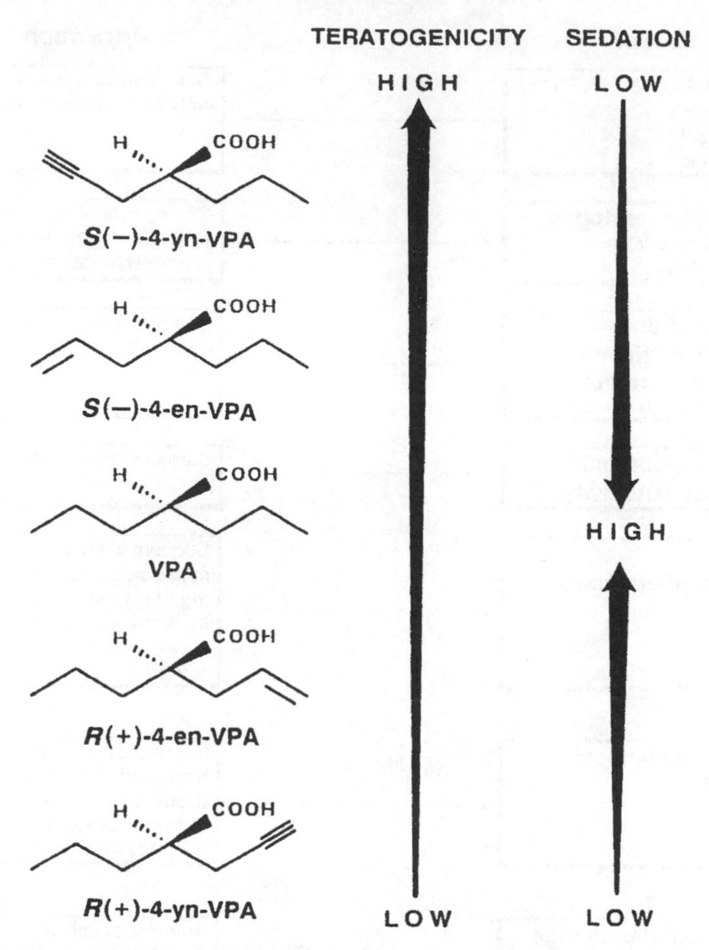

Figure 6.2 Structure of valproic acid (VPA) and derivatives, showing stereoselectivity and the separation of teratogenetic and anticonvulsant activities. Sedation is measured in adult mice as an index of anticonvulsant activity. Teratogenicity is measured as the induction of exencephaly in mice. Figure kindly provided by H. Nau (see review, Nau, 1994)

anticonvulsants are one of few classes of drug that women of child-bearing age take chronically. The pharmacological mechanisms vary widely across the chemical classes, but modulation of neurotransmitter levels is a feature in common. Some neurotransmitters function in other capacities during development, and indeed this may have been their primary role in early evolution (Lauder, 1993). The action of valproate may be a clue to such a process. However, it is clear that the pharmacological and teratological activities of VPA are separable (Figure 6.2).

Studies of the relationship between delivered dose (the amount that reaches the site of action – the embryo in this case) and response for VPA were instrumental in establishing the importance of pharmacokinetics in teratogenesis (Nau and Scott,

1987). In the case of VPA, it is peak plasma concentration (C_{max}), not duration of exposure (AUC), that correlates with teratogenic effect (see Nau and Scott, 1987). For other chemicals, cyclophosphamide for example, the opposite is true. The fact that not only dose, but also kinetics, determine teratogenic response has important implications for clinical management in pregnancy. For example, divided doses are preferable for VPA in women of child-bearing age. These studies also generated the general principle that much of the often observed variation in species sensitivity to teratogens is pharmacokinetic in origin, with wide variation in delivered dose after the same administered dose (reviewed by O'Flaherty and Clarke, 1994).

Small structural changes to VPA have a profound effect on teratogenicity (see Nau, 1994). Two aspects are particularly interesting: separation of pharmacological and teratological properties, and chirality (Figure 6.2). A metabolite of VPA, 2-ene-VPA, retains anticonvulsant activity but is not teratogenic and is a candidate replacement drug (Nau, 1994). Introducing a terminal triple bond in one of the carboxyl side-chains is one of the few modifications that enhances the teratogenicity of VPA. This produces a molecule with an asymmetric centre, and the two enantiomers (*R*-4-yn-VPA and *S*-4-yn-VPA) differ markedly in their teratogenic (and pharmacological) potency. This is due to intrinsic activity, since both enantiomers distribute equally into the embryonic compartment (Nau, 1994), and is suggestive of a receptor-mediated action, but this remains enigmatic. Short-chain carboxylic acids, in general, may share a common mechanism of teratogenicity (Coakley *et al.*, 1986). The teratogenicity of the glycol ethers, widely used industrial solvents, is mediated by their stable alkoxyacid metabolites. For example, methoxyacetic acid is responsible for the effects of ethyleneglycol monomethyl ether, and the structure–activity relationship of these alkoxyacids is reminiscent of VPA.

The critical stage for VPA induction of spina bifida in the mouse is gestational day 9 (Nau, 1994) and the initial dysmorphogenesis may involve the neural suture and presomitic mesoderm (Brown *et al.*, 1991). Initial histological changes have been described, including cell death in the neuroepithelium (Turner *et al.*, 1990), but the molecular mechanism of VPA teratogenicity remains unknown. Early effects on lipid synthesis, intracellular pH or zinc or neurotransmitter metabolism have been suggested (reviewed by Nau, 1994), but perhaps most plausible is an effect on folate metabolism. Supplementation with some folates can reduce the incidence of VPA-induced defects, and VPA alters folate metabolism, perhaps by inhibition of glutamate formyltransferase. However, folates also reduce the incidence of malformation from other genetic and chemical causes. This is clinically important and several programmes are under way to supplement food with folate to reduce the risk of neural tube defects, following the clear demonstration of its effectiveness in controlled trials (Hall and Solehdin, 1998). Nevertheless, the mechanism of folate protection against neural tube defects remains obscure (see Chapter 8) and, in the case of VPA-induced defects, has not been reproducible in all laboratories (Hansen and Grafton, 1991). Further research is clearly needed in this important area. Recently it has been suggested that VPA teratogenicity is related to inhibition of histone deacetylases (HDACs; Phiel *et al.*, 2001), perhaps involving the peroxisome proliferator-activated receptor delta (PPARδ; Lampen *et al.*, 2001; 2005) and activation of specificity protein 1 (Sp1)-related pathways (Kultima *et al.*, 2004).

Gene–teratogen interaction

The susceptibility to VPA teratogenesis varies markedly across different inbred mouse strains (Nau, 1994), a finding that has been utilized to map genes that confer susceptibility or resistance to the teratogenic effects of VPA (Lundberg *et al.*, 2004). Although the molecular basis of the genetic modulation of VPA teratogenicity remains unknown, it serves as an excellent example of a gene–teratogen interaction, which has an important general principle with implications for the entire field of teratology. We do not know the causes of most human birth defects, but it is clear that only a small proportion, perhaps 20%, are Mendelian genetic syndromes. It is a salutary thought that even when all the mutations responsible for McKusick's (1994) compendium are identified, it will not explain the vast majority of human malformations. It is often said that most birth defects are multifactorial, that is, the result of environmental action on a susceptible genotype. Some would say this is not profoundly helpful, since if we exclude genotype and the environment there is nothing left but chance, and it is too depressing to conclude that random developmental 'error' is responsible for most human malformations. On the other hand, the principle of gene–teratogen interaction has been formalized in the multifactorial/threshold hypothesis of Fraser (1977) and this provides a useful conceptual model (Figure 6.3).

Using the development of the palate as an example (Ferguson, 1988; see Chapter 10), Fraser (1977) suggested that the palatal shelves must become horizontal before a critical stage, otherwise they will be unable to fuse and a cleft will result. Any population of embryos will be distributed around a mean stage of shelf development, due to usual biological variation. Many aspects of head development (tongue motility, shelf growth, and so on) will contribute to this process, and each will be influenced by both genetic and environmental factors. The proportion of embryos that fall beyond the threshold depends upon this complex interaction. The search for genetic variations in the human population that determine sensitivity to particular environmental agents represents a major challenge for the future.

Teratogens and phenocopies

An understanding of the pathogenesis, that is, the sequence of cellular and tissue changes leading to a particular malformation, can help to design the best approach to corrective surgery, may suggest potential cellular and molecular changes, and can identify critical aspects of normal development. As discussed throughout this book, there are very many genetic animal models of human malformation, and new transgenic knock-out models are being generated rapidly (see Chapter 4). Chemical

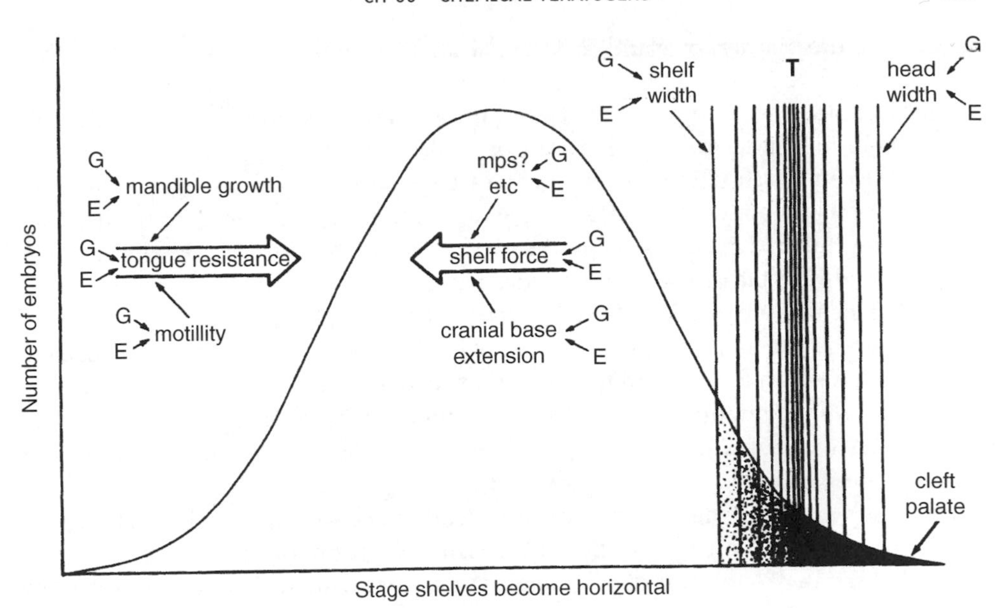

Figure 6.3 Multifactorial/threshold model of teratogenesis. Cleft palate is proposed to result if the palatal shelves do not become horizontal before a certain threshold (T) embryonic stage. A population of embryos is distributed Normally about a mean stage of shelf elevation, with a small proportion falling beyond the threshold. Many developmental processes (cranial base extension, extracellular matrix accumulation, etc.) shift either the distribution of stages or the position of the threshold. Each of these processes can be influenced by both genetic (G) and environmental (E) factors. After Fraser (1977)

teratogens can phenocopy the abnormal phenotypes of some of these genetic models and it is useful to be able to compare the pathogenesis of two different insults that lead to the same malformation.

Chemical phenocopies have a long history (Landauer, 1948) but a couple of examples are sufficient to illustrate the approach. The ideal phenocopy is one in which the chemical induces a single malformation in all treated embryos. As a model for the commonest human cardiac malformation, ventricular septal defect (VSD), the anticonvulsant trimethadione (TMD) comes very close to these criteria (Veuthey *et al.*, 1990). VSD can be induced in 98% of rat fetuses treated with TMD, and the critical changes appear to be in the proximal parts of the conotruncal ridges, particularly the septal ridge. Bisdiamine can also induce close to 100% incidence of cardiac malformation (Veuthey *et al.*, 1990) and, although more varied in morphology, this may be a phenocopy of the heart defects in the DiGeorge/velo-cardio-facial syndrome. Recent findings indicate that mutations in *Tbx1* are responsible for the heart defects in this syndrome and for other associated abnormalities (Yagi *et al.*, 2003).

The herbicide nitrofen (2,4-dichlorophenyl-4′-nitrophenyl ether) is an interesting experimental teratogen that phenocopies several malformations, including diaphragmatic hernia (Wickman *et al.*, 1993). Nitrofen can also induce a 100% incidence of absence of the Harderian gland, a lacrimal gland prominent in some species but rudimentary in man (reviewed by Manson, 1986). Its mechanism is not established but may involve thyromimetic activity and it should be a useful tool to study the role of the hypothalamic–pituitary–thyroid axis in development.

One area in which chemical phenocopies have been valuable is the gut atresias, both oesophageal and anorectal. These are amongst the most common life-threatening birth defects. Several chemical teratogens can induce gut atresias, including ethylenenitrourea. A rat model has been developed in which treatment with ethylenenitrourea leads to 80% incidence of anorectal malformations (Qi *et al.*., 2004). Recent work has shown that mouse embryos with defective hedgehog signalling have both foregut and hindgut abnormalities; for example, embryos lacking functional *Shh* (Litingtung *et al.*, 1998; Ramalho-Santos *et al.*, 2000) or *Gli* genes (Motoyama *et al.*, 1998; Mo *et al.*, 2001) that encode transcriptional effectors of Shh signalling.

One problem with the straight forward gene knock-out approach to studies of developmental mechanisms is that the normal function of the affected gene is prevented in all tissues and at all stages of development, which can considerably complicate the analysis of effects, since the resultant phenotype will include secondary and tertiary effects of gene elimination. There are, however, an increasing number of conditional approaches available that allow genes to be functionally inactivated in specific tissues and at specific times during development (Sauer, 1998; Metzger and Chambon, 2001). Antisense oligonucleotides also provide an alternative to complete knock-outs and may allow both spatial and temporal control over interference in gene function (Sadler and Hunter, 1994). RNA interference technologies may also prove to be useful to knock down genes in cultured mouse embryos (Calgeri *et al.*, 2004) and in chick embryos (Bron *et al.*, 2004; Chesnutt and Niswander, 2004).

Teratogens as manipulative tools

There is a distinguished history of advances in understanding mechanisms of development by following the consequences of induced abnormalities. Early studies, around the turn of the century, usually used surgical tools, like the cautery needle of Roux and the hair loop of Spemann (for historical perspectives, see Oppenheimer, 1967; Barrow, 1971), but chemical treatments were also common in those heydays of experimental embryology. Experiments using lithium to induce transformation of the germ layers of echinoderms, performed by Herbst in the 1890s and expanded by both Horstadius and von Ubisch in the 1920s, were seminal studies, as were 'animalization' treatments with cyanide and other respiratory inhibitors by Lindhal in the 1930s.

Clearly, early teratology contributed much to our understanding of development, but are chemicals useful tools today? I believe they can be, but with the usual caveat that one does not know all the consequences of treatment. When Roux killed a blastomere to examine its influence on the neighbouring cell, he did not know that the dead cell would have a mechanical constraint on further development. So even killing a cell, the easiest thing for an experimentalist to do, may induce an effect more complex than is immediately apparent.

Early mammalian development is a progressive hierarchy of regional specification by inductive interactions, rather than by autonomous cellular mechanisms. This enables mammalian embryos to be highly regulative and, at least in theory, able to repair damage. Very little is known of the mechanisms and capabilities of mammalian embryos to regulate, and it is here that chemical teratogens could be very useful tools but, ironically, are not being extensively utilized. In contrast, there are elegant studies of the embryonic response to physical damage and wounding in both mammalian and other embryos (Martin *et al.*, 1994; Redd *et al.*, 2004).

The only extensive studies of regulation following chemical insult concern the recovery of mouse embryos from mitomycin C (MMC) treatment (reviewed by Snow, 1987). This yielded important information on mechanisms of development and suggested a novel mechanism of teratogenesis. MMC is an alkylating agent that kills cells and arrests cell division. A single injection of MMC at primitive streak stages in mice results in massive cell death, so that 12 hours later the neural plate stage embryos contain only 10–15% of the normal number of cells. Despite this, most (> 85%) embryos survive and by the end of organogenesis are overtly normal and of usual size. This is remarkable, given that many populations of cells are specified by primitive streak stages, and suggests extensive respecification during recovery from damage. At the end of organogenesis, less than 10% of embryos show gross malformation, the most common defect being microphthalmia.

However, these apparently normal embryos harbour covert defects. Newborn animals have severe neurological defects and few (< 30%) survive to weaning. Even the healthy survivors have reduced fertility. Snow (1987) showed that the synchrony in development of individual organ systems was not normal during organogenesis, and suggested that this asynchrony is responsible for the subsequent abnormalities. Organs could be grouped into those that showed little or no retardation in their appearance, those with a moderate delay (5–6 hours), and those delayed by more than 10 hours. Derivatives of all three germ layers were found in all groups, but cells already committed to a particular tissue at the time of treatment (neural ectoderm, heart, germ cells, hindgut, allantois) appeared least delayed.

The precise relationship between the start of sensitivity of a tissue or organ to chemical disruption and its stage of development is another area where teratogens are underutilized as tools. In broad terms, the variation in sensitivity with gestational age has long been documented. For example, the peak sensitivity for thalidomide-induced defects was 21–27 days post-conception for external ears, 27–30 days for arms and 30–33 days for legs (Thomas *et al.*, 2004). The start of sensitivity must often relate to the allocation of cells to a particular fate, but this is poorly studied. An excellent example of what can be learned is the startling limb and lower body duplications following retinoic acid treatment at pre-implantation stages in the mouse, suggesting that aspects of body patterning may occur even before gastrulation (Rutledge *et al.*, 1994).

Programmed cell death is an important mechanism of morphogenesis. In many phases of development, too many progenitor cells are produced and cell numbers are subsequently regulated by programmed cell death. The numbers of neurones in the optic stalk are an example of this (Raff, 1992). It is possible to manipulate regions of cell death by chemical treatment. It has long been observed that cell death is a common feature in the pathogenesis of chemically induced malformation (Scott, 1977). Furthermore, several teratogens, such as ethanol and retinoic acid, increase the areas of normal programmed cell death (Sulik *et al.*, 1988). This provides an opportunity to vary, systematically, the proportion of cells in a particular region that die, then study subsequent development. Furthermore, retinoic acid treatment can ameliorate the interdigital webbing seen in the mouse hammertoe mutant by enhancing the cell death that normally serves to separate the digits (Ahuja *et al.*, 1997). On the other hand, diminution of physiological cell death can itself lead to congenital malformations (see Chapter 8), emphasizing the need during development for a precise balance between cell proliferation and cell loss.

The mechanisms of teratogen expansion of regions of cell death are unknown, but the view (Raff, 1992) that death is the fate of all cells, unless they receive sufficient survival factors, provides a potential explanation. Competition for limited quantities of survival factors from target cells may control the degree of 'programmed' death. Such conditions would involve a fine balance between the production of sufficient vs. insufficient factor by a group of signalling cells, with the population of responding cells on a knife-edge of survival. Any action that reduced the amount of factor would expand the proportion of responding cells that died. One can imagine many mechanisms by which a chemical teratogen could, rather unspecifically, reduce the amount of survival factor: metabolic or growth inhibition, or killing of signalling cells, for example. This would also provide a means whereby chemical teratogens of diverse mechanism might act additively at one site.

Teratogens as clues

When an exogenous chemical has an unexpected potent teratological effect, particularly when a reproducible syndrome of defects is induced with a high frequency, then suspicions should be raised that a fundamental developmental process is being disrupted. In several such cases, the search for the teratogenic mechanisms has provided valuable clues to normal mechanisms, whilst many others remain to be solved.

Retinoids

No chemical has received more attention, or contributed more to understanding development over the past 20 years, than vitamin A (retinol) and its derivatives (see Chapters 7, 11 and 14). The first demonstration that mammalian development could be profoundly affected by an environmental manipulation was Hale's (1933) observation of pigs born without eyeballs to sows on a vitamin A-deficient diet. The very wide range of defects that could be caused by deficiency were beautifully described by the fathers of modern teratology, Wilson and Warkany (see Wilson *et al.*, 1953), and they unknowingly provided phenocopies of recent multiple retinoic acid receptor (RAR) and retinoid X receptor (RXR) targeted mutations. Subsequently, vitamin A excess was shown also to be a teratogen with a remarkable spectrum of dysmorphic effects. Synthetic retinoids, isotretinoin and etretinate are perhaps the most effective human teratogens known (Thomas *et al.*, 2004).

It is now appreciated that retinoids play a widespread and critical role in developmental control. Genes that encode enzymes that metabolize retinoids have been identified and are expressed in embryos. Several of the genes encoding these enzymes have been knocked out in mice and this leads to abnormalities. Embryos in which the gene encoding Raldh2, retinaldehyde dehydrogenase 2, an enzyme that controls retinoic acid synthesis from retinaldehyde, has been functionally activated, do not undergo turning, are shortened with respect to the main body axis and die at mid-gestation (Niederreither *et al.*, 1999). Interestingly, it has been suggested that the teratogenic effects of ethanol may be mediated by inhibiting alcohol dehydrogenase catalysis of retinol to retinaldehyde (Deltour *et al.*, 1996). Mouse embryos deficient in the *Cyp26a1* gene, which encodes a cytochrome P450 enzyme that catabolizes retinoic acid, also die in mid-gestation and have major defects in several aspects of body patterning (Abu-Abed *et al.*, 2001). *Cyp26a1*$^{-/-}$ mouse embryos can be phenotypically rescued by heterozygous disruption of *Raldh2* (Niederreither *et al.*, 2002). This elegant genetic experiment shows that it is maintenance of the proper balance of retinoic acid levels that is critical for development. In addition, the fact that *Cyp26a1*$^{-/-}$, *Raldh2*$^{+/-}$ mice are normal and can even survive to adulthood shows that the oxidative derivatives of retinoic acid are not involved in retinoid signalling.

It is still not clear exactly how exogenous retinoids induce some of their teratogenic effects. For example, the archetypal morphological effect of retinoid treatment at late gastrulation/early neurulation is abnormal branchial arch development (Webster *et al.*, 1986). Treatment at this time can cause anterior shifts in the expression domains of *Hox* genes (see Chapters 11 and 14) and the most profound result of this seems to be the homeotic transformation of hindbrain rhombomeres (Marshall *et al.*, 1992). However, this does not seem to be involved in the development of abnormal arches, since the critical stage for their induction is slightly later, when no shifts in Hox expression are seen. The roles of extended cell death and inhibited cell migration, which do occur in specific cell populations (Sulik *et al.*, 1988), are not clear.

Jervine alkaloids/cyclopamine

A very good example of how understanding of developmental and of teratogenic mechanisms has converged is seen in holoprosencephaly, which is characterized by a partial or complete absence of forebrain division into telencephalic vesicles, severe skull defects and loss of midline facial structures. Holoprosencephaly in humans is associated with haplo-insufficiency for a number of genes, among which is Sonic hedgehog, one of the family of vertebrate Hedgehogs involved in mediating cell–cell interactions. Holoprosencephaly is also associated with ingestion of jervine alkaloids, and the reason why jervine alkaloids phenocopy genetic lesions has now been nicely explained.

Mouse embryos in which *Shh* has been functionally inactivated show holoprosencephaly (Chiang *et al.* 1996), while holoprosencephaly in humans can also be caused by mutations in *SHH* (Belloni *et al.*, 1996; Roessler *et al.*, 1996), in *Patched*, which encodes the Sonic hedgehog receptor (Ming *et al.*, 2002), and in *Gli2*, which encodes one of the transcriptional effectors of Sonic hedgehog signalling (Roessler *et al.*, 2003). Hedgehog proteins undergo cholesterol modification and this is important for their biological activity (Beachy *et al.*, 1997; Lewis *et al.*, 2001). Defects in genes encoding enzymes involved in cholesterol biosynthesis are also found in some human patients; thus, for example, mutation of the gene encoding 7-dehydrocholesterol reductase is the major defect in Smith–Lemli–Opitz syndrome, which has holoprosencephaly as part of its phenotype (Fitzky *et al.*, 1998; Cooper *et al.*, 2003).

Holoprosencephaly can be phenocopied by cholesterol inhibitors. Sheep born to ewes which ingested *Veratrum californicum* during pregnancy were found to suffer severe cyclopia. This plant produces jervine alkaloids and these have been shown to inhibit cholesterol biosynthesis (Cooper *et al.*, 1998). It has emerged, however, that jervine alkaloids do not act by perturbing cholesterol modification of Hedgehog proteins, as might be expected, but instead antagonize signalling through smoothened, the key cell-surface transducer of Hedgehog signals (Taipale *et al.*, 2000).

Dioxins

2,3,7,8-Tetrachlorodibenzo-*p*-dioxin (TCDD; dioxin) is the most potent of the halogenated aromatic hydrocarbons. It is a contaminant of many industrial mixtures, most famously the Agent Orange herbicide sprayed on Vietnam. Concern over the developmental effects of TCDD began with the demonstration that birth defects induced by 2,4,5-T (a component of Agent Orange) in rats and mice were actually caused by contaminating TCDD (reviewed by Peterson *et al.*, 1993). The usual model of TCDD teratogenicity is cleft palate induction in mice, but kidney, brain and other organs are also affected. It is thought that the ectodermal dysplasia syndrome in offspring of women from Yusho and Yu-Cheng who consumed contaminated rice oil was caused by TCDD, but many other contaminants were present (Peterson *et al.*, 1993).

The extraordinary potency of TCDD led to studies of molecular mechanisms that have provided one of the most complete descriptions of chemical teratogenesis. Postnatal behavioural and neuroendocrine functions are among the most TCDD-sensitive developmental processes. Significant effects on reproductive function have been found in male rat offspring after a single dose of 64 ng/kg on day 15 of gestation (Mably *et al.*, 1992). The offspring of rhesus monkeys exposed to less than 1 ng/kg/day before pregnancy were reported to have measurable behavioural changes (Schantz and Bowman, 1989).

TCDD produces cleft palate by an unusual cellular mechanism (Abbott and Bimbaum, 1989, 1990a, 1991). The palatal shelves of treated mice grow and make contact normally, but the subsequent loss of periderm, shelf adhesion and medial epithelium–mesenchyme transformation does not occur. Rather than transforming, TCDD-treated medial epithelium cells proliferate and differentiate into a stratified epithelium. This occurs in palate cultures from mouse, rat and human embryos, although the mouse is most sensitive. It is possible that this effect is mediated by an interference with epidermal growth factor (EGF) or transforming growth factor (TGF) functions (Abbott and Birnbaum, 1990b), and an effect on the regulation of EGF receptors may also be involved in TCDD actions on kidney development (Abbott and Bimbaum, 1990c). It is clear that palate epithelium cells have a high-affinity receptor for TCDD, the aryl hydrocarbon receptor (AhR; Abbott *et al.*, 1994a).

The mouse AhR gene encodes an 89 kDa transcription factor of the basic helix–loop–helix (bHLH) family (Figure 6.4; reviewed by Whitlock, 1993). It has a DNA-binding domain, a glutamate-rich activation domain, and a 'PAS' domain, named after homology with Per (encoding the *Drosophila* circadian rhythm protein), Arnt (encoding a protein that dimerizes with Ah, see below) and Sim (encoding a CNS development regulator in *Drosophila*), which may be involved in ligand binding. When unbound by ligand, AhR resides in the cytoplasm and is translocated to the nucleus on TCDD binding. Arnt (Ah receptor nuclear translocator, also known as hypoxia inducible factor 1α) is an 86 kDa nuclear protein, which also has bHLH and PAS domains (Whitelaw *et al.*, 1993). It has no affinity for TCDD or the unbound AhR, but forms a heterodimer with activated AhR. Neither activated AhR nor Arnt have substantial DNA-binding activity as monomers. By analogy with other bHLH proteins and other classes of transcription factor, it is possible that the diversity in biological effects of TCDD is the result of differential gene regulation, mediated by heterodimers of AhR with as-yet uncharacterized proteins. AhR associates with the 90 kDa heat-shock protein (Hsp90) in cytoplasm, which is thought to maintain the receptor in a conformation optimal for ligand binding and to prevent the inappropriate binding of the unliganded receptor to DNA (Pongratz *et al.*, 1992).

TCDD is a strong inducer of the expression of the cytochrome P450 lAl isozyme (*CYPIAI* gene). Studies of this gene have identified the dioxin-responsive element (DRE), a hexanucleotide core recognition sequence that is present in multiple copies (probably six) within the enhancer region (Saatcioglu *et al.*, 1990; Wu and Whitlock, 1993). DREs have been identified upstream of other TCDD-inducible genes (Pimental *et al.*, 1993).

The link between TCDD-AhR-mediated changes in gene expression and abnormal development has yet to be established. AhR mRNA and protein are expressed in

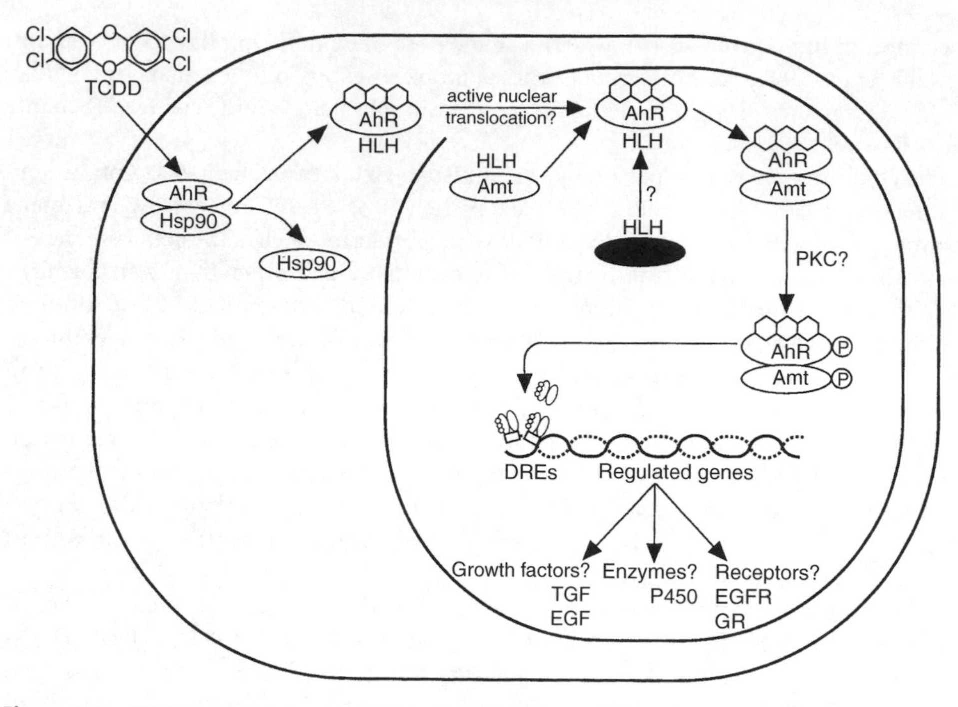

Figure 6.4 TCDD (2,3,7,8-tetrachlorodibenzo-*p*-dioxin) regulation of gene expression. TCDD binds to the aryl hydrocarbon receptor (AhR), which is held in an accessible conformation in the cytoplasm by a 90 kDa heat shock protein (Hsp90). Binding activates the AhR, releasing Hsp90 and translocating into the nucleus. Here it dimerizes with other helix–loop–helix (HLH) proteins, including the Arnt protein and other unidentified factors. The dimer is phosphorylated, possibly by protein kinase C (PKC), before binding to dioxin-responsive elements (DREs) in the enhancer regions of regulated genes. TGF, transforming growth factor; EGF, epidermal growth factor; GR, glucocorticoid receptor. After Whitlock (1993)

mouse palate, particularly in the epithelium (Abbott *et al.*, 1994a). There is tissue and spatial variation in intracellular distribution of the protein: perinuclear in the mesenchyme, and cytoplasmic and nuclear in the medial fusing epithelium. Changes in the levels of TGFα and TGFβ, and of EGF and its receptor, have been observed following TCDD exposure, presumably mediated by AhR regulation (Abbott and Birnbaum, 1989, 1990b). These growth factor changes are consistent with the abnormal proliferation of medial epithelial cells and the importance of these signalling pathways can be assessed using knock-out mice (Abbott *et al.*, 2003).

TCDD synergizes with excess glucocorticoids in the induction of cleft palate in the mouse, and it has been proposed that there is a cycle of mutual induction involving the AhR and glucocorticoid receptors (Abbott *et al.*, 1994b). Because TCDD affects neuroendocrine development, there may also be interactions between the AhR and oestrogen and/or androgen receptor-mediated regulation (Whitlock, 1993). The study of a recently developed transgenic mouse model in which the mouse *Ahr* gene has been replaced by a human *AHR* gene may provide more relevant information about the basis of teratogenic effects of TCCD in humans (Moriguchi *et al.*, 2003).

The natural ligand for AhR is not known. There are naturally occurring chemicals with a high affinity for the receptor, particularly in plants, so it is possible that the receptor evolved to induce the enzymes responsible for the metabolism of some ingested lipophilic chemicals. If this were the case, however, it is not clear why the receptor would be expressed during embryogenesis. An attractive hypothesis is that there is an unidentified natural ligand which has an important role in normal development. Consistent with this, the phenotype of AhR-deficient mice reveals that the receptor is critical for liver formation and development of the immune system (reviewed Carlson and Perdew, 2002). Arnt$^{-/-}$ mouse embryos die *in utero* due primarily to failure of placental differentiation (Kozak *et al.*, 1997), suggesting an essential role for Arnt in angiogenesis. Further evidence for members of this pathway being involved in normal development is the recent finding that CYP1A1 activity can be detected in early mouse embryos, although it is not clear whether expression is regulated through the Ah receptor (Campbell *et al.*, 2005).

Cytochrome P450s comprise one of the most important families of xenobiotic metabolizing enzymes. Another P450, CYP1B1, originally cloned as a dioxin-responsive cDNA, is now linked to abnormal eye development in humans (reviewed Stoilov, 2001). Indeed, there is increasing evidence for endogenous roles in development for other P450s, e.g. CYP26 (see above) in connection with retinoic acid signalling. Furthermore, functional inactivation in mice of P450 oxidoreductase (POR), the electron donor that is necessary for activity of all microsomal cytochrome P450 enzymes, leads to embryonic lethality (Otto *et al.* 2003), reinforcing the importance of these enzymes for normal development.

Xeno-oestrogens

Diethylstilboestrol (DES) was synthesized as a synthetic oestrogen and used for almost 30 years, until the 1970s, for the prevention of threatened miscarriage and other complications of pregnancy. DES acts by binding to the oestrogen receptor, a cytoplasmic protein that translocates to the nucleus on ligand-induced activation and acts as a transcription factor. The discovery of a rare form of cancer in the reproductive tracts of women who were exposed to DES *in utero* is now well known. However, vaginal adenocarcinoma was a rare (perhaps 1/1000) outcome of prenatal DES exposure, whilst structural and functional defects of both the male and female reproductive organs were much more common (reviewed by Edelman, 1986).

Human exposure to DES is no longer a problem, but there are a very large number of man-made and natural chemicals that have oestrogenic activity. Many of these bear no overt structural resemblance to oestradiol-17β (Figure 6.5), the natural ligand for the oestrogen receptor, but nevertheless have sufficient affinity to activate the receptor. There is currently concern that the complex mix of xeno-oestrogens in

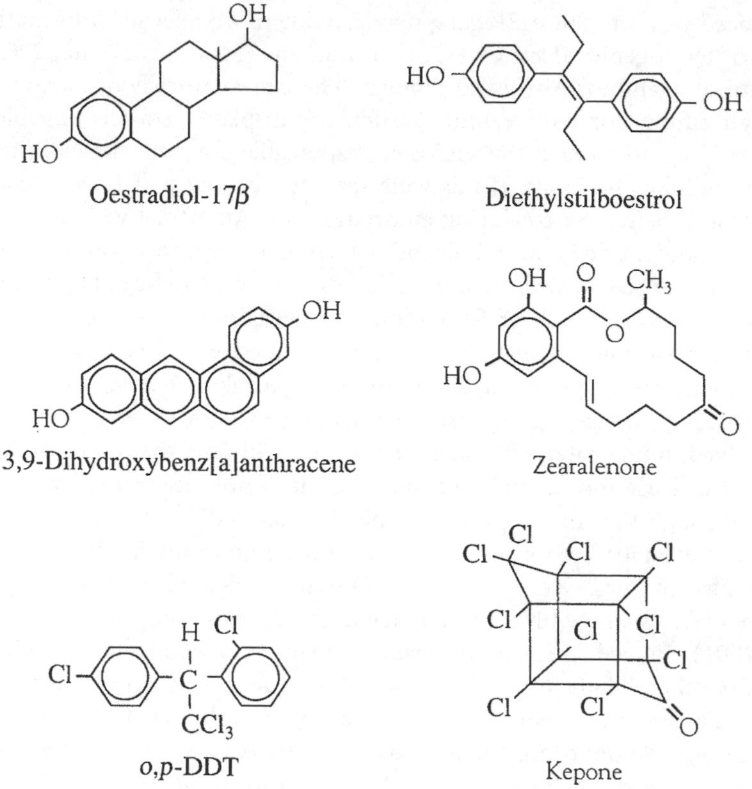

Oestradiol-17β

Diethylstilboestrol

3,9-Dihydroxybenz[a]anthracene

Zearalenone

o,p-DDT

Kepone

Figure 6.5 Diverse chemical structures that bind to the oestrogen receptor: oestradiol, natural ligand; diethylstilboestrol, synthetic oestrogen; dihydrobenzanthracene, metabolite of environmental combustion product; zearalenone, mycotoxin product; DDT, pesticide; kepone, flame retardant. After McLachlan (1993)

the environment may be affecting the reproductive health of wild animals and humans alike. For example, polychlorinated biphenyls (PCBs) can cause gonadal sex reversal in animals that exhibit temperature-dependent sex determination (Bergeron *et al.*, 1994) and the weakly oestrogenic bisphenol A, which is found in some food and drinks as a contaminant from polycarbonate plastics, can produce effects on prostate development in mice (Timms *et al.*, 2005). It has been suggested that the apparent decline in sperm counts, and increases in gonadal abnormalities, of men in Europe and the USA over the past 50 years is caused by environmental oestrogens (Sharpe and Skakkebaek, 1993). The studies of DES teratogenicity in mouse provide a model for xeno-oestrogens and have revealed interesting features of normal reproductive tract development.

Because the development of the reproductive tract is hormone-dependent, it is not surprising that DES is disruptive. Structural abnormalities of the uterus and oviducts in females, testicular and epididymal defects in males, and reproductive dysfunction in both sexes are consequences of prenatal DES exposure in mice (reviewed by Mori and Nagasawa, 1988). The molecular correlates of these actions are now being characterized. Lactoferrin is the major oestrogen-inducible uterine protein in mice, and is not normally expressed in the male. Prenatal DES exposure results in constitutive and inducible expression of lactoferrin in the seminal vesicle epithelium of adult offspring, without affecting the normal response to androgens (Beckman *et al.*, 1994). This is not an effect on circulating hormones, but appears to be due to an alteration in the differentiation of epithelial cells by oestrogen imprinting. Similarly, in the uterus of DES-exposed female mice, there is a permanent upregulation of lactoferrin and EGF expression, independent of normal ovarian oestrogen induction (Nelson *et al.*, 1994). These permanent changes are the molecular analogues of structural birth defects – 'molecular teratogenesis'.

At certain critical stages of gestation, prenatal DES 'masculinizes' the behaviour of female animals (reviewed by Newbold, 1993). This paradoxical effect illustrates the potential impact of chemical teratogens on functional brain development. The explanation for the paradox is that oestradiol is the normal mediator of testosterone imprinting of the developing male brain. Testosterone synthesized by the fetal testis is metabolized to oestradiol within cells of the brain. The brain is normally protected from circulating oestrogens, of maternal or fetal ovarian origin, by α-fetoprotein (AFP), which has a high affinity for oestradiol but not testosterone. DES and other xeno-oestrogens do not bind to AFP and so gain access to the brain, subsequently activating oestrogen receptors in an androgenizing manner. Whether there is an equivalent effect of DES in humans is contentious. Although several reports suggest changes in behaviour patterns associated with prenatal exposure to DES, current evidence is not convincing (Newbold, 1993). The relationship of DES developmental effects to the wider topic of endocrine disrupters has been well reviewed (Newbold, 2004).

Final comments

What does the future hold for chemical teratogenesis? Inventive chemists will continue to synthesize new molecules with surprising effects on development, providing clues for the astute investigator. In the right hands, chemical tools will help to unlock developmental mechanisms. The burgeoning molecular basis of development will make it much easier to identify the initial molecular insults inflicted by chemical teratogens, and to define the genetic susceptibility factors for each teratogenic mechanism. As these are characterized, we will be better able to detect and predict environmental hazards for the developing embryo and fetus. We can now reasonably conclude that a chemical with affinity for retinoid, oestrogen, glucocorticoid, androgen or Ah receptors is a potential teratogen. In time, this array of potential targets will expand and simple reporter-construct tests will be devised to screen new chemicals (McLachian, 1993). And we will begin to unravel the real contribution of the environment to birth defects.

References

Abbott, B.D. and Birnbaum, L.S. (1989) TCDD alters medial epithelial cell differentiation during palatogenesis. *Toxicol. Appl. Pharmacol.* **99**: 276–286.

Abbott, B.D. and Birnbaum, L.S. (1990a) Rat embryonic palatal shelves respond to TCDD in organ culture. *Toxicol. Appl. Pharmacol.* **103**: 441–451.

Abbott, B.D. and Birnbaum, L.S. (1990b) TCDD-induced altered expression of growth factors may have a role in producing cleft palate and enhancing the incidence of clefts after co-administration of retinoic acid and TCDD. *Toxicol. Appl. Pharmacol.* **106**: 418–432.

Abbott, B.D. and Birnbaum, L.S. (1990c) Effects of TCDD on embryonic ureteric epithelial EGF receptor expression and cell proliferation. *Teratology* **41**: 71–84.

Abbott, B.D. and Birnbaum, L.S. (1991) TCDD exposure of human embryonic palatal shelves in organ culture alters the differentiation of medial epithelial cells. *Teratology* **43**: 119–132.

Abbott, B.D., Perdew, G.H. and Birnbaum, L.S. (1994a) Ah receptor in embryonic mouse palate and effects of TCDD on receptor expression. *Toxicol. Appl. Pharmacol.* **126**: 16–25.

Abbott, B.D., Perdew, G.H., Buckalew, A.R. and Birnbaum, L.S. (1994b) Interactive regulation of Ah and glucocorticoid receptors in the synergistic induction of cleft palate by 2,3,7,8-tetrachlorodibenzo-*p*-dioxin and hydrocortisone. *Toxicol. Appl. Pharmacol.* **128**: 138–150.

Abbott, B.D., Buckalew, A.R., DeVito, M.J., Ross, D. *et al.* (2003) EGF and TGFα expression influence the developmental toxicity of TCDD: dose–response and AhR phenotype in EGF, TGFα, and EGF + TGFα knock-out mice. *Toxicol Sci.* **71**: 84–95.

Abu-Abed, S., Dolle, P., Metzger, D., Beckett, B. *et al.* (2001) The retinoic acid-metabolizing enzyme, CYP26A1, is essential for normal hindbrain patterning, vertebral identity, and development of posterior structures. *Genes Dev.* **15**: 226–240.

Ahuja, H.S, James, W. and Zakeri, Z. (1997) Rescue of the limb deformity in hammertoe mutant mice by retinoic acid-induced cell death. *Dev. Dyn.* **208**: 466–481.

Barrow, M.V. (1971) A brief history of teratology to the early 20th century. *Teratology* **4**: 119–130.

Beachy, P.A., Cooper, M.K., Young, K.E., von Kessler, D.P. *et al.* (1997) Multiple roles of cholesterol in hedgehog protein biogenesis and signaling. *Cold Spring Harb. Symp. Quant. Biol.* **62**: 191–204.

Beckman, W.C., Newbold, R.R., Teng, C.T. and McLachlan, J.A. (1994) Molecular feminization of mouse seminal vesicle by prenatal exposure to diethylstilbestrol: altered expression of messenger RNA. *J. Urol.* **151**: 1370–1378.

Belloni, E., Muenke, M., Roessler, E., Traverso, G. *et al.* (1996) Identification of sonic hedgehog as a candidate gene responsible for holoprosencephaly. *Nat. Genet.* **14**: 353–356.

Bergeron, J.M., Crews, D. and McLachlan, J.A. (1994) PCBs as environmental estrogens: turtle sex determination as a biomarker of environmental contamination. *Environ. Health Perspect.* **102**: 780–781.

Bron, R., Eickholt, B.J., Vermeren, M., Fragale, N., and Cohen, J. (2004) Functional knockdown of neuropilin-1 in the developing chick nervous system by siRNA hairpins phenocopies genetic ablation in the mouse. *Dev. Dyn.* **230**: 299–308.

Brown, N.A., Kao, J. and Fabro, S. (1980) Teratogenic potential of valproic acid. *Lancet* **1**: 660–661.

Brown, N.A., Clarke, D.O. and McCarthy, A. (1991) Adaptation of post-implantation embryos to culture: membrane lipid synthesis and response to valproate. *Reprod. Toxicol.* **5**: 245–253.

Calegari, F., Haubensak, W., Yang, D., Huttner, W.B., Buchholz, F. (2004) Tissue-specific RNA interference in postimplantation mouse embryos with endoribonuclease-prepared short interfering RNA. *Proc. Natl Acad. Sci. USA.* **99**: 14236–14240.

Campbell, S.J., Henderson, C.J., Anthony, D.C., Davidson, D. *et al.* (2005) The murine *Cyp1a1* gene is expressed in a restricted spatial and temporal pattern during embryonic development. *J. Biol. Chem.* **280**: 5828–5835.

Carlson, D.B. and Perdew, G.H. (2002) A dynamic role for the Ah receptor in cell signaling? Insights from a diverse group of Ah receptor interacting proteins. *J. Biochem. Mol. Toxicol.* **16**: 317–325.

Chesnutt, C. and Niswander, L. (2004) Plasmid-based short-hairpin RNA interference in the chicken embryo. *Genesis* **39**: 73–78.

Chiang, C., Litingtung, Y., Lee, E., Young, K.E. *et al.* (1996) Cyclopia and defective axial patterning in mice lacking sonic hedgehog gene function. *Nature* **383**: 407–413.

Coakley, M.E., Rawlings, S.J. and Brown, N.A. (1986) Short chain carboxylic acids, a new class of teratogens: studies of potential biochemical mechanisms. *Environ. Health Perspect.* **70**: 105–111.

Cooper, M.K., Porter, J.A., Young, K.E. and Beachy, P.A. (1998) Teratogen-mediated inhibition of target tissue response to Shh signaling. *Science* **280**: 1603–1607.

Cooper, M.K., Wassif, C.A., Krakowiak, P.A., Taipale, J. *et al.* (2003) A defective response to Hedgehog signaling in disorders of cholesterol biosynthesis. *Nat. Genet.* **33**: 508–513.

Deltour, L., Ang, H.L. and Duester, G. (1996) Ethanol inhibition of retinoic acid synthesis as a potential mechanism for fetal alcohol syndrome. *FASEB J.* **10**: 1050–1057.

Edelman, D.A. (1986) *DES/Diethylstilbestrol: New Perspective.* MTP Press: Boston, MA.

Ferguson, M.J.W. (1988) Palate development. *Development* **103**(suppl): 41–60.

Fitzky, B.U., Witsch-Baumgartner, M., Erdel, M., Lee, J.N. *et al.* (1998) Mutations in the Delta7-sterol reductase gene in patients with the Smith–Lemli–Opitz syndrome. *Proc. Natl Acad. Sci. USA.* **95**: 8181–8186.

Fraser, F.C. (1977) Relation of animal studies to the problem in man. In *Handbook of Teratology*, vol 1, Wilson, J.G. and Clarke-Fraser, F. (eds). Plenum: New York; 75–96.

Hale, F. (1933) Pigs born without eyeballs. *J. Hered.* **24**: 105–106.

Hall, J. and Solehdin, F. (1998) Folic acid for the prevention of congenital anomalies. *Eur. J. Pediatr.* **157**: 445–450.

Hansen, D.K. and Grafton, T.F. (1991) Lack of attenuation of valproic acid-induced effects by folinic acid in rat embryos *in vitro. Teratology* **43**: 575–582.

Jones, K.K. (2005) *Smith's Recognizable Patterns of Human Malformation*, 6th edn. Saunders: Philadelphia, PA.

Kao, J., Brown, N.A., Schmidt, B., Goulding, E.H. and Fabro, S. (1981) Teratogenicity of valproic acid: *in vivo* and *in vitro* investigations. *Teratogen. Carcinogen. Mutagen.* **1**: 376–382.

Kozak, K.R., Abbott, B. and Hankinson, O. (1997) ARNT-deficient mice and placental differentiation. *Dev. Biol.* **191**: 297–305.

Kultima, K., Nystrom, A.M., Scholz, B., Gustafson, A.L. *et al.* (2004) Valproic acid teratogenicity: a toxicogenomics approach. *Envir. Health Perspect.* **112**: 1225–1235.

Lampen, A., Grimaldo, P.A. and Nau, H. (2005) Modulation of peroxisome proliferator-activated receptor delta activity affect NCAM and ST8Sia IV (PST1) induction by teratogenic VPA-analogues in F9 cell differentiation. *Mol. Pharmacol.* (13 April, Epub ahead of print).

Lampen, A., Carlberg, C. and Nau, H. (2001) Peroxisome proliferator-activated receptor delta is a specific sensor for teratogenic valproic acid derivatives. *Eur. J. Pharmacol.* **431**: 25–33.

Landauer, W. (1948) Hereditary abnormalities and their chemically induced phenocopies. *Growth* **12**: 171–200.

Lauder, J.M. (1993) Neurotransmitters as growth regulatory signals: role of receptors and second messengers. *Trends Neurosci.* **16**: 233–240.

Lewis, P.M., Dunn, M.P., McMahon, J.A., Logan, M. *et al.* (2001) Cholesterol modification of sonic hedgehog is required for long-range signaling activity and effective modulation of signaling by Ptc1. *Cell* **105**: 599–612.

Litingtung, Y., Lei, L., Westphal, H. and Chiang, C. (1998) Sonic hedgehog is essential to foregut development. *Nat. Genet.* **20**: 58–61.

Lundberg, Y.X.W., Cabrera, R.M., Greer, K.A., Zhao, J. *et al.* (2004). Mapping a chromosomal locus for valproic acid-induced exencephaly in mice. *Mamm. Genome* **15**: 361–369.

Mably, T.A., Bjerke, D.L., Moore, R.W., Gendron-Fitzpatrick, A. and Peterson, R.E. (1992) *In utero* and lactational exposure of male rats to 2,3,7,8-tetrachlorodibenzo-*p*-dioxin: (3) effects on spermatogenesis and reproductive capability. *Toxicol. Appl. Pharmacol.* **114**: 118–126.

Manson, J.M. (1986) Mechanism of nitrofen teratogenicity. *Environ. Health Perspect.* **70**: 137–147.

Marshall, H., Nonchev, S., Sham, M.H. *et al.* (1992) Retinoic acid alters hindbrain *Hox* code and induces transformation of rhombomeres 2/3 into a 4/5 identity. *Nature* **260**: 756–759.

Martin, P., Nobes, C., McCluskey, J. and Lewis, J. (1994) Repair of excisional wounds in the embryo. *Eye* **8**: 155–160.

McKusick, V.A. (1994) *Mendelian Inheritance in Man*, 11th edn. Johns Hopkins University Press: Baltimore, MD.

McLachlan, J.A. (1993) Functional toxicology: a new approach to detect biologically active xenobiotics. *Environ. Health Perspect.* **101**: 386–387.

Metzger, D. and Chambon, P. (2001) Site- and time-specific gene targeting in the mouse. *Methods* **24**: 71–80.

Ming, J.E., Kaupas, M.E., Roessler, E., Brunner, H.G. *et al.* (2002) Mutations in PATCHED-1, the receptor for SONIC HEDGEHOG, are associated with holoprosencephaly. *Hum. Genet.* **110**: 297–301.

Mo, R., Kim, J.H., Zhang, J., Chiang, C. *et al.* (2001) Anorectal malformations caused by defects in sonic hedgehog signaling. *Am. J. Pathol.* **159**: 765–774.

Mori, T. and Nagasawa, H. (eds) (1988) *Toxicity of Hormones in Perinatal Life*. CRC Press: Boca Raton, FL.

Moriguchi, T., Motohashi, H., Hosoya, T., Nakajima, O. *et al.* (2003) Distinct response to dioxin in an arylhydrocarbon receptor (AHR)-humanized mouse. *Proc. Natl Acad. Sci. USA* **100**: 5652–5657.

Motoyama, J., Liu, J., Mo, R., Ding, Q. *et al.* (1998) Essential function of Gli2 and Gli3 in the formation of lung, trachea and oesophagus. *Nat. Genet.* **20**: 54–57.

Nau, H. (1994) Valproic acid-induced neural tube defects. In *Neural Tube Defects*. Ciba Foundation Symposium 181. Wiley: Chichester; 144–160.

Nau, H. and Scott, W.J. (eds) (1987) *Pharmacokinetics in Teratogenesis*, vols 1 and 2. CRC Press: Boca Raton, FL.

Niederreither, K., Subbarayan, V., Dolle, P. and Chambon, P. (1999) Embryonic retinoic acid synthesis is essential for early mouse post-implantation development. *Nat. Genet.* **21**: 444–448.

Niederreither, K., Abu-Abed, S., Schuhbaur, B., Petkovich, M. *et al.* (2002) Genetic evidence that oxidative derivatives of retinoic acid are not involved in retinoid signaling during mouse development. *Nat. Genet.* **31**: 84–88.

Nelson, K.G., Sakai, Y., Eitzman, B., Steed, T. and McLachlan, J. (1994) Exposure to diethylstilbestrol during a critical developmental period of the mouse reproductive tract leads to persistent induction of two estrogen-regulated genes. *Cell Growth Differ.* **5**: 595–606.

Newbold, R.R. (2004) Lessons learned from perinatal exposure to diethylstilbestrol. *Toxicol. Appl. Pharmacol.* **199**: 142–150.

Newbold, R.R. (1993) Gender-related behaviour in women exposed prenatally to diethylstilbestrol. *Environ. Health Perspect.* **101**: 208–213.

O'Flaherty, E.J. and Clarke, D.O. (1994) Pharmacokinetic/pharmacodynamic approaches for developmental toxicology. In *Developmental Toxicology*, 2nd edn, Kimmel, C.A. and Buelke-Sam, J. (eds). Raven: New York; 215–244.

Oppenheimer, J.M. (1967) *Essays in the History of Embryology and Biology.* MIT Press: Cambridge, MA.

Otto, D.M., Henderson, C.J., Carrie, D., Davey, M. *et al.* (2003) Identification of novel roles of the cytochrome p450 system in early embryogenesis: effects on vasculogenesis and retinoic acid homeostasis. *Mol. Cell Biol.* **23**: 6103–6116.

Peterson, R.E., Theobald, H.M. and Kimmel, G.L. (1993) Developmental and reproductive toxicity of dioxins and related compounds. *Crit. Rev. Toxicol.* **23**: 283–335.

Phiel, C.J., Zhang, F., Huang, E.Y., Guenther, M.G. *et al.* (2001) Histone deacetylase is a direct target of valproic acid, a potent anticonvulsant, mood stabilizer, and teratogen. *J. Biol. Chem.* **276**: 36734–36741.

Pimental, R.A., Liang, B., Yee, G.K. *et al.* (1993) Dioxin receptor and C/EBP regulate the function of the glutathione S-transferase Ya gene xenobiotic response element. *Mol. Cell. Biol.* **13**: 4365–4373.

Pongratz, I., Mason, G.G.F. and Poellinger, L. (1992) Dual roles of the 90 kDa heat shock protein hsp90 in modulating functional activities of the dioxin receptor. *J. Biol. Chem.* **267**: 13728–13734.

Qi, B.Q., Beasley, S.W. and Arsic, D. (2004) Abnormalities of the vertebral column and ribs associated with anorectal malformations. *Pediatr. Surg. Int.* **20**: 529–533.

Raff, MC. (1992) Social controls on cell survival and cell death. *Nature* **356**: 397–400.

Ramalho-Santos, M., Melton, D.A. and McMahon, A.P. (2000) Hedgehog signals regulate multiple aspects of gastrointestinal development. *Development* **127**: 2763–2772.

Redd, M.J., Cooper, L., Wood, W., Stramer, B. and Martin, P. (2004) Wound healing and inflammation: embryos reveal the way to perfect repair. *Phil. Trans. R. Soc. Lond. B Biol. Sci.* **359**: 777–784.

Roessler, E., Belloni, E., Gaudenz, K., Jay, P. *et al.* (1996) Mutations in the human Sonic Hedgehog gene cause holoprosencephaly. *Nat. Genet.* **14**: 357–360.

Roessler, E., Du, Y.Z., Mullor, J.L., Casas, E. *et al.* (2003) Loss-of-function mutations in the human *GLI2* gene are associated with pituitary anomalies and holoprosencephaly-like features. *Proc. Natl Acad. Sci. USA* **100**: 13424–13429.

Rutledge, J.C., Shourbaji, A.G., Hughes, L.A. *et al.* (1994) Limb and lower-body duplications induced by retinoic acid in mice. *Proc. Natl Acad. Sci. USA* **91**: 5436–5440.

Saatcioglu, F., Perry, D.J., Pasco, D.S. and Fagan, J.B. (1990) Multiple DNA-binding factors interact with overlapping specificities at the aryl hydrocarbon response elements of the cytochrome *P4501A1* gene. *Mol. Cell Biol.* **10**: 6408–6416.

Sadler, T.W. and Hunter, E.S. (1994) Principles of abnormal development: past, present and future. In *Developmental Toxicology*, 2nd edn, Kimmel C.A. and Buellce-Sam J. (eds). Raven: New York; 53–56.

Sauer, B. (1998) Inducible gene targeting in mice using the Cre/lox system. *Methods* **14**: 381–392.

Schantz, S.L. and Bowman, R.E. (1989) Learning in monkeys exposed perinatally to 2,3,7,8-tetrachlorodibenzo-*p*-dioxin (TCDD). *Neurotox. Teratol.* **11**: 13–19.

Schardein, J.L. (2000) *Chemically Induced Birth Defects*, 3rd edn. Marcel Dekker: New York.

Scott, W.J. (1977) Cell death and reduced proliferated rate. In *Handbook of Teratology*, vol 2, Wilson, J.G. and Clarke-Fraser, F. (eds). Plenum: New York; 81–98.

Sharpe, R. and Skakkebaek, N.E. (1993) Are oestrogens involved in falling sperm counts and disorders of the male reproductive tract? *Lancet* **341**: 1392–1395.

Shepard, T.H. (2002) Annual commentary of human teratogens. *Teratology* **66**: 275–277.

Snow, M.H.L. (1987) Uncoordinated development of embryonic tissue following cytotoxic damage. In *Approaches to Elucidate Mechanisms in Teratogenesis*, Welsch, F. (ed.). Hemisphere: London; 83–107.

Stoilov, I. (2001) Cytochrome P450s: coupling development and environment. *Trends Genet.* **17**: 629–632.

Sulik, K.K., Cook, C.S. and Webster, W.S. (1988) Teratogens and craniofacial malformations: relationships to cell death. *Development* **103**(suppl): 213–232.

Taipale, J., Chen, J.K., Cooper, M.K., Wang, B. *et al.* (2000) Effects of oncogenic mutations in *Smoothened* and *Patched* can be reversed by cyclopamine. *Nature* **406:** 1005–1009.

Thomas, H., Shepard, M.D. and Lemire, R.J. (2004) *Catalog of Teratogenic Agents*, 11th edn. Johns Hopkins University Press: Baltimore, MD.

Timms, B.G., Howdeshell, K.L., Barton, L., Bradley, S. *et al.* (2005) Estrogenic chemicals in plastic and oral contraceptives disrupt development of the fetal mouse prostate and urethra. *Proc. Natl Acad. Sci. USA* **102**: 7014–7019.

Turner, S., Sucheston, M.E., de Philip, R. and Paulson, R.B. (1990) Teratogenic effects on the neuroepithelium of the CD-1 mouse embryo exposed *in utero* to sodium valproate. *Teratology* **41**: 421–442.

Veuthey, S., Pexieder, T. and Scott, W.J. (1990) Pathogenesis of cardiac anomalies induced by trimethadione in the rat. In *Developmental Cardiology: Morphogenesis and Function*, Clark, E.B. and Takao, A. (eds). Futura: New York; 453–465.

Webster, W.S., Johnston, M.C., Lammer, E.J. and Sulik, K.K. (1986) Isotretinoin embryopathy and the cranial neural crest: an *in vivo* and *in vitro* study. *J. Craniofac. Genet. Dev. Biol.* **6**: 211–222.

Whitelaw, M., Pongratz, I., Wilhelmsson, A., Gustafsson, J.A. and Poellinger, L. (1993) Ligand-dependent recruitment of the Arnt co-regulator determines DNA recognition by the dioxin receptor. *Mol. Cell. Biol.* **13**: 2504–2514.

Whitlock, J.P. (1993) Mechanistic aspects of dioxin action. *Chem. Res. Toxicol.* **6**: 754–763.

Wickman, D.S., Siebert, J.R. and Benjamin, D.R. (1993) Nitrofen-induced congenital diaphragmatic defects in CD1 mice. *Teratology* **47**: 119–125.

Wilson, J.G., Roth, C.B. and Warkany, J. (1953) An analysis of the syndrome of malformations induced by vitamin A deficiency: effects of restoration of vitamin A at various times during gestation. *Am. J. Anat.* **92**: 189–217.

Wu, L. and Whitlock, J.P. (1993) Mechanism of dioxin action: receptor–enhancer interaction in intact cells. *Nucleic Acids Res.* **21**: 119–125.

Yagi, H., Furutani, Y., Hamada, H., Sasaki, T. *et al.* (2003) Role of TBX1 in human del22q11.2 syndrome. *Lancet* **362**: 1366–1373.

7

The Limbs

Patrizia Ferretti and **Cheryll Tickle**

Developmental anatomy of the human limb

Both upper and lower limbs originate from the lateral plate mesoderm, which thickens in discrete regions of the flank of the embryo. The primordium of the upper limb is first apparent 26 days after fertilization (stage 12, according to Moore, 1988) at the level of the cervical somites, while the lower limb bud appears 1–2 days later, at the beginning of stage 13, opposite the lumbar and upper sacral somites (Figure 7.1). Apart from this delay in the appearance of the lower limb anlage and its subsequent development, early stages of lower and upper limb development are fundamentally the same. For simplicity, the stages discussed here will refer to development of the upper limb.

The emerging limb bud consists of a rapidly proliferating mass of mesenchymal cells covered by an epithelium which will thicken at the tip of the bud along the anteroposterior ('thumb to little finger') axis to form the *apical ectodermal ridge.* Interactions between the apical ectodermal ridge and the underlying mesenchyme are of fundamental importance for proper progression of development and will be discussed in detail in the section concerned with the cellular and molecular interactions underlying normal development of the limb. Limb buds grow rapidly and in a coordinated way. A few days after the formation of buds (Carnegie stages 13–14), spinal nerves start to grow into the mesenchyme of the bud and innervation follows a segmental pattern. Initially the limb bud is supplied by a capillary network which, through processes which are still poorly understood, will transform into a main stem artery and its branches, which drain into a marginal vein. The limb bud continues to grow and elongate and, 32–34 days after fertilization (Carnegie stage 14), a paddle-shaped hand plate has formed. Prechondrogenic condensation of mesenchyme becomes apparent in the regions where the cartilaginous skeletal elements will

Embryos, Genes and Birth Defects, Second Edition Edited by Patrizia Ferretti, Andrew Copp, Cheryll Tickle and Gudrun Moore

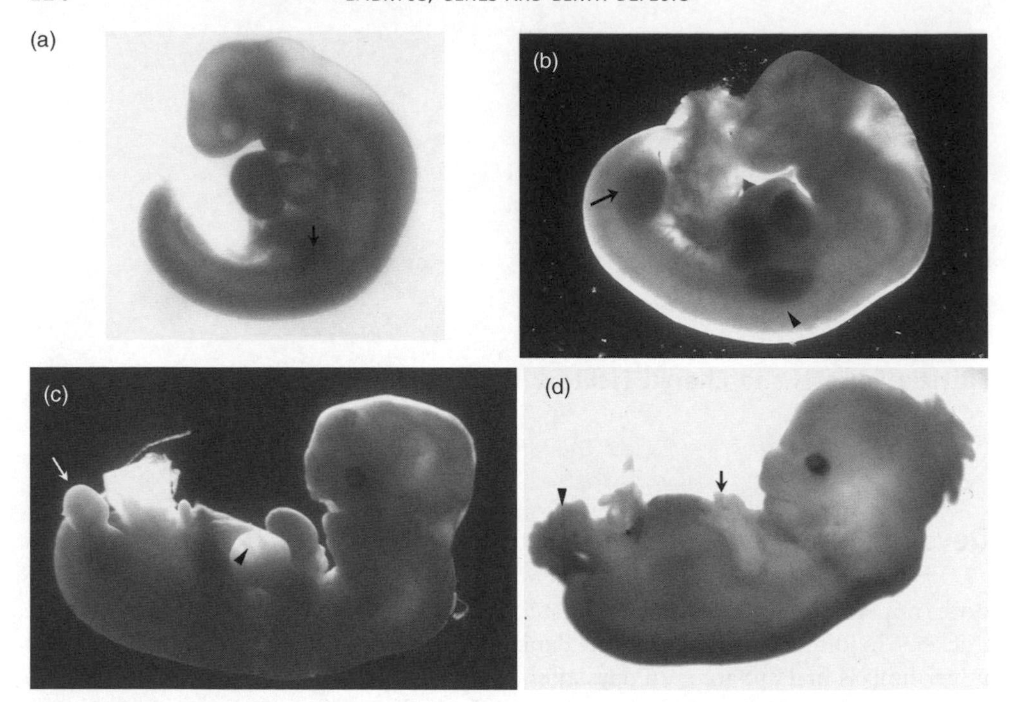

Figure 7.1 Limb development in human embryos at different times after fertilization. (a) Forelimb buds (arrow) are apparent in a 28 day-old embryo, but the hindlimb bud is hardly visible. (b) In a 33 day-old embryo the developing forelimb is paddle-shaped, but the hindlimb is still at the bud stage. (c) Finger rays are visible (arrowhead) in the forelimb and a foot plate (arrow) has formed in a 38 day-old embryo. (d) Short webbed fingers are present (arrow) and notches (arrowhead) are apparent between the digital rays of the foot of a 53 day-old embryo

form in a proximal to distal ('shoulder to digit') direction and, towards the end of the fifth week after fertilization (Carnegie stage 15), overt *chondrogenesis* is in progress. Myoblasts aggregate to form two *muscle masses* dorsal and ventral to the developing skeletal elements. The muscle masses will split to give rise to extensor muscles in the dorsal part of the limb and to flexors ventrally. At the end of the sixth week (stage 17), all the limb skeletal structures have been laid down in cartilage, digital rays are present, notches appear at the tip of the inter-digital ray mesenchyme, and the hand plate starts to assume a webbed appearance (44–50 days, stages 18–19). At the same time (48–50 days, stage 19), rotation of the upper limbs begins and bending of the elbow occurs. The limb will rotate 90° laterally on the longitudinal axis to assume the adult position with the palm of the hand facing anteriorly. Appropriate positioning of the lower limb, with the knee facing anteriorly and the foot downwards, is achieved through a medial rotation of 90° a few days later. In the meanwhile, ossification begins and digital separation is accomplished by destruction of interdigital tissue, probably through a process of programmed cell death, as described in other species. By the end of the eighth week, both upper and lower limbs appear as miniatures of the adult limb, but not all of the centres of primary *ossification* have yet formed, and ossification will continue throughout fetal life.

Main classes of limb defects

Given the numerous events which must be spatio-temporally synchronized in order to develop a normal limb, it is not surprising that limb abnormalities are frequently encountered. Gradually we are learning how impairment of specific cellular and molecular interactions can result in different limb abnormalities (see later). Limb defects can be caused by environmental factors (either chemical or mechanical), gene mutations and chromosomal abnormalities or a combination of such factors. As the most important events in limb development occur between the fourth and eighth week post-fertilization, this is also the period of higher susceptibility to teratogens or defective expression of developmentally regulated genes. However, limb defects can also occur once development of all of the limb structures has been accomplished, as a consequence of trauma. For example, secondary destruction, so called 'intrauterine amputation', is thought to be caused by constriction of the developing limb by amniotic bands. There have been several studies over the last 10 years that have suggested that an increased frequency of limb deficiencies, including 'amniotic band deformities', is associated with chorionic villus sampling carried out early in pregnancy (see e.g. Firth *et al.*, 1994; reviewed by Holmes, 2002).

Although many of the limb defects observed are quite minor and easily corrected by surgery, major limb abnormalities are observed in two of every 1000 births (Moore, 1988). *Amelia*, which is total absence of the limb, can occur, but it is a fairly rare condition. In contrast, partial absence of one or more of the limbs, *meromelia*, is frequently observed. Meromelia is often associated with other types of limb defects, such as *oligosyndactyly* (fusion of digits), *club foot* (deformity of the ankle), and bowed limbs. A large number of limb deficiencies occurred in the 1960s as a consequence of the anti-nausea drug thalidomide, which produced a wide range of bilateral limb deficiencies, including extreme cases of quadruple amelia. More commonly, prenatal exposure to thalidomide resulted in various deficiencies of long bones.

Limb defects are often associated with other malformations affecting, for example, craniofacial, kidney, cardiac and skin development (Stevenson and Meyer, 1993). Limb defects associated with other abnormalities are often heritable, either as autosomal recessive or autosomal dominant conditions. X-linked inheritance of limb defects is observed only in a few syndromes. The various causes believed to lead to each syndrome displaying limb abnormalities, including rare ones, can be found in the review by Stevenson and Meyer (1993). An extrapolation of these data for syndromes where either meromelias or synostoses (fusion of various bones) are observed is shown in Table 7.1.

Since the classification of limb abnormalities is rather complex and has been extensively covered in more specialized textbooks (Stevenson and Meyer, 1993;

Table 7.1 Summary of causation of limb abnormalities in syndromes displaying meromelias and synostosis

	Meromelias* (%)	Synostosis* (%)
Autosomal recessive	25.8	18.6
Autosomal dominant	25.8	52.1
X-linked dominant	3.2	4.3
Chromosomal abnormalities	4.8	1.4
Unknown/uncertain	27.4	12.8
Sporadic/maternal diabetes	9.6	5.7
Drug (thalidomide, alcohol)	1.6	2.9
Trauma	1.6	1.4

*These values have been extrapolated from Stevenson and Meyer (1993). Rare syndromes are also included.

Winter *et al.*, 1993; Gupta *et al.*, 2000), we will give only a few examples of the types of defects which can occur (Table 7.2). As already mentioned, limb deficiencies – amelias and meromelias – represent an important group of limb abnormalities. In other cases, instead of deletions of structures, the presence of supernumerary structures, such as fingers or toes (polydactylies), is observed. *Polydactyly* occurs fairly frequently and the extra digits formed are usually incomplete. In contrast,

Table 7.2 Classification of limb defects

Limb defects	Prominent features	
Amelia	Complete absence of a limb	
Meromelia	Partial absence of a limb	Terminal Intercalary Transverse Longitudinal
Brachydactyly	Shortening of digits	Single digit (one or more bones involved) Multiple digit (one or more bones involved)
Polydactyly	Supernumerary digits	Incomplete extra finger/toe Complete extra finger/toe Mirror hand/foot
Synostosis	Fusion of bones	Bones normally separated by joint space Bones of different rays Bones of different limbs
Syndactyly	Fusion of digits	Cutaneous Osseous (synostosis)
Skeletal dysplasia	Abnormal growth, organization and density of cartilage and bone	Epiphysis Metaphysis Diaphysis

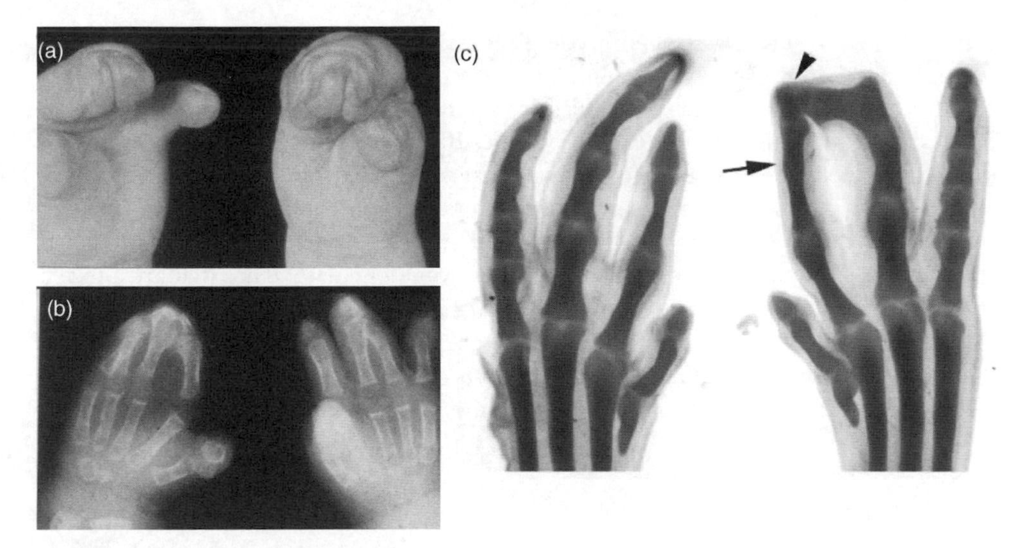

Figure 7.2 Human congenital limb abnormalities (a, b) and experimentally manipulated chick limbs (c) can have similar phenotypes. These examples were chosen to illustrate the close resemblance between human limb abnormalities and the results of experimental manipulations in chick embryos. (a, b) Fusion of digits (syndactyly) in Apert syndrome. Note also distal bone synostosis. (c) Chick limb with fused digits (arrowhead) produced by implanting a bead soaked in Fgf (arrow) in the interdigital space (Reproduced from Sonz-Ezquerro and Tickle, 2003, with permission)

mirror hands and feet (i.e. mirror-imaging of digits on either side of a proximo-distally orientated midline of the digital array), which are usually unilateral, are very rare. Most polydactylies (both isolated and associated with other anomalies) are heritable. They are mainly inherited as an autosomal trait, but a few recessive X-linked cases have also been reported. Other limb defects are the consequence of fusion of structures, either bones (synostosis) and/or cutaneous (Figure 7.2). Soft tissue *syndactyly* is a defect in which the web between the digital rays does not break down during development, and is usually an autosomal dominant abnormality. Another set of defects comprise the brachydactylies, where shortening of digits occurs; different digits, phalanges, metacarpals and metatarsals can be affected. *Brachydactyly* can be familial and is present in many syndromes and skeletal dysplasias. The latter are skeletal abnormalities originating from abnormal growth, organization and density of cartilage and bone.

Contemporary studies on mechanisms of limb development

Many of the contemporary studies that have shed light on mechanisms involved in normal development of the limb are based on experimental analysis of chick embryos. There is an extensive body of information on the effects of removing and transplanting specific pieces of tissue in developing chick limb buds (reviewed by

Saunders, 1977). Such classical embryological investigations have defined the cell–cell interactions that are involved in limb patterning. Considerable progress has been recently made in identifying molecules involved in these interactions (reviewed by Capdevila and Izpisúa-Belmonte, 2001; Tickle, 2003; Niswander, 2003). The potential importance of some of these molecules was first suggested from analysis of patterns of expression of genes encoding growth factors and of homeobox-containing genes in developing mouse embryos (see below). Other developmentally important genes expressed in the limb have been found by their homology with genes that are affected in developmental mutants of insects, e.g. *Hedgehog*, *Engrailed* and *Wnt* (see Chapter 1). Insights into the roles of these genes have been gained from the effects of gene overexpression, limb bud manipulations and application of defined chemicals in chick embryos, and from creating transgenic mice with specific mutations.

Cell–cell interactions

Work on chick embryos has revealed three major sets of cell–cell interactions which operate in the distal region of the developing limb bud (Figure 7.3); an epithelial–mesenchymal interaction between the *apical ectodermal ridge* and the underlying

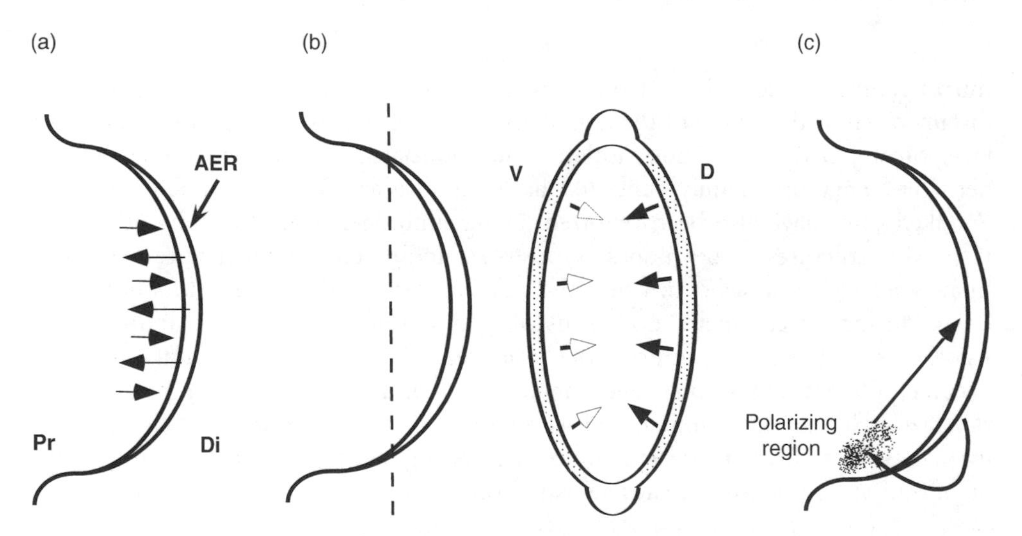

Figure 7.3 Diagram to illustrate the major interactions in the developing limb bud. (a) Arrows indicate reciprocal interactions between the apical ectodermal ridge (AER), the thickened epithelium at the tip of the limb bud and underlying mesenchyme. Pr, proximal; Di, distal. (b) Limb bud sectioned along the dotted line and then shown in cross-section. Arrows indicate potential signals from dorsal and ventral ectoderm (stippled). V, ventral; D, dorsal (c). Straight arrow indicates the interaction between the polarizing region (stippled) and the mesenchyme at the tip of the limb bud. Curved arrow indicates maintenance of the polarizing region by the apical ectodermal ridge

mesenchyme; a second epithelial–mesenchymal interaction involving the ectodermal covering of the limb bud and underlying mesenchyme; and a mesenchymal–mesenchymal interaction between the *polarizing region* (a region of mesenchyme cells at the posterior margin of the limb bud) and other mesenchyme cells at the tip of the limb bud.

The epithelial–mesenchymal interaction between the apical ridge and the underlying mesenchyme is required for bud outgrowth (Figure 7.3a). When the apical ridge is cut away from an early chick limb bud, outgrowth is halted and truncated limbs develop. Conversely, when an apical ridge is grafted to the surface of a bud near the tip, a second outgrowth is induced. Limb structures are laid down along the long axis of the limb in a proximo-distal sequence and removal of the apical ectodermal ridge at later stages of development gives less severe truncations than removal at early stages. The zone of mesenchyme immediately below the apical ridge consists of undifferentiated proliferating cells. A long-standing model suggests that the identity of structures being laid down is controlled by a timing mechanism operating autonomously in this zone, which has become known as the *progress zone* (Summerbell *et al.*, 1973). Cells that leave the progress zone early form proximal structures, whereas cells that leave it later form distal ones. Recently, this model has been challenged; it has been suggested that all the limb structures have already been specified in the early bud and that the primordia of these limb structures expand in a proximo-distal sequence as the limb bud grows out (Dudley *et al.*, 2002).

Signals from the ectoderm covering the tip of the bud appear to control pattern across the dorso-ventral axis (Figure 7.3b). When the ectodermal jacket of a left limb bud is replaced by the ectodermal jacket of a right limb bud, the dorso-ventral axis of the part of the limb laid down after the operation (i.e. the distal part of the limb) is reversed, as judged by muscle pattern, joint flexure and skin appendages (MacCabe *et al.*, 1974; Akita, 1996). In addition, when an apical ridge is grafted to the surface of a chick limb bud, the outgrowth has a symmetrical pattern, either double-dorsal, when the outgrowth arises from the dorsal surface, or double-ventral, when outgrowth arises from ventral surface (Saunders and Errick, 1976). It has also been suggested that the ectoderm may play a mechanical role in controlling bud shape. The apical ridge might provide a stiffened rim to the tip of the bud and help to keep the bud dorso-ventrally flattened (Figure 7.3b).

The mesenchymal–mesenchymal interaction involves signalling by the polarizing region to mesenchyme cells at the tip of the limb bud, and this determines the pattern of structures that develop across the antero-posterior axis of the limb (Figure 7.3c). Signalling of the polarizing region was discovered by grafting the posterior margin from one chick wing bud to the anterior margin of a second bud. In response to a signal from the graft, anterior mesenchyme cells in the host limb bud were found to form an additional set of digits (4, 3, 2) in mirror-image symmetry with the normal set of digits (2, 3, 4; Saunders and Gasseling, 1968). The strength of the polarizing region signal can be assayed by the character of the additional digits and was found to be dose-dependent. With a very small number of polarizing region cells grafted, just an additional digit 2 develops (Tickle, 1981). The polarizing region appears to provide a long-range positional signal and the structures that form depends on distance from the polarizing region. When the polarizing region is grafted more posteriorly, closer to the host polarizing region, anterior digits do not form and the digit pattern that results is 4, 3, 3, 4. Experiments in which blocks of leg tissue were interposed between graft and responding cells show that the polarizing signal can operate over about 10–30 cell diameters (Honig, 1981).

Signalling by the polarizing region also leads to maintenance of the apical ridge; this property of posterior mesenchyme was discovered even before the patterning effects of the polarizing region were identified and was postulated to be due to production of an apical ectodermal ridge maintenance factor (Zwilling and Hansborough, 1956). The length of the ridge is related to the number of digits that will form; thus, the maintenance of the apical ectodermal ridge links limb bud patterning and outgrowth. This link is further strengthened because the apical ectodermal ridge, in turn, has been shown to maintain the polarizing region at the posterior margin of the limb bud tip. Assays for polarizing activity after the posterior part of the ridge has been removed showed that the ability of posterior mesenchyme to induce digit duplications has been reduced (Vogel and Tickle, 1993).

Molecules implicated in signalling

Apical ectodermal ridge signals Genes encoding fibroblast growth factors (Fgfs) are expressed in the apical ectodermal ridge and can substitute for the apical ridge in mediating bud outgrowth (Niswander *et al.*, 1993; Fallon *et al.*, 1994). When beads soaked in Fgf are placed at or near the margin of a chick limb bud following removal of the apical ridge, outgrowth continues and distal structures are laid down. Even though Fgf beads can effectively promote outgrowth, the limb buds become bulbous rather than being dorsoventrally flattened. This change in bud shape could account for the bunching of digits that frequently occurs in Fgf-treated limbs.

Several members of the Fgf family are expressed in early chick and mouse limb buds. *Fgf8* is expressed throughout the apical ectodermal ridge from very earliest stages in limb bud development in mouse and chicken embryos and persists until the tips of the digits are formed, while *Fgf4*, *Fgf9* and *Fgf17* expression is initiated later, at limb bud stages, in the posterior part of the ridge, and *Fgf4* and *Fgf17* expression switches off prior to digit formation (reviewed by Martin, 1998). Mice have been created in which these genes have been conditionally knocked out – singly and in combination – in the apical ectodermal ridge of the limb. Some of the limb phenotypes are rather complex but several general points emerge. One is that if FGF signalling is completely deleted from the earliest stages of limb development, no limbs form (Sun *et al.*, 2002). A second general point is that there seems to be considerable redundancy between the *Fgfs* expressed in the ridge. Thus, for example, deletion of just one of the posteriorly expressed *Fgfs*, *Fgf4*, has no effect on limb development (Sun *et al.*, 2000).

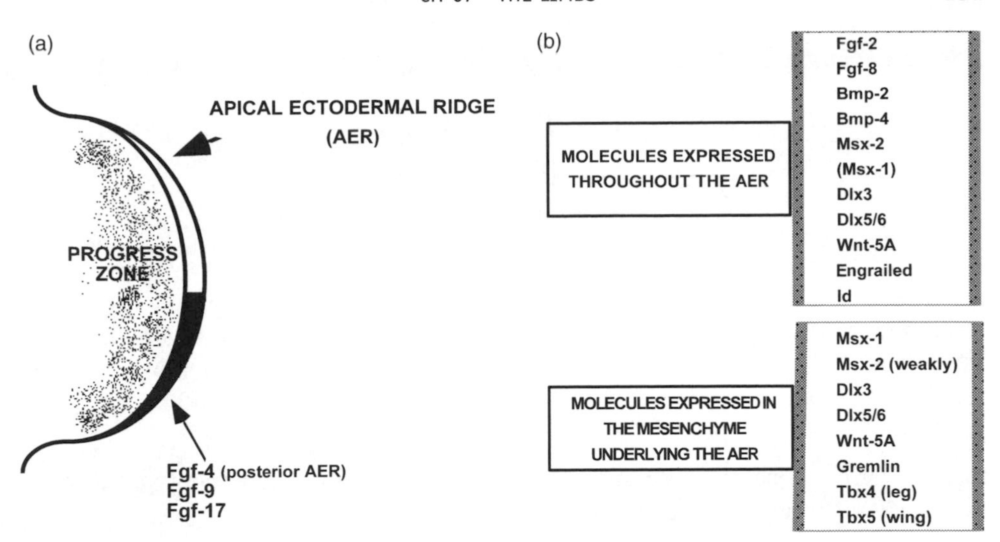

Figure 7.4 Molecules expressed in tissues at the tip of the limb bud that could be involved in mediating interactions between the apical ectodermal ridge and the underlying progress zone mesenchyme. (a) Limb bud showing the apical ectodermal ridge and progress zone and the pattern of expression of *Fgf4* transcripts. (b) Selected lists of molecules expressed in the two tissues

Apical ridge cells express genes encoding other signalling molecules in addition to Fgfs, including *bone morphogenetic proteins* (Bmps) and various members of the *Wnt* family. *Wnt* genes are a family of genes that comprise vertebrate homologues of the *Drosophila wingless* gene and an *int* gene which is involved in induction of mammary tumours in mice. Genes encoding Bmps and Wnt5A are also expressed in the mesenchyme at the tip of the limb bud (see below; Figure 7.4). Since Fgfs appear to be able to substitute for the apical ridge signal, these other growth factors and other molecules expressed in the apical ridge may be involved in either regulating Fgf expression or maintaining mechanical integrity of the ridge. In chicken limb development, *Wnt3a* has been shown to be involved in initiating *Fgf8* expression in the apical ridge (Kengaku *et al.*, 1998). In mice, this function seems to be served by *Wnt3* (Barrow *et al.*, 2003).

Several different transcription factors are expressed by apical ectodermal ridge cells. For example, *Distal-less* and *Engrailed* are expressed in ridge, while *Msx* genes (see later) and the *ld* gene are expressed in both ridge and mesenchyme. The *ld* gene is affected in the mouse limb deformity mutants and the location of the gene affected was characterized by identifying the site of an insertional mutation in a transgenic mouse. One gene in this region encodes a member of a previously unknown family of proteins called 'formins'. Formins are localized in the nucleus and may be involved in gene regulation (Woychik *et al.*, 1990). Another gene in this region encodes gremlin, an antagonist of Bmp signalling (see later), and it is now clear that it is the *gremlin*,

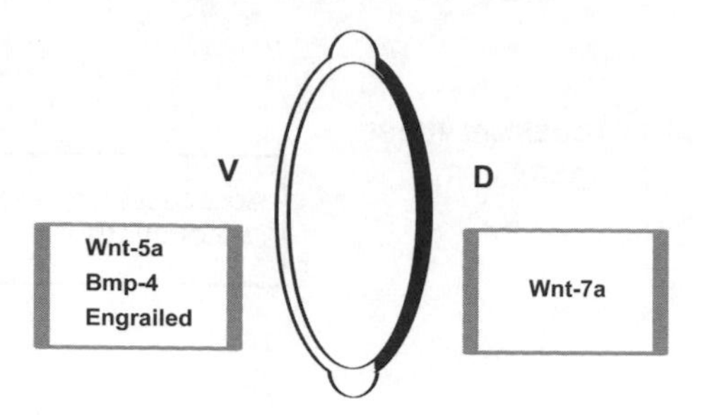

Figure 7.5 Molecules expressed in the dorsal and ventral ectoderm. Those expressed in the ventral ectoderm are later expressed in the apical ectodermal ridge (see Figure 7.4). V, ventral; D, dorsal

not the *formin* gene, that is responsible for the limb deformity phenotype (Khokha *et al.*, 2003; Zuniga *et al.*, 2004).

***Ectodermal signals* (Figure 7.5)** Several molecules are known to be expressed in either dorsal or ventral ectoderm. A striking dorsoventral ectodermal restriction is that of transcripts of the gene *Wnt7a*, which are localized in dorsal ectoderm (Parr *et al.*, 1993; Dealy *et al.*, 1993). Functional inactivation of *Wnt7a* in mice leads to ventralization of distal limb pattern (Parr and McMahon, 1995), showing that Wnt7a acts as a dorsalizing signal. At very early limb bud stages, a number of genes that are later restricted to the apical ridge are expressed in ventral ectoderm, including *Engrailed1* (Figure 7.5). There is evidence from studies in both chick and mouse embryos that Bmps act upstream of *Engrailed1* in apical ridge formation (Pizette *et al.*, 2001; Ahn *et al.*, 2001). Functional inactivation of *Engrailed1* results in dorsal transformation of ventral paw structures, suggesting a specific role for this gene, in addition to *Wnt7a*, in dorsoventral patterning of the limb (Loomis *et al.*, 1996).

Polarizing signals The first molecule to be identified that can mimic signalling of grafts of the polarizing region was a vitamin A derivative, *retinoic acid* (Tickle *et al.*, 1982). More recently, it has been shown that expression of the *sonic hedgehog* gene (*Shh*) maps to the polarizing region and the product of the *Sonic hedgehog* gene can provide a polarizing region signal (Riddle *et al.*, 1993). Beads soaked in retinoic acid or grafts of fibroblast cells transfected with the *Sonic hedgehog* gene or Shh-soaked beads placed at the anterior margin of a chick wing bud lead to digit duplications (Tickle *et al.*, 1985; Riddle *et al.*, 1993; Lopez-Martinez *et al.*, 1995; Yang *et al.*, 1997). Application of retinoic acid to the anterior margin induces expression of *Shh*, suggesting that the Shh signal acts downstream of retinoic acid (Riddle *et al.*, 1993; Niswander *et al.*, 1994). Shh is a good candidate for the long-range positional signal produced by the polarizing region. Diffusion of Shh across the limb bud has been demonstrated by immunohistochemistry (Gritli-Linde *et al.*, 2001) and by an indirect bioassay (Zeng *et al.*, 2001). The effects of Shh application are dose-dependent (Yang

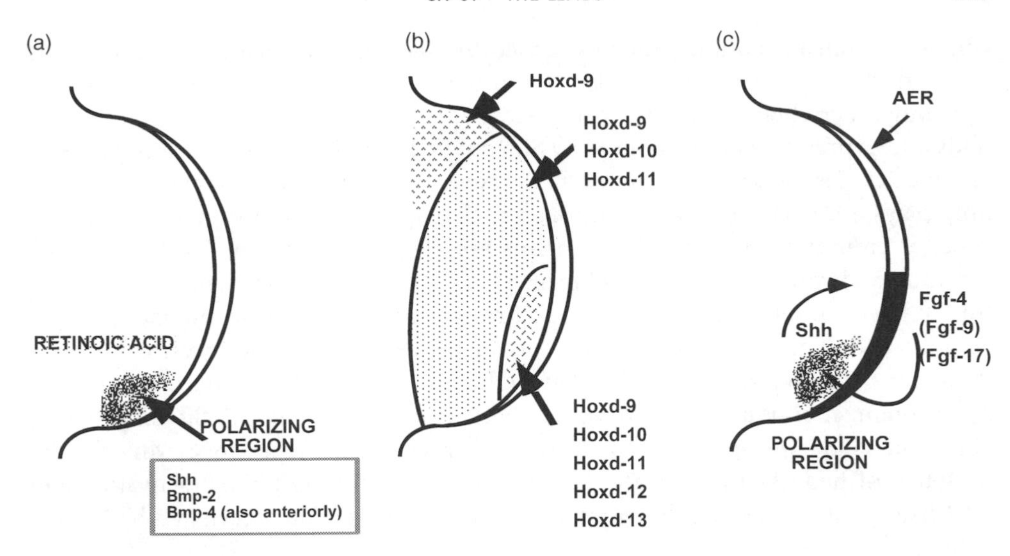

Figure 7.6 Molecules involved in the interaction between the polarizing region and the progress zone mesenchyme at the tip of the bud. (a) Potential signalling molecules; retinoic acid – high to low concentration indicated by large arrow; genes expressed in the polarizing region. (b) Pattern of expression of gene members of *HoxD* cluster across the antero-posterior axis of the limb bud. Changes in the expression pattern can be brought about by polarizing signals. (c) Feedback loop between signalling by the polarizing region and signalling by the posterior ridge. Arrows do not necessarily imply direct action

et al., 1997) but there has been considerable debate about whether Shh directly controls specified digit pattern or acts indirectly (see later). Recent work has also raised the possibility that the length of time cells are exposed to Shh in addition to the level of signalling may be important in determining which digit cells form (Harfe *et al.*, 2004; Ahn and Joyner, 2004).

There is also evidence that endogenous retinoic acid signalling plays an important role in limb development. Retinoic acid has been extracted from early chick and mouse limb buds and is estimated to be present in nanomolar concentrations (reviewed by Hofmann and Eichele, 1994). Chick limb bud mesenchyme cells express a range of molecules that mediate a retinoid response (Figure 7.6), including nuclear retinoic acid receptors and retinoid-binding proteins (reviewed by Mangelsdorf *et al.*, 1994; Kastner *et al.*, 1995). In addition, the genes encoding enzymes that metabolize retinoic acid – both those that are required to generate retinoic acid from retinol, e.g. retinaldehyde dehydrogenase (Raldh2), and those that break it down, e.g. Cyp26B1 – have been identified and are expressed in developing limb buds. In *Raldh2*$^{-/-}$ mice, forelimb development is not initiated, while in *Cyp26B1*$^{-/-}$ mouse embryos there are distal limb defects (Niederreither *et al.*, 2002; Mic *et al.*, 2004; Yashiro *et al.*, 2004).

The details of how cells respond to Hedgehog signalling and the components of the intracellular signal transduction machinery were first unravelled in *Drosophila*, in which the transmembrane protein encoded by the segment polarity gene *patched* is the hedgehog receptor and is also an immediate downstream regulator of Shh signalling. Vertebrate *patched* is expressed in the posterior part of vertebrate limb buds and expression can be induced by Shh applied ectopically (Goodrich *et al.*, 1996; Hahn *et al.*, 1996a; Marigo *et al.*, 1996; Pearse *et al.*, 2001). Interestingly, *patched* not only plays a crucial developmental role but also appears to be a tumour suppressor gene, as mutations of *patched* in humans are frequently associated with basal cell carcinoma (Hahn *et al.*, 1996b; Johnson *et al.*, 1996). When Shh binds to patched, this relieves inhibition of signalling through another transmembrane protein, smoothened. The transcriptional effectors of hedgehog signalling in vertebrates are the three bifunctional Gli proteins, Gli1, Gli2 and Gli3 (in *Drosophila,* there is only one protein, cubitus interruptus). In the absence of hedgehog ligand, the Gli proteins, Gli2 and Gli3, are processed to generate transcriptional repressors; while in the presence of hedgehog ligand, the Gli proteins act as transcriptional activators and *Gli1* itself is an immediate downstream target of Shh signalling (reviewed by Cohen, 2003).

Analysis of the limb phenotypes of mouse embryos in which *Shh*, *Gli3* and both *Shh* and *Gli3* together have been knocked out has revealed that the main function of Shh signalling appears to be to relieve Gli3 repressor activity in the posterior part of the limb bud. The limbs of $Shh^{-/-}$ mouse embryos are very reduced distally, and this can be understood in terms of high levels of Gli3 repressor being present throughout the limb bud and virtually abolishing limb outgrowth. In contrast, when *Gli3* is functionally inactivated, the limbs are polydactylous and all the digits look the same. A similar phenotype is seen in mouse embryos in which both *Shh* and *Gli3* have been functionally inactivated ($Shh^{-/-}$ $Gli3^{-/-}$), indicating that formation of morphologically identical digits from both posterior and anterior parts of the limb bud is independent of Shh signalling and depends, instead, on absence of Gli3 repressor (te Welscher *et al.*, 2002a; Litingtung *et al.*, 2002).

Shh expression is maintained by the apical ridge (Niswander *et al.*, 1994). The three *Fgf* genes expressed in the posterior part of the apical ridge are good candidates for fulfilling this role, but even in triple knock-outs (mouse embryos in which *Fgf4*, *Fgf9* and *Fgf17* have been functionally inactivated in the ridge) *Shh* expression is maintained, suggesting that *Fgf8* can suffice. Shh, in turn, maintains expression of the three *Fgf* genes in the posterior apical ridge, creating a positive feedback loop (at first thought to involve only *Fgf4*; Niswander *et al.*, 1994; Laufer *et al.*, 1994). The feedback loop from mesenchyme to ridge is now known to be accomplished via Shh, maintaining expression of *gremlin* (the limb deformity gene), which thus acts as the ridge maintenance factor (Zuniga *et al.*, 1999). In addition to the positive feedback loop between Fgfs and Shh, there is evidence that Wnt7a is also necessary for the maintenance of *Shh* expression (Parr and McMahon, 1995; Yang and Niswander, 1995). Thus, there is coordination of signalling along all three limb axes.

Bone morphogenetic proteins (e.g. *Bmp2, Bmp4*; Francis *et al.*, 1994) are signalling molecules of the *transforming growth factor* (TGF-β) superfamily (reviewed by

Wozney *et al.*, 1993) involved in cartilage and bone development. BMP transcripts are found in mesenchyme cells at the posterior margin of the early limb bud. In a *Drosophila* signalling pathway, *hedgehog* induces expression of a gene called *dpp*, which has a patterning function (Basler and Struhl, 1994). *Dpp* encodes a molecule very closely related to Bmp2 and Bmp4 and application of Shh can lead to *Bmp2* expression in anterior cells of chick limb buds (Yang *et al.*, 1997). There is also evidence from experiments on chick limb development that manipulating Bmp signalling in cells previously exposed to Shh can alter digit patterning (Drossopolou *et al.*, 2000). One recent model suggests that the most anterior digit of the mouse limb might be specified by Bmps, the next three digits by Shh and Bmps, and the most posterior digit by Shh alone (Lewis *et al.*, 2001) but this remains to be tested directly and is still very controversial.

Bmps seem to have other roles in limb development. For example, the gene encoding Bmp7 is also expressed in developing limbs, and functional inactivation of this gene in mice results in polydactyly (Luo *et al.*, 1995; Dudley *et al.*, 1995). There is also evidence that Bmp4, which is expressed more strongly anteriorly in chick limb buds, opposes Shh signalling posteriorly and that Shh, in turn, can inhibit *Bmp4* gene expression (Tümpel *et al.*, 2002). At later stages, Bmp4 signalling seems to regulate the programmed cell death that occurs interdigitally and serves to separate the individual digits. Thus, inactivation of a BMP receptor in chick limb buds leads to interdigital webbing (Zou and Niswander, 1996).

Recently it has been shown that digit morphology is surprisingly plastic at late stages in chick limb development. Thus, for example, when interdigital tissue from between two chick toes is grafted to a new location between two different toes, this can affect the adjacent toes and longer toes with additional phalanges or shorter toes with a reduced number of phalanges develop (Dahn and Fallon, 2000). Beads soaked in Shh can also induce the formation of additional phalanges in chick toes, when implanted at late stages when *Shh* is no longer expressed. There is evidence that the effects of Shh on digit length and induction of an extra phalange are mediated via Bmp signalling (Dahn and Fallon, 2000) and also that the duration of Fgf signalling in the apical ridge overlying the forming digits is prolonged (Sanz-Ezquerro and Tickle, 2003). When an Fgf bead is placed at the tip of the developing toe, this can also lead to digit elongation but tip formation is prevented and fusion of adjacent digits can result (Figure 7.2c). In contrast, cessation of Fgf signalling triggers a special developmental programme for making the digit tip (Sanz-Ezquerro and Tickle, 2003).

***Molecules expressed at the tip of the limb bud* (Figure 7.4)** Genes known to be expressed in mesenchyme at the tip of the limb bud could encode molecules that play roles in controlling cell proliferation, in maintaining cells in an undifferentiated state and in timing mechanisms. Among genes with this expression pattern are those encoding transcription factors and short-range signalling molecules.

Transcripts of a *Wnt* gene family member, *Wnt5A*, are found at high levels at the tip of the limb bud. In mice in which *Wnt5a* is functionally inactivated, the limbs are very short, due to progressive reduction of structures along the proximo-distal axis, culminating in complete loss of the distal-most structures, the phalanges of the digits (Yamaguchi *et al.*, 1999). Reduced outgrowth and patterning of the limb buds of *Wnt5a*$^{-/-}$ mouse embryos are associated with a reduction in cell proliferation.

Two related homeobox-containing genes, *Msx1* and *Msx2* (formerly known as *Hox7* and *Hox8*) could be important in maintaining cells at the tip of the limb bud in an undifferentiated state (Hill *et al.*, 1989; Robert *et al.*, 1991). Various grafting experiments show that, in chick limb buds, mesenchymal expression of *Msx1* is regulated by a signal from the ridge, which can be substituted by Fgf4 (Davidson *et al.*, 1991; Robert *et al.*, 1991; Vogel *et al.*, 1995a). The potential role of *Msx1* in maintaining an undifferentiated cell state was first suggested directly by experiments with a potentially myogenic cell line which, when transfected with the *Msx1* gene, can no longer be induced to differentiate into muscle (Song *et al.*, 1992). More recently it has been shown that *Msx1* transfection can even induce dedifferentiation of terminally differentiated mouse myotubes (Odelberg *et al.*, 2000). Functional inactivation of *Msx1* in a transgenic mouse has no apparent effect on limb development (Satokata and Maas, 1994) but it is possible that expression of *Msx2* compensates lack of *Msx1* in the limb.

Distal mesenchyme cells have also been shown to be linked by gap junctions (Kelley and Fallon, 1978). The presence of gap junctions between distal cells is dependent on ridge signalling (Green *et al.*, 1994) and mediated by Fgfs (Makarenkova *et al.*, 1997). There is evidence from functional blocking studies using antibodies that gap junctional communication could be important in limb patterning (Allen *et al.*, 1990) and, furthermore, that when expression of the gene encoding connexin43, a member of the family of proteins that make up gap junctions, is knocked down in chick limb buds using an antisense strategy, bud outgrowth is reduced and the limbs are short or have distal deletions (Law *et al.*, 2002).

Recently it has been shown that the gene *Hairy1*, which encodes a member of the Hairy/Enhancer of split family of transcriptional repressors, is expressed at the tip of developing chick limb buds (Vasiliauskas *et al.*, 2003). This is of particular interest because *Hairy1* can act as a transcriptional effector of Notch signalling and there is evidence that both Hairy and Notch are components of the timing mechanism that controls somite segmentation along the main body axis (reviewed by Rida *et al.*, 2004). It is not yet clear whether *Hairy1* is involved in measuring time in the developing limb. When *Hairy1* is overexpressed in wing buds in chick embryos, all the structures form normally but the wing is reduced in size (Vasiliauskas *et al.*, 2003).

Molecular response to signalling Genes involved in response to cell–cell signalling in the developing limb bud include gene members of the *HoxD* and *HoxA* clusters. Genes in the part of the *HoxD* cluster from *Hoxd9* to *Hoxd13* and in the part of *HoxA* cluster from *Hoxa9* to *Hoxa13* are expressed in overlapping domains in vertebrate limb buds, with transcripts of genes located more 5′ in the cluster being found more distally and, in the case of *Hoxd* genes, also more posteriorly (Dolle *et al.*, 1989; Nelson *et al.*, 1996). These nested patterns of gene expression are established in very early limb buds, with expression of 3′ genes appearing before that of 5′ genes. *Hoxc10* and *Hoxc11* are also expressed in early hindlimbs.

An ectopic sequence of *Hoxd* gene expression can be induced at the anterior of early chick wing buds by grafting a polarizing region or cells constitutively expressing *Shh* or by implanting retinoic acid-soaked beads (Nohno *et al.*, 1991; Izpisúa-Belmonte *et al.*, 1991; Riddle *et al.*, 1993). The establishment of an ectopic set of overlapping domains of expression of *HoxD* genes in anterior cells of early limb buds requires cooperation with FGF signalling from the apical ridge (Izpisúa-Belmonte *et al.*, 1992a; Niswander *et al.*, 1994) and it has also been shown that maintenance of *Hoxa13* expression at the tip of chick wing buds depends on FGF signalling (Hashimoto *et al.*, 1999; Vargesson *et al.*, 2001).

By the time the digits are forming in developing limbs, complex changes in the pattern of *Hox* gene expression have occurred with, for example, *Hoxd10* to *Hoxd13* and *Hoxa13* being expressed in the distal region of the limb and *Hoxa11* and *Hoxa10* more proximally (Nelson *et al.*, 1996; Kmita *et al.*, 2002). A global control region that contains an enhancer responsible for the late phase of *Hoxd-10* to *Hoxd-13* expression in the digits has been identified by a series of sophisticated genetic manipulations in transgenic mice (Spitz *et al.*, 2003). Another control region is responsible for the *Hox* gene expression in the early limb bud. Genetic manipulations around this 'early' region in transgenic mice lead to 5′ genes being expressed in the pattern of 3′ genes, e.g. the most 5′ genes, such as *Hoxd13* and *Hoxd12*, are expressed throughout the early limb bud. These transgenic mouse embryos have extra digits and this is associated with ectopic expression of *Shh* at the anterior margin of the limb bud (Zakany *et al.*, 2004; see also Knezevic *et al.*, 1997). This finding provides a new explanation for the extra digits observed after overexpressing a *HoxD* gene in chick limb buds (Morgan *et al.*, 1992).

There is now substantial evidence consistent with the idea that the pattern of *Hox* gene expression encodes position in the limb and is required for patterning different limb 'segments'. Creation of double and even triple knock-outs has been needed to reveal the roles of *Hox* genes in patterning the limb because there is considerable functional redundancy between corresponding genes (paralogues) from different clusters. Analysis of the limb phenotypes of mouse triple mutants for *Hox10* and *Hox11* paralogues, together with analysis of double mutants, such as mutants that lack *Hoxd13* and *Hoxa13*, suggests that *Hox10* (and *Hox9*, in the case of the forelimb) paralogues are required for development of the proximal 'segment' of the limb, femur and humerus. Similarly, *Hox11* paralogues are needed for development of the middle segment, tibia/fibula and radius/ulna, and *Hox13* paralogues are required for development of the distal segment, digits (Wellik and Capecchi, 2003).

Genes involved in specifying the proximal part of the limb bud in response to retinoic acid signalling have also been identified. Expression of the homeobox genes, *Meis1* and *Meis2*, is at first widespread in limb buds and then is restricted to the proximal regions of the limb bud. The importance of these genes in limb patterning has been shown by overexpression experiments in chick embryos. When either *Meis1* or *Meis2* is overexpressed in chick limb buds, this disrupts distal development and produces distal to proximal transformations (Mercarder *et al.*, 1999; Capdevila *et al.*,

1999). Genetic and biochemical studies have shown that Meis proteins can act as co-factors for transcription factors, including Hox proteins, and this may explain their role in limb development.

With respect to encoding dorso-ventral position, it has been shown that the transcription factor *Lmx1* can be induced by Wnt7a signalling in the dorsal chick limb mesenchyme (Riddle *et al.*, 1995; Vogel *et al.*, 1995b). Ectopic expression of *Lmx1* in the ventral mesenchyme of chick limb buds using retroviruses leads to formation of ectopic dorsal structures. Furthermore, when *Lmx1* targets are repressed by overexpressing the *Lmx1* DNA-binding domain fused to the *Engrailed* repressor, the dorsal pattern of chick wing buds is ventralized (Rodriguez-Esteban *et al.*, 1998).

Antero-posterior patterning in vertebrate limbs appears to mirror a signalling cascade identified in *Drosophila* wing development (see earlier). Vertebrate orthologues of target genes, which encode transcription factors, have been shown to be involved in responding to this cascade in insects. *Omb* orthologues, *Tbx3* and *Tbx2,* and *Spalt* orthologues, *Sall1* (mouse), *cSal1* and *cSal2* (chick), are expressed in limb buds. There is also experimental evidence that members of the *Tbx* gene family, *Tbx2* and *Tbx3,* may contribute to encoding antero-posterior position in the limb bud. *Tbx2* and *Tbx3* are expressed in posterior and anterior stripes, and expression in the posterior stripe requires Shh signalling (Tümpel *et al.*, 2002). When *Tbx3* and *Tbx2* are overexpressed throughout chick leg buds, this has been reported to lead to changes in digit morphology – anterior toes develop extra phalanges, consistent with a change to a more posterior identity (Suzuki *et al.*, 2004).

Initiation of limb bud development

The development of two pairs of limbs at different axial levels is a central feature of the vertebrate body plan. What controls limb number, limb position and limb type (i.e. forelimb vs. hindlimb)? Recent work suggests that homeobox-containing genes and growth factors may be involved in controlling the initiation of limb bud development and *Tbx* genes in encoding limb identity.

Limb position could be related to the pattern of expression of homeobox-containing genes along the body axis. In vertebrates, genes of the four *Hox* clusters are expressed in a series of overlapping domains, with all members of each cluster generally being expressed posteriorly at the 'tail' end of the embryo and anterior limits of expression near the 'head' end being staggered, with more 3′ genes being expressed more anteriorly. In mouse embryos, forelimb buds arise at the anterior limit of expression of *Hoxb8*. Transgenic mice in which the anterior limit of expression of *Hoxb8* has been shifted by linking the coding region of the gene to the promoter of the *RAR*β gene, a gene which is expressed in anterior regions of the embryo, have duplicated forelimb patterns (Charité *et al.*, 1994). The duplicated skeletal pattern is preceded in early limb buds by mirror-image expression patterns of *Shh* and *Fgf4*. Another transcription factor, D-Hand, also seems to be required for *Shh* expression in the posterior region of the limb bud (te Welscher *et al.*, 2002b).

Signalling by Fgfs and Wnts can initiate the formation of limb buds. The first clue that Fgf signalling can initiate limb formation came from mouse chimeras containing cells that constitutively express *Fgf4* (Abud *et al.*, 1996); the chimeric embryos develop ectopic 'limb buds' in the flank or interlimb region (the region lying between fore- and hindlimb buds). In chick embryos, a single bead soaked in Fgf placed in the presumptive flank can induce development of an ectopic bud that can then give rise to a complete extra limb (Cohn *et al.*, 1995; Ohuchi *et al.*, 1995; Crossley *et al.*, 1996). The type of extra limb induced is related to bead position along the flank: beads placed anteriorly tend to give wings, while more posteriorly placed beads give legs.

An interesting feature of the additional limbs is that they have reversed polarity and this is correlated with a reversed pattern of *Shh* expression in ectopic buds. It seems likely that the reversal of polarity is due to an anterior to posterior gradient in polarizing potential in cells of the flank (Hornbruch and Wolpert, 1991). This polarizing potential was assayed by systematically grafting pieces of flank to the anterior margin of chick wing buds and showing that additional digits could be specified. Flank cells will be recruited into an ectopic bud when Fgf is applied, and cells with the highest polarizing potential will be at the anterior of the bud. Ectopic buds also acquire an apical ectodermal ridge and once both signalling regions are established, the bud can then autonomously develop into a limb (Cohn *et al.*, 1995). The flank of mouse embryos also has polarizing potential, and implanting an Fgf bead can induce an *Fgf8*-expressing apical ridge all along the side of a mouse embryo (Tanaka *et al.*, 2000). Studies in chick embryos have also revealed that activation of Wnt signalling in the flank by, for example, implanting *Wnt*-expressing cells can also induce formation of additional limbs (Kawakami *et al.*, 2001).

There is now evidence that Fgfs and Wnts could be involved in the normal process of limb initiation in chicken and mouse embryos. *Fgf10* is expressed in presumptive limb-forming regions in both chick and mouse embryos. Moreover, mice in which *Fgf10* is functionally inactivated fail to develop limbs (Sekine *et al.*, 1999; Min *et al.*, 1998). Furthermore, mouse embryos in which the *Fgfr2IIIb* gene, expressed in limb ectoderm, is disrupted also lack limb buds, suggesting that *Fgf10* induces limb formation via Fgfr2IIIb (Xu *et al.*, 1998). In chick embryos, *Wnt2b* is expressed in the wing-forming region, while *Wnt8c* is expressed in the leg-forming region (Kawakami *et al.*, 2001) but transcripts of neither of these *Wnt* genes can be detected in limb-forming regions of early mouse embryos.

Two members of the *Tbx* gene family have been implicated in specifying limb identify and have striking limb type-specific expression patterns. *Tbx5* expression is associated with developing forelimbs and *Tbx4* with hindlimbs in embryos of a wide range of vertebrates, including teleost and cartilaginous fish, chickens and mice (Tamura *et al.*, 1999; Tanaka *et al.*, 2002; Gibson Brown *et al.*, 1996; reviewed by Logan, 2003). Furthermore, when *Tbx4* is ectopically expressed in the wing-forming regions of chick embryos, using retroviruses, changes towards a more leg-like morphology are induced, and more wing-like characteriztics are seen when *Tbx5* is ectopically expressed in leg-forming regions (Rodrizuez-Esteban *et al.*, 1999; Takeuchi *et al.*, 1999; Logan and Tabin, 1999). The importance of these genes for limb development is further shown by mouse knock-outs. When *Tbx5* is functionally inactivated, the forelimbs are completely absent (Rallis *et al.*, 2003), while in *Tbx4*$^{-/-}$ mouse embryos, limb buds form but fail to develop (Naiche and Papaioannou, 2003). Another gene, *Ptx-1* (*Pitx-1*), is specifically expressed in regions that will develop into hindlimbs. This is a homeobox gene, identified almost simultaneously in a screen for genes encoding DNA-binding proteins (and named *backfoot*) and as a transcription factor involved in pituitary gland development. *Pitx-1* seems to lie upstream of *Tbx4* in hindlimb development. When *Pitx1* is ectopically expressed in wing-forming regions in chick embryos, *Tbx4* is induced and leg-like transformations result (Logan and Tabin, 1999). In mouse embryos that lack *Pitx1* function, there are abnormalities in the hindlimbs (Marcil *et al.*, 2003). These defects are often asymmetrical, with the right leg being more affected than the left. This appears to be due to compensation by the related homeobox-containing gene, *Pitx2*, a gene known to be involved in determining laterality (left–right asymmetry) and which is expressed in the hindlimb-forming region on the left-hand side of vertebrate embryos (Marcil *et al.*, 2003). Genes encoding other transcription factors are also expressed in limb-forming regions, including *snail* and *twist* (Isaac *et al.*, 2000). In *twist*$^{-/-}$ mouse embryos, limb buds form but do not grow out (Chen and Behringer, 1995).

Limb regeneration

Only a very few adult vertebrates are able to regenerate their limbs, although some regenerative capability is present in the developing limbs of several species, including mammals. In the mouse, significant regeneration has been observed in the digit tip of the fetus, and it appears to be restricted to levels in which the amputation plane is within the distal region expressing *Msx1* (Reginelli *et al.*, 1995). The neonate can also regenerate its digit tips, although not always perfectly, and such capability is restricted to the nail bed, where both *Msx1* and *Msx2* are expressed. The nail organ has indeed been shown to have inductive ability on bone re-growth (Zhao and Neufeld, 1995). It has been reported that even young children can regenerate their last phalange, including the nail (Illingworth and Barker, 1974), but it is only urodele (tailed) amphibians, such as newts and the axolotl, which can regenerate functionally and morphologically perfect limbs in adulthood. The regenerating urodele limb therefore represents a valuable model for tackling the complex issue of what mechanisms underlie limb regeneration and why some animals are able to regenerate in adulthood and others are not. Numerous cellular and molecular approaches have been developed over the last few years which are proving very useful for tackling molecular mechanisms underlying regeneration, such as the availability of specific antibodies (Ferretti *et al.*, 1989; Kintner and Brockes, 1984, 1985), the establishment of long-term culture systems (Ferretti and Brockes, 1988), the isolation of urodele genes (Casimir *et al.*, 1988; Corcoran and Ferretti, 1997; Ferretti *et al.*, 1991; Géraudie and Ferretti, 1998; Khrestchatisky *et al.*, 1988; Onda *et al.*, 1991) and the development of transfection techniques (Brockes, 1994; Burns *et al.*, 1994; Kumar *et al.*, 2000;

Roy *et al.*, 2000). Crucially, urodele amphibian genome resources are now becoming available (Putta *et al.*, 2004)and the development of expressed sequence tags (ESTs) and of the genome map of the axolotl can be monitored at: http://salamander .uky.edu/about_sgp.htm

Limb regeneration proceeds by formation of a *blastema*, a mound of undifferentiated mesenchymal cells (blastemal cells) which accumulate at the stump surface after amputation and start to proliferate after 4–5 days. Innervation and the presence of a specialized *wound epidermis*, which lacks a distinct basement membrane, are both essential for regeneration in the newt, and they appear to control blastemal growth in the regenerating limb (reviewed by Niazi and Saxena, 1978; Stocum, 1985; Thornton, 1968; Wallace, 1981). Once a critical mass of blastemal cells has accumulated, differentiation and morphogenesis begin and, in about 10 weeks, all the structures distal to the plane of amputation are faithfully replaced. The original pattern, however, can be altered by administration of a class of putative morphogens, vitamin A and its derivatives, among which retinoic acid has been the most widely studied (see above). Retinoic acid induces formation of extra limb segments in a dose-dependent manner and has been shown to affect all three axes of the regenerating limb under certain experimental conditions (Brockes, 1990; Bryant and Gardiner, 1992; Maden, 1982; Niazi and Saxena, 1978; Stocum, 1991). Retinoic acid and retinoic acid receptors are indeed present in the regenerating limb and the different retinoic acid receptors expressed in the blastema have been shown to mediate different functions (Pecorino *et al.*, 1994, 1996). Expression of the recently identified newt homologue of *CD59*, *Prod 1*, which is regulated by location along the limb proximodistal axis, is increased by retinoic acid. Prod 1 is a surface molecule involved in modulating cellular interactions that underlie positional identity of blastemal cells (da Silva *et al.*, 2002).

Do developing and regenerating limbs use the same patterning mechanisms?

There is evidence from both classical tissue manipulation and, more recently, from analysis of gene expression, to suggest that developing and regenerating limbs largely use the same patterning mechanisms.

The specialized wound epidermis of the regenerating limb is believed to be homologous to the apical ectodermal ridge in developing limbs (reviewed by Stocum, 1985). Removal of the wound epidermis, like removal of the ectodermal ridge during development, has an inhibitory effect on further development of the regenerate. When formation of wound epidermis is impeded, for example by covering the wound with a skin flap, regeneration does not occur. Regeneration is also impaired in an axolotl mutant, *short-toes*, where, although blastemal cells accumulate following amputation, a thick and convoluted basement membrane forms. This is likely to adversely affect interactions between the wound epidermis and the underlying mesenchyme (Del Rio Tsonis *et al.*, 1992). In the frog, which loses regenerative ability following metamorphosis, a wound epidermis forms after limb amputation in the tadpole but not in the adult (see review by Thornton, 1968). These observations demonstrate the importance of epithelial–mesenchymal interactions during regeneration, and equivalent interactions in mammals are probably impaired by the presence of the basement membrane which rapidly forms after wounding mammalian skin. Interestingly, partial blastema formation has been induced in amputated toes of adult mice, in which the presence of a wound epithelium was maintained by repeated surgical skin removal and treatment with sodium chloride (Neufeld, 1980). In addition, some cases of regeneration of fingertips in young children were observed when the wound was not sutured after injury (Illingworth and Barker, 1974).

During limb development, the polarizing region plays a fundamental role in patterning. However, the regeneration blastema is in direct contact with the mature tissues of the stump, which have 'fixed' positional values. Therefore, it had been suggested that re-establishment of a polarizing region in order to trigger the chain of events which will lead to correct patterning of the regenerate may not be necessary, and that patterning could instead be determined by mesenchymal–mesenchymal interactions between the blastemal cells and the distal cells of the stump. Signalling molecules normally expressed by the polarizing zone, such as Shh, however, have been recently found to be upregulated in the posterior region of the regenerating limb (Imokawa and Yoshizato 1997; Torok *et al.*, 1999). On the other hand, following limb amputation, *Shh* is not expressed before the blastema has reached the medium bud stage, hence at a later stage than in the developing limb bud and possibly in a more restricted area. *Shh* expression in the regenerating limb, and the fact that its ectopic expression anteriorly induces formation of extra digits (Roy *et al.*, 2000), has been taken to indicate the existence of a zone of polarizing activity also in regenerating limbs.

Expression of transcription factors associated with limb development has been reported also in regenerating limbs, although certain genes appear to be deployed in a somewhat different fashion. In the case of *Tbx* gene expression, regenerating forelimbs and hindlimbs selectively express *Tbx5* and *Tbx4*, respectively, as the developing mammalian limb. In contrast, during development of the urodele limb *Tbx5* and *Tbx4* are expressed in both anterior and posterior limb buds (Khan *et al.*, 2002).

All of the developmentally regulated homeobox genes that are known to be expressed during limb development are also expressed in regenerating limbs. *HoxD* genes are expressed in the same spatio-temporal fashion as in development (Brown and Brockes, 1991; Simon and Tabin, 1993; Torok *et al.*, 1998), whereas the *HoxA* genes are not re-expressed in a co-linear fashion (Gardiner *et al.*, 1995). Interestingly, expression of *Hoxa6* and *Hoxc10* is not switched off in the adult newt limb, unlike in vertebrates, which cannot regenerate their limbs, suggesting a possible relationship between expression of these genes and maintenance of regenerative ability in adulthood (Savard *et al.*, 1988; Savard and Tremblay, 1995; Simon and Tabin, 1993). Furthermore, some of these genes, such as *Hoxa13* and *dlx3*, another transcription factor involved in determining positional identity, are downregulated by retinoic acid in the regenerating limb (Gardiner *et al.*, 1995; Mullen *et al.*, 1996), supporting their putative role in patterning it.

The presence of several Fgfs and at least three Fgf receptor variants in the limb regeneration blastema (Boilly *et al.*, 1991; Christen and Slack, 1997; Christensen *et al.*, 2002; Giampaoli *et al.*, 2003; Poulin and Chiu, 1995; Poulin *et al.*, 1993; Yokoyama *et al.*, 2001) further supports the view that the same key molecules are used to build both embryonic and adult limbs. Even in this case, however, some differences in deployment of various members of the Fgf family have been reported (e.g. *Fgf4* does not appear to be expressed in the regenerating urodele limb bud). The fact that some regeneration of the embryonic chick limb can be induced by Fgf2 and Fgf4 (Kostakopoulou *et al.*, 1996; Taylor *et al.*, 1994), which can substitute for the apical ectodermal ridge in developing chick limbs (Fallon *et al.*, 1994; Niswander *et al.*, 1993), indicates that the regenerative potential in vertebrates is higher than previously thought, and that, at least in the embryo, it can be stimulated when the right factor(s) is provided. Therefore, it will be of fundamental importance to achieve a full understanding of the basic mechanisms underlying limb development and regeneration if we are to devise strategies aimed at increasing regenerative potential in higher vertebrates, including humans.

Differences between developing and regenerating limbs

There are clear differences between development and regeneration concerning the origin of the cells and the control of their division, in particular regarding the role of nerves. While the developing limb bud starts to grow in the absence of innervation, the initial growth of the blastema requires an adequate level of nervous supply (Fekete and Brockes, 1988; Sicard, 1985; Singer, 1974; Wang *et al.*, 2000). If the limb is denervated and amputated, blastemal cells can accumulate but do not proliferate. However, if the limb is denervated after a blastema has formed, regeneration will progress but the regenerated limb will be smaller in size. Therefore, regeneration depends on the presence of the nerve only during the phase of rapid proliferation of blastemal cells. It has been suggested that the newt type III neuregulin and Fgf2 might be among the factors secreted by the nerve, which either directly or indirectly control blastemal cell proliferation (Brockes and Kintner, 1986; Ferretti and Brockes, 1991; Mullen *et al.*, 1996; Wang *et al.*, 2000), but more work will have to be carried out to fully define the molecular basis of nerve dependency. A number of factors that are not secreted by the nerve, but whose expression depends on the presence of innervation, such as Fgf8 and Fgf10, have been reported. Interestingly, some degree of limb regeneration has been induced in young opossums by transplantation of nervous tissue (Mizell and Isaacs, 1970), indicating that the nervous system can also play an important role in limb regeneration in higher vertebrates. It will therefore be extremely important to fully elucidate the mechanisms underlying the neural control of limb regeneration.

The other fundamental difference between development and regeneration is the origin of limb progenitor cells in the embryo and in the adult. As discussed earlier, the limb bud originates from the lateral plate mesoderm, whereas the limb blastema originates from the distal tip of the stump. The issue of whether these limb progenitor cells are 'equivalent' and share the same phenotype was neatly addressed when the monoclonal antibody technique became available and antibodies against blastemal antigen and markers of the differentiated state were developed. Such antibodies (reviewed by Ferretti and Brockes, 1991; Géraudie and Ferretti, 1998) have allowed the identification of a number of regeneration-associated molecules, analysis of their cellular distribution in developing and regenerating limbs, and isolation of the genes encoding them by screening blastema expression libraries. Two main findings have emerged from these studies. First, a difference in the phenotype of limb progenitor cells in embryos and adults has been revealed, since molecular markers such as 22/18 and the simple epithelial keratins 8 and 18, all of which have been shown to be expressed in the mesenchyme of regenerating limbs, are not detectable during limb development (Corcoran and Ferretti, 1997; Fekete and Brockes, 1987). Second, it has become apparent that blastemal cells do not comprise a homogeneous population, as previously believed on the basis of their morphological appearance, but are heterogeneous.

Blastemal cells are believed to originate from the mature tissues of the stump through a process of dedifferentiation. It is generally agreed that, while neither epidermis nor subepidermal glands contribute cells to the blastema, there is a contribution from mesodermal tissues of the stump and from Schwann cells (Maden, 1977). The most extensively studied tissue that contributes to blastema formation is the muscle. Some elegant labelling experiments have demonstrated the high plasticity of the urodele muscle (Kumar *et al.*, 2000; Lo *et al.*, 1993). When cultured myotubes are injected with a tracing dye and implanted in the blastema *in vivo*, labelled mononucleate cells can be found in the blastema, confirming previous work which suggested that muscle fibres contribute to blastema formation through a process of dedifferentiation (Hay, 1959). The molecular mechanisms underlying dedifferentiation of myoblasts are becoming better understood. *Msx1* has been shown to play an important role in myofibre dedifferentiation, and a decrease in its expression in myonuclei results in inhibition of myofibre fragmentation and formation of mononucleate cells (Kumar *et al.*, 2004). Another key step for the re-entry of dedifferentiated cells in the cell cycle is activation of thrombin at the injury site (Tanaka *et al.*, 1999). This is believed to induce cleavage of an as yet unidentified serum component and lead to phosphorylation of the retinoblastoma protein, with consequent cell cycle re-entry (Tanaka *et al.*, 1997).

Cells equivalent to mammalian satellite cells are also present in the newt muscle (post-satellite cells) and these cells proliferate and differentiate into myotubes *in vitro* (Cameron *et al.*, 1986). Although these cells might contribute to the blastema, there is evidence to suggest that their main role is in the repair of stump muscle, whereas the blastema contains cells produced through muscle dedifferentiation (Corcoran and Ferretti, 1999).

The possibility of inducing at least partial regeneration of limb structures in humans is still in its infancy, and the most formidable challenge ahead of us is to fully elucidate the molecular mechanisms underlying the remarkable plasticity of urodele limb tissues and controlling re-entry in the cell cycle in response to limb amputation.

How, when, and where experimental studies elucidate abnormal development

The results of embryological manipulations suggest that the cellular and molecular basis of limb development is conserved between vertebrates. For example, the polarizing region from embryonic limb buds of a wide range of vertebrates, including man, was shown to lead to additional digit formation in chick wing buds (Fallon and Crosby, 1977). This suggests that the signalling mechanism has been evolutionarily conserved (see Chapter 1) and this conclusion has since been confirmed at the molecular level. For instance, *Shh* transcripts have been detected in mouse and chick limb bubs and fish fin buds (Echelard *et al.*, 1993; Krauss *et al.*, 1993; Riddle *et al.*, 1993) and the human *SHH* gene has been identified (Marigo *et al.*, 1995). These and other data suggest that principles of limb patterning that emerge from experimental analysis of chick and mouse embryos can probably be applied directly to consideration of abnormal development and congenital limb abnormalities in humans. In addition, there have been considerable advances in clinical genetics and progress in pinpointing the genetic basis of limb defects in humans (Table 7.3), and the genes have often turned out to be those identified by basic research into limb development in model vertebrates.

Limb deficiencies

Amelia and meromelia Experimental analysis shows that the apical ectodermal ridge is central to bud outgrowth, and this has important implications for interpreting how amelia and terminal meromelic limbs (limbs that lack distal structures) could arise. If the apical ridge does not form at all, limbs will be completely absent and amelia will result. Absence of limbs could result from lack of appropriate initiation signals or failure to respond to these signals. Failure of limb bud outgrowth leading to truncated limbs could result from defective signalling; for example, absence of the apical ectodermal ridge and/or production of Wnts or Fgfs. Changes in apical ridge signalling could either have a genetic basis or be due to damage to the ridge. Another possibility is that there could be a failure to respond to ridge signals and, as a consequence, correct patterns of *Hox* gene expression, for example, are not established. In this respect it is interesting that the gene that is affected in the hypodactyly mutant mouse, which has only a single digit on each paw, is *Hoxa13* (Mortlock *et al.*, 1996).

Table 7.3 Gene defects associated with limb malformations

Gene	Human abnormality	Effects on limbs	Reference
ARL6 (Ras superfamily of small GTP-binding proteins)	Bardet–Biedl syndrome	Polydactyly	Chiang *et al.*, 2004 Fan *et al.*, 2004
ATPSK2 (ATP sulphurylase/ APS kinase)	Spondyloepimetaphyseal dysplasia	Bowed long bones, hemimelia, brachydactyly, enlarged knee joints, joint degeneration (early onset)	ul Haque *et al.*, 1998
BMPR1B	Brachydactyly type A2	Brachydactyly	Lehmann *et al.*, 2003
C7orf2	Acheiropodia	Hemimelia	Ianakiev *et al.*, 2001
(Lmbr1) Intron	Preaxial polydactyly	Polydactyly	Lettice *et al.*, 2002
CBP (CBEB binding protein gene)	Rubinstein–Taybi syndrome	Broad duplicated distal phalanges of thumbs and halluces	Petrij *et al.*, 1995
CDMP1 (GDF5)	Chondrodysplasia Grebe type	Brachydactyly, polydactyly, hemimelia, hypoplasia, aplasia	Thomas *et al.*, 1997
Col2A1	Spondyloepiphyseal dysplasia		Williams *et al.*, 1993
DHCR7 (human sterol $\Delta 7$ reductase)	Smith Lemli–Opitz syndrome	Syndactyly	Wassif *et al.*, 1998
EVC	Ellis van Creveld	Postaxial, polydactyly	Ruiz-Perez *et al.*, 2000
	Marfan syndrome	Arachnodactyly	Lee *et al.*, 1991
Fibrillin			
FGFR1	Pfeiffer syndrome	Syndactyly (soft tissue), broad digit 1, brachydactyly	Muenke *et al.*, 1994
FGFR2	Pfeiffer syndrome	As above	Meyers *et al.*, 1996
	Apert syndrome	Syndactyly (synostotic)	Muenke *et al.*, 1994
	Jackson–Weiss syndrome	Syndactyly (synostotic)	Wilkie *et al.*, 1995
FGFR3	Achondroplasia	Brachydacydactyly, hemimelia	Bellus *et al.*, 1995 Rousseau *et al.*, 1994
	Hypochondroplasia	Milder form of the above	Shiang *et al.*, 1994
GPC3 (glypican-3)	Simpson–Golabi–Behmel syndrome	Overgrowth, postaxial, polydactyly	Pilia *et al.*, 1996

GLI3	Greig cephalopolysyndactyly	Polydactyly, syndactyly	Vortkamp *et al.*, 1991
	Pallister–Hall syndrome		Wild *et al.*, 1997
	Postaxial polydactyly type A	Polydactyly	Kang *et al.*, 1997
			Radhakrishna *et al.*, 1997
HOXA13	Hand–foot–genital syndrome	Hemimelia/hypoplasia, syndactyly (synostotic), carpal fusion, delayed ossification	Mortlock and Innis, 1997
HOXD13	Type II syndactyly	Syndactyly (synostotic), polydactyly, meromelia, hemimelia	Goodman *et al.*, 1997
			Muragaki *et al.*, 1996
			Goodman *et al.*, 1998
HOXD13	Combination of brachydactyly and central polydactyly		Caronia *et al.*, 2003
HOXD13	Brachydactyly types D and E		Johnson *et al.*, 2003
IHH	Brachydactyly	Type A1	McCready *et al.*, 2002
			Gao *et al.*, 2001
LMX1B	Nail–patella syndrome	Meromelia, nail hypoplasia or dysplasia	Dreyer *et al.*, 1998
			Vollrath *et al.*, 1998
MID1	X-linked Opitz syndrome	Syndactyly	Quaderi *et al.*, 1997
MKKS (putative chaperonin)	McKusick–Kaufman	Postaxial, polydactyly	Stone *et al.*, 2000
MSX2	Autosomal dominant craniosynostosis	Brachydactyly, finger-like thumb	Jabs *et al.*, 1993
NIPBL	Cornelia de Lange syndrome	Limb reduction defects	Tonkin *et al.*, 2004
Noggin	Multiple synostoses syndrome	Symphalangism	Gong *et al.*, 1999
OFD1 (CXORF5)	Orofaciodigital syndrome 1	Syndactyly, brachydactyly	Ferrante *et al.*, 2001
p63	Split hand–split foot		van Bokhoven *et al.*, 2001
ROR2	Robinow syndrome	Brachydactyly	Afzal *et al.*, 2000
			van Bokhoven *et al.*, 2000
SALL1	Townes–Brockes syndrome	Polydactyly, finger-like thumb	Kohlhase *et al.*, 1998
SALL4	Okhiro syndrome	Preaxial meromelia	Kohlhase *et al.*, 2003

(*continued*)

Table 7.3 (*continued*)

Gene	Human abnormality	Effects on limbs	Reference
SHOX	Leri–Weill dyschondrosteosis	Meromelia, brachydactyly	Belin *et al.*, 1998
	Langer mesomelic dysplasia		Shears *et al.*, 1998
SOX9	Campomelic dysplasia	Bowed long bones	Foster *et al.*, 1994
TBX3	Ulnar–mammary syndrome	Meromelia, nail duplicated ventrally, hypoplasia, carpal fusion	Bamshad *et al.*, 1997
TBX5	Holt–Oram syndrome	Ecrodactyly, finger-like thumb, meromelia	Basson *et al.*, 1997
			Li *et al.*, 1997
TWIST	Saethre–Chotzen syndrome	Brachydactyly, syndactyly (soft tissue)	el Ghouzzi *et al.*, 1997
			Howard *et al.*, 1997
WNT3	Tetra-amelia	All four limbs absent	Niemann *et al.*, 2004

*Although only limb defects are described here, other malformations may be associated with these mutations.

In some cases of limb meromelia (intercalary or transverse), it is proximal rather than distal structures that are absent. A model for this class of defect is provided by X-irradiation of chick limb buds (Wolpert *et al.*, 1979). As the dose of X-irradiation is increased, proximal limb structures are deleted, whereas distal structures develop relatively normally. This result can be understood by reference to the progress zone model. X-irradiation kills cells in the progress zone and the number of cells at the tip of the limb will be reduced. Surviving cells will proliferate to fill the progress zone and as they do so, they will spend a longer time at the tip and hence give rise to distal rather than proximal structures. Therefore, death or killing of mesenchyme cells could be a mechanism that leads to proximal defects. Destruction of mesenchyme cells could be caused directly by cytotoxic drugs or indirectly by interference with the vascular supply. This second mechanism was suggested some time ago as the mechanism of action of thalidomide (Poswillo, 1975) and, more recently, thalidomide has been shown to inhibit angiogenesis, the growth and remodelling of blood vessels (D'Amato *et al.*, 1994). Roberts syndrome, a genetically inherited (autosomal recessive) limb defect, is phocomelia-like.

Polydactylies Experimental analysis shows that the polarizing region is central to anterior–posterior patterning. Therefore, defects in anterior–posterior patterning could either be due to changes in distribution and/or strength of the polarizing signal or to changes in cellular response to the signal. It could be that both signalling and response are abnormal. In polydactylous limbs, one possibility is that the polarizing signal might be more widespread and produced anteriorly and posteriorly. This could account for extra digits pre-axially or mirror hands/feet but does not explain additional post-axial digits.

In *talpid* (polydactylous) chicken mutants, an increased number of morphologically similar digits develop. The limb buds are abnormally broad, and the apical ectodermal ridge is correspondingly extended (Hinchliffe and Ede, 1967). *Shh* expression is restricted as normal to the posterior margin of the broadened buds (Francis-West *et al.*, 1995) but there are uniform expression patterns of *HoxD* genes across the tip of early buds, instead of the normal posterior restriction of expression, both in *talpid*^{3} (Izpisúa-Belmonte *et al.*, 1992b) and in a morphologically similar polydactylous chicken mutant, *talpid*^{2} (Coelho *et al.*, 1992). In *talpid*^{3}, both *Bmps* and *Fgf4* are also uniformly expressed (Francis-West *et al.*, 1995), while expression of *Gli* and high level *Ptc* that normally occurs in response to Shh signalling is absent (Lewis *et al.*, 1999). In contrast, in *talpid*^{2}, *Ptc* and *Gli* are ectopically expressed and Gli3 exists predominantly in the activator form throughout the limb bud (Caruccio *et al.*, 1999).

The mouse mutant *extra toes*, *Xt*, is characterized by preaxial digit duplications in the hindlimbs, and the gene affected has been identified as *Gli3* (Hui and Joyner, 1993). *Gli3* is one of the family of genes that encode transcriptional effectors of Shh signalling (see earlier). In this mouse mutant, ectopic expression of both *Shh* and *Fgf4* has been detected in the anterior region of the limb bud (Masuya *et al.*, 1995). As already mentioned, it has been shown that the limb phenotypes of *Shh*$^{-/-}$*Gli3*$^{-/-}$ mouse embryos are the same as those of *Gli3*$^{-/-}$ embryos. This means that the extra digits in the mouse mutant are not due to the ectopic domain of *Shh* expression but instead are due to the absence of Gli3 repressor anteriorly (Litingtung *et al.*, 2002; te Welscher *et al.*, 2002a) and, in this respect, there are similarities with the *talpid*^{2} chicken mutant discussed above. The homologous human syndrome, Greig cephalopolysyndactyly, is also due to mutations in *Gli3* (Hui and Joyner, 1993).

There are several other mouse mutants with polydactyly that have been shown to have ectopic *Shh* or, in one case, *Ihh* expression at the anterior of the limb buds. One such mutant, *Sasquatch*, was caused by an insertion of a transgene into an intron of the *Lmbr1* gene. Even though this site is 1Mb away from the *Shh* gene, it seems to contain an element that acts as an *cis*-acting enhancer driving *Shh* expression specifically in the limb. The equivalent region of the genome is also implicated in pre-axial polydactyly in human patients (Lettice *et al.*, 2002). Furthermore, other human limb abnormalities map near the same region. Patients with one of these conditions, known as acheiropodia, have meromelic limbs, lacking distal structures. This limb phenotype is similar to that of the limbs of *Shh*$^{-/-}$ mouse embryos, which have defects in many other systems, including holoprosencephaly, but in the acheiropodia patients only the limbs are affected. This can be explained by the idea that there is another gene regulator in this region, in this case driving normal *Shh* expression in the limb, and that this regulator is defective in acheiropodia patients (Hill *et al.*, 2003). Finally, it is also worth noting that synpolydactyly can be caused by mutations in *Hoxd13* (Muragaki *et al.*, 1996).

Synostosis Synostosis refers to the fusion between successive or adjacent skeletal elements. This could result from defects of patterning or be due to abnormalities in later events, such as growth and shaping. Fusion between successive elements and lack of elbow/knee joints have been reported in chick and mouse embryos following retinoic acid treatment at a time when chondrogenesis has begun (e.g. Kochhar, 1977). In the mouse mutant limb deformity (*ld*, see earlier), there is fusion between adjacent skeletal elements, for example between radius and ulna, and the early buds in the mutant are narrower than normal buds (Zeller *et al.*, 1989). Thus, it appears that Fgfs, as well as Bmps, are factors operating not only at the earliest stages in limb development (see earlier) but also in controlling skeletal form and growth at later stages. Mutations in Fgfrs (fibroblast growth factor receptors) are now known to be the cause of a number of human syndromes involving limb abnormalities (Wilkie *et al.*, 1995; Rutland *et al.*, 1995). For example, specific missense mutations in *Fgfr2* are found in patients with Apert syndrome, in which both soft tissue and digital fusions are seen, together with craniosynostosis (reviewed Wilkie, 2003; Figure 7.2a, b).

Syndactyly Syndactylies involving soft tissue fusions between adjacent digits are normally thought to be due to a failure of programmed cell death. In chick embryos, treatment with Janus green impairs mitochondrial function and leads to the absence of interdigital cell death. The result is soft tissue webbing between the digits. Interdigital cell death can also be inhibited by locally removing the apical ridge between presumptive digits. This leads not only to the persistence of interdigital mesenchyme but also cartilage differentiation occurs to give ectopic digit-like structures (Hurle and Ganan, 1986). There is increasing information about the control of cell death in the interdigital regions; not only do Bmps seem to be involved but also retinoids. There is currently considerable interest in the mechanisms and genetic basis of programmed cell death or apoptosis. From work in a nematode worm, genes that control programmed death have been identified and vertebrate homologues have also been found (Yuan *et al.*, 1993). When interdigital cell death in chick limb buds is blocked by overexpression of *bc12*, a vertebrate gene related to the *Ced9* nematode gene that is known to negatively regulate cell death in development of these worms, soft tissue webbing can result (Sanz-Ezquerro and Tickle, 2000). Interdigital cell death can also be suppressed in developing mouse paws by treatement with zVAD-fmk, a cell-permeable inhibitor of the Ced-3/ICE family of proteases (caspases), which are conserved between nematode worms and vertebrates and are involved in executing programmed cell death (Jacobsen *et al.*, 1996).

Skeletal dysplasias There are a large number of abnormalities in which either the size or shape of parts of the limb are abnormal. It is now emerging that closely related members of the TGF-β family, including bone morphogenetic proteins and growth/differentiation factors, play central roles in controlling skeletal form (reviewed Erlebacher *et al.*, 1995). Mutations in these genes are now known to underlie some skeletal dysplasias. The *brachypodism* mouse mutation affects a gene encoding growth/differentiation factor 5 (GDF-5). Transcripts of this gene are associated with developing skeletal elements in the limbs and the mutant phenotype is characterized by the limb skeleton being very reduced in length and the toes lacking the most distal elements (Storm *et al.*, 1994). Mutations in this gene, also known as the *CDMP1* (cartilage-derived morphogenetic protein-1) gene, have been identified in human patients with brachydactyly type C, in which there is a shortening of proximal phalanges in specific digits, together with 'hypersegmentation' (Thomas *et al.*, 1997). Genes associated with other types of brachydactyly (see Table 7.3) encode the bone morphogenetic protein receptor 1B (type A2), Indian hedgehog (type A1), ROR2, an orphan receptor tyrosine kinase (type B; reviewed by Gao and He, 2004) and Hoxd13 (types D and E; Johnson *et al.*, 2003). Fgfs have also been implicated in skeletal growth. *Achondroplasia*, in which skeletal elements are very reduced in length, is now known to be due to a mutation in a gene encoding a receptor for fibroblast growth factors (Shiang *et al.*, 1994; Rousseau *et al.*, 1994).

Agenda for the future

Three main areas look set for a rapid increase in understanding. One area is the further elucidation of signalling pathways that set up tissue patterns in the early limb bud. Most, if not, all of the steps that have so far been identified are probably indirect and it will be some time before the full details of the signalling pathways will be elucidated. The importance of knowing the details is that mutations might occur in any component of the pathway and affect coding or regulatory elements. A second area in which new insights are to be expected is in the control of limb development in the context of the vertebrate body plan, and such knowledge will have important evolutionary implications. Finally, there will be the continuing elucidation of the genetic basis of human limb abnormalities. Here human clinical and molecular genetics will combine with experimental embryology in a potentially powerful and productive way.

The progress in analysing the early pathways in establishing tissue pattern has been rapid. However, there is a large conceptual gap between, say, the pattern of expression of a homeobox gene *per se* and the development of a recognizable skeletal element with its characteristic shape and growth, etc. The morphogenesis of an individual skeletal element is a complex problem that is based on spatial control of cell behaviour. It remains a considerable challenge to understand how gene expression is translated into form.

References

Abud, H.E., Skinner, J.A., McDonald, F.J., Bedford, M.T. *et al.* (1996) Ectopic expression of FGF4 in chimaeric mouse embryos induces the expression of early markers of limb development in the lateral ridge. *Dev. Genet.* **19**: 51–65.

Afzal, A.R., Rajab, A., Fenske, C.D., Oldridge, M. *et al.* (2000) Recessive Robinow syndrome, allelic to dominant brachydactyly type B, is caused by mutation of ROR2. *Nat. Genet.* **25**: 419–422.

Ahn, K., Mishina, Y., Hanks, M.C., Behringer, R.R. and Crenshaw, E.B. (2001) BMPR-IA signalling is required for the formation of the apical ectodermal ridge and dorsal-ventral patterning of the limb. *Development* **128**: 4449–4461.

Ahn, S. and Joyner, A.L. (2004) Dynamic changes in the response of cells to positive hedgehog signalling during mouse limb patterning. *Cell* **118**: 505–516.

Akita, K. (1996). The effect of the ectoderm on the dorsoventral pattern of epidermis, muscles and joints in the developing chick leg – a new model. *Anat. Embryol.* **193**: 377–386.

Allen, F., Tickle, C. and Warner, A. (1990) The role of gap junctions in patterning of the chick limb bud. *Development* **108**: 623–634.

Bamshad, M., Lin, R.C., Law, D.J., Watkins, W.C. *et al.* (1997) Mutations in human TBX3 alter limb, apocrine and genital development in ulnar–mammary syndrome. *Nat. Genet.* **16**: 311–315.

Barrow, J.R., Thomas, K.R., Boussadia-Zahui, O., Moore, R. *et al.* (2003) Ectodermal Wnt3/beta-catenin signalling is required for the establishment and maintenance of the apical ectodermal ridge. *Genes Dev.* **17**: 394–409.

Basler, K. and Struhl, G. (1994) Compartment boundaries and the control of *Drosophila* limb pattern by hedgehog protein. *Nature* **368**: 208–214.

Basson, C.T., Bachinsky, D.R., Lin, R.C., Levi, T. *et al.* (1997) Mutations in human TBX5 cause limb and cardiac malformation in Holt–Oram syndrome. *Nat. Genet.* **15**: 30–35.

Belin, V., Cusin, V., Viot, G., Girlich, D. *et al.* (1998) SHOX mutations in dyschondrosteosis (Leri–Weill syndrome). *Nat. Genet.* **19**: 67–69.

Bellus, G.A., Hefferon, T.W., Ortiz de Luna, R.I., Hecht, J.T. *et al.* (1995) Achondroplasia is defined by recurrent G380R mutations of FGFR3. *Am. J. Hum. Genet.* **56**: 368–373.

Boilly, B., Cavanaugh, K.P, Thomas, D., Hondermarck, H. *et al.* (1991) Acidic fibroblast growth factor is present in regenerating limb blastemas of axolotls and binds specifically to blastema tissues. *Dev. Biol.* **145**: 302–310.

van Bokhoven, H., Celli, J., Kayserili, H., van Beusekom, E. *et al.* (2000) Mutation of the gene encoding the ROR2 tyrosine kinase causes autosomal recessive Robinow syndrome. *Nat. Genet.* **25**: 423–426.

van Bokhoven, H., Hamel, B.C., Bamshad, M., Sangiorgi, E. *et al.* (2001) p63 Gene mutations in eec syndrome, limb–mammary syndrome, and isolated split hand–split foot malformation suggest a genotype–phenotype correlation. *Am. J. Hum. Genet.* **69**: 481–492.

Brockes, J.P. (1990) Retinoic acid and limb regeneration. *J. Cell. Sci.* **13**(suppl): 191–198.

Brockes, J.P. (1994) New approaches to limb regeneration. *Trends Genet.* **10**: 169–173.

Brockes, J.P. and Kintner, C.R. (1986) Glial growth factor and nerve dependent proliferation in the regeneration blastema of urodele amphibians. *Cell* **45**: 301–306.

Brown, R. and Brockes, J.P. (1991) Identification and expression of a regeneration-specific homeobox gene in the newt limb blastema. *Development* **111**: 489–496.

Bryant, S.V. and Gardiner, D.M. (1992) Retinoic acid, local cell–cell interactions, and pattern formation in vertebrate limbs. *Dev. Biol.* **152**: 1–25.

Burns, J.C., Matsubara T., Lozinski G., Yee J.K. Friedmann T., Washabaugh C.H. and Tsonis P.A. (1994) Pantropic retroviral vector-mediated gene transfer integration and expression in cultured newt limb cells. *Dev. Biol.* **165**: 285–289.

Cameron, J.A., Hilgers, A.R. and Hinterberger, T.J. (1986) Evidence that reserve cells are a source of regenerated adult newt muscle *in vitro*. *Nature* **321**: 607–610.

Capdevila, J. and Izpisúa-Belmonte, J.C. (2001) Patterning mechanisms controlling vertebrate limb development. *Annu. Rev. Cell Dev. Biol.* **17**: 87–132.

Capdevila, J., Tsukui, T., Rodriguez Esteban, C., Zappavigna, V. and Izpisúa-Belmonte, J.C. (1999) Control of vertebrate limb outgrowth by the proximal factor Meis2 and distal antagonism of Bmps by Gremlin. *Mol. Cell* **4**: 839–849.

Caronia, G., Goodman, F.R., McKeown, C.M., Scambler, P.J. and Zappavigna, V. (2003) An I47L substitution in the HOXD13 homeodomain causes a novel human limb malformation by producing a selective loss of function. *Development* **130**: 1701–1712.

Carrucio, N.C., Martinez-Lopez, A., Harris, M., Dvorak, L. *et al.* (1999) Constitutive actiovation of sonic hedgehog signalling in the chicken mutant talpid(2): Shh-independent outgrwoth and polarizing activity. *Dev. Biol.* **212**: 137–149.

Casimir, C.M, Gates, P.B, Ross-Macdonald, P.B, Jackson, J.F. *et al.* (1988) Structure and expression of a newt cardio-skeletal myosin gene. Implications for the C value paradox. *J. Mol. Biol.* **202**: 287–296.

Charité, J., de Graaff, W. and Deschamps, J. (1994) Ectopic expression of *Hoxb-8* causes duplication of the ZPA in the forelimb and homeotic transformation of axial structures. *Cell* **78**: 589–601.

Chen, Z.F. and Behringer, R.R. (1995) Twist is required in head mesenchyme for cranial neural tube morphogenesis. *Genes Dev.* **15**: 686–699.

Chiang, A.P., Nishimura, D., Searby, C., Elbedour, K. *et al.* (2004) Comparative genomic analysis identifies an ADP-ribosylation factor-like gene as the cause of Bardet–Biedl syndrome (BBS3). *Am. J. Hum. Genet.* **75**: 475–484.

Christen, B. and Slack, J.M.W. (1997) FGF8 is associated with anteroposterior patterning and limb regeneration in *Xenopus*. *Dev. Biol.* **192**: 455–466.

Christensen, R.N., Weinstein, M. and Tassava, R.A. (2002) Expression of fibroblast growth factors 4, 8, and 10 in limbs, flanks, and blastemas of *Ambystoma*. *Dev. Dyn.* **223**: 193–203.

Coelho, C.N., Upholt, W.B. and Kosher, R.A. (1992) Role of the chicken homeobox-containing genes GHox4.6 and GHox8 in the specification of positional identities during the development of normal and polydactylous chick limb buds. *Development* **115**: 629–637.

Cohn, M.J., Izpisúa-Belmonte, J.C., Abud, H., Heath, J. and Tickle, C. (1995) Fibroblast growth factors induce additional limb development from the flank of chick embryos. Cell **80**: 739–746.

Cohen, M.M. Jr (2003) The hedgehog signalling network. *Am. J. Med. Genet.* **15**: 5–28.

Corcoran, J.P. and Ferretti, P. (1997) Keratin 8 and 18 expression in mesenchymal progenitor cells of regenerating limbs is associated with cell proliferation and differentiation. *Dev. Dyn.* **210**: 355–370.

Corcoran, J.P. and Ferretti, P. (1999) RA regulation of keratin expression and myogenesis suggests different ways of regenerating muscle in adult amphibian limbs. *J. Cell Sci.* **112**: 1385–1394.

Crossley, P.H. and Martin, G.M. (1995) The mouse *Fgf8* gene encodes a family of polypeptides and is expressed in regions that direct outgrowth and patterning in the developing embryo. *Development* **121**: 439–451.

Crossley, P.H., Minowada, G., Macarthur, C.A. and Martin, G.R. (1996) Roles for FGF8 in the induction, initiation, and maintenance of chick limb development. *Cell* **84**: 127–136.

Dahn, R.D. and Fallon, J.F. (2000) Interdigital regulation of digit identity and homeotic transformation by modulated BMP signalling. *Science* **289**: 438–441.

D'Amato, R.J., Loughnan, M.S., Flynn, E. and Folkman, J. (1994) Thalidomide is an inhibitor of angiogenesis. *Proc. Natl Acad. Sci. USA* **91**: 4082–4085.

da Silva, S.M., Gates, P.B. and Brockes, J.P. (2002) The newt orthologue of CD59 is implicated in proximodistal identity during amphibian limb regeneration. *Dev. Cell* **3**: 547–555.

Davidson, D.R., Crawley, A., Hill, R.E. and Tickle, C. (1991) Position-dependent expression of two related homeobox genes in developing vertebrate limbs. *Nature* **352**: 429–431.

Dealy, C.N., Roth, A., Ferrari, D., Brown, A.M.C. and Kosher, R.A. (1993) Wnt5a and Wnt7a are expressed in the developing chick limb bud in a manner suggesting roles in pattern formation along the proximodistal and dorsoventral axes. *Mech. Dev.* **43**: 175–186.

Del Rio Tsonis, K., Washabaugh, C.H. and Tsonis, P.A. (1992) The mutant axolotl Short toes exhibits impaired limb regeneration and abnormal basement membrane formation. *Proc. Natl Acad. Sci. USA* **89**: 5502–5506.

Dollé, P., Izpisúa-Belmonte, J.C., Falkenstein, H., Renucci, A. and Duboule, D. (1989) Coordinate expression of the murine HOX5 complex homeobox-containing genes during limb pattern formation. *Nature* **342**: 767–772.

Dreyer, S.D., Zhou, G., Baldini, A., Winterpacht, A. *et al.* (1998) Mutations in LMX1B cause abnormal skeletal patterning and renal dysplasia in nail patella syndrome. *Nat. Genet.* **19**: 47–50.

Drossopoulou, G., Lewis, K.E., Sanz-Ezquerro, J.J., Nikbakht, N. *et al.* (2000) A model for anteroposterior patterning of the vertebrate limb based on sequential long- and short- range shh signalling and Bmp signalling. *Development* **127**: 1337–1348.

Dudley, A.T., Lyons, K.M. and Robertson, E.J. (1995) A requirement for bone morphogenetic protein-7 during development of the mammalian kidney and eye. *Gene Dev.* **9**: 2795–2807.

Dudley, A.T., Ros, M.A. and Tabin, C.J. (2002) A re-examination of proximodistal patterning during vertebrate limb development. *Nature* **418**: 539–544.

Echelard, Y., Epstein, D., St-Jacques, B., Shen, L. *et al.* (1993) Sonic Hedgehog, a member of a family of putative signalling molecules, is implicated in the regulation of CNS polarity. *Cell* **75**: 1417–1430.

Erlebacher, A., Filvaroff, E.H., Gitelman, S.E. and Derynck, R. (1995) Toward a molecular understanding of skeletal development. *Cell* **80**: 371–378.

Fallon, J.F. and Crosby, G.M. (1977) Polarizing zone activity in limb buds of amniotes. In *Vertebrate Limb and Somite Morphogenesis*, Ede, D.A., Hinchliffe, J.R. and Balls, M. (eds). Cambridge University Press, Cambridge; 55–69.

Fallon, J.F., López, A., Ros, M.A., Savage, M.P. *et al.* (1994) FGF2: apical ectodermal ridge growth signal for chick limb development. *Science* **264**: 104–107.

Fan, Y., Esmail, M.A., Ansley, S.J., Blacque, O.E. *et al.* (2004) Mutations in a member of the Ras superfamily of small GTP-binding proteins causes Bardet–Biedl syndrome. *Nat. Genet.* **36**: 989–993.

Fekete, D.M. and Brockes, J.P. (1987) A monoclonal antibody detects a difference in the cellular composition of developing and regenerating limbs of newts. *Development* **99**: 589–602.

Fekete, D.M. and Brockes, J.P. (1988) Evidence that the nerve controls molecular identity of progenitor cells for limb regeneration. *Development* **103**: 567–573.

Ferrante, M.I., Giorgio, G., Feather, S.A., Bulfone, A. *et al.* (2001) Identification of the gene for oral–facial–digital type I syndrome. *Am. J. Hum. Genet.* **68**: 569–576.

Ferretti, P. and Brockes, J.P. (1988) Culture of newt cells from different tissues and their expression of a regeneration-associated antigen. *J. Exp. Zool.* **247**: 77–91.

Ferretti, P., Fekete, D.M., Patterson, M. and Lane, E.B. (1989) Transient expression of simple epithelial keratins by mesenchymal cells of regenerating newt limb. *Dev. Biol.* **133**: 415–424.

Ferretti, P. and Brockes, J.P. (1991) Cell origin and identity in limb regeneration and development. *Glia* **4**: 214–224.

Ferretti, P., Brockes, J.P. and Brown, R. (1991) A newt type II keratin restricted to normal and regenerating limbs and tails is responsive to retinoic acid. *Development* **111**: 497–507.

Firth, H.V., Boyd, P.A., Chamberlain, P.F., MacKenzie, I.Z. *et al.* (1994) Analysis of limb reduction defects in babies exposed to chorionic villus sampling. *Lancet* **343**: 1069–1071.

Foster, J.W., Dominguez-Steglich, M.A., Guioli, S., Kowk, G. *et al.* (1994) Campomelic dysplasia and autosomal sex reversal caused by mutations in an SRY-related gene. *Nature* **372**: 525–530.

Francis P.H., Richardson, M.K., Brickell, P.M. and Tickle, C. (1994) Bone morphogenetic proteins and a signalling pathway that controls patterning in the developing chick limb. *Development* **120**: 209–218.

Francis-West, P.H., Robertson, K.E., Ede, D.A., Rodriguez, C. *et al.* (1995) Expression of genes encoding bone morphogenetic proteins and sonic hedgehog in talpid [ta(3)] limb buds: their relationships in the signalling cascade involved in limb patterning. *Dev. Dyn.* **203**: 187–197.

Gao, B., Guo, J., She, C., Shu, A. *et al.* (2001) Mutations in IHH, encoding Indian hedgehog, cause brachydactyly type A-1. *Nat. Genet.* **28**: 386–388.

Gao, B. and He, L. (2004) Answering a century old riddle: brachydactyly type A1. *Cell Res.* **14**: 179–187.

Gardiner, D.M., Blumberg, B., Komine, Y. and Bryant, S.V. (1995) Regulation of HoxA expression in developing and regenerating axolotl limbs. *Development* **121**: 1731–1741.

Géraudie, J. and Ferretti, P. (1998) Gene expression during amphibian limb regeneration. *Int. Rev. Cytol.* **180**: 1–50.

el Ghouzzi, V., Le Merrer, M., Perrin-Schmitt, F., Laheunie, E. *et al.* (1997) Mutations of the TWIST gene in the Saethre–Chotzen syndrome. *Nat. Genet.* **15**: 42–46.

Giampaoli, S., Bucci, S., Ragghianti, M., Mancino, G. and Ferretti, P. (2003) Expression of FGF2 in the limb of two Salamandridae correlates with their regenerative capability. *Proc. R. Soc. Lond. B Biol. Sci.* **270**: 2197–2205.

Gibson-Brown, J.J., Agulnik, S.I., Chapman, D.L., Alexiou, M. *et al.* (1996) Evidence of a role for T-box genes in the evolution of limb morphogenesis and the specification of forelimb/hindlimb identity. *Mech. Dev.* **56**: 93–101.

Gong, Y., Krakow, D., Marcelino, J., Wilkin, D. *et al.* (1999) Heterozygous mutations in the gene encoding noggin affect human joint morphogenesis. *Nat. Genet.* **21**: 302–304.

Goodman, F.R., Mundlos, S., Muragako, Y., Donnai, D. *et al.* (1997) Synpolydactyly phenotypes correlate with size of expansions in HOXD13 polyalanine tract. *Proc. Natl Acad. Sci. USA* **94**: 7458–7463.

Goodman, F., Giovannucci-Uzielli, M.L., Hall, C., Reardon, W. *et al.* (1998) Deletions in HOXD13 segregate with an identical, novel foot malformation in two unrelated families. *Am. J. Hum. Genet.* **63**: 992–1000.

Goodrich, L.V., Johnson, R.L., Milenkovic, L., McMahon, J.A. and Scott, M.P. (1996) Conservation of the *hedgehog/patched* signalling pathway from flies to mice: induction of a mouse *patched* gene by Hedgehog. *Genes Dev.* **10**: 301–312.

Green, C.R., Bowles, L., Crawley, A. and Tickle, C. (1994) Expression of the connexin 43 gap junctional protein in tissues at the tip of the chick limb bud is related to the epithelial–mesenchymal interactions that mediate morphogenesis. *Dev. Biol.* **161**: 12–21.

Gritli-Linde, A., Lewis, P., McMahon, A.P. and Linde, A. (2001) The whereabouts of a morphogen: direct evidence for short- and graded long-range activity of hedgehog signalling peptides. *Dev. Biol.* **236**: 364–386.

Gupta, A., Kay, S.P.J. and Scheker, L.R. (2000) *The Growing Hand.* Harcourt: London.

Hahn, H., Christiansen, J., Wicking, C., Zaphiropoulos, P.G. *et al.* (1996a) A mammalian patched homologue is expressed in target tissues of sonic hedgehog and maps to a region associated with developmental abnormalities. *J. Biol. Chem.* **271**: 12125–12128.

Hahn, H., Wicking, C., Zaphiropoulos, P.G., Gailani, M.R. *et al.* (1996b) Mutations of the human homologue of *Drosophila patched* in the nevoid basal cell carcinoma syndrome. *Cell* **85**: 841–851.

Harfe, B.D., Scherz, P.J., Nissim, S., Tian, H. *et al.* (2004) Evidence for an expansion-based temporal Shh gradient in specifying vertebrate digit identities. *Cell* **118**: 517–528.

Hashimoto, K., Yokouchi, Y., Yamamoto, M. and Kuroiwa, A. (1999) Distinct signalling molecules control Hoxa11 and Hoxa13 expression in the muscle precursor and mesenchyme of the chick limb bud. *Development* **126**: 2771–2783.

ul Haque, M.F., King, L.M., Krakow, D., Cantor, R.M. *et al.* (1998) Mutations in orthologueous genes in human spondyloepimetaphyseal dysplasia and the brachymorphic mouse. *Nat. Genet.* **20**: 157–162.

Hay, E. D. (1959) Electron microscopic observations of muscle dedifferentiation in regenerating *Amblystoma* limbs. *Dev. Biol.* **1**: 555–585.

Hill, R.E., Jones, P.F., Rees, A.R., Sime, C.M. *et al.* (1989) A new family of mouse homeobox containing genes: molecular structure, chromosomal location and developmental expression of Hox7.1. *Genes Dev.* **3**: 26–37.

Hill, R.E., Heaney, S.J.H. and Lettice, L.A. (2003) Sonic hedgehog: restricted expression and limb dysmorphologies. *J. Anat.* **202**: 13–20.

Hinchliffe, J.R. and Ede, D.A. (1967) Limb development in the poldactylous talpid-3 mutant of the fowl. *J. Embryol. Exp. Morph.* **17**: 385–404.

Hofmann, C. and Eichele, G. (1994) Retinoids in development. In *The Retinoids: Biology, Chemistry and Medicine*, Sporn, M.B., Roberts, A.B. and Goodman, D.S. (eds). Raven: New York; 387–441.

Holmes, L.B. (2002) Teratogen-induced limb defects. *Am. J. Med. Genet.* **112**: 297–303.

Honig, L. (1981) Positional signal transmission in the developing chick limb. *Nature* **291**: 72–73.

Hornbruch, A. and Wolpert, L. (1991) The spatial and temporal distribution of polarizing activity in the flank of the pre-limb-bud stages in the chick embryo. *Development* **111**: 725–731.

Howard, T.D., Paznekas, W.A., Green, E.D., Chiang, L.C. *et al.* (1997) Mutations in TWIST, a basic helix–loop–helix transcription factor, in Saethre–Chotzen syndrome. *Nat. Genet.* **15**: 36–41.

Hui, C.-C. and Joyner, A.L. (1993) A mouse model of Greig cephalopolysyndactyly syndrome: the *extra-toes* mutation contains an intragenic deletion of the *Gli3* gene. *Nat. Genet.* **3**: 241–246.

Hurle, J.M. and Ganan, Y. (1986) Interdigital tissue chondrogenesis induced by surgical removal of the apical ectodermal ridge of the chick embryo leg bud in the stages previous to the onset of interdigital cell death. *Anat. Embryol.* **176**: 393–399.

Ianakiev, P., van Baren, M.J., Daly, M.J., Toledo, S.P. *et al.* (2001) Acheiropodia is caused by a genomic deletion in C7orf2, the human orthologue of the Lmbr1 gene. *Am. J. Hum. Genet.* **68**: 38–45.

Illingworth, C.M. and Barker, A.T. (1974) Trapped fingers and amputated finger tips in children. *J. Pediat. Surg.* **9**: 853–858.

Imokawa, Y. and Yoshizato, K. (1997) Expression of *Sonic hedgehog* gene in regenerating newt limb blastemas recapitulates that in developing limb buds. *Proc. Natl Acad. Sci. USA* **94**: 9159–9164.

Isaac, A., Cohn, M.J., Ashby, P., Ataliotis, P. *et al.* (2000) FGF and genes encoding transcription factors in early limb specification. *Mech. Dev.* **93**: 41–48.

Izpisúa-Belmonte, J.C., Tickle, C., Dollé, P., Wolpert, L. and Duboule, D. (1991) Expression of the homeobox Hox4 genes and the specification of pattern in chick wing development. *Nature* **350**: 585–589.

Izpisúa-Belmonte, J.C., Brown, J.M., Duboule, D. and Tickle, C. (1992a) Expression of Hox4 genes in the chick wing links pattern-formation to the epithelial mesenchymal interactions that mediate growth. *EMBO J.* **11**: 1451–1457.

Izpisúa-Belmonte, J.C., Ede, D.A., Tickle, C. and Duboule, D. (1992b) Misexpression of posterior *Hox4* genes in *talpid* (ta^3) mutant wings correlates with the absence of antero-posterior polarity. *Development* **114**: 959–963.

Jabs, E.W., Müller, U., Li, X., Ma, L. *et al.* (1993) A mutation in the homeodomain of the human *Msx2* gene in a family affected with autosomal dominant craniosynostosis. *Cell* **75**: 443–450.

Jacobsen, M.D., Weil, M. and Raff, M.C. (1996) Role of Ced-3/ICE-family proteases in staurosporine-induced programmed cell death. *J. Cell Biol.* **133**: 1041–1051.

Johnson, D., Kan, S.H., Oldridge, M., Trembath, R.C. *et al.* (2003) Missense mutations in the homeodomain of HOXD13 are associated with brachydactyly types D and E. *Am. J. Hum. Genet.* **72**: 984–997.

Johnson, R.L., Rothman, A.L., Xie, J.W., Goodrich, L.V. *et al.* (1996) Human homologue of *patched*, a candidate gene for the basal cell nevus syndrome. *Science* **272**: 1668–1671.

Kang, S., Graham, J.M. Jr, Olney, A.H. and Biesecker, L.G. (1997) GLI3 frameshift mutations cause autosomal dominant Pallister–Hall syndrome. *Nat. Genet.* **15**: 266–268.

Kastner, P., Mark, M. and Chambon, P. (1995) Nonsteroid nuclear receptors: what are genetic studies telling us. *Cell* **83**: 859–869.

Kawakami, Y., Capdevila, J., Buscher, D., Itoh, T. *et al.* (2001) WNT signals control FGF-dependent limb initiation and AER induction in the chick embryo. *Cell* **104**: 891–900.

Kengaku, M., Capdevila, J., Rodriguez-Esteban, C., De La Pena, J. *et al.* (1998) Distinct WNT pathways regulating AER formation and dorsoventral polarity in the chick limb bud. *Science* **280**: 1274–1277.

Kelley, R.O. and Fallon, J.F. (1978) Identification and distribution of gap junctions in the mesoderm of the developing chick limb bud. *J. Embryol. Exp. Morph.* **46**: 99–110.

Khan, P., Linkhart, B. and Simon, H.G. (2002) Different regulation of T-box genes Tbx4 and Tbx5 during limb development and limb regeneration. *Dev. Biol.* **250:** 383–392.

Khokha, M.K., Hsu, D., Brunet, L.J., Dionne, M.S. and Harland, R.M. (2003) Gremlin is the BMP antagonist required for maintenance of Shh and Fgf signals during limb patterning. *Nat. Genet.* **34**: 303–307.

Khrestchatisky, M., Djabali, M., Thouveny, Y. and Fontes, M. (1988) Expression of muscle actin genes in early differentiation stages of tail regeneration of the urodele amphibian *Pleurodeles waltlii*. *Cell Differ. Dev.* **25**: 203–212.

Kintner, C.R. and Brockes, J.P. (1984) Monoclonal antibodies identify blastemal cells derived from dedifferentiating muscle in newt limb regeneration. *Nature* **308**: 67–69.

Kintner, C.R. and Brockes, J.P. (1985) Monoclonal antibodies to the cells of a regenerating limb. *J. Embryol. Exp. Morphol.* **89**: 37–51.

Kmita, M., Fraudeau, N., Herault, Y. and Duboule, D. (2002) Serial deletions and duplications suggest a mechanism for the co-linearity of Hoxd genes in limbs. *Nature* **420**: 145–150.

Knezevic, V., De Santo, R., Schughart, K., Huffstadt, U. *et al.* (1997) Hoxd-12 differentially affects preaxial and postaxial chondrogenic branches in the limb and regulates sonic hedgehog in a positive feedback loop. *Development* **124**: 4523–4536.

Kochhar, D.M. (1977) Cellular basis of congenital limb deformity induced in mice by vitamin A. In *Morphogenesis and Malformation of the Limb. Birth Defects* original article series **13**: 11–123, Bergsma, D. and Lenz, W. (eds). Alan R Liss: New York.

Kohlhase, J., Wischermann, A., Reichenbach, H., Froster, U. and Engel, W. (1998) Mutations in the SALL1 putative transcription factor gene cause Townes–Brocks syndrome. *Nat. Genet.* **18**: 81–83.

Kohlhase, J., Schubert, L., Liebers, M., Rauch, A. *et al.* (2003) Mutations at the SALL4 locus on chromosome 20 result in a range of clinically overlapping phenotypes, including Okihiro syndrome, Holt–Oram syndrome, acro–renal–ocular syndrome, and patients previously reported to represent thalidomide embryopathy. *J. Med. Genet.* **40**: 473–478.

Kostakopoulou, K., Vogel, A., Brickell, P. and Tickle, C. (1996) 'Regeneration' of wing bud stumps of chick embryos and reactivation of Msx1 and Shh expression in response to FGF4 and ridge signals. *Mech. Dev.* **55**: 119–131.

Krauss, S., Concordet, J.P. and Ingham, P.W. (1993) A functionally conserved homologue of the *Drosophila* segment polarity gene hh is expressed in tissues with polarizing activity in zebrafish embryos. *Cell* **75**: 1431–1444.

Kumar, A., Velloso, C.P., Imokawa, Y. and Brockes, J.P. (2000) Plasticity of retrovirus-labelled myotubes in the newt limb regeneration blastema. *Dev. Biol.* **218**: 125–136.

Kumar, A., Velloso, C.P., Imokawa, Y. and Brockes, J.P. (2004) The regenerative plasticity of isolated urodele myofibres and its dependence on MSX1. *Publ. Libr. Sci. Biol.* **2**: E218.

Law, L.Y., Lin, J.S., Becker, D.L. and Green, C.R. (2002) Knockdown of connexin43-mediated regulation of the zone of polarizing activity in the developing chick limb leads to digit truncation. *Dev. Growth Differ.* **44**: 537–547.

Laufer, E., Nelson, C.E., Johnson, R.L., Morgan, B.A. and Tabin, C. (1994) Sonic hedgehog and FGF4 act through a signalling cascade and feedback loop to integrate growth and patterning of the developing limb bud. *Cell* **79**: 993–1003.

Lee, B., Godfrey, M., Vitale, E., Hori, H. *et al.* (1991) Linkage of Marfan syndrome and a phenotypically related disorder to two different fibrillin genes. *Nature* **352**: 330–334.

Lehmann, K., Seemann, P., Stricker, S., Sammar, M. *et al.* (2003) Mutations in bone morphogenetic protein receptor 1B cause brachydactyly type A2. *Proc. Natl Acad. Sci. USA* **100**: 12277–12282.

Lettice, L.A., Horikoshi, T., Heaney, S.J., van Baren, M.J. *et al.* (2002) Disruption of a long-range *cis*-acting regulator for Shh causes preaxial polydactyly. *Proc. Natl Acad. Sci. USA* **99**: 7548–7553.

Lewis, K.E., Drossopoulou, G., Paton, I.R., Morrice, D.R. *et al.* (1999) Expression of ptc and gli genes in talpid3 suggests bifurcation in Shh pathway. *Development* **126**: 2397–2407.

Lewis, P.M., Dunn, M.P., McMahon, J.A., Logan, M. *et al.* (2001) Cholesterol modification of sonic hedgehog is required for long-range signalling activity and effective modulation of signalling by Ptc1. *Cell* **105**: 599–612.

Li, Q.Y., Newbury-Ecob, R.A., Terrett, J.A., Wilson, D.I. *et al.* (1997) Holt–Oram syndrome is caused by mutations in TBX5, a member of the Brachyury (T) gene family. *Nat. Genet.* **15**: 21–29.

Litingtung, Y., Dahn, R.D., Li, Y.N., Fallon, J.F. and Chiang, C. (2002) Shh and Gli3 are dispensable for limb skeleton formation but regulate digit number and identity. *Nature* **418**: 979–983.

Lo, D.C., Allen, F. and Brockes, J.P. (1993) Reversal of muscle differentiation during urodele limb regeneration. *Proc. Natl Acad. Sci. USA* **90**: 7230–7234.

Logan, M. and Tabin, C.J. (1999) Role of Pitx1 upstream of Tbx4 in specification of hindlimb identity. *Science* **283**: 1736–1739.

Logan, M. (2003) Finger or toe: the molecular basis of limb identity. *Development* **130**: 6401–6410.

Loomis, C.A., Harris, E., Michaud, J., Wurst, W. *et al.* (1996) The mouse *engrailed-1* gene and ventral limb patterning. *Nature* **382**: 360–363.

Lopez-Martinez, A., Chang, D.T., Chiang, C., Porter, J.A. *et al.* (1995) Limb patterning activity and restricted posterior localization of the amino-terminal product of *sonic hedgehog* cleavage. *Curr. Biol.* **5**: 791–796.

Luo, G., Hofmann, C., Bronckers, A.L.J.J., Sohocki, M. *et al.* (1995) BMP7 is an inducer of nephrogenesis, and is also required for eye development and skeletal patterning. *Genes Dev.* **9**: 2808–2821.

MacCabe, J.A., Errick, J. and Saunders, J.W. (1974) Ectodermal control of the dorsoventral axis in the leg bud of the chick embryo. *Dev. Biol.* **39**: 69–82.

Maden, M. (1977) The role of Schwann cells in paradoxical regeneration in the axolotl. *J. Embryol. Exp. Morph.* **41**: 1–13.

Maden, M. (1982) Vitamin A and pattern formation in the regenerating limb. *Nature* **295**: 672–675.

Mangelsdorf, D.J., Umesono, K. and Evans, R.M. (1994) The retinoid receptors. In *The Retinoids: Biology, Chemistry, and Medicine*, Sporn, M.B., Roberts, A.B. and Goodman, D.S. (eds). Raven: New York; 319–349.

Marcil, A., Dumontier, E., Chamberland, M., Camper, S.A. and Drouin, J. (2003) Pitx1 and Pitx2 are required for development of hindlimb buds. *Development* **130**: 5–55.

Marenkova, H., Becker, D.L., Tickle, C. and Warner, A.E. (1997) Fibroblast growth factor 4 directs gap junction expression in the mesenchyme of the vertebrate limb bud. *J. Cell Biol.* **138**: 1125–1137.

Marigo, V., Roberts, D. J., Lee, S. M., Tsukurov, O. *et al.* (1995) Cloning, expression and chromosomal location of SHH and IHH: two human homologues of the *Drosophila* segment polarity gene hedgehog. *Genomics* **28**: 44–51.

Marigo, V., Scott, M.P., Johnson, R.L., Goodrich, L.V. and Tabin, C.J. (1996) Conservation in hedgehog signalling. Induction of a chicken patched homologue by sonic hedgehog in the developing limb. *Development* **122**: 1225–1233.

Martin, G.R. (1998) The roles of Fgfs in early development of vertebrate limbs. *Genes Dev.* **12**: 1571–1586.

Masuya, H., Sagai, T., Wakana, S., Moriwaki, K. and Shiroishi, T. (1995) A duplicated zone of polarizing activity in polydactylous mouse mutants. *Genes Dev.* **9**: 1645–1653.

McCready, M.E., Sweeney, E., Fryer, A.E., Donnai, D. *et al.* (2002) A novel mutation in the IHH gene causes brachydactyly type A1: a 95-year-old mystery resolved. *Hum. Genet.* **111**: 368–375.

Mercarder, N., Leonardo, E., Azpiazu, N., Serrano, A. *et al.* (1999) Conserved regulation of proximodistal limb axis development by Meis1/Hth. *Nature* **402**: 425–429.

Meyers, G.A., Day, D., Goldberg, R., Daentl, D.L. *et al.* (1996) FGFR2 exon IIIa and IIIc mutations in Crouzon, Jackson–Weiss, and Pfeiffer syndromes: evidence for missense changes, insertions, and a deletion due to alternative RNA splicing. *Am. J. Hum. Genet.* **58**: 491–498.

Mic, F.A., Sirbu, I.O. and Duester, G. (2004) Retinoic acid synthesis controlled by Raldh2 is required early for limb bud initiation and then later as a proximodistal signal during apical ectodermal ridge formation. *J. Biol. Chem.* **279**: 26698–27706.

Min, H., Danilenko, D.M., Scully, S.A., Bolon, B. *et al.* (1998) Fgf10 is required for both limb and lung development and exhibits striking functional similarity to *Drosophila* branchless. *Genes Dev.* **12**: 3156–3161.

Mizell, M. and Isaacs, J.J. (1970) Induced regeneration of hindlimbs in the newborn opossum. *Am. Zool.* **10**: 141–155.

Moore, K.L. (1988) *The Developing Human. Clinically Oriented Embryology*, 4th edn. Saunders: Philadelphia, PA.

Morgan, B.A., Izpisua-Belmonte, J.C., Duboule, D. and Tabin, C.J. (1992) Targeted misexpression of Hox4.6 in the avian limb bud causes apparent homeotic transformations. *Nature* **358**: 236–239.

Mortlock, D.P. and Innis, J.W. (1997) Mutation of HOXA13 in hand–foot–genital syndrome. *Nat. Genet.* **15**: 179–180.

Mortlock, D.P., Post, L.C. and Innis, J.W. (1996) The molecular basis of hypodactyly (*Hd*): a deletion in *Hoxa13* leads to arrest of digital arch formation. *Nat. Genet.* **13**: 284–289.

Muenke, M., Schell, U., Hehr, A., Robin, N.H. *et al.* (1994) A common mutation in the fibroblast growth fFactor receptor 1 gene in Pfeiffer syndrome. *Nat. Genet.* **8**: 269–274.

Muragaki, Y., Mundlos, S., Upton, J. and Olsen, B.R. (1996) Altered growth and branching patterns in synpolydactyly caused by mutations in HOXD13. *Science* **272**: 548–551.

Mullen, L.M., Bryant, S.V., Torok, M.A., Blumberg, B. and Gardiner, D.M. (1996) Nerve dependency of regeneration: the role of *Distal-less* and FGF signalling in amphibian limb regeneration. *Development* **122**: 3487–3497.

Naiche, L.A. and Papaioannou, V.E. (2003) Loss of Tbx4 blocks hindlimb development and affects vascularization and fusion of the allantois. *Development* **130**: 2681–2693.

Nelson, C.E., Morgan, B.A., Burke, A.C., Laufer, E. *et al.* (1996) Analysis of Hox gene expression in the chick limb bud. *Development* **122**: 1449–1466.

Niazi, I.A. and Saxena, S. (1978) Abnormal hindlimb regeneration in tadpoles of the toad, *Bufo andersoni*, exposed to excess vitamin A. *Folia Biol. (Krakow)* **26**: 3–11.

Niederreither, K., Vermot, J., Schuhbaur, B., Chambon, P. and Dolle, P. (2002) Embryonic retinoic acid synthesis is required for forelimb growth and anteroposterior patterning in the mouse. *Development* **129**: 3563–3574.

Niemann, S., Zhao, C., Pascu, F., Stahl, U. *et al.* (2004) Homozygous WNT3 mutation causes tetra-amelia in a large consanguineous family. *Am. J. Hum. Genet.* **74**: 558–563.

Neufeld, D.A. (1980) Partial blastema formation after amputation in adult mice. *J. Exp. Zool.* **212**: 31–36.

Niswander, L. (2003) Pattern formation: old models out on a limb. *Nat. Rev. Genet.* **4**: 133–143.

Niswander, L., Jeffrey, S., Martin, G.R. and Tickle, C. (1994) A positive feedback loop coordinates growth and patterning in the vertebrate limb. *Nature* **371**: 609–612.

Niswander, L., Tickle, C., Vogel, A., Booth, I. and Martin, G.R. (1993) FGF4 replaces the apical ectodermal ridge and directs outgrowth and patterning of the limb. *Cell* **75**: 579–587.

Nohno, T., Noji, S., Koyama, E., Ohyama, K. *et al.* (1991) Involvement of the *cHox4* chicken homeobox genes in determination of anteroposterior axial polarity during limb development. *Cell* **64**: 1197–1205.

Odelberg, S.J., Kollhoff, A. and Keating, M.T. (2000) Dedifferentiation of mammalian myotubes induced by msx1. *Cell* **103**: 1099–1109.

Onda, H., Poulin, M.L., Tassava, R.A. and Chiu, I.M. (1991) Characterization of a newt tenascin cDNA and localization of tenascin mRNA during newt limb regeneration by *in situ* hybridization. *Dev. Biol.* **148**: 219–232.

Ohuchi, H.T.N., Yamauchi, T., Ohata, T., Yoshioka, H. *et al.* (1995) An additional limb can be induced from the flank of the chick embryo by FGF4. *Biochem. Biophys. Res. Commun.* **209**: 809–816.

Parr, B.A. and McMahon, A.P. (1995). Dorsalizing signal Wnt7a required for normal polarity of d–v and a–p axes of mouse limb. *Nature* **374**: 350–353.

Parr, B.A., Shea, M.J., Vassileva, G. and McMahon, A.P. (1993) Mouse wnt genes exhibit discrete domains of expression in the early embryonic cns and limb buds. *Development* **119**: 247–261.

Pearse, R.V. II, Vogan, K.J. and Tabin, C.J. (2001) Ptc1 and Ptc2 transcripts provide distinct readouts of Hedgehog signalling activity during chick embryogenesis. *Dev. Biol.* **239**: 15–29.

Pecorino, L.T., Entwistle, A. and Brockes, J.P. (1996) Activation of a single retinoic acid receptor isoform mediates proximodistal respecification. *Curr. Biol.* **6**: 563–569.

Pecorino, L.T., Lo, C.D. and Brockes, J.P. (1994) Isoform-specific induction of a retinoid-responsive antigen after biolistic transfection of chimaeric retinoic acid/thyroid hormone receptors into a regenerating limb. *Development* **120**: 325–333.

Petrij, F., Giles, R.H., Dauwerse, H.G., Saris, J.J. *et al.* (1995) Rubinstein–Taybi syndrome caused by mutations in the transcriptional co-activator CBP. *Nature* **376**: 348–351.

Pilia, G., Hughes-Benzie, R.M., MacKenzie, A., Baybayan, P. *et al.* (1996) Mutations in GPC3, a glypican gene, cause the Simpson–Golabi–Behmel overgrowth syndrome. *Nat. Genet.* **12**: 241–247.

Pizette, S., Abate-Shen, C. and Niswander, L. (2001) BMP controls proximodistal outgrowth, via induction of the apical ectodermal ridge, and dorsoventral patterning in the vertebrate limb. *Development* **128**: 4463–4474.

Poswillo, D. (1975) Haemorrage in development of the face. *Birth Defects* **11**: 61–67.

Poulin, M.L. and Chiu, I.M. (1995) Re-programming of expression of the KGFR and *bek* variants of fibroblast growth factor receptor 2 during limb regeneration in newts (*Notophthalmus viridescens*). *Dev. Dyn.* **202**: 378–387.

Poulin, M.L., Patrie, K.M., Botelho, M.J., Tassava, R.A. and Chiu, I.M. (1993) Heterogeneity in the expression of fibroblast growth factor receptors during limb regeneration in newts (*Notophthalmus viridescens*). *Development* **119**: 353–361.

Putta, S., Smith, J.J., Walker, J.A., Rondet, M. *et al.* (2004) From biomedicine to natural history research: EST resources for ambystomatid salamanders. *BMC Genomics* **5**: 54.

Quaderi, N.A., Schweiger, S., Gaudenz, K., Franco, B. *et al.* (1997) Opitz G/BBB syndrome, a defect of midline development, is due to mutations in a new RING finger gene on Xp22. *Nat. Genet.* **17**: 285–291.

Radhakrishna, U., Wild, A., Grzeschik, K.H. and Antonarakis, S.E. (1997) Mutation in GLI3 in postaxial polydactyly type A. *Nat. Genet.* **17**: 269–271.

Rallis, C., Bruneau, B.G., Del Buono, J., Seidman, C.E. *et al.* (2003) Tbx5 is required for forelimb bud formation and continued outgrowth. *Development* **130**: 2741–2751.

Reginelli, A.D., Wang, Y.Q., Sassoon, D. and Muneoka, K. (1995) Digit tip regeneration correlates with regions of *Msx1* (*Hox7*) expression in fetal and newborn mice. *Development* **121**: 1065–1076.

Rida, P.C., Le Minh, N. and Jiang, Y.J. (2004) A Notch feeling of somite segmentation and beyond. *Dev. Biol.* **265**: 2–22.

Riddle, R.D., Ensini, M., Nelson, C., Tsuchida, T. *et al.* (1995) Induction of the lim homeobox gene LMX1 by WNT7a establishes dorsoventral pattern in the vertebrate limb. *Cell* **83**: 631–640.

Riddle, R., Johnson, R., Laufer, E. and Tabin, C. (1993) Sonic Hedgehog mediates the polarizing activity of the ZPA. *Cell* **75**: 1401–1416.

Robert, B., Lyons, G., Simandl, B.K., Kuroiwa, A. and Buckingham, M. (1991) The apical ectodermal ridge regulates Hox7 and Hox8 gene expression in developing chick limb buds. *Genes Dev.* **5**: 2363–2374.

Rodriguez-Esteban, C., Schwabe, J.W., Pena, J.D., Rincon-Limas, D.E. *et al.* (1998) Lhx2, a vertebrate homologue of apterous, regulates vertebrate limb outgrowth. *Development* **125**: 3925–3934.

Rodriguez-Esteban, C., Tsukui, T., Yonei, S., Magallon, J. *et al.* (1999) The T-box genes Tbx4 and Tbx5 regulate limb outgrowth and identity. *Nature* **398**: 814–818.

Rousseau, F., Bonaventure, J., Legeai-Mallet, L., Pelet, A. *et al.* (1994) Mutations in the gene encoding fibroblast growth factor receptor-3 in achondroplasia. *Nature* **371**: 252–254.

Roy, S., Gardiner, D.M. and Bryant, S.V. (2000) Vaccinia as a tool for functional analysis in regenerating limbs: ectopic expression of Shh. *Dev. Biol.* **218**: 199–205.

Ruiz-Perez, V.L., Ide, S.E., Strom, T.M., Lorenz, B. *et al.* (2000) Mutations in a new gene in Ellis–van Creveld syndrome and Weyers acrodental dysostosis. *Nat. Genet.* **24**: 283–286.

Rutland, P., Pulleyn, L.J., Reardon, W., Baraister, M. *et al.* (1995) Identical mutations in the FGFR2 gene cause both Pfeiffer and Crouzon syndrome phenotypes. *Nat. Genet.* **9**: 173–176.

Sanz-Ezquerro, J.J. and Tickle, C. (2000) Autoregulation of Shh expression and Shh induction of cell death suggest a mechanism for modulating polarising activity during chick limb development. *Development* **127**: 4811–4823.

Sanz-Ezquerro, J.J. and Tickle, C. (2003) Fgf signalling controls the number of phalanges and tip formation in developing digits. *Curr. Biol.* **13**: 1830–1836.

Satokata, I. and Maas, R. (1994) Msx1-deficient mice exhibit cleft palate and abnormalities of craniofacial and tooth development. *Nat. Genet.* **6**: 348–356.

Saunders, J.F. (1977) The experimental analysis of chick limb bud development. In *Vertebrate Limb and Somite Morphogenesis*, Ede, D.A., Hinchliffe, J.R. and Balls, M. (eds). Cambridge University Press: Cambridge; 1–24.

Saunders, J.F. and Gasseling, M.T. (1968) Ectodermal–mesodermal interactions in the origin of limb symmetry. In *Epithelial–Mesenchymal Interactions*, Fleischmajer, R. and Billingham, R.E. (eds). Williams & Wilkins: Baltimore, MD; 78–97.

Saunders, J.F. and Errick, J. (1976) Inductive activity and enduring cellular constitution of a supernumerary apical ectodermal ridge grafted to the limb bud of the chick embryo. *Dev. Biol.* **50**: 16–25.

Savard, P. and Tremblay, M. (1995). Differential regulation of Hox C6 in the appendages of adult urodeles and anurans. *J. Mol. Biol.* **249**: 879–889.

Savard, P., Gates, P.B. and Brockes, J.P. (1988) Position-dependent expression of a homeobox gene transcript in relation to amphibian limb regeneration. *EMBO J.* **7**: 4275–4282.

Sekine, K., Ohuchi, J., Fujiwara, M., Yamasaki, M. *et al.* (1999) Fgf10 is essential for limb and lung formation. *Nat. Genet.* **21**: 138–141.

Shears, D.J., Vassal, H.J., Goodman, F.R., Palmer, R.W. *et al.* (1998) Mutation and deletion of the pseudoautosomal gene SHOX cause Leri–Weill dyschondrosteosis. *Nat. Genet.* **19**: 70–73.

Shiang, R., Thompson, L.M., Zhu, Y.Z., Church, D.M. *et al.* (1994) Mutations in the transmembrane domain of FGFR3 cause the most common genetic form of dwarfism, achondroplasia. *Cell* **78**: 335–342.

Sicard, E.R. (1985) *Regulation of Vertebrate Limb Regeneration*. Oxford University Press: New York.

Simon, H.G. and Tabin, C.J. (1993) Analysis of Hox4.5 and Hox3.6 expression during newt limb regeneration: differential regulation of paralogueous Hox genes suggest different roles for members of different Hox clusters. *Development* **117**: 1397–1407.

Singer, M. (1974) Neurotrophic control of limb regeneration in the newt. *Ann. N.Y. Acad. Sci.* **228**: 308–322.

Song, K., Wang, Y. and Sassoon, D. (1992) Expression of Hox7.1 in myoblasts inhibits terminal differentiation and induces cell transformation. *Nature* **360**: 477–481.

Spitz, F., Gonzalez, F. and Duboule, D. (2003) A global control region defines a chromosomal regulatory landscape containing the HoxD cluster. *Cell* **113**: 405–417.

Stevenson, R.E. and Meyer, L.C. (1993) The limbs. In *Human Malformations and Related Anomalies*, Stevenson, R.E., Hall, J.G. and Goodman, R.M. (eds). Oxford Monographs on Medical Genetics, No. 27. Oxford University Press: Oxford; 699–804.

Stocum, D. (1985). Role of the skin in urodele limb regeneration. In *Regulation of Vertebrate Limb Regeneration*, Sicard, R.E. (ed.). Oxford University Press: New York.

Stocum, D.L. (1991) Retinoic acid and limb regeneration. *Semin. Dev. Biol.* **2**: 199–210.

Stone, D.L., Slavotinek, A., Bouffard, G.G., Banerjee-Basu, S. *et al.* (2000) Mutation of a gene encoding a putative chaperonin causes McKusick–Kaufman syndrome. *Nat. Genet.* **25**: 79–82.

Storm, E.E., Huynh, T.V., Copeland, N.G., Jenkins, N.A. *et al.* (1994) Limb alterations in brachypodism mice due to mutations in a new member of the TGFβ superfamily. *Nature* **368**: 639–643.

Summerbell, D., Lewis, J. and Wolpert, L. (1973) Positional information in chick limb morphogenesis. *Nature* **244**: 492–496.

Sun, X., Lewandoski. M., Meyers, E.M., Liu, Y.H. *et al.* (2000) Conditional inactivation of Fgf4 reveals complexity of signalling during limb bud development. *Nature* **418**: 501–508.

Sun, X., Mariani, F.V. and Martin, G.R. (2002) Functions of FGF signalling from the apical ectodermal ridge in limb development. *Nature* **418**: 501–508.

Suzuki, T., Takeuchi, J., Koshiba-Takeuchi, K. and Ogura, T. (2004) Tbx genes specify posterior digit identity through Shh and BMP signalling. *Dev. Cell* **6**: 43–53.

Takeuchi, J.K., Joshiba-Takeuchi, K., Matsumoto, K., Vogek-Hopker, A. *et al.* (1999) Tbx5 and Tbx4 genes determine the wing/leg identity of limb buds. *Nature* **398**: 810–814.

Tamura, K., Yonei-Tamura, S. and Belmonte, J.C. (1999) Differential expression of *Tbx4* and *Tbx5* in zebrafish fin buds. *Mech. Dev.* **87**: 181–184.

Tanaka, E.M., Drechsel, D.N. and Brockes, J.P. (1999) Thrombin regulates S-phase re-entry by cultured newt myotubes. *Curr. Biol.* **9**: 792–799.

Tanaka, E.M., Gann, A.A.F., Gates, P.B. and Brockes, J.P. (1997) Newt myotubes reenter the cell cycle by phosphorylation of the retinoblastoma protein. *J. Cell Biol.* **136**: 155–165.

Tanaka, M., Cohn, M.J., Ashby, P., Davey, M. *et al.* (2000) Distribution of polarizing activity and potential for limb formation in mouse and chick embryos and possible relationships to polydactyly. *Development* **127**: 4011–4021.

Tanaka, M., Münsterberg, A., Anderson, W.G., Prescott, A.R. *et al.* (2002) Fin development in a cartilaginous fish and the origin of vertebrate limbs. *Nature* **416**: 527–531.

Taylor, G.P., Anderson, R., Reginelli, A.D. and Muneoka, K. (1994) FGF2 induces regeneration of the chick limb bud. *Dev. Biol.* **163**: 282–284.

Thomas, J.T., Kilpatrick, M.W., Lin, K., Erlacher, L. *et al.* (1997) Disruption of human limb morphogenesis by a dominant negative mutation in CDMP1. *Nat. Genet.* **17**: 58–64.

Thornton, C.S. (1968). Amphibian limb regeneration. *Adv. Morphogen.* **7**: 205–249.

Tickle, C. (1981) The number of polarizing region cells required to specify additional digits in the developing chick wing. *Nature* **289**: 295–298.

Tickle, C. (2003) Patterning systems – from one end of the limb to the other. *Dev. Cell* **4**: 449–458.

Tickle, C., Alberts, B.M., Wolpert, L. and Lee, J. (1982) Local application of retinoic acid to the limb bud mimics the action of the polarizing region. *Nature* **296**: 564–565.

Tickle, C., Lee, J. and Eichele, G. (1985) A quantitative analysis of the effect of all-*trans*-retinoic acid on the pattern of chick wing development. *Dev. Biol.* **109**: 82–95.

Tonkin, E.T., Wang, T.J., Lisgo, S., Bamshad, M.J. *et al.* (2004) NIPBL, encoding a homologue of fungal Scc2-type sister chromatid cohesion proteins and fly Nipped-B, is mutated in Cornelia de Lange syndrome. *Nat. Genet.* **36**: 636–641.

Torok, M.A., Gardiner, D.M., Izpisúa-Belmonte, J.C. and Bryant, S.V. (1999) Sonic hedgehog (shh) expression in developing and regenerating axolotl limbs. *J. Exp. Zool.* **284**: 197–206.

Torok, M.A., Gardiner, D.M., Shubin, N.H. and Bryant, S.V. (1998) Expression of HoxD genes in developing and regenerating axolotl limbs. *Dev. Biol.* **200**: 225–233.

Tümpel, S., Sanz-Ezquerro, J.J., Isaac, A., Eblaghie, M.C. *et al.* (2002) Regulation of Tbx3 expression by anteroposterior signalling in vertebrate limb development. *Dev. Biol.* **250**: 251–262.

Vargesson, M., Kostakopoulou, K., Drossopoulou, G., Papageorgiou, S. and Tickle, C. (2001) Characterization of hoxa gene expression in the chick limb bud in response to FGF. *Dev. Dyn.* **220**: 87–90.

Vasiliauskas, D., Laufer, E. and Stern, C.D. (2003) A role for hairy1 in regulating chick limb bud growth. *Dev. Biol.* **262**: 94–106.

Vogel, A., Roberts-Clarke, D. and Niswander, L. (1995a) Effect of FGF on gene expression in chick limb bud cells *in vivo* and *in vitro*. *Dev. Biol.* **171**: 507–520.

Vogel, A., Rodriguez, C., Warnken, W. and Belmonte, J. (1995b) Dorsal cell fate specified by chick *Lmx-1* during vertebrate limb development. *Nature* **378**: 716–720.

Vogel, A. and Tickle, C. (1993) FGF4 maintains polarizing activity of posterior limb bud cells in vivo and *in vitro*. *Development* **119**: 199–206.

Vollrath, D., Jaramillo-Babb, V.L., Clough, M.V., McIntosh, I. *et al.* (1998) Loss-of-function mutations in the LIM-homeodomain gene, LMX1B, in nail–patella syndrome. *Hum. Mol. Genet.* **7**: 1091–1098.

Vortkamp, A., Gessler, M. and Grzeschik, K.-H. (1991) *GLI3* zinc-finger gene interrupted by translocation in Greig syndrome families. *Nature* **352**: 539–540.

Wallace, H. (1981). *Vertebrate Limb Regeneration*. Wiley: Chichester.

Wang, L., Marchionni, M.A. and Tassava, R.A. (2000) Cloning and neuronal expression of a type III newt neuregulin and rescue of denervated, nerve-dependent newt limb blastemas by rhGGF2. *J. Neurobiol.* **43**: 150–158.

Wassif, C.A., Maslen, C., Kachilele-Linjewile, S., Lin, D. *et al.* (1998) Mutations in the human sterol delta7-reductase gene at 11q12–13 cause Smith–Lemli–Opitz syndrome. *Am. J. Hum. Genet.* **63**: 55–62.

Wellik, D.M. and Capecchi, M.R. (2003) Hox10 and Hox11 genes are required to globally pattern the mammalian skeleton. *Science* **301**: 363–367.

te Welscher, T.T., Zuniga, A., Kuijper, S., Drenth, T. *et al.* (2002a) Progression of vertebrate limb development through SHH-mediated counteraction of GLI3. *Science* **298**: 827–830.

te Welscher, P., Fernandez-Teran, M., Ros, M.A. and Zeller, R. (2002b) Mutual genetic antagonism involving GLI3 and dHAND prepatterns the vertebrate limb bud mesenchyme prior to SHH signalling. *Genes Dev.* **16**: 421–426.

Wild, A., Kalff-Suske, M., Vortkamp, A., Bornholdt, D. *et al.* (1997) Point mutations in human GLI3 cause Greig syndrome. *Hum. Mol. Genet.* **6**: 1979–1984.

Wilkie, O.M., Slaney, S.F., Oldridge, M., Poole, M.D. *et al.* (1995) Apert syndrome (craniosynostosis/syndactyly) results from localised mutations of FGFR2 and is allelic with Crouzon syndrome. *Nat. Genet.* **9**: 165–172.

Wilkie, A.O. (2003) Why study human limb malformations? *J. Anat.* **202**: 27–35.

Williams, C.J., Considine, E.L., Knowlton, R.G., Reginato, A. *et al.* (1993) Spondyloepiphyseal dysplasia and precocious osteoarthritis in a family with an Arg75-Cys mutation in the procollagen type II gene. *Hum. Genet.* **92**: 499–505.

Winter, R.M., Schoroer, R.J. and Meyer, L.C. (1993) Hands and feet. In *Human Malformations and Related Anomalies*, Stevenson, R.E., Hall, J.G. and Goodman, R.M. (eds). Oxford Monographs on Medical Genetics, No. 27. Oxford University Press: Oxford; 805–843.

Wolpert, L., Tickle, C. and Sampford, M. (1979) The effect of cell killing by X-irradiation on pattern formation in the chick limb. *J. Embryol. Exp. Morph.* **50**: 175–198.

Woychik, R.P., Maas, R.L., Zeller, R., Vogt, T.F. and Leder, P. (1990) 'Formins': proteins deduced from the alternative transcripts of the *limb deformity* gene. *Nature* **346**: 850–853.

Wozney, J.M., Capparella, J. and Rosen, V. (1993) The bone morphogenetic proteins in cartilage and bone development. In *Molecular Basis of Morphogenesis*, Bernfield, M. (ed.). Wiley-Liss: New York; 221–230.

Xu, X., Weinstein, M., Li, C., Naski, M. *et al.* (1998) Fibroblast growth factor receptor 2 (FGFR2) – mediated reciprocal regulation loop between FGF8 and FGF10 is essential for limb induction. *Development* **125**: 753–765.

Yamaguchi, T.P., Bradley, A., McMahon, A.P. and Jones, S. (1999) A Wnt5a pathway underlies ougrowth of multiple structures in the vertebrate embryo. *Development* **126**: 1211–1223.

Yang, Y. and Niswander, L. (1995) Interaction between the signalling molecules WNT7 and SHH during vertebrate limb development: dorsal signals regulate anteroposterior patterning. Cell **80**: 939–947.

Yang, Y., Drossopoulou, G., Chuang, P.T., Duprez, D. *et al.* (1997) Relationship between dose, distance and time in Sonic Hedgehog-mediated regulation of anteroposterior polarity in the chick limb. *Development* **124**: 4393–4404.

Yashiro, K., Zhao, X., Uehara, M., Yamashita, K. *et al.* (2004) Regulation of retinoic acid distribution is required for proximodistal patterning and outgrowth of the developing mouse limb. *Dev. Cell* **6**: 411–422.

Yokoyama, H., Ide, H. and Tamura, K. (2001) FGF10 stimulates limb regeneration ability in *Xenopus laevis*. *Dev. Biol.* **233**: 72–79.

Yuan, J., Shaham, S., Ledoux, S., Ellis, H.M. and Horvitz, H.R. (1993) The *C. elegans* cell death gene *ced-3* encodes a protein similar to mammalian interleukin-1-beta converting enzyme. *Cell* **75**: 641–652.

Zakany, J., Kmita, M. and Duboule, D. (2004) A dual role for Hox genes in limb anterior-posterior asymmetry. *Science* **304**: 1669–1772.

Zeller, R., Jackson-Grusby, L. and Leder, P. (1989) The limb deformity gene is required for apical ectodermal ridge differentiation and antero-posterior limb pattern formation. *Genes Dev.* **3**: 1481–1492.

Zeng, X., Goetz, J.A., Suber, L.M., Scott, W.J. Jr *et al.* (2001). A freely diffusible form of Sonic hedgehog mediates long-range signalling. *Nature* **411**: 716–720.

Zhao, W.G. and Neufeld, D.A. (1995) Bone regrowth in young mice stimulated by nail organ. *J. Exp. Zool.* **271**: 155–159.

Zou, H. and Niswander, L. (1996) Requirement for BMP signalling in interdigital apoptosis and scale formation. *Science* **272**: 738–741.

Zuniga, A., Haramis, A.P., McMahon, A.P. and Zeller, R. (1999) Signal relay by BMP antagonism controls the SHH/FGF4 feedback loop in vertebrate limb buds. *Nature* **401**: 598–602.

Zuniga, A., Michos, O., Spitz, F., Haramis, A.P. *et al.* (2004) Mouse limb deformity mutations disrupt a global control region within the large regulatory landscape required for Gremlin expression. *Genes Dev.* **18**: 1553–1564.

Zwilling, E. and Hansborough, L. (1956) Interaction between limb bud ectoderm and mesoderm in the chick embryo. III. Experiments with polydactylous limbs. *J. Exp. Zool.* **132**: 219–239.

8

Brain and Spinal Cord

Andrew J. Copp

Abstract

The neural tube is the embryonic progenitor of the entire brain and spinal cord. It constitutes one of the most active areas of research in developmental biology today, not only because of the key importance of the central nervous sysem (CNS) in the body plan of all vertebrates, but also because of the wide range and varying severity of the clinical defects that arise from abnormal development of the human CNS. In this chapter, the main categories of congenital CNS defect are discussed in relation to the developmental processes that are principally involved in their generation. For example, disturbance of early neural inductive events in the forebrain is responsible for the severe defect holoprosencephaly, the morphogenesis of the neural plate, which closes dorsally to from the neural tube, is implicated in the neural tube defects anencephaly and open spina bifida, and disturbance of the events of neuronal migration in the developing cerebral cortex leads to the broad category of 'neuronal migration' defects, which includes lissencephaly and neuronal heterotopias. The rate at which new knowledge has accumulated in the field of CNS development and defects has accelarated over the past ten years, aided in large part by the identification of a number of human CNS disease genes and by the construction of mouse genetic models for many CNS defects. The mouse models enable an experimental approach to the molecular mechanisms underlying congenital CNS defects, and raise the possibility of, ultimately, developing new methods for treatment or even primary prevention of CNS birth defects.

Keywords

central nervous system, congenital defect, neural induction, neurulation, neuronal migration, neuronal differentiation, axon guidance, holoprosencephaly, neural tube defects, lissencephaly, agenesis of the corpus callosum

Introduction

There can be few parts of the mammalian embryo that play such a pivotal role in development as the neural tube. This dorsal midline structure runs the entire length

Embryos, Genes and Birth Defects, Second Edition Edited by Patrizia Ferretti, Andrew Copp, Cheryll Tickle and Gudrun Moore

of the embryo, giving rise to all of the neurons and most of the glia of the central nervous system (CNS). Moreover, its derivative, the neural crest, contributes to the peripheral nervous system and to a variety of other organ and body systems, including the craniofacial skeleton, pigment cells of the skin, thymus, thyroid and parathyroid and important vascular and cardiac structures. In addition to these cellular contributions, the neural tube is critically important as an inducer of the formation of other organ systems, for example, the mesoderm-derived vertebrae and the ectoderm-derived inner ear primordium (the otic vesicle).

Defects of the CNS arise when the processes of neural tube development become disturbed, particularly during the embryonic and fetal periods. The abnormalities may be structural, as when the neural tube fails to close during the 3rd and 4th weeks of human development, leading to the malformations anencephaly and myelomeningocele (spina bifida). CNS defects are of major clinical importance, both as a cause of death around birth (perinatal mortality) and as a source of disability in children and adults. They affect 0.5–1% of liveborn children (Table 8.1), but with a higher prevalence amongst the embryos and fetuses of early pregnancy, many of which are lost due to spontaneous abortion (miscarriage). Disturbance of later nervous system development yields functional, rather than gross structural deficits, causing conditions ranging from severe disorders, such as mental retardation and epilepsy, to milder conditions, including speech/language disorders and dyslexia. Up to 20% of all individuals are affected by functional defects of this type, one of the commonest and most challenging disabilities faced by members of society today.

In order to diagnose, effectively manage and ultimately prevent congenital nervous system defects, it is essential that we understand the genetic, molecular and cellular mechanisms of nervous system development and the ways in which these processes

Table 8.1 Varying frequencies of some CNS birth defects[1]

CNS defect	Frequency in late pregnancies and live births (per 10 000)[2]
Cerebral palsy	22
Dandy–Walker syndrome	3
Holoprosencephaly	0.6*
Hydrocephaly	14
Microcephaly	16
Neural tube defects	10
Schizencephaly	<0.01
Total CNS defects	50–100

[1]Frequency figures are approximate, and are intended only as a general guide to relative prevalence of CNS defects. Reported frequencies vary considerably between studies reflecting differences in ascertainment and also indicating variation in frequency of defects in different human populations (for further details see Myrianthopoulos, 1979; Ming and Muenke, 2002; Russman and Ashwal, 2004).
[2]Includes pregnancies undergoing therapeutic termination of pregnancy after prenatal diagnosis.
*Reported to be as frequent as 1/250 conceptuses during early pregnancy.

can be disturbed. This chapter reviews the main events of nervous system development, in each case discussing the principal types of congenital defect that result when development is disturbed, as well as recent insights into the cellular and molecular mechanisms that regulate nervous system development, both normal and abnormal. Evidence from sub-mammalian species is presented where particular advances have been made in these systems, but the main emphasis is on mammalian systems, which hold most promise for an ultimate understanding of human development.

Overview of nervous system development

Neural induction

Nervous system development is conventionally considered to begin with neural induction, in which the neural plate, the immediate precursor of the neural tube, forms as a result of cellular interactions that occur during emergence of the primary germ layers at gastrulation, particular involving a key region of the embryo termed the 'organizer'. As ectodermal cells become distinct from mesoderm and endoderm, so embryonic specification of neural, as opposed to non-neural, ectoderm gets under way (Stern, 2002). The following developmental events occur concurrently with the first appearance of cells committed to a neural programme of development.

Regional patterning

The neural plate becomes regionally patterned along each of its axes: rostro-caudal, dorso-ventral and medio-lateral. This patterning represents a latent potential for differentiation that can be visualized at this early stage as patterns of differential gene expression (Figure 8.1a–c). The patterning foreshadows the later development of morphological and functional subdivisions of the neural tube, for example, the distinct regions of the brain and regionalisation of neuronal types in the spinal cord.

Morphogenesis: neural tube closure

While regional patterning is under way, the neural plate also undergoes morphogenesis. The gastrulation stage embryo rapidly changes its overall shape through 'convergent extension', a process of medio-lateral narrowing and rostro-caudal lengthening, which generates a neural plate that is broad rostrally, as the future brain, but narrower caudally, as the future spinal cord. Subsequently, in the process of neurulation, folds arise at the edges of the neural plate, approach one other in the dorsal midline and fuse (Figure 8.1d). Once neural fold fusion is complete, epithelial remodelling occurs so that the inner aspect of the neural folds becomes a continuous

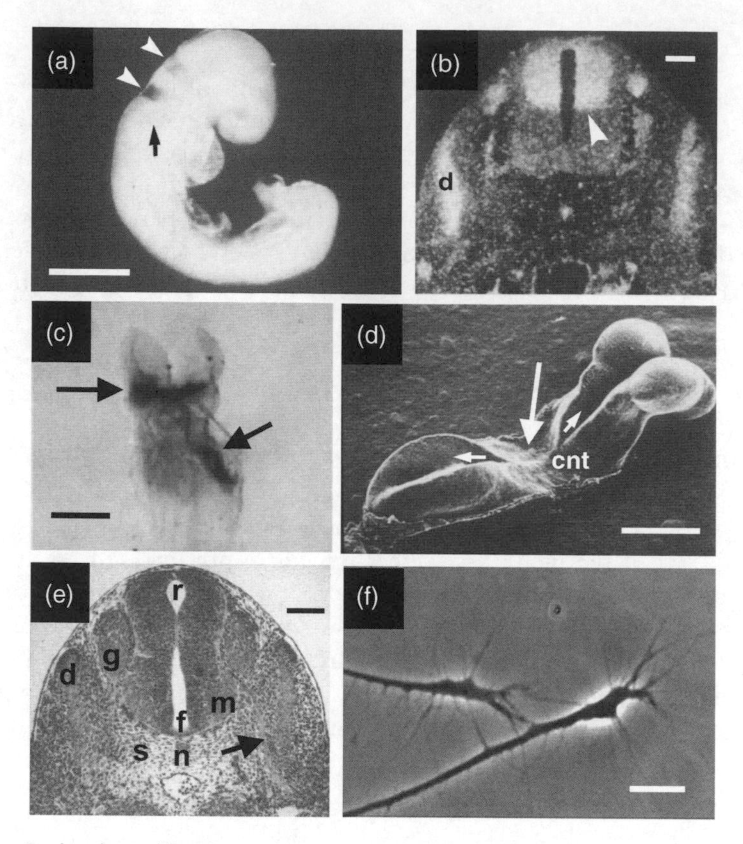

Figure 8.1 Regional specification and morphogenesis of the mouse nervous system. (a–c) Gene activity domains in the rostro-caudal (a), dorso-ventral (b) and left–right (c) embryonic axes, detected using *in situ* hybridization of mRNA. (a) *Krox-20* expression in the embryonic day (E) 9.5 day mouse embryo shows expression specifically in rhombomeres 3 (weaker expression, more rostral arrowhead) and 5 (stronger expression, more caudal arrowhead) of the hindbrain. Black arrow shows position of otocyst (not stained) which marks the boundary of rhombomeres 5 and 6. (b) *Pax3* expression in a transverse section through the E11.5 mouse trunk region. A sharp boundary (arrowhead) exists between *Pax3*-positive dorsal and *Pax3*-negative ventral neural tube regions (silver grains indicate sites of mRNA localization). Dermamyotome (d) and dorsal root ganglia (adjacent to neural tube) also express *Pax3*. (c) *Pitx2* expression in an E8.5 mouse embryo viewed from the ventral surface (the closing head folds are at the top). Note symmetrical *Pitx2* expression in the forebrain (large arrow) but left-sided in the lateral plate mesoderm (small arrow). (d–f) Morphogenetic events of nervous system development. (d) Scanning electron micrograph of E8.5 embryo showing an early stage of neural tube closure. Neural tube fusion has just initiated at the boundary (large arrow) of the future cervical and hindbrain regions (cnt). Closure is progressing rostrally into the hindbrain (small arrow to right) and caudally along the spine (small arrow to left). (e) Transverse section of E11.5 mouse trunk region (haematoxylin and eosin-stained) showing dorsal root ganglia (g) which are derived from ventrally migrating neural crest cells, and motor neurons (m) derived from neuroepithelial cells which have ceased proliferation and migrated away from the ventricular zone. Arrow points to axon tract arising from the motor horn. Floor plate (f), notochord (n), roof plate (r) and sclerotome (s) are also visible. (f) Two axon growth cones migrating on a tissue culture substratum, as viewed by phase contrast optics. Long filopodia emerge particularly from the expanded growth cone structure. Parts (c) and (d) are reproduced with permission from Henderson *et al.* (2001) and Copp *et al.* (1990), respectively. Scale bars: (a) 0.5 mm; (b) 0.1 mm; (c) 0.3 mm; (d) 0.25 mm; (e) 0.1 mm; (f) 10 μm

layer across the midline, comprising the roof-plate of the neural tube. Similarly, the outer aspects of the neural folds form the continuous mid-dorsal surface ectoderm. In this way, the neural tube takes up its internal, dorsal, midline position in the embryo, a process that is complete by the end of the 4th week of human development (day 10 of mouse gestation).

Rostro-caudal progression of CNS development

An important factor in the development of higher vertebrates, including birds and mammals, is that neurulation does not occur simultaneously at all levels of the cranio-caudal axis, in contrast to lower vertebrates, as exemplified by the amphibian *Xenopus laevis*. In mammals, neural induction initially defines a region of neural plate whose developmental fate is confined to forming the brain and upper spinal regions of the future nervous system. Lower spinal regions form progressively (thoracic, lumbar, sacral and caudal) by emergence of cells from the primitive streak as it regresses towards the caudal end of the embryo, marking the later stages of gastrulation.

Neural crest migration

Origin and emigration of the neural crest correlates temporally with closure of the neural tube. In mammals, neural crest cells migrate from the tips of the neural folds just prior to closure of the cranial neural tube and then migrate entirely beneath the surface ectoderm (i.e. the dorso-lateral route). In contrast, neural crest cells in the spinal region emerge from the roof of the neural tube after its closure is completed. Newly emerged spinal neural crest cells migrate either along the dorso-lateral route or between the neural tube and somite (ventro-medial route). Neural crest cells from both cranial and spinal regions subsequently give rise to differentiated derivatives in the peripheral nervous system and in several other organs and body structures (Gammill and Bronner-Fraser, 2003) (Figure 8.1e; see also Chapters 9–14).

Cell proliferation in the neural tube

The period prior to and immediately following completion of neural tube closure is marked by rapid cell proliferation within the neuroepithelium, with a relatively short cell cycle (e.g. 8–10 hours at 10.5 days of mouse development). The proliferative neuroepithelium is ‘pseudostratified’ with nuclei occupying varying positions from basal (i.e. inner) to apical within the epithelium (Figure 8.2a). Despite this varying nuclear position, all neuroepithelial cells (except those in mitosis) maintain contact with both surfaces. The passage of nuclei from basal to apical, and then back to the basal surface again (‘interkinetic nuclear migration’), is linked to progression through the cell cycle: nuclei in S phase are located at the basal surface, while mitotic nuclei are positioned apically. Cells in phases G_1 and G_2 have nuclei at intermediate positions.

(a)

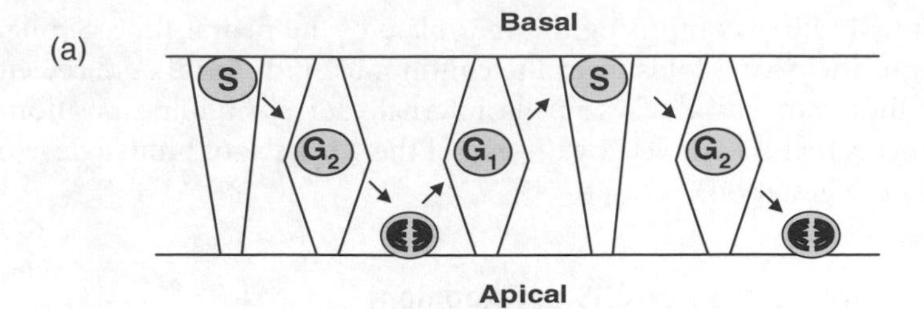
Basal
S
G2
G1
S
G2
Apical

(b)

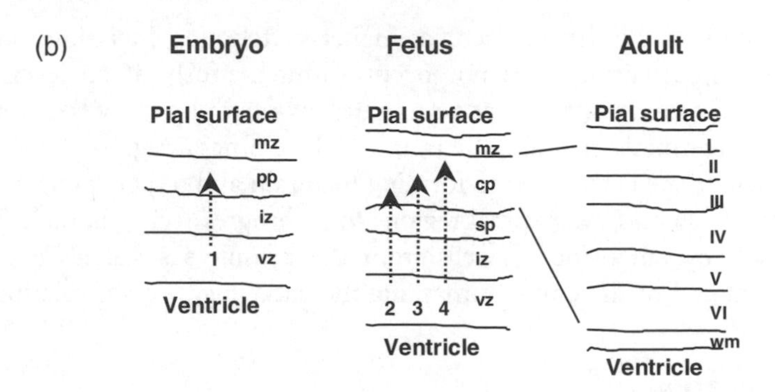
Embryo
Fetus
Adult
Pial surface
mz
pp
iz
vz
1
Ventricle
Pial surface
mz
cp
sp
iz
vz
2 3 4
Ventricle
Pial surface
I
II
III
IV
V
VI
wm
Ventricle

(c)

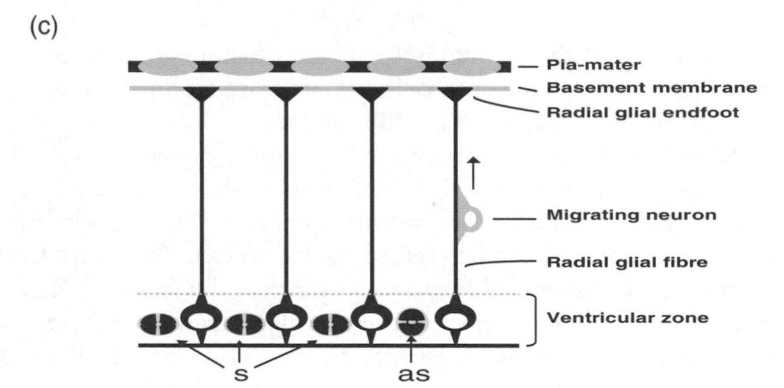
Pia-mater
Basement membrane
Radial glial endfoot
Migrating neuron
Radial glial fibre
Ventricular zone
s
as

(d)

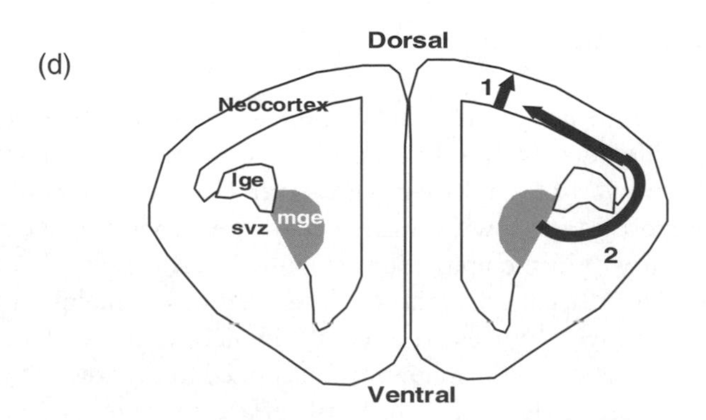
Dorsal
Neocortex
lge
svz
mge
1
2
Ventral

Onset of neuronal and glial cytodifferentiation

In the early neural tube, neuroepithelial cells divide symmetrically with both daughter cells continuing to form part of the proliferative population (the 'ventricular zone'). As development proceeds, however, an increasing proportion of neuroepithelial cells undergo asymmetric divisions, in which one daughter cell remains proliferative while the other leaves the cell cycle and embarks upon a pathway of cytodifferentiation. Early in CNS development, neuroepithelial differentiation generates mainly neurons, whereas, later in development, glia also arise from the neuroepithelial precursor population. The onset of differentiation towards neuronal and glial lineages coincides with the loss of contact of neuroepithelial cells with the lumen of the neural tube, and their cessation of proliferation. The only exceptions to this rule are the ependyma, a population of glial cells that differentiates *in situ* at the luminal border of the neural tube, and cells of the subventricular zone of the cerebral hemispheres that continue to proliferate for a time after migrating away from the neural tube lumen. Subventricular zone cells also cease proliferation before they migrate to the cortical plate.

Regionalization of neuronal differentiation

An important principle, is that the neuroepithelial cells at a particular cranio-caudal level of the neural tube differentiate into only those cell types that will characterize the mature nervous system at that level. Thus, the anterior hindbrain neural tube uniquely contains cells destined to form the cerebellar Purkinje neurons, whereas the forebrain neuroepithelium contains cells destined to develop as pyramidal neurons. Transplantation studies in the chick show that, soon after neural tube closure, cells of the future cerebellar region are already irreversibly committed to particular pathways of differentiation (Alvarado-Mallart *et al.*, 1990), presumably as a result of intracellular events that occurred during the phase of regional patterning.

Figure 8.2 (a) Interkinetic nuclear migration, as observed in the early neural tube and ventricular zone. (b) The stages of development of cerebral cortex. An initial wave of 'early' migrating post-mitotic neuroblasts (1) originates in the marginal zone (mz), passes through the intermediate zone (iz) and forms the pre-plate (pp). The latter is subsequently split by waves of 'later' arriving neuroblasts (2, 3, 4) which form the cortical plate (cp), splitting the pre-plate into a superficial marginal zone (mz) and deep sub-plate (sp). The cortical plate subsequently gives rise to the six-layered cerebral cortex (layers I–VI) overlying the white matter (wm). (c) Mechanism of radial glia-guided neuronal migration. Note that neuroepithelial cell division can be symmetrical (s) in the plane of the ventricular zone, or asymmetrical (as) to generate a migrating neuroblast and a proliferative daughter cell. (d) Two modes of neuroblast migration to the cerebral cortex. Radial glia-guided migration (1) generates the pyramidal neurons, while tangential migration from the medial (mge) and lateral (lge) ganglionic eminence (2), via the subventricular zone (svz) generates the GABAergic interneurons

Neuronal migration

As neuroepithelial cells leave the cell cycle, they migrate outwards, away from the ventricular zone. In the spinal cord, the neuronal and glial precursors move only short distances to take up a position in the intermediate and outer, marginal, layers where they differentiate. There is dorso-ventral specificity in this migration: for example, post-mitotic motor neurons differentiate only in the ventro-lateral portions of the spinal cord (Figure 8.1e). In the forebrain, neuronal precursors migrate much larger distances, in a series of waves across the intermediate zone, to form the layered structure of the cerebral cortex (Figure 8.2b). Early migrating neuroblasts form a transient structure, the pre-plate, that subsequently becomes split into inner (sub-plate) and outer (marginal zone) regions by the arrival of later migrating neuroblasts which form the cortical plate. Deep layers of the cortical plate are formed first, with later arriving cells populating progressively more superficial layers. The cortical plate thus expands as development proceeds, giving rise to the adult cortex.

Neuroblasts destined to form the glutaminergic pyramidal neurons of the cortex migrate outwards along the cellular processes of radial glia (Figure 8.2c). The latter are cells within the ventricular zone that maintain cellular contacts with the outer (pial) surface of the developing cortex. In contrast, neuroblasts that are destined to form the GABAergic interneurons of the cerebral cortex originate in the ganglionic eminence of the ventral forebrain and migrate tangentially to take up their positions within the laminar cortical structure (Figure 8.2d; Parnavelas, 2000).

Programmed cell death

Somewhat surprisingly, at this time of intensive growth and differentiation, there is also considerable programmed cell death (apoptosis) in the early nervous system (Nijhawan *et al.*, 2000). This is regionally restricted, for example characterizing the cortical sub-plate, a transient structure seemingly removed by apoptosis (Allendoerfer and Shatz, 1994), and spinal motor neurons (Sendtner *et al.*, 2000), where apoptosis may serve the function of pruning ineffective axonal connections. However, other instances of programmed cell death are not easily explained and may serve functions that are not yet understood.

Formation of nerve connections

Differentiating neurons extend multiple dendritic processes, one of which subsequently specializes as an axon, in some cases growing long distances to form nerve connections. For example, the axons of motor neurons grow away from the CNS to contact muscle fibres at neuromuscular junctions, while the bipolar axons of sensory neurons connect peripheral sense organs with interneurons in the CNS. The motile organ of the axon is the 'growth cone', a dynamic structure whose function is regulated by remodelling of the cytoskeleton (Figure 8.1F). Growth cones are guided by complex environmental cues during the growth of the axon towards its target site (Dickson, 2002). Once connections have formed, glial cells (oligodendrocytes in the CNS; Schwann cells in the peripheral nervous system) create an insulating myelin sheath around the axon, enabling the passage of nerve impulses. Axons group together ('fasciculate'), forming macroscopic nerves in the periphery and fibre tracts within the CNS. Of these, the massive corpus callosum, whose fibres connect the two cerebral hemispheres, the corticospinal tracts that connect the motor cortex to the spinal motor neurons, and the optic chiasm, the site where optic nerves cross the midline, are often implicated in CNS birth defects.

Onset of nervous system function

The human embryo first shows evidence of CNS function at 6 weeks (de Vries, 1992), correlating with the onset of peptide neurotransmitter production, which can be detected in neuroblasts during development of the cerebral cortical plate (Allendoerfer and Shatz, 1994). Synaptic connections develop progressively during the remainder of embryonic and fetal development, and well as into post-natal life. Upon these connections depend the majority of body functions and all of our mental activities.

Defects of CNS development: towards a genetic and developmental understanding

Defective nervous system development results in a diverse group of diseases, ranging from major malformations that are incompatible with post-natal life to disabilities that only slightly affect the physical or mental function of the individual. Although an over-simplification, it can be useful to consider the main categories of CNS birth defect in relation to the event or process of nervous system development that appears to be primarily disturbed. The following pages review the main categories of CNS defects, relating them where possible to the underlying genetic, cellular and molecular developmental mechanisms. A detailed description of the clinical and neuropathological aspects of these defects is beyond the scope of the present chapter, but can be found in Harding and Copp (2002).

Holoprosencephaly: a defect of neural induction and dorso-ventral forebrain patterning

Global defects of neural induction are expected to arrest development at an early stage, leading to lethal abnormalities that are unlikely to come to the attention of the clinician, except as cases of early pregnancy loss. On the other hand, a localized abnormality of forebrain induction has long been suggested to underlie the severe brain malformation holoprosencephaly. In recent years, mutations in the gene encoding sonic hedgehog, a morphogen of critical importance in neural induction and dorso-ventral patterning, have been identified in certain patients with holoprosencephaly, confirming the early prediction that holoprosencephaly is a disorder of neural induction.

Holoprosencephaly spectrum and its causes Subdivision of the early embryonic forebrain (prosencephalon) to form the bilateral telencephalic vesicles (Figure 8.3a), the precursors of the cerebral hemispheres, is defective in holoprosencephaly. At its most severe, the forebrain is completely undivided (alobar holoprosencephaly), a defect that may be accompanied by failure of separation of the optic vesicles and eye primordia, yielding the birth defect cyclopia (Figure 8.3b). Holoprosencephaly may also be associated with lack of olfactory bulb development (arrhinencephaly), and

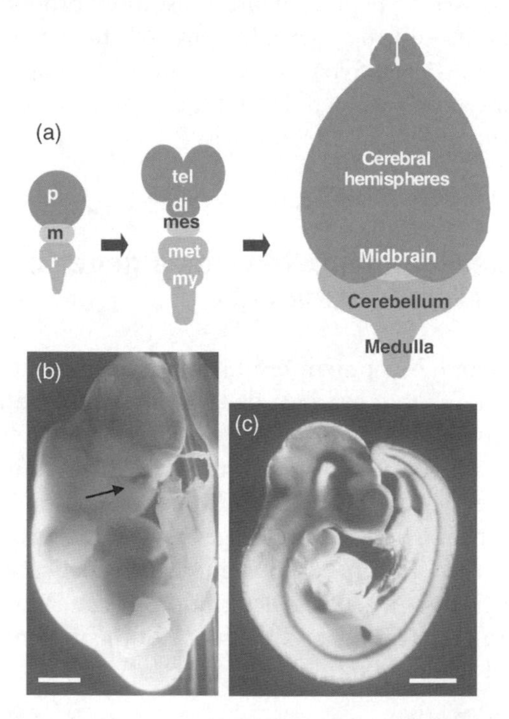

Figure 8.3 (a) Formation of the paired telencephalic vesicles (tel) by division of the initially single prosencephalic vesicle (p) in the midline. The telencephalic vesicles, with their lateral ventricles, form the cerebral hemispheres connected in the midline by the corpus callosum (not shown). (b) Human fetus (47 days post-fertilization) with cyclopia (arrow) and holoprosencephaly. (c) Mouse embryo at E10 showing expression of *sonic hedgehog* (*Shh*) mRNA detected by *in situ* hybridization. Note expression of *Shh* particularly beneath the ventral forebrain, a requirement for normal formation of the telencephalic vesicles. Abbreviations: di, diencephalon; m/mes, mesencephalon; met, metencephalon; my, myencephalon; r, rhombencephalon. Scale bars: (b) 2 mm; (c) 0.5 mm

Table 8.2 Genes implicated in holoprosencephaly*

Holoprosencephaly locus**	Gene	Gene function	Chromosome	Mouse model
HPE1	ND***	—	21q22.3	—
HPE2	*SIX3*	Transcription factor	2p21	—
HPE3	*SHH*	Signalling molecule	7q36	*Shh* knock-out
HPE4	*TGIF*	TGFβ-associated transcription factor	18p11.3	—
HPE5	*ZIC2*	Transcription factor	13q32	*Zic2* knock-out
HPE6	*ND*	—	2q37.1–q37.3	—
HPE7	*PATCHED*	SHH receptor	9q22.3	*Ptc1* knock-out
—	*GLI2*	Transcription factor	2q14	—
—	*TDGF1*	Nodal associated protein	3p23–p21	—

*For further details, see Ming and Muenke (2002) and Online Mendelian Inheritance in Man (OMIM; http://www.ncbi.nlm.nih.gov).
**HPE loci are sites of human genetic linkage detected originally from familial and/or cytogenetic studies. Although several have been subsequently associated with specific gene mutations, the HPE1 and HPE6 loci have not yet been associated with particular genes.
***Not determined.

with absence of midline structures such as the corpus callosum. In less severe forms (e.g. semi-lobar holoprosencephaly), the most noticeable abnormalities are often a reduction of midline craniofacial features, with close-set eyes (hypotelorism) or a single central incisor tooth.

Several chromosomal regions have been linked to holoprosencephaly, emphasising the heterogeneity of the defects, in keeping with the variable severity and nature of the clinical phenotype. During the past 10 years, mutations in specific genes have been identified in holoprosencephaly patients and families, with the result that the genetic basis of most of the chromosomal linkages in holoprosencephaly has now been elucidated (Table 8.2). Mutations at other genetic loci are being identified in specific groups of patients with holoprosencephaly. Several of the causative genes are transcription factors, for example, *ZIC2*, *SIX3* and *TGIF* (Ming and Muenke, 2002). These genes are known to be expressed in the embryonic brain and, in the case of *ZIC2*, holoprosencephaly is also found in the mouse loss-of-function mutant (Nagai *et al.*, 2000). Little is known, however, of the developmental mechanisms by which mutation of these genes leads to holoprosencephaly. In contrast, the finding of holoprosencephaly-causing mutations in genes that participate in the sonic hedgehog (SHH) signalling pathway has shed considerable light on the developmental mechanism of holoprosencephaly.

Neural induction and the role of SHH in dorso-ventral specification of the CNS An early demonstration of neural induction came from the identification of the amphibian 'organizer'. A region of the embryo situated at the dorsal lip of the blastopore was shown to induce a secondary body axis when transplanted to a non-midline region of a host embryo. The chick and mouse equivalents of the organizer ('Hensen's node' and the 'node', respectively) are situated at the cranial end of the primitive streak, and have similar inducing properties following grafting (Hemmati-Brivanlou and Melton, 1997). The organizer/node is a source of proteins, including noggin and follistatin, antagonists of bone morphogenetic proteins (BMPs). In amphibia, the main requirement for neural induction appears to be inhibition of BMP action, with noggin and follistatin diverting cell fate from the default pathway of non-neural ectoderm towards a neural fate (Hemmati-Brivanlou and Melton, 1997). Although the BMP system also applies to higher vertebrates, there is an earlier role for other neural inducing factors, in particular fibroblast growth factors (FGFs) and Wnts (Stern, 2002).

During gastrulation, the node retreats beneath the midline of the newly induced neural plate, leaving in its wake the notochord, a rod-like mesodermal structure. The notochord is a potent source of SHH (Figure 8.3c), a diffusible signalling protein that rapidly becomes established in a gradient of declining concentration from ventral to dorsal. SHH induces the midline floor plate of the newly formed neural tube, and subsequently is instrumental in patterning gene expression along the dorso-ventral axis of the neural tube (Jessell, 2000). In the forebrain, other genes, particularly bone morphogenetic proteins (BMPs), cooperate with SHH in this inductive process (Dale *et al.*, 1997). A series of genes, mostly encoding homeodomain-containing transcription factors, are either positively or negatively regulated by SHH at specific concentration thresholds, giving rise to a complex pattern of gene expression domains along the dorso-ventral axis. This patterning is subsequently transformed into a sequence of neuronal, and later glial, types that differentiate in response to the gene expression 'pre-pattern' in the neural tube (Jessell, 2000).

A lack of SHH influence (e.g. in the *Shh* knock-out mouse embryo) causes a lack of ventral midline structures in the neural tube. Genes normally expressed in the dorsal neural tube show expression domains that extend ventrally. In the embryonic forebrain, the lack of SHH secretion from the prechordal plate, the rostral 'continuation' of notochord-like cells, has a particularly early effect, detectable well before neural tube closure. The lack of ventral forebrain specification appears to adversely affect forebrain growth, so that *Shh* null embryos are characterized by an abnormally narrow forebrain territory (Chiang *et al.*, 1996). After neural tube closure, this narrow forebrain is incompatible with full telencephalic separation. Moreover, the absence of ventral forebrain structures precludes the separation of the prospective optic fields, leading to cyclopia (Figure 8.3b). Recent work has demonstrated that mutations in *GLI2*, a gene functioning downstream of *SHH*, are also associated with defects within the holoprosencephaly spectrum in humans (Roessler *et al.*, 2003), emphasizing the need for normal SHH signalling for full telencephalic development.

Neural tube defects: failure of the embryonic process of neurulation

Failure of neural tube closure results in malformations termed neural tube defects (Copp, 1999). In craniorachischisis (Figure 8.4a), the most severe type of neural tube

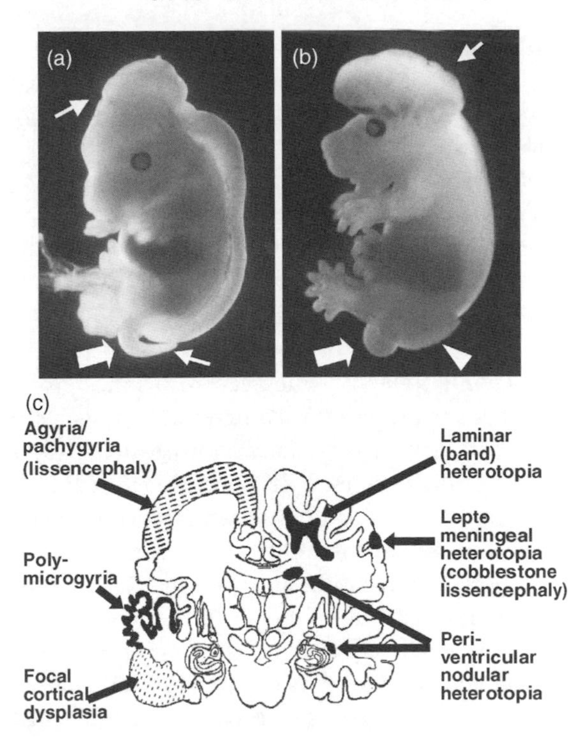

Figure 8.4 Neural tube defects and neuronal migration disorders. (a,b) Mouse fetuses at E15.5 to illustrate the appearance of craniorachischisis, in a *Celsr1* mutant (a) and exencephaly and open spina bifida, in a *curly tail* mutant (b). In craniorachischisis, the neural tube is open from midbrain to low spine (between the thin arrows in A). Exencephaly in the *curly tail* fetus is restricted to the midbrain (thin arrow in b), while the spina bifida affects the lumbosacral region (arrowhead in b). Note the presence of a curled tail in both fetuses (thick arrows in a and b). (c) The various types of neuronal migration disorder displayed diagrammatically on a coronal section of a postnatal human brain. The left side shows large-scale defects, while the right side shows typical focal lesions. See text for description of the different types of neuronal migration disorder. Parts (a) and (b) are reproduced with permission from Copp *et al.* (2003b) and part (c) from Copp and Harding (1999)

defect, the neural tube fails to close along most of the body axis, although the forebrain usually closes normally. If the neural tube fails to close specifically in the future brain, exencephaly results (i.e. exteriorization of the brain folds), which is converted to anencephaly (i.e. absence of brain) owing to neurodegeneration in later gestation. In contrast, if the low spine is primarily affected, this leads to open spina bifida (myelomeningocele; Figure 8.4b).

Related to these open CNS defects are a series of 'closed' defects, in which the neural tube and/or meninges herniate through an opening in the skull or in the neural arches of the vertebral column. Brain herniation yields a defect called encephalocele, while herniation of the spinal cord is termed meningocele. A further category of neural tube defects are so-called 'occult' spina bifida (also called spinal dysraphism), which mainly affect the low spinal region and are skin-covered lesions in which the spinal cord may be split or tethered to the surrounding tissues, often in association with a bony spur or a fatty mass (lipoma). While covered lesions are protected from the potentially toxic amniotic environment, open neural tube defect lesions undergo erosion of the exposed neuroepithelium so that, by the late stages of gestation, the region of affected nervous system is largely degenerate, leading to severe disability or death after birth. Surgery on the human fetus during pregnancy, with the aim of covering the neural tube defect lesion with muscle and skin, has shown that this process of degeneration can be halted (Johnson *et al.*, 2003), minimizing damage to the exposed CNS but not recovering function.

Genetic basis of neural tube defects The high recurrence risk in siblings and in close relatives of individuals with neural tube defects suggests a strong genetic basis, although there is a marked lack of large families with neural tube defects, arguing against causation based on single genes. It has been suggested, therefore, that neural tube defects have a multifactorial causation, with many genetic variants interacting to determine individual risk of neural tube defect, and with a marked contribution of environmental factors, both exacerbating and preventive. Since neural tube defects are relatively common malformations (Table 8.1), the predisposing genetic variants themselves seem likely to be relatively common, or else there may be many different combinations of rare genetic variants that can predispose to neural tube defect. In mice, more than 80 different mutant genes cause non-closure of the mouse neural tube, with different mutations affecting different rostro-caudal levels of the body axis, thereby mimicking the human situation (Copp *et al.*, 2003b). In contrast, only a few of the mouse mutants exhibit closed neural tube defects, for example resembling encephalocele. Despite the many genetic loci that have been implicated in mouse neural tube defects, few human genes have so far been definitively shown to predispose to human neural tube defects. The best known of these is the gene encoding 5,10-methylene tetrahydrofolate reductase (MTHFR), an enzyme of folic acid metabolism. MTHFR catalyses the reaction that produces 5-methyl tetrahydrofolate, a methyl donor for homocysteine during its conversion to methionine. A polymorphic, thermolabile variant of the *MTHFR* gene (the C677T variant) exhibits a higher frequency among neural tube defect cases and their families than among normal controls in several populations (Van der Put *et al.*, 1997) and seems responsible for imparting an increased risk of neural tube defect, especially in combination with a low folate and/or vitamin B_{12} level during pregnancy.

Environmental effects on occurrence of neural tube defects Many environmental factors have been demonstrated, in mice, to interact with the genotype to either increase or decrease the risk of neural tube defect (Copp *et al.*, 1990). Teratogenic (i.e. malformation increasing) influences range from physical factors such as hyperthermia to biologically active molecules such as retinoids (vitamin A derivatives). In humans, several of these agents are also suspected of increasing neural tube defect risk and the anti-epileptic drug sodium valproate, taken early in pregnancy, has conclusively been demonstrated to predispose to spina bifida (Lammer *et al.*, 1987). In contrast to these teratogenic influences, folic acid is well known to diminish the risk of neural tube defects in a proportion of predisposed human pregnancies (Wald *et al.*, 1991) and in several mouse mutants with neural tube defect. While studies in mice have demonstrated that folic acid acts directly on the developing embryo, its precise mode of action remains elusive. Genetic predisposition to neural tube defect via the C677T variant of *MTHFR* leads to elevated levels of homocysteine, a trend that is reversed by administration of exogenous folic acid. Homocysteine has not been found to directly cause neural tube defects in experimental animals, however, suggesting that other embryonic defects, such as diminished embryonic cell proliferation or excessive cell death, may be the primary target of folic acid in preventing neural tube defects. A proportion of neural tube defects in both humans and mice do not respond to folic acid therapy. In one folate-resistant mouse mutant, *curly tail*, administration of the vitamin-like molecule inositol can prevent the great majority of cases of spina bifida, through a molecular mechanism involving activation of specific isoforms of the enzyme family protein kinase C (Cogram *et al.*, 2004). It remains to be determined whether inositol will also prove to exert a preventive effect against human neural tube defects.

Embryonic mechanisms of neural tube defects In mice, the great majority of neural tube defects arise from non-closure of the neural tube during neurulation. Analysis of the types of mutant gene (especially gene knock-outs) that lead to mouse neural tube defects has highlighted several embryonic mechanisms that appear essential for closure of the neural tube. In some cases, experimental studies have confirmed the importance of these developmental mechanisms in neurulation.

Craniorachischisis. In this most severe neural tube defect, both cranial and spinal regions of the neural tube remain open (Figure 8.4a). The defect arises when the initial event of neural tube closure ('closure 1') fails at the hindbrain–cervical boundary. A small group of mouse mutant genes give rise to this neural tube defect and recent work has implicated these genes in the so-called 'planar cell polarity' signalling pathway, in which Wnt/frizzled signals are transduced by a β-catenin-independent mechanism (Copp *et al.*, 2003a). Hence, loss of function of *Vangl2* (also called *strabismus*), *Celsr1*, *Scrb1*, *Ptk7* and double mutants for *dishevelled-1* and *-2*, produce craniorachischisis in homozygous form. The planar cell polarity pathway is required for 'convergent extension', a net medially-directed movement of cells, with intercalation and rostro-caudal extension in the midline. Convergent extension fails in mice with planar cell polarity mutations, leading to short, broad embryos in which the neural folds are spaced widely apart. This wide spacing of the neural folds prevents closure 1 and causes craniorachischisis (Greene *et al.*, 1998).

Exencephaly and anencephaly. Many mutant genes and a large number of teratogens cause cranial neural tube defects in the mouse, with the neural tube failing to close in the future brain (Figure 8.4b). Analysis of the genetic models has revealed several critical events in cranial neurulation that are required for successful brain closure (Copp *et al.*, 2003b). The initial elevation of the cranial neural folds requires expansion of the cranial mesenchyme, as a result of cell proliferation and increase in extracellular space. This causes the elevating neural folds to adopt a bi-convex appearance, particularly in the midbrain. Mice with loss of function of the *Twist* or *Cart1* genes have cranial neural tube defects in which the principal defect is a reduction in the proliferation and expansion of the cranial mesenchyme (Chen and Behringer, 1995; Zhao *et al.*, 1996).

Once the bi-convex neural folds have formed, a second phase of cranial neurulation occurs in which the dorsolateral aspects of the neural fold bend medially, allowing the folds to adopt a bi-concave morphology and approach the dorsal midline for fusion. This second phase is highly dependent on the actin cytoskeleton, as illustrated by mice mutant for *shroom*, a gene involved in generating actin microfilaments, which fail to close their brains (Hildebrand and Soriano, 1999). The initiation of cranial neural crest emigration from the neural fold apices is also required, as shown by mice overexpressing *connexin 43,* which exhibit defects of cranial neural crest emigration and exencephaly (Ewart *et al.*, 1997). A third requirement for cranial closure is precise regulation of programmed cell death (apoptosis). Knock-out mice with either increased (e.g. *AP-2α*, *bcl10* and *Tulp1*) or decreased (e.g. *Apaf-1*, *caspase 9* and *p53*) apoptosis exhibit cranial neural tube defects (Copp *et al.*, 2003b). Apoptosis appears to synergize with neural crest cell emigration, to enable the conversion from bi-convex to bi-concave morphology. In addition, apoptosis at the neural fold tips may be necessary for midline epithelial remodelling, once the neural folds have met in the midline, since inhibition of apoptosis in the chick embryo prevents midline remodelling (Weil *et al.*, 1997). Cranial closure also requires precisely coordinated cell proliferation in the neural tube: mice mutant for *RBP-Jκ*, *Hes1* and *Numb* show premature differentiation of the neuroepithelium and failure of brain closure (Copp *et al.*, 2003b).

Open spina bifida. A number of mouse mutants exhibit low spinal neurulation defects leading to open spina bifida (Figure 8.4b). Here, the critical event appears to be regulation of dorsolateral bending of the neural plate. In contrast to the cranial region, dorsolateral bending in the spine does not require emigration of the neural crest (which begins after neurulation in the spine) or function of the actin cytoskeleton (Ybot-Gonzalez and Copp, 1999). Instead, sonic hedgehog (Shh) signalling appears critical for regulation of dorsolateral bending. Shh is produced by the notochord underlying the ventral neural plate and inhibits dorsolateral bending (Ybot-Gonzalez *et al.*, 2002). In the absence of Shh, for example in the *Shh* mutant mouse, dorsolateral bending occurs as a default mechanism that ensures spinal closure. Overstimulation of the Shh signalling pathway, on the other hand, is incompatible with spinal closure. Hence in the *Ptc1* and *Opb* mouse mutants, dorsolateral bending is absent and homozygotes fail to close their low spinal neural tube (Eggenschwiler and Anderson, 2000; Goodrich *et al.*, 1997). In contrast, the *curly tail* mutant does not lack dorsolateral bending but apposition of the neural folds is hampered, owing to ventral curvature of the caudal body axis (Brook *et al.*, 1991), secondary to an underproliferation of ventral cell types (Copp *et al.*, 1988), so that a proportion of homozygotes exhibit open spina bifida.

Regional brain disorders: rostro-caudal specification and divergence of CNS cell types

Early development of the nervous system is characterized by differentiation of the various cell types that are functionally appropriate for each level of the body axis. In some disorders of nervous system development there is reduction or even absence (aplasia) of whole structures, indicative of rostro-caudal specification defects, or maldevelopment of specific neuronal and glial cell types. These anomalies are seen particularly in some human cerebellar syndromes and has been described in a number of mouse genetic mutations affecting the cerebellum.

Rostro-caudal patterning of the nervous system During neural induction, the specific regional character of the neuroepithelium along the cranio-caudal axis is induced by an interaction between mesoderm and overlying ectoderm. In both amphibia and mice, anterior mesoderm can induce the expression of genes characterizing anterior neuroepithelium when co-cultured with posterior ectoderm that would not normally express these genes (Ang and Rossant, 1993). Regional specification has been studied in detail in the mouse hindbrain where, just prior to neural tube closure, neuroepithelial cells begin to express genes belonging to the *Hox* family of homeobox-containing genes. In general, *Hox* genes are expressed along much of the body axis but they have differing rostral boundaries of expression, with the boundary situated in some cases within the hindbrain. The hindbrain consists of six segments or rhombomeres, with each expressing a different combination of *Hox* genes. This has led to the suggestion that regional specification within the hindbrain may be determined by a '*Hox* code' (Hunt and Krumlauf, 1992). Evidence in support of this idea comes from the analysis of gastrulation stage mouse embryos treated with retinoic acid. The pattern of *Hox* gene expression is altered, causing cells of rhombomeres 2 and 3 to express Hoxb1, which is normally expressed only in rhombomeres 4 and 5. The regional character of rhombomeres 2 and 3 is altered so that they now give rise to a nerve resembling cranial nerve VII (facial) rather than cranial nerve V (trigeminal), as would normally occur. Thus, the facial nerve is duplicated in the retinoic acid-treated embryos (Marshall *et al.*, 1992). Further evidence for a *Hox* code comes from studies in which *Hox* genes are inactivated in transgenic mice. The most prominent abnormalities in these mice affect the neural crest and skeletal derivatives (see Chapter 15), but nervous system defects, particularly involving the cranial nerves, have also been described in mutations of *Hox* genes whose anterior expression boundary is located within the hindbrain (Capecchi, 1997).

Defects of rostro-caudal neural patterning in humans may be responsible for malformations such as cerebellar agenesis, in which an element in the rostro-caudal sequence of CNS structures is diminished or absent. This defect is similar to the agenesis of the midbrain and cerebellum observed in mice with a null mutation in the *Wnt-1* gene (McMahon and Bradley, 1990). Likewise, the gene *Krox-20* (Figure 8.1a) is needed for correct development of the cranial nerves originating in the hindbrain (Swiatek and Gridley, 1993). Another hindbrain anomaly, Dandy–Walker syndrome (absence of the cerebellar vermis and cystic dilatation of the fourth ventricle), has recently been shown to be related in some cases to genetic disorders of two members of the *ZIC* gene family, which encodes zinc finger transcription factors (Grinberg *et al.*, 2004). Joubert syndrome (absence of the cerebellar vermis combined with breathing and eye movement disorders) may also represent a developmental defect of rostro-caudal CNS patterning; a causative gene was recently identified (Ferland *et al.*, 2004).

Divergence of neuronal and glial cell lineages Much research effort has gone into defining the process of divergence of the neuronal and glial lineages in the developing cerebral cortex and cerebellum. Single neuroepithelial cells have been labelled by infection with a replication-defective retroviral vector containing a reporter gene such as bacterial β-galactosidase (*LacZ*). Correlation of the subsequent development of the labelled cell and its clonal descendants with their pattern of differentiation, as discerned by staining with antibodies specific for different neuronal and glial subsets, have shown that labelled clones most often contain only a single neuronal or glial cell type, suggesting that lineage specification occurs prior to the final mitotic division of neuroepithelial stem cells (Grove *et al.*, 1993; Walsh and Cepko, 1993). A recent surprising finding, however, has been that radial glial cells, previously thought to be already committed to a pathway of glial differentiation with an astrocytic fate, can in fact serve as early neuronal stem cells, although the extent of this stem cell function depends on brain region (Malatesta *et al.*, 2003). This suggests that neuroepithelial cells, upon withdrawing from the cell cycle, may pass through a radial glial stage, before embarking upon neuronal differentiation or becoming definitive glia. Clearly, there is still much to be learned about the fundamental aspects of cell lineage specification in the developing CNS.

Chimeric analysis of cell type-specific CNS defects in the mouse Several mouse mutations have been described in which particular neuronal cell types develop abnormally in the cerebellum. For example, cerebellar Purkinje cells are defective or degenerate in the *lurcher* and *Purkinje cell degeneration* (*pcd*) mutants, while the cerebellar granule cells are defective in the *staggerer* and *weaver* mutants (Mullen *et al.*, 1997). These defects characteriztically produce an ataxic gait, a phenotype that is readily identified in mice. An experimental approach to identifying the cell type in which the mutant gene acts in these disorders is provided by chimera analysis. Chimeras are individuals containing cells of two distinct genotypes as a result of experimental manipulation, for example the transplantation of cells of different genotypes. This is to be contrasted with mosaics, in which cells of two or more genotypes coexist as a result of spontaneous events during development, including mitotic recombination or X-inactivation. A useful approach is to create chimeras containing both mutant and wild-type cells, with the additional use of a marker that can distinguish between the two cell types. The brains of chimeras can then be assessed to determine whether the cellular abnormality is corrected in genetically mutant cells (indicating a non-cell-autonomous defect) or whether mutant cells persist in developing abnormally in chimeras despite a partially wild-type environment (indicating a cell-autonomous defect). Studies of this sort identified a cell-autonomous defect of Purkinje cells in the *pcd* mutation, whereas aberrant migration of Purkinje cells in *reeler* homozygotes was rescued in chimeras, suggesting a non-autonomous defect involving trophic support from neighbouring cells (Mullen and Herrup, 1979). Abnormal migration of Purkinje and granule cells to regions outside the cerebellum in *Unc5h3* mutants, which lack a receptor for the netrin-1 ligand, were shown in chimeras to depend on the pioneering influence of the granule cells, not the Purkinje neurons (Goldowitz *et al.*, 2000).

Neurocristopathies: origin and migration of the neural crest

The neural crest differentiates as the dorsal-most cell type of the CNS. A complex set of molecular interactions, involving BMPs, Wnts and other signalling molecules, specifies the neural crest as a 'boundary' cell type at the junction between the neural and non-neural ectoderm (Knecht and Bronner-Fraser, 2002). Shortly thereafter, cells of the neural crest begin to emigrate from the neural fold, migrating ventrolaterally or ventromedially to participate in the formation of a variety of tissue derivatives. Non-neural structures formed by the neural crest include many of the skeletal elements of the head (Chapter 12), the aorto-pulmonary septum of the cardiac outflow tract (Chapter 13), the pigmentary cells of the skin and inner ear (Chapter 10) and the thymus, thyroid and parathyroid glands. Abnormalities of these and other neural crest-derived body structures are termed 'neurocristopathies' and are exemplified by Waardenburg syndrome, one type of which is caused by mutations of the *PAX3* gene (Read and Newton, 1997). Neural crest cells are deficient in heterozygotes, leading to craniofacial anomalies, pigmentary defects and sensorineural hearing loss. Mice with *Pax3* mutations show a similar range of defects with, in addition, failure of septation of the cardiac outflow tract, producing the defect 'common arterial trunk' in homozygotes (Conway *et al.*, 1997).

Much of the peripheral nervous system is also derived from the neural crest, including the cranial and spinal ganglia, sympathetic chain, and the 'enteric nervous system', the network of neuronal ganglia and connections that control the motility of the gut. Defects of neural crest colonization of the gut, which give rise to the congenital defect Hirschsprung's disease, are described in Chapter 10.

Neuronal migration disorders

The neuronal migration disorders are a heterogeneous group of congenital brain defects (Figure 8.4c). Their variable time of clinical presentation makes it difficult to gain an accurate estimate of incidence but, in total, they are likely to be relatively

Table 8.3 Genes implicated in neuronal migration disorders*

Neuronal migration disorder	Gene	Gene function	Chromosome
Lissencephaly (Miller–Dieker and isolated lissencephaly sequence)	*LIS1*	Microtubule activating protein	17p13.3
X-linked lissencephaly (band heterotopia)	*DCX* (Double-cortin)	Microtubule stabilizing protein	Xq22.3–q23
X-linked lissencephaly with ambiguous genitalia	*ARX*	Homeobox transcription factor	Xp22.13
Autosomal recessive lissencephaly	*RELN* (Reelin)	Extracellular matrix signalling protein	7q22
Cobblestone (Type 2) lissencephaly (Fukuyama muscular dystrophy)	*FCMD* (Fukutin)	Enzyme modifying cell surface glycoproteins	9q31
Cobblestone (Type 2) lissencephaly (Walker–Warburg syndrome)	*POMT1*	*o*-Mannosyl transferase enzyme	9q34.1
Nodular periventricular heterotopia	*FLNA* (Filamin-A)	Actin binding protein	Xq28

*For further details, see Mochida and Walsh (2004) and Online Mendelian Inheritance in Man (OMIM; http://www.ncbi.nlm.nih.gov).

common defects. Table 8.3 summarizes current knowledge of the genetic causation of these disorders (see also (Barkovich *et al.*, 2001; Mochida and Walsh, 2004)).

Brain lamination defects – large-scale disorders of neuronal migration Large-scale disorders of neuronal migration are typified by 'lissencephaly' (smooth brain), in which the normal six-layered cerebral cortical structure is absent, with a thickened four-layered structure present instead. The lissencephalic brain lacks the normal folded (gyral) surface of the postnatal brain, exhibiting pachygyria (few, thick folds) or even agyria (no folds). This disorder is likely to result from severe disruption of the migration of neuroblasts to the developing cerebral cortex, with failure of the sequential population of the cortical plate, which generates the normal six-layered structure. Miller–Diecker lissencephaly is the commonest of these large-scale neuronal migration disorders and is an important cause of mental retardation and severe epilepsy. Many cases result from mutation of the *LIS1* gene, which encodes a protein that functions as part of a multi-protein complex, together with its binding partners NUDE-L and dynein heavy chain, to regulate the function of microtubules within the migrating neuroblasts (Shu *et al.*, 2004).

A second type of large-scale lamination disorder is X-linked lissencephaly, which is caused by mutations of the gene *Doublecortin* (Gleeson *et al.*, 1998; Des Portes *et al.*, 1998). In this condition, which is often familial, affected females exhibit apparent duplication of the cerebral cortex (so-called 'double cortex'), whereas the brain of affected male siblings is typically lissencephalic. This sex difference in phenotype is due to the X-chromosome-linked nature of the *Doublecortin* gene. Females inactivate one X-chromosome at random during early development, so females heterozygous for a *Doublecortin* mutation will have on average 50% of neuroblasts expressing a normal copy of the *Doublecortin* gene, while the other 50% express the mutated form. These genetically different populations of neuroblasts are thought to behave 'cell autonomously': those with normal *Doublecortin* function migrate to form a normal six-layered cerebral cortex, while those with mutant function migrate abnormally, forming an inner defective cortical layer. Male brains, in contrast, are formed solely by neuroblasts expressing mutant *Doublecortin*, and so develop an entirely abnormal, lissencephalic brain. The *Doublecortin* gene appears to function in the stabilization of microtubules at the extremities of growing neuronal processes, possibly via interaction with the LIS-1 protein (Schaar *et al.*, 2004; Tanaka *et al.*, 2004).

Several other disorders of cortical lamination are recognized in humans, some of which are becoming understood in terms of causation (Table 8.3). In mice, a particularly well-studied abnormality is the disturbed lamination of the cerebral cortex and cerebellum seen in homozygotes for the *reeler* mutation, in which the mutant phenotype involves an apparent reversal of the polarity of the normal cortical layers. *Reeler* neuroblasts begin their centrifugal migration along radial glial fibres but are unable to pass post-migratory neurons in the deeper cortical layers (Pinto-Lord *et al.*, 1982). The *reeler* mutation affects an extracellular matrix molecule, 'reelin', which exhibits similarities to molecules involved in cell-matrix adhesion, such as tenascin (D'Arcangelo *et al.*, 1995). Reelin is expressed by neuroblasts but not radial glia, supporting the idea that the reeler phenotype results from a defect in adhesion between early post-migratory neurons. Mice with mutations in the *mdab1* gene, and in genes encoding the very low density lipoprotein (VLDL) receptor and apolipoprotein E (ApoE) receptor 2, all show very similar phenotypes to reeler. Molecular studies demonstrate that VLDL receptor and ApoE receptor 2 bind the reelin protein, whereas the mdab1 protein acts intracellularly to transduce the reelin signal (D'Arcangelo *et al.*, 1999; Hiesberger *et al.*, 1999). Hence, reelin appears to regulate an important genetic pathway controlling neuroblast migration during CNS development.

Neuronal heterotopias – localized defects of neuronal migration At the mild end of the spectrum of neuronal migration disorders are conditions in which groups of neurons are situated in abnormal positions, particularly within the cerebral hemispheres, hippocampus and cerebellar cortex. They can comprise relatively discrete islands containing small numbers of neurons, which may be asymptomatic or can be associated with epileptic foci: surgical specimens of excised brain tissue from individuals with intractable seizure disorders often contain heterotopic neurons. Heterotopias may also be larger-scale defects and are often associated with complex maldevelopments of the cerebral cortex, such as polymicrogyria (increased numbers of poorly formed cortical folds) or with lissencephaly.

Neurons may fail to initiate migration, as in 'periventricular' heterotopia, in which groups of neurons are abnormally positioned close to the ventricular cavities. X-linked periventricular heterotopia results from mutations in the gene encoding Filamin-A, an actin-cross-linking phosphoprotein that interacts with Filamin-B and Filamin-A-interacting protein (FILIP) to regulate the actin reorganization necessary for neuronal migration (Sheen *et al.*, 2002; Nagano *et al.*, 2004). When the function of this protein complex is disturbed, neuroblasts are unable to initiate migration.

Neuroblasts may also 'over-migrate', as in conditions characterized by the presence of heterotopic neurons in the marginal zone, or even beneath the meninges on the surface of the brain (so-called 'cobblestone' lissencephaly or type II lissencephaly). Overmigration of neurons normally destined for layers II and III appears to arise from disturbance of the interaction between the end-feet of the radial glial fibres and the glial limiting membrane, an extracellular matrix layer that marks the edge of the cortex structure. Two severe human conditions with 'overmigration' heterotopias are Fukuyama muscular dystrophy and Walker–Warburg syndrome. In the former, apart from the muscular defects, patients also exhibit abnormalities of the cortical glial-limiting membrane (Yamamoto *et al.*, 1997). The gene mutated in Fukuyama muscular dystrophy, *fukutin*, appears to be an enzyme modifying cell surface glycoproteins and glycolipids (Muntoni *et al.*, 2004). Similarly, a disorder of protein glyosylation has been identified in families with Walker–Warburg syndrome (Beltran-Valero *et al.*, 2002).

Animal models provide further evidence for a role of the glial-limiting membrane in the pathogenesis of marginal zone heterotopias. Breaches of the glial-limiting membrane have been described in mice with autoimmune conditions predisposing to marginal zone heterotopias (Sherman *et al.*, 1990) and homozygotes for a null mutation of the *Marcks* gene (myristoylated alanine-rich kinase C substrate), in which leptomeningeal heterotopias are observed (Blackshear *et al.*, 1997). In the *dreher* mouse, mutation of the *LIM* homeobox gene *Lmx1a*, leads to a phenotype resembling cobblestone lissencephaly. Birth-dating studies have demonstrated that heterotopic neurons in the marginal zone are indeed overmigrated layer II and III cells in *dreher* brains (Costa *et al.*, 2001).

Microcephaly and megalencephaly: regulation of CNS cell number

Brain size is closely similar in all members of a species, implying a close regulation of brain cell number through cell proliferation and programmed cell death (apoptosis). A small brain is a common finding in adult neurological conditions, usually because of neurodegeneration secondary to a disease process. Reduced brain size can also be a congenital defect, however, as in microcephaly (strictly 'micrencephaly'), where reduced brain mass is often accompanied by mental retardation. Some cases of micrencephaly have been suggested to result from a reduction in the number of cell divisions undergone within the ventricular zone of the neural tube, with early onset of neuronal differentiation and premature formation of the ependyma. While the full complement of neuronal types is generated, total neuronal cell number is reduced (Woods *et al.* 2005). Conversely, a small brain could be generated by excessive apoptosis in the presence of normal neural tube cell proliferation, although there is little evidence to implicate this mechanism. Reduced apoptosis, on the other hand, seems likely to be an important factor in the generation of congenital defects characterized by excessive brain tissue. For example, in the condition megalencephaly, a major part or the whole of the brain is of massive size, possibly as a result of reduced programmed cell death during neurogenesis. In support of this idea, the targeted inactivation of genes involved in programmed cell death, for example intracellular 'caspase' enzymes, has led to the generation of mutant mice with increased amounts of brain tissue (Kuan *et al.*, 2000; Kuida *et al.*, 1996).

Molecular regulation of cell proliferation and cell death in the developing CNS Major decisions facing cells of the early neural tube are first, whether to continue proliferating or to embark upon neuronal/glial differentiation, and second, whether to survive or undergo programmed cell death. In both cases, considerable information is now available on the molecular regulatory mechanisms.

The continuation or cessation of proliferation in neuroepithelial cells is regulated in part by members of the *Notch* gene family, encoding a group of cell surface receptors that are activated by binding their ligands, encoded by members of the *Delta* gene family (Gaiano and Fishell, 2002). Both Notch and Delta proteins are expressed throughout the early neural tube, consistent with the almost entirely proliferative nature of the early neuroepithelium. As development proceeds, however, additional sets of genes become expressed which encourage neural tube cells to leave the cell cycle and embark upon a programme of neuronal, or later glial, differentiation. These 'proneural' genes include *NeuroD*, *neurogenin* and *Mash1*, with further groups of genes, for example, *Hes1*, acting to transduce the signals from the proneural genes (Kageyama *et al.*, 2005). Different sets of genes promote glial differentiation, for example, *Olig2* and *Nkx-2.1*, which promote oligodendrocyte differentiation, sequential to motor neuron producton, in the ventral spinal cord (Kessaris *et al.*, 2001). The function of proneural gene signalling is to encourage neuroepithelial cells to undergo an asymmetric division, to generate one post-mitotic, differentiating daughter cell as well as a proliferative daughter that continues in the cell cycle (Figure 8.2c). This asymmetric type of division contrasts with earlier neural tube cell proliferation, in which all divisions are symmetrical, with both daughter cells continuing to proliferate. The post-mitotic daughters rapidly lose contact with the ventricular zone of the neural tube, moving to the marginal zone, where they undergo further differentiative decisions that determine their ultimate fate as terminally differentiated neurons or glial cells. Hence, while the early neural tube is 'Notch-dominated' and largely proliferative, the influence of 'proneural' genes gradually increases until, by the end of CNS development, the great majority of neural tube cells have left the cell cycle and differentiated.

All cells are probably programmed to die by apoptosis, being kept alive only by the constant presence of survival factors in the extracellular environment

(Jacobson *et al.*, 1997). Teratogenic agents such as ethanol, which are capable of inducing CNS defects (e.g. the fetal alcohol syndrome), appear to trigger the apoptotic pathway by inhibiting neurotransmitter receptors (Ikonomidou *et al.*, 2000). The decision, in normal development, of whether to die or survive is regulated by opposing intracellular molecular events involving pro-apoptotic and anti-apoptotic genes (Hengartner, 2000). Neuronal survival factors, particularly nerve growth factor (NGF), brain-derived neurotrophic factor (BDNF) and neurotrophin-3, are extracellular molecules that bind to and activate members of the tyrosine kinase receptor (Trk) family (Huang and Reichardt, 2001). Intracellular signalling cascades are initiated that culminate in the upregulation of anti-apoptotic members of the *bcl* gene family, such as *bcl-2* and *bcl-x*, while inhibiting expression of pro-apoptotic genes, such as *Bax* and *Bad*. This intracellular regulation appears largely mediated at the level of gene transcription, involving the interplay of many regulatory elements within the cell. Withdrawal of neurotrophic support for neurons reverses the balance, so that pro-apoptotic influences become dominant, activating downstream caspase enzymes that initiate a molecular cascade culminating in apoptosis.

Clearly, there are many opportunities within this vast array of regulatory interactions where cell proliferation/differentiation and survival/death can be disrupted. Global effects are likely to be early embryonic lethal, as has been demonstrated in mouse knock-outs of many of the genes mentioned above. More subtle, region-specific alterations may ultimately be implicated in the generation of specific anomalies of CNS development in humans.

Malformations of fibre tracts in the CNS: the development of neuronal connections

Abnormalities of the large CNS fibre tracts are well-known birth defects recognized by virtue of their neurological sequelae or, in some cases, by their structural effects, for example on magnetic resonance imaging (MRI). More subtle anomalies of neuronal connection may also be common defects, but difficulty in detection prevents their true frequency from being realized.

Agenesis of the corpus callosum, the large white matter tract that connects the cerebral hemispheres, is a common congenital anomaly (Barkovich, 2002) that, by itself, has only minor effects on neurocognitive function. Its main significance is to alert the clinician to the likely presence of other brain malformations, which are frequently associated with this defect. Agenesis of the optic chiasm, the site of midline crossing of sensory fibres that connect the retina to the midbrain and visual cortex, is a further well-recognized congenital CNS defect (Guillery *et al.*, 1995). Stereoscopic vision depends on the fibre crossing at the chiasm, and total agenesis of the optic chiasm is associated with visual disturbance, particularly nystagmus (rhythmic eye oscillations). The functionally most significant anomaly of CNS fibre tracts involves the corticospinal tracts, which carry fibres from the motor cortex through the brainstem, where midline crossing occurs, to synapse on the motor neurons of the spinal cord. Defects of the corticospinal tracts, although often not structurally detectable on MRI, are a common cause of cerebral palsy, in which motor control of the legs (diplegia), legs and arms (quadriplegia) or one side of the body (hemiplegia) is severely compromised. Although cerebral palsy has been traditionally considered an injury caused by prolonged hypoxia at birth, there is considerable evidence that a sizeable proportion of cases have their origin in developmental defects during pregnancy (Ferriero, 2004).

Molecular regulation of axonal pathfinding and the establishment of neuronal connections Recent years have seen major advances in our understanding of the molecular mechanisms regulating the growth and guidance of axons towards their targets and the establishment of synaptic connections. The axon growth cone (Figure 8.1f) is subject to a myriad influences, both attractive and repulsive, either long- or short-range, which guide it during its journey to its target.

Considering first long-range influences, a series of ligand–receptor systems have been described that function in particular axon guidance systems, but not in others (Dickson, 2002). For example, netrin-1 is a diffusible protein that binds a receptor, 'deleted in colorectal cancer' (DCC), which is expressed on axon growth cones. This interaction was originally described as mediating the chemoattraction of commissural axons approaching the floor plate of the spinal cord. However, it was subsequently found that exposure to netrins within an extracellular environment rich in laminin converts the netrin influence to chemorepulsion (Höpker *et al.*, 1999), providing versatility in the action of this axon guidance ligand in different cellular contexts. Similarly, interaction of the intracellular portion of the DCC receptor with that of another netrin receptor, UNC5, also converts the influence of netrin-1 to chemorepulsion (Hong *et al.*, 1999). Further work has identified additional ligand–receptor interactions that mediate chemoattraction and chemorepulsion, depending on cellular context. For example, Slit proteins are diffusible ligands that bind members of the Robo family of cell surface receptors. Slit–Robo interactions can be chemoattractive or chemorepulsive in the guidance of olfactory and other axons (Dickson, 2002).

Local interactions also play an important role in guiding axon growth cones. The cellular and extracellular environment through which the axon migrates is rich in molecular cues that provide guidance information. For example, analysis of retinal ganglion cell axon guidance in the chick optic tectum (equivalent to the inferior colliculus of the mammalian visual pathway) has identified a key chemorepulsive role for interactions between Eph receptors expressed on ganglion cell axons and their ephrin ligands, which are expressed on cells of the optic tectum (McLaughlin *et al.*, 2003). Ganglion cells originating from the temporal retina express Eph receptors to a greater extent than ganglion cells from the nasal retina. Moreover, cells of the posterior optic tectum express ephrins more strongly than anterior cells. As retinal ganglion cell axons arrive in the tectum, temporal axons terminate preferentially in the anterior tectum, owing to their repulsion from the posterior tectum, whereas nasal axons are not similarly repelled and are able to synapse in the posterior tectum. This 'sorting out' of retinal axon terminations according to their site of origin generates a precise topographical map of the retina on the brain, a prerequisite for detailed vision. Many other analogous examples of axon guidance by local chemoattractive and chemorepulsive cues have now been described, giving rise to the idea that the complexity of connections in the CNS may, to some extent, reflect the diversity of guidance cues confronting growing axons.

Molecular basis of fibre tract malformations in the CNS Mouse genetic models are available for the analysis of CNS fibre tract malformations. For example, mice with null mutations in the genes *Emx1* (*Drosophila* empty spiracle orthologue), *Hesx1*, *Mrp* (Marcks-related protein) and *Nfia* (nuclear factor I-A) all exhibit agenesis of the corpus callosum (Wu *et al.*, 1996; Qiu *et al.*, 1996; Dattani *et al.*, 1998; Shu *et al.*, 2003). However, although some progress has been made in defining the neuronal populations that send out callosal fibres, there is as yet no definitive evidence on the molecular mechanisms underlying agenesis of the corpus callosum anomaly.

More progress has been made in understanding the guidance cues that specify the position, and the nature of axon crossing or non-crossing of the midline at the optic chiasm (Rasband *et al.*, 2003). The site of the chiasm appears to be specified by a combination of gene expression domains in the ventral diencephalon, particularly involving the Slit–Robo signalling system. Optic tract fibres arising from the ventrotemporal part of the retina express EphB receptors and are repelled from crossing by the expression of ephrinBs at the chiasm, ensuring ipsilateral projections. On the other hand, retinal ganglion cells in other parts of the retina do not express EphB receptors and their axons are able to cross, forming the contralateral projection. Expression of *Zic2* by retinal ganglion cells is also required for the ipsilateral projection (Herrera *et al.*, 2003), perhaps suggesting regulation of EphB expression on retinal ganglion cells by Zic2. Hence, a set of interactions between the retinal ganglion cell axons and the local environment of the chiasm determines the normal pattern of crossing and non-crossing fibres. Finally, in relation to the corticospinal tracts, a requirement for an interaction between midline ephrinB3 and EphA4 expressed on cortical motor neurons has been defined in preventing recrossing of the midline by corticospinal fibres (Yokoyama *et al.*, 2001; Kullander *et al.*, 2001; Dottori *et al.*, 1998). This ensures that the crossing (decussation) of fibres in the brainstem is total, and so enables unilateral, crossed motor control to be established. Determining the molecular basis of congenital disturbance of the corticospinal tracts, which leads to cerebral palsy, is a major challenge for future research.

Agenda for the future

In the previous edition of this book, completed in 1995, I ended the chapter with the words: 'there is hardly a single topic in the development of nervous system malformations that does not require further in-depth study'. Nearly 10 years on, and notwithstanding the enormous advances that have been made in understanding many fundamental aspects of the molecular regulation of CNS development, I am forced to draw the same conclusion. There has been an explosion of information on the genetic causation of CNS birth defects, in particular holoprosencephaly and neuronal migration disorders. Moreover, mouse models are now widely available, mostly as a result of gene targeting, to facilitate the study of CNS malformations. We are still lacking basic information, however, of the early development of most types of congenital CNS pathology and, in particular, we need to apply findings from the molecular regulation of normal development to CNS birth defects. Transferring this knowledge to the human embryo and fetus will be a difficult step to take, in view of the ethical and practical limitations on the use of early stages of human development. However, with careful use of the limited resources available for study of early human CNS development (Lindsay and Copp, 2005), such studies are certainly feasible. This work should provide important new information on the underlying mechanisms of CNS defects, which may in turn open new avenues for improved diagnosis and, most importantly, therapeutic interventions to prevent these disabling malformations.

Acknowledgements

I am extremely grateful to the following colleagues who allowed me to reproduce their unpublished material: Dr Nicholas Greene (Figure 8.1a), Dr Simon Conway (Figure 8.1b), Dr Andrew Stoker (Figure 8.1f) and Dr Cristina Costa (Figures 8.2c, 8.2d and 8.3a).

References

Allendoerfer, K.L. and Shatz, C.J. (1994) The subplate, a transient neocortical structure: its role in the development of connections between thalamus and cortex. *Annu. Rev. Neurosci.* **17**: 185–218.

Alvarado-Mallart, R.-M., Martinez, S. and Lance-Jones, C.C. (1990) Pluripotentiality of the 2-day-old avian germinative neuroepithelium. *Dev. Biol.* **139**: 75–88.

Ang, S.-L. and Rossant, J. (1993) Anterior mesendoderm induces mouse *Engrailed* genes in explant cultures. *Development* **118**: 139–149.

Barkovich, A.J. (2002) Magnetic resonance imaging: role in the understanding of cerebral malformations. *Brain Dev.* **24**: 2–12.

Barkovich, A.J., Kuzniecky, R.I., Jackson, G.D., Guerrini, R. and Dobyns, W.B. (2001) Classification system for malformations of cortical development: update 2001. *Neurology* **57**: 2168–2178.

Beltran-Valero, D.B., Currier, S., Steinbrecher, A., Celli, J. *et al.* (2002). Mutations in the O-mannosyltransferase gene *POMT1* give rise to the severe neuronal migration disorder Walker–Warburg syndrome. *Am. J. Hum. Genet.* **71**: 1033–1043.

Blackshear, P.J., Silver, J., Nairn, A.C., Sulik, K.K. *et al.* 1997. Widespread neuronal ectopia associated with secondary defects in cerebrocortical chondroitin sulfate proteoglycans and basal lamina in MARCKS-deficient mice. *Exp. Neurol.* **145**: 46–61.

Brook, F.A., Shum, A.S.W., Van Straaten, H.W.M. and Copp, A.J. (1991) Curvature of the caudal region is responsible for failure of neural tube closure in the curly tail (*ct*) mouse embryo. *Development* **113**: 671–678.

Capecchi, M.R. (1997) *Hox* genes and mammalian development. *Cold Spring Harbor Symp. Quant. Biol.* **62**: 273–281.

Chen, Z.-F. and Behringer, R.R. (1995) *twist* is required in head mesenchyme for cranial neural tube morphogenesis. *Genes Dev.* **9**: 686–699.

Chiang, C., Litingtung, Y., Lee, E., Young, K.E. *et al.* (1996) Cyclopia and defective axial patterning in mice lacking *Sonic hedgehog* gene function. *Nature* **383**: 407–413.

Cogram, P., Hynes, A., Dunlevy, L.P.E., Greene, N.D.E. and Copp, A.J. (2004) Specific isoforms of protein kinase C are essential for prevention of folate-resistant neural tube defects by inositol. *Hum. Mol. Genet.* **13**: 7–14.

Conway, S.J., Henderson, D.J., Anderson, R.H., Kirby, M.L. and Copp, A.J. (1997) Development of a lethal congenital heart defect in the *splotch* (*Pax3*) mutant mouse. *Cardiovasc. Res.* **36**: 163–173.

Copp, A.J. (1999) Neural tube defects. In *Encyclopaedia of Life Sciences*. Nature Publishing Group: London; www.els.net

Copp, A.J., Brook, F.A., Estibeiro, J.P., Shum, A.S.W. and Cockroft, D.L. (1990) The embryonic development of mammalian neural tube defects. *Prog. Neurobiol.* **35**: 363–403.

Copp, A.J., Brook, F.A. and Roberts, H.J. (1988) A cell-type-specific abnormality of cell proliferation in mutant (*curly tail*) mouse embryos developing spinal neural tube defects. *Development* **104**: 285–295.

Copp, A.J., Greene, N.D.E. and Murdoch, J.N. (2003a) Dishevelled: linking convergent extension with neural tube closure. *Trends Neurosci.* **26**: 453–455.

Copp, A.J., Greene, N.D.E. and Murdoch, J.N. (2003b) The genetic basis of mammalian neurulation. *Nat. Rev. Genet.* **4**: 784–793.

Copp, A.J. and Harding, R.N. (1999) Neuronal migration disorders in humans and in mouse models – an overview. *Epilepsy Research* **36**: 133–141.

Costa, C., Harding, B.N. and Copp, A.J. (2001) Neuronal migration defects in the *dreher* (*Lmx1a*) mouse mutant: role of disorders of the glial limiting membrane. *Cereb. Cortex* **11**: 498–505.

D'Arcangelo, G., Homayouni, R., Keshvara, L., Rice, D.S. *et al.* (1999) Reelin is a ligand for lipoprotein receptors. *Neuron* **24**: 471–479.

D'Arcangelo, G., Miao, G.G., Chen, S.-C., Soares, H.D. *et al.* (1995) A protein related to extracellular matrix proteins deleted in the mouse mutant reeler. *Nature* **374**: 719–723.

Dale, J.K., Vesque, C., Lints, T.J., Sampath, T.K. *et al.* (1997) Cooperation of BMP7 and SHH in the induction of forebrain ventral midline cells by prechordal mesoderm. *Cell* **90**: 257–269.

Dattani, M.T., Martinez-Barbera, J.P., Thomas, P.Q., Brickman, J.M. *et al.* (1998) Mutations in the homeobox gene *HESX1/Hesx1* associated with septo-optic dysplasia in human and mousc. *Nat. Genet.* **19**: 125–133.

de Vries, J.I.P. (1992) The first trimester. In *Fetal Behaviour, Developmental and Perinatal Aspects*, Nijhuis, J.E. (ed.). Oxford University Press: Oxford, 1992; 3–16.

Des Portes, V., Pinard, J.M., Billuart, P., Vinet, M.C. *et al.* (1998) A novel CNS gene required for neuronal migration and involved in X-linked subcortical laminar heterotopia and lissencephaly syndrome. *Cell* **92**: 51–61.

Dickson, B.J. (2002) Molecular mechanisms of axon guidance. *Science* **298**: 1959–1964.

Dottori, M., Hartley, L., Galea, M., Paxinos, G. *et al.* (1998) EphA4 (Sek1) receptor tyrosine kinase is required for the development of the corticospinal tract. *Proc. Natl Acad. Sci. USA* **95**: 13248–13253.

Eggenschwiler, J.T. and Anderson, K.V. (2000) Dorsal and lateral fates in the mouse neural tube require the cell-autonomous activity of the *open brain* gene. *Dev. Biol.* **227**: 648–660.

Ewart, J.L., Cohen, M.F., Meyer, R.A., Huang, G.Y. *et al.* (1997) Heart and neural tube defects in transgenic mice overexpressing the Cx43 gap junction gene. *Development* **124**: 1281–1292.

Ferland, R.J., Eyaid, W., Collura, R.V., Tully, L.D. *et al.* (2004) Abnormal cerebellar development and axonal decussation due to mutations in *AHI1* in Joubert syndrome. *Nature Genet.* **36**: 1008–1013.

Ferriero, D.M. (2004) Neonatal brain injury. *N. Engl. J. Med.* **351**: 1985–1995.

Gaiano, N. and Fishell, G. (2002) The role of Notch in promoting glial and neural stem cell fates. *Annu. Rev. Neurosci.* **25**: 471–490.

Gammill, L.S. and Bronner-Fraser, M. (2003) Neural crest specification: migrating into genomics. *Nat. Rev. Neurosci.* **4**: 795–805.

Gleeson, J.G., Allen, K.M., Fox, J.W., Lamperti, E.D. *et al.* (1998) *Doublecortin*, a brain-specific gene mutated in human X-linked lissencephaly and double cortex syndrome, encodes a putative signaling protein. *Cell* **92**: 63–72.

Goldowitz, D., Hamre, K.M., Przyborski, S.A. and Ackerman, S.L. (2000) Granule cells and cerebellar boundaries: Analysis of *Unc5h3* mutant chimeras. *J. Neurosci.* **20**: 4129–4137.

Goodrich, L.V., Milenkovic, L., Higgins, K.M. and Scott, M.P. (1997) Altered neural cell fates and medulloblastoma in mouse *patched* mutants. *Science* **277**: 1109–1113.

Greene, N.D.E., Gerrelli, D., Van Straaten, H.W.M. and Copp, A.J. (1998) Abnormalities of floor plate, notochord and somite differentiation in the *loop-tail* (*Lp*) mouse: a model of severe neural tube defects. *Mech. Dev.* **73**: 59–72.

Grinberg, I., Northrup, H., Ardinger, H., Prasad, C. *et al.* (2004) Heterozygous deletion of the linked genes *ZIC1* and *ZIC4* is involved in Dandy-Walker malformation. *Nat. Genet.* **36**: 1053–1055.

Grove, E.A., Williams, B.P., Li, D.-Q., Hajihosseini, M. *et al.* (1993) Multiple restricted lineages in the embryonic rat cerebral cortex. *Development* **117**: 553–561.

Guillery, R.W., Mason, C.A. and Taylor, J.S. (1995) Developmental determinants at the mammalian optic chiasm. *J. Neurosci.* **15**: 4727–4737.

Harding, B.N. and Copp, A.J. (2002) Congenital malformations. In *Greenfield's Neuropathology, 7th edn*, Graham, D.I., Lantos, P.L. (eds). Arnold: London, 2002; 357–483.

Hemmati-Brivanlou, A. and Melton, D. (1997) Vertebrate neural induction. *Annu. Rev. Neurosci.* **20**: 43–60.

Henderson, D.J., Conway, S.J., Greene, N.D.E., Gerrelli, D. *et al.* (2001) Cardiovascular defects associated with abnormalities in midline development in the *loop-tail* mouse mutant. *Circ. Res.* **89**: 6–12.

Hengartner, M.O. (2000) The biochemistry of apoptosis. *Nature* **407**: 770–776.

Herrera, E., Brown, L., Aruga, J., Rachel, R.A. *et al.* (2003) *Zic2* patterns binocular vision by specifying the uncrossed retinal projection. *Cell* **114**: 545–557.

Hiesberger, T., Trommsdorff, M., Howell, B.W., Goffinet, A. *et al.* (1999) Direct binding of Reelin to VLDL receptor and ApoE receptor 2 induces tyrosine phosphorylation of disabled-1 and modulates tau phosphorylation. *Neuron* **24**: 481–489.

Hildebrand, J.D. and Soriano, P. (1999) Shroom, a PDZ domain-containing actin-binding protein, is required for neural tube morphogenesis in mice. *Cell* **99**: 485–497.

Hong, K.S., Hinck, L., Nishiyama, M., Poo, M.M. *et al.* (1999) A ligand-gated association between cytoplasmic domains of UNC5 and DCC family receptors converts netrin-induced growth cone attraction to repulsion. *Cell* **97**: 927–941.

Höpker, V.H., Shewan, D., Tessier-Lavigne, M., Poo, M.M. and Holt, C. (1999) Growth-cone attraction to netrin-1 is converted to repulsion by laminin-1. *Nature* **401**: 69–73.

Huang, E.J. and Reichardt, L.F. (2001) Neurotrophins: roles in neuronal development and function. *Annu. Rev. Neurosci.* **24**: 677–736.

Hunt, P. and Krumlauf, R. (1992) *Hox* codes and positional specification in vertebrate embryonic axes. *Annu. Rev. Cell Biol.* **8**: 227–256.

Ikonomidou, C., Bittigau, P., Ishimaru, M.J., Wozniak, D.F. *et al.* (2000) Ethanol-induced apoptotic neurodegeneration and fetal alcohol syndrome. *Science* **287**: 1056–1060.

Jacobson, M.D., Weil, M. and Raff, M.C. (1997) Programmed cell death in animal development. *Cell* **88**: 347–354.

Jessell, T.M. (2000) Neuronal specification in the spinal cord: inductive signals and transcriptional codes. *Nat. Rev. Genet.* **1**: 20–29.

Johnson, M.P., Sutton, L.N., Rintoul, N., Crombleholme, T.M. *et al.* (2003) Fetal myelomeningocele repair: short-term clinical outcomes. *Am. J. Obstet. Gynecol.* **189**: 482–487.

Kageyama, R., Ohtsuka, T., Hatakeyama, J. and Ohsawa, R. (2005) Roles of bHLH genes in neural stem cell differentiation. *Exp. Cell Res.* **306**: 343–348.

Kessaris, N., Pringle, N. and Richardson, W.D. (2001) Ventral neurogenesis and the neuron–glial switch. *Neuron* **31**: 677–680.

Knecht, A.K. and Bronner-Fraser, M. (2002) Induction of the neural crest: a multigene process. *Nat. Rev. Genet.* **3**: 453–461.

Kuan, C.Y., Roth, K.A., Flavell, R.A. and Rakic, P. (2000) Mechanisms of programmed cell death in the developing brain. *Trends Neurosci.* **23**: 291–297.

Kuida, K., Zheng, T.S., Na, S.Q., Kuan, C.Y. *et al.* (1996) Decreased apoptosis in the brain and premature lethality in CPP32-deficient mice. *Nature* **384**: 368–372.

Kullander, K., Croll, S.D., Zimmer, M., Pan, L. *et al.* (2001) Ephrin-B3 is the midline barrier that prevents corticospinal tract axons from recrossing, allowing for unilateral motor control. *Genes Dev.* **15**: 877–888.

Lammer, E.J., Sever, L.E. and Oakley, G.P. (1987) Teratogen update: valproic acid. *Teratology* **35**: 465–473.

Lindsay, S. and Copp, A.J. (2005) MRC–Wellcome Trust Human Developmental Biology Resource: enabling studies of human developmental gene expression. *Trends Genet.* **210**: 986–990.

Malatesta, P., Hack, M.A., Hartfuss, E., Kettenmann, H. *et al.* (2003) Neuronal or glial progeny: regional differences in radial glia fate. *Neuron* **37**: 751–764.

Marshall, H., Nonchev, S., Sham, M.H., Muchamore, I. *et al.* (1992) Retinoic acid alters hindbrain *Hox* code and induces transformation of rhombomeres 2/3 into a 4/5 identity. *Nature* **360**: 737–741.

McLaughlin, T., Hindges, R. and O'Leary, D.D. (2003) Regulation of axial patterning of the retina and its topographic mapping in the brain. *Curr. Opin. Neurobiol.* **13**: 57–69.

McMahon, A.P. and Bradley, A. (1990) The *Wnt-1* (*int-1*) proto-oncogene is required for development of a large region of the mouse brain. *Cell* **62**: 1073–1085.

Ming, J.E. and Muenke, M. (2002) Multiple hits during early embryonic development: digenic diseases and holoprosencephaly. *Am. J. Hum. Genet.* **71**: 1017–1032.

Mochida, G.H. and Walsh, C.A. (2004) Genetic basis of developmental malformations of the cerebral cortex. *Arch. Neurol.* **61**: 637–640.

Mullen, R.J., Hamre, K.M. and Goldowitz, D. (1997) Cerebellar mutant mice and chimeras revisited. *Perspect. Dev. Neurobiol.* **5**: 43–55.

Mullen, R.J. and Herrup, K. (1979) Chimeric analysis of mouse cerebellar mutants. In *Neurogenetics: Genetic Approaches to the Nervous System*, Breakfield, X.O. (ed.). Elsevier: New York; 173–196.

Muntoni, F., Brockington, M., Torelli, S. and Brown, S.C. (2004) Defective glycosylation in congenital muscular dystrophies. *Curr. Opin. Neurol.* **17**: 205–209.

Myrianthopoulos, N.C. (1979) Our load of central nervous system malformations. *Birth Defects* **15**: 1–18.

Nagai, T., Aruga, J., Minowa, O., Sugimoto, T. *et al.* (2000) Zic2 regulates the kinetics of neurulation. *Proc. Natl Acad. Sci. USA* **97**: 1618–1623.

Nagano, T., Morikubo, S. and Sato, M. (2004) Filamin A and FILIP (filamin A-interacting protein) regulate cell polarity and motility in neocortical subventricular and intermediate zones during radial migration. *J. Neurosci.* **24**: 9648–9657.

Nijhawan, D., Honarpour, N. and Wang, X.D. (2000) Apoptosis in neural development and disease. *Annu. Rev. Neurosci.* **23**: 73–87.

Parnavelas, J.G. (2000) The origin and migration of cortical neurones: new vistas. *Trends Neurosci.* **23**: 126–131.

Pinto-Lord, M.C., Evrard, P., Caviness, V.S. Jr. (1982) Obstructed neuronal migration along radial glial fibers in the neocortex of the reeler mouse: a Golgi–EM analysis. *Dev. Brain Res.* **4**: 379–393.

Qiu, M.S., Anderson, S., Chen, S.D., Meneses, J.J. *et al.* (1996) Mutation of the *Emx-1* homeobox gene disrupts the corpus callosum. *Dev. Biol.* **178**: 174–178.

Rasband, K., Hardy, M. and Chien, C.B. (2003) Generating x. Formation of the optic chiasm. *Neuron* **39**: 885–888.

Read, A.P. and Newton, V.E. (1997) Waardenburg syndrome. *J. Med. Genet.* **34**: 656–665.

Roessler, E., Du, Y.Z., Mullor, J.L., Casas, E. *et al.* (2003) Loss-of-function mutations in the human *GLI2* gene are associated with pituitary anomalies and holoprosencephaly-like features. *Proc. Natl Acad. Sci. USA* **100**: 13424–13429.

Russman, B.S. and Ashwal, S. (2004) Evaluation of the child with cerebral palsy. *Semin. Pediatr. Neurol.* **11**: 47–57.

Schaar, B.T., Kinoshita, K. and McConnell, S.K. (2004) Doublecortin microtubule affinity is regulated by a balance of kinase and phosphatase activity at the leading edge of migrating neurons. *Neuron* **41**: 203–213.

Sendtner, M., Pei, G., Beck, M., Schweizer, U. and Wiese, S. (2000) Developmental motoneuron cell death and neurotrophic factors. *Cell Tissue Res.* **301**: 71–84.

Sheen, V.L., Feng, Y.Y., Graham, D., Takafuta, T. *et al.* (2002) Filamin A and Filamin B are co-expressed within neurons during periods of neuronal migration and can physically interact. *Hum. Mol. Genet.* **11**: 2845–2854.

Sherman, G.F., Morrison, L., Rosen, G.D., Behan, P.O. and Galaburda, A.M. (1990) Brain abnormalities in immune defective mice. *Brain Res.* **532**: 25–33.

Shu, T.Z., Ayala, R., Nguyen, M.D., Xie, Z.G. *et al.* (2004) Ndel1 operates in a common pathway with LIS1 and cytoplasmic dynein to regulate cortical neuronal positioning. *Neuron* **44**: 263–277.

Shu, T.Z., Butz, K.G., Plachez, C., Gronostajski, R.A. and Richards, L.J. (2003) Abnormal development of forebrain midline glia and commissural projections in *Nfia* knock-out mice. *J. Neurosci.* **23**: 203–212.

Stern, C.D. (2002) Induction and initial patterning of the nervous system – the chick embryo enters the scene. *Curr. Opin. Genet. Dev.* **12**: 447–451.

Swiatek, P.J. and Gridley, T. (1993) Perinatal lethality and defects in hindbrain development in mice homozygous for a targeted mutation of the zinc finger gene *Krox20*. *Genes Dev.* **7**: 2071–2084.

Tanaka, T., Serneo, F.F., Higgins, C., Gambello, M.J. *et al.* (2004) Lis1 and doublecortin function with dynein to mediate coupling of the nucleus to the centrosome in neuronal migration. *J. Cell Biol.* **165**: 709–721.

Van der Put, N.M.J., Eskes, T.K.A.B. and Blom, H.J. (1997) Is the common 677C → T mutation in the methylenetetrahydrofolate reductase gene a risk factor for neural tube defects? A meta-analysis. *Qu. J. Med.* **90**: 111–115.

Wald, N., Sneddon, J., Densem, J., Frost, C. and Stone, R. (1991) MRC Vitamin Study Research Group. Prevention of neural tube defects: results of the Medical Research Council Vitamin Study. *Lancet* **338**: 131–137.

Walsh, C. and Cepko, C.L. (1993) Clonal dispersion in proliferative layers of developing cerebral cortex. *Nature* **362**: 632–635.

Weil, M., Jacobson, M.D. and Raff, M.C. (1997) Is programmed cell death required for neural tube closure? *Curr. Biol.* **7**: 281–284.

Woods, C.G., Bond, J. and Enard, V. (2005) Autosomad recessive primary microcephaly (MCPH): a review of clinical, molecular and evolutionary findings. *Am. J. Hum. Genet.* **76**, 717–728.

Wu, M., Chen, D.F., Sasaoka, T. and Tonegawa, S. (1996) Neural tube defects and abnormal brain development in F52-deficient mice. *Proc. Natl Acad. Sci. USA* **93**: 2110–2115.

Yamamoto, T., Toyoda, C., Kobayashi, M., Kondo, E. *et al.* (1997) Pial–glial barrier abnormalities in fetuses with Fukuyama congenital muscular dystrophy. *Brain Dev.* **19**: 35–42.

Ybot-Gonzalez, P., Cogram, P., Gerrelli, D. and Copp, A.J. (2002) Sonic hedgehog and the molecular regulation of neural tube closure. *Development* **129**: 2507–2517.

Ybot-Gonzalez, P. and Copp, A.J. (1999) Bending of the neural plate during mouse spinal neurulation is independent of actin microfilaments. *Dev. Dyn.* **215**: 273–283.

Yokoyama, N., Romero, M.I., Cowan, C.A., Galvan, P. *et al.* (2001) Forward signaling mediated by ephrin-B3 prevents contralateral corticospinal axons from recrossing the spinal cord midline. *Neuron* **29**: 85–97.

Zhao, Q., Behringer, R.R. and De Crombrugghe, B. (1996) Prenatal folic acid treatment suppresses acrania and meroanencephaly in mice mutant for the *Cart1* homeobox gene. *Nat. Genet.* **13**: 275–283.

9

Birth Defects Affecting the Eye

Jane C. Sowden

The eye

The eye is an exquisitely complex sense organ, capable of gathering a wealth of information in the form of refracted light. Working in conjunction with visual centres of the brain, our eyes provide a rich visual experience. They are specialized to provide high visual acuity and colour vision as well as the ability to judge the relative distance of objects in space and to track moving objects. Light passes through the transparent cornea and the iris pupil in the anterior segment of the eye and is focused by the lens onto the retina at the back of the eye. Photoreceptor cells in the retina detect photons of light. These cells connect to interneurons, which partially process and then transmit visual information to ganglion cells projecting axons along the optic nerve to the brain. Intra-ocular pressure, important for stabilizing the shape of the eye, is controlled by the flow of aqueous humour through the drainage structures in the anterior chamber angle between the iris and the cornea.

In the newborn infant the eye is structurally complete, although visual acuity is low. Postnatal development consists of growth of the optical elements of the eyes (the cornea, lens, anterior chamber and axial length of the globe) to optimize focus, as well as maturation of the cone photoreceptors responsible for colour vision. The critical events that determine the structural integrity of the eye globe take place early in gestation, during the early stages of nervous system development, whereas differentiation of retinal neurons and maturation of the anterior segment occur in the second and third trimesters and are only completed around or after birth. Disruption of the process of eye development, either by genetic changes, environmental factors or a combination of the two, leads to the presentation of congenital eye defects in the newborn. The large number of different causes of eye defects reflects the complexity of eye development; the online database of human genetic disease, Online Mendelian

Embryos, Genes and Birth Defects, Second Edition Edited by Patrizia Ferretti, Andrew Copp, Cheryll Tickle and Gudrun Moore

Inheritance in Man (OMIM), lists more than 1000 diseases that include abnormalities of the structure of the eye. In this chapter the genetic causes of birth defects affecting the eye will be reviewed, together with a discussion of our current understanding of the molecular basis of normal eye development and how this is disrupted in cases of congenital eye malformations.

Development of the eye

The laboratory mouse has proved to be an invaluable model for understanding human eye development, as the morphological process of mouse eye development is very similar to that in humans. In addition to the high level of DNA sequence conservation between the mouse and human genomes, the ease of genetic analysis in the mouse has led to the characterization of many important genes. In this section, the key events in eye development are compared in humans and mouse (Cvekl and Tamm, 2004; Kaufman, 1992; Larsen, 2001; Mann, 1964; O'Rahilly, 1975; Table 9.1).

Eye development begins in the fourth week of life in a human embryo (equivalent to embryonic day (E) 8.5 in the mouse embryo). In the folding neural plate at the rostral end of the post-gastrulation embryo, two small optic pits appear in the neuroepithelium, one on either side of the midline (Figure 9.1a). By the time the

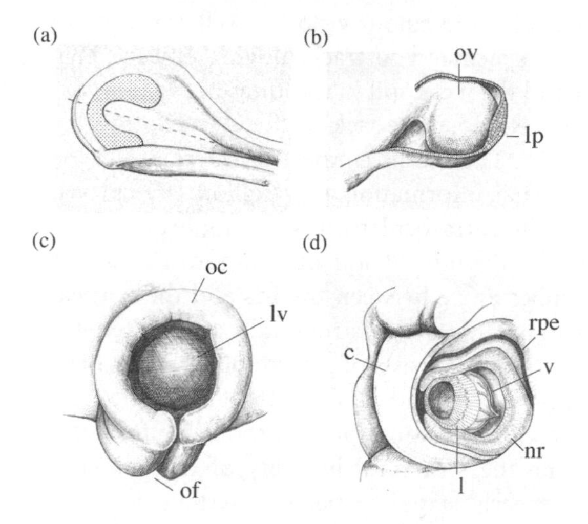

Figure 9.1 Development of the rudimentary eye in human embryos. (a) The rostral end of the folding neural plate in a 4 week-old human embryo. The optic vesicles extend from the eye field region, which is indicated by a dotted area. The first sign of the optic vesicles are two depressions (the optic pits) within the left and right regions of the eye field, respectively. The broken line indicates the mid-line of the embryo. (b) The optic vesicle (ov) contacting the overlying and thickened surface ectoderm, the lens placode (lp). (c) Frontal view of the optic cup showing the ventral optic fissure (of) and the lens vesicle (lv) within the optic cup (oc). (d) View of the optic cup at 6 weeks with side cut-away to show retina, primitive vasculature and lens. c, surface of future cornea; l, lens; nr, neural retina; rpe, retinal pigmented epithelium; v, vasculature of the lens. Original illustration by John Chilton

Table 9.1 Comparison of timing of key events in eye development in human and mouse

Human gestation	Mouse embryonic day (E), postnatal day (P)	(1) Specification of the eye field and optic vesicle morphogenesis	(2) Growth, patterning and closure of the optic cup	(3) Development of the anterior segment	(4) Lens development	(5) Development of retinal neurons and the optic nerve
3 Weeks	E8.5	Optic pit				
4 Weeks	E9.5	Optic vesicle			Lens placode	
5 Weeks	E11.5	Optic cup/lens vesicle	Optic cup growth, RPE/NR specified		Lens vesicle develops	
Day 37	E13.5		Optic fissure closes			Retinal neurogenesis under way, ganglion cells born
7 Weeks	E14.5			Anterior chamber forms between lens and cornea	Lens fibres	
8 Weeks						Optic nerve
8th Month	P11					Retinal neurogenesis completed
9th Month	P21			Anterior segment angle structures mature		

cranial neural tube closes, the optic pits have expanded to form paired optic vesicles (in the prosencephalon). These vesicles each extend laterally towards the overlying surface ectoderm (Figure 9.1b). Following contact with the surface ectoderm, a process of invagination of the distal neuroepithelium gives rise to a bi-layered optic cup (Figure 9.1c). The proximal neuroepithelium of the optic vesicle forms the optic stalk, which connects the optic cup to the forebrain (diencephalon). Concomitant with the morphogenetic process of optic cup formation, the surface ectoderm apposed to the optic vesicle thickens to form the lens placode (Figure 9.1b). The lens placode then invaginates and separates from the surface ectoderm to form the primitive lens vesicle (Figure 9.1c). The rudimentary eye is thus formed during the fifth week of development (by E11.5 in the mouse) (Figure 9.1d). The inner layer of the optic cup forms the presumptive neural retina and the outer layer forms the presumptive retinal pigmented epithelium (Figures 9.1d, 9.2a). The optic cup and optic stalk are initially incomplete along their ventral surfaces and this optic (choroidal) fissure (Figure 9.1c) allows entry of the primitive vasculature of the lens (a branch of the ophthalmic artery called the hyaloid artery) before it closes (Figure 9.1d). The optic fissure closes at around 6 weeks of human development (by day 37; E13.5 in the mouse) when the basement membranes abut and fuse.

The next phase of eye development is dependent upon the differentiation of the rudimentary embryonic structures (the lens and retina) and the formation of additional specialized structures through the coordinated integration and differentiation of tissues with different embryonic origins. In the latter case, neural crest cells migrating towards the anterior of the eye, together with cells from the surrounding peri-ocular mesenchyme, play an important role.

Development of the anterior segment (the cornea and iris) is dependent upon interactions between these neural crest/mesenchymal cells and cells derived from the neuroepithelium of the optic cup and from the surface ectoderm (Figure 9.2a). The corneal epithelium develops from the surface ectoderm overlying the lens vesicle. Secretion of extracellular matrix molecules from the surface ectoderm facilitates the migration of neural crest/mesenchymal cells. These cells form the corneal endothelium, the keratocytes of the corneal stroma lying between the corneal endothelium and the outer corneal epithelium. By the seventh week (E14.5 in the mouse) the fluid-filled anterior chamber has formed as the differentiating cornea separates from the lens (Figure 9.2b). At this stage differentiation of the anterior edge of the optic cup begins. Neural crest cells populate the developing anterior iris and other structures in the anterior segment angle, including the trabecular meshwork and Schlemm's canal, which is important for the maintenance of intra-ocular pressure. Neural crest cells also contribute to the ciliary muscle, which focuses the lens, and to the extra-ocular muscles, which facilitate eye movement. The posterior layer of the iris and the ciliary body epithelium is derived from neuroepithelium at the peripheral rim of the optic cup. Maturation of the structures in the anterior segment angle is complete by post-natal day (P) 21 in the mouse and around birth in humans (Figure 9.2c).

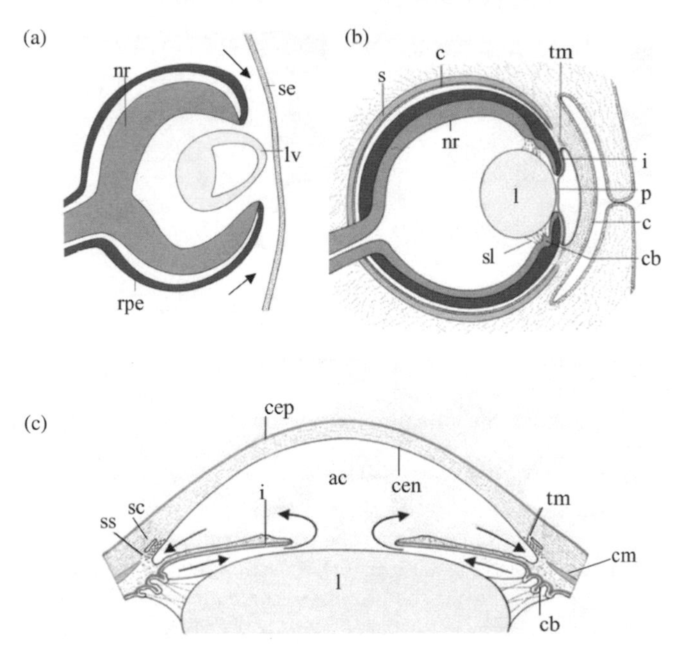

Figure 9.2 Development of the anterior segment of the human eye. (a) The optic cup at 6 weeks. Arrows show the direction of migration of neural crest cells towards the anterior region of the optic cup, between the surface ectoderm and the lens vesicle. lv, lens vesicle; nr, neural retina; rpe, retinal pigmented epithelium; se, surface ectoderm. (b) By 20 weeks the posterior iris and the ciliary body (cb) epithelium has formed from the periphery of the optic cup. Neural crest cells contribute to development of the anterior iris, the trabecular meshwork primordia (tm) and the cornea (c). c, choroid; i, iris; l, lens; nr, neural retina; p, pupil; s, sclera; sl, suspensory ligament of ciliary body attached to lens. (c) The anterior segment at birth. Arrows indicate the flow of aqueous. ac, anterior chamber; cb, ciliary body; cen, corneal endothelium; cep, corneal epithelium; cm, ciliary muscle; i, iris. l, lens; sc, Schlemm's canal; ss, scleral spur; tm, trabecular meshwork. Original illustration by Terry Tarrant

The lens vesicle loses its central cavity as the posterior cells of the lens vesicle elongate to form lens fibres orientated in an anteroposterior direction by the seventh week. Differentiating posterior lens fibres cells lose their mitochondria and nucleus and synthesize the lens protein, crystallin, whereas the cells at the anterior side of the lens vesicle remain as epithelial cells. Secondary fibre cells differentiate from proliferating epithelial cells in the equatorial zone at the margin of the lens epithelium. After birth, new fibres continue to be added at a lower rate throughout life.

The neuroepithelium of the presumptive neural retina undergoes a process of neurogenesis, starting by the end of the 6th week (E12.5 in mouse), to generate the mature retinal architecture (Figure 9.3). Multipotential retinal progenitor cells of the neural retina give rise to Muller glial cells and six distinct types of retinal neuron (Figure 9.3a). These are the cone and rod photoreceptor cells, the bipolar, amacrine and horizontal cells of the inner nuclear layer and the ganglion cells, which project

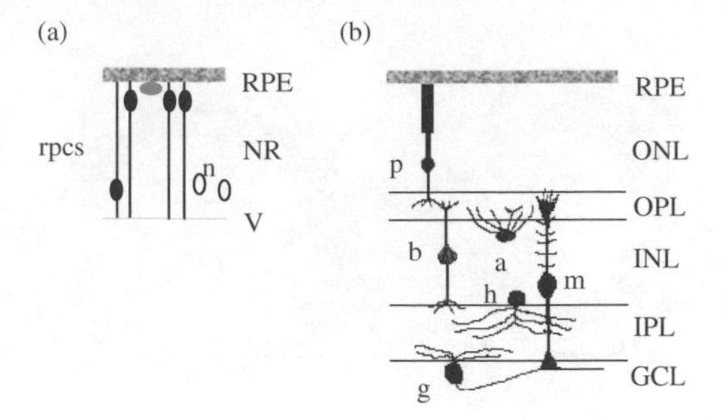

Figure 9.3 Retinal neurogenesis. (a) Retinal progenitor cells (rpcs) proliferate in the immature neural retina (NR); rpc nuclei, shown filled in black, migrate up and down between the retinal pigmented epithelium (RPE) and the vitreal surface (V) during the cell cycle. The mitotic phase of the rpc cell cycle takes place adjacent to the retinal pigmented epithelium. An rpc in mitosis is shown in grey. Newly born retinal neurons (post-mitotic rpcs), starting to differentiate (n), are shown in white. (b) The mature neural retina shows a laminated organization. It has three cellular layers, the outer nuclear layer (ONL) containing the photoreceptor cells (p), the inner nuclear layer (INL) containing the interneurons, bipolar cells (b), amacrine cells (a) and horizontal cells (h), as well as Muller glial cells (m) and the ganglion cell layer containing ganglion cells (g) which project their axons along the optic nerve. Synaptic connections between neurons in the respective nuclear layers form the outer plexiform layer (OPL) and the inner plexiform layer (IPL). RPE, retinal pigmented epithelium

axons along the optic nerve. The photoreceptor cells express the photo-pigment opsin genes, which bind the chromophore, 11-*cis*-retinal (derived from vitamin A) and provide photosensitivity to the retina. The characteristic laminated organization of the mature retina develops by the 8th month and in mouse neurogenesis is only completed by P11 (Figure 9.3b). The outer neuroepithelial layer of the optic cup develops into the non-neuronal retinal pigmented epithelium.

The cuboidal pigmented cells of the retinal pigmented epithelium form a characteristic epithelial monolayer structure adjacent to the photoreceptor cells. Ganglion cells project their axons into the optic stalk and promote the optic stalk neuroepithelium to develop as astroglia. By the 8th week the optic stalk has transformed into the optic nerve.

Congenital eye defects and paediatric blindness

Congenital malformation of the eye is one of the most common causes of blindness in children in the UK and is a significant cause worldwide. Of the 1.4 million blind children worldwide, 17% are due to congenital eye defects. In the UK about 10/10 000 children are visually impaired or blind (usually considered as a corrected visual acuity of less than 6/60 in the better eye). Around 21% of these cases in the UK are caused by congenital eye defects, mainly structural malformations of the globe. Other frequent causes are optic nerve disorders (~20%), retinal dystrophies and albinism (~13%)

and congenital cataracts (~4%) (reviewed by (Rahi and Dezateux, 2001). A higher proportion of paediatric blindness is caused by congenital malformation in the UK compared to the worldwide figures, reflecting the lower impact of infection and environmental factors (Muhit and Gilbert, 2003). The majority of congenital eye defects have complex uncharacterized causes and do not show a typical Mendelian pattern of dominant or recessive inheritance. Peri-natal and pre-natal factors have often been implicated, although currently there is little understanding of the interaction between environmental influences and genetic changes. Genetic causes have so far been identified in only a minority of cases (see Table 9.2). Congenital eye defects often affect multiple components of the eye, making their clinical diagnosis and classification difficult. Many observed clinical features may be secondary to a primary tissue defect at the site of action of a specific gene. However, with increased knowledge of the molecular basis of the normal process of eye development, it is possible to describe the developmental origin of different types of malformation resulting from single gene mutations. Several key events in eye development are dependent upon the function of specific genes. These are:

1. Specification of the eye field and optic vesicle morphogenesis.

2. Growth, patterning and closure of the optic cup.

3. Development of the anterior segment.

4. Lens development.

5. Development of retinal neurons and the optic nerve.

Disruption of these key events cause different types of eye malformation and these are summarized in Table 9.2, together with the human genes whose mutation is associated with each condition; conditions are only included where causative gene mutations have been identified. In several cases a single gene affects more than one developmental event.

In the next section, the genetic causes of congenital eye defects are reviewed and knowledge of gene function in normal eye development and in relation to malformation is discussed.

Gene mutations underlying congenital eye defects

Molecular genetic analysis of patients with congenital eye defects has led to the identification of single-gene mutations as the cause of many congenital eye defects. Several conditions have been found to be genetically heterogeneous and also different mutations in the same gene can cause clinically distinct phenotypes (allelic heterogeneity). Both recessive and dominant mutations have been identified and biochemical

Table 9.2 Molecular and cellular basis of congenital eye defects

Embryonic process Key events	Consequence of disruption of developmental process (clinical condition)	Genes implicated in this process	MIM Number	Chromosome location	Eye defects associated with gene mutation
(1) Specification of the eye field and optic vesicle morphogenesis	Abnormal morphogenesis of optic vesicle. Optic cup fails to form (anophthalmia)	*PAX6**	607108	11p13	Anophthalmia (recessive)
		*SOX2**	184429	3q26.3–q27	Anophthalmia/microphthalmia (dominant)
		*RAX**	601881	18q21.31	Anophthalmia/sclerocornea (recessive)
		SHH	600725	7q36	Holoprosencephaly/cyclopia/coloboma (dominant)
		*SIX3**	603714	2p21	Holoprosencephaly/microphthalmia/iris coloboma (dominant)
(2) Growth, patterning and closure of the optic cup	Growth of optic cup is abnormal (microphthalmia)	*CHX10**	142993	14q24.3	Microphthalmia (recessive)
		*BCOR**	300485	Xp11.4	Microphthalmia/congenital cataracts (oculofaciocardiodental, OFCD; X-linked dominant)
		*BCOR**	300485	Xp11.4	Microphthalmia (Lenz syndrome; X-linked recessive)
	Optic fissure in ventral optic cup fails to close (coloboma)	*MAF**	177075	16q22–q23	Congenital cataracts/microphthalmia/iris coloboma/Peters anomaly (dominant)
		*PAX2**	167409	10q25	Coloboma/microphthalmia (dominant)
(3) Development of the anterior segment	Abnormal differentiation of anterior tissues derived from neural crest mesenchymal cells (anterior segment dysgenesis)	*FOXC1**	601090	6p25	Axenfeld–Rieger/Peters anomaly (dominant)
		*PITX2**	601542	4q25–q27	Axenfeld–Rieger/Peters anomaly (dominant)
		*PAX6**	607108	11p13	Aniridia/cataracts/Peters anomaly (dominant)
		CYP1B1	601771	2p21	Primary congenital glaucoma/Peters anomaly (recessive)

(4) Lens development	Failure of lens development, signalling and differentiation (cataract)	*MAF* *	177075	16q22–q23	Congenital cataracts/microphthalmia/iris coloboma/Peters anomaly
		FOXE3 *	601094	1p32	Congenital cataracts/ASD
		PITX3 *	602669	10q25	Congenital cataracts/ASD
		EYA1 *	601653	8q13.3	Congenital cataracts/ASD/branchio-oto-renal (BOR) syndrome
(5) Development of retinal neurons and the optic nerve	Abnormal differentiation of neural retina	*GUCY2D*	600179	17p13.1	Leber congenital amaurosis/cone rod dystrophy (recessive)
		RPE65,	180069	1p31	Leber congenital amaurosis (recessive)
		AIPL1,	604392	17p13.1	Leber congenital amaurosis (recessive)/retinitis pigmentosa (dominant)
		RPGRIP	605446	14q11	Leber congenital amaurosis (recessive)
		CRB1	604210	1q31–q32.1	Leber congenital amaurosis (recessive)
		CRX *	602225	19q13.3	Leber congenital amaurosis/cone rod dystrophy (dominant)
	Abnormal optic nerve development	*PAX6* *	607108	11p13	Optic nerve aplasia, foveal hypoplasia (dominant)
		HESX1 *	601802	3p21.1–p21.2	Optic nerve hypoplasia, septo-optic dysplasia (recessive or dominant)

*Transcription factor.

characterization of the normal and altered proteins is providing a better understanding of genotype–phenotype correlations. Of interest is the prevalence of transcription factors as disease-causing genes. These DNA-binding proteins act to regulate expression of other genes during development and their coordinated activities are essential. In many cases mouse strains exist that carry mutations in orthologues of human disease genes. These are either naturally occurring or genetically engineered mutations. These disease models are described where their analysis has enhanced understanding of the human condition.

Anophthalmia and holoprosencephaly

Failure of event (1) – specification of the eye field and optic vesicle morphogenesis – causes the profound and distressing condition of absence of the eye, anophthalmia, which may be unilateral or bilateral. Epidemiological studies estimate incidence rates of around 0.3/10 000 for anophthalmia (Stoll *et al.*, 1992; Warburg, 1993).

Compound heterozygous mutations in the *RAX* gene (Voronina *et al.*, 2004) and *SIX6* hemizygosity (Gallardo *et al.*, 1999) have been associated with anophthalmia. *SOX2* mutation also causes anophthalmia and identified mutations were *de novo* and dominant (Fantes *et al.*, 2003); *Sox2* plays a role in lens specification and regulation of crystallin expression, as well as playing an early role in regulation of neural progenitor cells (Graham *et al.*, 2003; Kamachi *et al.*, 1995). Heterozygous mutation of *OTX2* is a newly indentified cause of anophthalmia (Ragge *et al.*, 2005). A special case of failure of event (1) is found in some cases of holoprosencephaly (HPE) (MIM 600725) associated with dominant mutation of the secreted signalling protein, *SHH* (Nanni *et al.*, 1999; Roessler *et al.*, 1997). Rarely, HPE patients have only a single eye, located at the midline (cyclopia), resulting from abnormal patterning of the eye field at the midline.

Microphthalmia

Disruption of event (2) causes microphthalmia (small eyes) and/or coloboma when the optic fissure fails to close. In children with these conditions, vision is variably affected and a single eye (unilateral) or both eyes (bilateral) can be affected. The microphthalmic eye has an axial length of less that 19.3 mm at 1 year of age and less than 20.9 mm in adulthood (at least 2 SD below the mean; Weiss *et al.*, 1989a, 1989b). Severe microphthalmia, often found with other ocular anomalies, is a common cause of childhood blindness. Incidence rates are estimated to be around 1.8/10 000 live births (Stoll *et al.*, 1992; Warburg, 1993).

Mutation of the *CHX10* gene was the first identified cause of isolated bilateral microphthalmia (Bar-Yosef *et al.*, 2004; Ferda Percin *et al.*, 2000). The condition is recessively inherited and the patient phenotype is similar to the eye phenotype of a mutant mouse strain with a recessive *Chx10* mutation, the *ocular retardation* mouse. The phenotype also includes optic nerve hypoplasia and cataracts; both these features are secondary to the defect in the neural retina, where the *Chx10* gene is specifically expressed. Mutations in the transcriptional co-repressor *BCOR* cause microphthalmia in two distinct clinical syndromes, oculofaciocardiodental syndrome (OFCD) and Lenz syndrome, thought to result from distinct protein mutations (Ng *et al.*, 2004).

Coloboma

Failure of the process of closure and fusion of the basement membranes of the optic fissure in the ventral optic cup results in the persistence of a fissure, or coloboma, in the globe. The coloboma can affect any part of the globe traversed by the fissure from the iris to the optic nerve (Onwochei *et al.*, 2000). Epidemiological studies estimate incidence rates of around 0.7/10 000 for coloboma (Stoll *et al.*, 1992; Warburg, 1993). Ocular (uveoretinal) colobomas represent a significant cause of congenital poor vision (estimated to account for 3.2–11.2% of childhood blindness; Fraser 1967) but in some cases coloboma can be asymptomatic and only visible with ophthalmic investigation.

OMIM lists over 100 conditions of coloboma, usually as part of a syndrome and often associated with microphthalmia, but largely without known causes. Mutations in *PAX2* are found in cases of coloboma of the retina and optic nerve that occur with renal anomalies as part of the renal–coloboma syndrome (MIM 120330; 167409). A similar phenotype of optic nerve coloboma together with renal hypoplasia is found in the *Krd* (*kidney and retinal defects*) mutant mouse, which lacks *Pax2* (Favor *et al.*, 1996; Sanyanusin *et al.*, 1995). In *Krd* heterozygous mice, which lack one copy of the normal *Pax2* gene, *Pax2*-positive cells show abnormal morphogenetic movements, causing misrouting of ganglion cell axons and a malformation of the optic disc (Otteson *et al.*, 1998). Colobomas are also part of the phenotypic range associated with HPE and *SHH* mutation (Schimmenti *et al.*, 2003) and rarely with *SIX3* gene mutation (Wallis *et al.*, 1999)(MIM 157170). Mutation in the retinol-binding protein *RBP4*, causing retinol deficiency, has also been associated with iris coloboma (Seeliger *et al.*, 1999) (MIM 180250).

To help understand their aetiology, the conditions anophthalmia, microphthalmia and coloboma have been considered separately here, although in many patients these malformations occur together. A recent study grouped these conditions together and estimated a live birth prevalence of 19/100 000 in Scotland of microphthalmia, anophthalmia and coloboma, with 70% of patients having coloboma (Morrison *et al.*, 2002). Clinical analysis also reveals the close relationship between events (1) and (2) in development. For example, patients with a *SOX2* mutation usually have bilateral anophthalmia but can present with unilateral anophthalmia and contralateral microphthalmia (Fantes *et al.*, 2003; Ragge *et al.*, 2005). Such observations suggest an early function of the gene in event (1) that in some cases can be compensated. The mechanism for such phenotypic variation is not fully understood, and suggests compensatory mechanisms that rely on relative levels of different factors (stoichiometric effects). The frequent observation of asymmetry in the ocular malformation, such that one eye has fared better despite their identical genetic backgrounds, indicates that anophthalmia/microphthalmia should be considered as a clinical spectrum.

Anterior segment dysgenesis

Failure of the normal development of the structures of the anterior segment of eye (event 3) – the cornea, the iris, and the anterior segment angle between the iris and the cornea – causes a complex range of malformations, termed anterior segment dysgenesis (ASD), which are associated with developmental glaucoma. These conditions include overt ocular features, such as iris hypoplasia, irregular-shaped pupils (corectopia) or additional pupils (polycoria) and adhesions between the iris and the lens (peripheral anterior synechiae). Abnormal development of structures of the anterior segment angle (trabecular meshwork, Schlemm's canal, ciliary muscle) can cause elevated intra-ocular pressure and associated optic nerve damage (glaucoma).

A range of clinical conditions concerning malformation of the anterior segment have been grouped under the heading *Axenfeld–Rieger syndrome* (ARS), based on their phenotypic similarities, and their often common genetic basis (Alward, 2000; Lines *et al.*, 2002). Clinical sub-types of ARS include Axenfeld anomaly, iridogoniodysgenesis and Rieger syndrome or anomaly. Deletions and point mutations of two genes, *PITX2* and *FOXC1*, cause autosomal dominant ARS (Mears *et al.*, 1998; Semina *et al.*, 1996; Lines *et al.*, 2004). These genes are expressed within the developing angle and other anterior tissue but are not expressed in the lens or retina. A spectrum of similar anterior segment phenotypes are associated with *PITX2* and *FOXC1* mutations, likely indicating that these genes act in a common pathway which is essential for differentiation of the anterior ocular tissues. *PITX2* and *FOXC1* have occasionally been identified as causes of *Peters anomaly* (Honkanen *et al.*, 2003; Perveen *et al.*, 2000). This phenotype involves a central corneal opacity (leukoma) often associated with adhesion of the lens to the back of the corneal opacity. The lens may also show anterior polar cataract. The reason for the wide spectrum of phenotypes, both between and within families, may indicate interactions with modifier genes as well as the variable function of different mutant proteins (Kozlowski and Walter, 2000).

Study of *FOXC1* and *PITX2* and the mouse homologues *Foxc1* and *Pitx2* has demonstrated the critical importance of these genes for the development of neural crest/mesenchymal cells of the anterior segment. Specifically *Foxc1* and *Pitx2* are essential for conversion of mesenchymal/neural crest cells to an endothelial phenotype in the developing cornea. Analysis of mice lacking these genes showed that the corneal endothelium layer does not develop and the outer corneal epithelium (derived from the surface ectoderm) is hypercellular and undifferentiated (Gage *et al.*, 1999; Kidson *et al.*, 1999; Kitamura *et al.*, 1999; Kume *et al.*, 2000). Heterozygous mice lacking one functional gene show a phenotype resembling ASD patient phenotypes. The commonly observed iris–corneal adhesions are likely to be caused by the abnormalities in the corneal endothelium. Both *Pitx2* and *Foxc1* also play important roles in the development of tissues other than the eye, and this is reflected in the finding of systemic malformations in patients with anterior segment malformations. For example, patients with *PITX2* mutation often show dental, cranio-facial and umbilical abnormalities (Semina *et al.*, 1996).

Aniridia is the absence or hypoplasia of the iris. Heterozygous mutation of the *PAX6* gene commonly causes human aniridia (Jordan *et al.*, 1992). Less commonly, *PAX6* mutation causes cataracts, macular hypoplasia, keratitis and Peters anomaly. The *PAX6* mutation database provides genotype/phenotype information on human *PAX6* mutations (http://www.hgu.mrc.ac.uk/Softdata/PAX6). Both the iris and the cornea are sensitive to *Pax6* gene dosage, as the phenotypes resembling aniridia, Peters anomaly and microphthalmia are present in heterozygous *Small eye* (*Sey*) mice, which carry a *Pax6* gene mutation. *Pax6* heterozygous eyes also show defects in angle differentiation that are associated the spectrum of anterior eye segment abnormalities (Baulmann *et al.*, 2002).

While the analysis of mouse models of ASD is providing insight into the underlying embryological malformations, the variability of clinical phenotypes resulting from mutations in the same gene still makes it difficult to predict the likely genetic cause in each patient. Variations in individual genetic background likely contribute to modifying patient phenotypes and, in addition, local environmental/stochastic factors, such as levels of important regulatory factors at specific time-points in development, are also likely to affect the phenotype. This is acutely demonstrated by the different phenotypes often seen in the two eyes of the same patient, for example, Peters anomaly in one eye and ARS in the other eye. Indeed, the sensitivity of eye development to levels of transcription factors has been elegantly demonstrated in several experiments in which genetic manipulation has been used to create mouse models with variable numbers of gene copies. For example, *Sey* heterozygous mice show microphthalmia, cataracts and Peter anomaly, whereas mice and humans carrying extra copies of the *Pax6* gene (up to five copies of the gene) also show eye malformation (Aalfs *et al.*, 1997; Glaser *et al.*, 1994; Schedl *et al.*, 1996). Likewise, deletion and duplication, as well as single amino acid substitution of the *FOXC1* gene, cause anterior segment malformations (Lehmann *et al.*, 2000; Nishimura *et al.*, 2001).

Primary congenital glaucoma (PCG) can be considered as an anterior segment dysgenesis. In this condition, drainage of the aqueous is impeded and children typically show enlarged eyes, buphthalmos, before the age of 3 years. Their eyes otherwise appear normal. PCG is an aggressive form of glaucoma in children. Homozygous mutations of *CYP1B1* cause a substantial proportion of PCG and juvenile open angle glaucoma cases (Stoilov *et al.*, 1997, 1998) and have also been identified in cases of Peters anomaly (Vincent *et al.*, 2001). *CYP1B1* may act as a modifier of a second gene associated with juvenile and adult forms of glaucoma, *MYOC* (myocilin/trabecular meshwork-induced glucocorticoid response protein; TIGR; Vincent *et al.*, 2002). Both *CYP1B1* and the *MYOC* gene (MIM 601652) are expressed in the iris, trabecular meshwork and ciliary body of the eye.

Cataracts

Failure of normal lens development (event 4) causes congenital cataracts in around 3/10 000 children and accounts for 10% of cases of childhood blindness (Francis and Moore, 2004; Rahi and Dezateaux, 2001). Disease-causing mutations have been identified in more than 20 genes encoding a wide variety of different lens proteins, including structural lens proteins (crystallins), gap junction proteins (connexin), membrane proteins and transcription factors involved in lens development (Graw, 2004; Hejtmancik and Smaoui, 2003). These different mutations all reduce lens transparency and in some cases disrupt lens formation. In part because of the ease of identification, a large number of mutant mouse lines with cataract phenotypes have been identified, and these have assisted gene identification and provide models to understand the molecular basis of the abnormal lens pathology (Graw and Loster, 2003). In this section (and in Table 9.2), only cataracts caused by mutations in transcription factor genes involved in lens development will be considered. These genes are expressed in the developing lens and their products often have a wider impact on eye development, causing complex ocular phenotypes, including microphthalmia and anterior segement dysgenesis (ASD).

Mutation of the genes *FOXE3*, *MAF* and *PITX3* cause cataracts in association with a range of ASDs, including Peters anomaly (Jamieson *et al.*, 2002; Semina *et al.*, 1998, 2001). Missense mutations in *EYA1* have also been associated with congenital cataracts and ASD (Azuma *et al.*, 2000), although *EYA1* is more often associated with branchio-oto-renal (BOR) syndrome, which affects development of branchial arch, ear and kidney. Analysis of mouse models has added to understanding of the function of these genes, also revealing their interactions. A dominant mutation of *Maf* alters the DNA-binding property of the Maf protein and causes murine cataract (Lyon *et al.*, 2003), whereas homozygous null mutations of *Maf* show defective lens formation, decreased expression of crystallins in the lens and microphthalmia (Kim *et al.*, 1999). The *dysgenetic lens* (*dyl*) mouse mutant encodes a Foxe3 protein unable to bind DNA. The mouse phenotype is variable but typically consists of the equivalent of Peters anomaly in humans, with central corneal opacity, keratolenticular adhesion and, in some cases, anterior polar cataract (Ormestad *et al.*, 2002). Deletion of a region upstream of the *Pitx3* gene causes small eyes that lack a lens in the *aphakia* mouse mutant. The deleted DNA region contains binding sites for the AP-2 and Maf transcription factors, suggesting that these proteins regulate *Pitx3* (Semina *et al.*, 2000). Maf proteins bind as dimers to Maf response elements (MAREs) to regulate transcription of the crystallin genes and *Pitx3* (Ring *et al.*, 2000).

The anterior segment phenotypes, often found with congenital cataracts caused by *FOXE3*, *MAF* and *PITX3* mutation, overlap with phenotypes caused by *PITX2*/*FOXC1* mutation, but *FOXE3*, *MAF* and *PITX3* act via a different mechanism, as they are expressed in the developing lens rather than the neural crest/mesenchymal tissue. That anterior segment development is dependent on signals from the lens is demonstrated by these different sites of action of genes that cause ASD, e.g lens vs. peri-ocular mesenchyme. Analysis of *Foxe3*, *Maf* and *Pitx3* mutant mice has shown that a common mechanism of anterior segment malformation caused by these genes is disruption of the process of separating the lens vesicle from the overlying surface ectoderm. Normal differentiation of the cornea does not then occur. The typical resulting phenotype, referred to as Peters anomaly, is corneal opacity, lack of corneal endothelium and adhesions between lens and cornea.

Abnormal development of retinal neurons and the optic nerve

Abnormal development of the retinal neurons and the optic nerve (event 5), although present at birth, may not become apparent until later in childhood. The eye globe is usually normal in appearance. This kind of defect includes: early onset retinal dystrophies, Leber congenital amaurosis and colour blindness, where retinal neurons are abnormal as well as optic nerve aplasia, where the optic nerve does not form normally or albinism, where the routing of ganglion cell axons at the optic chiasm is abnormal.

Leber congenital amaurosis (LCA) is the most common genetic cause of congenital retinal disorders in infants and children. Its incidence is 2–3/100 000 births and it accounts for 10–18% of cases of congenital blindness. Mutations in at least seven genes cause LCA (*GUCY2D*, *RPE65*, *CRX*, *AIPL1*, *RPGRIP*, *CRB1*; Hanein *et al.*, 2004). All are expressed in photoreceptors or retinal pigmented epithelium but the encoded proteins appear to function in different cellular pathways. *Septo-optic dysplasia* is a rare birth defect characterized by optic nerve hypoplasia together with any combination of absent septum pellucidum and/or pituitary dysfunction. Patients may present with strabismus, nystagmus, reduced visual acuity and visual impairment. Mutations in the homeobox gene *HESX1*/*Hesx1* are associated with septo-optic dysplasia in humans and mouse (Dattani *et al.*, 1998). *Albinism* is a heterogeneous group of conditions having in common an inherited error of melanin metabolism, resulting in misrouting of optic nerve fibres during embryogenesis, underdevelopment of the neural retina and varying degrees of hypopigmentation of the eyes, skin, and hair (Oetting *et al.*, 2003; Russell-Eggitt, 2001). Mutations in several genes have been identified as causes of albinism, Oculocutaneous albinism type 1 (OCA1), resulting from mutations of the *tyrosinase* (*TYR*) gene, is genetically and biochemically the best understood (Oetting *et al.*, 2003) (MIM 606933).

Cellular and molecular mechanisms affecting eye development and how they elucidate the causes of abnormal development

Classical embryological studies in chick and amphibian embryos provided the framework for our current understanding of the interactions between tissues during eye formation. Over the last two decades, modern genetic studies have identified molecules regulating these processes. In many cases, characterization of the expression pattern of genes during eye development has helped to elucidate their role. The availability of mouse models of different types of human eye defects has facilitated both gene identification and understanding of gene function. More recently, zebrafish mutagenesis programmes and screening for mutants showing eye phenotypes have provided a powerful additional route for the characterization of genes essential for eye development. These resources, together with the molecular genetic analysis of families with eye malformations, are providing a fuller understanding of the molecular and cellular basis of congenital eye defects. In many cases the identification of new causative gene mutations in patients is guided by detailed knowledge of gene function in model organisms.

The eyes of all vertebrates share a common structure. Development of the eye is regulated by equivalent genes (homologues) in diverse species. One of the most remarkable discoveries is that many of these genes play a similar role in the development of the invertebrate eye (Callaerts *et al.*, 1997). As well as posing important questions about the evolution of eyes, these findings also mean that studies in the invertebrate model organism, the fruit fly *Drosophila melanogaster*, have provided new insights into the genetic regulation of human eye development.

In the following sections, selected studies in embryology and developmental biology using classical embryological techniques and modern molecular techniques are described. These have been selected as examples in each of the key events of eye development, which have led to increased understanding of the molecular basis of eye development and the function and interactions between disease genes.

Specification of the eye field and optic vesicle morphogenesis

The earliest event in formation of the eye is the specification of the eye field (or eye anlage), which is a population of cells in the anterior neural plate that are fated to give rise to the optic vesicle. This specification occurs as part of the anterior–posterior patterning of the neural ectoderm and involves the expression of specific genes. Several genes encoding DNA-binding transcription factors have been identified, *Pax6*, *Rax*, *Six3*, *Six6* and *Lhx2*, which are expressed in the eye field and are critical for early stages of optic vesicle development. Signalling from the midline of the embryo by the secreted protein *Shh* is essential for bisection of the eye field to give paired eye primordia, and subsequent *Shh* signalling from the ventral forebrain also influences patterning of the optic vesicle (Huh *et al.*, 1999; Zhang and Yang, 2001). The targeted or spontaneous mutation of *Pax6*, *Rax*, *Six3* and *Lhx2* in the mouse results in animals with grossly abnormal or no eyes (Hill *et al.*, 1991; Lagutin *et al.*, 2003; Mathers *et al.*, 1997; Porter *et al.*, 1997). In the absence of these genes, eye development does not progress beyond the optic vesicle stage and downstream developmental processes are prevented. These genes are considered as eye determination genes, as they sit at the top of the hierarchy regulating eye development. Further evidence of their dominance in the genetic cascade of eye development comes from the observation that forced expression of *Pax6*, *Rax* or *Six3*, (or the related gene *Six6*) in vertebrates promotes the formation of retinal tissue at ectopic locations

(Lagutin *et al.*, 2001; Loosli *et al.*, 1999). In the zebrafish (*Danio rerio*), *Rax* appears dispensable for eye field specification but is critical for evagination and growth of the optic vesicle. Such findings position *Rax* downstream of the other eye determination genes. *Six6* is also dispensible for optic vesicle development, as optic cups form in mice lacking *Six6* and instead the retinae are hypoplastic (Li *et al.*, 2002).

The prototypical eye determination gene is the *Pax6* gene. Extraordinarily expression of the mouse *Pax6* gene in the antennae of *Drosophila* induces the formation of new compound eyes, indicating that it alone can induce the cascade of genetic interactions necessary for eye formation (Halder *et al.*, 1995). This experiment demonstrates a high level of conservation of *Pax6* function between vertebrates and invertebrates. Several other *Drosophila* genes, such as *Eyeless* (the *Drosophila* homologue of *Pax6*), *Eyes absent* (*Eya*), *Sine oculis* (homologous to *Six3/6*) and *Dachshund* have a similar function. In *Drosophila* these genes appear to act as a network with multiple steps of feedback regulation, including functional interactions between the encoded proteins (Pignoni *et al.*, 1997; Wawersik and Maas, 2000). Inactivating any of these genes results in flies lacking eyes.

In *Xenopus laevis* embryos, misexpression of a single eye determination gene, *Pax6*, results in the formation of new ectopic eyes in the head (Chow *et al.*, 1999). Moreover, misexpression of a cocktail of transcription factors, normally expressed in the eye field of the anterior neural plate, is sufficient to induce ectopic eyes outside the nervous system at high frequency (Zuber, Harris 2003). These experiments were carried out by injecting the RNAs encoding different transcription factors, including *Pax6*, *Six3*, *Rax* and *ET* (*Xenopus Tbx3*) into two cell *Xenopus* embryos and then allowing the embryos to grow to tadpole stage. The ectopic eyes included lens, retinal pigmented epithelium, ganglion cells and photoreceptors.

Growth, patterning and closure of the optic cup

The undifferentiated neuroepithelium of the optic vesicle becomes patterned by the expression of combinations of transcription factor genes within discrete domains. The domains of transcription factor gene expression are regulated by gradients of secreted signalling molecules acting in concert with the eye field transcription factors. This process is the first step in committing regions of the optic vesicle to their different fates. Distinct identities are conferred to regions destined to become the neural retina, the retinal pigmented epithelium, and the optic stalk. Although the full details of the molecular pathways and their coordinated interactions and feedback loops are not yet known, the pivotal roles of many genes and some of their downstream effectors and upstream regulators have been characterized in model organisms.

One of the most important fate determination steps is defining the region of the optic vesicle that will become neural retina. Initially the neural progenitor cells of the optic vesicle can give rise to either non-neural retinal pigmented epithelium or to neural retina. The bi-potentiality of the early neuroepithelium has been demonstrated in experiments in which ablation of the neural retina in chick embryos can promote its regeneration from retinal pigmented epithelium (Park and Hollenberg, 1989). This capacity is generally lost in older embryos. The *Chx10* transcription factor is the earliest known transcription factor, expressed in the presumptive neural retina of the optic vesicle upon its close association with the surface ectoderm (Liu *et al.*, 1994), and its expression signifies a neural retina fate. *Chx10* is known to play an important role in promoting proliferation of neural progenitors in the presumptive neural retina; the retinae of mice lacking *Chx10* fail to grow and mice have small, microphthalmic eyes (Burmeister *et al.*, 1996).

Determination of the neural retina and retinal pigmented epithelium is mediated by signals emanating from surrounding tissues, outside the optic vesicle. Early embryological studies, mainly in amphibian embryos, and in explant culture concluded that contact between the distal tip of the optic vesicle and the pre-lens ectoderm is essential for the specification of the neural retina (Lopashov, 1963). Later work showed that the pre-lens ectoderm is a source of fibroblast growth factor (Fgf; de Iongh and McAvoy, 1993), and Fgf signalling can substitute for the pre-lens ectoderm in specifying the neural retina (Hyer *et al.*, 1998; Nguyen and Arnheiter, 2000; Pittack *et al.*, 1997). Signals from the extra-ocular mesenchyme also play an important role. In explant cultures of optic vesicles from chick embryos, it was shown that signals from the extra-ocular mesenchyme (possibly the growth factor Tgfβ) could induce expression of retinal pigmented epithelium-specific genes such as *Mitf* and repress expression of the neural retina-specific gene *Chx10* (Fuhrmann *et al.*, 2000). Both *Pax2* and *Pax6* appear to be important for retinal pigmented epithelium determination. Both genes are expressed throughout the optic vesicle during establishment of the neural retina, retinal pigmented epithelium and optic stalk progenitor domains. *In vitro*, the *Pax2* and *Pax6* transcription factors bind to and activate the promoter of the *Mitf* gene. In the absence of both *Pax2* and *Pax6* in mice, the optic vesicle lacks expression of *Mitf*, and neural retina markers are expressed in the presumptive retinal pigmented epithelium (Baumer *et al.*, 2003).

During determination of the neural retina and retinal pigmented epithelium regions, the developing optic vesicle becomes demarcated by other patterns of gene expression across the dorso-ventral and naso-temporal axes of the eye, which give regional identity to the top and bottom of the developing optic cup. The *T-box* genes *Tbx2*, *Tbx3* and *Tbx5* are expressed in the dorsal neural retina, the *Bf-1* forkhead transcription factor is expressed in the anterior (nasal) optic vesicle, whereas the ventral region of the developing optic cup expresses *Vax2* and *Pax2* (Barbieri *et al.*, 2002; Huh *et al.*, 1999; Sowden *et al.*, 2001). Loss of regionally expressed transcription factors disrupts the process of eye development, often causing coloboma. For example, mice homozygous for targeted deletion of *Vax2* and *Bf-1* show a coloboma resulting from incomplete closure of the optic fissure (Barbieri *et al.*, 2002; Huh *et al.*, 1999). *Pax2* is important in establishing a boundary within the optic vesicle that defines the position of the optic stalk. In both mice and zebrafish null for the *Pax2* gene, retinal pigmented epithelium cells extend abnormally into the optic stalk, disrupting formation of the optic disc, the site at which retinal ganglion cell axons normally leave the optic cup and cause optic nerve coloboma.

After the neural retina domain has been specified by regional gene expression, the optic vesicle invaginates to form the optic cup. The mechanism of formation of the optic cup is intriguing. One important question is whether the lens is absolutely required for formation of the optic cup. It appears that the lens is dispensable; instead, signalling from the early 'pre-lens' ectoderm is essential for the formation of the optic cup (Hyer *et al.*, 2003). In experiments in which the lens placode was ablated in the chick embryo, the optic cup formed without the concomitant formation of the lens, whereas ablation of the 'pre-lens' ectoderm abolished cup formation (Hyer *et al.*, 2003). Similarly, mice lacking the AP-2 transcription factor have defects in the early morphogenesis of the lens and exhibit a range of phenotypes, including optic cups without lens (West-Mays *et al.*, 1999). In mice in which *Pax6* has been specifically inactivated in the eye surface ectoderm, lens development arrested. Remarkably, independent, fully differentiated neural retinas developed from a single optic vesicle, demonstrating that in this system the developing lens is not necessary to instruct the differentiation of the neural retina but, rather, is required for the correct placement of a single retina in the eye (Ashery-Padan *et al.*, 2000).

In addition to the influence of signalling molecules emanating from surrounding tissues (Shh, Fgf, Tgfβ), other important signalling molecules are produced within the optic cup. For example, the bone morphogenetic protein 4 (Bmp4) and retinoic acid metabolizing enzymes are expressed asymmetrically across the developing optic cup. Disruption of these gradients of signalling molecules prevents normal eye development. For example, reducing retinoic acid availability causes abnormalities in the ventral eye and prevents fissure closure (Marsh-Armstrong *et al.*, 1994; Stull and Wikler, 2000).

The molecular basis of normal closure of the optic fissure in the ventral optic cup is not well understood. Apoptosis is anatomically closely associated with fissure closure (Ozeki *et al.*, 2000). Pax2 protein is normally localized to the fissure as it forms in the ventral optic cup and stalk, and then persists in a cuff of cells encircling the developing optic disc at the site where ganglion cell axons exit the retina (Otteson *et al.*, 1998). Recent progress has been made in elucidating a genetic pathway involved in closure and acting upstream of the *Pax2* gene. Mice that lack members of the c-Jun NH(2)-terminal kinase (JNK) group of mitogen activated protein kinases have coloboma (Weston *et al.*, 2003). *In vitro*, JNK initiates a signalling cascade of dorsally expressed *Bmp4* and ventrally expressed *Shh* that induces *Pax2*.

Formation of the lens vesicle and lens development

Study of the signals needed to induce formation of the lens from surface ectoderm has been a classical system for study of mechanisms of embryonic induction. Tissue grafting experiments performed over the last century, mainly in *Xenopus* embryos, have led to a multi-step model of lens determination (Saha *et al.*, 1989). Early pioneering work by Hans Spemann, suggesting that signals from the optic vesicle were sufficient for lens induction, has been convincingly refuted (Grainger *et al.*,

1992). Several studies have shown that lens-inducing signals also act prior to optic vesicle formation. Signals emanating from the adjacent anterior neural plate induce a lens-forming bias in an extended region of head ectoderm, which is essential for subsequent lens development. The optic vesicle plays a role only in the final stages of lens determination. Progress has been made on identifying the profile of gene expression during lens induction (Zygar *et al.*, 1998). *Pax6*, *Sox2* and *Six3* are expressed in the head ectoderm prior to lens differentiation, whereas *L-maf*, *Prox1* and crystallin genes are expressed at a later stage in the lens placode in a more restricted fashion.

Analysis of the regulation of crystallin gene expression and study of mouse models with impaired lens induction has further unravelled important molecular interactions. Lens induction is absolutely dependent upon several genes, such as *Pax6* and *Lhx2*, as the lens placode does not form in mice null for these genes (Hill, 1991; Porter *et al.*, 1997). *Lhx2* appears to be important for secretion by the optic vesicle of factors that induce the lens vesicle, whereas *Pax6* is required for the surface ectoderm to respond to these factors (Ashery-Padan *et al.*, 2000). Activation of the transcription factor *Sox2* (in mouse; *Sox2/Sox3* in chick) in the *Pax6*-expressing ectoderm is also essential for lens induction. The Sox proteins and Pax6 activate crystallin gene expression (Kamachi *et al.*, 1998) and the crystallin gene promoters contain DNA-binding sites for regulation of their expression by these transcription factors (Nishiguchi *et al.*, 1998). *L-Maf*, which is expressed in the lens placode and is restricted to lens cells, can also trigger lens induction and differentiation (Ogino and Yasuda, 1998).

Several secreted factors (Fgf1, Fgf2, Igf1, Igf2, Bmp7 and Pdgf-A) are known to be important for maturation of the lens vesicle. In the absence of Bmp4 or Bmp7 in mice, the lens fails to develop (Furuta and Hogan, 1998; Wawersik *et al.*, 1999). Exposure of the posterior side of the lens vesicle to secreted factors within the vitreous humour controls the regionalized development of the lens. For example, misexpression of Igf1 in the mouse lens expands the transitional zone and perturbs lens polarization (Shirke *et al.*, 2001).

The importance of normal lens formation for development of the eye globe is suggested by the microphthalmic phenotypes found in association with cataracts. Disruption of *Gja8* (*α8 connexin*) in mice leads to microphthalmia associated with retardation of lens growth and lens fibre maturation (Rong *et al.*, 2002) and *Cx50* (*connexin50*)-null mice exhibited microphthalmia and nuclear cataracts (Mackay *et al.*, 1999). Gap junction proteins, such as the connexins, are important for maintaining normal lens transparency (Gong *et al.*, 1997). *Sox1* deletion in mice also causes microphthalmia and cataract (Nishiguchi *et al.*, 1998).

Development of the anterior segment

Development of the anterior segment requires the coordinated morphogenesis and differentiation of cells originating from the surface ectoderm, the periphery of the optic cup neuroepithelium and the peri-ocular mesenchyme, including neural crest-derived cells.

The contribution of neural crest cells to the developing anterior segment was first demonstrated in tissue grafting experiments between quail and chick embryos, which allowed tracing of cell populations (Johnston *et al.*, 1979; Noden, 1986). Similar experiments using florescent dyes to label migrating cells have confirmed the contribution of neural crest to the mouse anterior segment (Trainor and Tam, 1995). Classical chick embryo transplantation experiments have also demonstrated that inductive signals from the lens are important for anterior segment development (Beebe and Coats, 2000; Genis-Galvez, 1966). A large-scale screen of patterns of gene expression provided molecular evidence for the induction of anterior segment structures by the developing lens (Thut *et al.*, 2001).

Many important genes have now been identified which are important for coordinating development of the anterior segment and which are expressed in the neural crest-derived tissue (*FOX1/PITX2*) or other tissues, such as the lens (*MAF*, *PITX3*, *FOXE3*). *PAX6/Pax6* is considered as a panocular gene, as its essential role in several tissues of the eye has been demonstrated. *Pax6* is by far the most heavily studied of the eye development genes and an extensive body of knowledge has been gathered regarding its biological function and role in disease (van Heyningen and Williamson, 2002). In contrast to the other genes identified as causing anterior segment malformation, *Pax6* is widely expressed during eye development and appears to act to coordinate different processes (Cvekl *et al.*, 2004). It expression in cells deriving from surface ectoderm and optic cup is required for expression of downstream lens transcription factors, such as *maf*, and for crystallin expression. *Pax6* also has a cell autonomous role in development of the trabecular meshwork and the corneal endothelium (Baulmann *et al.*, 2002; Collinson *et al.*, 2003).

In an effort to understand the clinical heterogeneity observed in anterior segment malformations caused by mutation in the same transcription factor gene, *in vitro* studies of mutant protein function have been carried out (Lines *et al.*, 2002). Comparison has been made of the ability of normal and mutant transcription factor proteins to bind to DNA, and of their ability to regulate transcription of reporter genes in cultured cell lines, where the activity of the reporter gene is easily assayed. This approach has demonstrated, for example, that in addition to mutant PITX2 proteins that lack the ability to bind to DNA, other mutant proteins bind DNA, but have a reduced ability to activate transcription (Kozlowski and Walter, 2000). We still know little about the target genes regulated by the anterior segment transcription factors, and their identification is an important step in understanding the aetiology of anterior segment malformation.

Glaucoma occurs in 50–70% of patients with anterior segment malformation and threatens what is often already limited vision. Investigations of glaucoma-related genes, such as *PITX2*, *FOXC1* and *CYP1B1*, in mouse disease models are enabling a better understanding of the relationship between anterior segment development and glaucoma. *CYP1B1*-deficient mice have ocular drainage structure abnormalities resembling those reported in human PCG patients (Libby *et al.*, 2003). The tyrosinase gene (*TYR*; MIM606933) was identified as a modifier of the drainage structure phenotype in *Cyp1b1*$^{-/-}$ mice, with tyrosinase deficiency increasing the magnitude of dysgenesis. Severe dysgenesis in eyes lacking both *Cyp1b1* and *Tyr* was alleviated by administration of the tyrosinase product dihydroxyphenylalanine (L-dopa). Intriguingly *Tyr* also modified the drainage structure dysgenesis in mice with a mutant *Foxc1* gene. These findings suggest that a common pathway involving *Tyr*, *Foxc1* and *Cyp1b1* is important for anterior segment development and raise the possibility that modification of a tyrosinase–L-dopa pathway could ameliorate developmental glaucoma.

Development of the neural retina

In the newly formed optic cup, the presumptive retina comprises a pool of neural progenitor cells, which ultimately give rise to the photoreceptors as well as the other retinal neurons and Muller glial cells of the adult retina. Rapid progress has been made over the last decade in unravelling the cell and molecular processes underlying differentiation of the neural retina. Many of the key factors that are important for differentiation of specific retinal neurons have now been identified (Baumer *et al.*, 2003; Livesey *et al.*, 2000). One of the most important breakthroughs in the study of retinal progenitors was the discovery (using retroviruses carrying marker genes or injection of horseradish peroxidase to trace cell lineages) that all retinal neurons and Muller glial cells derive from the same progenitor cells (Fekete *et al.*, 1994; Holt *et al.*, 1988; Turner and Cepko, 1987; Turner *et al.*, 1990). Each cell type develops, or is born, during a specific period of development (Young, 1985). The order of these birth dates is conserved between mouse and man. For example, retinal ganglion cells are early-born cells, whereas rod photoreceptors are late-born cells. A combination of intrinsic cellular factors, particularly the coordinated expression of specific transcription factors, in addition to extrinsic factors, such as secreted signalling factors, determine the generation of different types of retinal cell (Cepko *et al.*, 1996).

One of the transcription factor genes found to be important for photoreceptor development is the *Crx* gene, which is required for differentiation and maintenance of photoreceptors. Mice lacking *Crx* show a hypocellular outer nuclear layer, normally the location of the photoreceptor cells, which degenerates rapidly (Furukawa *et al.*, 1999). Experiments using gene microarrays have compared the profile of gene expression within the normal retina and the retina lacking *Crx* to identify genes whose expression is altered (Livesey *et al.*, 2000). These experiments have identified putative targets of Crx regulation. Many of these targets are themselves retinal disease genes. For example, Crx regulates expression of the photoreceptor gene *rhodopsin*, whose mutation is a common cause of the inherited retina dystrophy retinitis pigmentosa.

Agenda for the future

This chapter has discussed congenital eye defects for which inherited causes have been discovered. OMIM contains hundreds of examples of ocular defects for which the cause is unknown. The high number of birth defects that include eye malformation likely reflects the complexity of developmental genetics underlying eye development. Nevertheless, with current advances there is now a realistic expectation of a molecular level description of the embryological events that shape the eye and of retinal neurogenesis. Compared with the brain, the eye is a much more tractable part of the CNS for study, due to it accessibility and size. Such future progress relies on state-of-the-art research conducted using modern molecular, cell and developmental biology technologies. It requires the use of different animal model systems, the completion of genome sequences of model organisms, in addition to the human genome, and the synthesis of this knowledge with increasing use of computer-based data-modelling tools.

Eye malformations are a significant cause of childhood blindness. They show complex patterns of inheritance and where genetic causes have been identified they are often genetically heterogeneous. Knowledge of the cell and molecular basis of eye development increasingly informs the choice of candidate genes that are screened for mutations in children with congenital eye defects. A candidate gene approach is often the only option for identification of genetic changes in patients who do not show Mendelian patterns of inheritance and where the defect may result from *de novo* gene mutation. High-throughput DNA screening technology makes it likely that in the future, with sufficient investment, it will become feasible to screen patients for multiple gene mutations. Identification of the genetic basis of a congenital eye defect is immediately of value to families, as parents can then be advised of the likely risk of having a second affected child. With knowledge of the genetic basis of a congenital eye defect, it is possible to determine whether a child has a new mutation (and therefore a minimal risk of recurrence) or whether either or both parents are carriers of the mutation. In conditions where the structural defect predisposes the child to further visual loss, for example anterior segment malformation increases the risk of glaucoma, further study of genotype–phenotype correlations should lead to improved identification of at-risk children. In some instances where the condition is progressive, for example mutations causing loss of photoreceptors, gene replacement therapies may be an effective route to reduce visual loss. Considerable progress has been made in delivering genes to photoreceptor cells and promoting the generation of new photoreceptor outer segments in animal models of retinal degeneration (Ali *et al.*, 2000). This approach has been shown to effectively reduce the rate of visual loss in animal models and future transfer to the clinic looks feasible.

Genetic analysis has highlighted the critical role that transcription factors play in birth defects affecting the eye. Mutations in transcription factor genes are by far the most common identified cause of structural eye defects. The downstream target genes, which are regulated by each transcription factor are still largely unknown. The availability of post-genomic high-throughput methods for identification of downstream genetic pathways is now set to change this impasse. In particular, the use of gene microarrays to compare gene expression profiles in normal and abnormal tissues and the use of chromatin immunoprecipitation to identify transcription factor binding sites are two examples of fruitful approaches for the identification of target genes. Completion of a molecular level description of eye development, combined with further identification of disease genes in patients, will make clear the common pathways affected by mutations in different genes. It is hoped that this knowledge will present new opportunities for therapeutic intervention. It may be possible to identify susceptible genotypes or gene–environment interactions that increase the likelihood of ocular birth defects.

Study of the embryology and development of the eye is also contributing to the expanding field of stem cell research and its possible application for the treatment of blindness. The identification of stem cells within the ciliary epithelium of the adult eye offers a potential source of cells that could be used to develop therapies to repair diseased parts of the eye (Ahmad *et al.*, 2000). This approach offers promise for the replacement of neurons lost in retinal degeneration, for which there is currently no treatment (Ali and Sowden, 2003). The ciliary epithelial stem cells appear to be quiescent multipotential cells derived from the embryonic neuroepithelium at the margins of the optic cup. To use these cells therapeutically, their biology needs to be sufficiently well understood so that their numbers can be expanded *in vitro* before transplantation into diseased retina. Genetic and cellular factors regulating retinal development are being explored as therapeutic tools to induce transplanted cells to repair diseased retina by forming new retinal neurons and integrating with existing retinal circuitry. An alternative possibility is that stem cells in the retina (endogenous stem cells), rather than transplanted cells, could be induced to recapitulate their behaviour during development to regenerate diseased retina (Fischer and Reh, 2001).

This chapter has focused on congenital and inherited anomalies, which are among the main unavoidable causes of blindness in children and among the largest causes in affluent countries. However, almost half of all blindness in children is avoidable, with three-quarters of the world's blind children living in developing countries (Gilbert and Awan, 2003). While the future reduction of inherited birth defects affecting the eye will remain a significant research challenge, there is now a global initiative – Vision 2020: the Right to Sight– which aims to eliminate avoidable blindness by 2020 (www.v2020.org). The priorities of the initiative include elimination of corneal scarring due to vitamin A deficiency, and measles, and treatment of cataract (Gilbert and Awan, 2003).

References

MIM numbers from the Online Mendelian Inheritance in Man database are cited throughout this chapter and in Table 9.2, and can be used to access additional references. This database is a catalogue of human genes and genetic disorders authored and edited by Dr Victor A. McKusick and his colleagues at Johns Hopkins and elsewhere, and developed for the World Wide Web by NCBI, the National Center for Biotechnology Information. The web address is: http://www.ncbi.nlm.nih.gov/entrez/query.fcgi?db=OMIM

Aalfs, C.M., Fantes, J.A., Wenniger-Prick, L.J., Sluijter, S. *et al.* (1997) Tandem duplication of 11p12-p13 in a child with borderline development delay and eye abnormalities: dose effect of the *PAX6* gene product? *Am. J. Med. Genet.* **73**: 267–271.

Ahmad, I., Tang, L. and Pham, H. (2000) Identification of neural progenitors in the adult mammalian eye. *Biochem. Biophys. Res. Commun.* **270**: 517–521.

Ali, R.R., Sarra, G.M., Stephens, C., Alwis, M.D. *et al.* (2000) Restoration of photoreceptor ultrastructure and function in retinal degeneration slow mice by gene therapy. *Nat. Genet.* **25**: 306–310.

Ali, R.R. and Sowden, J.C. (2003) Therapy may yet stem from cells in the retina. *Br. J. Ophthalmol.* **87**: 1058–9.

Alward, W.L. (2000) Axenfeld–Rieger syndrome in the age of molecular genetics. *Am. J. Ophthalmol.* **130**: 107–115.

Ashery-Padan, R., Marquardt, T., Zhou, X. and Gruss, P. (2000) Pax6 activity in the lens primordium is required for lens formation and for correct placement of a single retina in the eye. *Genes Dev.* **14**: 2701–2711.

Azuma, N., Hirakiyama, A., Inoue, T., Asaka, A. and Yamada, M. (2000) Mutations of a human homologue of the *Drosophila* eyes absent gene (*EYA1*) detected in patients with congenital cataracts and ocular anterior segment anomalies. *Hum. Mol. Genet.* **9**: 363–366.

Bar-Yosef, U., Abuelaish, I., Harel, T., Hendler, N. *et al.* (2004) *CHX10* mutations cause non-syndromic microphthalmia/anophthalmia in Arab and Jewish kindreds. *Hum. Genet.* **115**: 302–309.

Barbieri, A.M., Broccoli, V., Bovolenta, P., Alfano, G. *et al.* (2002) *Vax2* inactivation in mouse determines alteration of the eye dorsal–ventral axis, misrouting of the optic fibres and eye coloboma. *Development* **129**: 805–813.

Baulmann, D.C., Ohlmann, A., Flugel-Koch, C., Goswami, S. *et al.* (2002) *Pax6* heterozygous eyes show defects in chamber angle differentiation that are associated with a wide spectrum of other anterior eye segment abnormalities. *Mech. Dev.* **118**: 3–17.

Baumer, N., Marquardt, T., Stoykova, A., Spieler, D. *et al.* (2003) Retinal pigmented epithelium determination requires the redundant activities of Pax2 and Pax6. *Development* **130**: 2903–2915.

Beebe, D.C. and Coats, J.M. (2000) The lens organizes the anterior segment: specification of neural crest cell differentiation in the avian eye. *Dev. Biol.* **220**: 424–431.

Burmeister, M., Novak, J., Liang, M. Y., Basu, S. *et al.* (1996) Ocular retardation mouse caused by *Chx10* homeobox null allele: impaired retinal progenitor proliferation and bipolar cell differentiation. *Nat. Genet.* **12**: 376–384.

Callaerts, P., Halder, G. and Gehring, W.J. (1997) *PAX-6* in development and evolution. *Annu. Rev. Neurosci.* **20**: 483–532.

Cepko, C.L., Austin, C.P., Yang, X., Alexiades, M. and Ezzeddine, D. (1996) Cell fate determination in the vertebrate retina. *Proc. Natl Acad. Sci. USA* **93**: 589–595.

Chow, R.L., Altmann, C.R., Lang, R.A. and Hemmati-Brivanlou, A. (1999) Pax6 induces ectopic eyes in a vertebrate. *Development* **126**: 4213–4222.

Collinson, J.M., Quinn, J.C., Hill, R.E. and West, J.D. (2003) The roles of Pax6 in the cornea, retina, and olfactory epithelium of the developing mouse embryo. *Dev. Biol.* **255**: 303–312.

Cvekl, A. and Tamm, E.R. (2004) Anterior eye development and ocular mesenchyme: new insights from mouse models and human diseases. *Bioessays* **26**: 374–386.

Cvekl, A., Yang, Y., Chauhan, B. K. and Cveklova, K. (2004) Regulation of gene expression by Pax6 in ocular cells: a case of tissue-preferred expression of crystallins in lens. *Int. J. Dev. Biol.* **48**: 829–844.

Dattani, M.T., Martinez-Barbera, J.P., Thomas, P.Q., Brickman, J.M. *et al.* (1998) Mutations in the homeobox gene *HESX1/Hesx1* associated with septo-optic dysplasia in human and mouse. *Nat. Genet.* **19**: 125–133.

de Iongh, R., and McAvoy, J.W. (1993). Spatio-temporal distribution of acidic and basic FGF indicates a role for FGF in rat lens morphogenesis. *Dev. Dyn.* **198**: 190–202.

Fantes, J., Ragge, N.K., Lynch, S.A., McGill, N.I. *et al.* (2003) Mutations in *SOX2* cause anophthalmia. *Nat. Genet.* **33**: 461–463.

Favor, J., Sandulache, R., Neuhauser-Klaus, A., Pretsch, W. *et al.* (1996) The mouse *Pax2(1Neu)* mutation is identical to a human *PAX2* mutation in a family with renal–coloboma syndrome and results in developmental defects of the brain, ear, eye, and kidney. *Proc. Natl Acad. Sci. USA* **93**: 13870–13875.

Fekete, D.M., Perez-Miguelsanz, J., Ryder, E.F. and Cepko, C.L. (1994) Clonal analysis in the chicken retina reveals tangential dispersion of clonally related cells. *Dev. Biol.* **166**: 666–682.

Ferda Percin, E., Ploder, L.A., Yu, J.J., Arici, K. *et al.* (2000) Human microphthalmia associated with mutations in the retinal homeobox gene *CHX10*. *Nat. Genet.* **25**: 397–401.

Fischer, A.J. and Reh, T.A. (2001) Transdifferentiation of pigmented epithelial cells: a source of retinal stem cells? *Dev. Neurosci.* **23**: 268–276.

Francis, P.J. and Moore, A.T. (2004) Genetics of childhood cataract. *Curr. Opin. Ophthalmol.* **15**: 10–15.

Fuhrmann, S., Levine, E.M. and Reh, T.A. (2000) Extra-ocular mesenchyme patterns the optic vesicle during early eye development in the embryonic chick. *Development* **127**: 4599–4609.

Furukawa, T., Morrow, E.M., Li, T., Davis, F.C. and Cepko, C.L. (1999) Retinopathy and attenuated circadian entrainment in Crx-deficient mice. *Nat. Genet.* **23**: 466–470.

Furuta, Y. and Hogan, B.L. (1998) BMP4 is essential for lens induction in the mouse embryo. *Genes Dev.* **12**: 3764–3775.

Gage, P.J., Suh, H. and Camper, S.A. (1999) Dosage requirement of Pitx2 for development of multiple organs. *Development* **126**: 4643–4651.

Gallardo, M.E., Lopez-Rios, J., Fernaud-Espinosa, I., Granadino, B. *et al.* (1999) Genomic cloning and characterization of the human homeobox gene *SIX6* reveals a cluster of *SIX* genes in chromosome 14 and associates *SIX6* hemizygosity with bilateral anophthalmia and pituitary anomalies. *Genomics* **61**: 82–91.

Genis-Galvez, J.M. (1966) Role of the lens in the morphogenesis of the iris and cornea. *Nature* **210**: 209–210.

Gilbert, C. and Awan, H. (2003) Blindness in children. *Br. Med J.* **327**: 760–761.

Glaser, T., Jepeal, L., Edwards, J.G., Young, S.R. *et al.* (1994) *PAX6* gene dosage effect in a family with congenital cataracts, aniridia, anophthalmia and central nervous system defects. *Nat. Genet.* **7**: 463–471.

Gong, X., Li, E., Klier, G., Huang, Q. *et al.* (1997) Disruption of *α3 connexin* gene leads to proteolysis and cataractogenesis in mice. *Cell* **91**: 833–843.

Graham, V., Khudyakov, J., Ellis, P. and Pevny, L. (2003) *SOX2* functions to maintain neural progenitor identity. *Neuron* **39**: 749–765.

Grainger, R.M., Henry, J.J., Saha, M.S. and Servetnick, M. (1992) Recent progress on the mechanisms of embryonic lens formation. *Eye* **6**(2): 117–122.

Graw, J. (2004) Congenital hereditary cataracts. *Int. J. Dev. Biol.* **48**: 1031–1044.

Graw, J. and Loster, J. (2003) Developmental genetics in ophthalmology. *Ophthalm. Genet.* **24**: 1–33.

Halder, G., Callaerts, P. and Gehring, W.J. (1995) Induction of ectopic eyes by targeted expression of the eyeless gene in *Drosophila*. *Science* **267**: 1788–1792.

Hanein, S., Perrault, I., Gerber, S., Tanguy, G. *et al.* (2004) Leber congenital amaurosis: comprehensive survey of the genetic heterogeneity, refinement of the clinical definition, and genotype–phenotype correlations as a strategy for molecular diagnosis. *Hum. Mutat.* **23**: 306–317.

Hejtmancik, J.F. and Smaoui, N. (2003) Molecular genetics of cataract. *Dev. Ophthalmol.* **37**: 67–82.

Hill, R.E., Favor, J., Hogan, B.L., Ton, C.C. *et al.* (1991) Mouse small eye results from mutations in a paired-like homeobox-containing gene. *Nature* **354**: 522–525.

Holt, C.E., Bertsch, T.W., Ellis, H.M. and Harris, W.A. (1988) Cellular determination in the *Xenopus* retina is independent of lineage and birth date. *Neuron* **1**: 15–26.

Honkanen, R.A., Nishimura, D.Y., Swiderski, R.E., Bennett, S.R. *et al.* (2003) A family with Axenfeld–Rieger syndrome and Peters anomaly caused by a point mutation (*Phe112Ser*) in the *FOXC1* gene. *Am. J. Ophthalmol.* **135**: 368–375.

Huh, S., Hatini, V., Marcus, R.C., Li, S.C. and Lai, E. (1999) Dorsal–ventral patterning defects in the eye of *BF-1*-deficient mice associated with a restricted loss of *shh* expression. *Dev. Biol.* **211**: 53–63.

Hyer, J., Kuhlman, J., Afif, E. and Mikawa, T. (2003) Optic cup morphogenesis requires pre-lens ectoderm but not lens differentiation. *Dev. Biol.* **259**: 351–363.

Hyer, J., Mima, T. and Mikawa, T. (1998) *FGF1* patterns the optic vesicle by directing the placement of the neural retina domain. *Development* **125**: 869–877.

Jamieson, R.V., Perveen, R., Kerr, B., Carette, M. *et al.* (2002) Domain disruption and mutation of the bZIP transcription factor, MAF, associated with cataract, ocular anterior segment dysgenesis and coloboma. *Hum. Mol. Genet.* **11**: 33–42.

Johnston, M.C., Noden, D.M., Hazelton, R.D., Coulombre, J.L. and Coulombre, A.J. (1979) Origins of avian ocular and peri-ocular tissues. *Exp. Eye Res.* **29**: 27–43.

Jordan, T., Hanson, I., Zaletayev, D., Hodgson, S. *et al.* (1992) The human *PAX6* gene is mutated in two patients with aniridia. *Nat. Genet.* **1**: 328–332.

Kamachi, Y., Sockanathan, S., Liu, Q., Breitman, M. *et al.* (1995) Involvement of SOX proteins in lens-specific activation of crystallin genes. *EMBO J.* **14**: 3510–3519.

Kamachi, Y., Uchikawa, M., Collignon, J., Lovell-Badge, R. and Kondoh, H. (1998) Involvement of Sox1, 2 and 3 in the early and subsequent molecular events of lens induction. *Development* **125**: 2521–2532.

Kaufman, M.H. (1992) *The Atlas of Mouse Development*. Academic Press: London.

Kidson, S.H., Kume, T., Deng, K., Winfrey, V. and Hogan, B.L. (1999) The forkhead/winged-helix gene, *Mf1*, is necessary for the normal development of the cornea and formation of the anterior chamber in the mouse eye. *Dev. Biol.* **211**: 306–322.

Kim, J.I., Li, T., Ho, I.C., Grusby, M.J. and Glimcher, L.H. (1999) Requirement for the c-Maf transcription factor in crystallin gene regulation and lens development. *Proc. Natl Acad. Sci. USA* **96**: 3781–3785.

Kitamura, K., Miura, H., Miyagawa-Tomita, S., Yanazawa, M. *et al.* (1999) Mouse *Pitx2* deficiency leads to anomalies of the ventral body wall, heart, extra- and peri-ocular mesoderm and right pulmonary isomerism. *Development* **126**: 5749–5758.

Kozlowski, K. and Walter, M.A. (2000) Variation in residual PITX2 activity underlies the phenotypic spectrum of anterior segment developmental disorders. *Hum. Mol. Genet.* **9**: 2131–2139.

Kume, T., Deng, K. and Hogan, B.L. (2000) Murine forkhead/winged helix genes *Foxc1* (*Mf1*) and *Foxc2* (*Mfh1*) are required for the early organogenesis of the kidney and urinary tract. *Development* **127**: 1387–1395.

Lagutin, O., Zhu, C.C., Furuta, Y., Rowitch, D.H. *et al.* (2001) Six3 promotes the formation of ectopic optic vesicle-like structures in mouse embryos. *Dev. Dyn.* **221**: 342–349.

Lagutin, O.V., Zhu, C.C., Kobayashi, D., Topczewski, J. *et al.* (2003) Six3 repression of Wnt signalling in the anterior neuroectoderm is essential for vertebrate forebrain development. *Genes Dev.* **17**: 368–379.

Larsen, W.J. (2001). *Human Embryology.* Churchill Livingstone: Philadelphia, PA.

Lehmann, O.J., Ebenezer, N.D., Jordan, T., Fox, M. *et al.* (2000) Chromosomal duplication involving the forkhead transcription factor gene *FOXC1* causes iris hypoplasia and glaucoma. *Am. J. Hum. Genet.* **67**: 1129–1135.

Li, X., Perissi, V., Liu, F., Rose, D.W. and Rosenfeld, M.G. (2002) Tissue-specific regulation of retinal and pituitary precursor cell proliferation. *Science* **297**: 1180–1183.

Libby, R.T., Smith, R.S., Savinova, O.V., Zabaleta, A. *et al.* (2003) Modification of ocular defects in mouse developmental glaucoma models by tyrosinase. *Science* **299**: 1578–1581.

Lines, M.A., Kozlowski, K., Kulak, S.C., Allingham, R.R. *et al.* (2004) Characterization and prevalence of *PITX2* microdeletions and mutations in Axenfeld–Rieger malformations. *Invest. Ophthalmol. Vis. Sci.* **45**: 828–833.

Lines, M.A., Kozlowski, K. and Walter, M.A. (2002) Molecular genetics of Axenfeld–Rieger malformations. *Hum. Mol. Genet.* **11**: 1177–1184.

Liu, I.S., Chen, J.D., Ploder, L., Vidgen, D. *et al.* (1994) Developmental expression of a novel murine homeobox gene (*Chx10*): evidence for roles in determination of the neuroretina and inner nuclear layer. *Neuron* **13**: 377–393.

Livesey, F.J., Furukawa, T., Steffen, M.A., Church, G.M. and Cepko, C.L. (2000) Microarray analysis of the transcriptional network controlled by the photoreceptor homeobox gene *Crx*. *Curr. Biol.* **10**: 301–310.

Loosli, F., Winkler, S. and Wittbrodt, J. (1999) *Six3* overexpression initiates the formation of ectopic retina. *Genes Dev.* **13**: 649–654.

Lopashov, G.V. (1963) *Developmental Mechanisms of Vertebrate Eye Rudiments.* Pergamon: New York.

Lyon, M.F., Jamieson, R.V., Perveen, R., Glenister, P.H. *et al.* (2003) A dominant mutation within the DNA-binding domain of the bZIP transcription factor Maf causes murine cataract and results in selective alteration in DNA binding. *Hum. Mol. Genet.* **12**: 585–594.

Mackay, D., Ionides, A., Kibar, Z., Rouleau, G. *et al.* (1999) *Connexin46* mutations in autosomal dominant congenital cataract. *Am. J. Hum. Genet.* **64**: 1357–1364.

Mann, I. (1964) *The Development of the Human Eye.* Cambridge University Press: London.

Marsh-Armstrong, N., McCaffery, P., Gilbert, W., Dowling, J.E. and Drager, U.C. (1994) Retinoic acid is necessary for development of the ventral retina in zebrafish. *Proc. Natl Acad. Sci. USA* **91**: 7286–7290.

Mathers, P.H., Grinberg, A., Mahon, K.A. and Jamrich, M. (1997) The *Rx* homeobox gene is essential for vertebrate eye development. *Nature* **387**: 603–607.

Mears, A.J., Jordan, T., Mirzayans, F., Dubois, S. *et al.* (1998) Mutations of the forkhead/winged-helix gene, *FKHL7*, in patients with Axenfeld–Rieger anomaly. *Am. J. Hum. Genet.* **63**: 1316–1328.

Morrison, D., FitzPatrick, D., Hanson, I., Williamson, K. *et al.* (2002) National study of microphthalmia, anophthalmia and coloboma (MAC) in Scotland: investigation of genetic aetiology. *J. Med. Genet.* **39**: 16–22.

Muhit, M. and Gilbert, C. (2003) A review of the epidemiology and control of childhood blindness. *Trop. Doct.* **33**: 197–201.

Nanni, L., Ming, J.E., Bocian, M., Steinhaus, K. *et al.* (1999) The mutational spectrum of the *sonic hedgehog* gene in holoprosencephaly: *SHH* mutations cause a significant proportion of autosomal dominant holoprosencephaly. *Hum. Mol. Genet.* **8**: 2479–2488.

Ng, D., Thakker, N., Corcoran, C.M., Donnai, D. *et al.* (2004) Oculofaciocardiodental and Lenz microphthalmia syndromes result from distinct classes of mutations in BCOR. *Nat. Genet.* **36**: 411–416.

Nguyen, M. and Arnheiter, H. (2000) Signalling and transcriptional regulation in early mammalian eye development: a link between FGF and MITF. *Development* **127**: 3581–3591.

Nishiguchi, S., Wood, H., Kondoh, H., Lovell-Badge, R. and Episkopou, V. (1998) *Sox1* directly regulates the gamma-crystallin genes and is essential for lens development in mice. *Genes Dev.* **12**: 776–781.

Nishimura, D.Y., Searby, C.C., Alward, W.L., Walton, D. *et al.* (2001) A spectrum of *FOXC1* mutations suggests gene dosage as a mechanism for developmental defects of the anterior chamber of the eye. *Am. J. Hum. Genet.* **68**: 364–372.

Noden, D.M. (1986) Origins and patterning of craniofacial mesenchymal tissues. *J. Craniofac. Genet. Dev. Biol.* **2**(suppl): 15–31.

O'Rahilly, R. (1975) The prenatal development of the human eye. *Exp. Eye Res.* **21**: 93–112.

Oetting, W.S., Fryer, J.P., Shriram, S. and King, R.A. (2003) Oculocutaneous albinism type 1: the last 100 years. *Pigment Cell Res* **16**: 307–311.

Ogino, H. and Yasuda, K. (1998) Induction of lens differentiation by activation of a bZIP transcription factor, L-Maf. *Science* **280**: 115–118.

Onwochei, B.C., Simon, J.W., Bateman, J.B., Couture, K.C. and Mir, E. (2000) Ocular colobomata. *Surv. Ophthalmol.* **45**: 175–194.

Ormestad, M., Blixt, A., Churchill, A., Martinsson, T. *et al.* (2002) *Foxe3* haploinsufficiency in mice: a model for Peters anomaly. *Invest. Ophthalmol. Vis. Sci.* **43**: 1350–57.

Otteson, D.C., Shelden, E., Jones, J.M., Kameoka, J. and Hitchcock, P.F. (1998) *Pax2* expression and retinal morphogenesis in the normal and Krd mouse. *Dev. Biol.* **193**: 209–224.

Ozeki, H., Ogura, Y., Hirabayashi, Y. and Shimada, S. (2000) Apoptosis is associated with formation and persistence of the embryonic fissure. *Curr. Eye Res.* **20**: 367–372.

Park, C.M. and Hollenberg, M.J. (1989) Basic fibroblast growth factor induces retinal regeneration *in vivo*. *Dev. Biol.* **134**: 201–205.

Perveen, R., Lloyd, I.C., Clayton-Smith, J., Churchill, A. *et al.* (2000) Phenotypic variability and asymmetry of Rieger syndrome associated with *PITX2* mutations. *Invest. Ophthalmol. Vis. Sci.* **41**: 2456–2460.

Pignoni, F., Hu, B., Zavitz, K.H., Xiao, J. *et al.* (1997) The eye-specification proteins So and Eya form a complex and regulate multiple steps in *Drosophila* eye development. *Cell* **91**: 881–891.

Pittack, C., Grunwald, G.B. and Reh, T.A. (1997) Fibroblast growth factors are necessary for neural retina but not pigmented epithelium differentiation in chick embryos. *Development* **124**: 805–816.

Porter, F.D., Drago, J., Xu, Y., Cheema, S.S. *et al.* (1997) *Lhx2*, a *LIM* homeobox gene, is required for eye, forebrain, and definitive erythrocyte development. *Development* **124**: 2935–2944.

Ragge, N.K., Lorenz, B., Schneider, A., Bushby, K. *et al.* (2005) *SOX2* anophthalmia syndrome. *Am. J. Med. Genet. A* **135**: 1–7; discussion 8.

Ragge, N.K., Brown, A.G., Poloschek, C.M., Lorenz, B., Henderson, R.A., Clarke, M.P., Russell-Eggitt, I., Fielder, A., Gerreli, D., Martinez-Barbera, J.P., Ruddle, P., Hurst, J., Collin, J.R. Salt, A. Copper, S.T., Thompson, P.J., Sisodiya, S.M., Williamson, K.A., Fitzpatrick, D.R., van Heyningen, V., Hamson, I.M. (2005) Heterozygous mutations of OTX2 cause severe ocular malformations. *Am. J. Hum. Genet.* **76**: 1008–1022.

Rahi, J. and Dezateux, C. (2001) Epidemiology of childhood visual impairment in the United Kingdom and Ireland. In *Recent Advances in Paediatrics*, vol 19, David, T. (ed.). Churchill Livingstone: Edinburgh; 97–114.

Rahi, J.S. and Dezateaux, C. (2001) Measuring and interpreting the incidence of congenital ocular anomalies: lessons from a national study of congenital cataract in the UK. *Invest. Ophthalmol. Vis. Sci.* **42**: 1444–1448.

Ring, B.Z., Cordes, S.P., Overbeek, P.A. and Barsh, G.S. (2000) Regulation of mouse lens fiber cell development and differentiation by the *Maf* gene. *Development* **127**: 307–317.

Roessler, E., Belloni, E., Gaudenz, K., Vargas, F. *et al.* (1997) Mutations in the C-terminal domain of *Sonic hedgehog* cause holoprosencephaly. *Hum. Mol. Genet.* **6**: 1847–1853.

Rong, P., Wang, X., Niesman, I., Wu, Y. *et al.* (2002) Disruption of *Gja8* (*α8 connexin*) in mice leads to microphthalmia associated with retardation of lens growth and lens fiber maturation. *Development* **129**: 167–174.

Russell-Eggitt, I. (2001) Albinism. *Ophthalmol. Clin. N. Am.* **14**: 533–546.

Saha, M.S., Spann, C.L. and Grainger, R.M. (1989) Embryonic lens induction: more than meets the optic vesicle. *Cell Differ. Dev.* **28**: 153–171.

Sanyanusin, P., McNoe, L.A., Sullivan, M.J., Weaver, R.G. and Eccles, M.R. (1995) Mutation of *PAX2* in two siblings with renal–coloboma syndrome. *Hum. Mol. Genet.* **4**: 2183–2184.

Schedl, A., Ross, A., Lee, M., Engelkamp, D. *et al.* (1996) Influence of *PAX6* gene dosage on development: overexpression causes severe eye abnormalities. *Cell* **86**: 71–82.

Schimmenti, L.A., de la Cruz, J., Lewis, R.A., Karkera, J.D. *et al.* (2003) Novel mutation in *Sonic hedgehog* in non-syndromic colobomatous microphthalmia. *Am. J. Med. Genet.* **116A**: 215–221.

Seeliger, M.W., Biesalski, H.K., Wissinger, B., Gollnick, H. *et al.* (1999) Phenotype in retinol deficiency due to a hereditary defect in retinol binding protein synthesis. *Invest. Ophthalmol. Vis. Sci.* **40**: 3–11.

Semina, E.V., Brownell, I., Mintz-Hittner, H.A., Murray, J.C. and Jamrich, M. (2001) Mutations in the human forkhead transcription factor FOXE3 associated with anterior segment ocular dysgenesis and cataracts. *Hum. Mol. Genet.* **10**: 231–236.

Semina, E.V., Ferrell, R.E., Mintz-Hittner, H.A., Bitoun, P. *et al.* (1998) A novel homeobox gene *PITX3* is mutated in families with autosomal-dominant cataracts and ASMD. *Nat. Genet.* **19**: 167–170.

Semina, E.V., Murray, J.C., Reiter, R., Hrstka, R.F. and Graw, J. (2000) Deletion in the promoter region and altered expression of *Pitx3* homeobox gene in aphakia mice. *Hum. Mol. Genet.* **9**: 1575–1585.

Semina, E.V., Reiter, R., Leysens, N.J., Alward, W.L. *et al.* (1996) Cloning and characterization of a novel bicoid-related homeobox transcription factor gene, *RIEG*, involved in Rieger syndrome. *Nat. Genet.* **14**: 392–399.

Shirke, S., Faber, S.C., Hallem, E., Makarenkova, H.P. *et al.* (2001) Misexpression of IGF-I in the mouse lens expands the transitional zone and perturbs lens polarization. *Mech. Dev.* **101**: 167–174.

Sowden, J.C., Holt, J.K., Meins, M., Smith, H.K. and Bhattacharya, S.S. (2001) Expression of *Drosophila omb*-related T-box genes in the developing human and mouse neural retina. *Invest. Ophthalmol. Vis. Sci.* **42**: 3095–3102.

Stoilov, I., Akarsu, A.N., Alozie, I., Child, A. *et al.* (1998) Sequence analysis and homology modeling suggest that primary congenital glaucoma on 2p21 results from mutations disrupting either the hinge region or the conserved core structures of cytochrome P4501B1. *Am. J. Hum. Genet.* **62**: 573–584.

Stoilov, I., Akarsu, A.N. and Sarfarazi, M. (1997) Identification of three different truncating mutations in cytochrome P4501B1 (*CYP1B1*) as the principal cause of primary congenital glaucoma (buphthalmos) in families linked to the *GLC3A* locus on chromosome 2p21. *Hum. Mol. Genet.* **6**: 641–647.

Stoll, C., Alembik, Y., Dott, B. and Roth, M.P. (1992) Epidemiology of congenital eye malformations in 131 760 consecutive births. *Ophthalm. Paediatr. Genet.* **13**: 179–186.

Stull, D.L. and Wikler, K.C. (2000) Retinoid-dependent gene expression regulates early morphological events in the development of the murine retina. *J. Comp. Neurol.* **417**: 289–298.

Thut, C.J., Rountree, R.B., Hwa, M. and Kingsley, D.M. (2001) A large-scale *in situ* screen provides molecular evidence for the induction of eye anterior segment structures by the developing lens. *Dev. Biol.* **231**: 63–76.

Trainor, P.A. and Tam, P.P. (1995). Cranial paraxial mesoderm and neural crest cells of the mouse embryo: co-distribution in the craniofacial mesenchyme but distinct segregation in branchial arches. *Development* **121**: 2569–2582.

Turner, D.L. and Cepko, C.L. (1987) A common progenitor for neurons and glia persists in rat retina late in development. *Nature* **328**: 131–136.

Turner, D.L., Snyder, E.Y. and Cepko, C.L. (1990) Lineage-independent determination of cell type in the embryonic mouse retina. *Neuron* **4**: 833–845.

van Heyningen, V. and Williamson, K.A. (2002) *PAX6* in sensory development. *Hum. Mol. Genet.* **11**: 1161–1167.

Vincent, A., Billingsley, G., Priston, M., Williams-Lyn, D. *et al.* (2001) Phenotypic heterogeneity of *CYP1B1*: mutations in a patient with Peters anomaly. *J. Med. Genet.* **38**: 324–326.

Vincent, A.L., Billingsley, G., Buys, Y., Levin, A.V. *et al.* (2002) Digenic inheritance of early-onset glaucoma: *CYP1B1*, a potential modifier gene. *Am. J. Hum. Genet.* **70**: 448–460.

Voronina, V.A., Kozhemyakina, E.A., O'Kernick, C.M., Kahn, N.D. *et al.* (2004) Mutations in the human *RAX* homeobox gene in a patient with anophthalmia and sclerocornea. *Hum. Mol. Genet.* **13**: 315–322.

Wallis, D.E., Roessler, E., Hehr, U., Nanni, L. *et al.* (1999) Mutations in the homeodomain of the human *SIX3* gene cause holoprosencephaly. *Nat. Genet.* **22**: 196–198.

Warburg, M. (1993). Classification of microphthalmos and coloboma. *J. Med. Genet.* **30**: 664–669.

Wawersik, S. and Maas, R.L. (2000) Vertebrate eye development as modeled in *Drosophila*. *Hum. Mol. Genet.* **9**: 917–925.

Wawersik, S., Purcell, P., Rauchman, M., Dudley, A.T. *et al.* (1999) *BMP7* acts in murine lens placode development. *Dev. Biol.* **207**: 176–188.

Weiss, A.H., Kousseff, B.G., Ross, E.A. and Longbottom, J. (1989a) Complex microphthalmos. *Arch. Ophthalmol.* **107**: 1619–1624.

Weiss, A.H., Kousseff, B.G., Ross, E.A. and Longbottom, J. (1989b) Simple microphthalmos. *Arch. Ophthalmol.* **107**: 1625–1630.

West-Mays, J.A., Zhang, J., Nottoli, T., Hagopian-Donaldson, S. *et al.* (1999) AP-2α transcription factor is required for early morphogenesis of the lens vesicle. *Dev. Biol.* **206**: 46–62.

Weston, C.R., Wong, A., Hall, J.P., Goad, M.E. *et al.* (2003) JNK initiates a cytokine cascade that causes *Pax2* expression and closure of the optic fissure. *Genes Dev.* **17**: 1271–1280.

Young, R.W. (1985) Cell differentiation in the retina of the mouse. *Anat. Rec.* **212**: 199–205.

Zhang, X.M. and Yang, X.J. (2001) Temporal and spatial effects of Sonic hedgehog signalling in chick eye morphogenesis. *Dev. Biol.* **233**: 271–290.

Zuber, M.E., Gestro, G., Viczian, A.S., Barsacchi, G., Harris, W.A. (2003) Specification of the vertebrate eye by a network of eye field transcription factors. *Development*, **130**: 5155–5167.

Zygar, C.A., Cook, T.L. and Grainger, R.M. Jr. (1998) Gene activation during early stages of lens induction in *Xenopus*. *Development* **125**: 3509–3519.

10

The Ear

Sarah L. Spiden and **Karen P. Steel**

Introduction

The ear is a complex structure, with components derived from all three germ layers. Many basic developmental processes and well-known genes contribute to its construction. The adult ear consists of an outer, middle and inner ear with associated cochlear and vestibular ganglia. The outer ear includes the pinna and the external ear canal (external auditory meatus), which serve to direct sound vibrations towards the tympanic membrane (ear drum; Figure 10.1). The tympanic membrane converts the sound pressure waves into mechanical movement and these vibrations are conducted through the air-filled middle ear to the inner ear by the three middle ear ossicles (the malleus, incus and stapes). The ossicular chain preserves the sound energy and amplifies it by its lever action, and the relative size of the large tympanic membrane compared with the smaller opening into the inner ear (the oval window) allows vibration to be passed from air to inner ear fluid with minimal loss of energy. Movement of the stapes footplate inserted into the oval window initiates a travelling wave along the length of the coiled cochlear duct, leading to up-and-down motion of the flexible basilar membrane, upon which lies the sensory epithelium of the cochlea, the organ of Corti (Figure 10.2). This motion leads to deflection of the hair bundles (ordered arrays of modified microvilli called stereocilia) at the top of each sensory hair cell in the organ of Corti. Extracellular links between adjacent stereocilia pull open transduction channels when the hair bundle is deflected, allowing cations to flood through the opened channels into the hair cell, depolarizing it and triggering synaptic activity at the base of the cell. The flow of cations is enhanced by the high resting potential (endocochlear potential) of the fluid (endolymph) bathing the tops of the hair cells, generated by the stria vascularis on the lateral wall of the cochlear duct, providing a large potential difference across transduction channels. Endolymph

Embryos, Genes and Birth Defects, Second Edition Edited by Patrizia Ferretti, Andrew Copp, Cheryll Tickle and Gudrun Moore

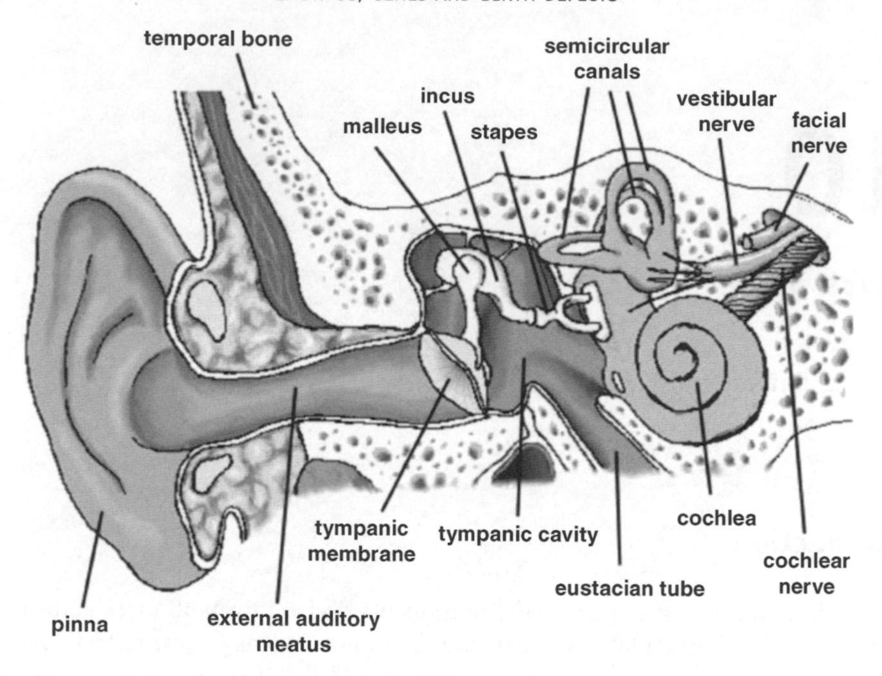

Figure 10.1 Diagram of the mammalian ear, showing the pinna, external ear canal, middle ear and inner ear

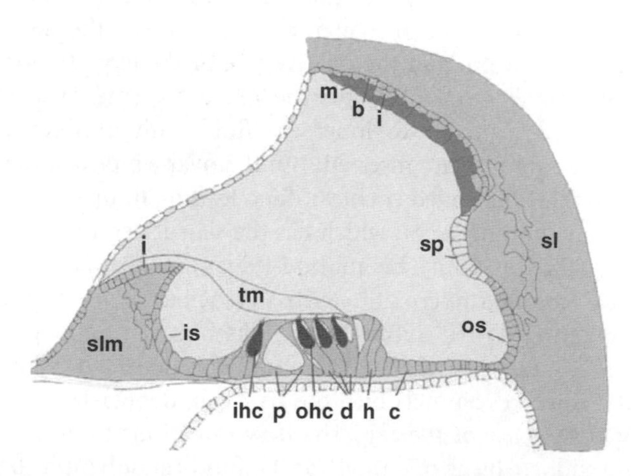

Figure 10.2 Diagram of the mammalian cochlear duct. b, basal cells of the stria vascularis; c, Claudius cells; d, Deiter's cells; h, Hensen's cells; i, intermediate cells of the stria vascularis; ihc, inner hair cell; is, inner sulcus; m, marginal cells of the stria vascularis; ohc, outer hair cell; os, outer sulcus; p, pillar cell; slm, spiral limbus (with i, interdental cells); sp, spiral prominance; sl, spiral ligament; tm, tectorial membrane. Figure kindly prepared by Sarah Holme and reproduced with permission from Holme and Steel (1999), copyright © Elsevier, 1999

has an unusual composition, high in potassium and low in sodium, so much of the transduction current is due to potassium ions flowing into the hair cells. The organ of Corti consists of two types of hair cell, inner and outer, together with a variety of specialized supporting cells, many of which contain dense collections of microtubules, giving the whole structure considerable rigidity. The inner hair cells are innervated primarily by afferent neurons and are the main receptor cells, while outer hair cells have mainly efferent innervation and respond to sound by rapidly changing shape, leading to enhanced motion of the whole organ of Corti. Amplification of the energy in sound by the middle ear mechanics, the large potential difference across the transduction channels and the active movement of outer hair cells, together with the rigid organization of the organ of Corti, which minimizes the dissipation of vibration, all contribute to our ability to detect vibration that is not much larger than Brownian motion at the threshold of hearing. Thus, it is not surprising that any minor anomaly in this finely-tuned process can lead to severe functional impairment.

The inner ear also contains the vestibular sense organs required for normal balance. Two sensory epithelia called maculae, in the saccule and utricle, act as gravity receptors. The sensory region contains hair cells and supporting cells, and stereocilia of hair cells are embedded in a gelatinous matrix containing dense crystals (otoliths), which deflects the stereocilia, depending upon the static position of the head in relation to gravity. Movement of the head is detected by three cristae located within ampulla at the ends of the three semicircular canals. As the head moves, the resulting fluid movement around these canals deflects the stereocilia of hair cells in the cristae, triggering synaptic activity.

Development of the outer and middle ear

The outer ear is formed from the first and second branchial arches, whose tissue types include arch ectoderm and mesoderm. The pinna is formed from the fusion and subsequent morphogenesis of six auricular hillocks that develop around the first branchial groove at the 6th week of development in humans (embryonic day (E)10.5 in mice). The first branchial groove itself deepens to become the external auditory meatus. The middle ear bones develop from neural crest cells from the branchial arches (Mallo, 2001). These condense to form the ossicles by 7 weeks of development (E13.5 in mice). Neural crest-derived cells of the first branchial arch (which give rise to Meckel's cartilage) form the major parts of the malleus and incus. Neural crest-derived cells of the second branchial arch (which give rise to Reichert's cartilage) form the stapes (Carlson, 1984). These ossicles develop further by ossifying the cartilaginous template (endochondral ossification). The air-filled space of the middle ear is formed from a pouch extending from the pharynx, and the connection to the pharynx is maintained as the Eustachian canal. Thus, the lining of the middle ear is largely derived from pharyngeal endoderm. This lining forms the inner layer of the tympanic membrane, with the outer layer formed of the branchial arch ectoderm and the middle fibrous layer formed from mesenchyme, by 9–10 weeks of development (E13.5–E16.5 in mice). The tympanic membrane is surrounded by a C-shaped ring, the tympanic ring, which is formed from first arch mesenchyme and eventually ossifies (Mallo and Gridley, 1996).

Development of the inner ear

The inner ear develops from otic placodes, which are ectodermal thickenings that form either side of rhombomeres 5 and 6 of the hindbrain at 4 weeks of development in humans (E8.5–E9.5 in mice). These placodes invaginate to form otic pits and then close completely to form otic vesicles or otocysts (Figures 10.3, 10.4a). The otocyst is a hollow ball of cells from which all of the core tissues and nerves of the inner ear are derived, apart from the neural-crest derived Schwann cells and melanocytes (Figure 10.4a). By the end of the 4th week (E9.5 in mice) the otocyst becomes surrounded with mesenchyme and the dorsal otocyst wall evaginates to form the endolymphatic duct and sac (Figure 10.3). The otocyst lengthens and widens to form the triangular vestibular pouch dorsally and the small flattened cochlear pouch ventrally. The semicircular canals form in week 5 of development (E13.5 in mice) in sequence from each edge of the vestibular pouch, with the anterior (superior) canal forming first, then the posterior and finally the lateral (horizontal) canal. In each case the edge expands outwards to form a flattened pocket. The centre of this pocket then collapses and the two layers fuse and resorb, so that an open lumen remains around the edge of the pocket (Martin and Swanson, 1993). The sensory regions associated with the semicircular canals are located within the ampullae at the end of each semicircular canal (Figure 10.4c). During the 6th week of development (E12.5 in

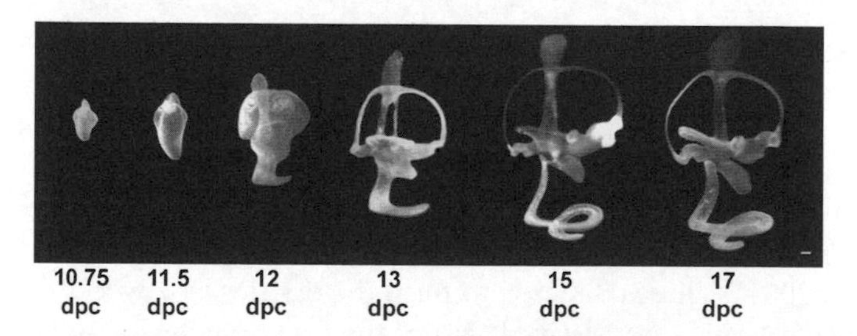

Figure 10.3 Development of the inner ear shown by paintfilled mouse labyrinths from E10.75 to E17. At E10.75, the endolymphatic duct projects dorsally and the cochlea anlage emerges as a ventral bulge. The cochlea expands ventrally till it reaches its mature one and three-quarter turns by E17. The semicircular canals start to develop at E11.5 from plates that form in the dorsolateral region of the otocyst. By E12 the anterior and posterior parts of the canal plates start to reabsorb, delineating the anterior and posterior semicircular canals. By E13, all three canals are formed. The utricle and saccule are distinguished by E15. Figure kindly provided by D. Wu and reproduced with permission from Morsli *et al.* (1998), copyright © The Society of Neuroscience, 1998

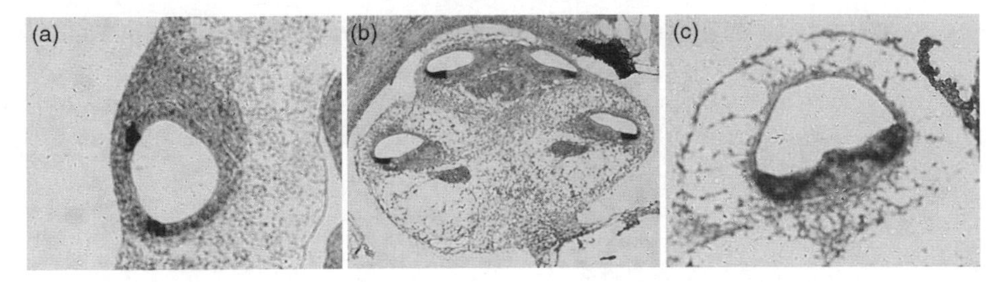

Figure 10.4 *Bmp4* expression in the developing inner ear and sensory epithelium. (a) *Bmp4* is expressed strongly in the otic vesicle at E10.5, shown by the darkly-stained patches. (b) At E16.5, expression becomes restricted to the sensory patches within the cochlea and (c) cristae of the vestibular system

mice) the utricle and saccule separate and the cochlear duct begins to separate from the saccule and lengthen, coiling as it grows (Figure 10.4b). The sensory epithelium within the cochlea, as in the other sensory patches within the inner ear, is immature and the cells are covered with tiny microvilli with a single kinocilium. The developing sensory patch in the cochlea forms two bands along the length of the cochlear duct, called the greater (modiolar side) and lesser (lateral side) epithelial ridges. As the membranous labyrinth develops, the sensory epithelium begins to differentiate and cells of the greater epithelial ridge near to the boundary start to differentiate as inner hair cells. A little later, outer hair cells develop from the lesser epithelial ridge and two rows of pillar cells eventually differentiate between inner and outer hair cells, along the boundary between the two ridges. Further towards the modiolar edge, the remaining cells of the greater epithelial ridge form the spiral limbus and Kölliker's organ. This latter structure secretes the major part of the tectorial membrane, a gelatinous extracellular matrix which attaches to the spiral limbus and extends over the sensory hair cell region. Kölliker's organ later regresses to a single cell layer lining the inner spiral sulcus. Meanwhile, the remaining cells of the lesser epithelial ridge develop into various specialized support cells, including Deiter's cells, Hensen's cells and Claudius' cells. Presumptive hair cells are not seen in the developing inner ear until 11–12 weeks of gestation (E16.5–E18.5 in mice).

The stria vascularis on the lateral wall of the cochlear duct plays a key role in secreting endolymph and generating the endocochlear potential. It forms from the ectodermal epithelial cells lining the otocyst (which differentiate into marginal cells), the mesenchymal cells that surround the otocyst (forming the basal cells) and the neural crest-derived melanocytes which migrate to the stria (later becoming intermediate cells), as well as a rich supply of blood vessels. These cell types interdigitate extensively during development, breaking down the basal lamina as they do so.

The vestibular and cochlear ganglia derive from cells that delaminate from the medio-ventral region of the otocyst. This delamination begins as the otocyst is closing and separating from the surface ectoderm of the head. The Schwann cells that myelinate these neurons migrate from the neural crest. At first both inner and outer hair cells are targeted by extending dendrites from the cochlear neurons, but later the afferent innervation of outer hair cells is replaced by efferent dendrites, leaving few if any afferent connections, and efferents extending to inner hair cells retract to contact only the afferent dendrites just below the hair cells, rather than synapsing directly onto the hair cell body (Pujol *et al.*, 1978).

Main classes of ear defects

Hearing impairment is the most common sensory disorder, estimated to affect ca. 1/850 live births (Fortnum *et al.*, 2001). Individuals with hearing impairment often show problems with social integration, speech development and quality of life, and early diagnosis is essential to provide early help with either hearing aids (for mild or moderate hearing impairment) or cochlear implants (for severe or profound deafness), the two major types of prosthesis currently available. A variety of environmental factors have been shown to cause deafness in infants, including infections such as meningitis (Eisenberg *et al.*, 1984), ototoxic drugs (Catlin, 1985) or low birth weight (Abramovich *et al.*, 1979). However, half of the cases documented are thought to have a primary genetic cause (Morton, 1991). Deafness in humans can be further subdivided into two main phenotypic classes: (a) syndromic, where the deafness is associated with other abnormalities; and (b) non-syndromic, where deafness is the only feature. Over 120 loci involved in non-syndromic deafness have been mapped in humans to date and 36 of the genes underlying the defects have been identified (Van Camp and Smith, 2005). Some of the genes involved specifically in the formation and maintenance of the inner ear sensory epithelia are listed in Table 10.1. Other genes involved in syndromic and non-syndromic deafness are listed in Table 10.2.

Much of our understanding of the pathology of deafness has come from studies of animal models such as the mouse, chick or zebrafish. There are several reasons for this. The main reason is the extreme difficulty in accessing and examining the primary defect in living affected human ears. Some limited access may be available to temporal bone tissue after a patient has died; however, it is difficult to tell whether the changes within the ear are the primary cause of the deafness in earlier life or a secondary consequence of long-term cochlear dysfunction (Schuknecht, 1974; Schuknecht and Gacek, 1993). Imaging is possible in humans, and has been successful in detecting major structural abnormalities, but it is not carried out routinely and cannot inform us of the developmental origin of the malformation (Jackler *et al.*, 1987; Phelps *et al.*, 1998). It is also possible to study mutations of genes that are critical for early development of the ear in an animal model which may prove lethal due to their role in the development of other structures, while this material will be very difficult to access in humans. The mouse in particular represents a very powerful tool in which to study ear development, due to the high degree of similarity between the human and mouse ear. Comparison between human deafness disorders and mouse models with mutations in orthologous genes have been shown to have largely similar phenotypes, given the limited human data that is available. The mouse is also

Table 10.1 Some genes involved in maintenance of the stereociliary bundle

Human gene	Human deafness locus	Human phenotype	Mouse mutant and phenotype
MYO7A	(a) USH1B	(a) Profound congenital deafness	*shaker1, sh1*
		Absent vestibular response	Deaf and vestibular defects
		Retinitis pigmentosa in 1st decade	Mild retinal anomalies
	(b) DFNB2	(b) Sensorineural hearing loss with variable onset	Disorganized stereociliary bundles
	(c) DFNA11	(c) Gradual bilateral hearing loss in 1st decade	Degeneration of neuroepithelia
CDH23	(a) USH1D	(a) Profound congenital deafness	*waltzer, v*
		Absent vestibular response	Deaf and vestibular defects
		Retinitis pigmentosa in 1st decade	Mild retinal anomalies
	(b) DFNB12	(b) Profound prelingual sensorineural deafness	Disorganized stereocilia bundles
			Degeneration of neuroepithelia
PCDH15	(a) USH1F	(a) Profound congenital deafness	*Ames waltzer, av*
		Absent vestibular response	Deaf and vestibular defects
		Retinitis pigmentosa in 1st decade	Disorganized stereocilia bundles
	(b) DFNB23	(b) Prelingual deafness	Degeneration of neuroepithelia
USH1C (Harmonin)	(a) USH1C	(a) Profound congenital deafness	*deaf circler, dfcr*
		Absent vestibular response	Deaf and vestibular defects
		Retinitis pigmentosa in 1st decade	Disorganized stereocilia bundles
	(b) DFNB18	(b) Sensorineural hearing loss, onset by 19 years	Degeneration of neuroepithelia
SANS	USH1G	Profound congenital deafness	*Jackson shaker, js*
		Absent vestibular response	Deaf and vestibular defects
		Retinitis pigmentosa in 1st decade	Disorganized stereocilia bundles
MYO6	(a) DFNB37	(a) Profound congenital deafness	*Snell's waltzer, sv*
	(b) DFNA22	(b) Progressive postlingual hearing loss with onset in childhood	Deaf and vestibular defects
			Disorganized stereocilia bundles
			Fusion of stereocilia
			Degeneration of neuroepithelia
MYO15	DFNB3	Profound congenital deafness	*shaker 2, sh2,*
			Deaf and vestibular defects
			Short stereocilia

Table 10.2 Some genes involved in human syndromic and non-syndromic deafness

Human gene	Human deafness locus	Human phenotype	Mouse mutant and phenotype
ACTG1	DFNA20/26	Progressive sensorineural hearing loss	None
CLDN14	DFNB29	Profound congenital hearing impairment	Targeted null Mice deaf Degeneration of neuroepithelium
COCH	DFNA9	Postlingual progressive hearing loss from 20 years	None
COL11A1	STL3, Stickler syndrome, type III	High-frequency progressive hearing loss associated with eye, skeletal and facial abnormalities	*Chondrodysplasia, cho* Mice deaf Cochlea under-developed Sensory hair cells missing
COL11A2	(a) DFNA13 (b) STL2, Stickler syndrome, type II	(a) Progressive sensorineural hearing loss, onset 2nd–4th decade (b) Early onset progressive hearing loss associated with eye, skeletal and facial anomalies	Targeted null Mice deaf Skeletal and craniofacial abnormalities
COL1A1	OI, osteogenesis imperfecta	Conductive and/or sensorineural hearing loss from late teens, associated with multiple fractures and blue sclerae of eye	*Mov13* transgene disruption Embryonic lethality Severely abnormal bone developement
COL2A1	STL1, Stickler syndrome I	Hearing impairment associated with eye, skeletal and facial anomalies	*Disproportionate micromelia, Dmm; spondyloepiphyseal dysplasia congentia, sedc;* mutant transgenes Mice deaf with malformed inner ears Otic and bone anomalies also present
COL4A3	Alport syndrome	Nephritis, with or without sensorineural hearing impairment	Targeted null Mice showed slight hearing loss Abnormalities in strial vessels Renal abnormalities
DFNA5	DFNA5	Progressive high-frequency hearing loss	None
DIAPH1	DFNA1	Progressive low-frequency hearing loss, onset 10 years	None

EDN3	WS4, Waardenburg–Shah syndrome	Sensorineural hearing impairment and pigmentation defects, with or without Hirschsprung's disease	*Lethal spotting, ls* Mice deaf, no hair cells present in cochlea. Pigmentation defects present, premature death due to colon defects
EDNRB	WS4, Waardenburg–Shah syndrome	Sensorineural hearing impairment and pigmentation defects, with or without Hirschsprung's disease	*piebald, s* Mice deaf, no hair cells present in cochlea. Pigmentation defects present, premature death due to colon defects
ESPN	DFNA36	Rapidly progressing sensorineural hearing loss, onset 5–10 years, deaf within 10–15 years	*jerker, je* Mice deaf and vestibular defects Hair cell stereocilia shorten and fuse Hair cells completely degenerate
EYA1	BOR, branchio-oto-renal syndrome	Severe hearing impairment with conductive and sensorineural component, in association with branchial and renal abnormalities	$Eya1^{bor}$ and targeted null Mice deaf and vestibular defects Shortened cochlea, no sensory epithelium. Absent or abnormal kidneys
EYA4	DFNA10	Progressive sensorineural hearing loss, onset 2nd–5th decade	None
GATA3	HDR syndrome, hypoparathyroidism, sensorineural deafness and renal dysplasia	Sensorineural hearing impairment in association with hypoparathyroidism and renal failure	Targeted null Abnormal inner ear with no semi-circular canals; abnormal nerve development to ear and face. Also brain and liver abnormalities.
GJB2 (CX26)	(a) DFNA3	(a) Sensorineural hearing impairment	Targeted conditional null Mice deaf
	(b) DFNB1	(b) Profound prelingual sensorineural hearing impairment	Degeneration of hair cells and neuroepithelia
GJB3 (CX31)	DFNA2	Progressive high-frequency sensorineural hearing loss, onset 1st–2nd decade	Targeted null 60 % Mutants died early embryonic stages. Those that survived showed no ear or skin defects

(*continued*)

Table 10.2 (*continued*)

Human gene	Human deafness locus	Human phenotype	Mouse mutant and phenotype
GJB6 (CX30)	(a) DFNA3 (b) DFNB1	(a) Sensorineural hearing impairment (b) Profound prelingual sensorineural hearing impairment	Targeted null Mice show progressive hearing loss Degeneration of neuroepithelium
KCNE1 (ISK)	JLNS2, Jervell and Lange–Nielsen syndrome, locus 2	Congenital hearing impairment in association with functional heart disease	Targeted null Deaf and vestibular defects Ataxia and heart defects
KCNQ1 (KVLQT1)	JLNS1, Jervell and Lange–Nielsen syndrome, locus 1	Congenital hearing impairment in association with functional heart disease	Targeted null Deaf and vestibular defects Inner ear malformed
KCNQ4	DFNA2	Progressive high-frequency sensorineural hearing loss, onset 1st–2nd decade	None
KIT	PBT, piebald trait	Sensorineural congenital hearing impairment in association with pigmentation defects	*Dominant spotting, W* Deaf Pigmentation defects
MASS1	USH2C, Usher syndrome, type IIC	Sensorineural hearing impairment in association with early onset retinitis pigmentosa	Frings and BUB/BnJ inbred strains, targeted null Early onset deafness Abnormal development of hair cells Degeneration of neuroepithelia
MITF	WS2A, Waardenburg syndrome, type IIA	Sensorineural hearing impairment in association with heterochromia iridum	*Microphthalmia, mi* Deaf, pigmentation defects, bone abnormalities
MYH9	DFNA17	Progressive high-frequency sensorineural hearing loss, onset 10 years	None
MYH14	DFNA4	Fluctuating but progressive sensorineural hearing loss, onset 2nd decade	None
MYO1A	DFNA48	Variable sensorineural hearing loss	None

MYO3A	DFNB30	Progressive sensorineural hearing loss, onset 2nd decade	None
NDP	ND, Norrie disease	Sensorineural heaing impairment in association with blindness and mental retardation	Targeted null Progressive hearing loss Abnormal stria vascularis Retinal defects
OTOA	DFNB22	Prelingual sensorineural hearing impairment	None
OTOF	DFNB9	Prelingual sensorineural hearing impairment	None
PAX2	Renal–coloboma syndrome	High-frequency neurosensory hearing impairment in association with renal and eye abnormalities	Targeted null Absent development of cochlea Abnormal development of optic nerve, retina, kidney, urogenital tract and mid-hindbrain
PAX3	(a) WS1, Waardenburg syndrome, type 1 (b) WS3, Waardenburg syndrome, type 1	(a and b) Sensorineural hearing impairment in association with facial and pigmentation defects	*Splotch, Sp* Normal hearing, pigmentation and craniofacial abnormalities
PMP22	CMT1A, Charcot–Marie–Tooth disease, type IA	Sensorineural hearing impairment in association with other neurological defects and progressive limb atrophy	*Trembler, Tr* Seizures and paralysis High juvenile mortality rate
POU3F4	DFN3	Profound sensorineural hearing impairment, with or with out a conductive component	Targeted null; *sex-linked fidget, slf* Deaf and vestibular defects
POU4F3	DFNA15	Progressive sensorineural hearing loss, onset 2nd–3rd decade	Targeted null; *dreidel, ddl* Deaf and vestibular defects Neuroepithelium fails to develop
SALL1	TBS, Townes–Brocks syndrome	Sensorineural hearing impairment in association with renal abnormalities, inperforate anus and radial dysplasia	Targeted null Heterozygotes have deafness with cystic kidneys. Homozygotes show renal agenesis, limb defects and exencephaly
SLC19A2	TRMA, thiamine-responsive megaloblastic anaemia syndrome	Sensorineural hearing impairment in association with diabetes and amino aciduria	Targeted null On thiamine-deficient diet was deaf, diabetic and megaloblastic anemia

(*continued*)

Table 10.2 *(continued)*

Human gene	Human deafness locus	Human phenotype	Mouse mutant and phenotype
SLC26A4	(a) DFNB4	(a) Profound sensorineural hearing impairment and developmental abnormalities of inner ear	Targeted null Deaf and vestibular defects Dilation of endolymphatic ducts
	(b) PDS, Pendred syndrome	(b) Hearing impairment as described above in association with goitre	Malformation of vestibular sensory epithelium Degeneration of neuroepithelia
SNAI2	WS2, Waardenburg syndrome, type II	Sensorineural hearing impairment in association with heterochromia iridum	Targeted null Pigmentation defects seen
SOX10	WS4, Waardenburg–Shah syndrome	Sensorineural hearing impairment in association with pigmentation defects	*Dominant megacolon, Dom* Pigmentation defects and megacolon
SPTBN4	CMT4F, Charcot–Marie–Tooth disease, type 4F	Neurosensory hearing impairment early in childhood	*quivering, qv* Deaf due to central neural defect Cochlear function normal, ataxia
STRC	DFNB16	Non-progressive sensorineural hearing loss, onset early childhood	None
TBX1	DGS, DiGeorge syndrome	Sensorineural hearing impairment in association with heart, thyroid and facial abnormalities	Targeted null Deaf and vestibular dysfunction Otitis media Cardiovascular malformations and thymus gland hypoplasia
TECTA	(a) DFNA8/12	(a) Prelingual sensorineural hearing impairment	Targeted null Impaired hearing
	(b) DFNB21	(b) Prelingual sensorineural hearing impairment	Tectorial membrane defect in cochlea
TFCP2L3	DFNA28	Post-lingual sensorineural hearing loss, onset from 7 years	None
TMC1	(a) DFNB7/B11	(a) Profound neurosensory hearing impairment	*Beethoven, Bth* Progressive deafness in heterozygote
	(b) DFNA36	(b) Rapidly progressing sensorineural hearing loss, onset 5–10 years, deaf within 10–15 years	Deaf in homozygote Neuroepithelium fails to mature and subsequently degenerates

TMIE	DFNB6	Sensorineural hearing impairment	*spinner, sr* Deaf and vestibular dysfunction Neuroepithelium degenerates
TMPRSS3	(a) DFNB10 (b) DFNB8	(a) Congenital sensorineural hearing impairment (b) Sensorineural hearing loss, onset in childhood	None
USH1G	USH1G, Usher syndrome, type IG	Profound congenital hearing impairment in association with prepubertal retinitis pigmentosa	*Jackson shaker, js* Deaf and vestibular dysfunction Disorganized stereocilia Degeneration of neuroepithelium
WFS1	(a) DFNA6/14/38 (b) WFS, Wolfram syndrome	(a) Low frequency sensorineural hearing loss, worsens over time (b) Hearing impairment in association with diabetes and optic atrophy	None
WHRN	DFNB31	Profound, prelingual sensorineural hearing impairment	*whirler, wi* Deaf and vestibular dysfunction Abnormal development and subsequent degeneration of neuroepithelium

DFNA loci are autosomal dominant, DFNB are autosomal recessive and DFN are X-linked. Adapted from Zheng and Johnson, 2005, and Van Camp and Smith, 2005.

very amenable to genetic manipulations that can abolish the function of a gene of interest completely (knock-out or null mutation), abolish the function of the gene within a specific cell type or at a specified stage (conditional knock-out mutation) or change just single base pairs within the gene that subtly alter its function (hypomorphs). Finally, with the completion of the mouse genome sequence and the high degree of conservation between mouse and human genes, any genes identified as being important for auditory function or development in the mouse can be readily identified as a candidate and the orthologous gene screened for mutations in DNA from hearing-impaired humans.

Using information from both human and mouse deafness, a number of broad categories of pathology of the ear can be defined (e.g. Steel *et al.*, 2002), but here we describe just four of the major classes of ear abnormalities:

1. Pinna and middle ear defects.

2. Malformation of the inner ear.

3. Neuroepithelial defects.

4. Abnormal endolymph homeostasis.

Pinna and middle ear defects

Genes involved in craniofacial development can affect development of the middle and outer ears as well as other features of the head, due to the shared embryological origins of these structures. Defects seen in the pinna and middle ear can be extremely variable and range from small or malformed pinnae, preauricular pits or fistules and slight malformations of individual ossicles, to absence of the external auditory meatus, agenesis of several ossicles or complete absence of the entire middle ear cavity and its components. An example of an ossicle defect in the hushpuppy mouse is shown in Figure 10.5. We discuss just two further examples here.

The first is *EYA1*, the human homologue of the *Drosophila eyes absent* gene, one copy of which has been shown to be mutated in patients with branchio-oto-renal (BOR) syndrome (Vincent *et al.*, 1997; Abdelhak *et al.*, 1997). BOR syndrome is characterized by craniofacial abnormalities, hearing impairment and kidney defects. Outer ear defects include malformed pinnae, malformation of the external ear canal and the presence of preauricular pits and cysts (Fraser *et al.*, 1978). Hearing impairment is present in 75 % of BOR patients and can be due to a conductive (30 %) or sensorineural (20 %) defect, or a mixture of both (50 %) (Cremers and Noord, 1980; Cremers *et al.*, 1981). A wide range of middle ear defects have been noted, including unconnected or fused stapes and incus (Cremers and Noord, 1980; Cremers *et al.*, 1981) and temporal bone changes (Fitch and Srolovitz, 1976). Imaging of the temporal bone by tomography in a large family of affected BOR individuals

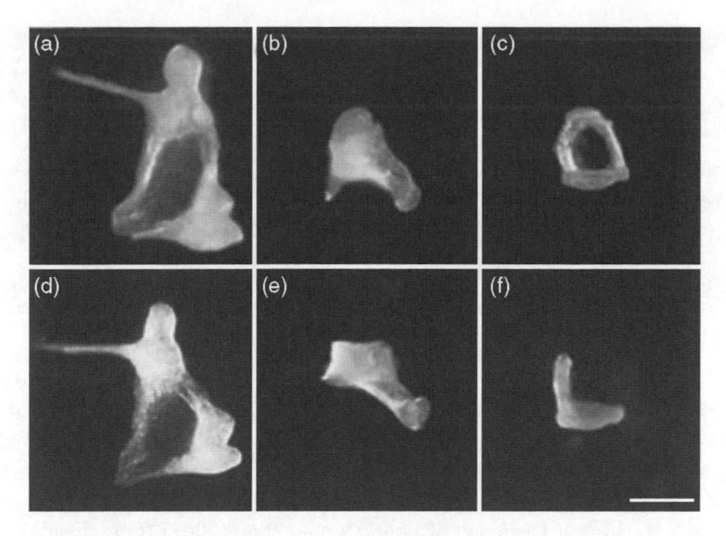

Figure 10.5 Ossicle defects seen in the hearing-impaired mouse mutant hushpuppy. Normal malleus, incus and stapes are shown in (a–c), respectively. Mutants have a normal malleus (d) but have various incus abnormalities, such as small bodies and reduced long and short processes (e). Mutants also show a characteristic stapes defect, with a reduced or absent posterior crus (f). Scale bar, 500 μm. Images reproduced with kind permission of Lippincott, Williams and Wilkins from Pau *et al.* (2005)

(14 ears) identified abnormalities of the internal auditory canal, dysplasia of the lateral semicircular canal and cochlear malformations (Cremers and Noord, 1980). Studies of *Eya1* heterozygote mice also showed deafness and middle ear defects, such as malformation of the ossicles and failure of the stapes to contact the oval window (Johnson *et al.*, 1999; Xu *et al.*, 1999).

A second gene in humans shown to affect development of the outer and middle ear is the POU homeodomain transcription factor, *POU3F4* (de Kok *et al.*, 1995). Mutations in *POU3F4* give rise to X-linked deafness with gusher (DFN3), where the internal auditory meatus is dilated and there is a conductive hearing impairment due to the fixation of the stapes in the oval window (de Kok *et al.*, 1995). The inner ear defect means that when surgery is performed to release the fixed stapes, a gush of fluid is released due to a failure of separation of the perilymph from the cerebrospinal fluid. Mouse mutants lacking the *Pou3f4* gene also show defects in the stapes and inner ear malformation (Phippard *et al.*, 1999, 2000).

Inner ear malformations

Defects in the development of the inner ear have been identified in both humans and mouse mutants, ranging from the truncation or thinning of one or more semicircular canal, failure of the semicircular canals to form leading to a cyst-like vestibular cavity, a shortened or malformed cochlear duct or a complete failure of the otic vesicle to

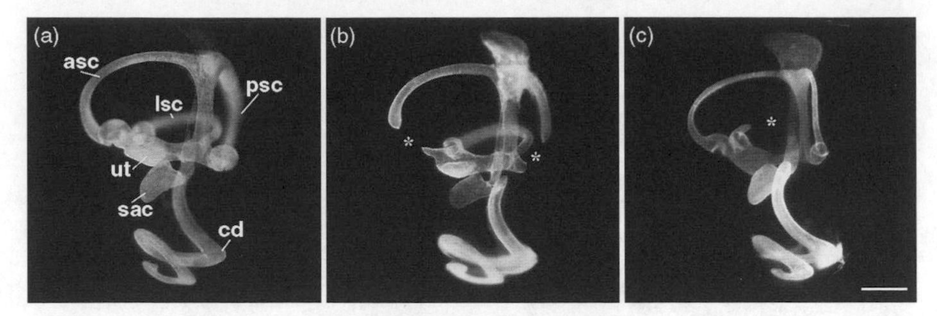

Figure 10.6 Semicircular defects seen in mouse mutants, shown in paintfilled E16.5 embryos. (a) Normal structure of the inner ear labyrinth. (b) Semicircular canal truncations (marked by asterix) of the anterior and posterior semicircular canals in the *headturner* mutant. (c) Lateral semicircular canal truncation in the *tornado* mouse mutant. asc, anterior semicircular canal; cd, cochlear duct; lsc, lateral semicircular canal; psc, posterior semicircular canal; sac, saccule; ut, utricle. Scale bar, 500 μm. Images reproduced with permission from Kiernan *et al.* (2001), copyright © National Academy of Sciences USA, 2001 (6A and 6B) and Kiernan *et al.* (2002) (6C), copyright © Springer-Verlag GmbH, 2002 (6c)

develop, leading to the production of an elongated cyst. These defects often result in deafness. However, in mice with mild defects in semicircular canal structure only, hearing can be near-normal (see e.g. Kiernan *et al.*, 2002). Examples of truncated semicircular canals in mice are shown in Figure 10.6. Imaging is not routinely carried out in deaf individuals, so we have limited knowledge about the full range of malformations in humans. Mutations in the human *EYA1* and *POU3F4* genes result in inner ear defects, such as truncations of the cochlea and enlarged internal auditory meatus (the canal through which the cochlear nerve passes), as mentioned earlier. Another inner ear malformation that has been seen in human patients is a Mondini malformation, in which the two apical turns of the cochlea are merged into a common cavity (Cremers *et al.*, 1998; Phelps *et al.*, 1998). Mondini defects are sometimes seen in Pendred syndrome, although enlarged endolymphatic duct and sac are more commonly seen in this disease, due to mutations of the *SLC26A4* (*PDS*) gene (Everett *et al.*, 1997; Cremers *et al.*, 1998). *PDS* is involved in non-syndromic deafness as well as Pendred syndrome (Li *et al.*, 1998) and appears to be one of the more common genes underlying deafness in the human population. Some of the largest studies of inner ear malformations suggest that defects of the lateral semicircular canal are the most common in the human population (Jackler *et al.*, 1987; Phelps, 1974; Sando *et al.*, 1984).

Neuroepithelial defects

Neuroepithelial defects are abnormalities within the six specialized sensory neuroepithelia within the inner ear, including the organ of Corti of the cochlea, the maculae of the saccule and utricle and the cristae of the ampulla at the end of each of the three semicircular canals. Neuroepithelial defects that have been identified

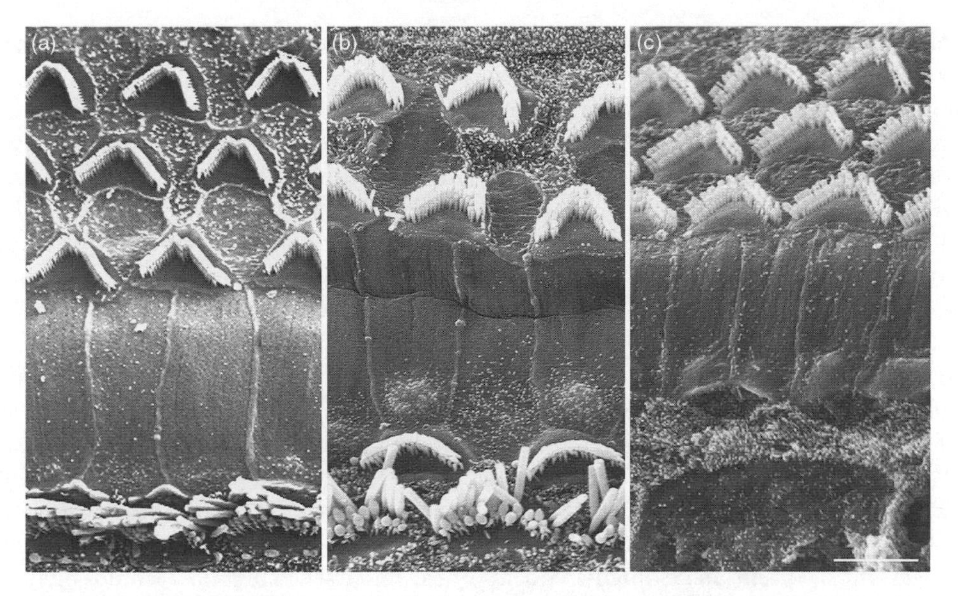

Figure 10.7 Examples of neuroepithelial defects seen in hearing-impaired mouse mutants. (a) Scanning electron micrographs showing the normal structure of the organ of Corti with three rows of outer hair cells (OHC) at the top and a single row of inner hair cells (IHC) near the bottom. (b) Organ of Corti from the *slalom* mutant with only two rows of OHC and atypical OHCs in the IHC row. (c) Organ of Corti from the *beethoven* mutant with three rows of normal OHC but the IHC row is absent in this region of the cochlear duct. Scale bar, 5 μm

range from complete failure of the sensory epithelium to develop, through abnormal differentiation of specific cell types, to failure to maintain sensory hair cells during the lifespan of the individual. Examples of neuroepithelial defects in mice are shown in Figure 10.7. It is not possible to detect primary neuroepithelial defects in humans by imaging and, using temporal bone specimens studied after death, it is often difficult to distinguish the initial defect leading to a functional impairment from the secondary degeneration of the whole organ of Corti that usually follows dysfunction. Nonetheless, careful study of temporal bones led to the proposal of several categories of age-related hearing loss by Schuknecht and Gacek (1993), including a common class called 'sensory defects' which probably correspond to neuroepithelial defects. Extensive studies in the mouse have allowed identification of many genes that are involved in the development and maintenance of the neuroepithelium and that are shown to be mutated in patients with both congenital and progressive forms of hearing loss. This will be discussed further below.

Abnormal endolymph homeostasis

Endolymph bathes the upper surface of all sensory hair cells in the inner ear. As mentioned earlier, it has an unusual ionic content, high in potassium and low in sodium, and in the cochlea the endolymph is maintained at a high resting potential (+ 100 mV in mouse). A number of genes are involved in maintaining endolymph homeostasis, as revealed by mutations in the mouse and/or human cases (for review, see Steel and Kros, 2001; Figure 10.3). The failure to maintain homeostasis can sometimes result in collapse of the endolymphatic compartments of the inner ear, which is a feature that can be detected in temporal bone sections (Schuknecht, 1974), or can lead to a reduced or absent endocochlear potential, as measured in mutant mice (e.g. Steel *et al.*, 1987; Minowa *et al.*, 1999). One example of abnormal fluid homeostasis in the inner ear is the *Pds* (*Slc26a4*) knock-out mouse, in which severe dilation of the endolympatic cavities is seen during development, leading ultimately to sensory hair cell degeneration and malformation of the inner ear (Everett *et al.*, 2001). Another example is the *Slc12a2* mouse mutation, which leads to early collapse of endolymphatic chambers and consequent malformation of semicircular canals (Dixon *et al.*, 1999). Schuknecht and Gacek (1993) proposed that the strial class of pathology of human age-related hearing loss showed a characteristic audiogram with a flat increase in thresholds across frequencies (in contrast to sensory pathology, which most often affects high-frequency hearing first). This class could include abnormal homeostasis caused by defects in other parts of the cochlear duct as well as the stria vascularis, as implied by the name.

Mechanisms involved in development of the outer and middle ear

Many genes have been implicated in the development of the outer and middle ear from studies in the mouse (Steel *et al.*, 2002; Mallo, 2003). Although in most cases the mutants die shortly after birth due to severe craniofacial abnormalities, the outer and middle ear abnormalities in these mice can be characterized and studied. Several of these genes have been associated with human deafness too, such as *POU3F4*, *EYA1* and *SIX1* (Abdelhak *et al.*, 1997; de Kok *et al.*, 1995; Ruf *et al.*, 2004).

Signalling molecules such as endothelin1 (Edn1), fibroblast growth factor 8 (Fgf8) and retinoic acid (RA) have been shown to act as mediators of epithelial–mesenchymal interactions and be involved in development of the branchial arches (Bee and Thorogood, 1980). Mutations in these factors also affect middle ear development. In mutants where *Edn1* (Clouthier *et al.*, 1998; Kurihara *et al.*, 1994) and *Fgf8* (Trumpp *et al.*, 1999) were inactivated, the malleus and incus were absent or underdeveloped and various stapedial defects seen. In mice carrying mutations in several RA receptors, the stapes was severely affected, while the incus was only slightly malformed and the malleus was normal (Lohnes *et al.*, 1994). The homeobox gene *Hoxa2* has also been shown to be essential for the proper formation of second branchial arch structures (Gendron-Maguire *et al.*, 1993; Mallo, 1997). In mutants of *Hoxa2*, a duplicate set of first branchial arch middle ear ossicles are formed. Mutation of a second homeobox gene, *Hoxa1,* has also been shown to lead to middle ear defects (Lufkin *et al.*, 1991; Gavalas *et al.*, 1998).

Several genes have also been shown to cause ossicular defects within the middle ear, due to defects in the process of skeletal condensation. Mutations in the *Distal-less*-related gene *Dlx5* causes absence of the stapes and the presence of an extra cartilaginous element attached to the malleus (Acampora *et al.*, 1999; Depew *et al.*, 1999). Inactivation of the *Prx1* gene causes the incus and stapes to be attached to abnormal cartilaginous structures within the branchial arches (Martin *et al.*, 1995). *Prx1* is expressed in mesenchymal cells (Cserjesi *et al.*, 1992) and may play a role in establishing the size or location of skeletogenic condensations within the post migratory neural crest (Mallo, 2001). Mutants in a second *distal-less*-related gene, *Dlx2*, show extra cartilaginous formations attached to the incus and a reduced stapes (Qiu *et al.*, 1995). It has been suggested that *Dlx2* may be involved in determination of the size of skeletogenic condensation (Hall and Miyake, 1995). *Dlx2* may also be implicated in mediating epithelial–mesenchymal factors required for skeletal development within the craniofacial area (Qiu *et al.*, 1997). In mutants of the *Goosecoid* gene, the tympanic ring, which is essential for the subsequent formation of the ear drum, fails to develop due to failure of skeletal condensations (Rivera-Perez *et al.*, 1995). Whilst these studies in the mouse have identified several of the key processes involved in outer and middle ear development, the exact roles and interactions between genes involved in these processes are not yet understood.

Mechanisms underlying inner ear development

Again, there are a large number of genes that have been implicated in inner ear development (e.g. Barald and Kelley, 2004; Kiernan and Steel, 2002). Studies from mouse mutants show that inner ear development can be affected by two types of gene action. The first is mediated by genes that are expressed outside the inner ear, such as those involved in hindbrain segmentation and definition of rhombomere identity, or those expressed in the mesenchyme surrounding the inner ear. These genes can be involved in both induction of the otic vesicle and formation and patterning of the otic vesicle. The second is mediated by genes that are expressed within the otocyst itself, such as transcription factors involved in establishing and maintaining patterning within the developing vesicle. The overlapping expression patterns of these genes in the otocyst, together with the phenotypes resulting from mutations, suggest that a complex combinatorial code is involved in defining the developmental fate of each region of the otic epithelium (Fekete and Wu, 2002). Again, many genes have been shown to affect inner ear development in the mouse, and some of these have been shown to be involved in human deafness too (Steel *et al.*, 2002; Zheng and Johnson, 2005).

Several fibroblast growth factor (FGF) genes, such as *Fgf2*, *Fgf3*, *Fgf8* and *Fgf10*, have been implicated as neural signals from the hindbrain that are involved in induction of the otic placode and vesicle in vertebrates (Adamska *et al.*, 2001; Leger and Brand, 2002; Lombardo *et al.*, 1998; Lombardo and Slack, 1998; Ohuchi *et al.*, 2000; Represa *et al.*, 1991; Vendrell *et al.*, 2000). The earliest candidate identified in this process was *Fgf3*, although mutations in this gene in mouse do not abolish otic vesicle formation altogether but lead to improper formation of the endolymphatic duct and sac (Mansour *et al.*, 1993). It was shown in zebrafish that, although loss of *Fgf3* function on its own moderately perturbed otic development, if *Fgf8* function was also lost, as in the *acerebellar* mutants, then the otic vesicles failed to develop at all (Phillips *et al.*, 2001). A double mouse knock-out for *Fgf3* and *Fgf10* was shown to have severely reduced otic vesicles, although early otic vesicle markers were expressed, suggesting that inner ear differentiation had occurred (Alvarez *et al.*, 2003). This perhaps suggests that, in the zebrafish, FGFs are sufficient for inner ear induction, whereas in the mouse FGFs act to reinforce or maintain early inductive signals (Alvarez *et al.*, 2003).

Several diffusible elements whose genes are expressed in the hindbrain, yet which influence development of the otocyst, have been identified. The *Krml* gene, which is mutated in the *kreisler* mouse mutant, is essential for hindbrain development and, when absent, formation of rhombomeres (r) 5 and 6 is abolished (Cordes and Barsh, 1994; McKay *et al.*, 1994). This in turn downregulates expression of *FGF3*, affecting inner ear development as described above (McKay *et al.*, 1996). Functional inactivation of the *Hoxa1* gene results in the formation of a cyst-like inner ear similar to that in the *kreisler* mutant, associated with a complete absence of r5 and a reduction in r4 (Chisaka *et al.*, 1992; Lufkin *et al.*, 1991). Mice with null mutations in sonic hedgehog (*Shh*) show ear induction, but the cochlear duct and cochleovestibular ganglion fail to develop (Liu *et al.*, 2002; Riccomagno *et al.*, 2002). This may be due to a reduction in the expression of the paired box gene *Pax2* (Liu *et al.*, 2002), which has been shown to be expressed early in otic placode epithelium (Ekker *et al.*, 1992). The homeobox genes *Prx1* and *Prx2* are also involved in the formation of the inner ear labyrinth, as in *Prx1/Prx2* mutants the otic capsule is reduced and the lateral semicircular canal is often absent (ten Berge *et al.*, 1998).

Genes that are expressed within the otocyst can be involved in the specification of particular regions of the inner ear labyrinth (Fekete and Wu, 2002). Mutations in the *Pax2* gene result in absence of the entire cochlear duct (Favor *et al.*, 1996; Torres *et al.*, 1996), whereas mutations in *Hmx3* result in severe disruption of the vestibular system (Hadrys *et al.*, 1998; Wang *et al.*, 1998), corresponding to the regions of maximum expression of these two genes during early otocyst development. Mutations in the *Dlx5* gene affect all regions of the inner ear, causing an absence of the anterior and posterior semicircular canals, truncation of the lateral semicircular canals and a shortening of the cochlear duct (Depew *et al.*, 1999). Mutations in *Otx1* result in milder defects of the vestibular system where only the lateral semi-circular canal is affected (Acampora *et al.*, 1996). Recent studies have identified a member of the *Six* family of homeobox genes, *Six1*, to act as a key regulator of otic vesicle patterning and to be involved in the control of the expression domains of downstream otic genes responsible for specific inner ear structures, such as *Pax2, Dlx5, Hmx3* and *Otx1* (Ozaki *et al.*, 2004). Inactivation of this gene leads to a fusion of the dorsal-most parts of the semicircular canals and the endolymphatic ducts and an absence of the rest of the vestibular and cochlear parts of the inner ear (Ozaki *et al.*, 2004).

Mechanisms underlying development of inner ear sensory epithelia

The sensory epithelia in the auditory and vestibular systems include mechanoreceptor cells called hair cells, which convert mechanical motion into electrochemical energy. The hair cells contain a highly ordered bundle of actin-rich stereocilia on their apical surfaces, which are essential for their function. Hair cells are normally separated from each other by supporting cells, which in the organ of Corti are highly specialized.

Immature hair cells and support cells arise from the same progenitor cells within a sensory-competent patch within the otocyst (Fekete *et al.*, 1998). We are beginning to understand some of the mechanisms involved in pattern formation in this sensory patch. Notch signalling appears to be involved in defining the boundaries of the sensory patch, as well as a presumed role in lateral inhibition, a mechanism that can lead to the precise mosaic arrangement of hair cells and supporting cells (Daudet and Lewis, 2005; Lewis, 1991; Corwin and Jones, 1991). This population of precursor cells in the sensory patch expresses both *Notch1* (Lanford *et al.*, 1999) and *Jag1* (Zine *et al.*, 2000; Morrison *et al.*, 1999), two of the genes involved in this signalling pathway. The *headturner* (Kiernan *et al.*, 2001) and *slalom* (Tsai *et al.*, 2001) mouse mutants, with missense mutations in *Jag1*, lack some sensory regions in the vestibular system and have abnormal boundaries in some of the remaining patches, suggesting that *Jag1* is involved in specification of the prosensory patch in the inner ear. Mice with mutations of *Jag2*, another gene involved in Notch signalling, also have been shown to have an increase in the number of sensory hair cells within the organ of Corti (Lanford *et al.*, 1999), suggesting abnormal boundary specification or a role in cell fate decisions by lateral inhibition within the prosensory patch. A model for lateral inhibition proposes that some cells in the prosensory epithelium produce slightly more Jag2 ligand, which activates the Notch1 receptor in adjacent cells and in turn leads to downregulation of Notch1 in the cells expressing more Jag2 (Lanford *et al.* 1999). The initial small imbalance is enhanced by this feedback loop, and the cells expressing Jag2 at higher levels develop as hair cells, while their neighbours become supporting cells. Progenitor cells destined to become hair cells express the basic helix–loop–helix (bHLH) transcription factor, mammalian atonal homologue 1 (*Math1, Atoh1*). In *Math1* mouse knock-outs, hair cells fail to form but supporting cells do form, suggesting that *Math1* is essential for sensory hair cell formation (Bermingham *et al.*, 1999). This is confirmed by the fact that overexpression of *Math1* in rat cochlear cultures induces the production of extra hair cells (Zheng and Gao, 2000). *Math1* is not required for establishing the sensory primordia in mammals, but is required for the differentiation of cells into sensory hair cells (Chen *et al.*, 2002). In cells that express *Math1*, there is an increase in the expression of the Notch ligand *Jag2* (Lanford *et al.*, 1999; Zine *et al.*, 2000). Two bHLH genes shown to be involved downstream of the Notch signalling pathway are *Hes1* and *Hes5*, homologues of the *Drosophila hairy* and *enhancer of split* genes, which have been shown to be negative regulators of inner ear hair cell differentiation (Zheng *et al.*, 2000; Zine *et al.*, 2001).

The transcriptional activity of *Hes1* and *Hes5* has been shown to repress the transcriptional activity of *Math1* (Akazawa *et al.*, 1995) and therefore may upregulate *Notch1* expression, reinforcing the non-sensory fate in these cells. The exact relationship between the various molecules involved in determining hair cell and supporting cell fate is complex and not yet fully resolved and is likely to involve feedback, rather than a linear cascade of gene activity (e.g. Woods *et al.*, 2004).

A further gene required for hair cell differentiation is the transcription factor gene *Pou4f3*, which has been shown to be expressed in both vestibular and auditory hair cells in mice (Xiang *et al.*, 1997). Mutants lacking this gene are deaf and have vestibular defects and show a complete absence of hair cells, with secondary loss of supporting cells and spiral ganglion neurones (Xiang *et al.*, 1997). Subsequent analysis showed that in mutants lacking *Pou4f3*, hair cells are specified and undergo some differentiation, as indicated by the expression of early hair cell markers, such as Myosin 6 and Myosin 7a (Xiang *et al.*, 1998). However, these differentiated cells fail to develop stereociliary bundles and undergo apoptotic cell death (Xiang *et al.*, 1998). Therefore, *Pou4f3* is essential for the survival of sensory hair cells. The human orthologue has been shown to be mutated in a non-syndromic form of human deafness, DFNA15 (Vahava *et al.*, 1998)).

Each sensory hair cell within the organ of Corti projects approximately 100 actin-packed stereocilia from its apical surface. These stereocilia are arranged in three parallel rows that increase in height from the inner row to the outer row and are organized in a V-shaped pattern (see Figure 10.7A). The base of each stereocilium is anchored in the cuticular plate and lateral cross-links between the stereocilia are thought to be involved in maintaining their precise arrangement. Analysis of many mouse mutants have shown that maintenance of this ordered arrangement of stereocilia is essential for normal hearing (e.g. Frolenkov *et al.*, 2004). One of the first families of genes shown to be involved in controlling the structure of the stereociliary bundle was the family of unconventional myosins. The first myosin gene implicated in hearing was *Myo7a*, which was shown to be mutated in shaker1 mouse mutants (Gibson *et al.*, 1995). These mice are deaf and, although they develop stereociliary bundles, these become progressively more disorganized as the mice develop (Self *et al.*, 1998). Snell's waltzer mutants have a mutation in another unconventional myosin gene, *Myo6*, and show disorganization and progressive fusion of stereocilia (Avraham *et al.*, 1995; Self *et al.*, 1999). *Myo15* mutations, present in the shaker2 mouse, do not affect the organization of the stereociliary bundle but the stereocilia appear shorter than normal (Probst *et al.*, 1998). *Myo1a*, a member of the myosin 1 family of proteins, has also been shown to be mutated in human deafness patients, although no mouse model has yet been identified (Donaudy *et al.*, 2003). Genes for other cytoskeletal proteins that have been shown to be essential for the integrity of the stereociliary bundle include: *Cdh23*, mutated in the waltzer mouse; (Di Palma *et al.*, 2001; Holme and Steel, 1999); *Pcdh15*, affected in the Ames waltzer mutant (Alagramam *et al.*, 2001); *Harmonin* (*Ush1c*), which is mutated in deaf circler (Johnson *et al.*, 2003); and *Sans*, which is mutated in the Jackson shaker mouse (Kikkawa *et al.*, 2003). Mutations in genes for some of these structural proteins have been shown to underlie forms of syndromic deafness, such as Usher syndrome, where affected individuals are born deaf and develop retinitis pigmentosa in childhood, as well as non-syndromic congenital or progressive hearing loss. A list of human and mouse deafness caused by these genes is summarized in Table 10.1.

Analysis of the genes involved in Usher syndrome suggest that there are direct protein interactions between harmonin and myosin 7a, cadherin 23, sans and protocadherin 15 (Boeda *et al.*, 2002). Myosin 7a may be involved in the transport of harmonin to the tip region of the stereocilia, as in myosin 7a mutants harmonin is absent from the stereocilia and is arranged in beadlike foci in the cuticular plate at its base. Cadherin 23 has been shown to interact directly with harmonin and is thought that harmonin anchors cadherin 23 to the actin core of the stereocilium. Cadherin 23 has been proposed to form transient links that interconnect the stereocilia from their emergence to maturation, presumably to maintain the structure of the developing bundle (Boeda *et al.*, 2002). However, recent findings suggest that cadherin 23 may also be a component of the tip link complex at the top of adjacent stereocilia, involved in opening the transduction channel (Sollner *et al.*, 2004; Siemens *et al.*, 2004). Myosin 6 protein is thought to play a role in anchoring the stereocilia to the cuticular plate, since in the absence of this protein the stereocilia fuse together (Self *et al.*, 1999). Mice with mutations in the *Sans* gene show progressive stereociliary disorganization (Kikkawa *et al.*, 2003). Based on the localization of sans protein at the base of stereocilia in the cuticular plate, its binding to myosin 7a and harmonin, and the presence of several predicted protein interaction domains, it has been suggested that sans protein may act as a scaffolding or anchoring protein for molecular complexes or may regulate trafficking of Usher proteins towards the stereocilia (Kikkawa *et al.*, 2003; Adato *et al.*, 2005).

Even if the hair cell develops normally, several factors are required for the survival and continued functioning of the cell. For example, *Barhl1*, the mouse homologue of the *Drosophila BarH* homeobox gene, has been shown to be required for the survival but not the specification of sensory hair cells. Mutations in this gene result in severe to profound hearing loss in mice and a progressive disorganization and degeneration of the hair cells within the cochlea, although the vestibular hair cells appear normal (Li *et al.*, 2002).

Mechanisms involved in endolymph homeostasis

The ionic balance of the endolymph, the fluid which surrounds the stereocilia on the apical surface of the hair cells, has been shown to be very important for normal functioning of the hair cells in the inner ear. Disruption of *Kcnq1 (Kvlqt1)* and *Kcne1* (*Isk*) genes, encoding proteins which associate to form potassium channels in marginal cells of the stria vascularis (Neyroud *et al.*, 1997; Sakagami *et al.*, 1991), cause deafness and vestibular dysfunction in mutant mice (Casimiro *et al.*, 2001; Lee *et al.*, 2000; Vetter *et al.*, 1996). Histological analysis showed collapse of the endolymphatic compartments and consequent degeneration of the hair cells in the adult cochlea. These two genes are proposed to mediate the secretion of potassium into endolymph, and mutations have been identified in the human orthologues in patients with Jervell and Lange–Nielsen disease, in which patients are deaf and have heart defects (Neyroud *et al.*, 1997; Tyson *et al.*, 1997).

Kcnq4 encodes another potassium channel shown to be expressed in outer hair cells by *in situ* hybridization analysis of the mouse cochlea (Kubisch *et al.*, 1999). Mutations in *KCNQ4* have been identified in families with progressive hearing loss in early childhood and it has been suggested that this channel may be involved in the removal of potassium from the base of the outer hair cell (Kubisch *et al.*, 1999). The hair cell pathology is unknown; however, it may be that an overload of potassium within the cell could lead to progressive degeneration and a decrease in hearing ability in these patients (Kubisch *et al.*, 1999).

The future

The future of hearing research is an extremely exciting one. Aided by the use of model organisms, major advances have been made in the understanding of the development of the inner ear and the sensory regions within it. The inner ear is crucial for normal hearing and mutations in genes involved the formation and function of the inner ear are likely to be a major cause of non-syndromic deafness, which accounts for the majority of cases of human deafness. However, there still remain over 50 loci involved in human non-syndromic deafness for which the genes have yet to be identified, many more loci not yet discovered, and even more mouse models for which no human form of deafness has yet been documented (VanCamp and Smith, 2005; Zheng and Johnson, 2005). Identifying the genes underlying these forms of deafness can only increase our understanding of the development and function of this complex sensory organ. Development of the outer and middle ear is perhaps not as well understood, and further work on mutants affecting the hindbrain and first and second branchial arches will allow the molecular mechanisms underlying the development of these regions of the ear to be elucidated. Despite the number of genes identified in deafness, most individuals with hearing impairment, whether congenital or progressive, have no molecular diagnosis, so there is still a need for large-scale gene discovery.

Identification of all the genes involved in the formation of the sensory patch will help us develop possibilities of therapy for deaf individuals. Studies have shown that non-mammalian vertebrates retain the ability to regenerate hair cells throughout their lives (Corwin and Cotanche, 1988; Ryals and Rubel, 1988). In contrast, in the mature mammalian cochlea these cells do not regenerate (Warchol *et al.*, 1993; Rubel *et al.*, 1995). In order to restore hearing, not only will hair cell development need to be triggered in the sensory epithelium, but all the supporting cells and normal ionic homeostasis will need to be regenerated to create the right environment for hair cells to function normally. A recent advance in the field of hair cell regeneration came when Li *et al.*, (2003) were able to induce proliferation of cells taken from the sensory region in the utricle. These cells were shown to express marker genes for sensory hair cells and their supporting cells *in vitro*, and when transplanted into the chick otocyst, a subset developed into sensory hair cells (Li *et al.*, 2003). Furthermore, transfection of Math1 in recently damaged sensory epithelia of guinea-pigs led to the reappearance of hair cells and some functional improvement (Izumikawa *et al.*, 2005). Whether or not sensory epithelia in the mammalian ear could be induced to undergo regeneration *in vivo* long after damage has occurred remains to be seen, but these findings are very encouraging. Despite the obstacles, the possibility of sensory patch regeneration remains a promising area of hearing research.

Several of the genes that underlie hereditary deafness from birth have been shown to be involved in progressive hearing loss in some families, and so may also play a role in predisposing individuals to age-related hearing loss at later stages. Examples of these include *MYO7A*, *USH1C* and *MYO6* (see Table 10.1 for details). Progressive hearing loss is a very common disorder in the population, with 60 % of people over age 70 have a hearing loss of 25 dB or greater, a significant level at which they would benefit from a hearing aid (Davis, 1989). Analysis of mice that show age-related hearing loss may allow more of the genes involved in this process to be identified and might allow more of the factors required for long-term survival of the sensory patches to be identified. This in turn may offer a way of preventing or slowing the sensory hair cell death in this condition and offer an alternative to hearing aids and cochlear implants, the only other options available for these individuals at the moment.

Acknowledgements

We thank Sarah Holme for Figure 10.2, Erika Bosman, Charlotte Rhodes and Alexandra Erven for unpublished images in Figures 10.4 and 10.7, Doris Wu for Figure 10.3 and Agnieszka Rzadzinska for help with figures. Supported by Defeating Deafness, the MRC and the Wellcome Trust.

References

Abdelhak, S., Kalatzis, V., Heilig, R., Compain, S. *et al.* (1997) A human homologue of the *Drosophila eyes absent* gene underlies branchio-oto-renal (BOR) syndrome and identifies a novel gene family. *Nat. Genet.* **15**: 157–164.

Abramovich, S.J., Gregory, S., Slemick, M. and Stewart, A. (1979) Hearing loss in very low birthweight infants treated with neonatal intensive care. *Arch. Dis. Child* **54**: 421–426.

Acampora, D., Mazan, S., Avantaggiato, V., Barone, P. *et al.* (1996) Epilepsy and brain abnormalities in mice lacking the Otx1 gene. *Nat. Genet.* **14**: 218–222.

Acampora, D., Merlo, G.R., Paleari, L., Zerega, B. *et al.* (1999) Craniofacial, vestibular and bone defects in mice lacking the *Distal-less*-related gene *Dlx5*. *Development* **126**: 3795–3809.

Adamska, M., Herbrand, H., Adamski, M., Kruger, M. *et al.* (2001) FGFs control the patterning of the inner ear but are not able to induce the full ear program. *Mech. Dev.* **109**: 303–313.

Adato, A., Michel, V., Kikkawa, Y., Reiners, J. *et al.* (2005) Interactions in the network of Usher syndrome type 1 proteins. *Hum. Mol. Genet.* **14**: 347–356.

Akazawa, C., Ishibashi, M., Shimizu, C., Nakanishi, S. and Kageyama, R. (1995) A mammalian helix–loop–helix factor structurally related to the product of *Drosophila* proneural gene *atonal* is a positive transcriptional regulator expressed in the developing nervous system. *J. Biol. Chem.* **270**: 8730–8738.

Alagramam, K.N., Murcia, C.L., Kwon, H.Y., Pawlowski, K.S. *et al.* (2001) The mouse Ames waltzer hearing-loss mutant is caused by mutation of *Pcdh15*, a novel protocadherin gene. *Nat. Genet.* **27**: 99–102.

Alvarez, Y., Alonso, M.T., Vendrell, V., Zelarayan, L.C. *et al.* (2003) Requirements for FGF3 and FGF10 during inner ear formation. *Development* **130**: 6329–6338.

Avraham, K.B., Hasson, T., Steel, K.P., Kingsley, D.M. *et al.* (1995) The mouse Snell's waltzer deafness gene encodes an unconventional myosin required for structural integrity of inner ear hair cells. *Nat. Genet.* **11**: 369–375.

Barald, K.F. and Kelley, M.W. (2004) From placode to polarization: new tunes in inner ear development. *Development* **131**: 4119–4130.

Bee, J. and Thorogood, P. (1980) The role of tissue interactions in the skeletogenic differentiation of avian neural crest cells. *Dev. Biol.* **78**: 47–62.

Bermingham, N.A., Hassan, B.A., Price, S.D., Vollrath, M.A. *et al.* (1999) *Math1*: an essential gene for the generation of inner ear hair cells. *Science* **284**: 1837–1841.

Boeda, B., El-Amraoui, A., Bahloul, A., Goodyear, R. *et al.* (2002) Myosin VIIa, harmonin and cadherin 23, three Usher I gene products that cooperate to shape the sensory hair cell bundle. *EMBO J.* **21**: 6689–6699.

Carlson, B.M. (1984) *Human Embryology and Developmental Biology.* Mosby: St Louis.

Casimiro, M.C., Knollmann, B.C., Ebert, S.N., Vary, J.C. Jr *et al.* (2001) Targeted disruption of the *Kcnq1* gene produces a mouse model of Jervell and Lange–Nielsen syndrome. *Proc. Natl Acad. Sci. USA* **98**: 2526–2531.

Catlin, F.I. (1985) Prevention of hearing impairment from infection and ototoxic drugs. *Arch. Otolaryngol.* **111**: 377–384.

Chen, P., Johnson, J.E., Zoghbi, H.Y. and Segil, N. (2002) The role of Math1 in inner ear development: uncoupling the establishment of the sensory primordium from hair cell fate determination. *Development* **129**: 2495–2505.

Chisaka, O., Musci, T.S. and Capecchi, M.R. (1992) Developmental defects of the ear, cranial nerves and hindbrain resulting from targeted disruption of the mouse homeobox gene *Hox-1.6. Nature* **355**: 516–520.

Clouthier, D.E., Hosoda, K., Richardson, J.A., Williams, S.C. *et al.* (1998) Cranial and cardiac neural crest defects in endothelin-A receptor-deficient mice. *Development* **125**: 813–824.

Cordes, S.P. and Barsh, G.S. (1994) The mouse segmentation gene *kr* encodes a novel basic domain-leucine zipper transcription factor. *Cell* **79**: 1025–1034.

Corwin, J. and Jones, J. (1991) In *Regeneration of Vertebrate Sensory Cells*, Bock, G. and Whelan, J. (eds). Wiley: Chichester; 103–130.

Corwin, J.T. and Cotanche, D.A. (1988) Regeneration of sensory hair cells after acoustic trauma. *Science* **240**: 1772–1774.

Cremers, C. and Noord, M.F. (1980) The ear-pits deafness syndrome. Clinical and genetic aspects. *Int. J. Paediat. Otorhinolaryngol.* **2**: 309–322.

Cremers, C., Thijssen, H., Fischer, A. and Marres, E. (1981) Otological aspects of the earpit-deafness syndrome. *J. Otorhinolaryngol. Relat. Spec.* **43**: 223–239.

Cremers, C.W., Admiraal, R.J., Huygen, P.L., Bolder, C. *et al.* (1998) Progressive hearing loss, hypoplasia of the cochlea and widened vestibular aqueducts are very common features in Pendred's syndrome. *Int. J. Pediatr. Otorhinolaryngol.* **45**: 113–123.

Cserjesi, P., Lilly, B., Bryson, L., Wang, Y. *et al.* (1992) MHox: a mesodermally restricted homeodomain protein that binds an essential site in the muscle creatine kinase enhancer. *Development* **115**: 1087–1101.

Daudet, N. and Lewis, J. (2005) Two contrasting roles for Notch activity in chick inner ear development: specification of prosensory patches and lateral inhibition of hair-cell differentiation. *Development* **132**: 541–551.

Davis, A.C. (1989) The prevalence of hearing impairment and reported hearing disability among adults in Great Britain. *Int. J. Epidemiol.* **18**: 911–917.

de Kok, Y.J., van der Maarel, S.M., Bitner-Glindzicz, M., Huber, I. *et al.* (1995) Association between X-linked mixed deafness and mutations in the POU domain gene *POU3F4*. *Science* **267**: 685–688.

Depew, M.J., Liu, J.K., Long, J.E., Presley, R. *et al.* (1999) Dlx5 regulates regional development of the branchial arches and sensory capsules. *Development* **126**: 3831–3846.

Di Palma, F., Holme, R.H., Bryda, E.C., Belyantseva, I.A. *et al.* (2001) Mutations in *Cdh23*, encoding a new type of cadherin, cause stereocilia disorganization in waltzer, the mouse model for Usher syndrome type 1D. *Nat. Genet.* **27**: 103–107.

Dixon, M.J., Gazzard, J., Chaudhry, S.S., Sampson, N. *et al.* (1999) Mutation of the Na–K–Cl co-transporter gene *Slc12a2* results in deafness in mice. *Hum. Mol. Genet.* **8**: 1579–1584.

Donaudy, F., Ferrara, A., Esposito, L., Hertzano, R. *et al.* (2003) Multiple mutations of *MYO1A*, a cochlear-expressed gene, in sensorineural hearing loss. *Am. J. Hum. Genet.* **72**: 1571–1577.

Eisenberg, L.S., Luxford, W.M., Becker, T.S. and House, W.F. (1984) Electrical stimulation of the auditory system in children deafened by meningitis. *Otolaryngol. Head Neck Surg.* **92**: 700–705.

Ekker, M., Akimenko, M.A., Bremiller, R. and Westerfield, M. (1992) Regional expression of three homeobox transcripts in the inner ear of zebrafish embryos. *Neuron* **9**: 27–35.

Everett, L.A., Belyantseva, I.A., Noben-Trauth, K., Cantos, R. *et al.* (2001) Targeted disruption of mouse *Pds* provides insight about the inner-ear defects encountered in Pendred syndrome. *Hum. Mol. Genet.* **10**: 153–161.

Everett, L.A., Glaser, B., Beck, J.C., Idol, J.R. *et al.* (1997) Pendred syndrome is caused by mutations in a putative sulphate transporter gene (*PDS*). *Nat. Genet.* **17**: 411–422.

Favor, J., Sandulache, R., Neuhauser-Klaus, A., Pretsch, W. *et al.* (1996) The mouse *Pax2* (*1Neu*) mutation is identical to a human *PAX2* mutation in a family with renal–coloboma syndrome and results in developmental defects of the brain, ear, eye, and kidney. *Proc. Natl Acad. Sci. USA* **93**: 13870–13875.

Fekete, D.M., Muthukumar, S. and Karagogeos, D. (1998) Hair cells and supporting cells share a common progenitor in the avian inner ear. *J. Neurosci.* **18**: 7811–7821.

Fekete, D.M. and Wu, D.K. (2002) Revisiting cell fate specification in the inner ear. *Curr. Opin. Neurobiol.* **12**: 35–42.

Fitch, N. and Srolovitz, H. (1976) Severe renal dysgenesis produced by a dominant gene. *Am. J. Dis. Child.* **130**: 1356–1357.

Fortnum, H.M., Summerfield, A.Q., Marshall, D.H., Davis, A.C. and Bamford, J.M. (2001) Prevalence of permanent childhood hearing impairment in the United Kingdom and implications for universal neonatal hearing screening: questionnaire-based ascertainment study. *Br. Med. J.* **323**: 536–540.

Fraser, F., Ling, D., Clogg, D. and Nogrady, B. (1978) Genetic aspects of the BOR syndrome – branchial fistulas, ear pits, hearing loss, and renal anomalies. *Am. J. Med. Genet.* **2**: 241–252.

Frolenkov, G.I., Belyantseva, I.A., Friedman, T.B. and Griffith, A.J. (2004) Genetic insights into the morphogenesis of inner ear hair cells. *Nat. Rev. Genet.* **5**: 489–498.

Gavalas, A., Studer, M., Lumsden, A., Rijli, F.M. *et al.* (1998) *Hoxa1* and *Hoxb1* synergize in patterning the hindbrain, cranial nerves and second pharyngeal arch. *Development* **125**: 1123–1136.

Gendron-Maguire, M., Mallo, M., Zhang, M. and Gridley, T. (1993) *Hoxa-2* mutant mice exhibit homeotic transformation of skeletal elements derived from cranial neural crest. *Cell* **75**: 1317–1331.

Gibson, F., Walsh, J., Mburu, P., Varela, A. *et al.* (1995) A type VII myosin encoded by the mouse deafness gene *shaker-1*. *Nature* **374**: 62–64.

Hadrys, T., Braun, T., Rinkwitz-Brandt, S., Arnold, H.H. and Bober, E. (1998) Nkx5–1 controls semicircular canal formation in the mouse inner ear. *Development* **125**: 33–39.

Hall, B.K. and Miyake, T. (1995) Divide, accumulate, differentiate: cell condensation in skeletal development revisited. *Int. J. Dev. Biol.* **39**: 881–893.

Holme, R.H. and Steel, K.P. (1999) Genes involved in deafness. *Curr. Opin. Genet. Dev.* **9**: 309–314.

Izumikawa, M., Minoda, R., Kawamoto, K., Abrashkin, K.A. *et al.* (2005) Auditory hair cell replacement and hearing improvement by *Atoh1* gene therapy in deaf mammals. *Nat. Med.* **11**: 271–276.

Jackler, R.K., Luxford, W.M. and House, W.F. (1987) Congenital malformations of the inner ear: a classification based on embryogenesis. *Laryngoscope* **97**: 2–14.

Johnson, K.R., Cook, S.A., Erway, L.C., Matthews, A.N. *et al.* (1999) Inner ear and kidney anomalies caused by IAP insertion in an intron of the *Eya1* gene in a mouse model of BOR syndrome. *Hum. Mol. Genet.* **8**: 645–653.

Johnson, K.R., Gagnon, L.H., Webb, L.S., Peters, L.L. *et al.* (2003) Mouse models of USH1C and DFNB18: phenotypic and molecular analyses of two new spontaneous mutations of the *Ush1c* gene. *Hum. Mol. Genet.* **12**: 3075–3086.

Kiernan, A. and Steel, K. (2002) Development of the mouse inner ear. In *Mouse Development. Patterning, Morphogenesis and Organogenesis*, Rossant, J. and Tam, P. (eds). Academic Press: New York; 539–566.

Kiernan, A.E., Ahituv, N., Fuchs, H., Balling, R. *et al.* (2001) The Notch ligand Jagged1 is required for inner ear sensory development. *Proc. Natl Acad. Sci. USA* **98**: 3873–3878.

Kiernan A.E., Erven, A., Voegeling, S., Peters, J. *et al.* (2002) ENU mutagenesis reveals a highly mutable locus on mouse chromosome 4 that affects ear morphogenesis. *Mamm. Genome* **13**: 142–148.

Kikkawa, Y., Shitara, H., Wakana, S., Kohara, Y. *et al.* (2003) Mutations in a new scaffold protein Sans cause deafness in Jackson shaker mice. *Hum. Mol. Genet.* **12**: 453–461.

Kubisch, C., Schroeder, B.C., Friedrich, T., Lutjohann, B. *et al.* (1999) *KCNQ4*, a novel potassium channel expressed in sensory outer hair cells, is mutated in dominant deafness. *Cell* **96**: 437–446.

Kurihara, Y., Kurihara, H., Suzuki, H., Kodama, T. *et al.* (1994) Elevated blood pressure and craniofacial abnormalities in mice deficient in endothelin-1. *Nature* **368**: 703–710.

Lanford, P.J., Lan, Y., Jiang, R., Lindsell, C. *et al.* (1999) Notch signalling pathway mediates hair cell development in mammalian cochlea. *Nat. Genet.* **21**: 289–292.

Lee, M.P., Ravenel, J.D., Hu, R.J., Lustig, L.R. *et al.* (2000) Targeted disruption of the *Kvlqt1* gene causes deafness and gastric hyperplasia in mice. *J. Clin. Invest.* **106**: 1447–1455.

Leger, S. and Brand, M. (2002) Fgf8 and Fgf3 are required for zebrafish ear placode induction, maintenance and inner ear patterning. *Mech. Dev.* **119**: 91–108.

Lewis, J. (1991) Rules for the production of sensory cells. In *Regeneration of Vertebrate Sensory Cells*, Bock, G. and Welan, J. (eds). Wiley: Chichester; 25–53.

Li, H., Liu, H. and Heller, S. (2003) Pluripotent stem cells from the adult mouse inner ear. *Nat. Med.* **9**: 1293–1299.

Li, S., Price, S.M., Cahill, H., Ryugo, D.K. *et al.* (2002) Hearing loss caused by progressive degeneration of cochlear hair cells in mice deficient for the *Barhl1* homeobox gene. *Development* **129**: 3523–3532.

Li, X.C., Everett, L.A., Lalwani, A.K., Desmukh, D. *et al.* (1998) A mutation in *PDS* causes non-syndromic recessive deafness. *Nat. Genet.* **18**: 215–217.

Liu, W., Li, G., Chien, J.S., Raft, S. *et al.* (2002) *Sonic hedgehog* regulates otic capsule chondrogenesis and inner ear development in the mouse embryo. *Dev. Biol.* **248**: 240–250.

Lohnes, D., Mark, M., Mendelsohn, C., Dolle, P. *et al.* (1994) Function of the retinoic acid receptors (RARs) during development (I). Craniofacial and skeletal abnormalities in RAR double mutants. *Development* **120**: 2723–2748.

Lombardo, A., Isaacs, H.V. and Slack, J.M. (1998) Expression and functions of *FGF-3* in *Xenopus* development. *Int. J. Dev. Biol.* **42**: 1101–1107.

Lombardo, A. and Slack, J.M. (1998) Postgastrulation effects of fibroblast growth factor on *Xenopus* development. *Dev. Dyn.* **212**: 75–85.

Lufkin, T., Dierich, A., LeMeur, M., Mark, M. and Chambon, P. (1991) Disruption of the *Hox-1.6* homeobox gene results in defects in a region corresponding to its rostral domain of expression. *Cell* **66**: 1105–1119.

Mallo, M. (1997) Retinoic acid disturbs mouse middle ear development in a stage-dependent fashion. *Dev. Biol.* **184**: 175–186.

Mallo, M. (2001) Formation of the middle ear: recent progress on the developmental and molecular mechanisms. *Dev. Biol.* **231**: 410–419.

Mallo, M. (2003) Formation of the outer and middle ear: molecular mechanisms. *Curr. Top. Dev. Biol.* **57**: 85–113.

Mallo, M. and Gridley, T. (1996) Development of the mammalian ear: coordinate regulation of formation of the tympanic ring and the external acoustic meatus. *Development* **122**: 173–179.

Mansour, S.L., Goddard, J.M. and Capecchi, M.R. (1993) Mice homozygous for a targeted disruption of the proto-oncogene *int-2* have developmental defects in the tail and inner ear. *Development* **117**: 13–28.

Martin, J.F. and Swanson, G.J. (1993) Descriptive and experimental analysis of the epithelial remodellings that control semicircular canal formation in the developing mouse inner ear. *Dev. Biol.* **159**: 549–558.

Martin, J.F., Bradley, A. and Olson, E.N. (1995) The paired-like homeobox gene *MHox* is required for early events of skeletogenesis in multiple lineages. *Genes Dev.* **9**: 1237–1249.

McKay, I.J., Lewis, J. and Lumsden, A. (1996) The role of *FGF-3* in early inner ear development: an analysis in normal and Kreisler mutant mice. *Dev. Biol.* **174**: 370–378.

McKay, I.J., Muchamore, I., Krumlauf, R., Maden, M. *et al.* (1994) The Kreisler mouse: a hindbrain segmentation mutant that lacks two rhombomeres. *Development* **120**: 2199–2211.

Minowa, O., Ikeda, K., Sugitani, Y., Oshima, T. *et al.* (1999) Altered cochlear fibrocytes in a mouse model of DFN3 nonsyndromic deafness. *Science* **285**: 1408–1411.

Morsli, H., Choo, D., Ryan, A., Johnson, R. and Wu, D. (1998) Development of the mouse inner ear and origin of its sensory organs. *J. Neurosci.* **18**: 3327–3335.

Morrison, A., Hodgetts, C., Gossler, A., Hrabe de Angelis, M. and Lewis, J. (1999) Expression of *Delta1* and *Serrate1* (*Jagged1*) in the mouse inner ear. *Mech. Dev.* **84**: 169–172.

Morton, N.E. (1991) Genetic epidemiology of hearing impairment. *Ann. N. Y. Acad. Sci.* **630**: 16–31.

Neyroud, N., Tesson, F., Denjoy, I., Leibovici, M. *et al.* (1997) A novel mutation in the potassium channel gene *KVLQT1* causes the Jervell and Lange–Nielsen cardioauditory syndrome. *Nat. Genet.* **15**: 186–189.

Ohuchi, H., Hori, Y., Yamasaki, M., Harada, H. *et al.* (2000) FGF10 acts as a major ligand for FGF receptor 2 IIIb in mouse multi-organ development. *Biochem. Biophys. Res. Commun.* **277**: 643–649.

Ozaki, H., Nakamura, K., Funahashi, J., Ikeda, K. *et al.* (2004) *Six1* controls patterning of the mouse otic vesicle. *Development* **131**: 551–562.

Pau, H., Fuchs, H., Hrabé De Angelis, M. and Steel, K.P. (2005) *Hush puppy*: a new mouse mutant with pinna, ossicle, and inner ear defects. *Laryngoscope* **115**: 116–124.

Phelps, P.D. (1974) Congenital legions of the inner ear, demonstrated by tomography. *Arch. Otolaryngol.* **100**: 11–18.

Phelps, P.D., Coffey, R.A., Trembath, R.C., Luxon, L.M. *et al.* (1998) Radiological malformations of the ear in Pendred syndrome. *Clin. Radiol.* **53**: 268–273.

Phillips, B.T., Bolding, K. and Riley, B.B. (2001) Zebrafish *fgf3* and *fgf8* encode redundant functions required for otic placode induction. *Dev. Biol.* **235**: 351–365.

Phippard, D., Lu, L., Lee, D., Saunders, J.C. and Crenshaw, E.B. III. (1999) Targeted mutagenesis of the POU-domain gene *Brn4/Pou3f4* causes developmental defects in the inner ear. *J. Neurosci.* **19**: 5980–5989.

Phippard, D., Boyd, Y., Reed, V., Fisher, G. *et al.* (2000) The sex-linked fidget mutation abolishes *Brn4/Pou3f4* gene expression in the embryonic inner ear. *Hum. Mol. Genet.* **9**: 79–85.

Probst, F.J., Fridell, R.A., Raphael, Y., Saunders, T.L. *et al.* (1998) Correction of deafness in *shaker-2* mice by an unconventional myosin in a *BAC* transgene. *Science* **280**: 1444–1447.

Pujol, R., Carlier, E. and Devigne, C. (1978) Different patterns of cochlear innervation during the development of the kitten. *J. Comp. Neurol.* **177**: 529–536.

Qiu, M., Bulfone, A., Ghattas, I., Meneses, J.J. *et al.* (1997) Role of the *Dlx* homeobox genes in proximodistal patterning of the branchial arches: mutations of *Dlx-1*, *Dlx-2*, and *Dlx-1* and *-2* alter morphogenesis of proximal skeletal and soft tissue structures derived from the first and second arches. *Dev. Biol.* **185**: 165–184.

Qiu, M., Bulfone, A., Martinez, S., Meneses, J.J. *et al.* (1995) Null mutation of *Dlx-2* results in abnormal morphogenesis of proximal first and second branchial arch derivatives and abnormal differentiation in the forebrain. *Genes Dev.* **9**: 2523–2538.

Represa, J., Leon, Y., Miner, C. and Giraldez, F. (1991) The *int-2* proto-oncogene is responsible for induction of the inner ear. *Nature* **353**: 561–563.

Riccomagno, M.M., Martinu, L., Mulheisen, M., Wu, D.K. and Epstein, D.J. (2002) Specification of the mammalian cochlea is dependent on *Sonic hedgehog*. *Genes Dev.* **16**: 2365–2378.

Rivera-Perez, J.A., Mallo, M., Gendron-Maguire, M., Gridley, T. and Behringer, R.R. (1995) Goosecoid is not an essential component of the mouse gastrula organizer but is required for craniofacial and rib development. *Development* **121**: 3005–3012.

Rubel, E.W., Dew, L.A. and Roberson, D.W. (1995) Mammalian vestibular hair cell regeneration. *Science* **267**: 701–707.

Ruf, R.G., Xu, P.X., Silvius, D., Otto, E.A. *et al.* (2004) *SIX1* mutations cause branchio–oto–renal syndrome by disruption of *EYA1–SIX1* DNA complexes. *Proc. Natl Acad. Sci. USA* **101**: 8090–8095.

Ryals, B.M. and Rubel, E.W. (1988) Hair cell regeneration after acoustic trauma in adult *Coturnix* quail. *Science* **240**: 1774–1776.

Sakagami, M., Fukazawa, K., Matsunaga, T., Fujita, H. *et al.* (1991) Cellular localization of rat Isk protein in the stria vascularis by immunohistochemical observation. *Hear. Res.* **56**: 168–172.

Sando, I., Takahara, T. and Ogawa, A. (1984) Congenital anomalies of the inner ear. *Ann. Otol. Rhinol. Laryngol. Suppl.* **112**: 110–118.

Schuknecht, H. (1974) *Pathology of the Ear.* Harvard University Press: Cambridge, MA.

Schuknecht, H.F. and Gacek, M.R. (1993) Cochlear pathology in presbyacusis. *Ann. Otol. Rhinol. Laryngol.* **102**: 1–16.

Self, T., Mahony, M., Fleming, J., Walsh, J. *et al.* (1998) *Shaker-1* mutations reveal roles for myosin VIIA in both development and function of cochlear hair cells. *Development* **125**: 557–566.

Self, T., Sobe, T., Copeland, N.G., Jenkins, N.A. *et al.* (1999) Role of myosin VI in the differentiation of cochlear hair cells. *Dev. Biol.* **214**: 331–341.

Siemens, J., Lillo, C., Dumont, R.A., Reynolds, A. *et al.* (2004) *Cadherin 23* is a component of the tip link in hair-cell stereocilia. *Nature* **428**: 950–955.

Sollner, C., Rauch, G.J., Siemens, J., Geisler, R. *et al.* (2004) Mutations in *cadherin 23* affect tip links in zebrafish sensory hair cells. *Nature* **428**: 955–959.

Steel, K.P., Barkway, C. and Bock, G.R. (1987) Strial dysfunction in mice with cochleo-saccular abnormalities. *Hear. Res.* **27**: 11–26.

Steel, K.P., Erven, A. and Kiernan, A.E. (2002) Mice as models for human hereditary deafness. In *Genetics and Auditory Disorders*, Keats, B.J.B., Popper, A.N. and Fay, R.R. (eds). Springer: New York.

Steel, K.P. and Kros, C.J. (2001) A genetic approach to understanding auditory function. *Nat. Genet.* **27**: 143–149.

ten Berge, D., Brouwer, A., Korving, J., Martin, J.F. and Meijlink, F. (1998) *Prx1* and *Prx2* in skeletogenesis: roles in the craniofacial region, inner ear and limbs. *Development* **125**: 3831–3842.

Torres, M., Gomez-Pardo, E. and Gruss, P. (1996) *Pax2* contributes to inner ear patterning and optic nerve trajectory. *Development* **122**: 3381–3391.

Trumpp, A., Depew, M.J., Rubenstein, J.L., Bishop, J.M. and Martin, G.R. (1999) Cre-mediated gene inactivation demonstrates that *FGF8* is required for cell survival and patterning of the first branchial arch. *Genes Dev.* **13**: 3136–3148.

Tsai, H., Hardisty, R.E., Rhodes, C., Kiernan, A.E. *et al.* (2001) The mouse *slalom* mutant demonstrates a role for *Jagged1* in neuroepithelial patterning in the organ of Corti. *Hum. Mol. Genet.* **10**: 507–512.

Tyson, J., Tranebjaerg, L., Bellman, S., Wren, C. *et al.* (1997) *IsK* and *KvLQT1*: mutation in either of the two subunits of the slow component of the delayed rectifier potassium channel can cause Jervell and Lange–Nielsen syndrome. *Hum. Mol. Genet.* **6**: 2179–2185.

Vahava, O., Morell, R., Lynch, E.D., Weiss, S. *et al.* (1998) Mutation in transcription factor POU4F3 associated with inherited progressive hearing loss in humans. *Science* **279**: 1950–1954.

VanCamp, G. and Smith, R. (2005). The hereditary hearing loss homepage: http://webhost.ua.ac.be/hhh/

Vendrell, V., Carnicero, E., Giraldez, F., Alonso, M.T. and Schimmang, T. (2000) Induction of inner ear fate by FGF3. *Development* **127**: 2011–2019.

Vetter, D.E., Mann, J.R., Wangemann, P., Liu, J. *et al.* (1996) Inner ear defects induced by null mutation of the *isk* gene. *Neuron* **17**: 1251–1264.

Vincent, C., Kalatzis, V., Abdelhak, S., Chaib, H. *et al.* (1997) BOR and BO syndromes are allelic defects of *EYA1*. *Eur. J. Hum. Genet.* **5**: 242–246.

Wang, W., Van De Water, T. and Lufkin, T. (1998) Inner ear and maternal reproductive defects in mice lacking the *Hmx3* homeobox gene. *Development* **125**: 621–634.

Warchol, M.E., Lambert, P.R., Goldstein, B.J., Forge, A. and Corwin, J.T. (1993) Regenerative proliferation in inner ear sensory epithelia from adult guinea pigs and humans. *Science* **259**: 1619–1622.

Woods, C., Montcouquiol, M. and Kelley, M.W. (2004) *Math1* regulates development of the sensory epithelium in the mammalian cochlea. *Nat. Neurosci.* **7**: 1310–1318.

Xiang, M., Gan, L., Li, D., Chen, Z.Y. *et al.* (1997) Essential role of POU-domain factor *Brn-3c* in auditory and vestibular hair cell development. *Proc. Natl Acad. Sci. USA* **94**: 9445–9450.

Xiang, M., Gao, W.Q., Hasson, T. and Shin, J.J. (1998) Requirement for *Brn-3c* in maturation and survival, but not in fate determination of inner ear hair cells. *Development* **125**: 3935–3946.

Xu, P.X., Adams, J., Peters, H., Brown, M.C. *et al.* (1999) *Eya1*-deficient mice lack ears and kidneys and show abnormal apoptosis of organ primordia. *Nat. Genet.* **23**: 113–117.

Zheng, J.L. and Gao, W.Q. (2000) Overexpression of Math1 induces robust production of extra hair cells in postnatal rat inner ears. *Nat. Neurosci.* **3**: 580–586.

Zheng, J.L., Shou, J., Guillemot, F., Kageyama, R. and Gao, W.Q. (2000) *Hes1* is a negative regulator of inner ear hair cell differentiation. *Development* **127**: 4551–4560.

Zheng, Q. and Johnson, K. (2005). Hereditary hearing impairment in mice website: http://www.jax.org/hmr/

Zine, A., Aubert, A., Qiu, J., Therianos, S. *et al.* (2001) *Hes1* and *Hes5* activities are required for the normal development of the hair cells in the mammalian inner ear. *J. Neurosci.* **21**: 4712–4720.

Zine, A., Van De Water, T.R. and de Ribaupierre, F. (2000) Notch signaling regulates the pattern of auditory hair cell differentiation in mammals. *Development* **127**: 3373–3383.

11

Development of the Enteric Nervous System in Relation to Hirschsprung's Disease

Heather M. Young, Donald F. Newgreen and **Alan J. Burns**

Introduction

The rapid progress of research into the major birth abnormality of the enteric nervous system (ENS), Hirschsprung's disease (HSCR), is one of the clearest illustrations of the meshing of clinical and basic research in delving into normal and abnormal early embryonic development. The advance of clinical genetic knowledge of the causes of HSCR has thrown light on the long-known variable inheritance modes and penetrance of HSCR. This has been combined with advances in basic developmental biology at the genetic, molecular, cellular and cell population levels. These advances have been facilitated by the relative simplicity of organization of the ENS, the conservative nature of its development across species, the variety of animal models and the availability of new techniques for answering early developmental questions. This combination makes this disease an outstanding model for the understanding of complex multigenic congenital dysmorphology syndromes.

Anatomy and function of the ENS

The ENS is the system of neurons and supporting glial cells within the wall of the gastrointestinal tract. There are many different morphological and functional types of

Embryos, Genes and Birth Defects, Second Edition Edited by Patrizia Ferretti, Andrew Copp, Cheryll Tickle and Gudrun Moore

neurons with complex connectivity patterns (Costa *et al.*, 1996; Furness *et al.*, 2004). Most of the neurotransmitters found in the CNS also occur in the ENS (Liu *et al.*, 1997; Galligan *et al.*, 2000; Furness, 2000; Brookes, 2001). Cell bodies of enteric neurons are located in myenteric (or Auerbach's) ganglia, between the circular and longitudinal muscle of the gut, and in submucous (or Meissner's) ganglia, internal to the circular muscle layer (Figure 11.1). Each enteric ganglion contains several different neuron types, and neighbouring ganglia in the same layer will contain similar types of neurons. The ENS therefore comprises units of neuronal circuitry repeated around and along the gastrointestinal tract.

The ENS mediates motility reflexes, and is also important in controlling water and electrolyte balance and intestinal blood supply (Furness and Costa, 1987; Vanner and Surprenant, 1996; Vanner and Macnaughton, 2004). The ENS shows considerable functional independence from the CNS, the degree varying with the region and species (Grider and Jin, 1994; Furness *et al.*, 1995). The functional independence is permitted by complete reflex circuits within the ENS, comprising motor neurons, intrinsic sensory and interneurons (Costa *et al.*, 1996; Furness *et al.*, 1998, 2004; Brookes, 2001).

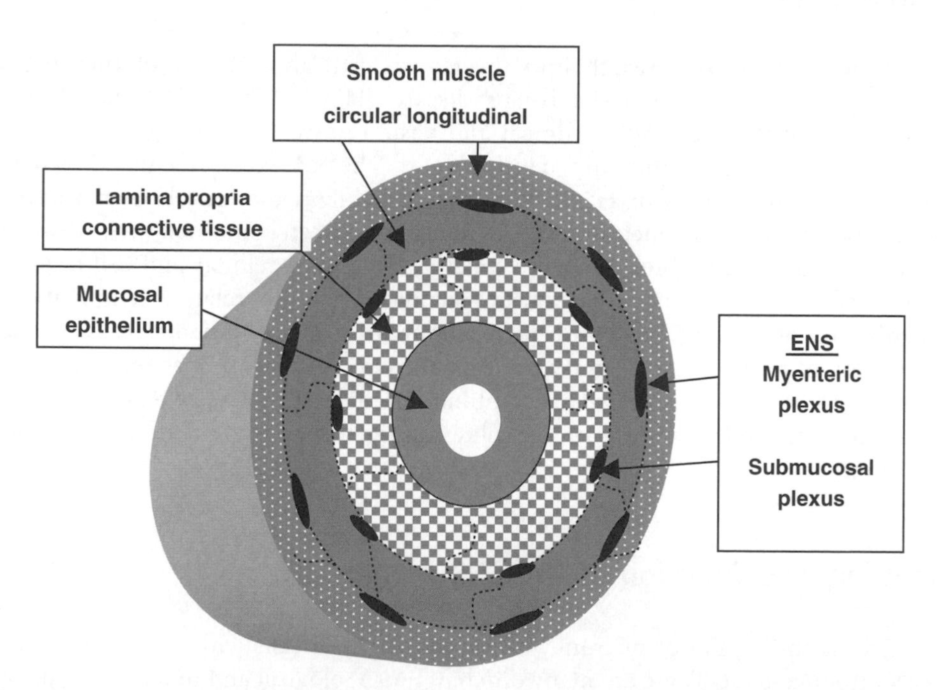

Figure 11.1 Diagram of the organization of the ENS. The intrinsic ganglia of the gastrointestinal tract form two concentric layers, the myenteric and submucosal plexuses, each consisting of small ganglia. Bundles of nerve fibres (dotted lines) connect the ganglia within and between each plexus. Nerve fibres also extend from the ganglia into the muscle layers and lamina propria

The ENS is crucial for post-natal life, as HSCR testifies. In contrast, the ENS is not essential at earlier embryonic and fetal stages, where nutrition is supplied by different routes. This means that abnormality of ENS development, by itself, will not cause pre-natal mortality or morbidity but will have powerful effects after birth.

The best-characterized developmental defect of the ENS – Hirschsprung's disease

HSCR is the name given to an intestinal motility disorder of the intestine first described in 1886 by the Danish physician Harald Hirschsprung (Holschneider and Puri, 2000). It was not until 1948 that the site of bowel functional abnormality and an effective surgical treatment were described by Swenson. The region involved in HSCR is always the most distal part of the bowel, of variable extent but usually the distal colon. Knowing the site of abnormality, the underlying cause, a congenital regional absence or gross reduction of enteric ganglia (aganglionosis), was soon identified. Subsequently, it was recognized that the ENS was established by proximal-to-distal migration of neural crest (NC) cells along the gastrointestinal tract (Yntema and Hammond, 1954) and it was therefore deduced that this regional absence of an ENS stemmed most likely from an early embryonic defect in NC cell migration. HSCR is diagnosed with certainty by a lack of intrinsic neurons in biopsy of intestinal submucosal tissues.

HSCR occurs in about 1/5000 live births. There is a pronounced sex bias (male:female ratio, 4:1), which is especially pronounced for short segment aganglionosis (Emison *et al.* 2005; for reviews, see Cass, 1986; Kapur, 1999b; Parisi and Kapur, 2000; Amiel and Lyonnet, 2001). HSCR also occurs in association with other syndromes and anomalies, such as congenital central hypoventilation syndrome (Ondine's curse), chromosome 22q11 deletion syndromes, and a variant of Shah–Waardenburg (WS4) syndrome (Amiel and Lyonnet, 2001). Down's syndrome patients are also at heightened risk of HSCR. Many of these conditions primarily involve abnormalities in systems developmentally related to the NC, collectively termed neurocristopathies (Parisi and Kapur, 2000).

Usually HSCR aganglionosis involves a short distal-most segment, the rectum and sigmoid colon, but rarely there is a greater length of the colon affected and this can even extend to the neighbouring ileum. Cases of essentially total intestinal aganglionosis have also been reported in humans (Shimotake *et al.*, 1997; Inoue *et al.*, 2000; Nemeth *et al.*, 2001). Functionally the disease is marked by intestinal obstruction, severe chronic constipation caused by the inability of the gut to transmit a peristaltic wave along the aganglionic segment, which is typically contracted and practically devoid of contents. In contrast, faecal accumulation causes the intestine proximal to the aganglionic region to become increasingly distended, a condition termed megacolon (Passarge, 2002). It is important to note that the grossly distended region contains a relatively normal ENS, and that the distension is an indirect effect of the blockage in the more distal regions lacking enteric neurons. Abnormalities of ENS structure or function are not always restricted to the aganglionic region, since in most cases, proximal to this there may be a variable length of intestine termed the 'transition zone' with either hypoganglionosis or even hyperganglionosis (Kapur, 1999b).

HSCR has a strong genetic component. The sibling-risk increases and the sex ratio difference decreases with the length of the aganglionic segment (Holschneider and Puri, 2000). Both long- and short-segment HSCR can be caused by the same mutation (Seri *et al.*, 1997). Mutations in at least eight genes are potentially involved in HSCR, but it should be noted that all of the known and tested mutations account for less than 50% of HSCR cases in humans. This suggests that some mutations may involve non-coding regulatory regions of known genes (Emison *et al.* 2005), about which little is known at present. Additionally, there may be more genes which, when mutated, predispose to HSCR. As with some other neurocristopathies, there may also be non-genetic risk factors.

Similar conditions of distal intestinal aganglionosis have occurred in animals, such as *lethal spotting* (*ls*), *Piebald lethal* (S^l) (Lane, 1966) and *Dominant megacolon* (*Dom*) (Lane and Liu, 1984) mutants in mice, *spotting lethal* (*sl*) in the rat (Dembowski *et al.*, 2000) and lethal white foal syndrome in horses (McCabe *et al.*, 1990). All of these mutations, unlike most human cases, involve prominent abnormalities in pigmentation, which involve another NC derivative, the melanocytes. Yet other examples have been engineered in mice by targeted gene mutation and silencing or other forms of gene manipulation (see below). In chickens, aganglionosis of the most distal intestine has also been induced by partial removal during early embryogenesis of a specific section of the NC at hindbrain level (Yntema and Hammond, 1954; Peters-van der Sanden *et al.*, 1993; Burns *et al.*, 2000).

Cell biology of ENS development

Axial origin of neural crest-derived ENS precursors

Vagal neural crest cells form the majority of the ENS A NC origin for the enteric nervous system (ENS) was initially proposed over 50 years ago by Yntema and Hammond (1954), following experiments whereby the entire vagal (post-otic hindbrain) neural tube was ablated in chick embryos, leading to severe gut aganglionosis. However, it was the establishment of the quail–chick interspecies grafting technique by Le Douarin, in the late 1960s, that first allowed the precise axial level of ENS precursor cells to be determined (Le Douarin and Teillet, 1973; Le Douarin, 1982). Using this grafting technique, Le Douarin and colleagues described the vagal NC adjacent to somites 1–7 as the major contributor of precursor cells to the ENS, with a minor contribution, originating in the lumbosacral NC caudal to the 28th pair of somites, providing a smaller number of cells to the hindgut (Le Douarin and Teillet, 1973, 1974), as shown in Figure 11.2.

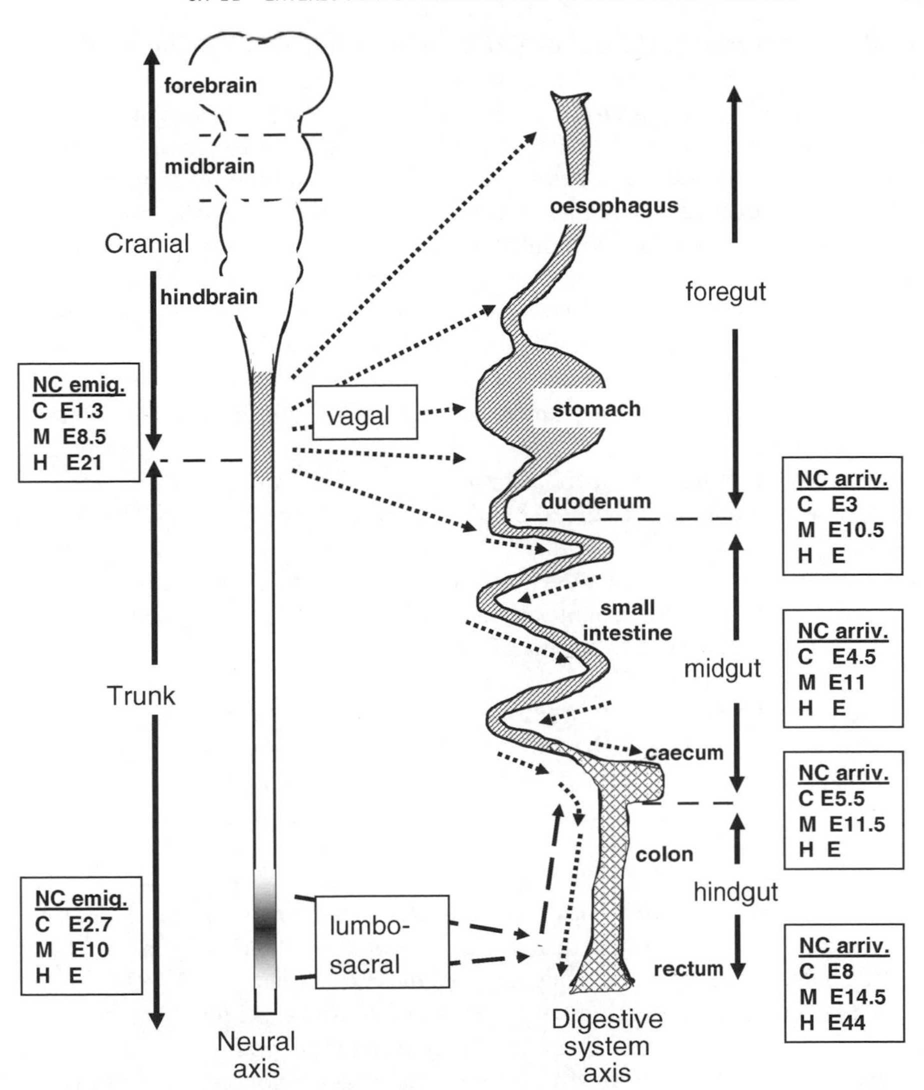

Figure 11.2 Diagram of the axial organization of the neural and digestive systems (not to scale), with the origin of ENS cells from the vagal level and lumbosacral levels of the NC shown. The approximate age of onset of NC emigration from the neural axis for chick, mouse and human (C, M, H) is shown in the boxes on the left side, with E being days of incubation (chick) or days post-fertilization (mouse and human). Note that the vagal level lies mostly in the caudal hindbrain but overlaps slightly the rostral trunk level. The vagal NC cells migrate directly to the foregut, then colonize the midgut and hindgut in a caudally directed wave. The approximate ages for colonization by vagal NC cells (vNC) are given in the boxes on the right. The lumbosacral NC cells colonize only the distal intestine (cross-hatched), but the number of ENS cells from this source is less than that from the vagal NC

In the chick embryo, the vagal contribution to the ENS has subsequently been confirmed following regional ablations of the neural tube (Peters-van der Sanden *et al.*, 1993), quail–chick grafting experiments (Burns and Le Douarin, 1998, 2001; Burns *et al.*, 2000), and cell labelling using injection of a virus vector containing the marker gene *lacZ* into the somites adjacent to the neural tube (Epstein *et al.*, 1994). More recently, quail–chick grafts, consisting of sub-regions of the vagal NC, have been employed to analyse the contribution of the vagal crest at a more precise axial level (Burns *et al.*, 2000). These studies demonstrated that the vagal crest, adjacent to somites 1–2, contributed cells mainly to the foregut, i.e. the oesophagus and stomach, whereas grafts adjacent to somites 6–7 labelled cells that were most numerous in the colon. When grafts adjacent to somites 3–5 or 3–6 were performed, labelled cells were present along the entire length of the gut, from the oesophagus to the distal colon.

Evidence confirming the contribution of the vagal NC to the ENS in mammalian embryos has been more difficult to obtain, due mainly to the technical difficulty of performing focal, long-term labelling of NC cells in embryos that normally develop *in utero*. Durbec *et al.* (1996) used the fluorescent lineage tracer DiI in conjunction with embryo culture to trace the fate of vagal NC cells. When DiI was applied to the neural tube of E8.5 mice adjacent to somites 1–4, labelled cells were subsequently found in the foregut and midgut. When DiI was injected into the neural tube adjacent to somites 6–7, labelled cells were subsequently found in the foregut. These results suggested that, in the mouse, the majority of the ENS is derived from a population of NC cells from the vagal level, adjacent to somites 1–5. In addition, the oesophagus ENS receives a contribution from a population of cells adjacent to somites 6–7 (Durbec *et al.*, 1996).

The lumbosacral NC contributes cells to the hindgut ENS Le Douarin and colleagues originally suggested that, in addition to the vagal NC, the lumbosacral NC also contributes cells to the hindgut (Le Douarin and Teillet, 1973, 1974). However, due to conflicting findings, these results remained controversial for a number of years. The issue was resolved by utilizing quail–chick interspecies grafting to selectively label subpopulations of NC cells, in conjunction with antibody double-labelling to identify quail cells and neuronal and glial phenotypes within chick enteric ganglia (Burns and Le Douarin, 1998; Burns *et al.*, 2000). These studies demonstrated that in the chick, lumbosacral NC cells initially form the nerve of Remak (an extramural nerve particular to avians that extends along the mesenteric border of the hindgut and midgut). Then, at E7, when the hindgut is colonized by vagal NC cells, nerve fibres project from the nerve of Remak into the hindgut. Lumbosacral crest cells then migrate along these fibres and colonize the hindgut in larger numbers from E10 onwards, i.e. 2–3 days after it has been colonized by vagal NC cells. This 'waiting period' for immigration of lumbosacral NC cells in the hindgut occurs independently of vagal crest cells, since lumbosacral cells colonized the gut and differentiated into neurons and glia in an apparently normal manner, even when the vagal-derived cells were absent from the hindgut following the ablation of the vagal NC (Burns *et al.*, 2000).

Evidence in the mouse embryo for a similarly later arrival of lumbosacral NC cells into the hindgut has also been obtained using *Wnt1–lacZ* transgene expression as an early marker of murine NC cells (Kapur, 2000). It appears that lumbosacral NC cells in both avians and mammals initially form extramural ganglia (i.e. the nerve of Remak in the chick; the pelvic ganglia in the mouse), then subsequently enter the hindgut after it has been colonized by vagal NC cells. However, in mammals, the spatiotemporal migration, extent of contribution and fate of lumbosacral NC cells has yet to be fully determined.

Are vagal and lumbosacral NCC prespecified as ENS precursors?

The ENS is derived from two specific regions of the neuraxis, the vagal and lumbosacral NC. Do NC cells at these axial levels have specific properties that allow them to colonize the gut, or is the environment favourable/permissive to gut colonization at these specific locations? Findings from heterotopic quail–chick transplantations have demonstrated that the fate of NC cells depends mainly on the signals they encounter along their migration pathways, rather than on their axial level of origin. For example, when vagal NC, which normally forms the ENS, was replaced with trunk (i.e. thoracic) crest, which normally gives rise to sensory and sympathetic but not ENS ganglia, the grafted cells colonized the gut and gave rise to neurons that displayed enteric phenotypes (Le Douarin and Teillet, 1974; Le Douarin *et al.*, 1975; Fontaine-Perus *et al.*, 1982; Rothman *et al.*, 1986). In contrast, when cephalic and vagal NC was transplanted into the thoracic axial level, transplanted cells gave rise to adrenergic cells in sympathetic ganglia and to adrenomedullary cells, which are typical of this level (Le Douarin and Teillet, 1974). These findings suggest that vagal crest cells are not restricted to an enteric fate, since they developed phenotypes typical of their new axial level following transplantation. When lumbosacral crest was transplanted to the thoracic level, and vice versa, these cells also behaved according to their new position rather than their site of origin. Therefore, transplanted thoracic level cells were able to colonize the gut, while lumbosacral cells, grafted to the thoracic level, did not reach the gut (Erickson and Goins, 2000). Hence it seems that lumbosacral NC cells do not possess an inherent ability to find their way to the gut, and that permissive pathways at the lumbosacral level allow cells from other axial levels to reach the gut (Erickson and Goins, 2000).

When these experiments are examined in more detail, however, differences can be seen. For example, the trunk NC transplanted to vagal levels colonized only the foregut and part of the midgut, and contributed ectopic melanocytes in the gut (Le Douarin and Teillet, 1974), whereas vagal NC cells populate the entire gut. Furthermore, the vagal transplants to the mid-trunk, as well as producing normal trunk derivatives, also produced cells that migrated into the gut and formed apparently normally placed enteric ganglia, a feat beyond the capability of the local trunk crest (Le Douarin and Teillet, 1974). When quail vagal crest was transplanted into the lumbosacral region of chick embryos, the transplanted vagal NC cells colonized the gut via lumbosacral pathways, in large numbers, early in development and ignoring the 'waiting period'. Conversely, when lumbosacral crest was transplanted to the vagal region, these cells also colonized the gut, but in much lower numbers than vagal cells (Burns *et al.*, 2002).

These results suggest a cell autonomous difference in the two NC cells populations, with vagal NC cells being more invasive of the gut than the lumbosacral (or other trunk) population, and being able to evade the colonic repulsive signals that delay lumbosacral NC cells.

Why are vagal NC cells more invasive of the gut than lumbosacral NC cells?

The reasons for the difference in invasive capacity of the vagal and lumbosacral populations are still unclear, but changes in gene expression along the rostro-caudal axis during development may result in changes in the expression of cell-signalling molecules involved in NC cell migration/proliferation. Suitable candidates are the homeobox-containing (*Hox*) genes, a group of highly conserved regulatory genes involved in patterning and specification during development (Krumlauf, 1994). Recent studies of *Hox* paralogous groups 4 and 5 have suggested that some genes within these groups may be organized to form a specific *Hox* code involved in ENS development (Pitera *et al.*, 1999). This theory is supported by the fact that *Hox* genes belonging to groups 4 and 5 are expressed in the developing hindbrain at the level of rhombomeres 6–8. Since rhombomeres 7 and 8 correspond to the anterior vagal NC, the region that contributes the majority of ENS precursors, it is feasible that *Hox* genes from this area may be implicated in vagal NC cell development. Indeed, the expression pattern of *Hoxb5* has recently been correlated with the migration and differentiation of NC cells within the developing human (Fu *et al.*, 2003) and mouse gut (Pitera *et al.*, 1999), and recently one element in the complex enhancer of *Hoxb3* has been found to specify expression to ENS-forming NC (Chan *et al.*, 2005). More 5′-located *Hox* genes, such as those from paralogous groups 10–11 that are expressed in the lumbosacral region of the NC (Lance-Jones *et al.*, 2001), may provide lumbosacral crest-derived enteric precursors with positional information.

Molecular biology of ENS development and Hirschsprung-like dysplasias

Many genes have been identified that, when mutated or deleted, interfere with ENS development.

GDNF/GFRα1/RET

RET is a receptor tyrosine kinase, and is activated by GDNF (see below). Heterozygous mutations in *RET* are the main known cause of HSCR, and about 50% of familial and some sporadic HSCR arise from mutations in *RET* (Robertson *et al.*, 1997; Parisi and Kapur, 2000; Amiel and Lyonnet, 2001; Belknap, 2002; Passarge, 2002). Parallel studies in mice revealed that *Ret* is expressed in specific regions of the developing central and peripheral nervous systems, including the ENS, and in the developing kidney (Pachnis *et al.*, 1993). Furthermore, mice with targeted disruption in the *Ret* gene show intestinal aganglionosis and renal agenesis, and die within 24 hours of birth (Schuchardt *et al.*, 1994).

Over 80 mutations in the *RET* gene have been described in HSCR patients (Amiel and Lyonnet, 2001). Interestingly, the same heterozygous mutation in a single family can result in very variable phenotypes, including long-segment HSCR, short-segment HSCR or no detectable defects (Romeo *et al.*, 1994; Edery *et al.*, 1994). This variability strongly suggests the involvement of other genes. As mentioned previously, HSCR shows a 4:1 male:female predominance, and in familial cases that are heterozygous for the same *RET* mutation, aganglionosis is more severe in male than in female siblings (Inoue *et al.*, 2000). Furthermore, the mean penetrance of *RET* mutations in familial HSCR is significantly higher in males than in females (Attie *et al.*, 1995). There is a rare form of HSCR, known as total intestinal aganglionosis, in which enteric neurons are absent from the entire small and large intestines. A homozygous RET mutation was found in a HSCR patient with total intestinal aganglionosis, while heterozygosity of the same mutation resulted in a less severe form of aganglionosis (extending distally from the jejunum) (Inoue *et al.*, 2000). In contrast to humans, mice with heterozygous mutations in *Ret* do not exhibit aganglionosis in any region of the gut (Schuchardt *et al.*, 1994; Gianino *et al.*, 2003). Although $Ret^{-/-}$ mice show severe kidney defects (Schuchardt *et al.*, 1994) and *RET* is expressed in the developing kidney of humans (Attie-Bitach *et al.*, 1998), HSCR patients only extremely rarely have renal abnormalities (Lore *et al.*, 2000).

Glial cell line-derived neurotrophic factor (Gdnf) is a secreted growth factor that belongs to the transforming growth factor-β (TGFβ) superfamily (Airaksinen and Saarma, 2002). Like $Ret^{-/-}$ mice, those with targeted disruption of the *Gdnf* gene showed renal agenesis and intestinal aganglionosis (Moore *et al.*, 1996; Pichel *et al.*, 1996; Sanchez *et al.*, 1996). Although Gdnf signals via Ret, Gdnf does not bind directly to Ret. Gdnf binds to a glycosylphosphatidylinsoitol (GPI) linked co-receptor, GFRα1 (Jing *et al.*, 1996; Treanor *et al.*, 1996), and the Gdnf–Gfrα1 complex then binds to Ret and triggers signalling. GFRα1 occurs both as a GPI-linked membrane-tethered form and as a released form (Paratcha *et al.*, 2001).

Although $Gdnf^{-/-}$ mice lack enteric neurons throughout most of the gastrointestinal tract, mutations in the *GDNF* gene alone have only rarely been found to be responsible for HSCR (Ivanchuk *et al.*, 1996; Salomon *et al.*, 1996; Amiel and Lyonnet, 2001; Eketjall and Ibanez, 2002). However, *GDNF* gene mutations may contribute to the severity of HSCR if the mutation coincides with mutations/polymorphisms in other HSCR genes (Angrist *et al.*, 1996; Hofstra *et al.*, 1997). Unlike $Ret^{+/-}$ mice, which have normal numbers of enteric neurons and exhibit no detectable defects (Gianino *et al.*, 2003), $Gdnf^{+/-}$ mice have reduced numbers of enteric neurons throughout the gastrointestinal tract, impaired intestinal motility and a higher incidence of post-natal death compared to wild-type mice (Shen *et al.*, 2002). Although there is no sex bias, there is significant variation between individual *Gdnf*$^{+/-}$ mice in the severity and the age of onset of the symptoms, supporting the data from human studies suggesting that *GDNF* is an important HSCR susceptibility locus (Shen *et al.*, 2002).

The RET co-receptors are members of the GFRα family. Mice with targeted disruption in the gene encoding *Gfr*α*1* have a similar phenotype to *Ret* and *Gdnf* null mice in that the ENS fails to develop in most regions of the gut (Cacalano *et al.*, 1998; Enomoto *et al.*, 1998; Tomac *et al.*, 2000). However, mutations in *GFR*α*1* have yet to be identified in HSCR patients (Angrist *et al.*, 1998; Myers *et al.*, 1999; Borrego *et al.*, 2003).

Role of GDNF/GFRα1/RET signalling in ENS development Gdnf plays multiple roles during ENS development. Gdnf is expressed by the gut mesenchyme prior to, and after, the entry of NC-derived cells (Trupp *et al.*, 1995; Hellmich *et al.*, 1996; Moore *et al.*, 1996; Suvanto *et al.*, 1996; Natarajan *et al.*, 2002). GFRα1 is expressed by vagal crest-derived cells within the gut and also by the gut mesenchyme (Chalazonitis *et al.*, 1998; Schiltz *et al.*, 1999), and Ret is expressed by vagal NC-derived cells prior to and after they have colonized the gut (Pachnis *et al.*, 1993; Lo and Anderson, 1995; Robertson and Mason, 1995; Iwashita *et al.*, 2003). In embryonic *Ret*$^{-/-}$ mice, vagal NC-derived cells die just before or just after they reach the foregut (Durbec *et al.*, 1996), indicating a role for Gdnf in survival and/or migration. Studies of enteric NC-derived cells *in vitro* have shown that Gdnf promotes survival, proliferation and neuronal differentiation (Taraviras *et al.*, 1999). An interesting aspect of the role of Gdnf in survival is that transfection of *Ret* into cells from an immortalized olfactory neuroblast cell line induces apoptosis, which can be prevented by the presence of Gdnf (Bordeaux *et al.*, 2000). The ability of Ret to induce apoptosis in the absence of a ligand suggests that the coordinated expression of Gdnf and Ret may be very important during development. Gdnf is also chemoattractive to vagal NC-derived cells and appears to play an important role in inducing the migration of crest cells into and along the gut, and also in retaining them within the gut (Young *et al.*, 2001; Natarajan *et al.*, 2002; Iwashita *et al.*, 2003).

Although Gdnf exerts multiple effects on enteric NC-derived cells – survival, proliferation, differentiation, migration – it is unclear how the different biological responses are controlled. For example, under what circumstances does Gdnf induce proliferation rather than differentiation? It is likely that the expression of a variety of intracellular molecules, particularly signalling molecules, will vary with age and will determine the response of a particular cell to Ret activation. Studies using cell lines and primary cell cultures have shown that Ret can activate a range of intracellular signalling pathways. In particular, activation of PI3K appears to be essential for a variety of Gdnf-induced responses. An *in vitro* study of mouse enteric NC-derived cells showed that both MAPK and PI3K are involved in Gdnf-induced migration and axon outgrowth (Natarajan *et al.*, 2002). PI3K activity is also necessary for Gdnf-induced proliferation of enteric crest cells (Focke *et al.*, 2001). However, MAPK activity is not required for Gdnf-induced proliferation of crest-derived cells

(Focke *et al.*, 2001). The circumstances under which PI3K activation induces different biological responses are unknown.

Other GDNF family members There are four known members of the GDNF family; GDNF, neurturin (NRTN), artemin and persephin. All GDNF family ligands signal through Ret, but each binds to a specific Gfrα: GDNF to Gfrα1, NRTN to Gfrα2, artemin to Gfrα3 and persephin to Gfrα4 (Airaksinen *et al.*, 1999). Human mutations in *NRTN* by themselves do not appear to result in aganglionosis, although *NRTN* mutations in combination with mutations in *RET* or other susceptibility loci can result in aganglionosis (Doray *et al.*, 1998; Inoue *et al.*, 2000). Mutations in *GFR*α*2*, *GFR*α*3* or *GFR*α*4* have not been found in any HSCR patients (Onochie *et al.*, 2000; Vanhorne *et al.*, 2001; Borrego *et al.*, 2003). As yet there have been no studies published in which patients with HSCR have been screened for mutations in artemin or persephin.

In mice, NRTN is expressed in all regions of the developing gastrointestinal tract, particularly in the circular muscle layer (Widenfalk *et al.*, 1997; Golden *et al.*, 1999; Xian *et al.*, 1999). Artemin is expressed in the oesophagus only (Enomoto *et al.*, 2001) and persephin is not expressed in any peripheral tissue (Milbrandt *et al.*, 1998). Mice lacking artemin or Gfrα3 are viable and have not been reported to show any ENS phenotype (Nishino *et al.*, 1999; Honma *et al.*, 2002). Mice lacking neurturin or Gfrα2 are viable and have similar neuron numbers to wild-type mice, indicating that *in vivo* it is not essential for the survival, proliferation or migration of enteric neuron precursors (Gianino *et al.*, 2003). However, there is a decrease in the density of excitatory nerve fibres in the circular muscle compared to wild-type mice (Heuckeroth *et al.*, 1999; Rossi *et al.*, 1999; Gianino *et al.*, 2003). The transit of contents through the small intestine is also 25% slower in *Gfr*α*2*$^{-/-}$ mice compared to wild-type mice (Rossi *et al.*, 2003), probably because of the lower density of excitatory nerve terminals in the circular muscle which mediate contraction during peristalsis. Thus, expression of NRTN by the circular muscle is important in inducing axon extension or branching by excitatory motor neurons. Although artemin is expressed by the oesophagus, artemin does not appear to induce neurite outgrowth or migration of crest-derived cells in the oesophagus or intestine of embryonic mice (Yan *et al.*, 2004).

Endothelin-3/endothelin receptor B

The endothelins (Et-1, Et-2 and Et-3) are secreted peptides that act via G-protein-coupled receptors (Ednra and Ednrb). In adults, endothelins play an important role in the cardiovascular system. During development, endothelins have important additional roles which have mostly been revealed by gene knock-out studies in mice (Gershon, 1995).

Around 5% of HSCR cases are due to mutations in *EDNRB* (Puffenberger *et al.*, 1994; Amiel *et al.*, 1996; Kusafuka *et al.*, 1996; Inoue *et al.*, 1998; Amiel and Lyonnet, 2001). *Ednrb*$^{-/-}$ mice have megacolon (a HSCR-like condition) as well as pigmentation defects (Hosoda *et al.*, 1994). The spontaneously occurring piebald lethal (S^l) mutant in the mouse (Hosoda *et al.*, 1994), spotting lethal (*sl*) mutant in the rat (Ceccherini *et al.*, 1995; Gariepy *et al.*, 1998; Dembowski *et al.*, 2000) and lethal white foal syndrome (Yang *et al.*, 1998) are also due to loss of endothelin function resulting from inactivation of the *Ednrb* gene.

Mutations in the *ET-3* gene in humans are responsible for 5% or less of HSCR cases (Svensson *et al.*, 1999; Amiel and Lyonnet, 2001). Mutation of the mouse *Et-3* gene, as in the lethal spotting mouse (*ls*), produces an ENS and pigmentation phenotype similar to that of *Ednrb* null mutations, but the length of the aganglionic segment of colon is shorter. This is thought to be due to partial compensation by ET-1, which also binds to Ednrb. Significantly, some changes in the numbers of interstitial cells of Cajal and submucosal neurons, and in the expression of neurotransmitters by enteric neurons, have been reported in the ileum and colon (proximal to the aganglionosis) in $ET\text{-}3^{-/-}$ mice compared to heterozygous mice (Sandgren *et al.*, 2002). These changes could contribute to the on-going dysmotility problems that can occur after surgical resection of the aganglionic segment in HSCR (Catto-Smith *et al.*, 1995).

The production of functional Ets requires endothelin converting enzyme (ECE-1). Mutations in *ECE-1* has been found in a patient with a group of complex neurocristopathies involving cardiac lesions, craniofacial defects, autonomic dysfunction and intestinal aganglionosis (Hofstra *et al.*, 1999). Mice lacking *Ece-1* have abnormalities that are seen in a combination of $Et\text{-}1^{-/-}$ and $Ednra^{-/-}$ mice and in $Et3^{-/-}$ and $Ednrb^{-/-}$ mice, including craniofacial and cardiac abnormalities, and an absence of epidermal melanocytes and enteric neurons in the distal gut (Yanagisawa *et al.*, 1998).

The effects of Et-3 on enteric NC-derived cells are complex, and the conclusions drawn from different studies have often been inconsistent.

Where and when are Et-3 and EdnrB expressed in relation to the ENS? Et-3 is produced by mesenchymal cells and in mice is first expressed in E10.0 midgut and hindgut (Leibl *et al.*, 1999; Barlow *et al.*, 2003). At E11 and E11.5, expression becomes markedly intensified in the caecal mesenchyme prior to and overlapping the arrival of NC cells. There is also Et-3 expression in the proximal colon from E11.5, but the distal hindgut shows very low expression at all stages of development (Barlow *et al.*, 2003). *Ednrb* is expressed by migrating NC cells in the gut (Nataf *et al.*, 1996; Brand *et al.*, 1998; Woodward *et al.*, 2000; Sidebotham *et al.*, 2002b; Lee *et al.*, 2003; Barlow *et al.*, 2003; McCallion *et al.*, 2003). *Ednrb* is also expressed by the gut mesenchyme (Barlow *et al.*, 2003).

In the absence of Et-3 signalling, crest migration is delayed In mice lacking components of ET-3/Ednrb signalling pathways, the migration of crest-derived cells through the gut is delayed, and aganglionosis of the terminal bowel occurs (Kapur *et al.*, 1995; Shin *et al.*, 1999; Lee *et al.*, 2003; Kruger *et al.*, 2003). Even in E10.5 mice (which is only around a day after NC cells enter the foregut) the number of vagal cells and the distance they had migrated was reported to be reduced in $Et\text{-}3^{-/-}$ mice compared to wild-type mice (Barlow *et al.*, 2003). Two fundamental questions arise that remain unanswered. First, why does an absence of ET-3 signalling retard the rate of migration? Second, if the migration of crest cells is delayed, why do they not reach the anal end at a later developmental stage, rather than fail to colonize the terminal bowel?

Does an absence of Et-3 signalling affect the environment of crest cells? When wild-type NC cells were co-cultured with segments of hindgut from either *Et-3*$^{-/-}$ or wild-type mice, neurons developed in the wild-type hindgut explants but not in the *Et-3*$^{-/-}$ hindgut explants (Jacobs-Cohen *et al.*, 1987). However, in the presence of exogenous Et-3 in the culture medium, wild-type or *Et-3*$^{-/-}$ crest cells will enter explants of *Et-3*$^{-/-}$ hindgut (Wu *et al.*, 1999). This suggests that, in the absence of Et-3 signalling, non-crest derived elements of the environment are abnormal. However, when labelled enteric NC stem cells are injected into the distal colon of embryonic *Ednrb*$^{-/-}$ rats, the stem cells migrate, survive and differentiate to a similar degree to cells injected into the colon of wild-type embryos (Kruger *et al.*, 2003). Therefore, the environment of the distal colon of *Ednrb*$^{-/-}$ rats does not appear to be deleterious to the survival and differentiation of wild-type crest cells (Kruger *et al.*, 2003).

At what developmental stage is Et-3 signalling required? Using an inducible system to modulate the expression of *Ednrb* in transgenic mice, it has been shown that, for normal ENS development, expression of *Ednrb* was only required from E10.5 to E12.5 (Shin *et al.*, 1999). The latter stage is interesting because complete colonization of the embryonic mouse gut is not achieved until E14.5 (Kapur *et al.*, 1992). This suggests either that the colonization of the distal colon does not require signalling via Ednrb or that *Ednrb* mRNA and/or protein persist for around 2 days.

In the absence of Et-3 signalling, does cell death contribute to the ENS phenotype? Programmed cell death (apoptosis) occurs during the development of most parts of the nervous system. However, there is no evidence for apoptosis occurring during the development of the ENS (Gianino *et al.*, 2003; Kruger *et al.*, 2003). Importantly, there is also no evidence for apoptosis amongst crest-derived cells in the gut of embryonic *Ednrb*$^{-/-}$ rats or *ET-3*$^{-/-}$ mice (Kruger *et al.*, 2003; Woodward *et al.*, 2003).

Effects of ET-3 signalling on proliferation Although one study showed that Et-3 signalling does not affect the percentage of crest-derived cells undergoing cell division *in vivo* (Woodward *et al.*, 2003), another study reported that the percentage of undifferentiated crest cells undergoing mitosis in dissociated intestine of *ET-3*$^{-/-}$ mice was significantly lower than that in wild-type mice (Barlow *et al.*, 2003). In mouse enteric crest-derived cells *in vitro*, exposure to Et-3 alone had little or no effect on proliferation (Wu *et al.*, 1999; Barlow *et al.*, 2003), but Et-3 enhanced the proliferative effects of Gdnf on undifferentiated enteric crest cells from embryonic mice and quail (Hearn *et al.*, 1998; Barlow *et al.*, 2003; see section on Interactions between GDNF and ET-3 signalling pathways, below).

Effects of Et-3 signalling on differentiation Et-3 decreases the number of neurons that develop from crest-derived cells immunoselected from the embryonic mouse and rat gut (Wu *et al.*, 1999; Kruger *et al.*, 2003). In crest-derived cells immunoselected from the gut of embryonic quail, Et-3 alone has little effect but Et-3 significantly decreases the differentiation of neurons induced by Gdnf (Hearn *et al.*, 1998). However, a recent study of crest-derived cells isolated from the embryonic rat gut showed that Et-3 does not inhibit the neuronal differentiation induced by BMP-4 (Kruger *et al.*, 2003).

It has been proposed that, in the absence of Et-3/Ednrb signalling, crest-derived cells in the gut differentiate into neurons too early, prior to colonizing the distal part of the hindgut, resulting in a deficit of proliferative and migratory cells to complete the colonization of the distal gut (Hearn *et al.*, 1998; Wu *et al.*, 1999). This would explain the absence of neurons in the distal hindgut in mice or humans with mutations in the genes encoding either Et-3 or Ednrb. However, there are also data suggesting that Et-3 signalling does not directly influence differentiation. One of the major subpopulations of enteric neurons, those that synthesize nitric oxide synthase (Nos), show a different regional pattern of appearance in $Et\text{-}3^{-/-}$ mice compared to wild-type mice – in wild-type mice Nos neurons first appear in the caecum, whereas in $Et\text{-}3^{-/-}$ mice they first appear in the distal small intestine (Woodward *et al.*, 2003). It was argued that this rostral shift in the appearance of Nos neurons indicates that Et-3 signalling affects migration but not differentiation (Woodward *et al.*, 2003). Moreover, it has also recently been shown that the Et-3-induced reduction in the number of neurons that develop from crest-derived cells immunoselected from the embryonic rat gut is not due to a general inhibition of differentiation, but rather to a promotion of myofibroblast differentiation at the expense of neuronal differentiation (Wu *et al.*, 1999; Kruger *et al.*, 2003). The implications of these *in vitro* data are unclear, as *in vivo*, crest-derived ENS cells do not differentiate into smooth muscle-like cells. Furthermore, the same study examined the number of NC-derived cells and the proportion of those cells that had differentiated into neurons in the ileum of E13 $Ednrb^{-/-}$ and wild-type rats (Kruger *et al.*, 2003). Although the density of crest-derived cells was similar, a higher proportion of crest-derived cells expressed neuronal markers in the gut of wild-type rats than in the ileum of *Ednrb* null rats. However, as the migration of NC cells is delayed in $Ednrb^{-/-}$ rats, cells in the ileum in $Ednrb^{-/-}$ rats will be at the migratory wavefront, whereas cells in the ileum of wild-type mice will be well behind (rostral to) the wavefront, and thus the higher proportion of NC cells showing a neuronal phenotype in wild-type compared to $Ednrb^{-/-}$ rats reported by Kruger *et al.* (2003) may be due to the differences in the location of the migratory wavefront.

Effects of Et-3 signalling on migration In the absence of Et-3 signalling, the migration of crest cells through the gut is delayed but it is unknown whether Et-3 has a direct or indirect effect on the migratory behaviour of crest cells. Unlike Gdnf, exogenous Et-3 is not chemoattractive to rat enteric crest-derived cells (Kruger *et al.*,

2003). Interestingly, in mice lacking Ednrb, some crest-derived cells are present outside the gut, in the mesentery between the mid- and hindgut (Kapur *et al.*, 1995; Lee *et al.*, 2003). This suggests that Et-3 signalling plays a role in retaining crest cells within the gut or in the migration through the caecum.

Overview of the effects of Et-3 signalling in ENS development Data from different studies on the effects of Et-3 on crest migration, differentiation and proliferation are inconsistent, although data are consistent in showing that Et-3 signalling does not affect ENS cell death. Shortly after crest-derived cells have entered the gut of embryonic mice (at E10.5), there are fewer crest cells in the gut of $Et\text{-}3^{-/-}$ mice than wild-type mice (Barlow *et al.*, 2003). This may be due to reduced proliferation of crest-derived cells (Barlow *et al.*, 2003), delayed migration of crest cells from the vagal-level hindbrain to the gut and/or reduced vagal crest cell production in the absence of Et-3 signalling. At E12.5, there are still fewer crest-derived cells in the small intestine of $Et\text{-}3^{-/-}$ mice than in wild-type mice, but by E15.5, the numbers of crest cells in the small intestine are similar in the null mutant and the wild-type (Barlow *et al.*, 2003). In adult mice, there is no difference between $Et\text{-}3^{-/-}$ and $Et\text{-}3^{+/-}$ mice in the density of myenteric neurons in the ileum and proximal colon, although there is a small but significant *increase* in the density of submucosal neurons in $Et\text{-}3^{-/-}$ mice compared to $Et\text{-}3^{+/-}$ mice (Sandgren *et al.*, 2002). Not surprisingly, there are significantly fewer myenteric and submucosal neurons in the distal colon (the transitional zone) of adult $Et\text{-}3^{-/-}$ mice compared to heterozygote mice (Sandgren *et al.*, 2002). This means that there are fewer enteric neurons overall in the gastrointestinal tract of $Et\text{-}3^{-/-}$ mice. It is unclear why a decreased number of crest-derived cells within the gut results in aganglionosis of the terminal bowel, rather than hypoganglionosis throughout the colon or entire gastrointestinal tract, as occurs in $Gdnf^{+/-}$ mice (see section on GDNF/GFRα1/RET, above). It seems that $Gdnf^{+/-}$ mice have reduced crest cell numbers but apparently not delayed migration (Shen *et al.*, 2002), but that $Et\text{-}3^{-/-}$ mice have delayed migration and probably reduced crest cell number, suggesting that crest cell number and migration speed are not tightly correlated (Figure 11.3). Additional factors to those discussed here could also contribute to the ENS phenotype observed in the absence of ET-3 signalling. For example, $Ednrb^{-/-}$ rats have less enteric NC stem cells, which is likely to have important functional consequences (Kruger *et al.*, 2003).

Interactions between GDNF and ET-3 signalling pathways

Studies in both humans and mice have shown that interactions between Ret- and Ednrb-mediated signalling appear to be extremely important during the development of the ENS. A particular HSCR individual who has heterozygous mutations in both *RET* and *EDNRB* was shown to have parents that were each heterozygous for one mutation, but neither had HSCR (Auricchio *et al.*, 1999). Subsequently, an important genetic study of a genetically isolated Mennonite population (in which the incidence of HSCR is 1/500)

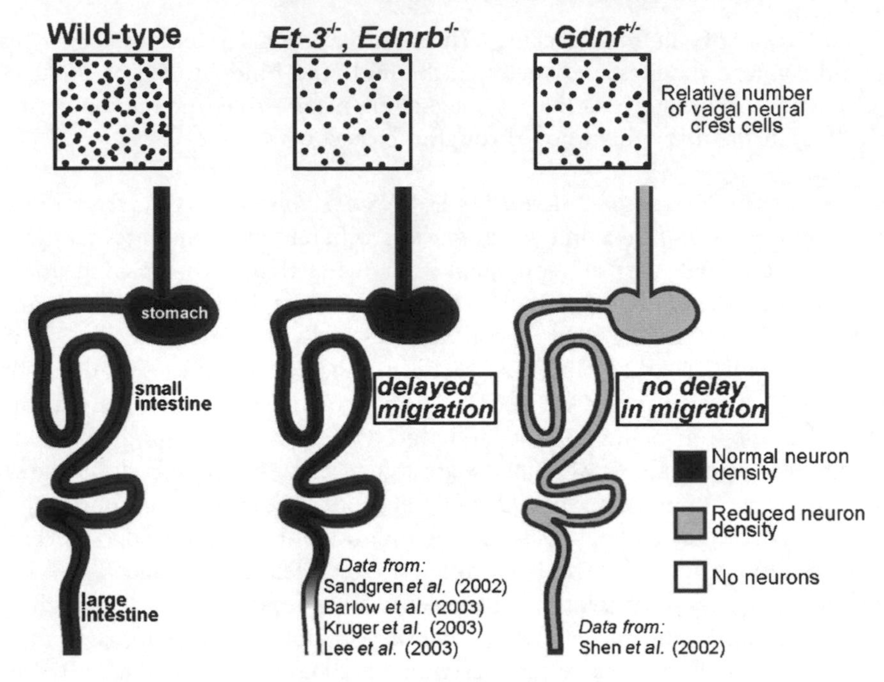

Figure 11.3 Diagram showing the relative number of vagal NC cells, rate of migration and density of neurons in different regions of the gastrointestinal tract of wild-type mice, in which there is no Et-3/ Ednrb signalling (*Et-3*$^{-/-}$ or *Ednrb*$^{-/-}$ mice), and *Gdnf*$^{+/-}$ mice. Compared to wild-type mice, *Et-3*$^{-/-}$ or *Ednrb*$^{-/-}$ mice have reduced numbers of vagal NC cells, delayed migration and aganglionosis of the distal regions of the gut, but relatively normal neuronal density in the normoganglionic regions. *Gdnf*$^{+/-}$ mice also have reduced vagal NC cells, but the timetable by which the gut is colonized is similar to wild-type mice and there is no aganglionic zone in the distal colon. However, the density of neurons throughout the gut of *Gdnf*$^{+/-}$ mice is lower than that of wild-type mice

showed that *RET* and *EDNRB* loci interact to govern the susceptibility to HSCR (Carrasquillo *et al.*, 2002). Furthermore, studies in mice have shown that genetic interactions between *Ret* and *Ednrb* or *ET-3* alleles determine the incidence and severity of HSCR-like symptoms in mice (Carrasquillo *et al.*, 2002; Barlow *et al.*, 2003; McCallion *et al.*, 2003). Although mice with single mutations do not exhibit any gender differences in the severity or penetrance of an HSCR-like phenotype, there is a sex bias in the penetrance yielded by the compound genotypes, which may in part account for the sex bias in the incidence of HSCR (McCallion *et al.*, 2003).

Some *in vitro* studies of enteric NC cells have shown that Et-3 enhances the Gdnf-induced proliferative effects and decreases the differentiation of neurons induced by Gdnf (Hearn *et al.*, 1998; Wu *et al.*, 1999; Barlow *et al.*, 2003). Et-3 has also been shown to inhibit the Gdnf -induced migratory and neurite outgrowth responses of enteric crest-derived cells in both embryonic mice and rats (Barlow *et al.*, 2003; Kruger *et al.*, 2003). Protein kinase A (PKA) is likely to be an important component in the interactions between Ret- or Ednrb-mediating signalling, as the effects of both Gdnf and ET-3 on the proliferation and migration of crest-derived cells are mediated by PKA (Barlow *et al.*, 2003).

Transcription factors

Phox2b Phox2b is a paired box homeodomain transcription factor (Brunet and Pattyn, 2002). *Phox2b*$^{-/-}$ mice lack an ENS along the entire gastrointestinal tract (Pattyn *et al.*, 1999). The presence of mutations or polymorphisms in *PHOX2B* that might contribute to HSCR has recently been examined (Garcia-Barcelo *et al.*, 2003a). One polymorphism showed a significantly higher incidence in HSCR patients compared to controls, although it is unclear whether this polymorphism directly contributes to HSCR (Garcia-Barcelo *et al.*, 2003a).

Phox2b is expressed by all developing autonomic neurons, including enteric neurons (Pattyn *et al.*, 1997). Sox10 (see section on *SOX10*, below) is required to induce *Phox2b* (Kim *et al.*, 2003), which can in turn regulate the expression of *Ret* (which is required for Gdnf signalling; Morin *et al.*, 1997; Lo *et al.*, 1998; Pattyn *et al.*, 1999). The absence of an ENS in *Phox2b*$^{-/-}$ mice is probably due to the absence of Ret (and hence Gdnf signalling).

Sox10 (dominant megacolon gene) Members of the Sox family of transcription factors are involved in a diverse range of developmental processes (Bowles *et al.*, 2000). *Sox10* is a member of the group E *Sox* genes, and it was identified as a transcriptional activator that is expressed in NC-derived cells and then later in glial cells (Kuhlbrodt *et al.*, 1998; Paratore *et al.*, 2001, 2002).

Mutations in *SOX10* can cause Waardenburg–Hirschsprung syndrome (Waardenburg–Shah syndrome, WS4) in humans (Pingault *et al.*, 1998; Southard-Smith *et al.*, 1999; Inoue *et al.*, 2002), and dominant megacolon in mice (Herbarth *et al.*, 1998; Southard-Smith *et al.*, 1998; Kapur, 1999a), both with defects in NC-derived melanocytes and ENS cells. Some WS patients who do not have HSCR have chronic intestinal pseudo-obstruction, and several of these patients have also been found to have *SOX10* mutations (Pingault *et al.*, 2002). Moreover, a patient who suffers from chronic intestinal pseudo-obstruction was found to have a heterozygous frameshift mutation in *SOX10* (Pingault *et al.*, 2000). Colon biopsies revealed the presence of both myenteric and submucosal plexuses in this patient, although it is unknown whether there are normal numbers of enteric neurons (Pingault *et al.*, 2000).

Sox10, alone or together with other transcription factors, regulates the expression of a number of different genes. For example, Sox10 and Pax3 interact to activate *Ret* (Lang *et al.*, 2000; Chan *et al.*, 2003; Lang and Epstein, 2003), which is of central importance for ENS formation. Sox10 is also required for the induction of *Phox2b* (Kim *et al.*, 2003), which is also essential for ENS development. As Phox2b can also regulate the expression of *Ret* (Morin *et al.*, 1997; Lo *et al.*, 1998; Pattyn *et al.*, 1999), Sox10 may be both directly and indirectly (via Phox2b) involved in the activation of *Ret*.

Sox10 is expressed in migrating NC cells and their derivatives in human (Bondurand *et al.*, 1998), mouse (Southard-Smith *et al.*, 1998) and chick (Cheng *et al.*, 2000). In *Sox10*$^{-/-}$ mice, vagal NC-derived cells die just prior to entering the foregut, and there are no neurons in any region of the gastrointestinal tract (Southard-Smith *et al.*, 1998; Kapur, 1999a). Studies *in vivo* and *in vitro* have shown that Sox10 is initially required for the survival of NC-derived cells prior to lineage segregation, and is later required for glial fate acquisition (Southard-Smith *et al.*, 1998; Kapur, 1999a; Britsch *et al.*, 2001; Paratore *et al.*, 2001, 2002; Kim *et al.*, 2003).

Pax3 Pax3 is a member of the paired-box-containing family of transcription factors (Goulding *et al.*, 1991). Heterozygous mutations in the *PAX3* gene are often found in patients with WS without HSCR (Baldwin *et al.*, 1992; Tassabehji *et al.*, 1992). Homozygous loss of function of *PAX3* is lethal (Ayme and Philip, 1995). To date, there are no reports of patients with mutations in *PAX3* that have HSCR or any other congenital ENS defects. In embryonic mice, *Pax3* is expressed by many NC-derived cells, including enteric neuron and melanocyte precursors (Goulding *et al.*, 1991; Lang *et al.*, 2000). *Pax3*$^{+/-}$ mice have a white belly spot (*Splotch* phenotype) but have no obvious ENS defects. In contrast, *Pax3*$^{-/-}$ mice die during mid-gestation with neural tube and cardiac defects and an absence of enteric neurons in the small and large intestines (Lang *et al.*, 2000). It appears that Pax3 interacts with Sox10 to initiate *Ret* expression (Lang *et al.*, 2000; Chan *et al.*, 2003; Lang and Epstein, 2003; see section on *Sox10*, above), and thus there is no expression of *Ret* caudal to the stomach in *Pax3*$^{-/-}$ mice (Lang *et al.*, 2000). However, the lack of any evidence for *PAX3* mutations in humans with ENS developmental defects suggest that PAX3 is not essential to initiate *RET* expression in human ENS precursors.

Mash1 (mammalian achaete–scute homologue 1) *Mash1* encodes a transcription factor that belongs to the basic helix–loop–helix (bHLH) family. In embryonic mice, *Mash1* is expressed by NC cells that colonize the gut (Johnson *et al.*, 1990; Lo *et al.*, 1991; Guillemot and Joyner, 1993; Durbec *et al.*, 1996). *Mash1*$^{-/-}$ mice die within 48 hours of birth; they have defects in sympathetic ganglia and lack enteric neurons in the oesophagus (Guillemot *et al.*, 1993). Enteric neurons are present in the stomach and intestine, although some classes of neurons, for example the serotonin-containing neurons, are absent in these regions (Blaugrund *et al.*, 1996). It appears that Mash1 is required for the differentiation of NC cells but not for their migration, because in *Mash1*$^{-/-}$ mice, NC cells migrate to their correct locations but fail to differentiate into neurons (Guillemot *et al.*, 1993; Sommer *et al.*, 1995). Mash1 indirectly activates Ret via the activation of Phox2b (Hirsch *et al.*, 1998; Lo *et al.*, 1998).

Hox11L1 The *Hox11* family of genes are expressed in non-overlapping ways in the developing nervous system and elsewhere in the mouse embryo (Roberts *et al.*, 1995). Mice with a null mutation of the *Hox11L1* gene (also known as *Tlx2*, *Enx* and *Ncx*) develop megacolon, and were reported to exhibit ENS hyperplasia in the colon and hypoplasia in the ileum (Hatano *et al.*, 1997; Shirasawa *et al.*, 1997). However, although a recent study confirmed the pseudo-obstruction phenotype, no detectable differences in the numbers of enteric neurons in *Hox11L1*-null mice was found (Parisi *et al.*, 2003). Interestingly, the penetrance of the pseudo-obstruction phenotype appears to be influenced by strain-specific differences in the genetic background of the mice (Parisi *et al.*, 2003). Two screens of patients diagnosed with intestinal neuronal dysplasia-B or HSCR have failed to find any mutations in *HOX11L1* associated with these conditions (Costa *et al.*, 2000; Fava *et al.*, 2002).

In mice, *Hox11L1* is expressed by a variety of NC-derived neurons, including enteric neurons (Hatano *et al.*, 1997; Shirasawa *et al.*, 1997). The generation of *Hox11L1lacZ* mice has shown that *Hox11L1* expression is not required for survival, as *lacZ*-expressing neurons are present in *Hox1L1*-null mice (Parisi *et al.*, 2003). *Hox11L1* is only expressed after NC-derived cells in the gut have started to differentiate into neurons, and hence it may be involved in their terminal differentiation (Parisi *et al.*, 2003).

SIP1 (ZFHX1B) The gene *ZFHX1B* encodes Smad-interacting protein-1 (SIP1). This is an adaptor protein for Smad proteins, which act as transducers for signals generated from TGFβ/BMP family growth factors (see section on Bone morphogenetic proteins, below). Some HSCR patients who also suffer from microencephaly, submucous cleft palate and short stature have mutations in *ZFHX1B* (Cacheux *et al.*, 2001; Wakamatsu *et al.*, 2001). The defects are thought to result from haplo-insufficiency of *ZFHX1B* caused by null mutations in one allele. In *Zfhx1b*$^{-/-}$ mice, the neural tube fails to close, vagal NC cells (which give rise to the ENS) do not form and there is an absence of the first branchial arch; they die around E9.5 (Van De Putte *et al.*, 2003). *Zfhx1b*$^{+/-}$ mice do not show any detectable defects (Van De Putte *et al.*, 2003).

SIP1 represses activity in target genes and can oppose the transcriptional activation produced via the Smad pathway (Tylzanowski *et al.*, 2001; Van De Putte *et al.*, 2003). In embryonic mice, *Sip1* is expressed in pre-migratory and migratory vagal NC cells (Van De Putte *et al.*, 2003). When the neuroepithelium differentiates from the ectoderm, E-cadherin is normally downregulated. However, in *Zfhx1b*$^{-/-}$ mice, E-cadherin expression persists (Van De Putte *et al.*, 2003). The association of deficits in SIP1 with some cases of HSCR therefore appears to be due to the role of SIP1 in NC cell formation.

The hedgehog signalling system

Members of the hedgehog family are secreted proteins that play crucial roles during development. Two members of the hedgehog family, Indian hedgehog (Ihh) and Sonic hedgehog (Shh), play a role in ENS development in mice. Both Ihh and Shh bind to the transmembrane protein, Patched (Ptc). Signalling via Ihh or Shh activates the transcription factor Gli1, and also induces expression of bone morphogenetic protein 4 (BMP4) (Roberts *et al.*, 1995, 1998; Marigo *et al.*, 1996; Narita *et al.*, 1998). *Shh* and *Ihh* are expressed by the gut endoderm (Bitgood and McMahon, 1995; Echelard *et al.*, 1993), while *Ptc*, *Gli* and *Bmp4* are expressed by the gut mesenchyme.

Ihh Fetal $Ihh^{-/-}$ mice exhibit a dilated colon, and enteric neurons are often missing from some parts of the small intestine and from the dilated regions of the colon (Ramalho-Santos *et al.*, 2000). However, in humans, a screen of 90 HSCR patients failed to detect any mutations in *IHH* (Garcia-Barcelo *et al.*, 2003b). As enteric neurons are present in non-dilated parts of the colon of fetal $Ihh^{-/-}$ mice, it appears that NC-derived cells migrate into the gut but fail to survive and/or differentiate.

Shh $Shh^{-/-}$ mice do not lack an ENS in any region of the gastrointestinal tract, but nerve cell bodies are present in ectopic locations – within the mucosa, under the endodermal epithelium and in the lamina propria of the villi (Ramalho-Santos *et al.*, 2000). Shh secreted by the gut endoderm induces *Ptc* and *Bmp4* expression in the neighbouring non-muscle mesenchyme and inhibits neuronal and smooth muscle differentiation (Sukegawa *et al.*, 2000). Thus, signalling via Shh may be important for the radial patterning of the gut tube, so that smooth muscle and neurons differentiate in the outer layers (distant from the endoderm).

Neurotrophins and growth factors

Members of the neurotrophin (NT) family of neurotrophic factors (nerve growth factor, brain-derived neurotrophic factor, NT-3 and NT4/5) play crucial roles in the survival, differentiation and growth of many parts of the nervous system, including the NC-derived sympathetic and dorsal root ganglion neurons. The actions of neurotrophins are mediated largely through Trk receptor tyrosine kinases. All neurotrophins except NT4/5, and all three Trk receptors (TrkA, TrkB and TrkC), are present in the developing ENS of a variety of species, including humans (Hoehner *et al.*, 1996; Sternini *et al.*, 1996; Facer *et al.*, 2001; Chalazonitis, 2004). However, only Nt-3 appears to play a role in ENS development (Chalazonitis, 2004).

Nt-3 promotes the differentiation of ENS precursors immunoselected from the embryonic gut (Chalazonitis *et al.*, 1994, 1998), and it also promotes neurite outgrowth and neuron differentiation in dissociated ganglia from post-natal rats (Saffrey *et al.*, 2000). Mice lacking Nt-3, or its receptor TrkC, have reduced numbers of both myenteric and submucosal neurons, and mice overexpressing Nt-3 have increased numbers of myenteric, but not submucosal, neurons (Chalazonitis *et al.*, 2001). It therefore seems likely that Nt-3 is required for the development of subpopulations of enteric neurons (Chalazonitis *et al.*, 2001).

The proportion of submucosal and myenteric neurons showing TrkC immunostaining has been examined in control and HSCR infants (in the aganglionic, transitional and normoganglionic regions; Facer *et al.*, 2001). The percentage of $TrkC^{+}$ submucosal neurons was significantly lower in the normoganglionic regions of HSCR patients compared to age-matched controls, and it was proposed that altered signalling via TrkC may contribute to the dysmotility problems that occur after resection of the aganglionic region (Facer *et al.*, 2001).

Bone morphogenetic proteins (BMPs)

TGFβ family members BMP-2 and BMP-4 and their receptors, BMPR-IA, BMPR-IB and BMPR-II, are all expressed by both crest-derived and non-crest-derived cells in the rat gut (Sukegawa *et al.*, 2000; Bixby *et al.*, 2002; Chalazonitis *et al.*, 2004). A variety of BMP antagonists, including noggin, chordin and follistatin, are also present in the embryonic rat gut (Chalazonitis *et al.*, 2004). BMP-2 and BMP-4 have been shown to have variable effects on the neuronal differentiation of enteric NC-derived cells *in vitro* (Sukegawa *et al.*, 2000; Pisano *et al.*, 2000; Bixby *et al.*, 2002; Kruger *et al.*, 2003). However, a recent detailed study showed that the effects of BMP-2 and BMP-4 on neuronal differentiation are concentration-dependent – at low concentrations both BMP-2- and BMP-4-promoted neuronal differentiation, but at higher concentrations, BMP-2 had no effect on differentiation and BMP-4 inhibited differentiation (Chalazonitis *et al.*, 2004). BMP-2 and BMP-4 also induce the expression of the neurotrophin receptor, TrkC, in cultured crest-derived cells and induce the cells to coalesce into ganglion-like clumpings. In mice overexpressing the BMP antagonist noggin, there are significantly more neurons in both the myenteric and submucosal ganglia and the external (circular and longitudinal) muscle layers are thicker than in control mice (Chalazonitis *et al.*, 2004). Thus, BMP signalling appears to regulate the number of crest-derived and external muscle cells in the developing gut, and may contribute to the specification of particular neuron types by inducing the expression of TrkC (Chalazonitis *et al.*, 2004).

Neural cell and axon guidance molecules

Axon guidance and directed neural migration (collectively called 'neuronal navigation') (Song and Poo, 2001) use common guidance molecules – netrins, Semaphorins, Slits and ephrins. To date there have been no reports of defects in neural guidance cues associated with human ENS developmental abnormalities.

Netrin/DCC Netrins are a conserved family of secreted proteins that can exert either attractive or repulsive effects (Dickson and Keleman, 2002). The repulsive effects of netrins mainly involve the UNC5 family of receptors, and the attractive effects of netrin are mediated through the deleted in colorectal cancer (DCC) family of receptors.

Netrin/DCC signalling appears to be important for the formation of the submucosal ganglia. NC cells initially occupy the myenteric region (between the longitudinal and circular muscle layers), and the colonization of the submucosal region by

NC cells does not occur until several days later, probably from a secondary migration of cells from the myenteric region (McKeown *et al.*, 2001). The centripetal migration of cells from the myenteric to the submucosal region appears to be mediated by netrins and DCC (Jiang *et al.*, 2003). The gut epithelium of embryonic chick and mice expresses netrins, and NC-derived cells within the gut express the netrin receptor, DCC (Seaman *et al.*, 2001). Netrin is chemoattractive to enteric NC cells *in vitro* and in mice lacking DCC there are no submucosal ganglia (Jiang *et al.*, 2003). The pancreas is colonized by NC cells that migrate from the small intestine (Kirchgessner *et al.*, 1992). Netrins expressed by the pancreas are also important for inducing the migration of crest-derived cells into the pancreas (Jiang *et al.*, 2003).

Slit/Robo Slits are large, secreted proteins and are primarily known for their role in neural repulsion (Wong *et al.*, 2002). Slit-induced repulsion is mediated via a family of receptors called roundabout (Robo), of which there are three known vertebrate members (Robo 1-3).

Trunk NC cells never migrate ventrally beyond the dorsal aorta and therefore do not enter the gut (Le Douarin and Teillet, 1973). Slit/Robo signalling may play a role in preventing trunk level NC cells from entering the gut. In chick embryos, Slit1, Slit2 and Slit3 are expressed in the splanchnic mesoderm, dorsal to the gut, and trunk NC cells express Robo receptors (De Bellard *et al.*, 2003). Slit2 is repulsive to trunk NC cells, and it is therefore likely that Slit proteins play a role in preventing trunk NC cells from entering the gut (De Bellard *et al.*, 2003). Vagal NC cells do not express Robo receptors, and hence it has been suggested that they ignore the repulsive effects of Slit2 and enter the gut (De Bellard *et al.*, 2003). Inhibitory molecules in the extracellular matrix may also contribute to preventing trunk NC cells from entering the gut (de Freitas *et al.*, 2003).

Collapsin-1/semaphorin3A It has been shown in the chick embryo that lumbosacral NC cells do not initially enter the gut but accumulate in the region adjacent to the gut wall, where they form the nerve of Remak (Burns and Le Douarin, 1998). Lumbosacral NC cells remain within this nerve until the hindgut is colonized by vagal NCC (see section on The lumbosacral NC, above). Prior to this inward migration of lumbosacral NC cells, the secreted glycoprotein collapsin-1 (Sema3A) is expressed throughout the rectal wall (Shepherd and Raper, 1999). However, once the hindgut is colonized by vagal NC cells, collapsin-1 expression retreats from the outer muscle layers and is confined to the inner submucosal and mucosal regions, thus allowing axons to project from the nerve of Remak into the gut, along which lumbosacral NC cells migrate into the hindgut. Collapsin-1 may not affect the migration of lumbosacral NCC directly, but instead repels the axons along which these cells migrate to gain entry to the gut. Collapsin-1 does not seem to play a role in patterning vagal NCC within the gut, as vagal cells colonize the submucosal region of the chick hindgut precisely when collapsin-1 expression is confined to this region, suggesting that the migration of these cells is not restricted by this signalling molecule.

Retinaldehyde dehydrogenase 2 (RALDH2)

Retinoic acids play important roles in development by acting at nuclear retinoic acid receptors and regulating the transcriptional activity of target genes. During early embryonic development, retinoic acids are mainly synthesized by retinaldehyde dehydrogenase 2 (RALDH2). *Raldh2*$^{-/-}$ mice die during mid-gestation (around E10) because of severe cardiovascular defects (Niederreither *et al.*, 1999). However, maternal retinoic acid supplementation prolongs the survival of *Raldh2*$^{-/-}$ embryos until late gestational stages (Niederreither *et al.*, 1999). RA-rescued *Raldh2*$^{-/-}$ embryos lack an ENS, probably due to defects in the posterior pharyngeal arches and vagal level hindbrain and consequent defects in vagal level NC cell migration (Niederreither *et al.*, 2003). There is currently no evidence for mutations in *RALDH2* underlying ENS defects in humans but, given the widespread use of retinoic acid signalling during development, these may be part of much more severe and complex abnormalities.

Neuregulin/ErbB2 signalling

Neuregulins are signalling proteins that bind to erbB3 or erbB4 receptor tyrosine kinases. *ErbB2-null* mutant mice die at around E10, due to heart defects (Britsch *et al.*, 1998). *ErbB2* and *erbB3* are normally strongly expressed in the mucosa of post-natal mice, and neuregulin is expressed by both the ENS and epithelial cells (Crone *et al.*, 2003). Conditional mutants, in which the *erbB2* gene is disrupted in colonic epithelium cells, are indistinguishable from wild-type littermates at birth, but grow slowly and die at 3–8 weeks of age (Crone *et al.*, 2003). The mutants exhibit a constricted distal colon and a distended (mega-) proximal colon, which is due to the post-natal apoptosis of enteric neurons in the colon. Conditional mutants in which *erbB2* is disrupted in the ENS do not show a phenotype, indicating that erbB2 is not acting cell autonomously to mediate survival. It has been proposed that neuregulin, produced by the ENS or the mucosa, binds to erbB2 and erbB3 receptors on the colonic mucosa, which results in the production of unidentified survival factors that are required for the post-natal survival of enteric neurons (Crone *et al.*, 2003). To date there are no reports of patients with HSCR or other ENS defects that have mutations in neuregulin-signalling molecules. Interestingly, there is also evidence that the ENS may be the source of a factor that regulates epithelial cell growth and repair (Bjerknes and Cheng, 2001). Thus, although epithelial cells are a considerable distance from enteric neurons (particularly in humans), there appears to be some cross-talk between the two cell types.

L1CAM

L1CAM is a cell adhesion molecule that is highly expressed in the nervous system and is involved in axon pathfinding and neural migration (Hortsch, 1996; Brummendorf *et al.*, 1998; Demyanenko and Maness, 2003). Mutations in the human *L1CAM* have been implicated in X-linked hydrocephalus (Rosenthal *et al.*, 1992), a neurological disorder which is also called MASA syndrome (mental retardation, aphasia, shuffling gait and adducted thumbs). A recent study has reported that some individuals with *L1CAM* mutations and X-linked hydrocephalus also have HSCR (Parisi *et al.*, 2002). It appears unlikely that the *L1CAM* mutation alone causes HSCR, but *L1CAM* may act as an X-linked modifier gene for the development of HSCR (Parisi *et al.*, 2002). To date there have been no reported studies that have examined the role of L1CAM during ENS development in laboratory animals.

HSCR: current and future treatments

Current diagnosis and treatment for Hirschsprung's disease

HSCR is usually suggested by failure to pass meconium in the 48 h period directly after birth. Most cases are detected in the first 6 weeks post-natally. HSCR is confirmed by suction rectal biopsies or muscular biopsy (Kapur, 1999b). Current treatment of HSCR involves surgical resection, preferably in the first weeks post-natally, of the aganglionic segment and re-anastomosis (Puri and Wester, 2000).

Future treatments for HSCR based on stem cell therapy

Stem cell transplantation as a treatment for developmental disorders of the ENS The ENS is derived from vagal and lumbosacral NC cells that can be thought as multipotent stem cells capable of giving rise to various neuronal and glial cell types within the gut, and to other cell types when transplanted to other regions of the embryo. Recent advances in the ability to identify, isolate, purify and transplant stem cells, obtained either from pre- or post-natal gut or from other sources, have created new possibilities for the treatment of ENS developmental disorders. However, each group of potential stem cells has particular advantages and disadvantages that may have consequences for their future therapeutic use in cell replacement strategies.

Stem cells in pre-natal gut Neural crest stem cells (NCSC) have been isolated from dissociated pre-natal gut using NC cell-specific antibodies (mainly anti-RET and p75) in conjunction with fluorescence activated cell sorting (FACS) or flow cytometry to retrieve immunopositive cells (Morrison *et al*, 1999; Bixby *et al*, 2002; Lo and Anderson, 1995; Natarajan *et al.*, 1999). Using such isolation methods, RET^+ cells, obtained from the gut of rat embryos, were found to generate mostly neurons in clonogenic assays (Lo and Anderson, 1995). Similar Ret^+ ENS precursor populations have been retrieved from the gut of mouse embryos which, when injected into fetal mouse gut grown in organ culture, gave rise to neuronal and glial progeny (Natarajan *et al*, 1999). Hence, NC-derived, RET^+ cells obtained from fetal gut have the potential to populate embryonic gut grown in organ culture. Cells isolated from the embryonic rat gut by flow cytometry that show high p75 and high α4 integrin expression were also shown to have the properties of gut NCSC because, in clonal analysis, 70% of the colonies formed by these cells contained neurons, glia and myofibroblasts (Bixby *et al.*, 2002). NCSC isolated from the embryonic gut show different properties from NCSC isolated from the sciatic nerve (Bixby *et al.*, 2002). There does not appear to be a difference in the capacity to colonize the hindgut between NC cells at the migration front and NC cells located within the proximal gut, suggesting that the entire pre-natal ENS contains multipotential ENS progenitors (Sidebotham *et al.*, 2002a).

In another study (Bondurand *et al.*, 2003), gut from embryonic and post-natal mice aged up to 2 weeks was dissociated and cultured in a medium designed to encourage growth of NCSC (Morrison *et al.*, 1999). After a number of days, neurosphere-like bodies (NLBs), which contained neurons and glial cells, were identified in the cultures. In addition to the differentiated cells, NLBs also contained proliferating progenitors that were capable of giving rise to colonies containing enteric neurons and glial cells. When pieces of NLBs were grafted into normal and aganglionic embryonic mouse gut, progenitors were able to colonize the gut and differentiate into appropriate enteric phenotypes at the appropriate locations. Interestingly, similar progenitors were isolated from the normoganglionic region of mice with colonic aganglionosis (Bondurand *et al*, 2003), thus raising the possibility of utilizing cells isolated from one region of the gut as autologous transplants to treat developmental defects prevalent in another, therefore eliminating the need for immunosuppression.

Stem cells in post-natal gut Aganglionic gut conditions, such as HSCR, are generally diagnosed post-natally. Thus, if stem cell replacement is to be used as a treatment for such conditions, it is important to determine whether cells with multipotent properties and self-renewal capacity persist in the post-natal gut. Such cells, obtained from normoganglionic regions of the bowel, could then be used in autotransplants to replace cells in affected regions of gut. Recently, such cells have been described in post-natal and adult rat gut (Kruger *et al.*, 2002). When integrin $\alpha 4/p75^{NTR}$-double-positive cells were isolated from rat gut, dissociated into single cell suspensions and plated in culture at clonal density, multilineage colonies containing neurons, glia and myofibroblasts were subsequently observed. These colonies were similar to those formed by embryonic NCSC, although they were smaller in size (Morrison *et al.*, 1999; Bixby *et al.*, 2002). The progenitor cells within the colonies also had the capacity to self-renew, as demonstrated by subcloning into secondary cultures, which in turn gave rise to multipotential daughter colonies (Kruger *et al.*, 2002). However, the capacity to self-renew decreased with the age of the gut from which the cells were isolated. The ability to generate certain subtypes of neurons also declined with age.

CNS stem cells The developmental potential of central nervous system-derived neural stem cells (CNS-NSC) has been explored (Micci *et al.*, 2001) as an alternative to using stem cells isolated from the gut. Neural stem cells, which have a broad developmental potential, including neuronal and glial cells, have the notional potential to treat a wide variety of neurodegenerative conditions (Yandava *et al.*, 1999; Gage *et al.*, 1995; Shihabuddin *et al.*, 1999), and thus may be useful in generating neurons within hypoganglionic or aganglionic gut. When CNS-NSC, isolated from the subventricular zone of E17 rat brain, were injected into the stomach of adult mice, these cells, which expressed RET, GFRα1 and neuronal nitric oxide synthase (nNOS), differentiated into neurons, continued to express nNOS and survived for at least 8 weeks, with little inflammation observed in the host tissue (Micci *et al.*, 2001). However, transplanted CNS-NSC did not migrate from the site of injection in the mouse stomach, a potential drawback, since precursor cells normally migrate extensively in order to form the ENS.

The data from these CNS-NSC, NCSC and NLB studies support the idea that cell transplantation approaches may be of potential benefit in the therapeutic treatment of gut conditions where lack of neurons is prevalent. The immediate challenges for developmental biologists are to determine the optimum source of ENS stem cells, bearing in mind the problem of immune responses, to determine whether the gut remains receptive to donor cells during pre-natal and post-natal development, and to determine whether methods can be devised to introduce sufficient numbers of stem cells into defective gut in order to form a functional ENS.

Conclusions

Research into ENS development and the pathogenesis of HSCR has grown explosively in the last decade. Mutations in known genes do not account for the bulk of HSCR cases. Moreover, the classification of a gene (such as RET) as a HSCR gene, but with the phenotypic variation ascribed to 'modifier genes', is no longer helpful. The effects of the many HSCR genes and molecules need to be integrated, with their cross-talk and interactions detailed. HSCR provides one of the most accessible test-beds for studying the efficacy of stem cell-based therapies for neural deficit disorders. The aims of stem cell therapies must not be simply to achieve appropriate numbers of neural cells and differentiation of a particular cell type, but also to generate multiple sublineages with appropriate positioning and connectivities. The development of the ENS and HSCR is likely to remain at the forefront of understanding of neurocristopathies and of complex multigenic birth defects in general.

References

Airaksinen, M.S. and Saarma, M. (2002) The gdnf family: signalling, biological functions and therapeutic value. *Nat. Rev. Neurosci.* **3**: 383–394.

Airaksinen, M.S., Titievsky, A. and Saarma, M. (1999) GDNF family neurotrophic factor signaling: four masters, one servant? *Mol. Cell Neurosci.* **13**: 313–325.

Amiel, J., Attie, T., Jan, D., Pelet, A. *et al.* (1996) Heterozygous endothelin receptor B (EDNRB) mutations in isolated Hirschsprung disease. *Hum. Mol. Genet.* **5**: 355–357.

Amiel, J. and Lyonnet, S. (2001) Hirschsprung disease, associated syndromes, and genetics: a review. *J. Med. Genet.* **38**: 729–739.

Angrist, M., Bolk, S., Halushka, M., Lapchak, P.A. *et al.* (1996) Germline mutations in glial cell line-derived neurotrophic factor (GDNF) and RET in a Hirschsprung disease patient. *Nat. Genet.* **14**: 341–344.

Angrist, M., Jing, S., Bolk, S., Bentley, K. *et al.* (1998) Human GFRA1: cloning, mapping, genomic structure, and evaluation as a candidate gene for Hirschsprung disease susceptibility. *Genomics* **48**: 354–362.

Attie, T., Pelet, A., Edery, P., Eng, C. *et al.* (1995) Diversity of RET proto-oncogene mutations in familial and sporadic Hirschsprung disease. *Hum. Mol. Genet.* **4**: 1381–1386.

Attie-Bitach, T., Abitbol, M., Gerard, M., Delezoide, A.L. *et al.* (1998) Expression of the RET proto-oncogene in human embryos. *Am. J. Med. Genet.* **80**: 481–486.

Auricchio, A., Griseri, P., Carpentieri, M.L., Betsos, N. *et al.* (1999) Double heterozygosity for a RET substitution interfering with splicing and an EDNRB missense mutation in Hirschsprung disease. *Am. J. Hum. Genet.* **64**: 1216–1221.

Ayme, S. and Philip, N. (1995) Possible homozygous Waardenburg syndrome in a fetus with exencephaly. *Am. J. Med. Genet.* **59**: 263–265.

Baldwin, C.T., Hoth, C.F., Amos, J.A., da-Silva, E.O. *et al.* (1992) An exonic mutation in the *HuP2* paired domain gene causes Waardenburg's syndrome. *Nature* **355**: 637–638.

Barlow, A., de Graaff, E. and Pachnis, V. (2003) Enteric nervous system progenitors are coordinately controlled by the G protein-coupled receptor EDNRB and the receptor tyrosine kinase RET. *Neuron* **40**: 905–916.

Belknap, W.M. (2002) The pathogenesis of Hirschsprung disease. *Curr. Opin. Gastroenterol.* **18**: 74–81.

Bitgood, M.J. and McMahon, A.P. (1995) Hedgehog and Bmp genes are coexpressed at many diverse sites of cell–cell interaction in the mouse embryo. *Dev. Biol.* **172**: 126–138.

Bixby, S., Kruger, G., Mosher, J., Joseph, N. *et al.* (2002) Cell-intrinsic differences between stem cells from different regions of the peripheral nervous system regulate the generation of neural diversity. *Neuron* **35**: 643.

Bjerknes, M. and Cheng, H. (2001) Modulation of specific intestinal epithelial progenitors by enteric neurons. *Proc. Natl Acad. Sci. USA* **98**: 12497–12502.

Blaugrund, E., Pham, T.D., Tennyson, V.M., Lo, L. *et al.* (1996) Distinct subpopulations of enteric neuronal progenitors defined by time of development, sympathoadrenal lineage markers and Mash-1-dependence. *Development* **122**: 309–320.

Bondurand, N., Kobetz, A., Pingault, V., Lemort, N. *et al.* (1998) Expression of the *SOX10* gene during human development. *FEBS Lett.* **432**: 168–172.

Bondurand, N., Natarajan, D., Thapar, N., Atkins, C. *et al.* (2003) Neuron and glia generating progenitors of the mammalian enteric nervous system isolated from foetal and post-natal gut cultures. *Development* **130**: 6387–6400.

Bordeaux, M.C., Forcet, C., Granger, L., Corset, V. *et al.* (2000) The RET proto-oncogene induces apoptosis: a novel mechanism for Hirschsprung disease. *EMBO J.* **19**: 4056–4063.

Borrego, S., Fernandez, R. M., Dziema, H., Niess, A. *et al.* (2003) Investigation of germline GFRα4 mutations and evaluation of the involvement of GFRα1, GFRα2, GFRα3, and GFRα4 sequence variants in Hirschsprung disease. *J. Med. Genet.* **40**: e18.

Bowles, J., Schepers, G. and Koopman, P. (2000) Phylogeny of the SOX family of developmental transcription factors based on sequence and structural indicators. *Dev. Biol.* **227**: 239–255.

Brand, M., Le Moullec, J.M., Corvol, P. and Gasc, J.M. (1998) Ontogeny of endothelins-1 and -3, their receptors, and endothelin converting enzyme-1 in the early human embryo. *J. Clin. Invest.* **101**: 549–559.

Britsch, S., Goerich, D. E., Riethmacher, D., Peirano, R. I. *et al.* (2001). The transcription factor Sox10 is a key regulator of peripheral glial development. *Genes Dev.* **15**: 66–78.

Britsch, S., Li, L., Kirchhoff, S., Theuring, F. *et al.* (1998) The ErbB2 and ErbB3 receptors and their ligand, neuregulin-1, are essential for development of the sympathetic nervous system. *Genes Dev.* **12**: 1825–1836.

Brookes, S.J. (2001) Classes of enteric nerve cells in the guinea-pig small intestine. *Anat. Rec.* **262**: 58–70.

Brummendorf, T., Kenwrick, S. and Rathjen, F.G. (1998) Neural cell recognition molecule L1: from cell biology to human hereditary brain malformations. *Curr. Opin. Neurobiol.* **8**: 87–97.

Brunet, J.F. and Pattyn, A. (2002) *Phox2* genes – from patterning to connectivity. *Curr. Opin. Genet. Dev.* **12**: 435–440.

Burns, A.J., Champeval, D. and Le Douarin, N.M. (2000) Sacral neural crest cells colonize aganglionic hindgut *in vivo* but fail to compensate for lack of enteric ganglia. *Dev. Biol.* **219**: 30–43.

Burns, A.J., Delalande, J.M. and Le Douarin, N.M. (2002) *In ovo* transplantation of enteric nervous system precursors from vagal to sacral neural crest results in extensive hindgut colonization. *Development* **129**: 2785–2796.

Burns, A.J. and Le Douarin, N.M. (1998) The sacral neural crest contributes neurons and glia to the post-umbilical gut: spatiotemporal analysis of the development of the enteric nervous system. *Development* **125**: 4335–4347.

Burns, A.J. and Le Douarin, N.M. (2001) Enteric nervous system development: analysis of the selective developmental potentialities of vagal and sacral neural crest cells using quail–chick chimeras. *Anat. Rec.* **262**: 16–28.

Cacalano, G., Farinas, I., Wang, L.C., Hagler, K. *et al.* (1998) GFRα1 is an essential receptor component for GDNF in the developing nervous system and kidney. *Neuron* **21**: 53–62.

Cacheux, V., Dastot-Le Moal, F., Kaariainen, H., Bondurand, N. *et al.* (2001) Loss-of-function mutations in SIP1 Smad interacting protein 1 result in a syndromic Hirschsprung disease. *Hum. Mol. Genet.* **10**: 1503–1510.

Carrasquillo, M.M., McCallion, A.S., Puffenberger, E.G., Kashuk, C.S. *et al.* (2002) Genome-wide association study and mouse model identify interaction between RET and EDNRB pathways in Hirschsprung disease. *Nat. Genet.* **32**: 237–244.

Cass, D. (1986) Hirschsprung's disease: an historical review. *Prog. Pediatr. Surg.* **20**: 199–214.

Ceccherini, I., Zhang, A.L., Matera, I., Yang, G. *et al.* (1995) Interstitial deletion of the endothelin-B receptor gene in the spotting lethal (sl) rat. *Hum. Mol. Genet.* **4**: 2089–2096.

Chalazonitis, A. (2004) Neurotrophin-3 in the development of the enteric nervous system. *Prog. Brain Res.* **146**: 243–263.

Chalazonitis, A., D'Autreaux, F., Guha, U., Pham, T.D. *et al.* (2004) Bone morphogenetic protein-2 and -4 limit the number of enteric neurons but promote development of a TrkC-expressing neurotrophin-3-dependent subset. *J. Neurosci.* **24**: 4266–4282.

Chalazonitis, A., Pham, T.D., Rothman, T.P., DiStefano, P.S. *et al.* (2001) Neurotrophin-3 is required for the survival–differentiation of subsets of developing enteric neurons. *J. Neurosci.* **21**: 5620–5636.

Chalazonitis, A., Rothman, T.P., Chen, J. and Gershon, M.D. (1998) Age-dependent differences in the effects of GDNF and NT-3 on the development of neurons and glia from neural crest-derived precursors immunoselected from the fetal rat gut: expression of GFRα-1 *in vitro* and *in vivo*. *Dev. Biol.* **204**: 385–406.

Chalazonitis, A., Rothman, T.P., Chen, J., Lamballe, F. *et al.* (1994) Neurotrophin-3 induces neural crest-derived cells from fetal rat gut to develop *in vitro* as neurons or glia. *J. Neurosci.* **14**: 6571–6584.

Chan, K.K., Wong, C.K., Lui, V.C., Tam, P.K. *et al.* (2003) Analysis of *SOX10* mutations identified in Waardenburg–Hirschsprung patients: differential effects on target gene regulation. *J. Cell Biochem.* **90**: 573–585.

Cheng, Y., Cheung, M., Abu-Elmagd, M.M., Orme, A. *et al.* (2000) Chick *sox10*, a transcription factor expressed in both early neural crest cells and central nervous system. *Brain Res. Dev. Brain Res.* **121**: 233–241.

Costa, M., Brookes, S.J., Steele, P.A., Gibbins, I. *et al.* (1996) Neurochemical classification of myenteric neurons in the guinea-pig ileum. *Neuroscience* **75**: 949–967.

Costa, M., Fava, M., Seri, M., Cusano, R. *et al.* (2000) Evaluation of the *HOX11L1* gene as a candidate for congenital disorders of intestinal innervation. *J. Med. Genet.* **37**: E9.

Crone, S.A., Negro, A., Trumpp, A., Giovannini, M. *et al.* (2003) Colonic epithelial expression of ErbB2 is required for post-natal maintenance of the enteric nervous system. *Neuron* **37**: 29–40.

De Bellard, M.E., Rao, Y. and Bronner-Fraser, M. (2003) Dual function of Slit2 in repulsion and enhanced migration of trunk, but not vagal, neural crest cells. *J. Cell Biol.* **162**: 269–279.

de Freitas, P.F., Ferreira Fde, F. and Faraco, C.D. (2003) PNA-positive glycoconjugates are negatively correlated with the access of neural crest cells to the gut in chicken embryos. *Anat. Rec.* **273A**: 705–713.

Dembowski, C., Hofmann, P., Koch, T., Kamrowski-Kruck, H. *et al.* (2000) Phenotype, intestinal morphology, and survival of homozygous and heterozygous endothelin B receptor-deficient (spotting lethal) rats. *J. Pediatr. Surg.* **35**: 480–488.

Demyanenko, G.P. and Maness, P.F. (2003) The L1 cell adhesion molecule is essential for topographic mapping of retinal axons. *J. Neurosci.* **23**: 530–538.

Dickson, B.J. and Keleman, K. (2002) Netrins. *Curr. Biol.* **12**: R154–155.

Doray, B., Salomon, R., Amiel, J., Pelet, A. *et al.* (1998) Mutation of the *RET* ligand, neurturin, supports multigenic inheritance in Hirschsprung disease. *Hum. Mol. Genet.* **7**: 1449–1452.

Durbec, P.L., Larsson-Blomberg, L.B., Schuchardt, A., Costantini, F. *et al.* (1996) Common origin and developmental dependence on c-ret of subsets of enteric and sympathetic neuroblasts. *Development* **122**: 349–358.

Echelard, Y., Epstein, D.J., St-Jacques, B., Shen, L. *et al.* (1993) Sonic hedgehog, a member of a family of putative signaling molecules, is implicated in the regulation of CNS polarity. *Cell* **75**: 1417–1430.

Edery, P., Pelet, A., Mulligan, L.M., Abel, L. *et al.* (1994) Long segment and short segment familial Hirschsprung's disease: variable clinical expression at the RET locus. *J. Med. Genet.* **31**: 602–606.

Eketjall, S. and Ibanez, C.F. (2002) Functional characterization of mutations in the *GDNF* gene of patients with Hirschsprung disease. *Hum. Mol. Genet.* **11**: 325–329.

Enomoto, H., Araki, T., Jackman, A., Heuckeroth, R.O. *et al.* (1998) GFRα1-deficient mice have deficits in the enteric nervous system and kidneys. *Neuron* **21**: 317–324.

Enomoto, H., Crawford, P.A., Gorodinsky, A., Heuckeroth, R.O. *et al.* (2001) RET signaling is essential for migration, axonal growth and axon guidance of developing sympathetic neurons. *Development* **128**: 3963–3974.

Epstein, M.L., Mikawa, T., Brown, A.M. and McFarlin, D.R. (1994) Mapping the origin of the avian enteric nervous system with a retroviral marker. *Dev. Dyn.* **201**: 236–244.

Erickson, C.A. and Goins, T.L. (2000) Sacral neural crest cell migration to the gut is dependent upon the migratory environment and not cell-autonomous migratory properties. *Dev. Biol.* **219**: 79–97.

Facer, P., Knowles, C.H., Thomas, P.K., Tam, P.K. *et al.* (2001) Decreased tyrosine kinase C expression may reflect developmental abnormalities in Hirschsprung's disease and idiopathic slow-transit constipation. *Br. J. Surg* **88**: 545–552.

Fava, M., Borghini, S., Cinti, R., Cusano, R. *et al.* (2002) *HOX11L1*: a promoter study to evaluate possible expression defects in intestinal motility disorders. *Int. J. Mol. Med.* **10**: 101–106.

Focke, P.J., Schiltz, C.A., Jones, S.E., Watters, J.J. *et al.* (2001) Enteric neuroblasts require the phosphatidylinositol 3-kinase pathway for GDNF-stimulated proliferation. *J. Neurobiol.* **47**: 306–317.

Fontaine-Perus, J.C., Chanconie, M. and Le Douarin, N.M. (1982) Differentiation of peptidergic neurones in quail–chick chimaeric embryos. *Cell Differ.* **11**: 183–193.

Fu, M., Chi Hang Lui, V., Har Sham, M., Nga Yin Cheung, A. *et al.* (2003) *HOXB5* expression is spatially and temporarily regulated in human embryonic gut during neural crest cell colonization and differentiation of enteric neuroblasts. *Dev. Dyn.* **228**: 1–10.

Furness, J.B. (2000) Types of neurons in the enteric nervous system. *J. Auton. Nerv. Syst.* **81**: 87–96.

Furness, J.B. and Costa, M. (1987) *The Enteric Nervous System.* Churchill Livingstone: Edinburgh.

Furness, J.B., Johnson, P.J., Pompolo, S. and Bornstein, J.C. (1995) Evidence that enteric motility reflexes can be initiated through entirely intrinsic mechanisms in the guinea-pig small intestine. *Neurogastroenterol. Motil.* **7**: 89–96.

Furness, J.B., Jones, C., Nurgali, K. and Clerc, N. (2004) Intrinsic primary afferent neurons and nerve circuits within the intestine. *Prog. Neurobiol.* **72**: 143–164.

Furness, J.B., Kunze, W.A., Bertrand, P.P., Clerc, N. *et al.* (1998) Intrinsic primary afferent neurons of the intestine. *Prog. Neurobiol.* **54**: 1–18.

Gage, F.H., Ray, J. and Fisher, L.J. (1995) Isolation, characterization, and use of stem cells from the CNS. *Annu. Rev. Neurosci.* **18**: 159–192.

Galligan, J.J., LePard, K.J., Schneider, D.A. and Zhou, X. (2000) Multiple mechanisms of fast excitatory synaptic transmission in the enteric nervous system. *J. Auton. Nerv. Syst.* **81**: 97–103.

Garcia-Barcelo, M., Sham, M.H., Lui, V.C., Chen, B.L. *et al.* (2003a) Association study of *PHOX2B* as a candidate gene for Hirschsprung's disease. *Gut* **52**: 563–567.

Garcia-Barcelo, M.M., Lee, W.S., Sham, M.H., Lui, V.C. *et al.* (2003b) Is there a role for the *IHH* gene in Hirschsprung's disease? *Neurogastroenterol. Motil.* **15**: 663–668.

Gariepy, C.E., Williams, S.C., Richardson, J.A., Hammer, R.E. *et al.* (1998) Transgenic expression of the endothelin-B receptor prevents congenital intestinal aganglionosis in a rat model of Hirschsprung disease. *J. Clin. Invest.* **102**: 1092–1101.

Gershon, M.D. (1995) Neural crest development. Do developing enteric neurons need endothelins? *Curr. Biol.* **5**: 601–604.

Gianino, S., Grider, J.R., Cresswell, J., Enomoto, H. *et al.* (2003) GDNF availability determines enteric neuron number by controlling precursor proliferation. *Development* **130**: 2187–2198.

Golden, J.P., DeMaro, J.A., Osborne, P.A., Milbrandt, J. *et al.* (1999) Expression of neurturin, GDNF, and GDNF family-receptor mRNA in the developing and mature mouse. *Exp. Neurol.* **158**: 504–528.

Goulding, M.D., Chalepakis, G., Deutsch, U., Erselius, J.R. *et al.* (1991) Pax-3, a novel murine DNA binding protein expressed during early neurogenesis. *EMBO J.* **10**: 1135–1147.

Grider, J.R. and Jin, J.G. (1994) Distinct populations of sensory neurons mediate the peristaltic reflex elicited by muscle stretch and mucosal stimulation. *J. Neurosci.* **14**: 2854–2860.

Guillemot, F. and Joyner, A.L. (1993) Dynamic expression of the murine *achaete–scute* homologue *Mash-1* in the developing nervous system. *Mech. Dev.* **42**: 171–185.

Guillemot, F., Lo, L.C., Johnson, J.E., Auerbach, A. *et al.* (1993) Mammalian *achaete–scute homolog 1* is required for the early development of olfactory and autonomic neurons. *Cell* **75**: 463–476.

Hatano, M., Aoki, T., Dezawa, M., Yusa, S. *et al.* (1997) A novel pathogenesis of megacolon in *Ncx/Hox11L.1* deficient mice. *J. Clin. Invest.* **100**: 795–801.

Hearn, C.J., Murphy, M. and Newgreen, D. (1998) GDNF and ET-3 differentially modulate the numbers of avian enteric neural crest cells and enteric neurons *in vitro*. *Dev. Biol.* **197**: 93–105.

Hellmich, H.L., Kos, L., Cho, E.S., Mahon, K.A. *et al.* (1996) Embryonic expression of glial cell-line derived neurotrophic factor (GDNF) suggests multiple developmental roles in neural differentiation and epithelial–mesenchymal interactions. *Mech. Dev.* **54**: 95–105.

Herbarth, B., Pingault, V., Bondurand, N., Kuhlbrodt, K. *et al.* (1998) Mutation of the Sry-related *Sox10* gene in Dominant megacolon, a mouse model for human Hirschsprung disease. *Proc. Natl Acad. Sci. USA* **95**: 5161–5165.

Heuckeroth, R.O., Enomoto, H., Grider, J.R., Golden, J.P. *et al.* (1999) Gene targeting reveals a critical role for neurturin in the development and maintenance of enteric, sensory, and parasympathetic neurons. *Neuron* **22**: 253–263.

Hirsch, M.R., Tiveron, M.C., Guillemot, F., Brunet, J.F. *et al.* (1998) Control of noradrenergic differentiation and Phox2a expression by *MASH1* in the central and peripheral nervous system. *Development* **125**: 599–608.

Hoehner, J.C., Wester, T., Pahlman, S. and Olsen, L. (1996) Alterations in neurotrophin and neurotrophin-receptor localization in Hirschsprung's disease. *J. Pediatr. Surg.* **31**: 1524–1529.

Hofstra, R.M., Osinga, J. and Buys, C.H. (1997) Mutations in Hirschsprung disease: when does a mutation contribute to the phenotype. *Eur. J. Hum. Genet.* **5**: 180–185.

Hofstra, R.M., Valdenaire, O., Arch, E., Osinga, J. *et al.* (1999) A loss-of-function mutation in the endothelin-converting enzyme 1 (ECE-1) associated with Hirschsprung disease, cardiac defects, and autonomic dysfunction. *Am. J. Hum. Genet.* **64**: 304–308.

Holschneider, A.M. and Puri, P. (2000) *Hirschsprung's Disease and Allied Disorders*. Harwood Academic: Amsterdam.

Honma, Y., Araki, T., Gianino, S., Bruce, A. *et al.* (2002) Artemin is a vascular-derived neurotropic factor for developing sympathetic neurons. *Neuron* **35**: 267–282.

Hortsch, M. (1996). The L1 family of neural cell adhesion molecules: old proteins performing new tricks. *Neuron* **17**: 587–593.

Hosoda, K., Hammer, R.E., Richardson, J.A., Baynash, A.G. *et al.* (1994) Targeted and natural (piebald-lethal) mutations of endothelin-B receptor gene produce megacolon associated with spotted coat color in mice. *Cell* **79**: 1267–1276.

Inoue, K., Shilo, K., Boerkoel, C.F., Crowe, C. *et al.* (2002) Congenital hypomyelinating neuropathy, central dysmyelination, and Waardenburg–Hirschsprung disease: phenotypes linked by *SOX10* mutation. *Ann. Neurol.* **52**: 836–842.

Inoue, K., Shimotake, T. and Iwai, N. (2000) Mutational analysis of *RET/GDNF/NTN* genes in children with total colonic aganglionosis with small bowel involvement. *Am. J. Med. Genet.* **93**: 278–284.

Inoue, M., Hosoda, K., Imura, K., Kamata, S. *et al.* (1998) Mutational analysis of the endothelin-B receptor gene in Japanese Hirschsprung's disease. *J. Pediatr. Surg.* **33**: 1206–1208.

Ivanchuk, S.M., Myers, S.M., Eng, C. and Mulligan, L.M. (1996) *De novo* mutation of *GDNF*, ligand for the RET/GDNFR-α receptor complex, in Hirschsprung disease. *Hum. Mol. Genet.* **5**: 2023–2026.

Iwashita, T., Kruger, G.M., Pardal, R., Kiel, M.J. *et al.* (2003) Hirschsprung disease is linked to defects in neural crest stem cell function. *Science* **301**: 972–976.

Jacobs-Cohen, R.J., Payette, R.F., Gershon, M.D. and Rothman, T.P. (1987) Inability of neural crest cells to colonize the presumptive aganglionic bowel of *ls/ls* mutant mice: requirement for a permissive microenvironment. *J. Comp. Neurol.* **255**: 425–438.

Jiang, Y., Liu, M.T. and Gershon, M.D. (2003) Netrins and DCC in the guidance of migrating neural crest-derived cells in the developing bowel and pancreas. *Dev. Biol.* **258**: 364–384.

Jing, S., Wen, D., Yu, Y., Holst, P.L. *et al.* (1996) *GDNF*-induced activation of the *ret* protein tyrosine kinase is mediated by GDNFR-α, a novel receptor for GDNF. *Cell* **85**: 1113–1124.

Johnson, J.E., Birren, S.J. and Anderson, D.J. (1990) Two rat homologues of *Drosophila achaete–scute* specifically expressed in neuronal precursors. *Nature* **346**: 858–861.

Kapur, R.P. (1999a) Early death of neural crest cells is responsible for total enteric aganglionosis in *Sox10(Dom)/Sox10(Dom)* mouse embryos. *Pediatr. Dev. Pathol.* **2**: 559–569.

Kapur, R.P. (1999b) Hirschsprung disease and other enteric dysganglionoses. *Crit. Rev. Clin. Lab Sci.* **36**: 225–273.

Kapur, R.P. (2000) Colonization of the murine hindgut by sacral crest-derived neural precursors: experimental support for an evolutionarily conserved model. *Dev. Biol.* **227**: 146–155.

Kapur, R.P., Sweetser, D.A., Doggett, B., Siebert, J.R. *et al.* (1995) Intercellular signals downstream of endothelin receptor-B mediate colonization of the large intestine by enteric neuroblasts. *Development* **121**: 3787–3795.

Kapur, R.P., Yost, C. and Palmiter, R.D. (1992) A transgenic model for studying development of the enteric nervous system in normal and aganglionic mice. *Development* **116**: 167–175.

Kim, J., Lo, L., Dormand, E. and Anderson, D.J. (2003) *SOX10* maintains multipotency and inhibits neuronal differentiation of neural crest stem cells. *Neuron* **38**: 17–31.

Kirchgessner, A.L., Adlersberg, M.A. and Gershon, M.D. (1992) Colonization of the developing pancreas by neural precursors from the bowel. *Dev. Dyn.* **194**: 142–154.

Kruger, G., Mosher, J., Bixby, S., Joseph, N. *et al.* (2002) Neural crest stem cells persist in the adult gut but undergo changes in self-renewal, neuronal subtype potential, and factor responsiveness. *Neuron* **35**: 657.

Kruger, G.M., Mosher, J.T., Tsai, Y.H., Yeager, K.J. *et al.* (2003) Temporally distinct requirements for endothelin receptor B in the generation and migration of gut neural crest stem cells. *Neuron* **40**: 917–929.

Krumlauf, R. (1994) *Hox* genes in vertebrate development. *Cell* **78**: 191–201.

Kuhlbrodt, K., Herbarth, B., Sock, E., Hermans-Borgmeyer, I. *et al.* (1998) Sox10, a novel transcriptional modulator in glial cells. *J. Neurosci.* **18**: 237–250.

Kusafuka, T., Wang, Y. and Puri, P. (1996) Novel mutations of the endothelin-B receptor gene in isolated patients with Hirschsprung's disease. *Hum. Mol. Genet.* **5**: 347–349.

Lance-Jones, C., Omelchenko, N., Bailis, A., Lynch, S. *et al.* (2001) Hoxd10 induction and regionalization in the developing lumbosacral spinal cord. *Development* **128**: 2255–2268.

Lane, P.W. (1966) Association of megacolon with two recessive spotting genes in the mouse. *J. Hered.* **57**: 29–31.

Lane, P.W. and Liu, H.M. (1984) Association of megacolon with a new dominant spotting gene (*Dom*) in the mouse. *J. Hered.* **75**: 435–439.

Lang, D., Chen, F., Milewski, R., Li, J. *et al.* (2000). Pax3 is required for enteric ganglia formation and functions with Sox10 to modulate expression of *c-ret*. *J. Clin. Invest.* **106**: 963–971.

Lang, D. and Epstein, J.A. (2003) Sox10 and Pax3 physically interact to mediate activation of a conserved c-RET enhancer. *Hum. Mol. Genet.* **12**: 937–945.

Le Douarin, N. (1982) *The Neural Crest*. Cambridge University Press: Cambridge, UK.

Le Douarin, N.M., Renaud, D., Teillet, M.A. and Le Douarin, G.H. (1975) Cholinergic differentiation of presumptive adrenergic neuroblasts in interspecific chimeras after heterotopic transplantations. *Proc. Natl Acad. Sci. USA* **72**: 728–732.

Le Douarin, N.M. and Teillet, M.A. (1973) The migration of neural crest cells to the wall of the digestive tract in avian embryo. *J. Embryol. Exp. Morphol.* **30**: 31–48.

Le Douarin, N.M. and Teillet, M.A. (1974) Experimental analysis of the migration and differentiation of neuroblasts of the autonomic nervous system and of neurectodermal mesenchymal derivatives, using a biological cell marking technique. *Dev. Biol.* **41**: 162–184.

Lee, H.O., Levorse, J.M. and Shin, M.K. (2003) The endothelin receptor-B is required for the migration of neural crest-derived melanocyte and enteric neuron precursors. *Dev. Biol.* **259**: 162–175.

Leibl, M.A., Ota, T., Woodward, M.N., Kenny, S.E. *et al.* (1999) Expression of endothelin 3 by mesenchymal cells of embryonic mouse caecum. *Gut* **44**: 246–252.

Liu, M.T., Rothstein, J.D., Gershon, M.D. and Kirchgessner, A.L. (1997) Glutamatergic enteric neurons. *J. Neurosci.* **17**: 4764–4784.

Lo, L. and Anderson, D.J. (1995) Postmigratory neural crest cells expressing *c-RET* display restricted developmental and proliferative capacities. *Neuron* **15**: 527–539.

Lo, L., Tiveron, M.C. and Anderson, D.J. (1998) *MASH1* activates expression of the paired homeodomain transcription factor Phox2a, and couples pan-neuronal and subtype-specific components of autonomic neuronal identity. *Development* **125**: 609–620.

Lo, L.C., Johnson, J.E., Wuenschell, C.W., Saito, T. *et al.* (1991). Mammalian *achaete–scute homolog 1* is transiently expressed by spatially restricted subsets of early neuroepithelial and neural crest cells. *Genes Dev.* **5**: 1524–1537.

Lore, F., Di Cairano, G. and Talidis, F. (2000) Unilateral renal agenesis in a family with medullary thyroid carcinoma. *N. Engl. J. Med.* **342**: 1218–1219.

Marigo, V., Johnson, R.L., Vortkamp, A. and Tabin, C.J. (1996) Sonic hedgehog differentially regulates expression of *GLI* and *GLI3* during limb development. *Dev. Biol.* **180**: 273–283.

McCabe, L., Griffin, L.D., Kinzer, A., Chandler, M. *et al.* (1990) Overo lethal white foal syndrome: equine model of aganglionic megacolon (Hirschsprung disease). *Am. J. Med. Genet.* **36**: 336–340.

McCallion, A.S., Stames, E., Conlon, R.A. and Chakravarti, A. (2003) Phenotype variation in two-locus mouse models of Hirschsprung disease: tissue-specific interaction between Ret and Ednrb. *Proc. Natl Acad. Sci. USA* **100**: 1826–1831.

McKeown, S.J., Chow, C.W. and Young, H.M. (2001) Development of the submucous plexus in the large intestine of the mouse. *Cell Tissue Res.* **303**: 301–305.

Micci, M.A., Learish, R.D., Li, H., Abraham, B.P. *et al.* (2001) Neural stem cells express RET, produce nitric oxide, and survive transplantation in the gastrointestinal tract. *Gastroenterology* **121**: 757–766.

Milbrandt, J., de Sauvage, F.J., Fahrner, T.J., Baloh, R.H. *et al.* (1998) Persephin, a novel neurotrophic factor related to GDNF and neurturin. *Neuron* **20**: 245–253.

Moore, M.W., Klein, R.D., Farinas, I., Sauer, H. *et al.* (1996) Renal and neuronal abnormalities in mice lacking *GDNF*. *Nature* **382**: 76–79.

Morin, X., Cremer, H., Hirsch, M.R., Kapur, R.P. *et al.* (1997) Defects in sensory and autonomic ganglia and absence of locus coeruleus in mice deficient for the homeobox gene *Phox2a*. *Neuron* **18**: 411–423.

Morrison, S.J., White, P.M., Zock, C. and Anderson, D.J. (1999) Prospective identification, isolation by flow cytometry, and *in vivo* self-renewal of multipotent mammalian neural crest stem cells. *Cell* **96**: 737–749.

Myers, S.M., Salomon, R., Goessling, A., Pelet, A. *et al.* (1999) Investigation of germline *GFRα-1* mutations in Hirschsprung disease. *J. Med. Genet.* **36**: 217–220.

Narita, T., Saitoh, K., Kameda, T., Kuroiwa, A. *et al.* (2000) BMPs are necessary for stomach gland formation in the chicken embryo: a study using virally induced BMP-2 and Noggin expression. *Development* **127**: 981–988.

Nataf, V., Lecoin, L., Eichmann, A. and Le Douarin, N.M. (1996) Endothelin-B receptor is expressed by neural crest cells in the avian embryo. *Proc. Natl Acad. Sci. USA* **93**: 9645–9650.

Natarajan, D., Grigoriou, M., Marcos-Gutierrez, C.V., Atkins, C. *et al.* (1999) Multipotential progenitors of the mammalian enteric nervous system capable of colonizing aganglionic bowel in organ culture. *Development* **126**: 157–168.

Natarajan, D., Marcos-Gutierrez, C., Pachnis, V. and de Graaff, E. (2002) Requirement of signalling by receptor tyrosine kinase RET for the directed migration of enteric nervous system progenitor cells during mammalian embryogenesis. *Development* **129**: 5151–5160.

Nemeth, L., Yoneda, A., Kader, M., Devaney, D. *et al.* (2001) Three-dimensional morphology of gut innervation in total intestinal aganglionosis using whole-mount preparation. *J. Pediatr. Surg.* **36**: 291–295.

Niederreither, K., Subbarayan, V., Dolle, P. and Chambon, P. (1999) Embryonic retinoic acid synthesis is essential for early mouse post-implantation development. *Nat. Genet.* **21**: 444–448.

Niederreither, K., Vermot, J., Le Roux, I., Schuhbaur, B. *et al.* (2003) The regional pattern of retinoic acid synthesis by *RALDH2* is essential for the development of posterior pharyngeal arches and the enteric nervous system. *Development* **130**: 2525–2534.

Nishino, J., Mochida, K., Ohfuji, Y., Shimazaki, T. *et al.* (1999) GFRα3, a component of the artemin receptor, is required for migration and survival of the superior cervical ganglion. *Neuron* **23**: 725–736.

Onochie, C.I., Korngut, L.M., Vanhorne, J.B., Myers, S.M. *et al.* (2000) Characterisation of the human GFRα-3 locus and investigation of the gene in Hirschsprung disease. *J. Med. Genet.* **37**: 669–673.

Pachnis, V., Mankoo, B. and Costantini, F. (1993) Expression of the *c-ret* proto-oncogene during mouse embryogenesis. *Development* **119**: 1005–1017.

Paratcha, G., Ledda, F., Baars, L., Coulpier, M. *et al.* (2001) Released GFRα1 potentiates downstream signaling, neuronal survival, and differentiation via a novel mechanism of recruitment of c-Ret to lipid rafts. *Neuron* **29**: 171–184.

Paratore, C., Eichenberger, C., Suter, U. and Sommer, L. (2002) *Sox10* haplo-insufficiency affects maintenance of progenitor cells in a mouse model of Hirschsprung disease. *Hum. Mol. Genet.* **11**: 3075–3085.

Paratore, C., Goerich, D.E., Suter, U., Wegner, M. *et al.* (2001) Survival and glial fate acquisition of neural crest cells are regulated by an interplay between the transcription factor Sox10 and extrinsic combinatorial signaling. *Development* **128**: 3949–3961.

Parisi, M.A., Baldessari, A.E., Iida, M.H., Clarke, C.M. *et al.* (2003) Genetic background modifies intestinal pseudo-obstruction and the expression of a reporter gene in *Hox11L1*$^{-/-}$ mice. *Gastroenterology* **125**: 1428–1440.

Parisi, M.A. and Kapur, R.P. (2000) Genetics of Hirschsprung disease. *Curr. Opin. Pediatr.* **12**: 610–617.

Parisi, M.A., Kapur, R.P., Neilson, I., Hofstra, R.M. *et al.* (2002) Hydrocephalus and intestinal aganglionosis: is *L1CAM* a modifier gene in Hirschsprung disease? *Am. J. Med. Genet.* **108**: 51–56.

Passarge, E. (2002) Dissecting Hirschsprung disease. *Nat. Genet.* **31**: 11–12.

Pattyn, A., Morin, X., Cremer, H., Goridis, C. *et al.* (1997) Expression and interactions of the two closely related homeobox genes *Phox2a* and *Phox2b* during neurogenesis. *Development* **124**: 4065–4075.

Pattyn, A., Morin, X., Cremer, H., Goridis, C. *et al.* (1999) The homeobox gene *Phox2b* is essential for the development of autonomic neural crest derivatives. *Nature* **399**: 366–370.

Peters-van der Sanden, M.J., Kirby, M.L., Gittenberger-de Groot, A., Tibboel, D. *et al.* (1993) Ablation of various regions within the avian vagal neural crest has differential effects on ganglion formation in the fore-, mid- and hindgut. *Dev. Dyn.* **196**: 183–194.

Pichel, J.G., Shen, L., Sheng, H.Z., Granholm, A.C. *et al.* (1996) Defects in enteric innervation and kidney development in mice lacking *GDNF*. *Nature* **382**: 73–76.

Pingault, V., Bondurand, N., Kuhlbrodt, K., Goerich, D.E. *et al.* (1998) *SOX10* mutations in patients with Waardenburg–Hirschsprung disease. *Nat. Genet.* **18**: 171–173.

Pingault, V., Girard, M., Bondurand, N., Dorkins, H. *et al.* (2002) *SOX10* mutations in chronic intestinal pseudo-obstruction suggest a complex physiopathological mechanism. *Hum. Genet.* **111**: 198–206.

Pingault, V., Guiochon-Mantel, A., Bondurand, N., Faure, C. *et al.* (2000) Peripheral neuropathy with hypomyelination, chronic intestinal pseudo-obstruction and deafness: a developmental 'neural crest syndrome' related to a *SOX10* mutation. *Ann. Neurol.* **48**: 671–676.

Pisano, J.M., Colon-Hastings, F. and Birren, S.J. (2000) Postmigratory enteric and sympathetic neural precursors share common, developmentally regulated, responses to BMP2. *Dev. Biol.* **227**: 1–11.

Pitera, J.E., Smith, V.V., Thorogood, P. and Milla, P.J. (1999) Coordinated expression of 3' *hox* genes during murine embryonal gut development: an enteric Hox code. *Gastroenterology* **117**: 1339–1351.

Puffenberger, E.G., Hosoda, K., Washington, S.S., Nakao, K. *et al.* (1994) A missense mutation of the endothelin-B receptor gene in multigenic Hirschsprung's disease. *Cell* **79**: 1257–1266.

Puri, P. and Wester, T. (2000) Experience with Swenson's operation. In *Hirschsprung's Disease and Allied Disorders*, Holschneider, A.M. and Puri, P. (eds). Harwood Academic: Amsterdam.

Ramalho-Santos, M., Melton, D.A. and McMahon, A.P. (2000) Hedgehog signals regulate multiple aspects of gastrointestinal development. *Development* **127**: 2763–2772.

Roberts, C.W., Sonder, A.M., Lumsden, A. and Korsmeyer, S.J. (1995) Development expression of *Hox11* and specification of splenic cell fate. *Am. J. Pathol.* **146**: 1089–1101.

Roberts, D.J., Smith, D.M., Goff, D.J. and Tabin, C.J. (1998) Epithelial–mesenchymal signaling during the regionalization of the chick gut. *Development* **125**: 2791–2801.

Robertson, K. and Mason, I. (1995) Expression of *ret* in the chicken embryo suggests roles in regionalisation of the vagal neural tube and somites and in development of multiple neural crest and placodal lineages. *Mech. Dev.* **53**: 329–344.

Robertson, K., Mason, I. and Hall, S. (1997) Hirschsprung's disease: genetic mutations in mice and men. *Gut* **41**: 436–441.

Romeo, G., Ronchetto, P., Luo, Y., Barone, V. *et al.* (1994) Point mutations affecting the tyrosine kinase domain of the *RET* proto-oncogene in Hirschsprung's disease. *Nature* **367**: 377–378.

Rosenthal, A., Jouet, M. and Kenwrick, S. (1992). Aberrant splicing of neural cell adhesion molecule L1 mRNA in a family with X-linked hydrocephalus. *Nat. Genet.* **2**: 107–112.

Rossi, J., Herzig, K.H., Voikar, V., Hiltunen, P.H. *et al.* (2003) Alimentary tract innervation deficits and dysfunction in mice lacking GDNF family receptor $\alpha 2$. *J. Clin. Invest.* **112**: 707–716.

Rossi, J., Luukko, K., Poteryaev, D., Laurikainen, A. *et al.* (1999) Retarded growth and deficits in the enteric and parasympathetic nervous system in mice lacking GFRα2, a functional neurturin receptor. *Neuron* **22**: 243–252.

Rothman, T.P., Sherman, D., Cochard, P. and Gershon, M.D. (1986) Development of the monoaminergic innervation of the avian gut: transient and permanent expression of phenotypic markers. *Dev. Biol.* **116**: 357–380.

Saffrey, M.J., Wardhaugh, T., Walker, T., Daisley, J. *et al.* (2000) Trophic actions of neurotrophin-3 on post-natal rat myenteric neurons *in vitro. Neurosci. Lett.* **278**: 133–136.

Salomon, R., Attie, T., Pelet, A., Bidaud, C. *et al.* (1996) Germline mutations of the RET ligand *GDNF* are not sufficient to cause Hirschsprung disease. *Nat. Genet.* **14**: 345–347.

Sanchez, M.P., Silos-Santiago, I., Frisen, J., He, B. *et al.* (1996) Renal agenesis and the absence of enteric neurons in mice lacking *GDNF. Nature* **382**: 70–73.

Sandgren, K., Larsson, L.T. and Ekblad, E. (2002) Widespread changes in neurotransmitter expression and number of enteric neurons and interstitial cells of Cajal in lethal spotted mice: an explanation for persisting dysmotility after operation for Hirschsprung's disease? *Dig. Dis. Sci.* **47**: 1049–1064.

Schiltz, C.A., Benjamin, J. and Epstein, M.L. (1999) Expression of the GDNF receptors Ret and GFRα1 in the developing avian enteric nervous system. *J. Comp. Neurol.* **414**: 193–211.

Schuchardt, A., D'Agati, V., Larsson-Blomberg, L., Costantini, F. *et al.* (1994) Defects in the kidney and enteric nervous system of mice lacking the tyrosine kinase receptor Ret. *Nature* **367**: 380–383.

Seaman, C., Anderson, R., Emery, B. and Cooper, H.M. (2001) Localization of the netrin guidance receptor, DCC, in the developing peripheral and enteric nervous systems. *Mech. Dev.* **103**: 173–175.

Seri, M., Yin, L., Barone, V., Bolino, A. *et al.* (1997) Frequency of *RET* mutations in long- and short-segment Hirschsprung disease. *Hum. Mutat.* **9**: 243–249.

Shen, L., Pichel, J.G., Mayeli, T., Sariola, H. *et al.* (2002) Gdnf haplo-insufficiency causes Hirschsprung-like intestinal obstruction and early-onset lethality in mice. *Am. J. Hum. Genet.* **70**: 435–447.

Shepherd, I.T. and Raper, J.A. (1999) Collapsin-1/semaphorin D is a repellent for chick ganglion of Remak axons. *Dev. Biol.* **212**: 42–53.

Shihabuddin, L.S., Palmer, T.D. and Gage, F.H. (1999) The search for neural progenitor cells: prospects for the therapy of neurodegenerative disease. *Mol. Med. Today* **5**: 474–480.

Shimotake, T., Iwai, N., Inoue, K., Kimura, T. *et al.* (1997) Germline mutation of the RET proto-oncogene in children with total intestinal aganglionosis. *J. Pediatr. Surg.* **32**: 498–500.

Shin, M.K., Levorse, J.M., Ingram, R.S. and Tilghman, S.M. (1999) The temporal requirement for endothelin receptor-B signalling during neural crest development. *Nature* **402**: 496–501.

Shirasawa, S., Yunker, A.M., Roth, K.A., Brown, G.A. *et al.* (1997) *Enx* (*Hox11L1*)-deficient mice develop myenteric neuronal hyperplasia and megacolon. *Nat. Med.* **3**: 646–650.

Sidebotham, E.L., Kenny, S.E., Lloyd, D.A., Vaillant, C.R. *et al.* (2002a) Location of stem cells for the enteric nervous system. *Pediatr. Surg. Int.* **18**: 581–585.

Sidebotham, E.L., Woodward, M.N., Kenny, S.E., Lloyd, D.A. *et al.* (2002b) Localization and endothelin-3 dependence of stem cells of the enteric nervous system in the embryonic colon. *J. Pediatr. Surg.* **37**: 145–150.

Sommer, L., Shah, N., Rao, M. and Anderson, D.J. (1995) The cellular function of MASH1 in autonomic neurogenesis. *Neuron* **15**: 1245–1258.

Song, H. and Poo, M. (2001) The cell biology of neuronal navigation. *Nat. Cell Biol.* **3**: E81–88.

Southard-Smith, E.M., Angrist, M., Ellison, J.S., Agarwala, R. *et al.* (1999) The *Sox10(Dom)* mouse: modeling the genetic variation of Waardenburg–Shah (WS4) syndrome. *Genome Res.* **9**: 215–225.

Southard-Smith, E.M., Kos, L. and Pavan, W.J. (1998) *Sox10* mutation disrupts neural crest development in Dom Hirschsprung mouse model. *Nat. Genet.* **18**: 60–64.

Sternini, C., Su, D., Arakawa, J., de Giorgio, R. *et al.* (1996) Cellular localization of Pan-trk immunoreactivity and trkC mRNA in the enteric nervous system. *J. Comp. Neurol.* **368**: 597–607.

Sukegawa, A., Narita, T., Kameda, T., Saitoh, K. *et al.* (2000) The concentric structure of the developing gut is regulated by Sonic hedgehog derived from endodermal epithelium. *Development* **127**: 1971–1980.

Suvanto, P., Hiltunen, J.O., Arumae, U., Moshnyakov, M. *et al.* (1996) Localization of glial cell line-derived neurotrophic factor (GDNF) mRNA in embryonic rat by *in situ* hybridization. *Eur. J. Neurosci.* **8**: 816–822.

Svensson, P.J., Von Tell, D., Molander, M.L., Anvret, M. *et al.* (1999) A heterozygous frameshift mutation in the endothelin-3 (*EDN-3*) gene in isolated Hirschsprung's disease. *Pediatr. Res.* **45**: 714–717.

Taraviras, S., Marcos-Gutierrez, C.V., Durbec, P., Jani, H. *et al.* (1999) Signalling by the RET receptor tyrosine kinase and its role in the development of the mammalian enteric nervous system. *Development* **126**: 2785–2797.

Tassabehji, M., Read, A.P., Newton, V.E., Harris, R. *et al.* (1992) Waardenburg's syndrome patients have mutations in the human homologue of the *Pax-3* paired box gene. *Nature* **355**: 635–636.

Tomac, A.C., Grinberg, A., Huang, S.P., Nosrat, C. *et al.* (2000) Glial cell line-derived neurotrophic factor receptor $\alpha 1$ availability regulates glial cell line-derived neurotrophic factor signaling: evidence from mice carrying one or two mutated alleles. *Neuroscience* **95**: 1011–1023.

Treanor, J.J., Goodman, L., de Sauvage, F., Stone, D.M. *et al.* (1996) Characterization of a multicomponent receptor for GDNF. *Nature* **382**: 80–83.

Trupp, M., Ryden, M., Jornvall, H., Funakoshi, H. *et al.* (1995) Peripheral expression and biological activities of GDNF, a new neurotrophic factor for avian and mammalian peripheral neurons. *J. Cell Biol.* **130**: 137–148.

Tylzanowski, P., Verschueren, K., Huylebroeck, D. and Luyten, F.P. (2001) SIP1 is a repressor of liver/bone/kidney alkaline phosphatase transcription in BMP-induced osteogenic differentiation of C2C12 cells. *J. Biol. Chem* **27**: 27.

Van De Putte, T., Maruhashi, M., Francis, A., Nelles, L. *et al.* (2003) Mice lacking *Zfhx1b*, the gene that codes for Smad-interacting protein-1, reveal a role for multiple neural crest cell defects in the etiology of Hirschsprung disease–mental retardation syndrome. *Am. J. Hum. Genet.* **72**: 2.

Vanhorne, J.B., Gimm, O., Myers, S.M., Kaushik, A. *et al.* (2001) Cloning and characterization of the human *GFRA2* locus and investigation of the gene in Hirschsprung disease. *Hum. Genet.* **108**: 409–415.

Vanner, S. and Macnaughton, W.K. (2004) Submucosal secretomotor and vasodilator reflexes. *Neurogastroenterol. Motil.* **16**(suppl 1): 39–43.

Vanner, S. and Surprenant, A. (1996) Neural reflexes controlling intestinal microcirculation. *Am. J. Physiol* **271**: G223–230.

Wakamatsu, N., Yamada, Y., Yamada, K., Ono, T. *et al.* (2001) Mutations in SIP1, encoding Smad interacting protein-1, cause a form of Hirschsprung disease. *Nat. Genet.* **27**: 369–370.

Widenfalk, J., Nosrat, C., Tomac, A., Westphal, H. *et al.* (1997) Neuturin and glial cell line-derived neurotrophic factor receptor-b, novel proteins related to GDNF and GDNFR-α with specific cellular patterns of expression, suggesting roles in the developing and adult nrevous system and in peripheral organs. *J. Neurosci.* **17**: 8506–8519.

Wong, K., Park, H.T., Wu, J.Y. and Rao, Y. (2002) Slit proteins: molecular guidance cues for cells ranging from neurons to leukocytes. *Curr. Opin. Genet. Dev.* **12**: 583–591.

Woodward, M.N., Kenny, S.E., Vaillant, C., Lloyd, D.A. *et al.* (2000) Time-dependent effects of endothelin-3 on enteric nervous system development in an organ culture model of Hirschsprung's disease. *J. Pediatr. Surg.* **35**: 25–29.

Woodward, M.N., Sidebotham, E.L., Connell, M.G., Kenny, S.E. *et al.* (2003) Analysis of the effects of endothelin-3 on the development of neural crest cells in the embryonic mouse gut. *J. Pediatr. Surg.* **38**: 1322–1328.

Wu, J.J., Chen, J.X., Rothman, T.P. and Gershon, M.D. (1999) Inhibition of *in vitro* enteric neuronal development by endothelin-3: mediation by endothelin B receptors. *Development* **126**: 1161–1173.

Xian, C.J., Huang, B.R. and Zhou, X.F. (1999) Distribution of neurturin mRNA and immuno-reactivity in the peripheral tissues of adult rats. *Brain Res.* **835**: 247–258.

Yan, H., Bergner, A.J., Enomoto, H., Milbrandt, J. *et al.* (2004) Neural cells in the esophagus respond to glial cell line-derived neurotrophic factor and neurturin, and are RET-dependent. *Dev. Biol.* **272**: 118–133.

Yanagisawa, H., Yanagisawa, M., Kapur, R.P., Richardson, J.A. *et al.* (1998) Dual genetic pathways of endothelin-mediated intercellular signaling revealed by targeted disruption of endothelin converting enzyme-1 gene. *Development* **125**: 825–836.

Yandava, B.D., Billinghurst, L.L. and Snyder, E.Y. (1999) 'Global' cell replacement is feasible via neural stem cell transplantation: evidence from the dysmyelinated shiverer mouse brain. *Proc. Natl Acad. Sci. USA* **96**: 7029–7034.

Yang, G.C., Croaker, D., Zhang, A.L., Manglick, P. *et al.* (1998) A dinucleotide mutation in the endothelin-B receptor gene is associated with lethal white foal syndrome (LWFS), a horse variant of Hirschsprung disease. *Hum. Mol. Genet.* **7**: 1047–1052.

Yntema, C.L. and Hammond, W.S. (1954) The origin of intrinsic ganglia of trunk viscera from vagal neural crest in the chick embryo. *J. Comp. Neurol.* **101**: 515–541.

Young, H.M., Hearn, C.J., Farlie, P.G., Canty, A.J. *et al.* (2001) GDNF is a chemoattractant for enteric neural cells. *Dev. Biol.* **229**: 503–516.

12

The Head

Gillian M. Morriss-Kay

Introduction

The human head represents an astonishing evolutionary achievement. It houses the brain and the special sense organs of vision, hearing, balance, taste and olfaction; it provides the structures and mechanisms for taking in food and oxygen; it mediates our social interactions through facial recognition, facial expression and speech. The anatomical and functional integration of all of these makes the head a uniquely complex part of the body. The evolutionary history of the vertebrate head is reflected in its development (for further details and references, see Morriss-Kay, 2001). The defining evolutionary change was the invention of neural crest, together with the potential for head neural crest to differentiate into connective tissue, cartilage and bone in addition to neural tissues. Prevertebrate chordate embryos, such as *Amphioxus*, have cells at the edge of the neural plate that express some genes characteristic of vertebrate neural crest, but lack the genes required to transform these cells into multipotent migratory cells (Holland and Holland, 2001). It was that transformation that enabled development of a skeletal protection for the brain and a new pharyngeal organization from which the face, teeth, tongue and ears are derived.

Our understanding of human craniofacial development is derived from three major sources: (a) morphological descriptions of human embryos, mainly analysed as serial sections; (b) experimental studies on animal models, most importantly the mouse; (c) genetic and phenotypic studies of human patients. Craniofacial abnormalities are rarely confined to a single structural component. An integrated approach to understanding development of the head is therefore essential for meaningful insights into the causes of dysmorphogenesis, and for distinguishing between abnormality and normal variation.

Embryos, Genes and Birth Defects, Second Edition Edited by Patrizia Ferretti, Andrew Copp, Cheryll Tickle and Gudrun Moore

The significance of the contribution of studies on rodent embryos to our current knowledge of the developmental basis of human birth defects cannot be underestimated. Long before the molecular revolution of the early 1980s, analysis of the phenotypes of many spontaneous mouse mutants by Grüneberg (1952) and others paved the way for modern mutation studies using transgenic technology (see Chapter 4). In the 1970s, the whole-embryo culture technique developed by New (1978) enabled rat and mouse embryos to be observed and manipulated *in vitro* during the period of cranial neurulation and neural crest formation. This technique facilitated experimental studies on morphogenetic mechanisms and cell lineage not previously possible in mammals because of their dependence on fetal membranes and the placenta. Rodent embryos undergo normal development *in vitro* during early morphogenesis stages because they use the less physiologically intimate yolk sac placenta during early developmental stages, until the discoid haemochorial placenta is formed. More recently, experimental manipulations on the developing head have been carried out *in vivo*, using an open uterus technique that allows development to continue until just before birth (Iseki *et al.*, 1997, 1999).

Avian embryos, which have contributed so much to our understanding of the mechanisms of limb development, have been less useful for craniofacial studies because significant mammalian–avian differences are established relatively early in this region. This reflects the very early evolutionary divergence of the reptilian lines leading to birds and mammals, affecting in particular the bone structure of the jaws and middle ear. There are also specializations in the avian head related to the supreme importance of vision, and to characteristic patterns of growth in the skull vault.

This chapter will describe the structure, development and defects of the human head, together with some observations in mouse embryos and fetuses. I will then explain how advances in mouse molecular genetics and developmental biology have been integrated with the discovery of mutations underlying human congenital abnormalities, to provide a developmental understanding of the molecular basis of craniofacial birth defects. The embryonic head includes not only the structures of the adult head but also some components of the anterior neck. These are the hyoid bone and larynx, together with their associated nerves, muscles and blood vessels, and the glands of the neck – thyroid, parathyroids and thymus. In addition, a subpopulation of cranial neural crest cells is essential for dividing the outflow tract of the heart into aortic and pulmonary trunks. Congenital defects of the head may therefore be combined with defects of these other structures; understanding such a complex system requires an appreciation of the developmental anatomy of the whole embryonic craniofacial region.

Developmental anatomy

By the time the embryo forms a structure composed of three germ layers (ectoderm, mesoderm and endoderm), the three body axes are clearly defined.

The future head region, together with the tissue that will form the heart and liver, lies anterior (rostral) to Hensen's node and the primitive streak. Major morphogenetic movements ('reversal', or 'folding') form the foregut, bringing the heart and liver to their ventral positions, while the cranial neural folds form the primitive brain tube (Figure 12.1). These processes are complete by 9–10 days in the mouse and 28 days in the human embryo. The embryonic head has now completed the first phase of morphogenesis: all of its essential components are in place and cranial neural crest cells have migrated to the branchial arches and heart. Segmentation is evident in the form of somites in the trunk; four occipital somites (of which the first is transient) are also present in the head. The head at this stage has two major components: (a) the developing brain and associated paired sense organ primordia, together with the surrounding mesenchyme; and (b) three branchial arches (eventually to become five) enclosing the embryonic pharynx. This second group of structures will form the organs associated with feeding, breathing, vocalization and sound conduction, as well as the endocrine glands of the neck. The period from early gastrulation (formation of the primitive streak) to the completion of embryonic folding takes only 2 days in the mouse and 9 days in the human embryo. It is during this remarkable period that some of the most serious birth defects have their developmental origin.

Origin and migration of the neural crest

Mammalian cranial neural crest cells emigrate during neurulation, except in the occiptal region, where they emerge at the time of neural tube closure, as in the trunk. Cranial neurulation is a slow process in mammals and more complex than neurulation in the trunk, probably because the cranial neural folds are broader than those of other vertebrates, reflecting the relatively large size of the brain (see Chapter 8). The cranial neural plate first forms convex neural folds (Figure 12.2a); the lateral edges then rise and the epithelial surface becomes concave as the edges approach each other in the dorsal midline and fuse, forming a closed brain tube. While the cranial neural folds are still open, forebrain, midbrain and hindbrain regions can be identified. The forebrain is divided into telencephalon ('end brain') and diencephalon. In the telencephalon, left and right optic sulci mark the future eyes and optic nerves; the diencephalon has a dorsal process, the pineal gland, and a ventral process, the infundibulum, which makes contact with Rathke's pouch, an upgrowth of the oral cavity just rostral to the tip of the notochord. The infundibulum and Rathke's pouch, respectively, form the neurohypophysis and adenohyphysis of the pituitary gland. By the time cranial neurulation is complete, the hindbrain is divided by a series of sulci and gyri into seven rhombomeres and an unsegmented occipital region (sometimes called rhombomere 8) alongside the four occipital somites (Figure 12.2b).

The first neural crest cells emigrate from the lateral edges of the convex neural folds of the four-somite stage embryo. At this stage a distinct transverse groove, the preotic sulcus, is present, and a second groove, the otic sulcus, is beginning to form (Figure 12.2a). These grooves divide the hindbrain into three prorhombomeres, A, B and C, from which three separate neural crest cell populations emerge in a rostrocaudal sequence (Figure 12.3). First, the trigeminal population migrates from a long stretch of the neural fold edges, from the diencephalic region of the forebrain to prorhombomere A of the hindbrain, forming the frontonasal mesenchyme (which covers the telencephalon and forms the nasal swellings) and the first branchial arch (maxillary and mandibular) mesenchyme, i.e. the region that will be innervated by the trigeminal nerve. Second, the hyoid population migrates from prorhombomere B to populate the second branchial arch. Third, the post-otic population migrates from prorhombomere C to the third branchial arch and forms the lower part of the hyoid

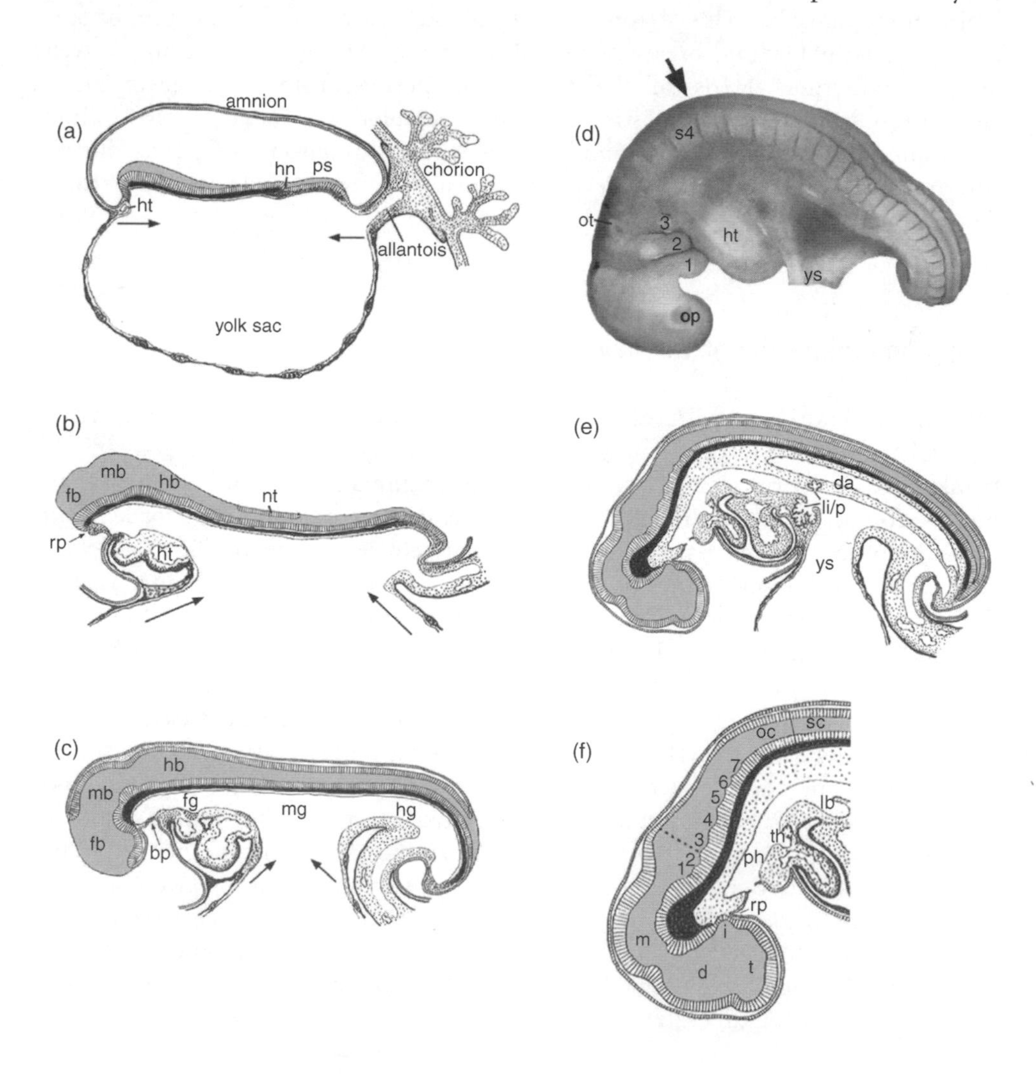

bone. Neural crest cells from the occipital part of the hindbrain migrate to the third, fourth and sixth branchial arches and include the cardiac crest.

These migration pathways have been elucidated by cell labelling studies in rat and mouse embryos (Tan and Morriss-Kay, 1985, 1986; Fukiishi and Morriss-Kay, 1992; Serbedzija *et al.*, 1992; Osumi-Yamashita *et al.*, 1994, 1997). After neural crest cell emigration, the preotic sulcus is transformed into the gyrus between rhombomeres 2 and 3, and the otic sulcus becomes rhombomere 5 (Ruberte *et al.*, 1997).The pattern of cranial neural crest migration in birds is the same as in mammals, except that migration begins as the neural tube closes, at which stage the rhombomeric divisions of the hindbrain are already clear (Kirby and Stewart, 1983; Köntges and Lumsden, 1996). A detailed account of the developmental biology of the neural crest is provided by Le Douarin and Kalcheim (1999).

Recently, a molecular cell lineage tracer for mouse neural crest has enabled elucidation of all of the contributions of these cells to the heart (Jiang *et al.*, 2000), jaws and teeth (Chai *et al.*, 2000), skull vault (Jiang *et al.*, 2002) and skull base (B. McBratney-Owen and G.M. Morriss-Kay, unpublished). Figure 12.3 shows embryos stained to show this lineage marker during neural crest cell migration stages. These

Figure 12.1 The period of embryonic folding in human embryos (16–28 days). (a) 16 days (late presomite stage), with Hensen's node (hn) and the primitive streak (ps), through which the embryonic axis is extended by addition of new notochord (black) and formation of new ectoderm, mesoderm and endoderm from undifferentiated epiblast. (b) 18 days (seven-somite stage): the cranial neural folds show division into forebrain (fb), midbrain (mb) and hindbrain (hb), and the neural tube has begun to close in the upper cervical region (nt). (c) 22 days (14-somite stage): the cranial neural tube is now closed except for the forebrain (rostral neuropore). (d–f) 28 days (25-somite stage). (d) An external view, showing the hindbrain–spinal cord boundary between somites 4 (s4) and 5 (arrowed); three of the five branchial (pharyngeal) arches have formed (1–3); the optic placode (op) lies over the optic outgrowth from the telencephalon, and the otic pit (ot) lies dorsal to the second branchial cleft. (e) Sagittal section, whole embryo, and (f), detail of the head: the forebrain has now divided into telencephalon (t) and diencephalon (d); seven rhombomeres are present in the hindbrain, rostral to the unsegmented occipital region (oc) and spinal cord (sc). The broken line in (f) represents the former position of the preotic sulcus (see text and Figure 12.2). During the whole of this period, the top of the yolk sac contracts (small arrows in a–c) bringing the heart (ht) and allantois (body stalk) beneath the embryo to form the foregut (fg) and hindgut (hg), with the yolk sac stalk (ys) at the level of the midgut (mg). By 28 days, major anatomical changes have occurred. The buccopharyngeal membrane (bp in c) has ruptured, joining the endoderm-lined embryonic pharynx (ph) to the ectoderm-lined stomodaeum; the thyroid gland (th) has begun to form, as a diverticulum in the floor of the pharynx; other foregut diverticula have initiated formation of the lungs (lb), liver and pancreas (li/p). Rathke's pouch (rp), which will form the adenohypophysis (anterior pituitary) is present as an ectodermal diverticulum just rostral to the buccopharyngeal membrane by the seven-somite stage; in (f) it can be seen to be adjacent to the infundibulum, a downgrowth of the diencephalon that will form the neurohypophysis (posterior pituitary). The apical (luminal) surface of the neuroepithelium is shown grey. All other structures are shown in sagittal section, except that the rhombomeres are in reality lateral structures; da, dorsal aorta. Compiled from various sources, including drawings by Bradley M. Patten and data from Morriss-Kay (1981)

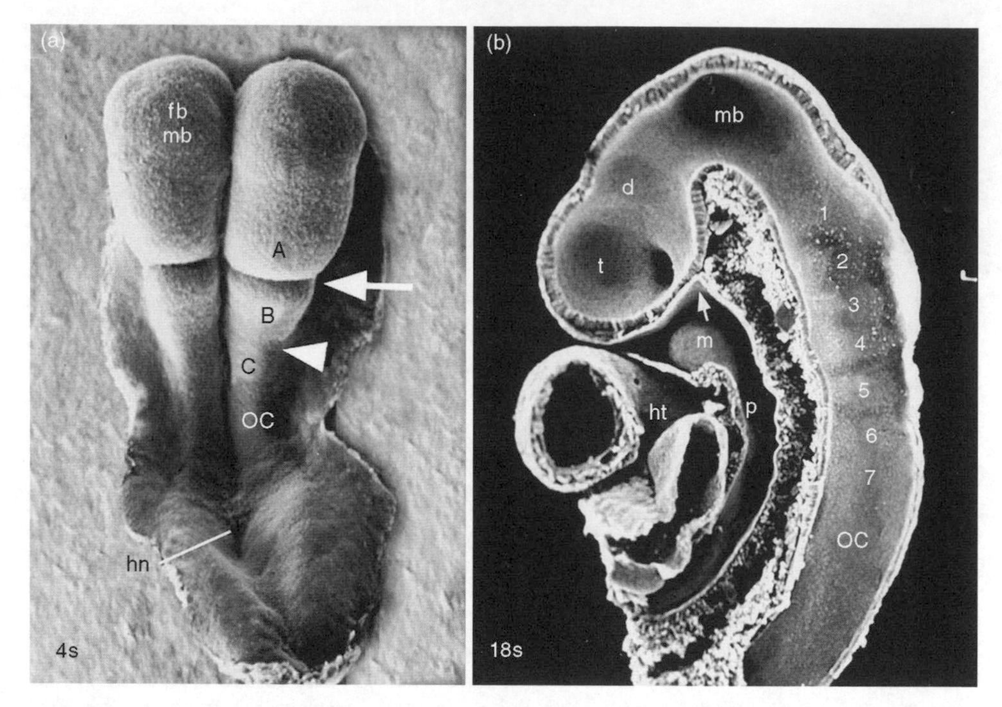

Figure 12.2 Scanning electron micrographs of (a) a whole four-somite stage (4s) and (b) a halved 18-somite stage (18s) mouse embryo (8.0 days and 9.5 days, equivalent to 20 and 26 days human). At the four-somite stage, the cranial neural folds are convex; the boundaries between the forebrain (fb), midbrain (mb) and hindbrain are still morphologically undefined, but the hindbrain is clearly divided into prorhombomeres A and B by the preotic sulcus (arrow) and the otic sulcus (arrowhead) has begun to form, separating prorhombomeres B and C. The preotic sulcus later becomes the rhombomere 2/3 boundary, and the otic sulcus forms rhombomere 5. Caudal to prorhombomere c, the occipital region (oc) overlies the first four somites (not visible), just rostral to Henson's node (hn) and the primitive streak. In the 18-somite stage embryo, the neural tube has closed and all seven rhombomeres have formed (numbers). The forebrain is divided into telencephalon (t) and diencephalon (d). Other structures visible include the heart (ht), mandibular arch (m) and pharynx (p); Rathke's pouch is arrowed. The plane of cut is slightly to the right of the midline, so the notochord is missing and only the lateral part of Rathke's pouch is present

studies have defined the contribution of neural crest to the mouse skull; this information, extrapolated to the human skull, is shown in Figure 12.4. Neural crest cells also contribute to the somatic sensory and parasympathetic cranial ganglia (Figures 12.4, 12.5), although the proportion that form glial rather than neuronal cells has not been determined. The motor components of cranial nerves are not neural crest-derived, but form from neuroblasts within the developing brain: these extend processes (neurites) to mesoderm-derived muscle in association with some (but not all) of the cranial nerve branches (Table 12.1). Neural crest cells also differentiate to form the Schwann cells that invest the larger neurons of both neural crest and neural tube origin; others migrate to the epidermis, where they differentiate into melanocytes.

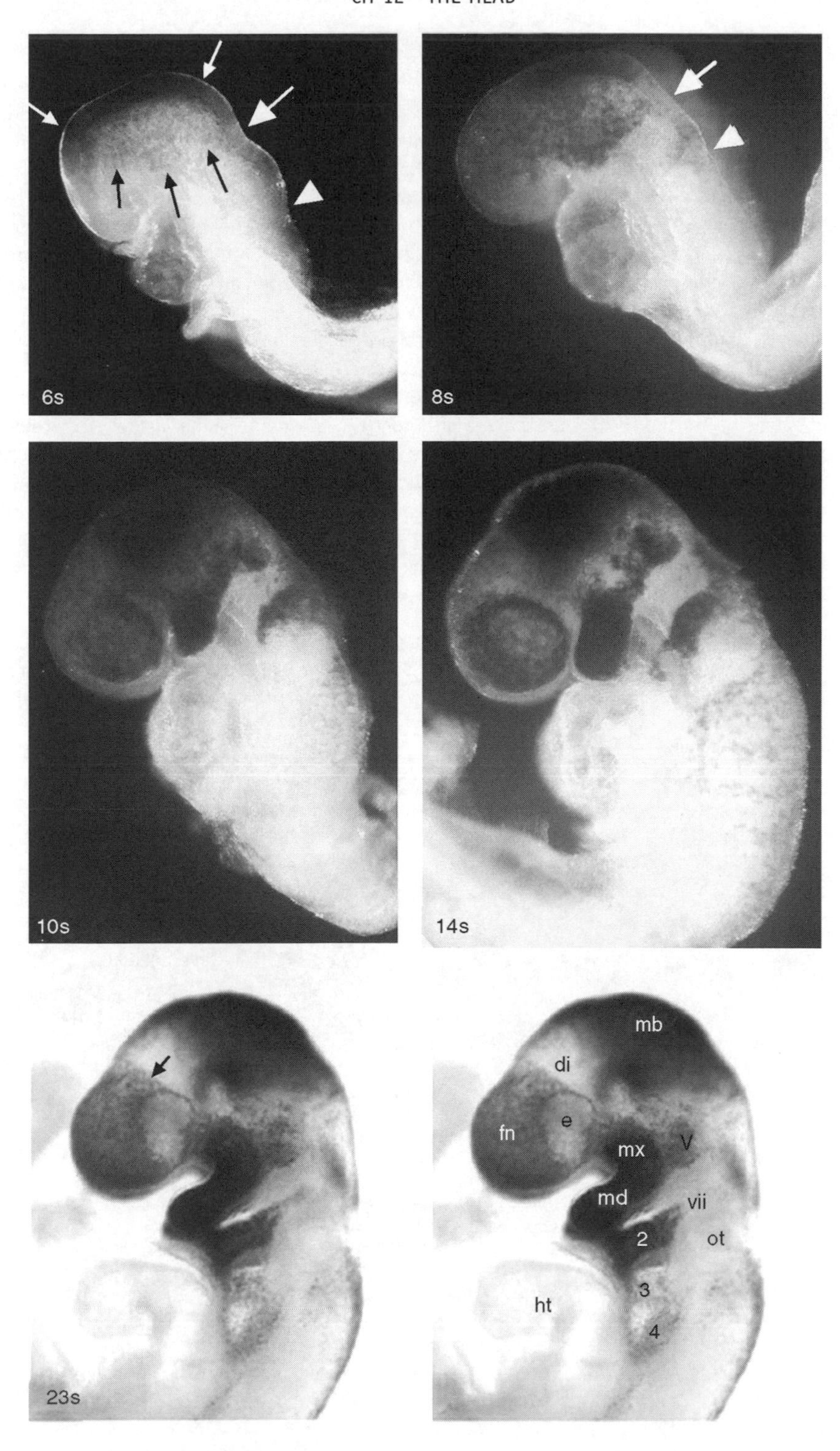
6s
8s
10s
14s
23s
mb
di
fn
e
mx
V
md
vii
2
ot
3
ht
4

Figure 12.3 Cranial neural crest cell migration in 6–23-somite (s) stage (E8.0–E9.5) mouse embryos, revealed by X-gal staining of *Wnt1-Cre/R26R* mutant embryos (Jiang *et al.*, 2002). In the six-somite stage image, small white arrows indicate the approximate forebrain–midbrain and midbrain–hindbrain boundaries of the neural folds, and small black arrows indicate the advancing edge of the migrating neural crest cells. The hindbrain is divided into prorhombomeres a, b and c by the preotic sulcus (large arrow) and the otic sulcus (arrowhead). The most rostral (trigeminal) neural crest cell population, which originates from the caudal forebrain, midbrain and prorhombomere A of the hindbrain, migrates to form the frontonasal mesenchyme (fn), which covers the telencephalon and forms a sharp boundary with the (unstained) mesodermal cranial mesenchyme. Cells from this population also migrate around the eye (e) and into the maxillary (mx) and mandibular (md) regions of the first branchial arch, as well as contributing to the trigeminal ganglion (v). Migration of this population is complete by the 23-somite stage; the frontonasal crest forms a clear boundary (arrowed) with adjacent mesoderm-derived cranial mesenchyme (unstained). From the 8s stage onwards, the hyoid population migrates from prorhombomere b into the second branchial arch. From the 10-somite stage, neural crest cells from prorhombomere c migrate as a diffuse population to arches 3 and 4, mixing with cells from the occipital region, which migrate into arches 3, 4 and 6 (not shown). Some of the neural crest cells migrating to arches 4–6 continue into the outflow tract of the heart (Fukiishi and Morriss-Kay, 1992). The midbrain (mb) and part of the diencephalon (di) and hindbrain are also X-gal-positive. In the 23-somite stage embryo, the otic pit (ot) is at the level of rhombomere 5 (formerly the otic sulcus)

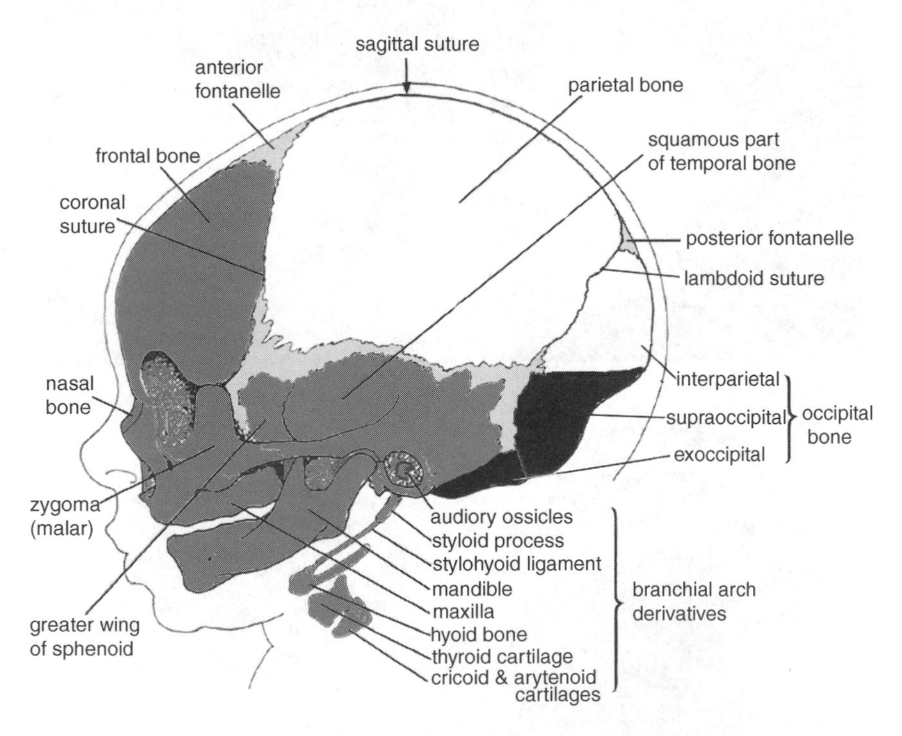

Figure 12.4 The human skull at full term, showing its tissue origins. The skull bones are derived from neural crest (dark grey), sclerotomal component of the occipital somites (black) and cranial mesoderal mesenchyme (unshaded). There is also a small neural crest contribution to the interparietal part of the occipital bone (not shown; see Jiang *et al.*, 2002). Reproduced from Morriss-Kay (1990), with permission from Oxford University Press

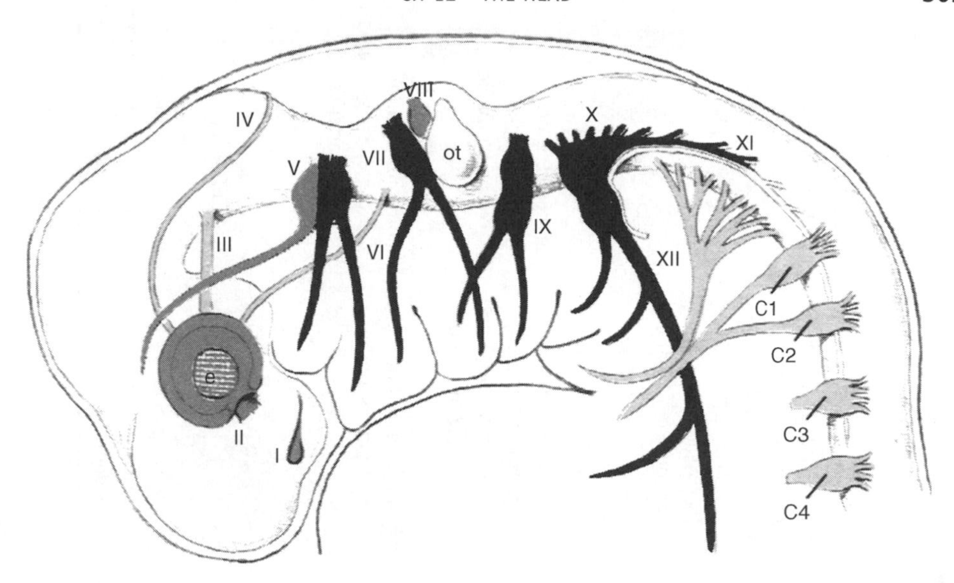

Figure 12.5 The cranial nerves in a 28-day human embryo (equivalent to mouse E9.5). Branchial arch nerves are shown in black (see Table 12.1). Reproduced from Morriss-Kay (1990), with permission from Oxford University Press

Placodes

Neural crest cells are not the only component of the cranial ganglia: they receive at least half of their neuronal cells from thickenings of the overlying ectoderm and the epibranchial. Placodes are regions of pseudostratified epithelium within the otherwise squamous epithelium that forms the ectodermal covering of the embryo. Cells from the epibranchial placodes undergo epithelial–mesenchymal transformation before joining the neural crest-derived cells forming the ganglia. There is no functional difference between the two components: both form bipolar neurons that extend centrally into the appropriate region of the developing brain, and peripherally to innervate the skin and tooth germs.

Other ectodermal placodes form parts of the cranial special sense organs. The nasal placodes form the olfactory epithelium: they are initially formed adjacent to the rostral neural plate (Bhattacharyya *et al.*, 2004). After neural tube closure, as the nasal swellings form around each deepening nasal pit, the olfactory epithelium remains adjacent to the olfactory lobe-forming region of the telencephalon, extending sensory neurites into it. Later, when the skull base cartilages differentiate (Figure 12.6), the mesenchyme around the many olfactory nerve bundles undergoes chondrogenesis to form the cribriform plates of the ethmoid bone. This process of chondrogenesis of the mesenchyme around existing cranial nerves and blood vessels is how all cranial foramina form.

The otic placodes form in surface ectoderm adjacent to the hindbrain, at the level of rhombomere 5. Each placode becomes concave, forming an otic pit; this soon

Table 12.1 Derivatives of the pharyngeal arches and their nerve supply

	Arch	Nerve	Pre-trematic branch (br.) Post-trematic branch		Muscles (ms)	Bones, cartilages, and ligaments (ligt)	Arteries
1	Maxillary		Maxillary division		–	Incus, tympanic, bones of face and palate	Terminal branch of maxillary artery
1	Mandibular	V	Mandibular division	→	Mm of mastication, ant. belly digastric, mylohyoid, tensor palati, tensor tympani	Malleus, mandible, Meckel's cartilage, ant. ligt malleus, sphenomandibular ligt	1st arch artery (transitory)
			Chorda tympani		–		
			Greater petrosal nerve		–		
2	Hyoid	VII	Facial	→	Stapedius, stylohyoid, post. belly digastric, mm of facial expression, auricle and scalp, buccinator	Stapes, styloid process, stylohyoid ligament, hyoid (lesser horn and upper part of body)	Stapedial artery (transitory); corticotympanic artery
			Tympanic br. IX		–		
3		IX	Glossopharangeal	→	Stylopharyngeus	Hyoid (greater horn and lower part of body)	Common carotid arteries, 1st part internal carotid arteries
			Pharyngeal br. X	→	Sup. and middle pharyngeal constrictors, palatal mm except tensor palati		
4			Sup. laryngeal br. X	→	Inf. pharyngeal constrictor, criothyroid		Arch of aorta, base of subclavian arteries
6		X	Recurrent laryngeal br. X with cranial XI	→	Laryngeal mm except cricothyroid	Laryngeal cartilages: thyroid, cricoid, arytenoid	Base of pulmonary trunk, ductus arteriosus (ligamentum arteriosum)

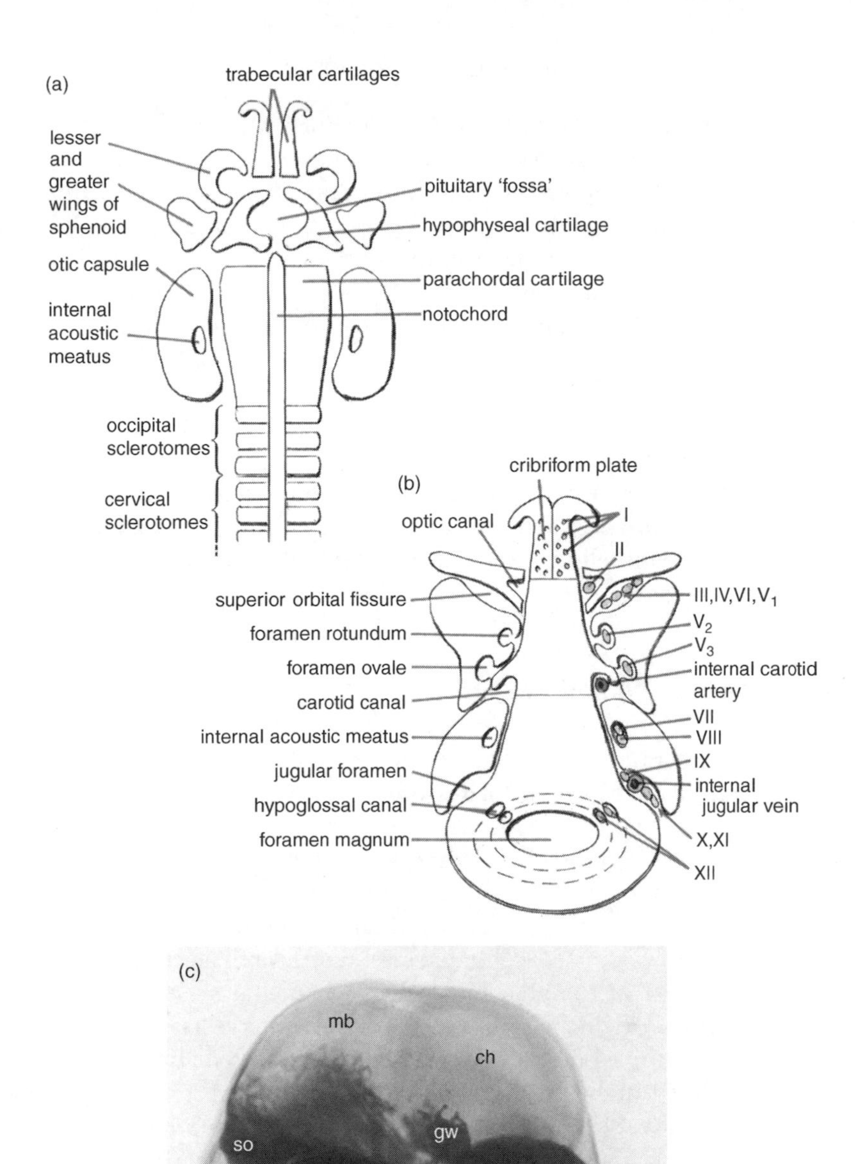

Figure 12.6 The cartilaginous skull base. (a, b) Scheme of the skull base at approximately 7 and 8 weeks. (b) shows foramina (labelled on the left) and the nerves and blood vessels that pass through them (labelled on the right). (c) Mouse cartilaginous skull in side view at E14.5, stained with Alcian blue av, atlas vertebra; ch, cerebral hemisphere; e, eye; gw, cartilagious part of greater wing of sphenoid; h, hyoid; lw, lesser wing of sphenoid; mb, midbrain; Mc, Meckel's cartilage; n, nasal capsule; ot, otic capsule; so, supraoccipital cartilage. (a) and (b) produced from Morriss-Kay (1990), with permission from Oxford University Press

closes to form the otic vesicle. This process involves microfilament contraction at the apical (outer) surface of the placodal cells (Morriss-Kay and Tuckett, 1985). Like the epibranchial placodes, the otic vesicle contributes cells to the adjacent neural crest-derived acousticovestibular ganglion. It undergoes a more complex morphogenesis than any other cranial placode derivative, forming the semicircular canals, utricle, saccule and cochlea of the inner ear.

The lens placode is the only cranial placode that does not have neural derivatives, although it forms part of a neural structure. It is induced by the optic cup, a lateral outgrowth of the telencephalon (see Chapter 9). Like the otic placode, it forms a lens pit, which closes to form a lens vesicle, separated from the surface ectoderm in which it formed. The ring of ectoderm around the lens placode covers the closed lens vesicle and forms the cornea. The lens vesicle epithelium close to the surface ectoderm remains as a thin epithelium, while the inner part forms an increasingly thick pseudostratified epithelium in which lens crystallins are expressed, forming the spherical, transparent lens.

The branchial arches and pouches

The embryonic viscerocranium is organized as a series of five paired branchial arches separated from each other by clefts (for reasons of comparison with more primitive vertebrates, the mammalian branchial arches are numbered 1, 2, 3, 4, 6). Internally, the shape of the pharynx mirrors the arches, forming a series of pouches opposite each cleft. In fishes, the tissue between the clefts and pouches is perforated to form gill slits, but this does not occur in the embryos of land vertebrates.

Each branchial arch has a cartilage, blood vessel (aortic arch artery), muscular and neural component. Most of the arch mesenchyme is neural crest-derived, but the muscle is of mesodermal origin (Noden, 1983). Each arch nerve has a main branch to 'its own' arch, which is mixed motor and sensory, and a pretrematic ('in front of the hole') branch, which is purely sensory. The ophthalmic branch of the trigeminal is not a branchial arch nerve, being equivalent to the separate profundus nerve of fishes. The components of each branchial arch are summarized in Table 12.1. The tongue forms from three swellings in the floor of the first arch and one in the floor of the third arch (Figure 12.7). This dual arch origin explains the innervation of the anterior two-thirds by the cranial nerves of the first arch (mandibular V for common sensation and the pretrematic branch of the facial nerve VII for taste) and the posterior one-third by the nerve of the third arch, the glossopharyngeal IX (for both common sensation and taste). The muscles of mastication and of facial expression are derived from first and second arch mesoderm, as their innervation by the trigeminal and facial nerves indicates. In contrast, the intrinsic muscles of the tongue migrate from the occipital somites (myotome), bringing with them their motor innervation from cranial nerve XII. It is interesting to note that there are often two foramina for this nerve on one or both sides of the occipital bone, recalling the origin of this bone from three fused sclerotomal components of the occipital somites (Figure 12.6).

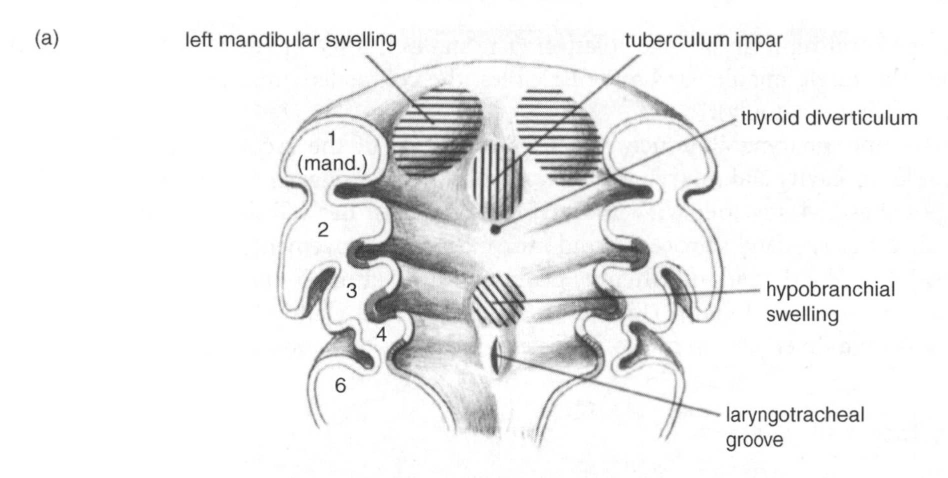

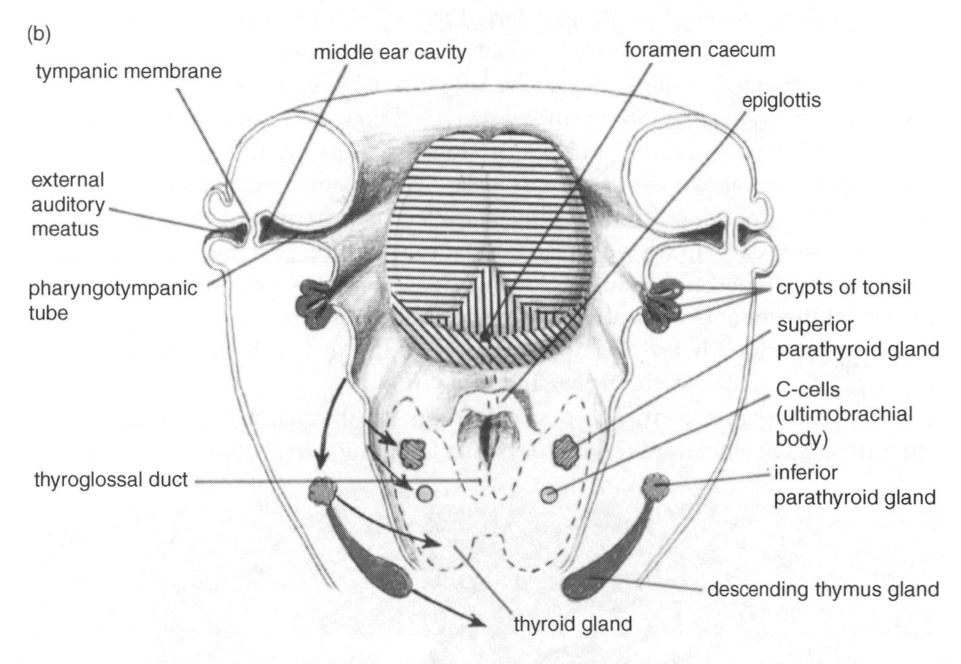

Figure 12.7 The embryonic pharynx and its derivatives. (a) Late 5th week; (b) seventh week. The tongue is derived from three swellings on the floor of the first arch and one from arch three (striped). The derivatives of the pharyngeal pouches, and their movements, are indicated. Reproduced from Morriss-Kay (1990), with permission from Oxford University Press

The external ears, composed of the pinnae and the external auditory meatus, are derived from six swellings, three on the first arch and two on the second, and the cleft between the two arches. The three ear ossicles also have a first and second arch origin. The malleus and incus are first arch cartilages, and their articulation is derived from the ancestral reptilian jaw articulation. The stapes is a second arch cartilage, derived from the single middle ear bone of reptiles, the columella auris (birds have retained the reptilian anatomy).

The first pharyngeal pouch, together with part of the second, gives rise to the middle ear cavity and pharyngotympanic tube. Thickenings of the endodermal lining of pouches 2–4, together with underlying neural crest-derived mesenchyme, undergo specific differentiation processes and morphogenetic movements to form the palatine tonsil, the parathyroid and thymus glands, and the ultimobranchial body (calcitonin cells) of the thyroid gland (Figure 12.7). The thyroid gland itself descends into the neck from a diverticulum in the midline of the pharyngeal floor, the foramen caecum.

The face and palate

The face and palate are formed by the growth and coalescence of the neural crest-derived, ectoderm-covered swellings referred to earlier. At E11 in the mouse, 6 weeks in human (Figure 12.8a), the medial and lateral nasal swellings surround the nasal pit. The nasolacrimal groove lies between the lateral nasal swelling and the upper border of the maxillary swelling; when these two swellings fuse, the groove forms the nasolacrimal duct, conveying lacrimal secretions from the eye to the nasal cavity, into which it opens just above the secondary palate. The medial nasal swellings fuse with the medial tip of each maxillary swelling, completing formation of the upper lip. This point of fusion is particularly vulnerable to failure, causing unilateral or bilateral cleft lip if one or both sides fail to fuse. The area between the two maxillary/nasal fusions forms the philtrum of the lip; it is continuous internally with the 'premaxilla' region of the maxilla, which bears the upper incisor teeth, and with the small triangular primary palate internal to the upper incisors.

The secondary palate forms from swellings on the internal aspect of the maxillae, which form at E12 in the mouse and 45 days after fertilization in the human (Figure 12.9a).

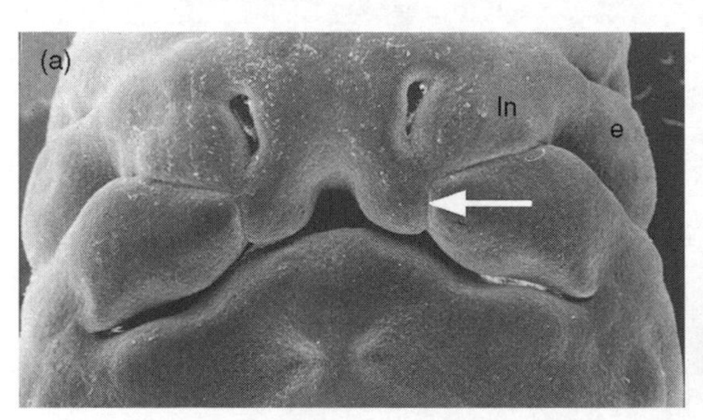

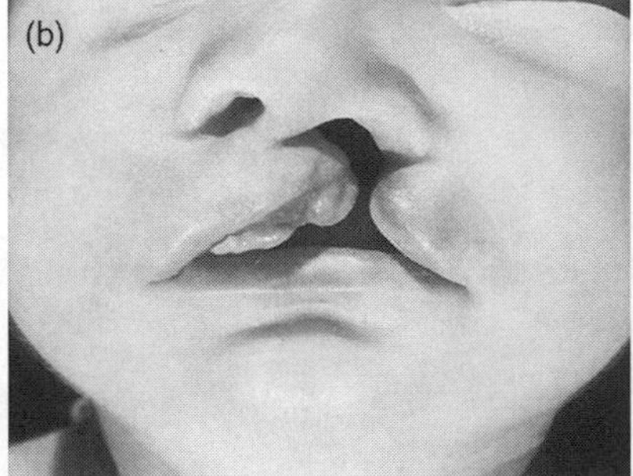

Figure 12.8 Fusion of the facial swellings. (a) Scanning electron micrograph of the face of a normal 6-week human embryo, showing the sites of fusion of each maxillary swelling (below the eye, e) with the lateral nasal swelling (ln) and medial nasal swellings. The arrow indicates the site of fusion of the maxillary and medial nasal swellings; failure of fusion here and between the lateral and medial nasal swellings causes cleft lip. (b) Unilateral cleft lip

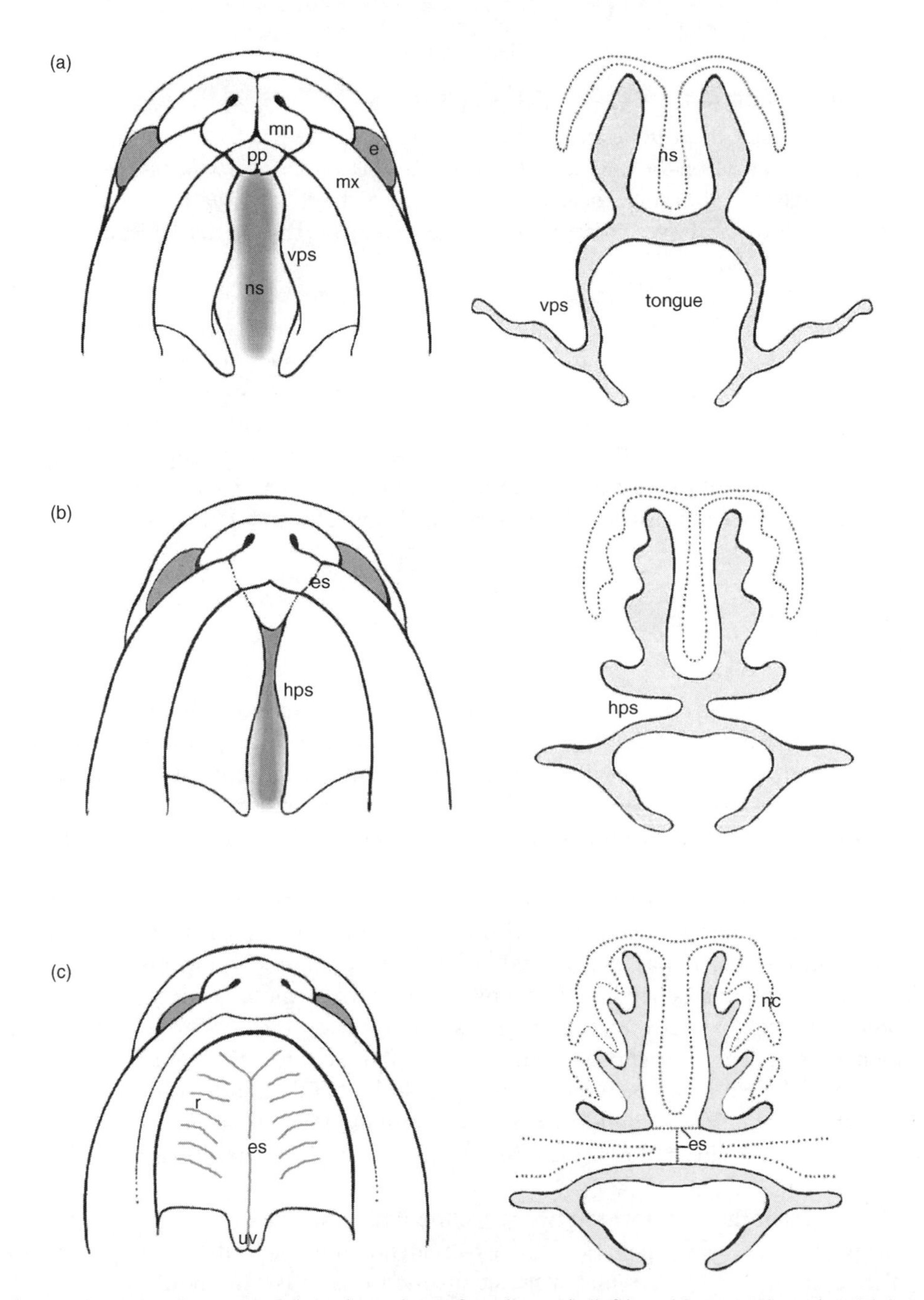

Figure 12.9 Development of the palate, shown from beneath (left) and in coronal section (right). (a) 7th week; (b) late 8th week; (c) 10 weeks. Swellings on the medial aspect of the maxillae grow downwards either side of the tongue (a), then swing medially to form horizontal palatal shelves (b); these fuse with each other and with the nasal septum by breakdown of the apposed epithelial seams (c). e, eye; es, epithelial seam; hps, horizontal palatal shelf; mn, medial nasal swelling; mx, maxillary process; nc, nasal conchae; ns, nasal septum; pp, primary palate; r, rugae; uv, uvula; vps, vertical palatal shelf

Each palatal swelling is at first rounded in form, then triangular in shape, projecting downwards beside the tongue and probably moulded by the shape of the available space. With growth of the face, the tongue descends and the palatal shelves swing into a horizontal position (Figure 12.9b). Their medial edges fuse with each other and with the nasal septum above them, forming the secondary palate and separating the left and right sides of the nasal cavity (Figure 12.9c).

After fusion of the facial processes, further development of the face involves growth and changes of proportion; at first, the gape of the mouth is relatively large, but growth of the cheeks results in the proportionately smaller mouth seen at birth. Occasionally this change in proportion is less than usual, leading to a wide mouth (macrostomia), sometimes referred to as a 'fetal face'. Normally, growth of the mandible is maintained at the same rate as the maxilla, enabling proper occlusion of the teeth. Growth of the temples without concomitant growth in breadth of the nose results in a change in position of the eyes, which move from a lateral position to the front of the head.

The skull

The skull has two functional components: the neurocranium (braincase), which surrounds and protects the brain, and the viscerocranium (face, palate and pharynx), which supports the functions of feeding, breathing and facial expression. The parts of the skull that protect the special sense organs of olfaction, vision, hearing and balance are intimately connected to the neurocranium, although the eyes and nose are clearly part of the face.

The osteogenic mesenchyme from which the skull differentiates is derived from two sources: cranial neural crest and cranial mesoderm (Figure 12.4). The mesodermal component is paraxial mesoderm, i.e. the somite-forming mesoderm that lies alongside the notochord in the trunk and occipital region; at more rostral levels of the head the paraxial mesoderm does not form epithelial somites, remaining mesenchymal. The notochord extends up to Rathke's pouch (Figure 12.1f), the position of the boundary between neural crest-derived and mesoderm-derived cranial mesenchyme (McBratney and Morriss-Kay, unpublished observations). The relationship between the tip of the notochord and the cartilaginous skull base is shown in Figure 12.6a.

The bones of the skull form in two ways: the skull base, the occipital region, the ear ossicles and the styloid process form by endochondral ossification, the process by which most of the extracranial skeleton forms. In contrast, the bones of the vault (calvaria) and face/palate form by direct ossification of cranial mesenchyme, i.e. by intramembranous ossification (the lateral part of the clavicle also forms in this way). Some bones form from both components: the occipital bone is formed from the sclerotomal component of the occipital somites, making this part of the skull equivalent to three fused vertebrae (Figure 12.4) and from the membranous interparietal bone. The sphenoid bone is also mainly endochondral but has a membranous component, the upper part of the greater wing (alisphenoid), which contributes to the side wall of the calvaria. The mandible forms initially as a membrane bone, but has later additions of endochondral bone.

The endochondral skull base and the neural crest-derived branchial arch cartilages of the skull (Meckel's cartilage, the styloid process and the ear ossicles) form as cartilage before any bone differentiates (Figure 12.6). The first-formed membrane bone (at E14.0 in the mouse) is the mandible; it differentiates from mesenchyme lateral to Meckel's cartilage, which then degenerates. The membrane bones of the skull vault – the frontal, parietal and interparietal bones – form last. They ossify within the skeletogenic membrane, the outermost layer of the mesenchyme that condenses around the brain soon after the neural tube closes (its inner layers form the meninges). The juxtaposed edges of these bones form the coronal, metopic, sagittal and lambdoid sutures, in which the major part of skull growth takes place, although appositional growth and remodelling within the bones is important for increasing skull thickness and for adjusting the curvature of the bones as the skull diameter increases. The meninges around the cerebral hemispheres are neural crest-derived; interaction between this layer and the overlying parietal mesoderm is required for ossification of the parietal bone (Jiang *et al.*, 2002).

Main classes of craniofacial defect

Detailed and illustrated information on the whole range of craniofacial defects is available in the classic textbook *Syndromes of the Head and Neck* (Gorlin *et al.*, 2001), in which additional supporting references for this section may be found. The following summary includes those for which we have some insight into the molecular and cellular mechanisms, which will be described in the next section of this chapter. In general, the more severe the defect, the earlier its developmental origin. Craniofacial defects with an origin during axis formation, neural induction and neurulation include holoprosencephaly, anencephaly and encephalocele. These are covered in Chapter 8.

Neural crest-related defects

Since the facial skeleton, frontal, nasal and ethmoid bones are derived from neural crest, it is clear that this tissue is involved in all abnormalities of facial skeletal structure. The following account is based on examples that illustrate defects of areas of the head and neck formed from distinct neural crest populations, beginning with the most rostral population, the frontonasal mesenchyme, and followed by the maxillary/mandibular crest and then the more caudal crest cell populations.

Frontonasal dysplasia (frontonasal malformation) is considered by Sedano and Gorlin (1988) to be a collection of related anomalies, not a unified syndrome. The documented cases have the following features in common: hypertelorism, broad nasal

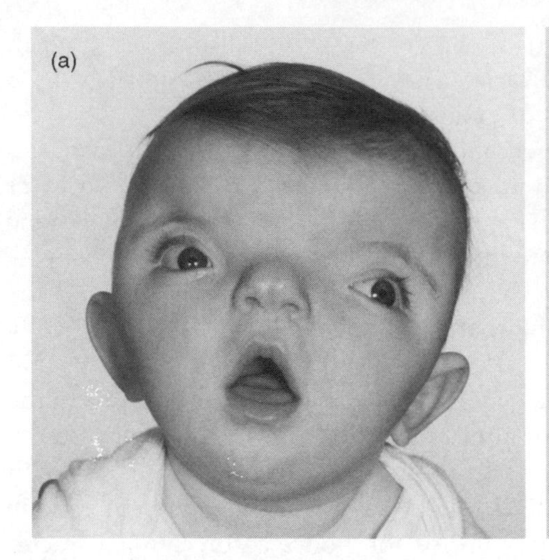

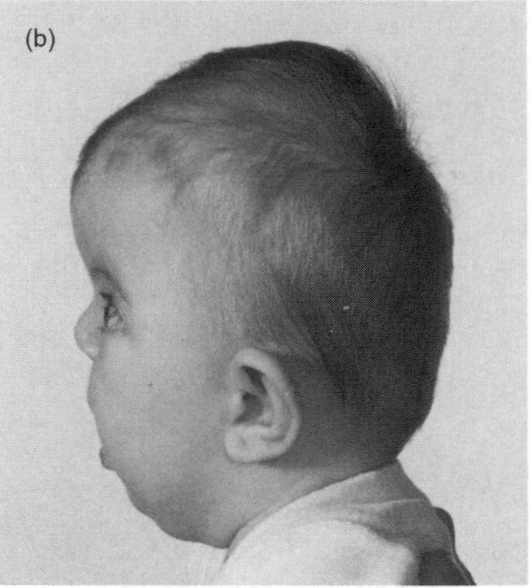

Figure 12.10 Craniofrontonasal dysplasia, 1 year-old infant with a deletion in *EFNB1* (Twigg *et al.*, 2004). (a) Facial view showing hypertelorism, divergent squint and nasal groove. (b) Side view showing brachycephaly due to coronal synostosis

root, lack of nasal tip, and anterior cranium bifidum occultum (i.e. the frontal bones have failed to oppose to form the metopic suture). In contrast, craniofrontonasal dysplasia (Figure 12.10) is an inherited syndrome, characterized by the features of frontonasal dysplasia together with craniosynostosis (see below) and splitting of the nails (Cohen, 1979).

Maxillofacial dysostosis is an autosomal dominantly inherited condition characterized by maxillary hypoplasia, downslanting palpebral fissures and minor abnormalities of the pinnae (Melnick and Eastman, 1977; Escobar *et al.*, 1977). An X-linked form of maxillofacial dysostosis additionally shows mild micrognathia and some hearing loss.

Mandibulofacial dysostosis (Franceschetti and Klein, 1949; Treacher Collins, 1960) has a prevalence of 1/50 000 live births and is dominantly inherited. It is characterized by symmetrical defects of the facial bones, including reduced mandible, incomplete zygomatic arch and orbits, associated with downward-sloping palpebral fissures. The pinna and external meatus of the ear are commonly small and malformed and the meatus may be absent; defects of the auditory ossicles are associated with conductive deafness.

Many other conditions involving increased or decreased size of the mandible have been described, including Pierre Robin sequence (a variable collection of facial defects) and hemifacial microsomia. Mandibular growth abnormality may be associated with other defects, including maxillary hypoplasia and/or craniosynostosis. Complete absence of the mandible (agnathia) is rare, and is associated with the absence of pharyngeal structures that are essential for life. Agnathia is also termed otocephaly, since the low-set ears lie beneath the maxillae and may even be fused to each other. It is clearly due to absence of the mandibular arch, suggesting a very early embryological origin; consistent with this conclusion, otocephaly is commonly associated with holoprosencephaly (see Gorlin *et al.*, 2001, for references).

The variable features of DiGeorge sequence are due to defective neural crest cell contributions to all the branchial arches. The condition is characterized by absence or hypoplasia of the thymus and/or parathyroid glands, and heart defects involving the outflow tract and/or defects of the aortic arch-derived great vessels (Lammer and Opitz, 1986; Thomas *et al.*, 1987, and references therein). Craniofacial anomalies are additionally present in 60% of patients. These include (variably) hypertelorism, cleft palate or bifid uvula, micrognathia, nasal, eye, ear and central nervous system abnormalities, potentially involving not only the whole range of cranial neural crest but also the neural tube.

Clefts

Clefts can occur at any site where fusion of two embryonic primordia (facial swellings) is required for normal development. Clefting of the upper lip may be unilateral (Figure 12.8b) or bilateral, and vary from a small notch in the lip (unilateral incomplete cleft lip) to a double cleft extending up into both nostrils (bilateral cleft lip). Cleft lip may be associated with cleft palate; cleft lip with or without cleft palate (CL/P) occurs in 0.5–3.6/1000 live births, with marked racial and geographical variation and a 2:1 male:female ratio.

Isolated cleft palate is genetically unrelated to CL/P; it has a frequency of 0.4/1000 births and is more common in females. The phenotype varies from failure of the vertical palatal shelves to elevate, to failure of the apposed horizontal palatal shelves to fuse. Bifid uvula is the least severe form of palatal clefting, although it is frequently accompanied by a submucous cleft palate, in which the bones have failed to fuse but the defect is covered by mucous membrane.

Formation of the cheeks requires fusion of the proximal regions of the maxillary and mandibular processes. The extent of fusion, together with subsequent differential growth, regulates the width of the mouth. Failure of, or insufficient, maxillary/mandibular process fusion results in a lateral facial cleft, which in extreme cases may extend as far as the ear. Compared to clefting of the lip and palate, lateral facial clefts are rare, variously estimated as 1/50 000–1/175 000 live births. Oblique facial clefts are even rarer. They extend from a unilateral or bilateral cleft lip to the eye, and may be due to failure of fusion of the lateral nasal and maxillary swellings to form the nasolacrimal duct.

Ossification defects

Ossification defects are of two main types, characterized by (a) excessive ossification leading to premature loss of sutural growth centres (craniosynostosis), and (b) deficient ossification, mainly affecting the membrane bones of the skull vault. The sutures and skull bones are shown in Figure 12.4.

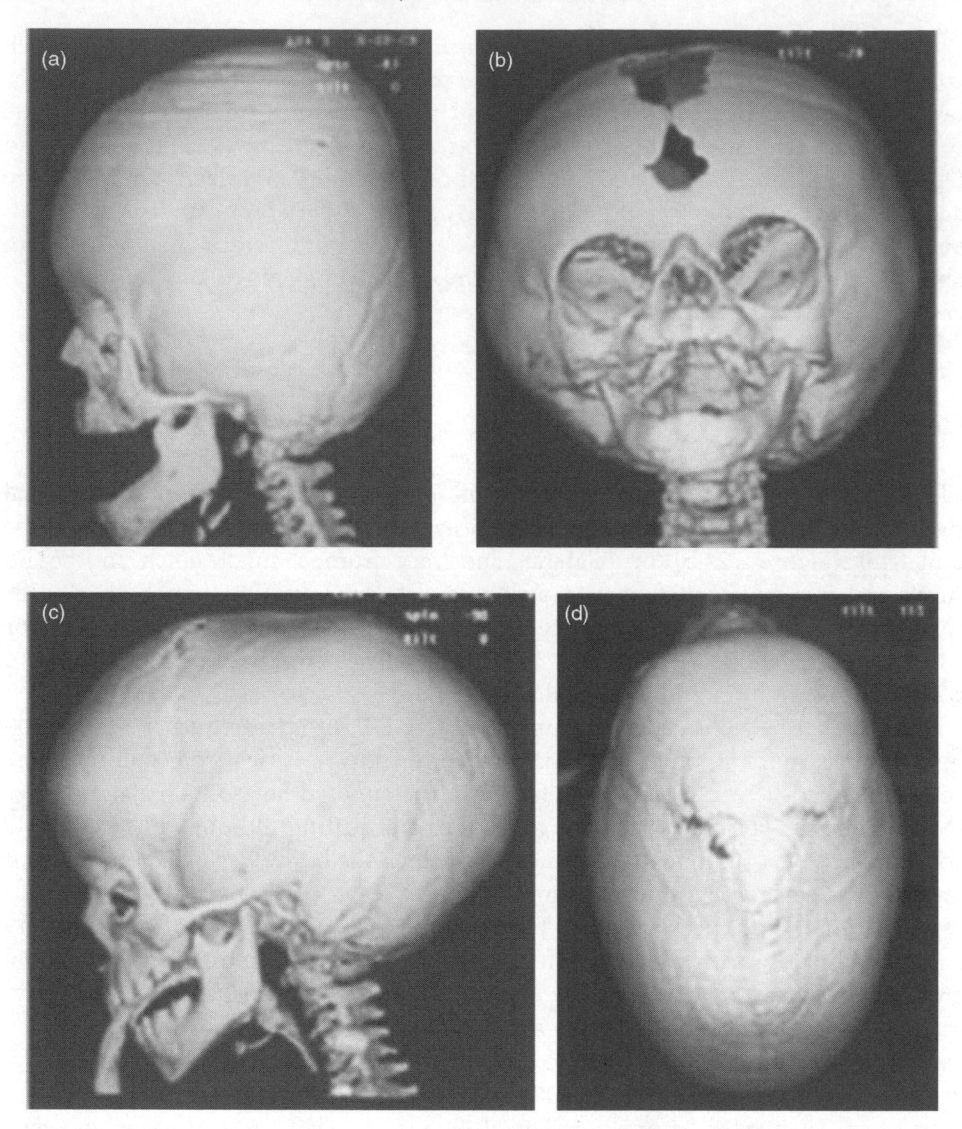

Figure 12.11 3-D reconstructions of CT scans showing craniosynostosis. (a, b) Side and face view of an infant with bicoronal synostosis. The cranial suture is fused, the metopic suture is wide open, and the lambdoid suture is normal. The skull is brachycephalic but there is compensatory growth in breadth. (c, d) Side and vertex views of a child with sagittal synostosis. Growth in breadth has been restricted and there is compensatory growth in the fronto-occipital plane

In craniosynostosis, one or more sutures are obliterated by bone before growth of the skull is complete. The effects range from mild to moderate skull asymmetry where a single suture is affected, to severe dysmorphism and intracranial restraint of the growing brain where multiple sutures are involved (Wilkie, 1997; Wilkie and Morriss-Kay, 2001). Growth deficiency of the facial bones leads to mid-face hypoplasia and often accompanies craniosynostosis. Fusion of the growth centres (synchondroses) of the skull base shortens the skull base, leading to secondary effects on the shape of the neurocranium. Examples of coronal and sagittal synostosis are shown in Figure 12.11. Since growth of the skull is perpendicular to each suture, compensatory growth in the unaffected sutures and through appositional growth contributes to the distortion of skull shape.

Ossification deficiency syndromes affecting the skull vault range from cranium bifidum, in which the parietal bones may be completely absent at birth, to small defects of the parietal bones (persistent parietal foramina; Figure 12.12). In cleido-cranial dysplasia the skull is short (brachycephalic) and broad, with many small Wormian bones in the widely patent sutures, which may remain wide open into adulthood. The lateral parts of the clavicles may also be deficient, enabling the shoulders to be brought forward, almost to make contact.

Cellular and molecular mechanisms

Our understanding of the mechanisms underlying craniofacial birth defects has undergone a major revolution during the past 12 or so years, during which many of the mutations that cause recognized syndromes have been identified. Broadly speaking, mutations are associated with either loss or gain of function. Most loss-of-function mutations are deletions or other alterations of DNA structure that prevent the gene and its RNA transcript from being translated to make a complete functional protein. Different loss-of-function mutations in the same gene may be associated with different degrees of severity of the birth defect. In contrast, gain-of-function mutations often result in the synthesis of a protein with different properties disturbing the balance between the many factors that control each developmental process. Since a single protein may be altered in many different ways, different mutations in a single gene may cause a range of related defects. Some recent reviews covering this area include Wilkie and Morriss-Kay (2001), Helms and Schneider (2003) and Santagati and Rijli (2003).

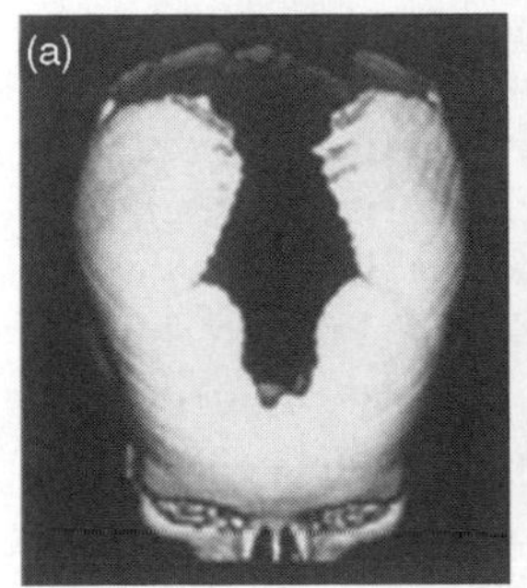

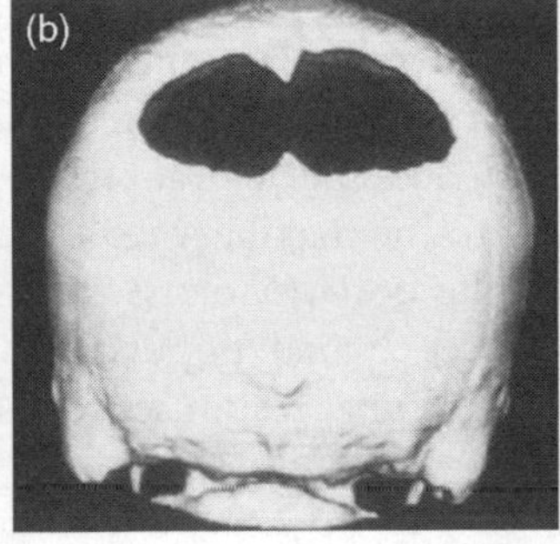

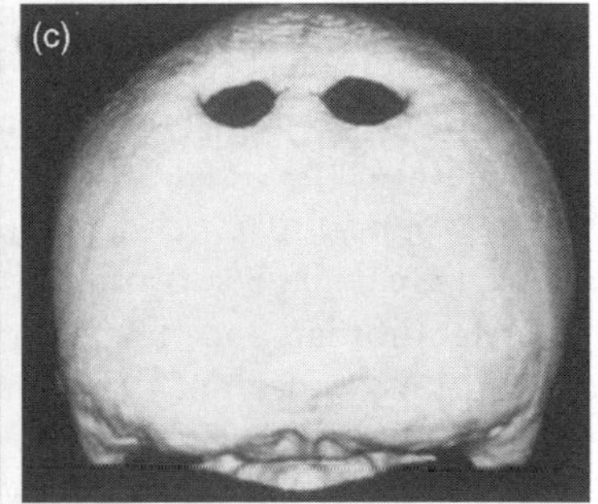

Figure 12.12 Cranium bifidum and parietal foramina 3-D CT scans. (a) Frontal view of a 1 year-old infant showing a wide parietal defect and broad sagittal and metopic sutures. (b) Occipital view of his mother, with a parietal defect, and (c) his grandfather, with persistent parietal foramina

Tissue interactions and craniofacial patterning

Normal developmental mechanisms during morphogenesis and organogenesis stages of development involve molecular signalling between adjacent tissues, usually as epithelial–mesenchymal interactions. In craniofacial development the mesenchymal source may be neural crest or mesoderm, and the epithelium may be surface or neural ectoderm, or pharyngeal endoderm. The mechanism of information exchange between the two adjacent tissues usually involves a ligand released by one of the tissues, which has an affinity for a receptor expressed on the cell surfaces of the adjacent tissue (paracrine signalling). The resulting receptor activation results in a cascade of molecular activation events from the cell surface receptors to the nucleus, ultimately affecting gene expression. The functional outcome of the transcriptional changes is an effect on cell proliferation (stimulation, inhibition or maintenance) or differentiation (initiation or inhibition). Differentiation is a multi-step process, involving a series of transcriptional events; this category includes apoptosis (physiological cell death).

Examples of intercellular signalling that are particularly important in craniofacial development include fibroblast growth factors and their receptors (Fgfs and Fgfrs), and the transforming growth factor-beta superfamily (Tgfβ), which includes bone morphogenetic proteins (Bmps). A different type of interaction, juxtacrine signalling, in which both receptors and ligands are attached to the cell surface, regulates adhesion and mixing between adjacent populations of similar cells. The interactions between Eph receptors and ephrin ligands are of particular interest in the craniofacial context; they are mainly inhibitory, preventing cell mixing and thereby establishing and maintaining clearly-defined boundaries between adjacent tissue domains (Xu *et al.*, 2000). Eph–ephrin interactions prevent mixing between adjacent streams of migrating neural crest cells and between the cell populations of adjacent rhombomeres in the developing hindbrain. Eph–ephrin interactions may also be instrumental in creating the sharp boundary between frontonasal neural crest and the adjacent cranial mesenchyme, and hence in the establishing the site of the future coronal and sagittal sutures (Twigg *et al.*, 2004).

Cranial neural crest cells migrating to different sites form different skeletal elements. Although local tissue interactions at the end point of migration are important, it has become clear that there is also some prepatterning in the different populations. When mouse or rat embryos are exposed to excess retinoic acid during neurulation, the hindbrain is shortened and neural crest cells normally designated for the mandibular arch migrate into the maxillary region, where they form ectopic Meckel's cartilage and mandible-like bone (Morriss-Kay, 1993). Prepatterning of the mandibular and maxillary mesenchymal populations has also been demonstrated on the basis of their transcriptional responses to the epithelial ligand FGF8 (Ferguson *et al.*, 2000).

Hox genes play major roles in the organization and fate of the hindbrain rhombomeres and the neural crest cells derived from them (Krumlauf, 1993; Trainor and Krumlauf, 2000). However, apart from transitory expression of Hoxa2 in rhombomere 2, *Hox* genes are not expressed in the first two rhombomeres, which contribute neural crest cells to the maxillary and mandibular mesenchyme. Transplantation experiments in avian embryos have shown that Fgf8 secreted by the midbrain–hindbrain isthmus is essential for inhibiting Hoxa2 expression in rhombomeres 1 and 2; Hox-expressing neural crest cells are unable to give rise to skeletal tissues, and in fact negatively regulate neural crest-mediated skeletogenesis (Couly *et al.*, 1998; Creuzet *et al.*, 2002). In contrast, mouse embryos lacking Hoxa2 expression in rhombomere 3 form a second lower jaw beneath the normal one (Barrow and Cappecchi, 1999). The absence of *Hox* gene expression from the first arch neural crest is therefore of developmental and evolutionary importance, and may explain why no *HOX* gene mutations have been identified in craniofacial syndromes.

Neural crest-related defects

Mandibulofacial dysostosis is due to mutation of the gene *TCOF1* (Treacher Collins–Franceschetti syndrome 1), identified by the Treacher Collins Syndrome Collaborative Group (1996). Most of the identified mutations introduce stop codons, bringing about the unusual situation of a dominantly inherited mutation that acts through a haplo-insufficiency (loss-of-function) mechanism. The *TCOF1* gene product, treacle, is localized to the nucleolus, where it is thought to contribute to ribosome processing. In the mouse, *Tcof1* is widely expressed, but particularly high levels of expression are seen at the edges of the neural folds and in the first and second branchial arches (Dixon *et al.*, 1997). The mouse model for the human mutation is not a perfect phenocopy, possibly because the level of conservation between the human mouse proteins is only 62%; nevertheless, its analysis provides evidence of potential roles in neural crest cell emigration, proliferation and survival (Dixon *et al.*, 2000).

Neural crest cell defects may also be caused by rupture of branchial arch blood vessels. Hemifacial microsomia, involving reduction of only one side of the mandible and ears, and accompanied by conductive deafness, occurred in babies exposed to thalidomide, and has been reproduced in an animal model in which thalidomide exposure resulted in vascular rupture of the second arch (stapedial) artery, causing a haematoma in the region of the developing middle ear and proximal part of the mandible (Poswillo, 1973). The stapedial artery is a transitory blood vessel that plays an essential role in development of the middle ear. Hemifacial microsomia can also be inherited as a syndrome genetically linked to human chromosome 14q32 (Kelberman *et al.*, 2000). A hemifacial microsomia (*Hfm*) mouse mutant similarly shows bleeding of the second arch artery at E9.5 (Naora *et al.*, 1994).

Clefts

The developmental mechanisms leading to lateral and oblique facial clefts are not understood. A large collection of patients with these clefts has been amassed for study at the Department of Plastic Surgery in Rotterdam (J.M. Vaandrager and I. Mathijssen, personal communication), but there is currently no animal model for this birth defect. In theory, failure of fusion of adjacent primordia for all orofacial clefts may be caused by insufficient growth, so that the edges are not closely apposed at the appropriate developmental stage, or by the failure of epithelial breakdown at the apposing surfaces. For facial clefts that do not coincide with embryonic lines of fusion, amniotic bands are the only plausible explanation (Keller *et al.*, 1978; Bagatin *et al.*, 1997).

Cleft lip and cleft palate have been much more extensively investigated. Human genetic studies on cleft lip with or without cleft palate (CL/P) show the characteristics of a complex genetic trait, compatible with either a multifactorial threshold trait (i.e. incorporating both genetic and environmental factors) or with an oligogenic cause (Mitchell and Risch, 1992). In a summary of genes associated with human orofacial clefting, Spritz (2001) listed growth factors [transforming growth factor-alpha (*TGFα*), *TGFβ1*, *TGFβ* and *TGFβ3*], transmembrane cell adhesion molecules (Nectin-1, Nectin-2) and transcription factors (*MSX1*, *AP2*) as well as a cell cycle regulator, an enzyme involved in folate metabolism, and endothelin-1. The functional breadth of this list confirms experimental evidence for multiple developmental processes that are vulnerable to perturbation during lip and palate formation. On the basis of previous mouse knock-out studies, Scapoli *et al.* (2002) carried out linkage disequilibrium and linkage analysis studies of five candidate genes. The results suggested a major role for the β3 subunit of the γ-aminobutyric acid receptor (GABRB3) and minor roles for retinoic acid receptor alpha (RARα) and transforming growth factor receptor beta 3 (TGFβ3).

The importance of folic acid and other B-group vitamins for prevention of CL/P has been established in both clinical and experimental studies, confirming the environmental component of these disorders. Tolarová (1987) found a reduction in the recurrence rate of CL/P after periconceptual supplementation with a multivitamin preparation including 10 mg folic acid. Czeizel *et al.* (1999), in a randomized, double-blind controlled trial of periconceptual vitamin supplementation, found the preventative effect to be dose-dependent. Unlike the requirements for reducing neural tube defects, supplementation was more effective at the stage of gestation during the period of facial process development than during the periconceptual period, suggesting a direct requirement by the embryonic tissue involved. Czeizel (2000) suggested that the mechanism might involve restoration of impaired mitosis caused by folate deficiency. Evidence for the specificity of folate deficiency comes from a mouse deficient in folic acid-binding protein-1 (Folbp1), in which a number of genetic markers known to be involved in face and palatal development were altered (Tang and Finnel, 2003). Schubert *et al.* (2002) tested two cleft palate-susceptible mouse strains for the effects of B-group vitamin deficiency and found an increased incidence from 3.8% to 25% in one strain and from 28% to 44% in the other. Other environmental factors associated with increased incidence of CL/P include alcohol, periconceptual cigarette smoking, steroids and anticonvulsants (Carinci *et al.*, 2003).

Cleft palate without cleft lip is genetically separate from CL/P (Dronamraju, 1971). It is morphogenetically heterogeneous, being found as a part of the phenotype of a great many malformation syndromes (Winter and Barraitser, 2000), suggesting that it may be secondary to developmental problems initiated at an earlier stage of facial development. This means that mutations that specifically cause cleft palate are likely to be rare, and the evidence for multifactorial underlying causes is as strong as that for CL/P. However, three genes have been identified that are associated with syndromes in which cleft palate is the predominant feature: *TBX22* (Braybrook *et al.*, 2001), *IRF6* (Kondo *et al.*, 2002), and *SATB2* (FitzPatrick *et al.*, 2003).

Secondary palate development begins when bilateral swellings arise on the medial aspect of the maxillary processes, associated with a high level of local cell proliferation (Burdett *et al.*, 1988). It is not clear why they grow vertically downwards, either side of the tongue, since in the alligator the palatal shelves simply grow horizontally towards the midline, where they fuse (Ferguson, 1981). The most likely explanation is that the shape of the oral cavity, in which the upper surface of the tongue lies against the skull base, imposes constraints on the direction of growth. Nevertheless, in spite of any shape constraints, the shelves subsequently reorientate to a horizontal position. This key event in palatogenesis has been theoretically associated with a number of extrinsic factors, including descent of the tongue, growth of the mandible and growth of the neck (allowing the mandible to descend; Ferguson, 1978). The best understood intrinsic factor is the accumulation of hydrated hyaluronan, which sets up a turgor pressure leading to rapid reorientation of the shelves when extrinsic factors permit movement. Production of hyaluronan is stimulated by Egf, Tgfα and Tgfβ. Ferguson (1978) estimated that delay or failure of shelf elevation accounts for many human cases of isolated cleft palate. It is not clear how a slight delay in shelf elevation could have this result without an accompanying growth defect of the shelves, unless palatal fusion is itself affected by the delay.

Growth deficiency of the palatal shelves, before and/or after elevation, is the most likely cause of cleft palate (Ferguson, 1988). Mesenchymal cell proliferation is affected by composition of the extracellular matrix, so defects in matrix production, composition or turnover may lead to cleft palate through both proliferation defects and elevation delay. Matrix metalloproteins (MMPs) are essential components of the mesenchymal matrix during palatal shelf growth. Their activity is regulated by epidermal growth factor receptor (Egfr) signalling, stimulated by the ligand Tgfα. Functional defects in Egfr are associated with cleft palate in mice, causing poor palatal shelf growth accompanied by greatly reduced mandibular growth (Miettinen *et al.*, 1999). Cleft palate is also associated with *Msx1* deficiency in mice. MSX1 is a transcription factor whose activity promotes growth and inhibits differentiation; deficiency results in defective cell proliferation through an inability to upregulate cyclin D1, leading to premature cessation of growth and differentiation and hence to palatal shelves of reduced size; the defect can be rescued by insertion of transgenic human *BMP4* in the mouse *Msx1* promoter (Zhang *et al.*, 2002a). *Pax9* is another transcription factor required for cell proliferation in the developing palate. It is regulated by Fgf8-induced Fgfr signalling. *Pax9* expression is downregulated as the shelves fuse, suggesting a possible mechanism for the reduced cell proliferation that is observed at this time (Hamachi *et al.*, 2003). FGF8 also induces *Lhx7* expression; deficiency of this gene is associated with cleft palate without any other defect (Zhang *et al.*, 2002b).

Defects in TGFβ signalling have been associated with failure of palatal shelf fusion (see below), but Ito *et al.* (2003) observed reduced cell proliferation as a cause of cleft palate in mice with conditional inactivation of the gene encoding a Tgfβ receptor, TGFBR2, specifically in neural crest cells. Neural crest cell migration was normal, and the abnormal cell proliferation of the postmigratory cells occurred specifically in the palate and in the dura mater, associated with downregulation of cyclin D1. Cyclin D1 downregulation cannot by itself be the cause of cleft palate in these or any other genetic defect with which it is associated, since cyclin D1-null mutant mice do not have cleft palate (Fantl *et al.*, 1995).

When the medial edges of the two horizontal palatal shelves make contact in the midline, the adjacent medial edge epithelia (MEE) fuse and then break down. Fusion involves an increase in sticky cell surface glycoproteins; cell adhesion molecules are also involved as fusion proceeds, leading to the formation of specialized junctions (desmosomes) between the apposing epithelia (Bittencourt and Bolognese, 2000). The interaction between the two MEE is specific – they will not fuse with the tongue. Palatal fusion does not occur in Tgfβ3-null mutant mice (Proetzel *et al.*, 1995; Kaartinen *et al.*, 1995). In this genetic defect, cleft palate is correlated with failure of formation of filopodia on the MEE cell surfaces; addition of recombinant Tgfβ3 to cultured Tgfβ3-null palates rescued filopodia formation (Taya *et al.*, 1999). These cell surface specializations increase surface area and are coated with a cell surface-associated glycocalyx that may be essential for adhesion of the two epithelial surfaces. Cell labelling studies have demonstrated that breakdown of the epithelial seam after fusion involves a combination of epithelial–mesenchymal transformation and apoptosis (Ferguson, 1988). Although failure of this process may not seem very significant for palate formation, it is important to bear in mind that without epithelial breakdown the two bony palatal processes would be unable to make contact and form a sutural growth centre.

Craniosynostosis

Because genes and their encoded proteins act in pathways or networks, mutations in two or more different genes may have a similar effect. For example, craniosynostosis involving only the coronal suture has two syndromic forms that were only distinguished from each other unambiguously when genetic analysis of affected patients revealed two underlying genetic defects. One of these is Saethre-Chotzen syndrome, which is due to heterozygous deletions or intragenic mutations of the gene encoding the transcription factor TWIST, abolishing its ability to bind to DNA (El Ghouzzi *et al.*, 2001). Muenke syndrome has a similar phenotype but is caused by the Pro250Arg mutation of *FGFR3*, which encodes fibroblast growth factor receptor type 3, a cell surface signalling molecule (Bellus *et al.*, 1996). The similarity of phenotype is likely to be due to the TWIST protein acting upstream of *FGFR* genes (Rice *et al.*, 2000).

Analysis of *FGFR* gene function and mutations has revealed that three FGFR proteins are involved in development and growth of the skeleton: FGFR1, FGFR2 and FGFR3. Each of these has two splice variants, 'b' and 'c', of which the 'c' form is the more important for skeletogenesis. Mouse mutants with loss of function (knock-out) of the two isoforms of Fgfr2 have been constructed. The *Fgfr2b* mutant lacks limbs and has multiple abnormalities of organs that form by branching morphogenesis (De Moerlooze *et al.*, 2000; Revest *et al.*, 2001). In contrast, the *Fgfr2c* mutant is viable and is of relatively normal appearance but reduced size; ossification is delayed and growth is slow, but the coronal suture and the skull base synchondroses of the occipital bone begin to synostose during late fetal stages (Eswarakumar *et al.*, 2002).

The three FGFR proteins and some of the mutations associated with craniostosis and dwarfism are shown diagrammatically in Figure 12.13. The three FGFRc splice

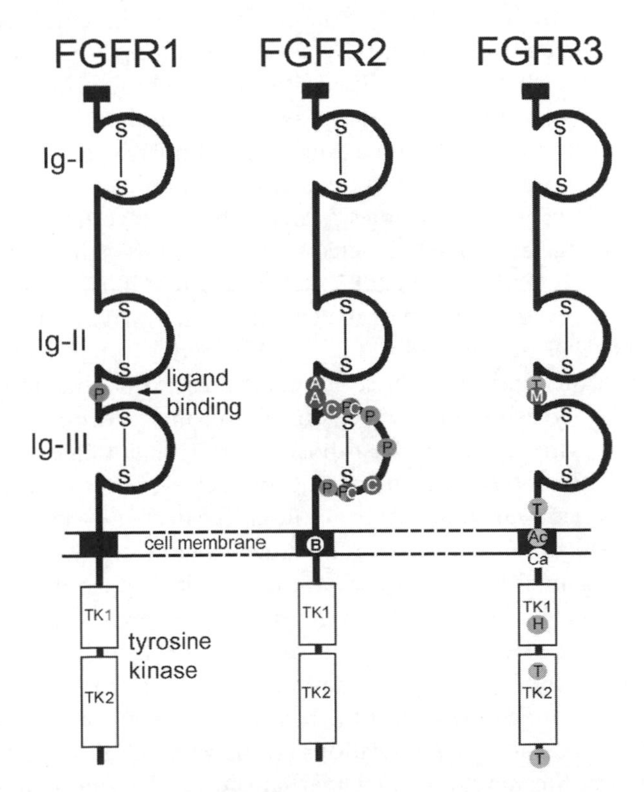

Figure 12.13 Diagrammatic representation of the three FGFR proteins, indicating some positions of the mutations associated with craniosynostosis and dwarfism syndromes. Three different craniosynostosis syndromes are associated with amino acid change at the equivalent position of each protein. Craniosynostosis mutations are commonest in *FGFR2*, particularly in the immunoglobulin-like (Ig)-IIIa/c domain. A, Apert syndrome; Ac, achondroplasia; B, Beare–Stevenson; C, Crouzon; Ca, Crouzon syndrome with acanthosis nigricans; M, Muenke; P, Pfeiffer, T, thanatophoric dysplasia type 1. See text for details and references

variants have distinct but overlapping functions in skeletal development and growth. Their distinct nature is illustrated by the different effects of an equivalent activating mutation of all three genes, resulting in a proline-to-arginine change at the same position within the IgII–IgIII linker domain in each of the three proteins. The Pro252Arg mutation of *FGFR1* causes a mild form of Pfeiffer syndrome, characterized by craniosynostosis together with broad first digits on both hands and feet; the Pro253Arg mutation of *FGFR2* causes the severe craniosynostosis phenotype, Apert syndrome, which is accompanied by bony syndactyly of all four limbs; the Pro250Arg mutation of *FGFR3* causes Muenke-type coronal synostosis, described above.

Different activating mutations in the same gene reveal the range of activities of the protein. Muenke-type coronal synostosis and a rare form of Crouzon syndrome are the only craniosynostosis syndromes to have been identified as an *FGFR3* mutation; this gene is more commonly associated with three forms of dwarfism: hypochondroplasia, achondroplasia and thanatophoric dysplasia. The mildest of these, hypochondroplasia, involves premature fusion of the epiphyses of the long bones with no effects on the craniofacial skeleton. Achondroplasia affects growth of the endochondral parts of the skull in addition to the long bones. Thanatophoric dysplasia affects both intramembranous and endochondral parts of the skull, as well as having the most severe effects on the long bones. Similarly, mutations in different parts of the *FGFR2* gene cause four craniosynostosis syndromes, Crouzon, Pfeiffer, Apert and Beare–Stevenson syndromes, which differ in severity of the craniofacial phenotype and in their associated defects of the limbs, skin and other organs (Reardon *et al.*, 1994; Wilkie *et al.*, 1995; Muenke and Wilkie, 2001). The majority of the mutations are located in the exons encoding the Ig-IIIa and Ig-IIIc domains, but in a screen of 259 patients Kan *et al.* (2002) found mutations associated with Pfeiffer and Crouzon syndromes in seven additional exons, including six distinct mutations in the tyrosine kinase region. Crouzon syndrome is occasionally associated with a skin condition, acanthosis nigricans, but these patients have a mutation of *FGFR3*, not *FGFR2*, providing further evidence of the similarity of function and/or functional cooperation of the different FGFR-signalling molecules.

These correlations have been made from clinical studies alone. To understand the mechanisms linking an altered genotype with the resulting phenotype, it is necessary to analyse the normal and altered developmental processes in embryos. Mouse models have been extensively used for this purpose and have been constructed, for example, for the Pro250Arg Pfeiffer mutation of *Fgfr1* (Zhou *et al.*, 2000) and the Cys342Tyr mutation of *Fgfr2*, which is the commonest Crouzon syndrome mutation (Eswarakumar *et al.*, 2004). The relationship between clinical and experimental studies in revealing the developmental mechanisms underlying craniosynostosis is explored in more detail by Morriss-Kay and Wilkie (2005). Gene expression studies in mouse embryos and fetuses have revealed a close relationship between gene expression domains of *Twist, Fgfr1, Fgfr2* and *Fgfr3* in both the coronal and sagittal sutures (Iseki *et al.*, 1997, 1999; Kim *et al.*, 1998; Johnson *et al.*, 2000; Rice *et al.*, 2000). In the coronal suture, *Fgfr2* is expressed in proliferating preosteoblasts, suggesting that signalling through Fgfr2 results in cell proliferation; Fgfr1 is expressed in differentiating osteoblasts but is downregulated when differentiation is complete, suggesting a role for Fgfr1 signalling in the differentiation process (Iseki *et al.*, 1999). Experimental studies, using an Fgf-soaked bead implanted onto the early coronal suture (Iseki *et al.*, 1997, 1999), have revealed the distinct but cooperative functions of Fgfr1 and Fgfr2 signalling. Stimulation of signalling by Fgf mimics the effects of an activating mutation. *Fgfr2* is downregulated and the cells leave the cell cycle; *Fgfr1* and *Fgfr3* are upregulated in these cells, which begin to express bone differentiation markers, such as *Spp1* and alkaline phosphatase. These observations suggest how mutations in all three receptors have the same outcome, namely craniosynostosis. According to this interpretation, an activation mutation of *FGFR2* leads to loss of cell proliferation and stimulation of osteogenic differentiation; an activating mutation of *FGFR1* or *FGFR3* increases the rate of differentiation, therefore having a similar effect (Figure 12.14). This model also explains how craniosynostosis can occur as a result of both loss- and gain-of-function mutations of *Fgfr2*, since both will affect the ability of sutural cells to proliferate.

The relatively mild form of coronal synostosis caused by the Pro250Arg mutation of *FGFR3* may be explained by the much lower level of expression of this gene than of *FGFR1* and *FGFR2* in the coronal suture. The mechanism of Saethre–Chotzen syndrome is haplo-insufficiency of *TWIST;* the phenotype is less severe in heterozygous *Twist* null mutant mice, but there is growth deficiency in the coronal suture (Bourgeois *et al.*, 1998).

The different phenotypes associated with the other craniosynostosis mutations are thought to be due to the specific nature of functional activation caused by each mutation. The mutations of Crouzon syndrome, which result in an unpaired cysteine residue, enable mutant receptors to cross-link and form activated dimers, so that signalling is not restricted by the availability of ligand (Neilson and Friesel, 1995; Robertson *et al.*, 1998). The amino acid substitutions of the Apert and Pfeiffer mutations cause increased or new FGF-binding affinities (Anderson *et al.*, 1998; Yu *et al.*, 2000, Ibrahimi *et al.*, 2004) or, in rare cases (as well as in a subtype of Pfeiffer syndrome), there is ectopic expression of alternative splice-forms of *FGFR2* (Oldridge *et al.*, 1999; Hajihosseini *et al.*, 2001).

The mechanism of normal sutural closure involves interaction with Tgfβ signalling from the underlying dura mater (Opperman *et al.*, 1997). Bmp4 is expressed in the proliferating sutural cells, where it is opposed by the Bmp antagonist, noggin, which is itself suppressed by Fgf2 and activating *Fgfr* mutations (Warren *et al.*, 2003). However, no mutations in these genes have yet been discovered in craniosynostosis patients.

The most recent syndromic form of craniosynostosis to be genetically understood is craniofrontonasal dysplasia. In a study of 20 families, Twigg *et al.* (2004) found deletions in the gene *EFNB1*, which encodes a cell surface (transmembrane) ligand, ephrin B1. This discovery supports the idea that Eph–ephrin interactions may be involved in formation of the neural crest–mesoderm boundary that forms the coronal suture (Figures 12.3, 12.4). The expression domain of *Efnb1* in mouse embryos makes it a strong candidate for involvement in this mechanism, raising the question of how formation of a boundary leads to initiation of the Twist–Fgfr signalling

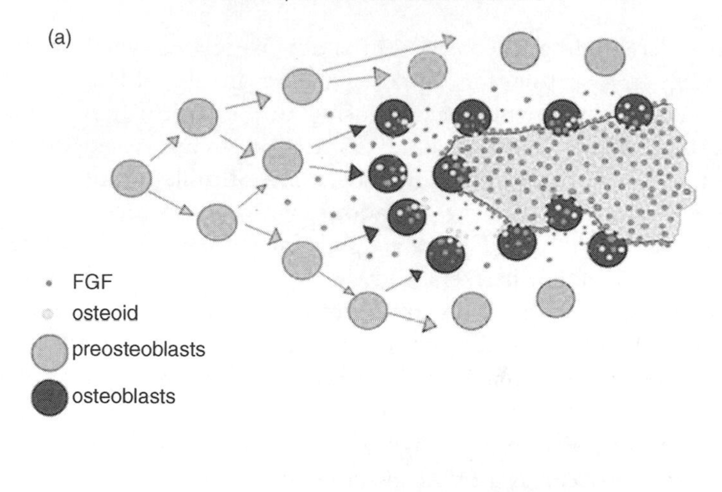

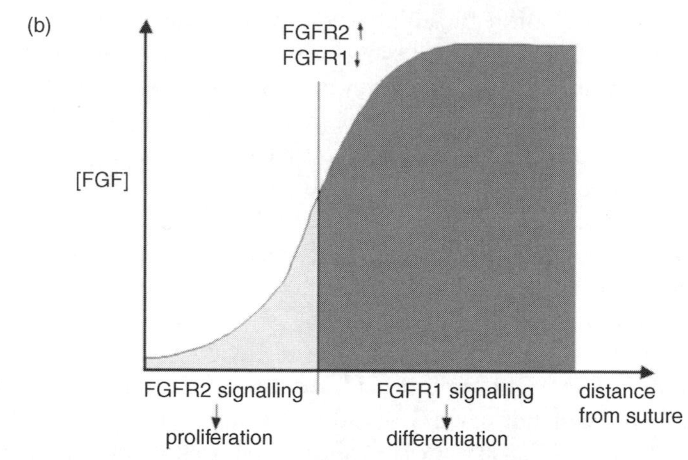

Figure 12.14 Models to explain the observed data on FGFR and FGF expression in relation to cell proliferation and differentiation in the coronal suture. (a) Section through the unmineralized edge of a growing calvarial bone, in which proliferating preosteoblasts (which express FGFR2) are continuously converted to differentiating osteoblasts (which express FGFR1). Osteoblasts secrete (among other products) FGF proteins and osteoid (unmineralized bone matrix). The concentration of FGF in the extracellular matrix is therefore high in the differentiated region and low where it diffuses to surround the proliferating cells. Cells are converted from proliferating to differentiating cells as they become closer to the most recently differentiated FGF-secreting cells. (b) The same information as a graph, indicating that where FGF levels are low, signalling through FGFR2 results in proliferation; where FGF levels are higher, FGFR2 is downregulated, FGFR1 upregulated, cells cease to proliferate and undergo differentiation. A notional threshold of FGF concentration at which this change takes place is suggested

mechanism that governs sutural growth and differentiation. Interactions between ephrin-B1 and Fgfr have been observed in other systems (Chong *et al.*, 2000; Moore *et al.*, 2004) and now need to be followed up for suture formation and function.

Ossification deficiency defects of the skull vault

Ossification deficiency defects of the skull vault, including cleidocranial dysplasia, cranium bifidum and persistent parietal foramina, are characterized by reduced ossification of the skeletogenic membrane. Cleidocranial dysplasia affects mineralization of the membrane bones of the skull vault and the intramembranous component of the clavicle. It is caused by haplo-insufficiency of the 'bone master gene' *RUNX2*, which encodes a transcription factor, CBFA1, required for osteoblast differentiation. Mice homozygous for loss of *Runx2* function show complete absence of the skull vault and failure of ossification of endochondral as well as intramembranous bones.

In persistent parietal foramina, the rate of ossification of the parietal bones is reduced, leading to a large midline defect (an expanded area across the sagittal suture) that may gradually resolve into a small foramen in each parietal bone (Figure 12.11). Loss-of-function mutations (deletions) in two genes, *MSX2* and *ALX4*, have been found in patients with persistent parietal foramina (Wilkie *et al.*, 2000; Wuyts *et al.*, 2000a, 2000b; Mavrogiannis *et al.*, 2001). Both of these genes encode transcription factors that are associated with the intramembranous ossification process. Mice homozygous for *Msx2* and *Alx4* loss of function do not exactly phenocopy the human defects, but show a greater effect on the frontal bones. Nevertheless, the decreased rate of ossification indicates that the mechanism is the same at the cellular level. Gene expression studies indicate that both *Msx2* and *Alx4* act downstream of *Runx2*, but upstream of *Fgfr1* and the bone differentiation markers *Spp1* and alkaline phosphatase (Antonopoulou *et al.*, 2004). Both transcription factors appear to influence the rate of osteoblast differentiation. This function may be considered as the converse of craniosynostosis, in which the balance between proliferation and differentiation is shifted in favour of differentiation. *MSX2* is particularly relevant in this context, since, in addition to loss-of-function mutations causing parietal foramina, an activating mutation causes sagittal synostosis (Boston-type synostosis); the mechanism involves enhanced DNA binding (Jabs *et al.*, 1993; Ma *et al.*, 1996).

In mice, deficiency of dura mater due to retinoic acid-induced neural crest cell deficiency led to reduced ossification of the parietal bone, suggesting that differentiation of this mesodermal bone requires interaction with the underlying neural crest-derived cells, whereas ossification of the neural crest-derived frontal bone is autonomous (Jiang *et al.*, 2002). However, Ito *et al.* (2003) reported failure of ossification of both frontal and parietal bones in the absence of dura mater due to conditional inactivation of *Tgfbr2* in neural crest cells, suggesting that both calvarial bones require Tgfβ signalling from the dura. An alternative possibility is that the frontal and parietal bone could be affected through different mechanisms; the parietal through absence of dura (Jiang *et al.*, 2002), and the frontal through failure of the Bmp signalling that is required for upgrowth from its initial position as a small primordium above the eye (Y. Yoshida, I. Ishikawa, K. Eto and S. Iseki, personal communication), since loss of the Tgfβ IIR may attenuate Bmp as well as Tgfβ signalling (Massague, 1990).

Agenda for the future

The past 12 years have seen tremendous advances in identifying mutations underlying syndromic craniofacial birth defects. This in turn had led to studies of the normal developmental functions of the genes involved. Before the fruits of this work can be applied in the clinic, we need animal models of the human genetic defects to be rigorously analysed to provide further insight into the molecular and morphogenetic basis of the defects. Many knock-out mice are now available for genes known to be important players in craniofacial development, and the next step will be to interbreed them in order to discover how genes and proteins act cooperatively in development. This has already been carried out for some gene pairs, e.g. *Msx1/Msx2* (Ishii *et al.* 2005), *Alx4/Msx2* (Antonopoulou *et al.*, 2004) and *Twist/Msx2* (Ishii *et al.*, 2003). Through these painstaking studies on pairs of genes, we will be able to build up a picture of the context of each gene's function within a network of genes and proteins. These studies must be carried out for each developing system, since it has become clear that molecular interactions and functions are context-specific.

We also need to understand better at what point in development an abnormality is initiated. The finding that the coronal and sagittal sutures form at the neural crest–mesoderm interface at mouse E9.5 (Jiang *et al.*, 2002) points to the need for further research into the mechanisms by which this boundary generates the system of molecular signalling that governs the proliferation–differentiation balance in the sutures. Similarly, cleft lip and palate may originate at an early stage of development of the facial primordia, and be due to minor changes in the timing of localized cell proliferation that affect the initial size of the affected parts. The fact that some defects can be unilateral, such as cleft lip and coronal synostosis, suggests chance elements in development that may reflect right–left differences in the proportions of normal and abnormal copies of the gene that have been synthesized. We know very little about this aspect of morphogenesis.

The importance of folic acid for the prevention of neural tube defects and cleft lip/palate suggests that there may be other simple ways of preventing recurrence of birth defects that have a multifactorial origin, but it is not immediately evident what these might be. More progress is likely to be made in treating defects of known genetic cause. The observation that cleft palate in mice can be rescued when the mutation is known (Taya *et al.*, 1999; Zhang *et al.*, 2002a) suggests possible applications in humans if the defect is diagnosed by ultrasound scanning at an early stage and the mutation identified.

Tissue engineering is an important new area that is applying the new developmental and genetic knowledge. Current experimental and clinical work indicates that improved repair of cranial ossification defects can be obtained by integrating stem cell biology, gene therapy and biopolymers (Chang *et al.*, 2003a, 2003b). Many genetic defects have long-term effects, e.g. craniosynostosis is not simply a problem that can be corrected by postnatal surgery, but an ongoing defect of growth and differentiation affecting skeletal and other systems. If the genetic defect is known, it should be possible to prevent or lessen the ongoing problems by modification of the functional consequences of the mutation. In the case of an activating *FGFR* mutation, for instance, the FGFR signalling overactivity could be inhibited by an antibody or antisense morpholino approach.

Although some of these possibilities have already moved into the clinic, it is essential for more work on potential new therapies to be carried out in the laboratory, using mouse models, tissue and organ culture (e.g. Erfani *et al.*, 2002). Smooth transition from laboratory observation to clinical application requires much better communication between basic scientists and clinicians than exists at present. In particular, trainee craniofacial surgeons should be offered the opportunity to spend properly funded and substantial amounts of time in an appropriate basic science laboratory, and more basic scientists should be welcomed into the clinic.

Acknowledgements

I thank Action Medical Research for supporting my work, and Chad Perlyn and Jonathan Bard for helpful comments on the manuscript.

References

Anderson, J., Burns, H.D. *et al.* (1998) Apert syndrome mutations in fibroblast growth factor receptor 2 exhibit increased affinity for FGF ligand. *Hum. Mol. Genet.* **7**: 1475–1483.

Antonopoulou, I., Mavrogiannis, L.M., Wilkie, A.O.M.W. and Morriss-Kay, G.M. (2004) Alx4 and Msx2 play phenotypically similar and additive roles in skull vault differentiation. *J. Anat.* **205** (in press).

Bagatin, M., Der Sarkissian R. and Larrabee W.F. Jr. (1997) Craniofacial manifestations of the amniotic band sequence. *Otolaryngol. Head Neck Surg.* **116**: 525.

Barrow, J.R. and Capecchi, M.R. (1999) Compensatory defects associated with mutations in *Hoxa1* restore normal palatogenesis to *Hoxa2* mutants. *Development* **126**: 5011–5026.

Bellus, G.A. Gaudenz, K., Zackai, E.H., Clarke, L.A. *et al.* (1996) Identical mutations in three different fibroblast growth factor receptor genes in autosomal dominant craniosynostosis syndromes. *Nature Genet.* **14**: 174–176.

Bhattacharyya, S., Bailey, A.P., Bronner-Fraser, M. and Streit, A. (2004) Bhattacharyya S., Bailey, A.P., Brown-Fraser, M. and Streit, A. (2004) Segregation of lens and olfactory precursors from a common territory: cell sorting and reciprocity of Dlx5 and Pax6 expression. *Dev. Biol.* **271**: 403–414

Bittencourt, M.A. and Bolognese, A.M. (2000) Epithelial alterations of secondary palate formation. *Braz. Dent. J.* **11**: 117–126.

Bourgeois, P., Bolcato-Bellemin, A.L. *et al.* (1998) The variable expressivity and incomplete penetrance of the twist-null heterozygous mouse phenotype resemble those of human Saethre–Chotzen syndrome. *Hum. Mol. Genet.* **7**: 945–957.

Braybrook, C., Doudney, K., Marcano, A.C., Arnason, A. *et al.* (2001) The T-box transcription factor gene *TBX22* is mutated in X-linked cleft palate and ankyloglossia. *Nat. Genet.* **29**: 179–183.

Burdett, D.N., Waterfield, J.D. and Shah, R.M. (1988) Vertical development of the secondary palate in hamster embryos following exposure to 6-mercaptopurine. *Teratology* **37**: 591–597.

Carinci, F. *et al.* (2003) Recent development in orofacial cleft genetics. *J. Craniofac. Surg.* **14**: 130–143.

Chai, Y. (2000) Fate of the mammalian cranial neural crest during tooth and mandibular morphogenesis. *Development* **127**: 1671–1679.

Chang, S.C., Chuangm, H.L., Chen, Y.R., Chen, J.K. *et al.* (2003a) *Ex vivo* gene therapy in autologous bone marrow stromal stem cells for tissue-engineered maxillofacial bone regeneration. *Gene Ther.* **10**: 2013–2019.

Chang, S.C., Wei, F.C., Chuang, H., Chen,Y.R. *et al.* (2003b) *Ex vivo* gene therapy in autologous critical-size craniofacial bone regeneration. *Plast. Reconstr. Surg.* **112**: 1841–1850.

Chong, L.D., Park, E.K., Latimer, E., Friesel, R. and Daar, I.O. (2000) Fibroblast growth factor receptor-mediated rescue of x-ephrin B1-induced cell dissociation in *Xenopus* embryos. *Mol. Cell Biol.* **20**: 724–734.

Cohen, M.M. Jr. (1979) Craniofrontonasal dysplasia. *Birth Defects* **15**(5B): 85–89.

Couly, G., Grapin-Botton, A., Coltey, P., Ruhin, B. and Le Douarin, N.M. (1998) Determination of the identity of the derivatives of the cephalic neural crest: incompatibility between *Hox* gene expression and lower jaw development. *Development* **125**: 3445–3459.

Creuzet, S., Couly, G., Vincent, C. and Le Douarin, N.M. (2002) Negative effect of *Hox* gene expression on the development of the neural crest-derived facial skeleton. *Development* **129**: 4301–4313.

Czeizel, A.E., Timar, L. and Sarkozi, A. (1999) Dose-dependent effect of folic acid on the prevention of orofacial clefts. *Pediatrics* **104**: e66.

Czeizel, A.E. (2000) Primary prevention of neural tube defects and some other major congenital abnormalities: recommendations for the appropriate use of folic acid during pregnancy. *Paediatr. Drugs* **2**: 437–449.

De Moerlooze, L., Spencer-Dene, B., Revest, J., Hajihosseini, M. *et al.* (2000) An important role for the IIIb isoform of fibroblast growth factor receptor 2 (FGFR2) in mesenchymal–epithelial signalling during mouse organogenesis. *Development* **127**: 483–492.

Dixon, J., Hovanes, K., Shiang, R. and Dixon, M. (1997) Sequence analysis, identification of evolutionary conserved motifs and expression analysis of murine *tcof1* provide further evidence for a potential function for the gene and its human homologue, *TCOF1*. *Hum. Mol. Genet.* **6**: 727–737.

Dixon, J., Brakebusch, C., Fässler, R. and Dixon, M. (2000) Increased levels of apoptosis in the prefusion neural folds underlie the craniofacial disorder, Treacher Collins syndrome. *Hum. Mol. Genet.* **9**: 1473–1480.

Dronamraju, K.R. (1971) Genetic studies of a cleft palate clinic population. *Birth Defects Orig. Artic. Ser.* **7**: 54–57.

El Ghouzzi, V., Legeai-Mallet, L., Benoist-Lasselin, C., Lajeunie, E. *et al.* (2001) Mutations in the basic domain and the loop–helix II junction of TWIST abolish DNA binding in Saethre–Chotzen syndrome. *FEBS Lett.* **492**: 112–118.

Erfani, S., Maldonado, T.S., Crisera, C.A., Warren, S.M. *et al.* (2002) Rescue of an *in vitro* palate non-fusion model using interposed embryonic mesenchyme. *Plast. Reconstr. Surg.* **109**: 2363–2372.

Escobar, V., Eastman, J., Weaver, D. and Melnick, M. (1977) Maxillofacial dysostosis. *J. Med. Genet.* **14**: 355–358.

Eswarakumar, V.P., Monsonego-Ornan, E., Pines, M., Antonopoulou, I. *et al.* (2002) The IIIc alternative of Fgfr2 is a positive regulator of bone formation. *Development* **129**: 3783–3793.

Eswarakumar, V.P., Horowitz, M.C., Locklin, R., Morriss-Kay, G.M. and Lonai, P. (2004) A gain-of-function mutation of *Fgfr2c* demonstrates the roles of this receptor variant in osteogenesis. *Proc. Natl Acad. Sci. USA* **101**: 12555–12560.

Fantl, V., Stamp, G., Andrews, A., Rosewell, I. and Dickson, C. (1995) Mice lacking cyclin D1 are small and show defects in eye and mammary gland development. *Genes Dev.* **9**: 2364–2372.

Ferguson, C.A., Tucker, A.S. and Sharpe, P.T. (2000) Temporospatial cell interactions regulating mandibular and maxillary arch patterning. *Development* **127**: 403–412.

Ferguson, M.W.J. (1978) Palatal shelf elevation in the Wistar rat fetus. *J. Anat.* **125**: 555–577.

Ferguson, M.W.J. (1981) The structure and development of the palate in *Alligator mississipiensis*. *Arch. Oral Biol.* **26**: 427–443.

Ferguson, M.W.J. (1988) Palate development. *Development* **103**(suppl): 41–60.

FitzPatrick, D.R., Carr, I.M., McLaren, L., Leek, J.P. *et al.* (2003) Identification of *SATB2* as the cleft palate gene on 2q32-q33. *Hum. Mol. Genet.* **12**: 2491–2501.

Franceschetti, A. and Klein, D. (1949) Mandibulo-facial dysostosis: new hereditary syndrome. *Acta Ophthalmol.* (*Kbh*) **27**: 143–224.

Fukiishi, Y. and Morriss-Kay, G.M. (1992) Migration of cranial neural crest cells to the pharyngeal arches and heart in rat embryos. *Cell Tissue Res* **268**: 1–8.

Gorlin, R.J., Cohen, M.M. and Hennekam, R.C.M. (2001) *Syndromes of the Head and Neck*, 4th edn. Oxford University Press: New York.

Grüneberg, H. (1952) *The Genetics of the Mouse*, 2nd edn. Martinus Nijhoff: The Hague.

Hamachi, T., Sasaki, Y., Hidaka, K. and Nakata, M. (2003) Association between palatal morphogenesis and Pax9 expression in CL/Fr embryos with clefting during palatal development. *Arch. Oral Biol.* **48**: 581–587.

Hajihosseini, M.K., Wilson, S., De Moerlooze, L. and Dickson, C.A. (2001) A splicing switch and gain-of-function mutation in *FgfR2-IIIc* hemizygotes causes Apert–Pfeiffer syndrome-like phenotypes. *Proc. Natl Acad. Sci. USA* **98**: 3855–3860.

Helms, J.A. and Schneider, R.A. (2003) Cranial skeletal biology. *Nature* **423**: 326–331.

Holland, L.Z. and Holland, N.D. (2001) Evolution of neural crest and placodes: *Amphioxus* as a model for the ancestral vertebrate? *J. Anat.* **199**: 85–98.

Ibrahimi, O.A., Zhang, F., Eliseenkova, A.V., Linhardt, R.J. and Mohammadi, M. (2004) Proline to arginine mutations in FGF receptors 1 and 3 result in Pfeiffer and Muenke craniosynostosis syndromes through enhancement of FGF binding affinity. *Hum. Mol. Genet.* **13**: 69–78.

Iseki, S., Wilkie, A.O.M., Heath, J.K., Ishimaru, T. *et al.* (1997) Fgfr2 and osteopontin domains in the developing skull vault are mutually exclusive and can be altered by locally applied FGF2. *Development* **124**: 3375–3384.

Iseki, S., Wilkie, A.O.M. and Morriss-Kay, G.M. (1999) Fgfr1 and Fgfr2 have distinct differentiation- and proliferation-related roles in the developing mouse skull vault. *Development* **126**: 5611–5620.

Ishii, M., Merrill, A.E., Chan, Y.S., Gitelman, I. *et al.* (2003) Msx2 and Twist cooperatively control the development of the neural crest-derived skeletogenic mesenchyme of the murine skull vault. *Development* **130**: 6131–6142.

Ishii, M. Han, J., Yen, H.Y., Sucov, H.M. *et al.* (2005) Combined deficiencies of Msx1 and Msx2 cause impaired patterning and survival of the cranial neural crest. *Development* **132**: 4937–4950.

Ito, Y., Yeo, J.Y., Chytil, A., Han, J. *et al.* (2003) Conditional inactivation of Tgfbr2 in cranial neural crest causes cleft palate and calvaria defects. *Development* **130**: 5269–5280.

Jabs, E.W. *et al.* (1993) A mutation in the homeodomain of the human *MSX2* gene in a family affected with autosomal dominant craniosynostosis. *Cell* **75**: 443–450.

Jiang, X., Rowitch, D.H., Soriano, P., McMahon, A.P. and Sucov, H.M. (2000) Fate of the mammalian cardiac neural crest. *Development* **127**: 1607–1616.

Jiang, X., Iseki, S., Maxson, R.E., Sucov, H.M. and Morriss-Kay, G.M. (2002) Tissue origins and interactions in the mammalian skull vault. *Dev. Biol.* **241**: 106–116.

Johnson, D., Iseki, I., Wilkie, A.O.M. and Morriss-Kay, G.M. (2000) Expression patterns of Twist and Fgfr1, -2 and -3 in the developing mouse coronal suture suggest a key role for Twist in suture initiation and biogenesis. *Mech. Dev.* **91**: 341–345.

Kaartinen, V. and Volcken, J.W. *et al.* (1995) Abnormal lung development and cleft palate in mice lacking TGF-β3 indicates defects of epithelial–mesenchymal interaction. *Nature Genet.* **11**: 425–421.

Kan, S.H., Elanko, N., Johnson, D. *et al.* (2002).Genomic screening of fibroblast growth-factor receptor 2 reveals a wide spectrum of mutations in patients with syndromic craniosynostosis. *Am. J. Hum. Genet.* **70**: 472–486.

Kelberman, D. *et al.* (2000) Mapping of a locus for autosomal dominant hemifacial microsomia. *J. Med. Genet.* **37**: S76.

Keller, H., Neuhauser, G., Durkin-Stamm, M.V., Kaveggia, E.G. *et al.* (1978) ADAM complex (amniotic deformity, adhesions, mutilations) – a pattern of craniofacial and limb defects. *Am. J. Med. Genet.* **2**: 81–98.

Kim, H.J., Rice, D.P., Kettunen, P.J. and Thesleff, I. (1998) FGF-, BMP- and Shh-mediated signalling pathways in the regulation of cranial suture morphogenesis and calvarial bone development. *Development* **125**: 1241–1251.

Kirby M.L. and Stewart, D.E. (1983) Neural crest origin of cardiac ganglion cells in the chick embryo: identification and extirpation. *Dev. Biol.* **97**: 433–443.

Kondo, S., Schutte, B.C., Richardson, R.J., Bjork, B.C. *et al.* (2002). Mutations in *IRF6* cause Van der Woude and popliteal pterygium syndromes. *Nat. Genet.* **32**: 285–289.

Köntges, G. and Lumsden, A. (1996) Rhombencephalic neural crest segmentation is preserved throughout craniofacial ontogeny. *Development* **122**: 3229–3242.

Krumlauf R. (1993) *Hox* genes and pattern formation in the branchial region of the vertebrate head. *Trends Genet.* **9**: 106–112.

Lammer, E.J. and Opitz, J.M. (1986) The DiGeorge anomaly as a developmental field effect. *Am. J. Med. Genet.* (*Suppl*) **2**: 113–127.

Le Douarin, N.M. and Kalcheim, C. (1999) *The Neural Crest*, 2nd edn. Cambridge University Press: Cambridge, UK.

Ma, L., Golden, S., Wu, L. and Maxson, R. (1996) The molecular basis of Boston-type craniosynostosis: the *Pro148-His* mutation in the N-terminal arm of the *MSX2* homeodomain stabilizes DNA binding without altering nucleotide sequence preferences. *Hum. Mol. Genet.* **5**: 1915–1920.

Massague, J. (1990) The transforming growth factor-β family. *Annu. Rev. Cell Biol.* **6**: 597–641.

Mavrogiannis, L.A., Antonopoulou, I., Baxova, A. *et al.* (2001) Haplo-insufficiency of the human homeobox gene *ALX4* causes skull ossification defects. *Nat. Genet.* **27**: 17–18.

Melnick, M. and Eastman, J.R. (1977) Autosomal dominant maxillofacial dysostosis. *Birth Defects* **13**(3B): 39–44.

Miettinen, P.J., Chin, J.R., Shum, L. *et al.* (1999) Epidermal growth factor receptor function is necessary for normal craniofacial development and palate closure. *Nat. Genet.* **22**: 69–73.

Mitchell, L.E. and Risch, N. (1992) Mode of inheritance of nonsyndromic cleft lip with or without cleft palate: a reanalysis. *Am. J. Hum. Genet.* **51**: 323–332.

Moore, K.B., Mood, K., Daar, I.O. and Moody, S.A. (2004) Morphogenetic movements underlying eye field formation require interactions between the FGF and ephrinB1 signalling pathways. *Dev. Cell* **6**: 55–67.

Morris, J.F. and MacKinnon, P.C.B. (1990) *Oxford Textbook of Functional Anatomy*, vol 3. Oxford University Press: Oxford.

Morriss-Kay, G.M. (1981) Growth and development of pattern in the cranial neural epithelium of rat embryos during neurulation. *J. Embryol. Exp. Morphol.* **65** (suppl): 225–241.

Morriss-Kay, G.M. and Tuckett, F. (1985) The role of microfilaments in cranial neurulation in rat embryos: effects of short-term exposure to cytochalasin D. *J. Embryol. Exp. Morphol.* **88**: 333–348.

Morriss-Kay, GM (1990) Development of the head and neck. In *Oxford Textbook of Functional Anatomy*, vol 3, Morris, J.F. and MacKinnon, P.C.B. (eds). Oxford University Press: Oxford; 17–23.

Morriss-Kay, G.M. (1993) Retinoic acid and craniofacial development: molecules and morphogenesis. *BioEssays* **15**: 9–15.

Morriss-Kay, G.M. (2001) Derivation of the mammalian skull vault. *J. Anat.* **199**: 143–151.

Morriss-Kay, G.M. and Wilkie, A.O.M. (2005) Growth of the normal skull vault and its alteration in craniosynostosis: insights from human genetics and experimental studies. *J. Anat.* **207**: 637–654.

Muenke, M. and Wilkie, A.O.M. (2001) In *The Metabolic and Molecular Bases of Inherited Disease*, 8th edn, Scriver, C.R. *et al.* (eds). McGraw-Hill: New York; 6117–6146.

Naora, H. *et al.* (1994) Transgenic mouse model of hemifacial microsomia: cloning and characterization of insertional mutation region on chromosome 10. *Genomics* **23**: 515–519.

New, D.A.T. (1978) Whole-embryo culture and the study of mammalian embryos during organogenesis. *Biol. Rev.* **53**: 81–122.

Neilson, K.M. and Friesel, R.E. (1995) Constitutive activation of fibroblast growth factor receptor-2 by a point mutation associated with Crouzon syndrome. *J. Biol. Chem.* **270**: 26037–26040.

Noden, D.M. (1983) The role of the neural crest in patterning of avian cranial skeletal, connective, and muscle tissues. *Dev. Biol.* **96**: 144–165.

Oldridge, M., McDonald-McGinn, D.M. *et al.* (1999) *De novo Alu* element insertions in *FGFR2* identify a distinct pathological basis for Apert syndrome. *Am. J. Hum. Genet.* **64**: 446–461.

Opperman, L.A., Nolen, A.A. and Ogle, R.C. (1997) TGF-β1, TGF-β2 and TGF-β3 exhibit distinct patterns of expression during cranial suture formation and obliteration *in vivo* and *in vitro*. *J. Bone Min. Res.* **12**: 301–310.

Osumi-Yamashita, N., Ninomiya, Y., Doi, H. and Eto, K. (1994) The contribution of both forebrain and midbrain crest cells to the mesenchyme in the frontonasal mass of mouse embryos. *Dev. Biol.* **164**: 409–419.

Osumi-Yamashita, N., Ninomiya, Y. and Eto, K. (1997) Mammalian craniofacial embryology *in vitro*. *Int. J. Dev. Biol.* **41**: 187–194.

Poswillo, D. (1973) The pathogenesis of the first and second branchial arch syndrome. *Oral Surg. Oral Med. Oral Pathol.* **35**: 302–328.

Proetzel, G., Pawlowski, S.A. *et al.* (1995). Transforming growth factor-β3 is required for secondary palate fusion. *Nat. Genet.* **11**: 409–414.

Reardon, W. *et al.* (1994) Mutations in the fibroblast growth factor receptor 2 gene cause Crouzon syndrome. *Nat. Genet.* **14**: 174–176.

Revest, J.M., Spencer-Dene, B., Kerr, K., De Moerlooze, L. *et al.* (2001) Fibroblast growth factor receptor 2-IIIb acts upstream of Shh and Fgf4 and is required for limb bud maintenance but not for the induction of Fgf8, Fgf10, Msx1, or Bmp4. *Dev. Biol.* **231**: 47–62.

Rice, D.P.C., Åberg, T., Chan, Y.-S., Tang, Z. *et al.* (2000) Integration of FGF and TWIST in calvarial bone and suture development. *Development* **127**: 1845–1855.

Robertson, S.T. *et al.* (1998) Activating mutations in the extracellular domain of the fibroblast growth factor receptor 2 function by disruption of the disulphide bond in the third immunoglobulin-like domain. *Proc. Natl Acad. Sci. USA* **95**: 4567–4572.

Ruberte E, Wood H.B. and Morriss-Kay G.M. (1997) Prorhombomeric subdivision of the-mammalian embryonic hindbrain: is it functionally meaningful? *Int. J. Dev. Biol.* **41**: 213–222.

Santagati, F. and Rijli, F.M. (2003) Cranial neural crest and the building of the vertebrate head. *Nat. Rev. Neurosci.* **4**: 806–818.

Scapoli, L., Martinelli, M., Pezzetti, F., Carinci, F. *et al.* (2002) Linkage disequilibrium between *GABRB3* gene and nonsyndromic familial cleft lip with or without cleft palate. *Hum. Genet.* **110**: 15–20.

Schubert, J., Schmidt R. and Syska, E. (2002) B group vitamins and cleft lip and cleft palate. *Int. J. Oral Maxillofac. Surg.* **31**: 410–413.

Sedano, H.O. and Gorlin, R.J. (1988) Frontonasal malformation as a field defect and in syndromal associations. *Oral Surg.* **65**: 704–710.

Serbedzija, G.N., Bronner-Fraser, M. and Fraser, S.E. (1992) Vital dye analysis of cranial neural crest cell migration in the mouse embryo. *Development* **116**: 297–307.

Spritz, R.A. (2001) The genetics and epigenetics of orofacial clefts. *Curr. Opin. Pediatr.* **13**: 556–560.

Tan, S.-S. and Morriss-Kay, G.M. (1985) The development and distribution of the cranial neural crest in the rat embryo. *Cell Tissue Res.* **240**: 403–416.

Tan, S.-S., and Morriss-Kay, G.M. (1986) Analysis of cranial neural crest cell migration and early fates in postimplantation rat chimaeras. *J. Embryol. Exp. Morph.* **98**: 21–58.

Tang, L.S. and Finnell, R.H. (2003) Neural and orofacial defects in Folp1 knock-out mice [corrected]. *Birth Defects Res. A Clin. Mol. Teratol.* **67**: 209–218 [erratum in: *Birth Defects Res. A Clin. Mol. Teratol.* **67**: 473].

Taya, Y., O'Kane, S. and Ferguson, M.W.J. (1999) Pathogenesis of cleft palate in TGF-β3 knock-out mice. *Development* **126**: 3869–3879.

Thomas, R.A., Landing, B.H., Wells, T.R. (1987) Embryologic and other developmental considerations of thirty-eight possible variants of the DiGeorge anomaly. *Am. J. Med. Genet. Suppl.* **3**: 43–66.

Tolarová, M. (1987) Orofacial clefts in Czechoslovakia. Incidence, genetics and prevention of cleft lip and palate over a 19-year period. *Scand. J. Plast. Reconstr. Surg. Hand Surg.* **21**: 19–25.

Trainor, P.A. and Krumlauf, R. (2000) Patterning the cranial neural crest: hindbrain segmentation and *Hox* gene plasticity. *Nat. Rev. Neurosci.* **1**: 116–124.

Treacher Collins, E. (1960) Cases with symmetrical congenital notches in the outer part of each lid and defective development of the malar bones. *Trans. Ophthalmol. Soc. UK* **20**: 190–192.

Treacher Collins Syndrome Collaborative Group (1996) Positional cloning of a gene involved in the pathogenesis of Treacher Collins syndrome. *Nat. Genet.* **12**: 130–136.

Twigg, S., Kan, R., Babbs, C. *et al.* (2004) Mutations of ephrin-B1 (*EFNB1*), a marker of tissue boundary formation, cause craniofrontonasal syndrome. *Proc. Natl Acad. Sci. USA* **101**: 8652–8657.

Warren, S.M., Brunet, L.J., Harland, R.M., Economides, A.N. and Longaker, M.T. (2003) The BMP antagonist noggin regulates cranial suture fusion. *Nature* **422**: 625–629.

Wilkie, A.O.M., Slaney, S.F., Oldridge, M. *et al.* (1995) Apert syndrome results from localized mutations of *FGFR2* and is allelic with Crouzon syndrome. *Nat. Genet.* **9**: 165–172.

Wilkie, A.O.M. (1997) Craniosynostosis: genes and mechanisms. *Hum. Mol. Genet.* **6**: 1647–1656.

Wilkie A.O.M. *et al.* (2000) Functional haplo-insufficiency of the human homeobox gene *MSX2* causes defects in skull ossification. *Nat. Genet.* **24**: 387–390.

Wilkie, A.O.M. and Morriss-Kay, G.M. (2001) Genetics of craniofacial development and malformation. *Nat. Rev. Genet.* **2**: 458–468.

Winter, R.M. and Barraitser, M. (2000). *London Dysmorphology Database.* Oxford University Press: Oxford.

Wuyts, W. *et al.* (2000a) Identification of mutations in the *MSX2* homeobox gene in families affected with foramina parietalia permagna. *Hum. Mol. Genet.* **9**: 1251–1255.

Wuyts, W. *et al.* (2000b) The *ALX4* homeobox gene is mutated in patients with ossification defects of the skull (foramina parietalia permagna, OMIM 168500). *J. Med. Genet.* **37**: 916–920.

Xu, Q., Mellitzer, G. and Wilkinson, D.G. (2000) Roles of Eph receptors and ephrins in segmental patterning. *Phil. Trans. R. Soc. Lond. B Biol. Sci.* **355**(1399): 993–1002.

Yu, K., Herr, A.B., Waksman, G. and Ornitz, D.M. (2000). Loss of fibroblast growth factor receptor 2 ligand-binding specificity in Apert syndrome. *Proc. Natl Acad. Sci. USA* **97**: 14536–14541.

Zhang, Z., Song, Y., Zhao, X., Fermin, C. and Chen, Y. (2002a) Rescue of cleft palate in *Msx1*-deficient mice by transgenic *Bmp4* reveals a network of BMP and Shh signaling in the regulation of mammalian palatogenesis. *Development* **129**: 4135–4146.

Zhang, Y., Mori, T., Takaki, H. *et al.* (2002b) Comparison of the expression patterns of two *LIM*-homeodomain genes, *Lhx6* and *L3/Lhx8*, in the developing palate. *Orthodont. Craniofac. Res.* **5**: 65–70.

Zhou, Y.X., Xu, X., Chen, L., Li, C. *et al.* (2000) A Pro250Arg substitution in mouse *Fgfr1* causes increased expression of Cbfa1 and premature fusion of calvarial sutures. *Hum. Mol. Genet.* **9**: 2001–2008.

13

The Heart

Deborah Henderson, Mary R. Hutson and **Margaret L. Kirby**

Developmental anatomy

The heart forms from the pair of cardiogenic fields (primary heart fields) in the anterior part of the lateral plate mesoderm (Figure 13.1). Angioblasts appear in clusters that later form vesicles, which join to create a network of channels; these channels enlarge to become two endothelial tubes, which fuse craniocaudally. The primary heart tube is formed when this endothelial tube is invested by myocardium, the cell layer destined to form the heart muscle. The myocardium secretes an expansive extracellular matrix between the endothelial and myocardial cell layers. Thus, the primitive heart tube consists of a homogeneous myocardial layer, several cells thick, and an endocardial layer separated from the myocardium by cardiac jelly (Figure 13.2a; Manasek, 1968). Since the heart begins to beat very early during its morphogenesis, it is important to establish a working arrangement to support the metabolic needs and vascular growth of the embryo while the transformation from the primitive tube into an adult heart with four chambers is taking place. The tube elongates and loops to the right, at the same time pivoting to the right on the anteroposterior axis (Figure 13.2b; Manasek *et al.*, 1972). The first functional segments of the heart are an inflow or descending limb and an outflow or ascending limb. The descending limb will give rise to the atria, atrioventricular canal and left ventricle, while the ascending limb will give rise to the right ventricle and conus. The convexity of the loop, called the bulboventricular or primary fold, demarcates the inflow from the outflow portion of the looped tube (Figure 13.2b). Elongation of the heart tube results not only from expansion of the tissue already in the tube but also from progressive addition of cells to both the outflow pole and, to a lesser extent, the inflow pole (Stalsberg and DeHaan, 1969). The truncus arteriosus (arterial trunks) is the most distal part of the tube invested with myocardium, and is the

Embryos, Genes and Birth Defects, Second Edition Edited by Patrizia Ferretti, Andrew Copp, Cheryll Tickle and Gudrun Moore

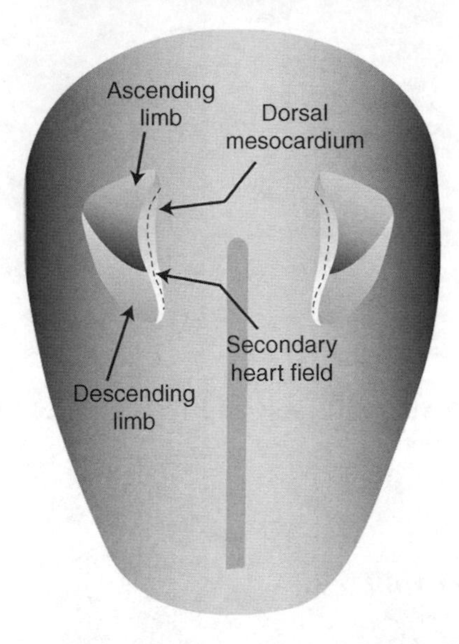

Figure 13.1 Schematic representation, using both chick and mouse data, of the cardiogenic fields at the early gastrula stage. The heart field is formed by continuous migration of the mesodermal cells through the primitive streak to bilateral positions in the anterior lateral plate mesoderm. The cardiogenic region is subdivided into the regions thought to provide cells to the ascending or outflow limb and the descending or inflow limb, which is partitioned from the presumptive secondary heart field region by the presumptive dorsal mesocardial cells

last portion of the heart to be added. Its junction with the aortic sac is the region where the aortic and pulmonary semilunar valves will form (Figure 13.2b).

A series of aortic arch arteries connect the aortic sac bilaterally with the left and right dorsal aortae. The aortic arch arteries traverse and develop from tissue located in the pharyngeal arches (Figure 13.3). In both the atrioventricular and conotruncal (outflow) regions, bulges called cardiac cushions form in the lumen of the heart tube. After looping, the heart continues rearrangement of the inflow and outflow tracts, such that they are aligned correctly with respect to the developing left and right ventricles (Figure 13.4). As the alignment is adjusted, various septation events divide the chambers and outflow vessels. Lengthening of the outflow seems to be required for the proper alignment of the outflow vessels.

While the sinus venosus and venous system are originally bilaterally symmetrical, early regression of specific veins causes a shift of the central venous return to the right side of the primitive atrium. The sinus venosus is incorporated into the nascent right atrium and interatrial septum, which divides the cavity of the primitive atrium into definitive right and left atria. The atrioventricular canal is converted into right and left channels by growth and fusion of endocardial cushions arising from the dorsal and ventral walls of the tube. The primitive ventricle is divided into right and left ventricular chambers by inward growth of a muscular ventricular septum. During the process of septation, the outflow tract (which can be divided into a proximal conus, a

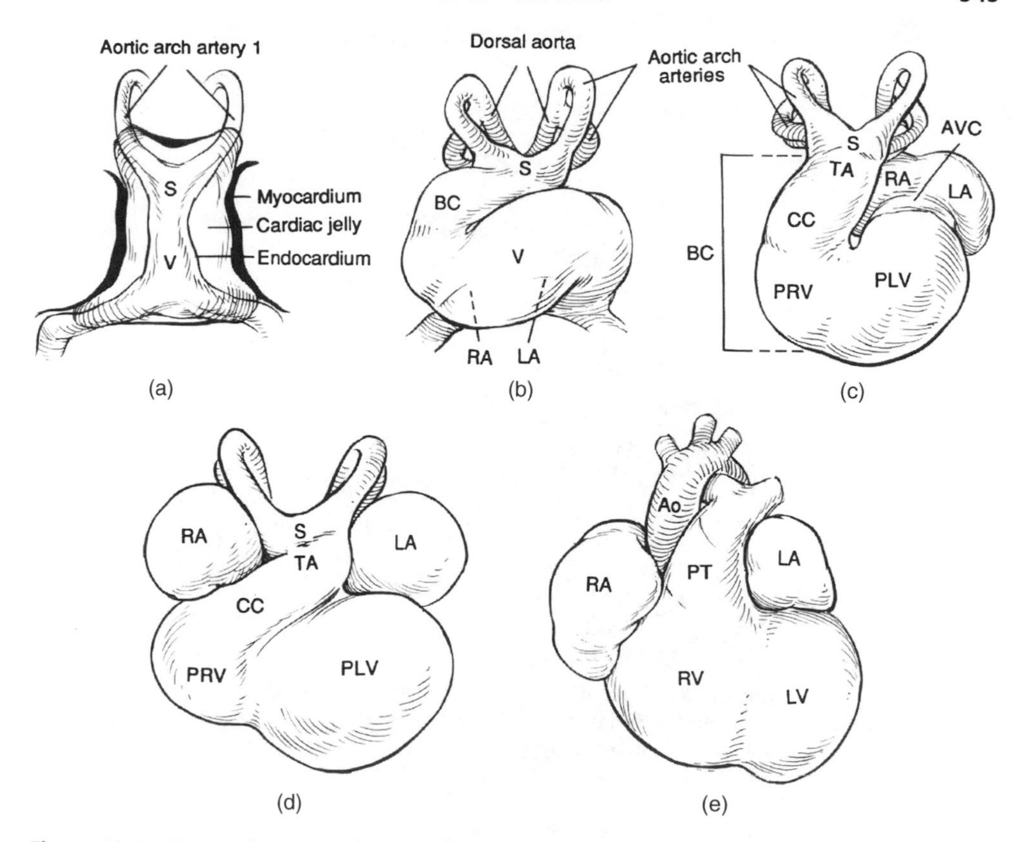

Figure 13.2 Progressive stages in heart development from the initial tubular heart (a) to a fully septated adult-type heart (e). In (a) the forming heart tube (V) connects distally with the aortic sac (S), from which arise a pair of aortic arch arteries that connect with the dorsal aorta. The heart tube has only three layers at early stages of development, which are designated endocardium, myocardium and cardiac jelly. Expansion and looping of the heart tube shown in (b) result in identifiable regions of the tubular heart. The regions that become right and left atria (RA and LA) are shown. (c) The ventricle (V) becomes the presumptive left ventricle, while the ascending limb of the looped tube gives rise to the right ventricle (PRV), conus (CC) and truncus (TA). The regions that form the right and left atria have shifted and absorption of the sinus venosus into the presumptive right atrium contributes to shifting that chamber toward the right, probably by expansion. AVC, atrioventricular canal. (d) The chambers are shown in their correct locations but septation is not complete, as can be seen in (e). The most prominent sign of septation externally is division of the aortic sac into the aorta (Ao) and pulmonary trunk (PT) above the valves, the truncus arteriosus into aortic and pulmonary semilunar valves, and the conus into the infundibulum and vestibule, which are portions of the ventricles just proximal to the semilunar valves

more distal truncus and the aortic sac) will be converted into the parts of the right and left ventricles just below the semilunar valves, the region of the valves, and the proximal parts of the aorta and pulmonary trunk (Figure 13.2c,d,e). Outflow septation begins in the aortic sac by growth of a partition between the fourth and sixth aortic arch arteries. This partition is continued into the distal truncal and

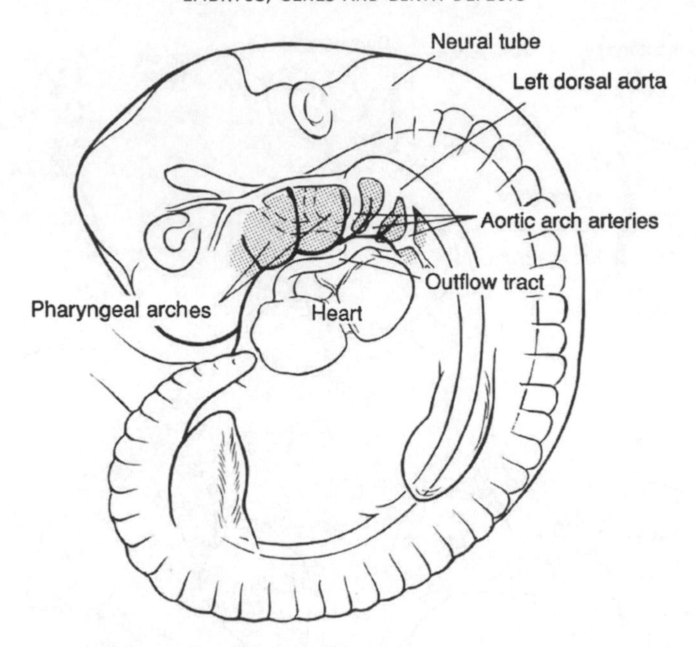

Figure 13.3 Diagrammatic representation of a human embryo viewed from the left side, showing the relationship of the heart with the outflow tract connecting to the dorsal aorta via a series of aortic arch arteries. Aortic arch arteries 1 and 2 located in pharyngeal arches 1 and 2 regress relatively early in development, while the caudal 3 arch arteries persist as the great arteries of the thorax

proximal conal cushions, progressively separating the pulmonary from the systemic circulation. A final fusion of the conal part of the septum with the ventricular septum and atrioventricular cushion tissue completes ventricular septation. The ventricular conduction system develops concomitant with ventricular septation to ensure simultaneous contraction of both ventricles (Chuck *et al.*, 1997; Moorman *et al.*, 1998).

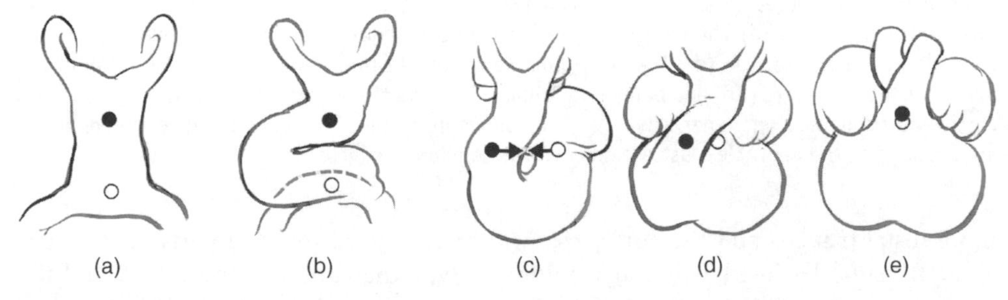

Figure 13.4 Convergence of the ends of the inflow (open circle) and outflow limbs (solid circle) of the heart occurs during looping. After convergence, the inflow and outflow become aligned with the ventricles

Major cell populations needed for heart development

Four major groups of cells are known to be necessary for normal structural development of the heart: myocardium, endocardium, epicardium and neural crest-derived ectomesenchymal cells. The myocardium is derived from lateral plate mesoderm (Figure 13.1). In the axolotl, the heart field forms as the anterior lateral plate mesoderm migrates over the underlying pharyngeal endoderm, and the mid-ventral and lateral walls of the pharyngeal cavity have been shown to have inductive capacity (Easton *et al.*, 1994). Studies in mice suggest that the myocardium of the entire future right ventricle and outflow tract are added to the elongating heart tube after fusion of the primary heart fields from a cardiogenic region of cells termed the anterior heart field (Kelly *et al.*, 2001). Studies in chick have shown that the entire myocardium of the truncus (distal outflow tract) is added during looping stages from a secondary cardiogenic region, located in the ventral pharyngeal mesenchyme caudal to the outflow and termed the secondary heart field (de la Cruz *et al.*, 1977; Figure 13.5). More recently, the secondary heart field has also been shown to contribute smooth muscle cells that form the proximal walls of the aorta and

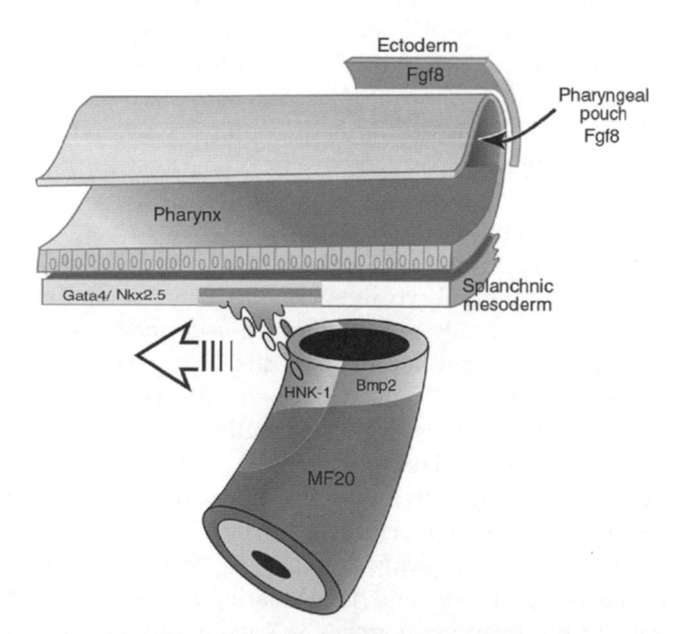

Figure 13.5 Secondary heart field. Diagram illustrating the temporally dynamic location of the secondary heart field. The view is from the right-hand side of the embryo. Rostral is to the right side of the page, caudal to the left. The outflow tract is displaced caudally across this mesenchyme (open arrow). The mesenchyme of the secondary heart field expresses Gata4 and Nkx2.5. As the outflow tract approaches the secondary heart field, Gata4/Nkx2.5-positive cells begin to express HNK1. The lateral walls of the pharynx express FGF8 and the distal outflow myocardium and secondary heart field express BMP2, both of which are thought to be involved in the myocardial induction

pulmonary trunk in the chick embryo (Waldo *et al.*, 2005). Whether similar additions of smooth muscle cells occur in mammalian embryos remains to be established, but seems likely.

While the origin of the myocardium is being more clearly understood, the derivation of the endocardium is unclear. Initial studies have not identified cardiogenic plate cells except in myocardium, so it is possible that atrial and ventricular endocardium are not derived from the cardiogenic plate, although it is more likely that the two lineages have a common progenitor but are separated at the time of gastrulation (Mikawa *et al.*, 1992). The endocardium of the outflow region has been mapped to the cephalic paraxial and lateral plate mesoderm underlying and slightly rostral and lateral to the otic placode (Noden, 1991), and these cells become interspersed with atrial and ventricular endocardium during the initial stages of cardiogenesis. Cell-marking studies have also shown that the ventral midline endoderm of the foregut also gives rise to endocardium (Kirby *et al.*, 2003).

The epicardium grows from mesothelial protrusions from the dorsal mesocardium on the right ventral wall of the sinus venosus (Hiruma and Hirakow, 1989; Ho and Shimada, 1978). The protrusions touch the dorsal wall of the atrioventricular groove, adhere and begin to form a sheet-like epicardium that ultimately invests the entire myocardium, with the exception of the outflow. The epicardium covering the outflow is derived from a pericardial epithelium near the aortic sac (Perez-Pomares *et al.*, 2003). Cells that accompany the epicardium from the liver form the cardiac vascular plexus, which is transformed into the adult coronary vessels (Poelmann *et al.*, 1993). The cardiac plexus extends towards the outflow tract and vessels grow into the aortic wall to form the main coronary arteries (Bogers *et al.*, 1989; Waldo *et al.*, 1990).

The last major contribution to heart development is made by cells derived from the neural crest (Kirby *et al.*, 1983). While the neural crest extends from the mid-diencephalon to the tail of the embryo and participates in craniofacial and peripheral nervous system development, only cells originating from the caudal rhombencephalon participate in structural development of the heart (Figure 13.6). The cells originate from rhombomeres 6, 7 and 8, located between the mid-optic placode and somite 3. The neural crest cells migrate from the neural folds and pause in the circumpharyngeal region while pharyngeal arches 3, 4 and 6 form, and then as each arch forms it is populated by cells migrating from the circumpharyngeal region (Kuratani and Kirby, 1991). These ectomesenchymal cells support development of the aortic arch arteries in the pharyngeal arches and form the tunica media of the persisting arch vessels (Le Liévre and Le Douarin, 1975). A population of cells continues migrating from pharyngeal arches 3, 4 and 6 into the outflow tract, where they will participate in aorticopulmonary septation. Although the cardiac neural crest cells in avians and mouse form only mesenchymal and neural derivatives in the heart, in zebrafish the cardiac neural crest also gives rise to some of the myocardium (Li *et al.*, 2003; Sato and Yost, 2003).

The pharynx is increasingly being implicated as a major participant in heart development. The pharyngeal mesenchyme is the source of the secondary heart field

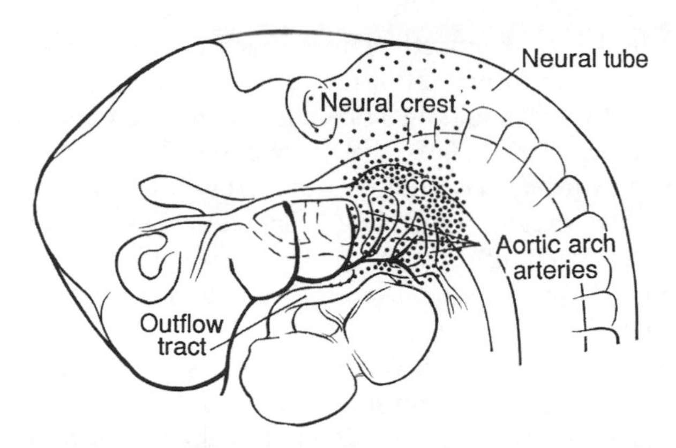

Figure 13.6 Diagrammatic representation of a human embryo showing the migration of neural crest cells into the circumpharyngeal region (CC) and then into the caudal pharyngeal arches, where they will form the tunica media of the arch arteries that give rise to the great arteries. A subpopulation of neural crest cells in the caudal arches migrate into the developing outflow tract of the heart, where they will participate in outflow septation

myocardium and the outflow epicardium. The pharyngeal endoderm and mesenchyme give rise to endocardium. Finally, the cardiac neural crest cells traverse the pharynx on their way to the outflow tract.

Molecular regulation of heart development

In recent years, developments in the field of molecular genetics have dramatically improved our knowledge of cardiogenesis. Despite marked differences between dorsal vessel formation in *Drosophila* and heart tube formation in vertebrates, related genes are involved in controlling these processes, suggesting that the process of 'heart' formation is highly evolutionarily conserved, and that much may be learned by analysing cardiovascular development in simpler organisms. Moreover, the advances in 'knock-out' and related technologies in mice, and more recently in zebrafish, have given us major insights into the molecular mechanisms regulating development of the heart, and the consequences of disruption of these processes.

Cardiac induction and formation of the heart tube

Prospective heart cells can be identified during gastrulation in the anterior part of the primitive streak, and are one of the first cell lineages to be established in the vertebrate embryo. At this stage the cardiac progenitors express CITED2, a transcriptional co-activator, which is thought to be one of the earliest markers of these cells (Schlange *et al.*, 2000), although it is unclear to what extent these cardiac progenitors are prespecified. Growth factors, particularly those of the TGFβ superfamily, have been implicated in this early inductive process as a consequence of their ability to induce cardiac mesoderm formation in the early embryo (Yatskievych *et al.*, 1997) and because their absence results in reductions or even absence of cardiac tissue (Griffin and Kimelman, 2002). After their formation, the pre-cardiac cells migrate laterally to form the cardiogenic fields of mesoderm on either side of the primitive streak. Closure of the foregut brings these bilateral fields to the midline to form a cardiac crescent. Movement of the cardiogenic cells toward the midline is dependent on the action of the transcription factors MesP1 and MesP2 (Saga *et al.*, 1999; Kitajima *et al.*, 2000).

Cardiac induction in the primary heart fields has been a subject of intensive investigation over recent years, but no single factor has been shown to be responsible for the differentiation of lateral plate mesoderm cells into myocardium, which is likely to be a multi-stage induction. A number of factors are produced by the endoderm and have been implicated in inducing the expression of myocardial-specific genes in the cardiac mesoderm. These include bone morphogenetic proteins (BMPs) and fibroblast growth factors (FGFs). Fibroblast growth factors, including FGF4 and FGF8, are expressed in the anterior endoderm (Zhu *et al.*, 1996; Alsan and Schultheiss, 2002) and are capable of inducing the expression of myocardial-specific markers in combination with BMP2 (Lough *et al.*, 1996, Barron *et al.*, 2000). Further evidence for a role for BMPs in cardiac induction comes from the finding that noggin, an inhibitor of BMP activity, blocks cardiac mesoderm formation when it is applied to the lateral plate mesoderm in chick embryos (Schultheiss *et al.*, 1997; Andree *et al.*, 1998). Synergy between FGF and BMP signalling therefore appears to be essential for myocardial differentiation in the primary heart fields. Other factors are thought to be inhibitory to myocardial differentiation. Wnt1 is produced by the neural ectoderm in chick embryos and is co-expressed with BMP2 in the lateral plate mesoderm. Ectopic expression of Wnt1 in the heart-forming fields inhibits myocardial differentiation and promotes blood formation (Marvin *et al.*, 2001). Moreover, antagonizing canonical Wnt signalling induces myocardial formation in posterior mesoderm, where it would not normally appear (Tzahor and Lassar, 2001). In mice, blocking Wnt signalling by ablation of the downstream mediator β-catenin results in multiple hearts being formed at the expense of endoderm (Liao *et al.*, 2002), further supporting the idea that regulated Wnt signalling is essential for proper designation of the heart-forming fields, perhaps as early as the endoderm–mesoderm lineage decision. Interestingly, Wnt11, which is expressed in the mesoderm at the posterior edge of the heart-forming field and which participates in a separate Wnt-activated pathway, can induce cardiogenesis in chick and *Xenopus* embryos. Wnt11 also induces myocardial-specific genes in a mouse embryonal carcinoma cell line (Eisenberg and Eisenberg, 1999; Pandur *et al.*, 2002). Complex interactions of positive and negative factors therefore appear to be involved in induction of the myocardial lineage in the heart-forming fields. Figure 13.7 gives an overview of the major factors thought to be involved in myocardial induction, and their spatial relationship to one another and to the forming cardiac mesoderm.

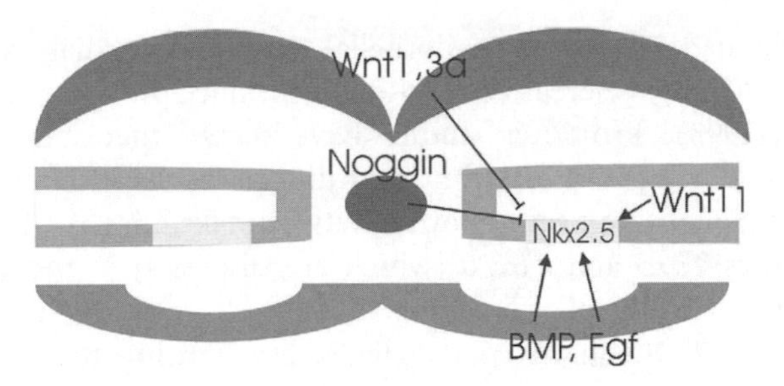

Figure 13.7 Spatial relationship between major factors regulating formation of the cardiac mesoderm. Synergy between BMP and FGF factors, from the pharyngeal endoderm (blue), induces formation of the (Nkx2.5-positive) cardiac mesoderm (yellow). Non-canonical Wnt signalling (Wnt11) from the posterior edge of the heart-forming field is also involved in the inductive process. In contrast, canonical Wnt signalling, from the neural tube (red), is inhibitory, and noggin, secreted by the notochord (purple), restricts the inductive effects of BMP signalling

Cardiac mesoderm formation is characterized by the expression of a number of cardiac-specific transcription factors, such as Nkx2.5, myocardin and GATA factors. The *Nkx2.5* gene is a vertebrate homologue of the *Drosophila* tinman gene, which is necessary, although not sufficient, for dorsal vessel formation in the fruitfly (Bodmer, 1993). Nkx2.5 is one of the earliest markers of the myogenic precursors in vertebrates, although it does not appear to be essential for heart formation and/or myocardial specification. Instead, it appears to be required for differentiation and morphogenesis of the developing heart, playing a crucial role in the development of the left ventricle (Lyons *et al.*, 1995; Yamagishi *et al.*, 2001). As several other *Nkx* genes are expressed in the early heart, it may be that functional redundancy between family members may compensate, in part, for loss of Nkx2.5, although it is more likely that myocardial transcription is directed by large multimeric complexes and that absence of a single member decreases transcriptional efficiency but does not block it entirely (Schwartz and Olson, 1999). Myocardin is another cardiac-specific transcription factor that is thought to be important for the early differentiation of myocardial cells, associating with serum response factor (SRF) to activate cardiac muscle-specific promoters (Wang *et al.*, 2001). SRF is regulated by a divergent homeodomain protein, Hop, which physically interacts with SRF to prevent binding of the complex to DNA. Both Hop and myocardin are regulated by Nkx2.5 (Chen *et al.*, 2002; Ueyama *et al.*, 2003), and similarly, the Nkx2.5 promoter has been shown to be activated by myocardin (Wang *et al.*, 2001). Complex feedback mechanisms are likely, therefore, to be involved in the regulation of myocardial development. GATA factors have also been shown to be important for early heart formation. In *Drosophila* a single *GATA* gene, called *pannier*, acts with *tinman* to induce cardiac-specific gene expression (Gajewski *et al.*, 1999). Three *GATA* genes (*GATA4–6*) are expressed in vertebrate hearts. *GATA4*-deficient mice have bilateral heart tubes (cardia bifida) and reduced numbers of cardiomyocytes, although the primary defect in these mice is thought to lie in the endoderm (Kuo *et al.*, 1997; Molkentin *et al.*, 1997). *GATA5*-deficient mice are viable but have specific deficiencies in endothelial and endocardial cell development (Molkentin *et al.*, 2000b), whereas *GATA6*-deficient mice die soon after implantation (Morrisey *et al.*, 1998). Promoter studies have shown that there is a regulatory network of factors during early vertebrate cardiogenesis, as in *Drosophila*. Mutually reinforcing interactions have been reported between *Nkx2.5*, *GATA* factors, *SRF* and other genes, such as *Tbx5* and *Tbx20* (which are discussed in more detail later) in *Drosophila* and mice (Molkentin *et al.*, 2000a; Gajewski *et al.*, 2001; Garg *et al.*, 2003; Sepulveda *et al.*, 2002; Stennard *et al.*, 2003), highlighting the complexity of the system and showing its conservation through evolution.

The secondary heart field cells are labelled by Nkx2.5, GATA4 and Nkx2.8 (Waldo *et al.*, 2001), as well as by β-galactosidase, in a transgenic line with an enhancer-trap of the *FGF10* gene (Kelly *et al.*, 2001). The secondary heart field cells therefore express a panel of genes similar to that of the primary heart field. The factors involved in inducing the secondary heart field are currently unknown, although it has been suggested that the presence of high levels of BMP antagonists in this location prevent the differentiation of the secondary heart field cells into cardiac cells at the time of primary heart field formation (Tzahor and Lassar, 2001). Further, it has been shown that BMP2 is likely to induce myocardial differentiation in the chick secondary heart field (See Figure 13.5 for an overview of factors in the secondary heart field; Waldo *et al.*, 2001).

Much less is known about the induction of the endocardial lineage in the heart-forming fields. Analysis of the zebrafish mutants faust (encoding the *GATA5* gene) and cloche (mutated gene currently unknown) has shown that vascular and endocardial endothelial cells have distinct origins (Reiter *et al.*, 1999; Liao *et al.*, 1998), as both mutants lack endocardial but not vascular endothelial cells. Similarly, the two lineages can be separated at the molecular level in mice (de la Pompa *et al.*, 1998; Puri *et al.*, 1999).

Formation of the ventral midline heart tube is dependent on the closure of the anterior intestinal portal, and prevention of this process results in failure of heart tube fusion (cardia bifida). Mutations in genes expressed in both the myocardium and the endocardium can result in the failure of heart tube fusion and the cardia bifida phenotype. Most notably, inactivation of *GATA4* in mice causes cardia bifida (Kuo *et al.*, 1997; Molkentin *et al.*, 1997). GATA4 is most highly expressed within the precardiogenic splanchnic mesoderm at the posterior lip of the anterior intestinal portal, corresponding to the region of the embryo that undergoes ventral fusion. It seems, then, that GATA4 is required for the migration or folding of the precardiogenic splanchnic mesodermal cells at the level of the anterior intestinal portal. The cardia bifida seen in the *GATA4* null mutants has recently been linked with downregulation of N-cadherin expression (Zhang *et al.*, 2003), which itself has been associated with failure of heart tube fusion (Nakagawa and Takeichi, 1997). These data suggest that N-cadherin in the middle part of the heart-forming region is essential for fusion of the cardiogenic cells and formation of the heart tube, and that this is regulated by GATA4. Several zebrafish mutants that lack endoderm formation, including casanova (encoding an HMG box-containing gene) and one-eyed pinhead (which encodes a CFC homologue) also display aberrant heart tube fusion, highlighting the importance of the endoderm in heart tube fusion (Alexander *et al.*, 1999). Finally, the paraxial mesoderm itself also seems to play a cell non-autonomous role in directing cardiac mesoderm to the midline, as mutation of the mesoderm-expressed sphingosine-1-phosphate receptor gene also results in cardia bifida in the zebrafish mutant miles apart (Kupperman *et al.*, 2000).

Left–right determination and cardiac looping

As the heart forms in the ventral midline it undergoes the process of looping morphogenesis. Regardless of species, among the vertebrates looping is always to the right. At present, there is no consensus about the mechanisms that initially cause looping, although a number of hypotheses exist. In vertebrates, the node is essential for left–right determination. Interestingly, despite asymmetrical expression of a number of genes in the node and lateral plate mesoderm of mouse and chick embryos, the details differ between species. For example, whereas sonic hedgehog (Shh) is restricted to the left side in the chick node (Levin *et al.*, 1995), it is symmetrically expressed in the mouse node (Meyers and Martin, 1999). Nevertheless, mice lacking functional SHH display abnormalities in left–right determination, including isomerism of the left lung and left atrial appendage, suggesting that a functional SHH signalling pathway is essential for left–right axis formation in mice (Tsukui *et al.*, 1999). Similarly, FGF8 is expressed on opposite sides of the lateral plate mesoderm in mouse and chick, but again appears to play a crucial role in both species (Boettger *et al.*, 1999; Meyers and Martin, 1999). Ultimately, in both species, the TGFβ family member Nodal is induced on the left side, in the lateral plate mesoderm. This, with the co-factor CFC, establishes the expression of Lefty in the midline, which limits Nodal expression and prevents spreading of left-sided signals to the right side (Schlange *et al.*, 2001). *Pitx2*, a homeobox gene, functions downstream of Nodal in the left lateral plate mesoderm and is essential for left–right axis formation, with both overexpression and absence of Pitx2 causing laterality defects (Logan *et al.*, 1998; Ryan *et al.*, 1998; Bruneau *et al.*, 1999b).

Whereas cardiac looping direction is frequently reversed in experimentally-induced situs inversus, cardiac looping can be uncoupled from other aspects of left–right determination. Pitx2-deficient mice have numerous cardiac defects, including right atrial isomerism, atrioventricular septal defects and abnormal arterioventricular connections, but cardiac looping is apparently normal (Kitamura *et al.*, 1999; Lu *et al.*, 1999). It may be that Pitx2 regulates the later aspects of asymmetrical morphogenesis in the inflow and outflow tract but that another, probably earlier-acting, molecule regulates looping morphogenesis. Nodal is a good candidate for this molecule, as ectopic Nodal can reverse the direction of cardiac looping in chick embryos (Patel *et al.*, 1999). Moreover, a Nodal hypomorph, in which Nodal is expressed at lower levels than in normal mouse embryos, displays aberrant cardiac looping (Levin *et al.*, 1995; Lowe *et al.*, 2001). A number of other genes are thought to be involved in regulating looping morphogenesis, on the basis that looping is prevented when they are disrupted, although we currently have no clear idea of the mechanism involved. This includes the *Tbx20* gene, where it has been shown that morpholino-abrogation of *Tbx20* expression results in apparently normal development until the cardiac looping stage, but the heart tube remains linear. *Tbx5* mutants also have an unlooped heart and, as it has been shown that Tbx20 negatively regulates Tbx5 expression, it may be that it is the deregulation of Tbx5 expression that is responsible for the looping defects (Szeto *et al.*, 2002; Garrity *et al.*, 2002). BMP signalling has also been implicated in cardiac looping morphogenesis, as both loss- and gain-of-function experiments result in defects of cardiac looping in zebrafish and *Xenopus* (Breckenridge *et al.*, 2001).

Incomplete looping has been associated with a range of congenital heart defects, principally involving the alignment of the great vessels with the ventricular chambers. These include defects such as double-outlet right ventricle (Figure 13.8c), transposition

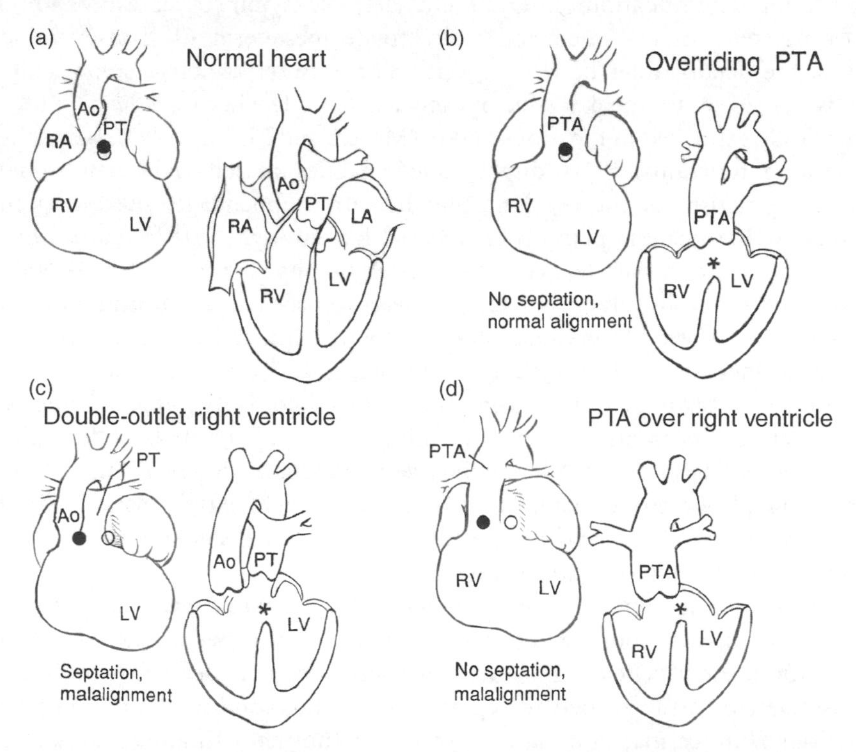

Figure 13.8 Diagrammatic representation of alignment and septation defects of the outflow tract. The open and closed circles show the position of the outflow and inflow regions as seen in Figure 13.5. (a) Normal alignment and septation. (b) Persisting truncus arteriosus (PTA) overriding the ventricular septum. In this case septation has not occurred but the alignment of the outflow tract is normal. Compare this configuration with (c), showing a PTA originating completely from the right ventricle, a malalignment. In (d), septation has occurred but the two outflow vessels, the aorta and pulmonary trunk originate from the right ventricle, a malalignment. RA and LA, right and left atrium; RV and LV, right and left ventricle; Ao, aorta; PT, pulmonary trunk

of the great arteries and ventricular septal defects (Bouman *et al.*, 1995; Bartram *et al.*, 2001). Such defects are seen in the mouse mutant *loop-tail* (*Lp*), which has abnormalities in axial rotation and cardiac looping and later develops a range of cardiac alignment defects that include double-outlet right ventricle and ventricular septal defects (Henderson *et al.*, 2001). Recently, it has been suggested that addition of myocardium to the outflow region of the heart tube is essential for cardiac looping. If this is prevented, then defects in cardiac looping and alignment of the aorta with the left ventricle result (Yelbuz *et al.*, 2002).

Cardiac cushion development

NF-ATc is a transcriptional regulator that has been shown to be essential for endocardial cushion development. Mice lacking functional NF-ATc develop valve and septal abnormalities (de la Pompa *et al.*, 1998; Ranger *et al.*, 1998), structures that are dependent on the endocardium for their formation. More recently, it has been shown that GATA5 and NF-ATc interact to regulate the endothelial/endocardial differentiation of cardiogenic cells, and synergistically activate endocardial transcription (Nemer and Nemer, 2002). Furthermore, these data suggest that, like NF-ATc, GATA5 might be important to the early stages of valvuloseptal development. GATA4 is also strongly expressed in the endocardium, and a recent knock-in mutation which affects the interaction of GATA4 with its co-factor, FOG2, suggests that it may also play a role in valve development (Crispino *et al.*, 2001). The recent development of assays for studying endocardial cell differentiation (Nemer and Nemer, 2002) has set the stage for an explosion in knowledge, and it will be interesting to see whether complex regulatory interactions operate in the endocardium, as they do in the myocardium.

Development and specialization of the chambers

At the time of cardiac looping, morphological and gene expression differences are readily observable between the developing atrial and ventricular chambers. Prior to this, however, fate-mapping studies have generated controversy as to whether the atrial and ventricular progenitors are organized in an antero-posterior pattern in the cardiac mesoderm, as they are in the primitive streak (Stalsberg and DeHaan, 1969; Garcia-Martinez and Schoenwolf, 1993; Redkar *et al.*, 2001). Retinoic acid has been shown to act as a morphogen, establishing posterior polarity within the heart tube. Retinoic acid is capable of posteriorizing the developing heart tube, such that excess levels result in truncation of the anterior part of the heart tube (the outflow component) but expansion of the posterior region (the inflow component; Yutzey *et al.*, 1994; Xavier-Neto *et al.*, 1999). Interestingly, mice lacking retinaldehyde dehydrogenase-2 (RALDH2), which is a critical enzyme involved in the biosynthesis of retinoic acid, have a severe truncation of the inflow (atria and sinus venosus) region of the heart tube. Moreover, although ventricular tissue is apparent, the trabeculae do not form and instead the ventricular wall consists of a thick layer of loosely attached cells that are prematurely differentiated (Niederreither *et al.*, 2001). This suggests that as well as playing roles in the development of the inflow region of the heart, retinoic acid is also essential for the specialization and maturation of the ventricular chamber. The observation that members of the homeotic selector (*Hox*) gene family play essential roles in defining anterior–posterior polarity in the *Drosophila* dorsal vessel has re-established interest in the possibility that *Hox* genes play similar roles in the vertebrate heart (Lo *et al.*, 2002; Lovato *et al.*, 2002; Ponzielli *et al.*, 2002; Lo and Frasch, 2003). However, despite several *Hox* genes being expressed in the vertebrate heart tube (Searcy and Yutzey, 1998), there is little evidence currently for essential roles for these genes in the vertebrate heart, although functional redundancy between related genes might explain the lack of cardiac phenotypes when these genes are inactivated in mice.

A recent model has suggested that the atria and ventricles develop from the primary heart tube by a process of ballooning (Christoffels *et al.*, 2000). In this model, the atria and ventricles originate from the outer curvature of the looped heart and undergo early differentiation to form the chamber or working myocardium. In contrast, the myocardium of the inner curvature of the looped heart retains features of the primary heart tube and forms the atrial midline, the atrioventricular canal, and with contributions from the secondary heart field, the outflow tract. This region contributes to the conduction system and is involved in cardiac septation (Lamers and Moorman, 2002). Several genes are expressed in the outer curvature of the heart tube that will form the chamber myocardium, including *atrial natriuretic factor, connexins* 40 and 43, and *Chisel*. These genes are not expressed in the inner curvature, however, suggesting that these populations are molecularly distinct at an early stage (Christoffels *et al.*, 2000). Many genes are expressed in a regionally restricted pattern by the time of cardiac looping. The hairy-related transcription factors Hey1 and Hey2, which are regulated by Notch signalling, are expressed in the atrial (posterior) and ventricular (anterior) precursors, respectively, at the linear heart tube stage (Leimeister *et al.*, 1999; Nakagawa *et al.*, 1999). Similarly, the Iroquois homeobox gene, *Irx4*, is restricted to the ventricular precursors at all stages of development (Bruneau *et al.*, 2000), whereas the T-box transcription factor Tbx5 is restricted to the presumptive atrial myocardium in the linear heart tube. Mouse embryos lacking Tbx5 manifest severe hypoplasia of the atria and left ventricle, whereas the right ventricle and outflow tract are unaffected, confirming the importance of this gene for development of the posterior regions of the heart tube (Bruneau *et al.*, 2001). The dHand (*Hand2*) and eHand (*Hand1*) genes display complementary patterns of expression in the mouse heart, with the *dHand* gene being expressed more abundantly in the developing right ventricle, whereas *eHand* is more abundant in the left ventricle. Each gene is essential for normal growth of the ventricle in which they are expressed, and the pattern of expression is highly dynamic as development proceeds. Inactivation of *Hand2* in mice has shown that this gene is essential for the survival and expansion of ventricular cells and, moreover, *Hand2* acts in a genetic pathway together with *Nkx2.5* to regulate development of the ventricular chambers (Srivastava *et al.*, 1997; Yamagishi *et al.*, 2001).

Specialization and maturation of the ventricular chambers requires the initially thin-walled vessel to acquire projections, the trabeculae, which are involved in nutrient transfer in the immature heart and thickening of the ventricular wall to form the mature pumping chamber of the heart. Neuregulin and its receptor, erbB2, have been implicated in regulating development of the trabeculae, as these do not form in mice where erbB2 has been inactivated (Lee *et al.*, 1995; Hertig *et al.*, 1999). Retinoic acid, as mentioned earlier, also appears to play essential roles in trabeculation and development of the compact myocardium. Other growth factors, including IGF1, neurotrophin-3, FGF1 and FGF4 (Zhu *et al.*, 1996; Hertig *et al.*, 1999; Lin *et al.*, 2000) have been associated with myocyte proliferation, and may play roles in development of the highly proliferative compact myocardium of the ventricular wall.

Chamber septation

The atrial and ventricular chambers originate as single chambers linked by the atrioventricular canal (Figure 13.2c). These individual chambers are then divided by the formation of the primary and secondary atrial septa and the ventricular septum, respectively. Atrial septation begins when a protrusion appears in the roof of the atrium. This thin muscular structure then grows down towards the atrioventricular canal, where it fuses with the superior atrioventricular cushion. Shortly before this fusion occurs, holes appear in the septum close to the atrial roof, allowing blood to continue to flow between the chambers. A second structure, the spina vestibuli or atrial spine, merges with the primary atrial septum to fuse with the atrioventricular cushion tissue. The origin of this structure is a matter of some controversy, but its importance in atrial septation now appears to be well established. Somewhat later, a fold in the atrial roof, to the right of the primary atrial septum, protrudes into the lumen of the atrium, although it does not fuse with the atrioventricular cushion tissue. This structure, the secondary atrial septum, is therefore not a true septum. This arrangement of atrial septa allows blood to flow between the chambers during fetal life. At birth, the flexible primary atrial septum is forced against the rigid secondary septum, preventing further communication between the two atrial chambers.

As alluded to earlier, the two atria arise from the caudal part of the heart-forming region and do not participate in cardiac looping. The two atria therefore retain their original left–right identity, and this is an important factor in atrial development. The primary and secondary atrial septa are a feature of the left atrium, as deduced by gene expression and their formation in hearts with atrial isomerism. Pitx2 isoform c (Pitx2c) is a left-sided signal in early embryogenesis, and is expressed exclusively in the left atrium. Importantly, both the primary and secondary atrial septa express Pitx2c, suggesting that they are left-sided structures. Moreover, in mice in which *Pitx2* has been inactivated by homologous recombination, right atrial isomerism results and there is a complete failure of atrial septation (Kitamura *et al.*, 1999). Similarly, there is frequently an association between right atrial isomerism and failure of atrial septation in human patients. Several other mouse mutants have been reported to have defects in atrial septation. These include mice in which the tolloid-like 1 (*tll-1*) gene was inactivated, resulting in complete absence of the primary septum in the most severe cases (Clark *et al.*, 1999), and mice lacking neurotrophin-3 (NT-3) and its receptor trkC (Donovan *et al.*, 1996; Tessarollo *et al.*, 1997). In each of these mouse mutants, atrial septal defects were accompanied by ventricular septal defects, suggesting that common mechanisms are likely to be involved in the formation of both the atrial and ventricular septa.

The ventricular chambers begin to grow out from the outer loop of the anterior region of the heart tube in a process referred to as ballooning. The muscular ventricular septum arises at the boundary between the two forming ventricles, with the expansive growth of the left and right ventricles forcing the septum towards the atrioventricular canal, where it eventually fuses (Lamers *et al.*, 1992; Anderson and Brown, 1996). In the linear heart tube, Tbx5 is restricted to the most posterior part of the heart tube which will give rise to the atria and sinus venosus, but by E9.0, when the heart has looped, Tbx5 transcripts are also found in the left ventricle, although they are completely excluded from the right ventricle (Bruneau *et al.*, 1999a). The ventricular septum is formed at the boundary between the Tbx5-positive and -negative cells, suggesting that it might play some role in the positioning of the septum. Indeed, when Tbx5 is expressed ubiquitously throughout the right and left ventricles the ventricular septum does not form, giving a single ventricle expressing left ventricle-specific markers (Takeuchi *et al.*, 2003). Moving the boundary of expression of Tbx5 into the presumptive right ventricle altered the position of the developing ventricular septum, resulting in a small right ventricle and an expanded ventricle, and creating a second boundary of Tbx5-positive and -negative cells in the right ventricle resulted in the formation of a second ectopic ventricular septum (Takeuchi *et al.*, 2003). The boundary of Tbx5-positive and -negative cells does, therefore, appear to be important for the positioning of the ventricular septum and thus for ventricular specification.

Tbx5 interacts with other genes to bring about chamber-specific patterns of gene expression. For example, *Tbx5* acts synergistically with *Nkx2.5* and *GATA4* to activate the atrial natiuretic factor (ANF) promoter (Durocher *et al.*, 1997; Hiroi *et al.*, 2001; Takeuchi *et al.*, 2003). Interestingly, mutations in these genes have also been implicated in human congenital heart defects.

Cardiovascular defects

Left–right patterning defects

Defects in left–right patterning (laterality defects) are found in humans in a variety of forms. In the most extreme cases there is complete reversal of symmetry resulting in so-called situs inversus. In other cases, the defect in laterality is less extreme, and may present as reversal or isomerism (mirror imagery) of particular organs, with the rest of the body showing normal left–right patterning. Cardiac defects are frequently associated with abnormalities in left–right patterning, and include abnormal position of the heart in the chest (malposition), atrial appendage isomerism, atrial septal defects, partial anomalous pulmonary venous return, ventricular septal defects and conotruncal abnormalities, including transposition of the great arteries and double-outlet ventricle. Several genes reported to be important for left–right determination in animal models have been found to be mutated in patients with laterality defects. These include *CRYPTIC, ZIC3, NODAL* and *LEFTY A* (Gebbia *et al.*, 1997; Kosaki *et al.*, 1999; Bamford *et al.*, 2000), in each case with the human defects closely mimicking those seen in the animal model (Carrel *et al.*, 2000; Gaio *et al.*, 1999; Yan *et al.*, 1999; Meno *et al.*, 1998; Lowe *et al.*, 2001). Genes that affect the development of the primary cilia found at the node also cause laterality defects in humans and animal models. Mutations in the dynein gene *DNAI1* have been found in patients with ciliary dyskinesia, also known as immotile cilia syndrome, who manifest laterality defects affecting the heart (Pennarun *et al.*, 1999; Zariwala *et al.*, 2001), and mutations in *left–right dynein* are responsible for similar defects in mice (Supp *et al.*, 1999). Similarly, mutations in subunits of another key cilia protein, kinesin, cause laterality defects in mice (Nonaka *et al.*, 1998; Marszalek *et al.*, 1999; Takeda *et al.*, 1999) although to date, mutations in these genes have not been reported in humans with laterality defects.

Transposition of the great vessels

In transposition of the great vessels, the aorta arises from the right ventricle and the pulmonary trunk from the left ventricle. This defect is almost immediately fatal at birth, unless a ventricular septal defect allows blood from the pulmonary circulation to mix with blood from the systemic circulation. Most mouse models of transposition are actually an incomplete form of transposition that is more closely related to overriding aorta, which falls under the purview of alignment defects. In contrast, the perlecan-null mouse shows a true transposition (Costell *et al.*, 2002), exhibiting a defect that is known clinically as 'common' or 'isolated' SDD transposition of the great arteries (formerly called D-transposition). This defect is characterized by discordance between the ventricles and the arteries, but with no discordance between the atria and the ventricles. In other words, the perlecan-null mouse has 'isolated' ventriculoarterial discordance.

Transposition can also occur as congenital, physiologically corrected transposition (designated SLL or IDD transposition). This happens when the heart loops to the wrong side or the atria are reversed. These physiological corrections are mostly associated with abnormal situs, or placement of the asymmetric organs. Several animal models of transposition with altered situs have been described, including one based on excess retinoic acid teratogenesis. Among genetic models, null mutations of the *type II activin receptor* (Oh and Li, 1997) and *cryptic* (Gaio *et al.*, 1999) show the clearest examples of transposition, although this phenotype is only one of several outflow defects in these models. Double-outlet right ventricle and persistent truncus arteriosus are also seen in these mutants (Figure 13.8). Retinoic acid shows a dose-dependent differential induction of transposition of the great arteries at high doses and dextroposition of the aorta at low doses (Yasui *et al.*, 1999). All of these perturbations alter left–right axis determination and ventricular septation, which are defects associated with physiologically corrected transposition. Importantly, in these models transposition occurs only infrequently in the spectrum of outflow malformations.

In contrast to the other models of transposition, which are physiologically corrected, the perlecan-null mouse presents common or uncorrected transposition. The high incidence of common transposition (11/15 perlecan-null embryos) with intact ventricular septum (10/11) is not found in any other animal models. Intact septum is found with some frequency in humans with transposition. Perlecan is a heparan sulphate proteoglycan (HSPG2) that is expressed in all basement membranes, in cartilage, and several other mesenchymal tissues during development. Perlecan binds growth factors and interacts with various extracellular matrix proteins and cell adhesion molecules. Since the heparan sulphate side-chains bind fibroblast growth factors (FGFs), perlecan may serve as a low-affinity receptor. If so, perlecan could modulate a number of other FGF-controlled processes (Aviezer *et al.*, 1994; Sharma *et al.*, 1998). Costell *et al.* (2002) described alterations in the mesenchyme of the outflow tract of perlecan-null mice that disrupted the formation of the outflow tract cushions. In normal development the cushions spiral in a counterclockwise manner when viewed from above. Disruption of the ridge pattern in the lumen of the outflow tract causes formation of a straight outlet septum rather than a spiralling septum, which results ultimately in transposed positions of the aorta and pulmonary trunk.

Defects of alignment

These defects include malalignment of either inflow or outflow portions of the heart. Inflow malalignments include straddling tricuspid valve and double-inlet left ventricle. Tricuspid atresia may also be classified with inflow malalignments. The outflow tract can also be malaligned. In double-outlet right ventricle (DORV), both the pulmonary trunk and aorta arise from the right ventricle (Figure 13.8c), while tetralogy of Fallot appears to be a milder form, in that the aorta overrides the ventricular septum to a lesser degree. Tetralogy of Fallot has four components, including a ventricular septal defect, pulmonary stenosis, an aorta that 'overrides' the ventricular septal defect and right ventricular hypertrophy. In humans, tetralogy of Fallot is often seen as a component of larger syndromes.

While there are probably multiple aetiologies for outflow malalignment defects, the best understood currently is failure of addition of the myocardium from the secondary heart field to the truncus (distal outflow tract) (Ward *et al.*, 2005). Many models have been described in which the right ventricle, conus (proximal outflow tract) and truncus fail to grow. These include mice with null mutations in *Tbx1* or *Fgf8*, both of which are associated with syndromic rather than isolated occurrence of cardiovascular defects (Abu-Issa *et al.*, 2002; Frank *et al.*, 2002; Vitelli *et al.*, 2002b). Mutations in *NKX2.5* in humans are associated with isolated tetralogy of Fallot, although the embryogenesis of the overriding aorta has not been described and mouse embryos with null mutations of *Nkx2.5* do not show such defects (Goldmuntz *et al.*, 2001; Harvey, 1996).

Defects of septation

Failure of development of any of the septa in the four-chambered heart can occur. Thus, there can be failure of the primary septa of the initial atrial cavity to make left and right atria (common atrium or atrial septal defect), of the atrioventricular canal to divide the region of the atrioventricular valves into right and left channels (atrioventricular canal defect), and of the muscular ventricular septum that divides the primitive ventricle into left and right ventricles (muscular ventricular septal defect). Septation of the outflow tract is complex and involves division of the aortic sac into the proximal parts of the major vessels emanating from the left and right ventricles, the aorta and pulmonary trunk. Defects in outflow septation result in ventricular septal defect, where there is a small to moderate-sized defect below the semilunar valves or persistent truncus arteriosus (common arterial trunk), where the entire outflow below the semilunar valves and above is a single large vessel (Figure 13.8).

Dominant mutations in *NKX2.5* have been found in patients with atrial and ventricular septal defects, tetralogy of Fallot and Ebstein's anomaly of the tricuspid valve (Schott *et al.*, 1998, Benson *et al.*, 1999; Goldmuntz *et al.*, 2001). More recently, mutations in the *GATA4* gene have been found in patients with atrial septal defects (Garg *et al.*, 2003). Interestingly, one of the reported mutations in *GATA4* disrupts interactions with the *TBX5* gene, suggesting that interactions between these genes might be essential to bring about correct atrial septation (Garg *et al.*, 2003).

Interruption, stenosis and atresia

Valves and the outflow portion of the ventricles can be narrowed or stenotic, and veins coming into the heart or arteries leaving the heart can undergo atresia. Interrupted aorta is a form of atresia in which a portion of the aorta has probably formed but then regressed abnormally (Figure 13.9). The specific term 'aortic atresia' is applied to a condition in which the aortic semilunar valve and base of the aorta have disappeared. Interrupted aorta is associated with inadequate ectodermal Fgf8 signalling in genetically engineered mice (Macatee *et al.*, 2003). *Tbx1* heterozygous

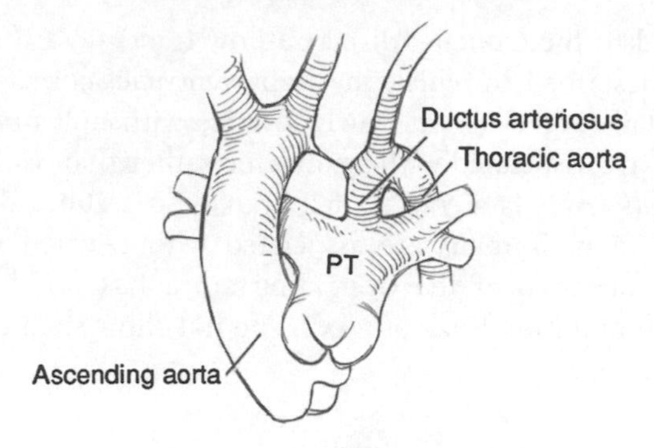

Figure 13.9 Interruption of the aorta (type B). The aorta is interrupted between two of its branches (the left carotid and left subclavian arteries). The remarkable ability of the cardiovascular system to make do with what is available is shown by the pulmonary trunk (PT), providing the major blood supply to the thoracic aorta via a patent ductus arteriosus in place of the missing piece of the aorta

and homozygous null mice have interrupted aorta also but, since Fgf8 expression is regulated by Tbx1, it is likely that the interrupted aorta in the context of *Tbx1* mutation is via a lack of or abnormal Fgf8 signalling (Vitelli *et al.*, 2002b).

Defects of ventricular growth/specification

Both right and left ventricles can be hypoplastic. In some cases, hypoplastic ventricles are secondary to stenotic or atretic outflow vessels while in others it is a primary failure of specification of the ventricular chambers. In the case of stenotic or atretic vessels or valves, it is easy to see the importance of haemodynamic factors in growth and remodelling of the cardiac chambers. However, almost nothing is known about the molecular mechanisms that underlie deficient ventricular growth in the absence of haemodynamic deficiency. The transcriptional network that controls ventricular development is conserved across all vertebrate species. *HAND* genes play a role in ventricular specification and growth (Srivastava, 1999). Epigenetic factors that regulate gene expression through chromatin remodelling appear to be important in ventricular growth. Bop is one such factor that acts through regulation of histone methylation. Mouse embryos lacking Bop expression have right ventricular hypoplasia, while atrial myocytes develop normally (Gottlieb *et al.*, 2002).

Cardiovascular defects in the context of syndromes

Many heart defects are present as part of a complex or syndrome. For example, in DiGeorge or velocardiofacial syndrome, the defects include persistent truncus arteriosus (common arterial trunk; a septation defect; Figure 13.8b, d) and interrupted aortic arch (Figure 13.9), accompanied by aplasia or hypoplasia of the thymus, parathyroids and/or thyroid glands with abnormal facies. Although these syndromes often occur in patients with microdeletions of chromosome 22q11, the gene from this region that appears most closely associated with abnormal cardiovascular development is *TBX1*. Mouse embryos with homozygous deletion of the *Tbx1* gene show a phenotype that resembles closely that seen in the DiGeorge syndrome (Jerome and Papaioannou, 2001; Lindsay *et al.*, 2001; Merscher *et al.*, 2001; Vitelli *et al.*, 2002a). Moreover, a bacterial artificial chromosome (BAC) carrying the human *TBX1* gene was shown to partially rescue the conotruncal defects generated by a 1.5 Mb deletion of the region corresponding to 22q11 in mice. These data, together with the expression pattern of the *Tbx1* gene, in the pharyngeal arches and outflow tract of the heart, suggest a major role for TBX1 in the molecular aetiology of DiGeorge syndrome. Three mutations of *TBX1* in two unrelated patients without the 22q11.2 deletion have been identified in association with a DiGeorge-like phenotype (Yagi *et al.*, 2003).

Chromosomal deletions or trisomies are also accompanied by heart defects. For example, Down's syndrome (trisomy 21) is characterized by craniofacial dysmorphology, mental retardation, hypotonia, short metacarpals and phalanges, and about 25% incidence of common atrioventricular canal (a septation defect). Some single gene mutations are associated with cardiovascular defects. Noonan syndrome results from a mutation in *PTPN11* and the defects include pulmonary stenosis and conduction anomalies (Tartaglia *et al.*, 2001).

Mutations in the transcription factor *TBX5* have been identified in patients with Holt–Oram syndrome. This syndrome is characterized by defects of the upper limb, the conduction system and atrial and ventricular septal defects. The clinical phenotype of patients with Holt–Oram syndrome can vary considerably, however, even within a single family, with some patients manifesting more complex lesions, such as tetralogy of Fallot and hypoplastic left heart syndrome (Basson *et al.*, 1997; Li *et al.*, 1997). The primary abnormality in tetralogy of Fallot is an antero-cephalad deviation of the ventricular septum, resulting in defects affecting the outflow region of the right ventricle. In contrast, hypoplastic left heart syndrome affects the left-sided structures of the heart, most notably resulting in a small left ventricular chamber. Both of these defects are, however, associated with abnormal positioning of the ventricular septum, supporting a role for TBX5 in this process. Direct evidence for a role for Tbx5 in positioning of the ventricular septum has come from studies in mice, where mis-expression of Tbx5 altered the position of the ventricular septum and in some cases resulted in the formation of a second septum (Takeuchi *et al.*, 2003)

Heart defects are also associated with environmental teratogens. Alcohol exposure during the time the neural crest cells populate the face and pharyngeal arches interferes with neural crest migration and survival (Chen *et al.*, 1996; Hassler and Moran, 1986) and results in fetal alcohol syndrome. The heart defects include tetralogy of Fallot, atrial and ventricular septal defects, in addition to many craniofacial anomalies. Cardiac and craniofacial defects also occur following early *in utero* exposure to retinoic acid, usually via administration of the drug Accutane, which is used to treat chronic acne. Interestingly, too much or to little retinoic acid is detrimental to normal development. In both cases, the range of cardiac defect include transposition of the great vessels, co-arctation of the aorta, aortic arch hypoplasia, tetralogy of Fallot, persistent truncus arteriosus and ventricular septal defects.

The future

It is important to understand the molecular and cellular biology of heart development in order to have a better understanding of congenital malformations that affect the heart. In addition, many adult cardiac diseases are established during development and knowledge of development is essential for counseling patients who are at risk for cardiac failure because of such problems. Moreover, interest is growing in treatment of cardiac failure in adults using stem cells. Recent reports using embryonic and adult stem cells to restore functional myocardium are encouraging but the extent to which these cells are incorporated as functional myocardium is limited (Jackson *et al.*, 2001). Thus, understanding the earliest steps of cardiogenesis will not only impact on the treatment of children with congenital heart defects, but may also provide new therapies for adults with cardiovascular disease. Although major advances have been made in identifying genes important for the early stages of heart specification, little is known about the molecules and the target genes that allocate and specify the precardiogenic cells. What are the molecular differences between a cell with cardiogenic potential and a fully determined cardiomyocyte? Complicating this progression from specified cell to working cardiomyocyte is the fact that the developing heart must function as a pump while it undergoes the intricacies of looping morphogenesis, chamber specification and formation and, finally, septation. Investigations will be facilitated by the availability of several animal models with fully characterized genomes. This, combined with emergent molecular, morphological and functional analyses, provides us with new ways to access the intrinsic programme of the developing pump, both in the context of its functional requirements and the extrinsic factors that influence the programme of heart development.

Acknowledgements

Our thanks to Karen Waldo for providing most of the illustrations in this chapter.

References

Abu-Issa, R., Smyth, G., Smoak, I., Yamamura, K.-I. and Meyers, E. (2002) Fgf8 is required for pharyngeal arch and cardiovascular development in the mouse. *Development* **129**: 4613–4625.

Alexander, J., Rothenberg, M., Henry, G.L. and Stainier, D.Y. (1999) Casanova plays an early and essential role in endoderm formation in zebrafish. *Dev. Biol.* **215**: 343–357.

Alsan, B.H. and Schultheiss, T.M. (2002) Regulation of avian cardiogenesis by Fgf8 signalling. *Development* **129**: 1935–1943.

Anderson, R.H. and Brown, N. (1996) The anatomy of the heart revisited. *Anat. Rec.* **246**: 1–7.

Andree, B., Duprez, D., Vorbusch, B., Arnold, H.H. and Brand, T. (1998) BMP-2 induces ectopic expression of cardiac lineage markers and interferes with somite formation in chicken embryos. *Mech. Dev.* **70**: 119–131.

Aviezer, D., Hecht, D., Safran, M., Eisinger, M. *et al.* (1994) Perlecan, basal lamina proteoglycan, promotes basic fibroblast growth factor-receptor binding, mitogenesis, and angiogenesis. *Cell* **79**: 1005–1013.

Bamford,. RN., Roessler, E., Burdine, R.D., Saplakoglu, U. *et al.* (2000) Requirement for BMP and FGF signalling during cardiogenic induction in non-precardiac mesoderm is specific, transient, and cooperative. *Dev. Dyn.* **218**: 383–393.

Barron, M., Gao, M. and Lough, J. (2000) Requirement for BMP and FGF signalling during cardiogenic induction in non-precardiac mesoderm is specific, transient, and cooperative. *Dev. Dyn.* **218**: 383–393.

Bartram, U., Molin D.G., Wisse, L.J., Mohamad, A. *et al.* (2001) Double-outlet right ventricle and overriding tricuspid valve reflect disturbances of looping, myocardialization, endocardial cushion differentiation, and apoptosis in TGF-β2-knockout mice. *Circulation* **103**: 2745–2752.

Basson, C.T., Bachinsky, D.R., Lin, R.C., Levi, T. *et al.* (1997) Mutations in human cause limb and cardiac malformation in Holt–Oram syndrome. *Nat. Genet.* **15**: 30–34.

Benson, D.W., Silberbach, G.M., Kavanaugh-McHugh, A., Cottrill, C. *et al.* (1999) Mutations in the cardiac transcription factor *NKX2.5* affect diverse cardiac developmental pathways. *Circulation* **104**: 1567–1573.

Bodmer, R. (1993) The gene tinman is required for specification of the heart and visceral muscles in *Drosophila*. *Development* **118**(3): 719–29. Erratum in: *Development* (1994) **119**: 969.

Boettger, T., Wittler, L. and Kessel, M. (1999) FGF8 functions in the specification of the right body side of the chick. *Curr. Biol.* **9**: 277–280.

Bogers, A.J.J.C., Gittenberger-de Groot, A.C., Poelmann, R.E. and Huysmans, H.A. (1989) Development of the origin of the coronary arteries, a matter of ingrowth or outgrowth? *Anat. Embryol.* **180**: 437–441.

Bouman, H.G., Broekhuizen, M.L., Baasten, A..M, Gittenberger-de Groot, A.C. and Wenink, A.C. (1995) Spectrum of looping disturbances in stage 34 chicken hearts after retinoic acid treatment. *Anat. Rec.* **243**: 101–108.

Breckenridge, R.A., Mohun, T.J. and Amaya, E. (2001) A role for BMP signalling in heart looping morphogenesis in *Xenopus*. *Dev. Biol.* **232**: 191–203.

Bruneau, B.G., Nemer, G., Schmitt, J.P., Charron, F. *et al.* (2001) A murine model of Holt–Oram syndrome defines roles of the T-box transcription factor Tbx5 in cardiogenesis and disease. *Cell* **106**: 709–721.

Bruneau, B.G., Bao, Z.Z., Tanaka, M., Schott, J.J. *et al.* (2000) Cardiac expression of the ventricle-specific homeobox gene *Irx4* is modulated by Nkx2–5 and dHand. *Dev. Biol.* **217**: 266–277.

Bruneau, B.G., Logan, M., Davis, N., Levi, T. *et al.* (1999a) Chamber-specific cardiac expression of Tbx5 and heart defects in Holt–Oram syndrome. *Dev. Biol.* **211**: 100–108.

Bruneau, B.G., Nemer, G., Schmitt, J.P., Charron, F. *et al.* (1999b) The homeobox gene *Pitx2*: mediator of asymmetric left–right signalling in vertebrate heart and gut looping. *Development* **126**: 1225–1234.

Carrel, T., Purandare, S.M., Harrison, W., Elder, F. *et al.* (2000) The X-linked mouse mutation Bent tail is associated with a deletion of the *Zic3* locus. *Hum. Mol. Genet.* **9**: 1937–1942.

Chen, F., Kook, H., Milewski, R., Gitler, A.D. *et al.* (2002) *Hop* is an unusual homeobox gene that modulates cardiac development. *Cell* **110**: 713–723.

Chen, S.Y., Yang, B., Jacobson, K. and Sulik, K.K. (1996) The membrane disordering effect of ethanol on neural crest cells *in vitro* and the protective role of GM1 ganglioside. *Alcohol* **13**: 589–595.

Christoffels, V.M., Habets, P.E., Franco, D., Campione, M. *et al.* (2000) Chamber formation and morphogenesis in the developing mammalian heart. *Dev. Biol.* **223**: 266–278. Erratum in: *Dev. Biol.* (2000) **225**: 266.

Chuck, E.T., Freeman, D.M., Watanabe, M. and Rosenbaum, D.S. (1997) Changing activation sequence in the embryonic chick heart – implications for the development of the His–Purkinje system. *Circ. Res.* **81**: 470–476.

Clark, T.G., Conway, S.J., Scott, I.C., Labosky, P.A. *et al.* (1999) The mammalian Tolloid-like 1 gene, *Tll1*, is necessary for normal septation and positioning of the heart. *Development* **126**: 2631–2642.

Costell, M., Carmona, R., Gustafsson, E., Gonzalez-Iriarte, M. *et al.* (2002) Hyperplastic conotruncal endocardial cushions and transposition of great arteries in perlecan-null mice. *Circ. Res.* **91**: 158–164.

Crispino J.D., Lodish M.B., Thurberg B.L., Litovsky S.H. *et al.* (2001) Proper coronary vascular development and heart morphogenesis depend on interaction of GATA-4 with FOG co-factors. *Genes Dev.* **15**: 839–44.

de la Cruz, M.V., Gomez, C.S., Arteaga, M.M. and Arguello, C. (1977) Experimental study of the development of the truncus and the conus in the chick embryo. *J. Anat.* **123**: 661–686.

de la Pompa, J.L., Timmerman, L.A., Takimoto, H., Yoshida, H. *et al.* (1998) Role of the NF-ATc transcription factor in morphogenesis of cardiac valves and septum. *Nature* **392**: 182–186.

Donovan, M.J., Hahn, R., Tessarollo, L. and Hempstead, B.L. (1996) Identification of an essential nonneuronal function of neurotrophin 3 in mammalian cardiac development. *Nat. Genet.* **14**: 210–213.

Durocher, D., Charron, F., Warren, R., Schwartz, R.J. and Nemer, M. (1997) The cardiac transcription factors Nkx2–5 and GATA-4 are mutual co-factors. *EMBO J.* **16**: 5687–5696.

Easton, H.S., Armstrong, J.B., and Smith, S.C. (1994) Heart specification in the Mexican axolotl (*Ambystoma mexicanum*). *Dev. Dyn.* **200**: 313–320.

Eisenberg, C.A. and Eisenberg, L.M. (1999) WNT11 promotes cardiac tissue formation of early mesoderm. *Dev. Dyn.* **216**: 45–58.

Frank, D., Fotheringham, L., Brewer, J., Muglia, L. *et al.* (2002) An *Fgf8* mouse mutant phenocopies human 22q11 deletion syndrome. *Development* **129**: 4591–4603.

Gaio, U., Schweickert, A., Fischer, A., Garratt, A.N. *et al.* (1999) A role of the cryptic gene in the correct establishment of the left–right axis. *Curr. Biol.* **9**: 1339–1342.

Gajewski, K., Fossett, N., Molkentin, J.D. and Schulz, R.A. (1999) The zinc finger proteins Pannier and GATA4 function as cardiogenic factors in *Drosophila*. *Development* **126**: 5679–5688.

Gajewski, K., Zhang, Q., Choi, C.Y., Fossett, N. *et al.* (2001) Pannier is a transcriptional target and partner of Tinman during *Drosophila* cardiogenesis. *Dev. Biol.* **233**: 425–436.

Garcia-Martinez, V. and Schoenwolf, G.C. (1993) Primitive-streak origin of the cardiovascular system in avian embryos. *Dev. Biol.* **159**: 706–719.

Garg, V., Kathiriya, I.S., Barnes, R., Schluterman, M.K. *et al.* (2003) *GATA4* mutations cause human congenital heart defects and reveal an interaction with *TBX5. Nature* **424**: 443–447.

Garrity, D.M., Childs, S. and Fishman, M.C. (2002) The heartstrings mutation in zebrafish causes heart/fin Tbx5 deficiency syndrome. *Development* **129**: 4635–4645.

Gebbia, M., Ferrero, G.B., Pilia, G., Bassi, M.T. *et al.* (1997) X-linked situs abnormalities result from mutations in *ZIC3. Nat. Genet.* **17**: 305–308.

Goldmuntz, E., Geiger, E. and Benson, D.W. (2001) *NKX2.5* mutations in patients with tetralogy of Fallot. *Circulation* **104**: 2565–2568.

Gottlieb, P.D., Pierce, S.A., Sims, R.J. III, Yamagishi, H. *et al.* (2002). *Bop* encodes a muscle-restricted protein containing MYND and SET domains and is essential for cardiac differentiation and morphogenesis. *Nat. Genet.* **31**: 25–32.

Griffin, K.J. and Kimelman, D. (2002) One-eyed Pinhead and Spadetail are essential for heart and somite formation. *Nat. Cell. Biol.* **4**: 821–825.

Harvey, R.P. (1996) *NK-2* homeobox genes and heart development. *Dev. Biol.* **178**: 203–216.

Hassler, J.A. and Moran, D.J. (1986) Effectiveness of ethanol on the cytoskeleton of migrating and differentiating neural crest cells: possible role in teratogenesis. *J. Craniofac. Genet. Dev. Biol.* **2**(suppl): 129–136.

Henderson, D.J., Conway, S.J., Greene, N.D., Gerrelli, D. *et al.* (2001) Cardiovascular defects associated with abnormalities in midline development in the Loop-tail mouse mutant. *Circ. Res.* **89**: 6–12.

Hertig, C.M., Kubalak, S.W., Wang, Y. and Chien, K.R. (1999) Synergistic roles of neuregulin-1 and insulin-like growth factor-I in activation of the phosphatidylinositol 3-kinase pathway and cardiac chamber morphogenesis. *J. Biol. Chem.* **274**: 37362–3739.

Hiroi, Y., Kudoh, S., Monzen, K., Ikeda, Y. *et al.* (2001) *Tbx5* associates with *Nkx2–5* and synergistically promotes cardiomyocytes differentiation. *Nat. Genet.* **28**: 276–280.

Hiruma, T., and Hirakow, R. (1989) Epicardium formation of chick embryonic heart: Computer-aided reconstruction, scanning, and transmission electron microscopic studies. *Am. J. Anat.* **184**: 129–138.

Ho, E., and Shimada, Y. (1978) Formation of the epicardium studied with the scanning electron microscope. *Dev. Biol.* **66**: 579–585.

Jerome, L.A. and Papaioannou, V.E. (2001) DiGeorge syndrome phenotype in mice mutant for the T-box gene, *Tbx1. Nat. Genet.* **27**: 286–291.

Jackson K.A., Majka S.M., Wang H., Pocius J. *et al.* (2001) Regeneration of ischemic cardiac muscle and vascular endothelium by adult stem cells. *J. Clin. Invest.* **107**: 1395–1402.

Kelly, R.G., Brown, N.A. and Buckingham, M.E. (2001) The arterial pole of the mouse heart forms from Fgf10-expressing cells in pharyngeal mesoderm. *Dev. Cell* **1**: 435–440.

Kirby, M.L., Gale, T.F. and Stewart, D.E. (1983) Neural crest cells contribute to aorticopulmonary septation. *Science* **220**: 1059–1061.

Kirby, M., Lawson, A., Stadt, H., Kumiski, D. *et al.* (2003) Hensen's node gives rise to the ventral midline of the foregut: implications for organizing head and heart development. *Dev. Biol.* **253**: 175–188.

Kitajima, S., Takagi, A., Inoue, T. and Saga, Y. (2000) MesP1 and MesP2 are essential for the development of cardiac mesoderm. *Development* **127**: 3215–3226.

Kitamura, K., Miura, H., Miyagawa-Tomita, S., Yanazawa, M. *et al.* (1999) Mouse Pitx2 deficiency leads to anomalies of the ventral body wall, heart, extra- and periocular mesoderm and right pulmonary isomerism. *Development* **126**: 5749–5758.

Kosaki, K., Bassi, M.T., Kosaki, R., Lewin, M. *et al.* (1999) Characterization and mutation analysis of human *LEFTY A* and *LEFTY B*, homologues of murine genes implicated in left–right axis development. *Am. J. Hum. Genet.* **64**: 712–721.

Kuo, C.T., Morrisey, E.E., Anandappa, R., Sigrist, K. *et al.* (1997) GATA4 transcription factor is required for ventral morphogenesis and heart tube formation. *Genes Dev.* **11**: 1048–1060.

Kupperman, E., An, S., Osborne, N., Waldron, S. and Stainier, D.Y. (2000) A sphingosine-1-phosphate receptor regulates cell migration during vertebrate heart development. *Nature* **406**: 192–195.

Kuratani, S.C. and Kirby, M.L. (1991) Initial migration and distribution of the cardiac neural crest in the avian embryo: an introduction to the concept of the circumpharyngeal crest. *Am. J. Anat.* **191**: 215–227.

Lamers, W.H., Wessels, A., Verbeek, F.J., Moorman, A.F. *et al.* (1992) New findings concerning ventricular septation in the human heart. Implications for maldevelopment. *Circulation* **86**: 1194–1205.

Lamers, W.H. and Moorman, A.F. (2002) Cardiac septation: a late contribution of the embryonic primary myocardium to heart morphogenesis. *Circ. Res.* **91**: 93–103.

Le Liévre, C.S. and Le Douarin, N.M. (1975) Mesenchymal derivatives of the neural crest. Analysis of chimaeric quail and chick embryos. *J. Morphol. Exp. Embryol.* **34**: 125–154.

Lee, K.F., Simon, H., Chen, H., Bates, B. *et al.* (1995) Requirement for neuregulin receptor erbB2 in neural and cardiac development. *Nature* **378**: 394–398.

Leimeister, C., Externbrink, A., Klamt, B. and Gessler, M. (1999) *Hey* genes: a novel subfamily of hairy- and Enhancer of split-related genes specifically expressed during mouse embryogenesis. *Mech. Dev.* **85**: 173–177.

Levin, M., Johnson, R.L., Stern, C.D., Kuehn, M. and Tabin, C. (1995) A molecular pathway determining left–right asymmetry in chick embryogenesis. *Cell* **82**: 803–814.

Li, Q.Y., Newbury-Ecob, R.A., Terrett, J.A., Wilson, D.I. *et al.* (1997) Holt–Oram syndrome is caused by mutations in *TBX5*, a member of the Brachyury (T) gene family. *Nat. Genet.* **15**: 21–35.

Li, Y.-X., Zdanowicz, M., Young, L., Kumiski, D. *et al.* (2003) Cardiac neural crest in zebrafish embryos contributes to myocardial cell lineage and early heart function. *Dev. Dyn.* **226**: 540–550.

Liao, E.C., Paw, B.H., Oates, A.C., Pratt, S.J. *et al.* (1998) *SCL/Tal-1* transcription factor acts downstream of cloche to specify hematopoietic and vascular progenitors in zebrafish. *Genes Dev.* **12**(5): 621–626.

Liao, W., Bisgrove, B.W., Sawyer, H., Hug, B. *et al.* (2002) Formation of multiple hearts in mice following deletion of β-catenin in the embryonic endoderm. *Dev. Cell* **3**: 171–181.

Lin, M.I., Das, I., Schwartz, G.M., Tsoulfas, P. *et al.* (2000) Trk C receptor signalling regulates cardiac myocyte proliferation during early heart development *in vivo*. *Dev. Biol.* **226**: 180–191.

Lindsay, E.A., Su, H., Morishima, M., Huynh, T. *et al.* (2001) *Tbx1* haploinsufficiency in the DiGeorge syndrome region causes aortic arch defects in mice. *Nature* **410**: 97–101.

Lo, P.C. and Frasch, M. (2003) Establishing A–P polarity in the embryonic heart tube: a conserved function of *Hox* genes in *Drosophila* and vertebrates? *Trends Cardiovasc. Med.* **13**: 182–187.

Lo, P.C., Skeath, J.B., Gajewski, K., Schulz, R.A. and Frasch, M. (2002) Homeotic genes autonomously specify the anteroposterior subdivision of the *Drosophila* dorsal vessel into aorta and heart. *Dev. Biol.* **251**: 307–319.

Logan, M., Pagan-Westphal, S.M., Smith, D.M., Paganessi, L. and Tabin, C.J. (1998) The transcription factor Pitx2 mediates site-specific morphogenesis in response to left–right asymmetric signals. *Cell* **94**: 307–317.

Lough, J., Barron, M., Brogley, M., Sugi, Y. *et al.* (1996) Combined BMP-2 and FGF-4, but neither factor alone, induces cardiogenesis in non-precardiac embryonic mesoderm. *Dev. Biol.* **178**: 198–202.

Lovato, T.L., Nguyen, T.P., Molina, M.R. and Cripps, R.M. (2002) The *Hox* gene abdominal-A specifies heart cell fate in the *Drosophila* dorsal vessel. *Development* **129**: 5019–5027.

Lowe, L.A., Yamada, S. and Kuehn, M.R. (2001) Genetic dissection of nodal function in patterning the mouse embryo. *Development* **128**: 1831–1843.

Lu, M.F., Pressman, C., Dyer, R., Johnson, R.L. and Martin, J.F. (1999) Function of Rieger syndrome gene in left–right asymmetry and craniofacial development. *Nature* **401**: 276–278.

Lyons, I., Parsons, L.M., Hartley, L., Li, R. *et al.* (1995) Myogenic and morphogenetic defects in the heart tubes of murine embryos lacking the homeobox gene *Nkx2–5*. *Genes Dev.* **9**: 1654–1666.

Macatee, T.L., Hammond, B.P., Arenkiel, B.R., Francis, L. *et al.* (2003) Ablation of specific expression domains reveals discrete functions of ectoderm- and endoderm-derived FGF8 during cardiovascular and pharyngeal development. *Development* **130**: 6361–6374.

Manasek, F.J. (1968) Embryonic development of the heart I: a light and electron microscope study of myocardial development in the early chick embryo. *J. Morphol.* **125**: 329–366.

Manasek, F.J., Burnside, M.B. and Waterman, R.E. (1972) Myocardial cell shape change as a mechanism of embryonic heart looping. *Dev. Biol.* **29**: 349–371.

Marszalek, J.R., Ruiz-Lozano, P., Roberts, E., Chien, K.R. and Goldstein, L.S. (1999) Situs inversus and embryonic ciliary morphogenesis defects in mouse mutants lacking the KIF3A subunit of kinesin-II. *Proc. Natl Acad. Sci. USA* **96**: 5043–5048.

Marvin, M.J., Di Rocco, G., Gardiner, A., Bush, S.M. and Lassar, A.B. (2001) Inhibition of Wnt activity induces heart formation from posterior mesoderm. *Genes Dev.* **15**: 316–327.

Meno, C., Shimono, A., Saijoh, Y., Yashiro, K. *et al.* (1998) lefty-1 is required for left–right determination as a regulator of lefty-2 and nodal. *Cell* **94**: 287–297.

Merscher, S., Funke, B., Epstein, J.A., Heyer, J. *et al.* (2001) TBX1 is responsible for cardiovascular defects in velo-cardio-facial/DiGeorge syndrome. *Cell* **104**: 619–629.

Meyers, E.N. and Martin, G.R. (1999) Differences in left–right axis pathways in mouse and chick: functions of FGF8 and SHH. *Science* **285**: 403–406.

Mikawa, T., Borisov, A., Brown, A.M.C. and Fischman, D.A. (1992) Clonal analysis of cardiac morphogenesis in the chick embryo using a replication-defective retrovirus. I. Formation of the ventricular myocardium. *Dev. Dyn.* **193**: 11–23.

Molkentin, J.D., Antos, C., Mercer, B., Taigen, T. *et al.* (2000a) Direct activation of a GATA6 cardiac enhancer by Nkx2.5: evidence for a reinforcing regulatory network of Nkx2.5 and GATA transcription factors in the developing heart. *Dev. Biol.* **217**: 301–309.

Molkentin, J.D., Tymitz, K.M., Richardson, J.A. and Olson, E.N. (2000b) Abnormalities of the genitourinary tract in female mice lacking GATA5. *Mol. Cell Biol.* **20**: 5256–5260.

Molkentin, J.D., Lin, Q., Duncan, S.A. and Olson, E.N. (1997) Requirement of the transcription factor GATA4 for heart tube formation and ventral morphogenesis. *Genes Dev.* **11**: 1061–1072.

Moorman A.F., de Jong F., Denyn M.M., Lamers W.H. (1998) Development of the cardiac conduction system. *Circ Res.* **82**: 629–44

Morrisey, E.E., Tang, Z., Sigrist, K., Lu, M.M. *et al.* (1998) GATA6 regulates HNF4 and is required for differentiation of visceral endoderm in mouse. *Development* **128**: 1019–1031.

Nakagawa, O., Nakagawa, M., Richardson, J.A., Olson, E.N. and Srivastava, D. (1999) HRT1, HRT2, and HRT3: a new subclass of bHLH transcription factors marking specific cardiac, somitic, and pharyngeal arch segments. *Dev. Biol.* **216**: 72–84.

Nakagawa, S. and Takeichi, M. (1997) N-cadherin is crucial for heart formation in the chick embryo. *Dev. Growth Differ.* **39**: 451–455.

Nemer, G. and Nemer, M. (2002) Cooperative interaction between GATA5 and NF-ATc regulates endothelial–endocardial differentiation of cardiogenic cells. *Development* **129**: 4045–4055.

Niederreither, K., Vermot, J., Messaddeq, N., Schuhbaur, B. *et al.* (2001) Embryonic retinoic acid synthesis is essential for heart morphogenesis in the mouse embryo. *Genes Dev.* **12**: 3579–3590.

Noden, D.M. (1991) Origins and patterning of avian outflow tract endocardium. *Development* **111**: 867–876.

Nonaka, S., Tanaka, Y., Okada, Y., Takeda, S. *et al.* (1998) Randomization of left–right asymmetry due to loss of nodal cilia generating leftward flow of extraembryonic fluid in mice lacking KIF3B motor protein. *Cell.* **95**: 829–37. Erratum in: *Cell* (1999) **99**: 117.

Oh, S.P. and Li, E. (1997) The signalling pathway mediated by the type IIB activin receptor controls axial patterning and lateral asymmetry in the mouse. *Genes Dev.* **11**: 1812–1826.

Pandur, P., Lasche, M., Eisenberg, L.M. and Kuhl, M. (2002) Wnt-11 activation of a non-canonical Wnt signalling pathway is required for cardiogenesis. *Nature* **418**: 636–641.

Patel, K., Isaac, A. and Cooke, J. (1999) Nodal signalling and the roles of the transcription factors SnR and Pitx2 in vertebrate left–right asymmetry. *Curr. Biol.* **9**: 609–612.

Pennarun, G., Escudier, E., Chapelin, C., Bridoux, A.M. *et al.* (1999) Loss-of-function mutations in a human gene related to *Chlamydomonas reinhardtii* dynein IC78 result in primary ciliary dyskinesia. *Am. J. Hum. Genet.* **65**: 1508–1519.

Perez-Pomares, J.M., Phelps, A., Sedmerova, M. and Wessels, A. (2003) Epicardial-like cells on the distal arterial end of the cardiac outflow tract do not derive from the proepicardium but are derivatives of the cephalic pericardium. *Dev. Dyn.* **227**: 56–68.

Poelmann, R.E., Gittenberger-de Groot, A.C., Mentink, M.M.T., Bokenkamp, R. and Hogers, B. (1993). Development of the cardiac coronary vascular endothelium, studied with anti-endothelial antibodies, in chicken–quail chimeras. *Circ. Res.* **73**: 559–568.

Ponzielli, R., Astier, M., Chartier, A., Gallet, A. *et al.* (2002) Heart tube patterning in *Drosophila* requires integration of axial and segmental information provided by the Bithorax complex genes and hedgehog signalling. *Development* **129**: 4509–4521.

Puri, M.C., Partanen, J., Rossant, J. and Bernstein, A. (1999) Interaction of the TEK and TIE receptor tyrosine kinases during cardiovascular development. *Development* **126**: 4569–4580.

Ranger, A.M., Grusby, M.J., Hodge, M.R., Gravallese, E.M. *et al.* (1998) The transcription factor NF-ATc is essential for cardiac valve formation. *Nature* **392**: 186–190.

Redkar, A., Montgomery, M. and Litvin, J. (2001) Fate map of early avian cardiac progenitor cells. *Development* **128**: 2269–2279.

Reiter, J.F., Alexander, J., Rodaway, A., Yelon, D. *et al.* (1999) Gata5 is required for the development of the heart and endoderm in zebrafish. *Genes Dev.* **13**: 2983–2995.

Ryan, A.K., Blumberg, B., Rodriguez-Esteban, C., Yonei-Tamura, S. *et al.* (1998) Pitx2 determines left–right asymmetry of internal organs in vertebrates. *Nature* **394**: 545–551.

Saga, Y., Miyagawa-Tomita, S., Takagi, A., Kitajima, S. *et al.* (1999) MesP1 is expressed in the heart precursor cells and required for the formation of a single heart tube. *Development* **126**: 3437–3447.

Sato, M. and Yost, H.J. (2003) Cardiac neural crest contributes to cardiomyogenesis in zebrafish. *Dev. Biol.* **257**: 127–139.

Schlange, T., Andree, B., Arnold, H. and Brand, T. (2000) Expression analysis of the chicken homologue of *CITED2* during early stages of embryonic development. *Mech. Dev.* **98**: 157–160.

Schlange, T., Schnipkoweit, I., Andree, B., Ebert, A. *et al.* (2001) Chick CFC controls Lefty1 expression in the embryonic midline and nodal expression in the lateral plate. *Dev. Biol.* **234**: 376–389.

Schott, J.J., Benson, D.W., Basson, C.T., Pease, W. *et al.* (1998) Congenital heart disease caused by mutations in the transcription factor *NKX2–5*. *Science* **281**: 108–111.

Schultheiss, T.M., Burch, J.B. and Lassar, A.B. (1997) A role for bone morphogenetic proteins in the induction of cardiac myogenesis. *Genes Dev.* **11**: 451–462.

Schwartz, R.J. and Olson, E.N. (1999) Building the heart piece by piece: modularity of *cis*-elements regulating Nkx2–5 transcription. *Development* **126**: 4187–4192.

Searcy, R.D. and Yutzey, K.E. (1998) Analysis of *Hox* gene expression during early avian heart development. *Dev. Dyn.* **213**: 82–91.

Sepulveda, J.L., Vlahopoulos, S., Iyer, D., Belaguli, N. and Schwartz, R.J. (2002) Combinatorial expression of GATA4, Nkx2–5, and serum response factor directs early cardiac gene activity. *J. Biol. Chem.* **277**: 25775–25782.

Sharma, B., Handler, M., Eichstetter, I., Whitelock, J.M. *et al.* (1998) Antisense targeting of perlecan blocks tumor growth and angiogenesis *in vivo*. *J. Clin. Invest.* **102**: 1599–1608.

Srivastava, D. (1999) HAND proteins: molecular mediators of cardiac development and congenital heart disease. *Trends Cardiovasc. Med.* **9**: 11–18.

Srivastava, D., Thomas, T., Lin, Q., Kirby, M.L. *et al.* (1997) Regulation of cardiac mesodermal and neural crest development by the bHLH transcription factor, dHAND. *Nat. Genet.* **16**: 154–60. Erratum in: *Nat. Genet.* (1997) **16**: 410.

Stalsberg, H. and DeHaan, R.L. (1969) The precardiac areas and formation of the tubular heart in the chick embryo. *Dev. Biol.* **19**: 128.

Stennard, F.A., Costa, M.W., Elliott, D.A., Rankin, S. *et al.* (2003) Cardiac T-box factor Tbx20 directly interacts with Nkx2–5, GATA4, and GATA5 in regulation of gene expression in the developing heart. *Dev. Biol.* **262**: 206–224.

Supp, D.M., Brueckner, M., Kuehn, M.R., Witte, D.P. *et al.* (1999) Targeted deletion of the ATP binding domain of left–right dynein confirms its mrole in specifying development of left–right asymmetries. *Development* **126**: 5495–5504.

Szeto, D.P., Griffin, K.J. and Kimelman, D. (2002) HrT is required for cardiovascular development in zebrafish. *Development* **129**: 5093–5101.

Takeda, S., Yonekawa, Y., Tanaka, Y., Okada, Y. *et al.* (1999) Left–right asymmetry and kinesin superfamily protein KIF3A: new insights in determination of laterality and mesoderm induction by $kif3A^{-/-}$ mice analysis. *J. Cell Biol.* **145**: 825–836.

Takeuchi, J.K., Ohgi, M., Koshiba-Takeuchi, K., Shiratori, H. *et al.* (2003) *Tbx5* specifies the left/right ventricles and ventricular septum position during cardiogenesis. *Development* **130**: 5953–5964.

Tartaglia, M., Mehler, E.L., Goldberg, R., Zampino, G. *et al.* (2001) Mutations in *PTPN11*, encoding the protein tyrosine phosphatase SHP-2, cause Noonan syndrome. *Nat. Genet.* **29**: 465–468.

Tessarollo, L., Tsoulfas, P., Donovan, M.J., Palko, M.E. *et al.* (1997) Targeted deletion of all isoforms of the *trkC* gene suggests the use of alternate receptors by its ligand neurotrophin-3 in neuronal development and implicates *trkC* in normal cardiogenesis. *Proc. Natl Acad. Sci. USA* **94**: 14776–14781.

Tsukui, T., Capdevila, J., Tamura, K., Ruiz-Lozano, P. *et al.* (1999) Multiple left–right asymmetry defects in *Shh*$^{-/-}$ mutant mice unveil a convergence of the Shh and retinoic acid pathways in the control of *Lefty-1*. *Proc. Natl Acad. Sci. USA* **96**: 11376–11381.

Tzahor, E. and Lassar, A.B. (2001) Wnt signals from the neural tube block ectopic cardiogenesis. *Genes Dev.* **15**: 255–260.

Ueyama, T., Kasahara, H., Ishiwata, T., Nie, Q. and Izumo, S. (2003) Myocardin expression is regulated by *Nkx2.5*, and its function is required for cardiomyogenesis. *Mol. Cell. Biol.* **23**: 9222–9232.

Vitelli, F., Morishima, M., Taddei, I., Lindsay, E.A. and Baldini, A. (2002a) *Tbx1* mutation causes multiple cadiovascular defects and disrupts neural crest and cranial nerve migratory pathways. *Hum. Mol. Genet.* **11**: 915–922.

Vitelli, F., Taddei, I., Morishima, M., Meyers, E. *et al.* (2002b) A genetic link between Tbx1 and fibroblast growth factor signalling. *Development* **129**: 4605–4611.

Waldo, K.L., Willner, W. and Kirby, M.L. (1990) Origin of the proximal coronary artery stem and a review of ventricular vascularization in the chick embryo. *Am. J. Anat.* **188**: 109–120.

Waldo, K.L., Kumiski, D.H., Wallis, K.T., Stadt, H.A. *et al.* (2001) Conotruncal myocardium arises from a secondary heart field. *Development.* **128**: 3179–3188.

Waldo, K.L., Hutson, M.R., Ward, C.C., Zdanowicz, M. *et al.* (2005) Secondary heart field contributes myocardium and smooth muscle to the arterial pole of the developing heart. *Dev. Biol.* **281**: 78–90.

Wang, D., Chang, P.S., Wang, Z., Sutherland, L. *et al.* (2001) Activation of cardiac gene expression by myocardin, a transcriptional co-factor for serum response factor. *Cell* **105**: 851–862.

Ward, C., Stadt, H., Hutson, M. and Kirby, M.L. (2005) Ablation of the secondary heart field leads to tetralogy of Fallot and pulmonary atresia. *Dev. Biol.*; 8 June, E-pub ahead of print.

Xavier-Neto, J., Neville, C.M., Shapiro, M.D., Houghton, L. *et al.* (1999) A retinoic acid-inducible transgenic marker of sino-atrial development in the mouse heart. *Development* **126**: 2677–2687.

Yagi, H., Furutani, Y., Hamada, H., Sasaki, T. *et al.* (2003) Role of *TBX1* in human del22q11.2 syndrome. *Lancet* **362**: 1366–1373.

Yamagishi, H., Yamagishi, C., Nakagawa, O., Harvey, R.P. *et al.* (2001) The combinatorial activities of Nkx2.5 and dHAND are essential for cardiac ventricle formation. *Dev. Biol.* **239**(2): 190–203.

Yan, Y.T., Gritsman, K., Ding, J., Burdine, R.D. *et al.* (1999) Conserved requirement for *EGF-CFC* genes in vertebrate left–right axis formation. *Genes Dev.* **13**: 2527–2537.

Yasui, H., Morishima, M., Nakazawa, M., Ando, M. and Aikawa, E. (1999) Developmental spectrum of cardiac outflow tract anomalies encompassing transposition of the great arteries and dextroposition of the aorta: pathogenic effect of extrinsic retinoic acid in the mouse embryo. *Anat. Rec.* **254**: 253–260.

Yatskievych, T.A., Ladd, A.N. and Antin, P.B. (1997) Induction of cardiac myogenesis in avian pregastrula epiblast: the role of the hypoblast and activin. *Development* **124**: 2561–2570.

Yelbuz, T.M., Waldo, K.L., Kumiski, D.H., Stadt, H.A. *et al.* (2002) Shortened outflow tract leads to altered cardiac looping after neural crest ablation. *Circulation* **106**: 504–510.

Yutzey, K.E., Rhee, J.T. and Bader, D. (1994) Expression of the atrial-specific myosin heavy chain AMHC1 and the establishment of anteroposterior polarity in the developing chicken heart. *Development* **120**: 871–883.

Zariwala M., Noone P.G., Sannuti A., Minnix S. *et al.* (2001) Germline mutations in an intermediate chain dynein cause primary ciliary dyskinesia. *Am. J. Resp. Cell Mol. Biol.* **25**: 577–583.

Zhang, H., Toyofuku, T., Kamei, J. and Hori, M. (2003) *GATA-4* regulates cardiac morphogenesis through transactivation of the *N-cadherin* gene. *Biochem. Biophys. Res. Commun.* **312**: 1033–1038.

Zhu, X., Sasse, J., McAllister, D. and Lough, J. (1996) Evidence that fibroblast growth factors 1 and 4 participate in regulation of cardiogenesis. *Dev. Dyn.* **207**: 429–438.

14

The Skin

Ahmad Waseem and **Irene M. Leigh**

Abstract

Skin consists of an outer layer, the epidermis, glued to the inner layer, the dermis. The epidermis and its appendages, such as hair and nail, exist in a dynamic equilibrium between the cells produced by the stem cells located the deepest layer and those lost by differentiation and apoptosis. This mechanism is supported and maintained by a host of proteins, including structural proteins, such as keratins, and proteins involved in cell–cell and cell–extracellular interactions. This chapter describes the basic keratinocyte biology, including the role of epidermal stem cells in normal homeostasis. The role of various genes in keratinocyte differentiation and apoptosis, in migration of epithelial and non-epithelial cells in structural proteins, mutations in enzymes involved in keratinocyte terminal differentiation and proteins involved in cell–cell and cell–extracellular matrix interactions lead to different types of genodermatoses. We have also described skin disorders associated with anomalous signalling, such as mosaicism and ectodermal dysplasia, and rare hereditory syndromes with predisposition to skin cancer, such as Muir–Torre and Gorlin syndromes. At the end of this chapter we have summarized a number of approaches that are being pursued to correct genetic lesions in skin and the challenges researchers face in this area.

Keywords

epidemis, cutaneous diseases, hair and nail, keratinocyte, differentiation, stem cells

Introduction

The skin is the largest organ, providing a protective barrier between the body and its environment. The adult human skin consists of an outer layer, the epidermis, a stratified epithelium derived from embryonic ectoderm, and an inner layer, the dermis, of mesodermal origin. The epidermis is made of several layers of keratinocytes (95%) which are infiltrated with non-epithelial cells, including melanocytes, Langerhans cells (dendritic cells) and Merkel cells (with sensory receptors). The epidermis and dermis

Embryos, Genes and Birth Defects, Second Edition Edited by Patrizia Ferretti, Andrew Copp, Cheryll Tickle and Gudrun Moore

are separated by a basement membrane made of extracellular matrix proteins, which act as a glue between components of the two different compartments. The interaction of epidermis with dermis through the basement membrane is thought to regulate all aspects of skin physiology.

Developmental anatomy

Epidermis development

During embryonic development the epidermis arises as a single layer of keratinocytes at the blastocyst stage, when the ectoderm and endoderm are morphologically defined. A second outer layer, called the periderm, arises at the end of 4 weeks EGA (estimated gestational age) in human and by E12 (day 12 of embryonic development) in mouse (Weiss and Zelickson, 1975). A third intermediate (two outer and one inner) layer appears at 4–9 weeks EGA in human and E13–E16 in mice. In humans, the periderm becomes three layered at 9–10 weeks EGA and begins stratification at 13–19 weeks EGA (Akiyama *et al.*, 2000). In humans it takes 24 weeks and in mice 17 days for all epidermal layers to develop. Mitotic activity is present in all layers, including the two- to three-layered periderm but, as the suprabasal layers begins to show signs of stratification, mitotic activity is drastically reduced in these layers, and in post-natal epidermis cell proliferation is restricted to the basal layer (Fuchs and Byrne, 1994).

Hair follicle development

Morphogenesis of specialized structures, such as hair and nails, begins at the onset of integumental stratification. The hair grows from hair follicles. In humans they begin to develop on the head, particularly on the eyebrows, lower and upper lip and gradually cover the entire body, with the exception of palms and soles. Roughly 5 million hair follicles cover the human body and no additional follicles are formed after birth, although the size of the follicles and hair may change with time, primarily under the influence of hormones (reviewed in Paus and Cotsarelis, 1999).

In embryogenesis the establishment of a dermal papilla is vital to the development of all hair follicles and associated structures, such as sebaceous glands (Figure 14.1). The dermal papilla begins to develop with the aggregation of a group of dermal fibroblasts just below the epidermis. In humans this initial aggregation begins at approximately 9 weeks of gestation and marks the site for the future development of a hair follicle. The keratinocytes above the dermal papilla develop into an epidermal plug, which grows into the dermis at an angle towards the dermal papilla. The communication between the dermal papilla and the epidermal plug results in proliferation and differentiation of epidermal cells into various sheath and hair

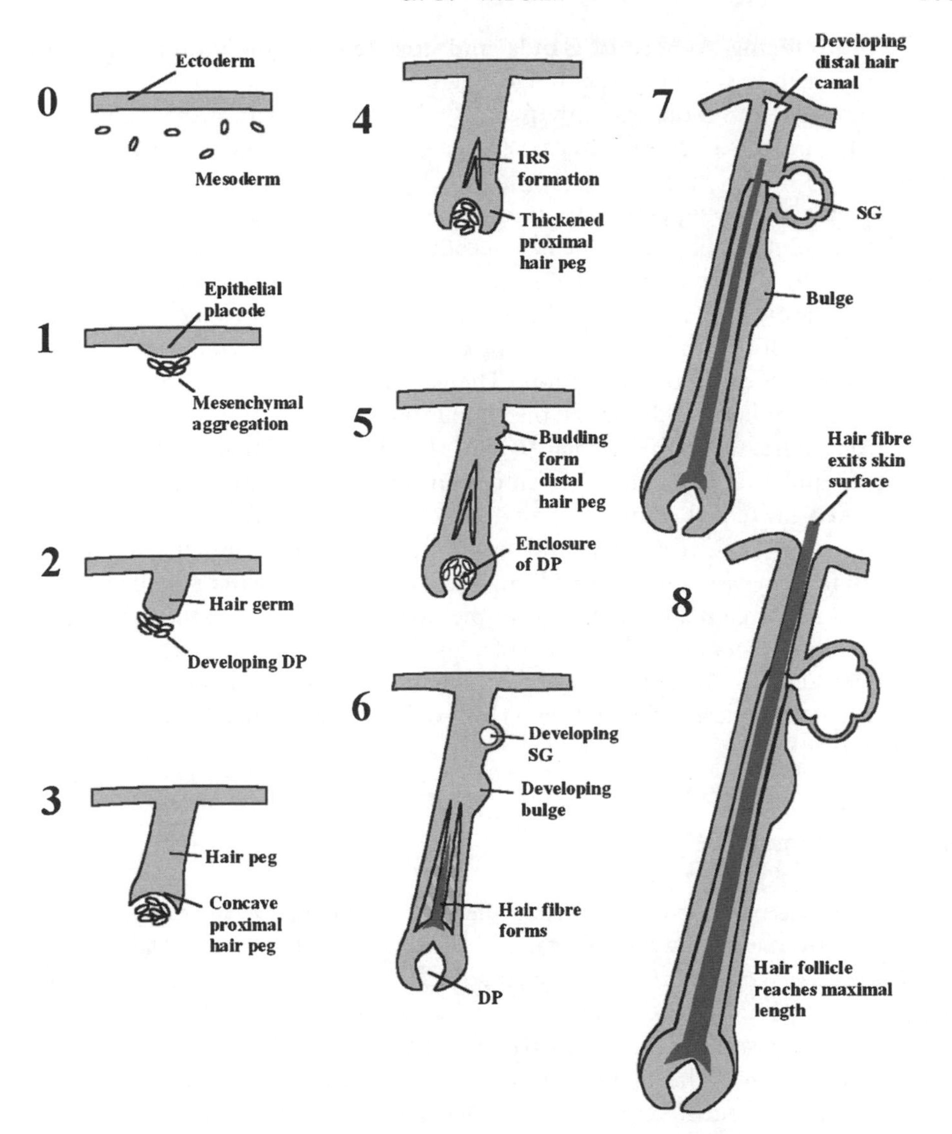

Figure 14.1 Stages of hair follicle development. Schematic diagrams to illustrate 8 distinct stages of hair follicle development. Stage 0–1 represents the induction phase, when coordinated signalling between the ectoderm and underlying mesenchymal cells induces placode formation. Stages 2–4 represent the morphogenesis phase, when the appendage elongates into a hair peg. Stages 5–8 illustrate further enlargement and the differentiation programme resulting in distinct compartments that make up a mature hair follicle. DP, dermal papilla; IRS, inner root sheath; SG, sebaceous gland

fibre structures (Holbrook and Minami, 1991). The gradual differentiation of the hair plug begins with the development of three distinct buds, one above the other. The one closest to the epidermis sometimes develops into sweat glands, but in most cases it regresses and disappears in mature follicles. The cells in the middle bud develop into oil-producing sebaceous glands and the cells in the lower bud form a structure which is known as the 'bulge' (Figure 14.1). A small muscle, the arrector pili, which develops independently from the hair follicle within the dermis, grows towards the bulge region and attaches the epidermis at one end to the follicle at the other.

As the epidermal plug penetrates into the dermis, dermal cells aggregate around it and develop into a fibrous follicular sheath, then a collagen capsule encased by epidermal cells. The epidermal plug seems to push this dermal papilla down into the dermis as it grows to its full size. The dermal papilla contains rounded, mostly non-proliferating cells, which secrete growth factors and matrix components vital for follicle development and maturation. The intercellular signalling between dermal papilla cells and plug cells leads the epidermal cells to arrange themselves into concentric layers above the dermal papilla. The layers eventually differentiate into hair fibres and into the inner and outer root sheaths encasing them. These layers begin to keratinize higher up the hair follicles, while the cells closer to the dermal papilla remain undifferentiated and continue to multiply. As the keratinocytes differentiate they are incorporated into the layers of the hair follicle, become keratinized and eventually are shed from the surface of the skin. Hence, development of a hair follicle requires a series of events, involving induction, initiation, elongation and differentiation. The basic structure of a hair follicle in human embryonic skin is complete by 160 days of gestation (Holbrook and Minami, 1991); reviewed in (Paus *et al.*, 1999).

Nail development

In humans, the nail apparatus starts to develop in the 9th week of gestation and is completed by the 20th week of fetal life (Telfer, 1991; Baran and Dawber, 1994). At 10 weeks, a rectangular area overlying the dorsal tip of the digit defines the nail field from which the entire nail apparatus will develop. The nail field is demarcated by a continuous shallow groove proximally, laterally and distally. The distal ridge, which later becomes the hyponychium, develops in close proximity to the distal groove. The nail matrix grows at the proximal part of the nail field as growing keratinocytes move downward into the dermis. By 11 weeks EGA the proximal and lateral nail folds appear, and the area between distal ridge, lateral nail folds and nail matrix becomes the nail bed (Fistarol and Itin, 2002). At the same time the nail bed begins to keratinize from the distal ridge and by 14 weeks the whole nail bed has developed a granular layer. The nail plate, an accumulation of keratinized 'dead' horn cells (onychocytes), emerges from the nail matrix beneath the proximal nail fold and grows distally. The granular layer of the nail bed gradually disappears and keratinocytes of the nail bed are integrated beneath the nail plate. At 17 weeks the nail plate covers most of the nail bed and at 22 weeks it grows over the distal ridge, now called the hyponychium (Figure 14.2; reviewed in Paus and Peker, 2003).

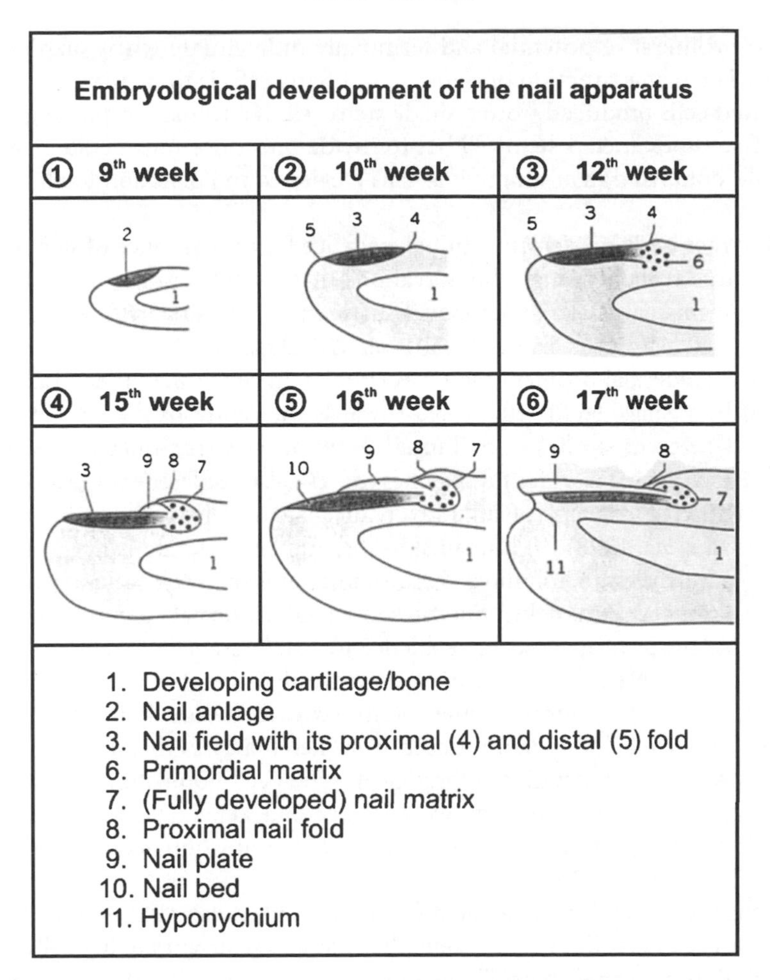

Figure 14.2 Developmental anatomy of nail apparatus. Schematic diagram to illustrate different stages of embryological development of the nail apparatus in primates. Reproduced with permission from Paus and Peker (2003)

Cellular and molecular mechanisms affecting skin development and how they contribute to elucidating the causes of abnormal development

Epidermal stem cell The epidermis is a dynamic structure that renews itself constantly throughout life. Its turnover is estimated at about a week in mice (Potten, 1981) and about 2 months in humans (Hunter *et al.*, 1995). This high turnover requires adult stem cells that are slow-cycling (hence less susceptible to DNA damage) but maintain high self-renewal capacity throughout the life of the individual (Lavker and Sun, 2000). Such cells are biochemically and morphologically primitive and are multipotent. The stem cells divide to produce transit-amplifying cells, which have finite proliferative potential and terminally differentiate when such potential is exhausted. The prime function of transit-amplifying cells is to increase the number of differentiated cells produced from a single stem cell. If a transit-amplifying cell is able to divide five times, then a stem cell has to divide only once to generate one stem cell and 32 differentiated transit-amplifying cells (reviewed in Fuchs and Raghavan, 2002).

Stem cell detection in epidermis Initial evidence for the existence of stem cells in the epidermis came from histological observations that squames in the upper layer of the murine epidermis have a large hexagonal surface area and are arranged into columns that are aligned with cells in the basal layer (Mackenzie, 1970). These columns are referred to as epidermal proliferation units (EPUs), in which the differentiated cells are maintained by a single stem cell located towards the centre of the cluster (Allen and Potten, 1974). Recent studies using human keratinocytes transduced with retrovirus-encoding *LacZ* gene have found patches of β-galactosidase expression, which is consistent with the concept of an EPU with a stem cell at its origin (Mackenzie, 1997; Kolodka *et al.*, 1998). In human skin the number of strata is far greater than in the mouse; nevertheless, a columnar organization similar to the EPU of mouse skin is found. The ability of primary keratinocytes to generate normal epidermis when grafted into a suitable recipient gave strong evidence for the presence of stem cells in culture (Compton *et al.*, 1998). Indeed, *in vitro* clonal analysis by Barrandon and Green has defined three categories of proliferating keratinocytes in culture, holoclones, paraclones and meroclones. Holoclones are produced from keratinocytes that are able to undergo 120–160 divisions and most likely represent the stem cell population. Clones in which keratinocytes have limited proliferative capacity and generate abortive colonies are called paraclones; meroclones contain cells that are in transition between holoclone and paraclone (Barrandon and Green, 1987).

One of the most reliable ways to identify stem cells in epidermis is to distinguish slow-cycling stem cells from the more frequently cycling transit-amplifying cells, using cell kinetic techniques. In this approach all epithelial cells, including the stem and transit-amplifying cells, are first labelled by continuously perfusing the tissue with tritiated thymidine or bromodeoxyuridine. In the subsequent 4–8 week chase period, the label in the rapidly dividing transit-amplifying cells is lost due to dilution and only cells which cycle slowly, and therefore are true stem cells, retain the label (Bickenbach *et al.*, 1986; Morris and Potten, 1994). Since the labelling strategy described above cannot be used on human skin, researchers have looked at molecular markers that can be used to identify stem cells in human epidermis. One of the most widely studied markers is $\beta 1$ integrin, a cell surface receptor that binds to extracellular matrix components including fibronectin and type IV collagen. It has been proposed that high levels of $\beta 1$ integrin expression make stem cells the most adhesive cells in the epidermis, and this characteristic has been used for their isolation (Adams and Watt, 1989; Jones and Watt, 1993). Epidermal stem cells have also been reported to express a high level of $\alpha 6$ integrin (Li *et al.*, 1998), p63 (Pellegrini *et al.*, 2001), keratins K19 (Stasiak *et al.*, 1989), K15 (Lyle *et al.*, 1998), β-catenin (Zhu and Watt, 1999) and melanoma-associated chondroitin sulphate proteoglycan (Legg *et al.*, 2003) and a low level of transferrin receptors (Tani *et al.*, 2000). However, none of these markers give satisfactory results when employed individually to identify a stem cell population.

Stem cells niche Localization of stem cells only in restricted areas of the adult epidermis suggests regulation of their local microenvironment via a combination of cellular activity and extracellular matrix components, in order to control all aspects of their behaviour. This has given rise to the concept of a stem cell 'niche' that supports and controls stemness (Spradling *et al.*, 2001).

Application of the pulse-chase approach to the hair follicle led to the discovery that label-retaining cells (LRCs, stem cells) were mostly localized in the bulge region, a specialized portion of the outer root epithelium defined as the insertion site of the arrector pili muscle (Figure 14.1; Cotsarelis *et al.*, 1990; Taylor *et al.*, 2000). Other evidence to support the hypothesis that the LRCs in the bulge are stem cells includes the formation of viable hair follicles by recombination of dermal papillae with hair follicle fragments containing bulge cells (Kobayashi and Nishimura, 1989). Furthermore, the bulge keratinocytes display very high growth capacity *in vitro* and undergo a transient burst of cell proliferation early in the first phase of the hair growth cycle (anagen), or after stimulation by hair plucking (Cotsarelis *et al.*, 1990). At the ultrastructural level they appear relatively undifferentiated. LRCs are scarce in interfollicular epidermis as compared to the bulge. This led some investigators to propose that in hairy skin the bulge may represent the main stem cell niche in the epidermis, supplying stem cells to both the interfollicular epidermis and the sebaceous glands. Long-term cell kinetic experiments have demonstrated multipotency of the bulge stem cells, which can differentiate not only into the hair lineage but also into interfollicular epidermis and sebaceous glands (Taylor *et al.*, 2000; Oshima *et al.*, 2001). These studies have been contradicted by other findings, demonstrating independent pools of stem cells located in the interfollicular epidermis, sebaceous gland and bulge region (Ghazizadeh and Taichman, 2001) that can replenish each other as required. Nonetheless, currently the most commonly accepted view is that the bulge region in the hair follicle is the major stem cell niche in mammalian epidermis.

Recently, Fuchs and colleagues used a fluorescent label to specifically tag LRCs in the bulge, allowing them to study these cells within their native microenvironment. These studies have finally confirmed many of the characteristics that stem cells are believed to possess. Using microarray hybridization, these investigators have identified more than 100 genes that are preferentially expressed in the niche keratinocytes. These include known stem cell markers, such as stem cell factor (kit ligand) Dab2, ephrin tyrosine kinase receptors (Ephs), tenasin C, interleukin-11 receptor, Id binding protein-2 (Idb-2), four-and-a-half lim domains (Fh11), CD34, S100A6 and growth arrest-specific protein. Many of these are surface receptors and secretory proteins. These data confirm the existence of a unique microenvironment within the bulge that allows stem cells to signal and respond to their surroundings (Fuchs *et al.*, 2004; Tumbar *et al.*, 2004). If stem cells can differentiate into hair matrix as well as interfollicular epidermis and sebaceous gland, then which factor(s) will decide the pathway to choose? The key factor in lineage decision making appears to be the Wnt

β-catenin signalling pathway (Huelsken *et al.*, 2001; Merrill *et al.*, 2001; Niemann *et al.*, 2002). Transgenic experiments have demonstrated *de novo* hair follicle production in postnatal interfollicular epidermis by overexpression of stabilized β-catenin (Gat *et al.*, 1998). Conversely, β-catenin knock-out or inhibition of its activity leads to the development of cysts of interfollicular epidermis with associated sebocytes instead of hair follicles. Thus, the level of β-catenin appears to control lineage selection in epidermal stem cells, with high levels favouring hair follicle formation and low levels stimulating the differentiation of interfollicular epidermis and sebocytes (DasGupta *et al.*, 2002; Niemann *et al.*, 2002).

Regulation of proliferation and differentiation of epidermal keratinocytes

The developing human epidermis at 5–7 weeks EGA contains two single cell layers, a basal layer and an upper periderm. The periderm first becomes three-layered and then gradually stratifies into several layers constituting the suprabasal compartment. The cell number in the two compartments must be regulated precisely to maintain tissue homeostasis, and these mechanisms must begin during embryonic development and continue to operate throughout life, because the epidermis is a self-renewing tissue. These mechanisms must regulate interactions among keratinocytes, interactions of keratinocytes with other cells such as melanocytes and Langerhans cells, adhesion of keratinocytes with basal lamina and regulation of terminal differentiation to produce the cornified envelope (CE) (reviewed in Fuchs and Byrne, 1994). One mechanism vital for tissue homeostasis is the adhesion of basal keratinocytes to the extracellular matrix of lamina lucida to form hemidesmosomes, the electron-dense plaques connecting keratin filaments to filaments of lamina lucida. Not only are these structures important for a stable epidermal–dermal association, they also generate the necessary signalling required for regulation of cell proliferation and differentiation in the epidermis (reviewed in Borradori and Sonnenberg, 1999). The hemidesmosome is assembled by three integral proteins, $\alpha 6$ and $\beta 4$ integrins, bullous pemphigoid 180 (also called type XVII collagen) and several peripheral proteins (Figure 14.3; reviewed in Jones *et al.*, 1998). Present knowledge of hemidesmosome structure is probably incomplete and more components have yet to be identified. Genetic mutations in hemidesmosome-associated proteins result in junctional epidermolysis bullosa (EB), a heterogeneous group of congenital skin-blistering disorders (see section on classes of skin defects).

The stem cells located in the basal layer divide to produce transit-amplifying cells that are committed to differentiation but remain in the basal layer. In response to signals mediated by the epithelial mesenchymal interactions, the committed keratinocytes exit the cell cycle and move up into the spinous layer (Janes *et al.*, 2002). Proliferation of keratinocytes in the basal layer is regulated by growth factors, including epidermal growth factor (EGF) and transforming growth factor α (TGFα) and in hyperproliferating epidermis, such as in psoriasis (see next section), TGFα expression is elevated. Exit from the cell cycle, accompanied by translocation

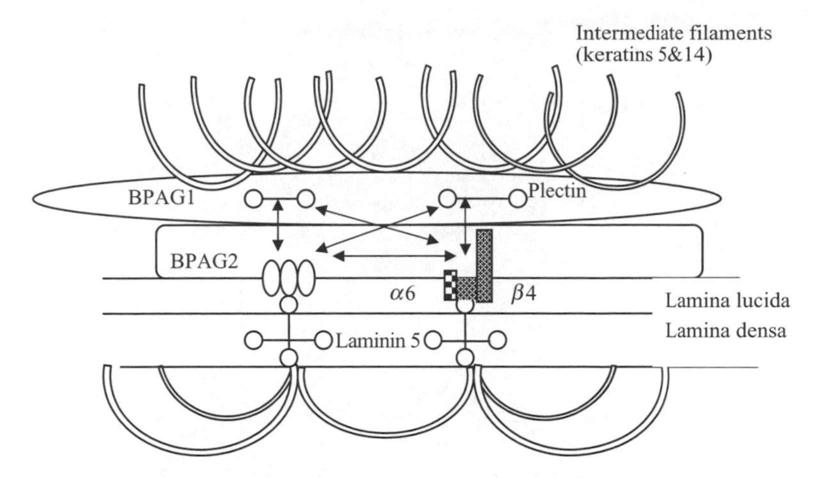

Figure 14.3 Structure of hemidesmosome. Schematic representation of the major proteins interactions involved in the association of hemidesmosomes is indicated. These associations have been identified using *in vitro* binding assays

from the basal to the suprabasal layer, is regulated by several genes, including c-myc, TGFβ and bone morphogenetic protein (BMP)/Smad family members (reviewed in Fuchs and Byrne, 1994)). This triggers a cascade of gene expression which governs the progression of suprabasal keratinocyte differentiation to corneocytes. Epidermal keratinocytes express pairs of type I (acidic) and type II (basic) keratins in a differentiation-specific fashion that allows basal keratinocytes to be distinguished from suprabasal ones (Steinert, 1993). In addition to their structural function, keratins may be involved in signalling pathways, as expression of K10 may play an indirect role in the arrest of the cell cycle and commitment to terminal differentiation via c-Myc.

In normal epidermis basal keratinocytes express keratins K5/K14, whereas suprabasal keratinocytes express keratins K1 and K10. Palmoplantar epidermis expresses K9 suprabasally and interfollicular epidermis expresses K2e in the high stratum spinosum. The epidermis and associated appendages also express other minor keratins in restricted sites. Keratin K15 is expressed in a subpopulation of basal epidermal and outer root sheath basal keratinocytes, where it is preferentially found in slowly cycling keratinocytes. This has led to the suggestion that K15 is a marker of pluripotential stem cells in the bulge region of the hair follicle. Keratin K19 (40 kDa) is also expressed in progenitor subpopulations of keratinocytes in the bulge region of the hair follicle, where it may temporarily stabilize keratin complexes.

Keratins K5, K14 and K15 in the adult epidermis are downregulated as soon as the keratinocytes move from the basal to the spinous layer (see Figure 14.4). The keratinocytes in the embryonic periderm express keratins K7, K13 and K19, but in fetal and adult epidermis they are replaced by K1, K10 and K2e (Moll *et al.*, 1982; Dale *et al.*, 1985). In addition, the keratinocytes in the spinous layer express involucrin, a precursor of the cornified envelope and transglutaminase 1 (TG1), the enzyme responsible for cross-linking involucrin, loricrin and other components

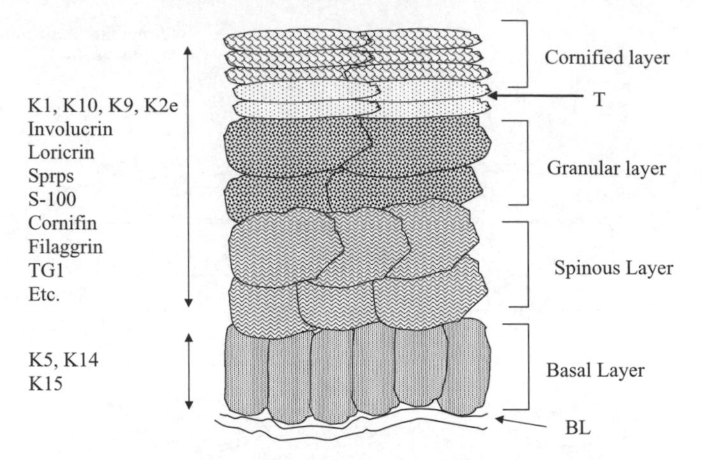

Figure 14.4 Epidermal keratinocyte differentiation. Schematic representation of the major morphologically distinct layers in the epidermis including basal, spinous, granular and cornified. The transition zone (T) separating the dead keratinocytes from the living cells and the basal lamina (BL) separating epidermis from the dermis is shown. The location of expression of marker proteins induced during keratinocyte differentiation is indicated. Redrawn from Eckert and Welter (1996)

of the cornified envelope, as early markers of differentiation. With migration of keratinocytes to the stratum granulosum characterized by the presence of large granules, expression of filaggrin, loricrin (another cornified envelope precursor) and keratin K2e is induced (Figure 14.4). The role of filaggrin during terminal differentiation is to laterally bundle keratin filaments. The different components of the cornified envelope (see Table 14.1) are cross-linked by TG1 to form an insoluble

Table 14.1 Cornified envelop (CE) precursor proteins*

Name	Gene locus	Size (kDa)	Relative abundance in human foreskin CE (%)	Cross-linking sites identified *in vivo*
Involucrin	1q21 (EDC)	65	2–5	Yes
Loricrin	1q21 (EDC)	26	80	Yes
Small proline-rich proteins (SPRs)	1q21 (EDC)	6–26	3–5	Yes
Cystatin A	3cen-q21	12	2–5	Yes
Proelafin	20q12-q13	10	<1	Yes
(Pro)filaggrin	1q21 (EDC)	>400	<1	Yes
Type II keratins	12q13	56–60	<1	Yes
Desmoplakin	6p21-ter	330–250	<1	Yes
Envoplakin	17q25	210	<1	Yes
Perplakin	16p 13.3	195	<1	Yes
S100 proteins	1q21 (EDC)	12	<1	No
Annexin I	9q12-q21.2	36	<1	No

Reproduced with permission from Nemes and Steinert (1999).

envelope just underneath the plasma membrane in the granular layer. These changes are accompanied by a high level of lipid synthesis, including acylated/glycosylated/hydroxylated ceramides, cholestrol and its acyl and sulphate esters and free fatty acids. In the transition zone (Figure 14.4) the cells lose their organelles and nuclei and the proteins are covalently attached to lipids to form the stratum corneum, the cornified layer providing a water-impermeable barrier to the skin (Eckert and Welter, 1996; Nemes and Steinert, 1999).

The normal expression pattern of differentiation markers changes in skin lesions, and the most significant changes are reported for keratin genes. For example, in epidermis undergoing wound repair the epidermal keratins K1 and K10 are suppressed (de Mare *et al.*, 1990; Kallioinen *et al.*, 1995) and additional keratins, K6, K16 and K17, are induced. These keratins are also expressed in the psoriatic epidermis and in the epidermis of hypertrophic and keloid scars (Leigh *et al.*, 1995; Machesney *et al.*, 1998). In squamous cell carcinoma, epidermal expression of differentiation-specific keratins, such as K1, K10 and K2e, is suppressed and simple epithelial keratins, such as K8, K18 and K19, are induced (Markey *et al.*, 1991; Bloor *et al.*, 2003).

One of the most important factors regulating keratinocyte differentiation is extracellular calcium, which is essential for the assembly of desmosomal and adherens junctions. In the epidermis the extracellular calcium concentration increases approximately four-fold between the basal and cornified layers. Exposure of cultured keratinocytes to increased levels of calcium in the medium results in cell cycle arrest and expression of differentiation markers, such as keratins K1 and K10, involucrin, loricrin and filaggrin. Some of these events are mediated by protein kinase C (PKC) (Denning *et al.*, 1995; Yang *et al.*, 2003).

Retinoids are another important regulator of epidermal differentiation. Among the retinoid receptors, RAR-γ1 is the major isoform expressed in epidermis which can mediate the effect of retinoids. The effects of retinoids on epidermal keratinocytes *in vivo* differ significantly from those in culture. *In vivo* exposure of human and murine epidermis to retinoic acid (RA) suppresses the expression of differentiation markers, including filaggrin, transglutaminase, loricrin and involucrin, but does not affect expression of K1 and K10 (Rosenthal *et al.*, 1992). However, keratin K2e transcription in human volunteers was suppressed by 100–1000-fold on exposure to RA (Virtanen *et al.*, 2000). A RA gradient with high levels in the basal layer and low levels in the suprabasal layers, as reported in the literature (Vahlquist *et al.*, 1987), would help to explain the effects of retinoids on the epidermis. *In vitro* studies of RA-treated epidermal keratinocytes have confirmed the *in vivo* data, with the exception that K1, K10 along with markers of cornification are suppressed (Kopan and Fuchs, 1989; Griffiths *et al.*, 1992).

The first evidence that mutations in keratin polypeptides could be involved in genodermatoses (hereditary genetic skin diseases) came from transgenic studies in which expression of a mutant K14 produced severe blistering defects (Vassar *et al.*, 1991). Since then a large number of genetic lesions in the keratin gene family have been associated with skin fragility syndromes (see next section). In addition, there are several birth defects linked to other markers of keratinocytes differentiation, for example, mutations in loricrin and TG1 genes are associated with a variety of congenital skin abnormalities (see next section). Other abnormalities in skin barrier function include X-linked ichthyosis, caused by cholesterol sulphate accumulation because of deficiency of the arylsulphatase C/cholesterol sulphatase gene (reviewed in (Nemes and Steinert, 1999).

Migration of epithelial and non-epithelial cells in the epidermis

Migration of non-epithelial cells into the skin is important. The melanocytes, which are implicated in disorders of impaired pigmentation, are neural crest derivatives that migrate into the epidermis early in development (80–90 days EGA). They also migrate to the hair matrix and outer root sheath of hair follicles. Neural crest migration is regulated by specific interaction between cell surface receptors (e.g. N-cadherin) and the extracellular matrix. The composition of the extracellular matrix changes in different regions of the embryo, and this modulates melanoblast migration and its directionality (reviewed in Bronner-Fraser, 1993). In the developing skin, homing of melanocytes to specific locations is regulated by specific interactions with keratinocytes, which also determine their regional distribution within the epidermis (Haake and Scott, 1991).

Recent studies have identified genes important in the migration of melanoblasts. These include transcription factor paired box 3 (*PAX3*), microphthalmia-associated transcription factor (*MITF*), cell surface transmembrane tyrosine kinase (*C-KIT*), embryonic stem cell growth factor (*STEEL*) and *endothelin-B receptor* (Tomita *et al.*, 2000). The transcription factor PAX3 induces MITF, which in turn induces C-KIT expression. The role of C-KIT is to ensure melanoblast survival during their migration from the neural tube to other regions of the developing embryo (Okura *et al.*, 1995). Mutations in *C-KIT* and *PAX3* are implicated in piebaldism, an autosomal dominant disorder of abnormal pigmentation, and Waardenburg syndrome, characterized by white forelock, deafness and heterochromia iridis (see next section).

Cellular proliferation and migration also plays an equally important role in adult skin, especially during wound healing, which is orchestrated by the interaction of several cell types and involves a wealth of growth factors and cytokines (reviewed in Coulombe, 1997). An injury to the epidermis activates a homeostatic response, leading to inflammation and re-epithelialization, followed by tissue remodelling (reviewed in Martin, 1997). Several studies have shown that damaged keratinocytes release interleukin-1 (IL-1), which induces dermal fibroblasts to migrate, proliferate and produce extracellular matrix components (Mauviel *et al.*, 1993). The released IL-1 also induces dermal endothelial cells to express selectins, which slow down circulating lymphocytes (Wyble *et al.*, 1997) and target them to the site of injury. IL-1 also acts as an autocrine signal that induces surrounding keratinocytes to proliferate, become migratory and express an activation-specific set of genes, including IL-3, IL-6, IL-8, granulocyte-macrophage colony-stimulating factor, tumour necrosis factor-α, transforming growth factor-α and keratins K6 and K16. (Kupper, 1990; Chen *et al.*, 1995). The changes in gene expression that accompany re-epithelialization are similar to those seen in other disorders associated with hyperproliferation, such as psoriasis, contact dermatitis and squamous cell carcinoma (SCC), suggesting considerable overlap in molecular signalling. The development of a normal scar is dependent on the reversal of gene expression at the wound site. However, in some cases the inflammatory and proliferative signals persist even after wound closure, resulting in pathological scars, such as hypertrophic and keloid scars. Although these scars are generally considered to be dermal phenomena (English and Shenefelt, 1999), abnormalities associated with epidermal keratinocytes in these scars have also been identified (Andriessen *et al.*, 1998; Bloor *et al.*, 2003).

Experiments using transgenic animals have identified the genes encoding keratin K6a (Wojcik *et al.*, 2000), fibroblast growth factor 2 (FGF2) (Ortega *et al.*, 1998), intermediate filament protein vimentin (Eckes *et al.*, 2000) and syndecan-1 (Stepp *et al.*, 2002) among the genes whose ablation does not affect skin development but causes a defective wound-healing response. These examples highlight that the significance of certain gene products becomes apparent only when a wound-healing response is induced. At present, birth defects in humans associated with delayed wound healing have not been identified. Genetic defects, however, do predispose certain individuals towards delayed wound healing, resulting in the development of pathological scars.

Role of transcription factors in developing and adult epidermis Identification of major transcription factors controlling gene expression in epidermis is vital to understanding the molecular mechanism of epidermal differentiation. One of the first transcription factors reported to be expressed in the epidermis was AP-2, a 52 kDa protein recognizing a palindromic sequence, GCCNNNGGC, which is functionally active in the promoter of keratin K14 (reviewed in Fuchs and Byrne, 1994). AP-2 has been shown to play a role in the expression of differentiation-specific genes, such as those encoding cystatin A (Takahashi *et al.*, 2000) and keratin K10 (Maytin *et al.*, 1999). There are five known isoforms, AP-2α, -β, -γ, -δ and -ε, and they control basal epidermal genes as well as regulating the expression of differentiation-specific genes (Oyama *et al.*, 2002). However, deficiency of AP-2α in mice does not affect normal expression of intermediate filament proteins, suggesting a considerable degree of functional redundancy amongst the isoforms (Talbot *et al.*, 1999).

The major downstream target of PKC and calcium-induced keratinocyte differentiation is activation protein 1 (AP-1), a family of transcription factors belonging to fos-related (Fra-1, c-fos, fos-B and Fra-2) and jun-related (c-jun, jun-D and jun-B) families. These transcription factors are abundantly expressed in different layers of the epidermis (reviewed in Eckert and Welter, 1996). Indeed, differentiation markers, including keratin K1 (Lu *et al.*, 1994), involucrin (Ng *et al.*, 2000), loricrin (Jang and Steinert, 2002) and TG1 (Jessen *et al.*, 2000) promoters, contain functional AP-1 binding sites. Another target of PKC is C/EBPβ, the CCAAT/enhancer-binding protein, which is implicated in keratinocyte differentiation and induction of differentiation markers including K1, K10 and involucrin (Agarwal *et al.*, 1999; Zhu *et al.*, 1999). Other transcription factors associated with regulation of differentiation markers in keratinocytes include SP-1 (Chen *et al.*, 1997), NF-κB (Ma *et al.*, 1997), STAT-1 (Komine *et al.*, 1996) and retinoid and thyroid hormone receptors (Tomic-Canic *et al.*, 1996).

One of the most important families of genes involved in embryonic development are homeobox genes, a class of transcriptional modulators that share a 183-nucleotide highly conserved region of DNA that encodes a 61-residue homeodomain capable of binding specific DNA sequences by virtue of helix–turn–helix motifs (Botas, 1993; Scott and Goldsmith, 1993; see also Chapter 1).

Evidence for the role of *HOX* genes in epidermal morphogenesis came to light from expression studies in which several members of this family have been detected in fetal and adult murine skin (Detmer *et al.*, 1993; Mathews *et al.*, 1993). These studies found differential expression of *Hox* genes in the developing murine epidermis (Kanzler *et al.*, 1994). Stelnicki and co-workers reported spatial and temporal changes in the expression of several *Hox* genes, including *HOXA4*, *-A5*, *-A7*, *-C4*, *-B7* and *-B4*, in the basal and suprabasal layers of developing human epidermis, but none of these genes was detected in the adult dermis (Stelnicki *et al.*, 1998). Although HOX homeoproteins have been proposed to function as transcription factors, both nuclear and cytoplasmic forms of HOX and transcription factor proteins have been detected in the epidermis (Komuves *et al.*, 2002). HOXC4 expression is closely linked to keratinocyte differentiation in normal epidermis and is also found in the differentiated areas of squamous cell carcinoma (Rieger *et al.*, 1994). Recently HOXA7 was shown to bind a regulatory element in the TG1 promoter and repress its expression. Furthermore, overexpression of HOXA7 inhibited TG1 expression in differentiating keratinocytes, suggesting a role for HOXA7 in the regulation of keratinocyte differentiation (La Celle and Polakowska, 2001). Anomalous expression of *HOX* genes has been associated with tumour development, indeed, overexpression of *HOXB4* in neonatal keratinocytes increases cell proliferation and reduced expression of cell adhesion molecules, the hallmark of cellular transformation. These observations, taken together with strong expression of *HOXB4* in psoriasis and basal cell carcinoma (BCC), suggest a role for this homeobox gene in cell proliferation (Komuves *et al.*, 2002).

In addition to *HOX* genes there are several transcription factors that are differentially expressed in developing epidermis. These include *POU* domain genes, such as *SKN-1a* and *OCT-6,* which contain a second conserved DNA-binding domain called the 'paired' box in addition to the homeobox, and *DLX3. SKIN-1a* and *DLX3* are expressed in differentiating epidermal mouse keratinocytes (Andersen *et al.*, 1997; Park and Morasso, 1999). In human keratinocytes SKN-1a has been shown to activate expression of the differentiation markers keratin K10 (Andersen *et al.*, 1993) and SPRR2A envelope protein (Fischer *et al.*, 1996), and non-hox gene *OCT-6* inhibits expression of basal-specific keratins K5 and K14 (Faus *et al.*, 1994), suggesting induction of differentiation. In addition, SKN-1a can also inhibit the human involucrin promoter. Thus, transcription factors have a dual role in keratinocytes, as they can either inhibit or activate expression of suprabasal markers and can also activate expression of basal cell markers.

At present there is no known human epidermal birth defect associated with genetic lesions in homeobox genes, but dysregulation of *SKN-1a* gene expression has been implicated in oncogenic transformation of keratinocytes (Svingen and Tonissen, 2003). It is also possible that genetic defects in members of this family predispose individuals to certain cutaneous lesions. More research is needed to understand the role of *HOX* genes in cutaneous diseases. Recent studies using transgenic animals suggest that deficiency and overexpression of *HOXC13* in mice cause severe hair growth and patterning defects leading to alopecia (Godwin and Capecchi, 1999; Tkatchenko *et al.*, 2001). Whether defects in this gene are associated with congenital alopecia in humans remains to be established.

Genes involved in hair morphogenesis

Most of our knowledge of hair follicle development comes from studies conducted on animals, particularly mouse. Although gene expression data in human skin are not sufficient for understanding all aspects of human follicle biology, they suggest that the basic mechanism of follicle development is similar in human and mouse (Holbrook *et al.*, 1993). The formation of hair follicles occurs at defined places during embryogenesis and relies heavily upon signalling between dermal cells and overlying epithelial stem cells. This interaction induces fate changes in both cell types and finally results in the differentiation of the hair shaft, root sheaths and dermal papillae (reviewed in Hardy, 1992). Evidence for the existence of these signals came from tissue recombination experiments using mouse and chick skin, because of the similarities in the early steps of hair and feather development. These experiments showed that dermis from a body region that will eventually develop hair or feather, when combined with epidermis from non-hair-bearing region, will direct the formation of appendages with characteristics of the region from which the dermis was derived. Therefore, the first signal for the development of a hair follicle at a particular skin site comes from the dermis and is defined as the 'first dermal signal' (Figure 14.5). There is evidence from both chick feather and murine hair development to suggest that this signal may be Wnt-mediated activation of β-catenin (Noramly *et al.*, 1999), but this has yet to be confirmed.

In response to the first dermal signal, the epithelial stem cells express inducers and inhibitors of placodal fate. Studies of mouse hair and chick feather development have suggested that inducers of placode formation include FGF and its receptor (FGFR), TGF-β and the homeobox-containing genes *MSX1* and *MSX2* (Millar, 2002). In addition in the placode there is induction of ectodysplasin (EDA), a molecule related to TNF, and its receptor, the ectodysplasin A receptor (EDAR) (Headon and Overbeek, 1999). Activation of *EDA* (*Tabby* in mouse) and *EDAR* (*Downless* in

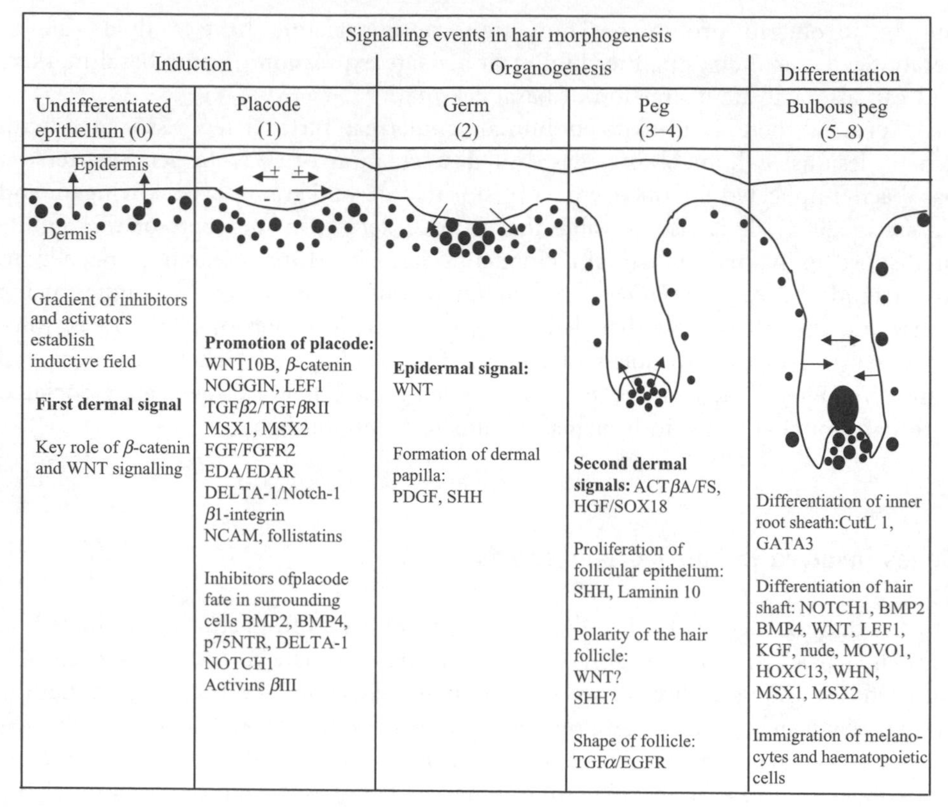

Figure 14.5 Molecular interactions in hair follicle development. Schematic representation of the different stages of murine hair follicle development with key intercellular signals indicated by arrows and candidate molecules important for that stage listed below. See text for details. Redrawn from Millar (2002)

mouse) in epithelial stem cells induces high-level expression of β-catenin, which activates LEF/TCF family of transcription factors. The expression of LEF/TCF transcription factors is suppressed by the activation of BMP-2 and BMP-4, which act as potent inhibitors of hair placode formation (Jung *et al.*, 1998). Inhibition of BMPs by noggin (Botchkarev *et al.*, 1999), a mesenchyme-derived factor, results in LEF/TCF activation at the site of placode formation. In turn, induction of LEF/TCF suppresses transcription of E-cadherin, an important component of adherens junctions. Absence of E-cadherin disrupts their assembly, allowing epithelial stem cells to reposition themselves to form the placode (Jamora *et al.*, 2003). Other factors thought to play a role in placode development include the transmembrane protein Notch and its ligand Delta1 (Figure 14.5).

Following the establishment of the hair placode, the epithelial cells signal to the dermal tissue, which induces the formation of a dermal condensate. Platelet-derived growth factor-A (PDGFA) has been suggested to play a role in the signalling between follicle epithelium and mesenchyme (Karlsson *et al.*, 1999). The expression of β-catenin in the dermal condensate and follicular epithelium (DasGupta and Fuchs, 1999) and the fact that dermal condensate does not develop in the absence of β-catenin suggest an important role for Wnt signalling in its formation (Huelsken *et al.*, 2001). Sonic hedgehog (SHH), which lies downstream of Wnt, also plays a major role in epithelial–mesenchymal signalling (Chiang *et al.*, 1999). The genes encoding Patched (*PTC1*), a receptor for SHH, and Gli1, a transcriptional effector of SHH signalling, are expressed in the follicular epithelium and in the dermal condensate, consistent with the requirement for SHH in the development of both follicle components (Ghali *et al.*, 1999). Mutations in *PTC1* play a causative role in the aetiology of Gorlin's syndrome (see later section), a rare condition featuring the development of several basal cell carcinomas (reviewed in Saldanha *et al.*, 2003). Putative targets of SHH signalling in the developing placode and dermal condensate include Wnt5a and TGFβ2 (Reddy *et al.*, 2001). Other factors that are important at this stage of follicle development include neurotrophin receptors (TrkC, p75) and the gene encoding neurotrophin 3 (reviewed in Millar, 2002).

Following assembly of the dermal condensate, the epithelial germ cells proliferate and grow downward in response to a 'second dermal signal', which is activated by the SHH signalling. Although the identity of this signal is not defined, it may involve activin βA (ActβA), a secreted signalling molecule expressed in the dermal condensate (Feijen *et al.*, 1994). Mice lacking ActβA display defective morphogenesis of vibrissae follicles (Matzuk *et al.*, 1995). Other factors likely to be important in the growth of epithelial germs include hepatocyte growth factor (HGF) and its receptor, MET (Lindner *et al.*, 2000), SOX18, a member of the Sry high mobility group box transcription factor family, (Pennisi *et al.*, 2000), α-catenin (Vasioukhin *et al.*, 2001) and laminin 10 (Li *et al.*, 2003). These molecules appear to promote proliferation of cells in the invaginating epithelial bud, which differentiates into the hair bulb and surrounds the dermal condensate to develop into dermal papilla. As the hair follicle begins to take shape, at least seven different epithelial cell layers arise from the differentiation of matrix cells. One of the key factors determining differentiation of matrix cells into inner root sheath and hair shaft is Notch1 and its ligands Serrate 1 and Serrate 2 (Lin *et al.*, 2000a). The proliferation and differentiation of matrix cells is regulated by a complex mechanism involving members of the BMP family, components of WNT signalling, MOVO1 transcription factor, winged-helix/forkhead transcription factor (FOXN1) and homeobox genes *HOXC13*, *MSX1* and *MSX2* (Figure 14.5). Inhibitors of BMP signalling also inhibit *HOXC13*, *FOXN1*, *MSX1* and *MSX2*, suggesting that these genes require BMP signalling (reviewed in Millar, 2002).

Hair follicles always grow at an angle in relation to the skin surface, pointing from anterior to posterior. There is evidence to suggest that this polarity is regulated by SHH and WNT signalling (Gat *et al.*, 1998; Widelitz *et al.*, 1999). The hair follicles also have a characteristic morphology, which is tightly regulated during development. Mutations in genes encoding TGFα, the EGF receptor and the transcription factor ETS2 have been shown to cause altered hair follicle architecture and wavy hair, indicating that these factors play critical roles in regulating the shape of hair follicle

(Luetteke *et al.*, 1994; Yamamoto *et al.*, 1998) and may determine the variations in hair texture that we see in the human population.

Programmed cell death

Programmed cell death or apoptosis is a process that allows cells to actively remove themselves. Its role in the removal of interdigital mesoderm between developing fingers and toes, in deleting male organs from a female fetus and vice versa and in removing the tails of tadpoles are well known in developmental biology (Hinchcliff, 1981). In adults, apoptosis plays a crucial role in tissue homeostasis (balancing cell production with cell loss) and in the prevention of tumour development by removing cells that are damaged due to repeated exposure to genotoxic substances (Evan and Littlewood, 1998).

In normal epidermis, the number of apoptotic cells is very low. However, this increases drastically following exposure to genotoxic agents such as UV-B (ultraviolet radiation B), which causes 'sunburn'. The cells in 'sunburn' skin patches display the characteristic features of apoptosis, including clumping of chromatin, condensation of nucleus and cytoplasm and formation of membrane blebs containing fragments of nucleus and cell organelles. Cells harbouring damaged DNA should either be repaired or removed from the system. The DNA damage that occurs to transit-amplifying cells is believed to be harmless, since these cells are short-lived and are lost during differentiation. However, if the DNA damage occurs in epidermal stem cells, this must be corrected or the frequency of deleterious somatic mutations will increase and may eventually lead to cancer development (Brash *et al.*, 1996). To remove damaged cells, the cell cycle must be arrested and the DNA repaired before the next round of DNA replication. However, under acute DNA damage a programmed cell death response is initiated to eliminate the badly damaged cells.

One of the key molecules rapidly responding to DNA damage is p53, a multifunctional protein designed to sense the extent of DNA damage. In normal cells the half-life of p53 is very short, as the protein is rapidly degraded by ubiquitinylation. However, in response to DNA damage cellular levels of p53 increase rapidly and induce either cell cycle arrest to permit DNA repair, or apoptosis. Exactly how p53 performs this role is not understood but it may involve oligomerization, which stabilizes p53 and induces its transcriptional activity. These changes lead to activation of p53-responsive genes, such as *p21*, which inhibits cell cycle progression and also induces DNA repair genes. A number of p53-activated pro-apoptotic genes have been identified to date, including *Pig3*, *Killer/DR5*, *Noxa* and *PUMA* (reviewed in Storey, 2002). The *Bax* gene, which is a p53-inducible pro-apoptotic gene in other systems, is not induced in the epidermis in response to UV-B (Qin *et al.*, 2002).

Several p53-independent apoptotic pathways also protect the epidermis from potentially cancer-causing cells. These include mechanisms mediated by Fas/APO-1 receptor and its ligand FasL, and also those involving Bak protein. There are other caspase-independent apoptotic pathways involving AIF (apoptosis-inducing factor),

Smac/Diablo and Omi (reviewed in Storey, 2002). However, these apoptotic pathways have not been characterized in epidermal keratinocytes.

Main classes of skin defects

Skin diseases are traditionally characterized by clinical morphological appearance (macules, papules, nodules, etc.) and clinico-pathological correlations. Recent advances in developmental, molecular and cell biology have aided in understanding the molecular basis of skin diseases. There are a large number of genodermatoses, which manifest from birth and result from defined genetic lesions, especially affecting structural proteins. In addition to such *monogenic* diseases there is genetic predisposition to common inflammatory diseases, such as psoriasis and eczema. Some of the loci responsible for predisposition to *polygenic* diseases have also been identified. Genetic or acquired lesions of specific developmental signalling pathways are involved in skin carcinogenesis and have led to further understanding of the function of these proteins in skin development and organogenesis. Defence mechanisms to ultraviolet radiation, particularly DNA repair and apoptosis, may also be abolished in skin carcinogenesis.

Monogenic disease

Disorders of keratinocyte integrity: skin and hair keratin disorders The integrity of the epidermal keratinocytes depends on the presence and normal function of the keratin intermediate filament cytoskeleton. Mutations in either member of a keratin pair produce a disease characterized by abnormalities in the structure or function of the cells normally expressing that keratin; thus, the disease phenotype mirrors the tissue distribution (Table 14.2; Irvine and McLean, 1999). Most lesions are autosomal dominantly inherited missense mutations within the helix initiation or termination motifs causing severe disease, although milder mutations can occur elsewhere in the keratin molecule. As a result, the keratinocytes show cytolytic changes and increasing fragility to trauma, with increasing hyperkeratosis (epidermolytic hyperkeratosis) if suprabasal keratinocytes are affected.

The first keratin disorder to be analysed was epidermolysis bullosa simplex (EBS) (Horn and Tidman, 2000), a hereditary blistering condition induced by mild trauma, which includes three subtypes: the severe EBS Dowling–Meara (EBS-DM; OMIM 131760) the milder EBS Weber–Cockayne (EBS-WC; OMIM 131800) and the variable EBS Kobner (EBS-K; OMIM 131900). Animal data, combined with human studies, showed that mutations in the genes encoding basal keratin as K5 and K14 cause EBS. Further studies have shown that the severe EBS results from mutations in the helix initiation and termination motifs, whereas mutations outside these regions generally are associated with the milder form of the disease. In a handful of families with EBS Kobner homozygous, mutations leading to premature termination codons result in loss of keratin K14 expression.

Table 14.2 Molecular basis of hereditary skin diseases

Disorders	Defective gene product
Keratin disorders	
Epidermolytic hyperkeratosis	Keratins K1, K10
Ichthyosis hystrix of Curth–Macklin	Keratin K1
Ichthyosis bullosa of Siemens	Keratin K2e
Epidermolytic palmoplantar keratoderma	Keratins K1, K9
Non-epidermolytic palmoplantar keratoderma	Keratins K1, K16
Pachyonchia congenita type 1 (Jadassohn–Lewandowsky)	Keratins K6a, K16
Pachyonchia congenita type 2 (Jackson–Lawlor)	Keratins K6b, K17
Steatocystoma multiplex	Keratins K17
Monilethrix	Keratins Hb1, Hb6
White sponge naevus	Keratins K4, K13
Epidermolysis bullosa simplex	Keratins K5, K14
Connexin disorders	
Erythrokeratoderma variabilis (Mendes da Costa)	Connexin 31
	Connexin 30.3
Hydrotic ectodermal dysplasia (Cloustons)	Connexin 30
Non-syndromic hearing loss (A	Connexin 30
Keratitis–ichthyosis–deafness (KID) syndrome	Connexin 26
Non-syndromic hearing loss (AD, AR)	Connexin 26
Vohkinkels's syndrome	Connexin 26
Neuropathy and hearing loss	Connexin 31
Non-syndromic hearing loss (AD, AR)	Connexin 31
Desmosome disorders	
Striate palmoplantar keratoderma (AD)	Desmoplakin 1, desmoglein 1
Carvajal syndrome (AR)	Desmoplakin 1
Naxos syndrome (AR)	Plakoglobin

The discovery of aggregates of keratins K1 and K10 in the suprabasal epidermis of patients with congenital bullous ichthyosiform erythroderma (BCIE/EHK, OMIM 1138000) was followed by the identification of mutations in keratins K1/K10. Ichthyosis bullosa of Siemens, a milder bullous ichthyosis with superficial skin peeling (IBS; OMIM 146800), which is characterized by epidermolytic hyperkeratosis confined to the high suprabasal layer, was found to result from mutations in K2e, predominantly in the helix termination motif. K9 mutations were found in epidermolytic palmoplantar keratoderma, characterized by hyperkeratosis of palm and sole skin, resulting in diffuse thickening and cracking with no other cutaneous involvement (OMIM 144200). Palmoplantar keratoderma is found in pachyonychia congenital (PC), a group of autosomal dominant syndromes characterized by hypertrophic nail dystrophy, associated with either oral mucosal changes ((PC type I, PC-1; Jadahsson–Lewandowsky syndrome; OMIM 167200) or hair/cystic changes (PC type-2, PC-2; Jackson–Lawler syndrome; OMIM 167210). Keratin 6a and 16 mutations give rise to PC-1 and K6b and K17 mutations to PC-2.

The induction and development of hair follicles during embryonic development and the hair cycle are regulated by recently established developmental pathways (see above). During hair fibre formation, outer root sheath keratinocytes give rise to trichocytes (hair cells) which synthesize different type I and type II hair keratins in a defined pattern (Langbein *et al.*, 2001). Defects in these structural proteins give fragile broken hair in monilethrix (OMIM 158000). Variations in hair shape structure or quality could result from polymorphisms in these structural genes or in keratin-associated small proteins (KAPS; Rogers *et al.*, 2001). At present, woolly or frizzy hair appears to result only from mutations in K17 or desmosomal proteins.

Most keratin mutations occur in hot spots in helix boundary motifs, the most common being arginine 125 in EBS-DM, and result in severe disease phenotypes. There are still epithelial and hair keratins that have not yet been linked to human disorders.

Diseases of intercellular connections There are a number of diseases in which mutations in non-keratin genes that encode cell junction proteins or extracellular matrix receptors cause disorders by disrupting intercellular and intracellular connections with the cytoskeleton. Mutations in the plectin gene with loss of plectin protein disrupts anchorage of intermediate filaments to the hemidesmosome in the skin, and in the muscle sarcomere in EBS with muscular dystrophy. This occurs in the plectin 1a isoform, which is expressed in keratinocytes and co-localizes with the hemidesmosome (McLean *et al.*, 1996). BP180/Collagen XVII and integrin $\beta 4$ are also components of the hemidesmosome, and mutations of the cytoplasmic domain of these proteins also cause skin blistering like EBS (Huber *et al.*, 2002).

Desmosomes provide intercellular adhesion plaques between adjacent keratinocytes in the epidermis, with a complex structure involving plaque proteins and desmosomal cadherins. Mutations in desmoplakin, the major desmosomal plaque protein, result in loss of intercellular cohesion as well as defects in intermediate filament attachment in skin, hair and cardiac muscle. Recessive mutations in desmoplakin lead to palmoplantar keratoderma and woolly hair with cardiomyopathy (Norgett *et al.*, 2000), whereas haplo-insufficiency in desmoplakin due to heterozygous premature termination codons has a skin restricted phenotype of striate palmoplantar keratoderma (Armstrong *et al.*, 1999). Naxos disease with the association of arrhythmogenic right ventricular cardiomyopathy (ARVC), keratoderma and woolly hair results from homozygous deletion of plakoglobin (McKoy *et al.*, 2000). In addition to playing an important role in cell–cell and cell–matrix adhesion, plakoglobin has a function in cell signalling. Loss of function of plakoglobin probably leads to fragility of keratinocytes, trichocytes and myocytes, particularly under physical stress.

Connexins are the constituent proteins of gap junctions. There are at least 10 different connexins in the human epidermis, which are assembled within the endoplasmic reticulum into hexameric complexes called connexons. They migrate to the cell membrane, where they dock to adjacent connexons to form gap junctions, which allow the passage of ions and small proteins of less than 100 Da. Mutations in human connexins have been found to cause not only skin diseases, such as Vohwinkel's keratoderma and erythrokeratoderma variabilis (Richard *et al.*, 1998), but also to be the major cause of hereditary sensorineural deafness. Mutations in the connexins do not segregate to particular regions. There are functional differences between connexin mutations causing deafness, where connexins can traffic to the cell membrane but do not form functional gap junctions, and skin disease mutations, where they fail to traffic and are associated with epidermal cell death.

Basement membrane and collagen disorders The epidermal basement membrane separates the epidermis from the underlying dermis, and can be divided ultrastructurally into four regions: the basal keratinocyte plasma membrane including hemidesmosomes, the lamina lucida, the lamina densa and the sublamina region of the dermis (fibroreticularis). The biochemical composition of these regions has been extensively characterized over the last 10–15 years. They contain protein complexes that anchor the epidermis to the dermis, provide a substrate for cell migration, and also influence tissue differentiation. The major proteins forming the hemidesmosome are integrins ($\alpha6$, $\beta4$), bullous pemphigoid antigen 1 and 2 (Type XVII collagen) and plectin (Figure 14.3). The lamina lucida and densa contain laminins 5, 6 and 10 and type IV collagen. The anchoring fibrils of the sublamina densa are formed by lateral aggregation of type VII collagen dimers. Hereditary defects of specific components of the basement membrane zone give rise to blistering diseases, whereas acquired immunity to the same components gives autoimmune bullous diseases.

All forms of junctional epidermolysis bullosa (JEB; Fine, 1999) are inherited as autosomal recessive traits and are predominantly due to mutations in laminin 5. The most severe form, Herlitz JEB, results from homozygous or compound heterozygous mutations resulting in premature termination codons in α, β or γ chains of laminin 5. The loss of this protein causes ultrastructural loss of anchoring filaments and blistering through the lamina lucida. The skin and mucous membranes are both affected and the resulting widespread loss of skin and mucosae, without scarring, results in death in infancy. Milder forms of JEB are rare and may result from mutations in $\alpha6$, $\beta4$ (Jonkman *et al.*, 2002) integrins or type XVII collagen, as well as from milder laminin 5 mutations.

Dystrophic epidermolysis bullosa (DEB) may be either recessively (RDEB) or dominantly (DDEB) inherited and is characterized by a loss of anchoring fibrils, caused by dominant negative mutations in the type VII collagen gene. In RDEB, premature termination codons result in loss of type VII collagen and affected children not only blister severely following minor trauma but also develop scarring, which results in mitten hand and foot deformities (pseudosyndactyly). Although through greatly improved nursing care children with RDEB survive into adult life, a major complication of the disease is the greatly increased risk of squamous cell carcinomas of the skin, with a cumulative risk of 77% by the age of 60. This may also be a limiting factor for the use of *ex vivo* gene therapy using transduced autologous keratinocytes in these patients.

Other collagen genes are mutated in Ehlers–Danlos syndrome, particularly collagens I, III and V (COL1 A1/A2, COL3 A1, COL5 A1/2). Alterations in dermal collagen produce the characteristic hyperelastic skin and hypermobile joints. Changes in the vascular intima are more serious and can lead to life-threatening vascular events. Mild defects in collagen proteins are probably fairly common within the normal population.

Cornification disorders: ichthyoses The ichthyoses, literally meaning fish (*ichthys*), are characterized by dry scaly skin, which may be associated with altered epidermal differentiation (Traupe, 1989). Hereditary ichthyoses are usually present at birth and show abnormal differentiation or cornification on biopsy. The most common form of ichthyosis, ichthyosis vulgaris, has two forms: X-linked recessive (XLRI) and autosomal dominant inheritance (ADIV). The molecular mechanism of XLRI was one of the first to be described following the observation that women with an affected fetus had low levels of urinary oestriol, due to steroid sulphatase (STS) deficiency (Webster *et al.*, 1978). This is caused by deletion or mutation of the *STS* gene on chromosome Xp22.32. The failure to hydrolyse cholesterol sulphate in the high epidermis leads to failure of desquamation, with an accumulation of corneocytes, a thick stratum corneum and a normal granular cell layer. Clinically this presents as large plate like scales, which are brown in colour, and affects mainly boys. In contrast, the mechanisms underlying ADIV are not clearly understood, although this form is much more common, with finer scaling, hyperlinear palms, flexural sparing and an association with atopy. Genetic studies have shown that abnormalities in profilaggrin expression, suggested by diminution in keratohyalin granules and stratum granulosum, are secondary to the primary defect.

Lamellar ichthyosis is an autosomal recessive form of ichthyosis with some overlaps with non-bullous ichthyosiform erythroderma (NBIE). This results from deficiency in transglutaminase-1. In this severe disorder, at birth, babies (collodion babies) are encased by a collodion membrane (sausage skin, shiny, tight inelastic scale, resembling oiled parchment), which is subsequently shed and replaced by large plate-like brown scales with no erythroderma. Secondary changes include ectropion (stretching of the eyelid) and alopecia (hair loss), nail plate changes, heat intolerance and keratoderma. NBIE is a biologically heterogenous group of conditions, with a collodion membrane at birth followed by a generalized erythroderma and brawny scaling. Multiple genetic loci have been described.

Erythrokeratodermas are sometimes confused with other syndromes because of the presence of palmoplantar keratoderma with fixed red scaly plaques and some more transient erythemas in early life. Some cases have been found to result from mutations in connexin genes *GBJ3* and *GBJ4*, encoding the gap junction proteins connexins 31 and 30.3.

Animal models of keratin, connexin and junctional protein defects There are several animal models, largely transgenic mice with gene knockouts, that provide models for the genodermatoses. However, as many of the human diseases have dominant disruptor mutations, the parallels are not exact. The ability to readily culture human epidermal keratinocytes has allowed the development of disease-associated cell lines and transduction of normal keratinocytes with mutant genes, which can provide insight into the biological effects of structural gene dysfunction (Morley *et al.*, 2003). A major goal for clinicians is to develop strategies for the treatment of affected individuals. Gene therapy is particularly difficult in the case of diseases caused by a dominant negative mutation. Approaches to treatment have included targeting the mutant DNA or RNA by chimeraplasts (small chimeric RNA/DNA oligonucleotides) and the use of ribozymes to specifically cleave target mRNA sequences, but the use of short inhibitory RNA (siRNA) to inactivate mutant RNA (Elbashir *et al.*, 2001) offers the most promising route to date.

Pigmentation disorders Newborn skin is not fully pigmented and darkens in neonatal life. Melanocytes reside in the basal epidermis (1 in 10 basal cells) and their dendrites are in contact with several keratinocytes (30–40). Despite the wide range of skin colour, the density of melanocytes is similar in all races: the differences in skin colour result from the amount and type of pigment produced by the melanocyte. The genetic basis of skin colour appears to result from polymorphisms of the MCIR (melanocortin 1 receptor). The cytoplasm of the melanocyte contains a unique organelle, the melanosome, enclosing a scaffold-binding melanin and enzymes regulating its biosynthesis. The key enzyme is tyrosinase, which catalyses the formation of DOPA quinone from DOPA. Melanins quench oxidative free radicals generated by UVR, so when the skin is irradiated the melanosomes increase in size and are transferred into the keratinocytes. Delayed tanning is visible 48–72 hours after UVR and represents new tyrosine-mediated pigment production.

Oculocutaneous albinism (OCA) is characterized by a genetic loss of melanin pigment in the skin, hair and eyes, which is usually autosomal recessively inherited. In African patients the extreme sensitivity to ultraviolet radiation and high risk of skin cancer of the white skin is the major cause of death. Four different types of albinism have been defined genetically. OCA1 results from mutations in the tyrosinase gene, leading to reduced or absent tyrosinase activity and hence loss of melanogenesis, although the number of melanocytes is normal (Toyofuku *et al.*, 2001). OCA2 results from mutations in the *P* gene, OCA3 from mutations in the tyrosinase-related protein 1 gene and OCA4 is heterogenous. Photophobia results from loss of retinal pigment and misrouting of optic fibres results in nystagmus, strabismus and monocular vision. The classical OCA1 is associated with white hair, milky skin and blue grey eyes at birth, but other variants may have more subtle pigmentary changes and develop large pigmented naevi.

Piebaldism is a rare autosomal dominant disorder where there are fixed areas of loss of pigment in skin and hair, particularly a white forelock. The leukoderma is characteristically in the midline from the front of scalp, forehead and trunk. There may be islands of pigmented macules within well-circumscribed milk-white areas where there are no melanocytes. This syndrome results from mutations in the KIT gene on chromosome 4p12, which encodes a tyrosine kinase transmembrane receptor on melanocytes essential for melanoblast migration from the neural crest development.

Waardenburg's syndrome is a very rare condition resulting from mutations in PAX3 and MITF transcription factors (see Table 14.2, Chapter 10). A white forelock, heterochromia of the iris, congenital deafness and facial dysmorphism are characteristic of this syndrome (Spritz, 1997).

Signalling disorders

Mosaicism Cutaneous mosaic traits follow characteristic whorled patterns on the skin, which are called Blaschko's lines, defined as a stereotyped pigmented pattern assumed by many naevoid lesions (Figure 14.6). Mosaicism may occur through several mechanisms. Two distinct cell clones may arise early in embryogenesis from X chromosome inactivation, somatic mutation, gametic half-chromatid mutation or

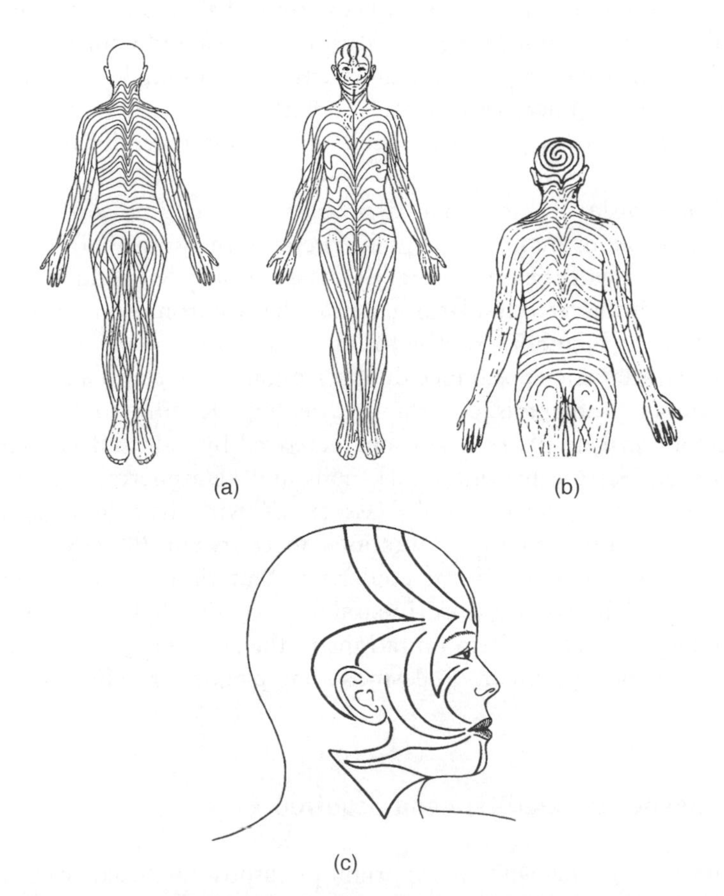

Figure 14.6 Blaschko's lines, (a) as originally described, (b, c) as modified by Happle (1985) and Bolognia *et al.* (1994), respectively. Reprinted with permission from Bolognia *et al.* (1994)

true chimerism. These may occur in X-linked disorders from lyonization (process of inactivation of one X chromosome), but may also result from somatic mutation or chimerism. It was assumed that the patterning reflected the migration of cells during embryogenesis, but this has not been definitively investigated (Paller, 2001). An example of an X-linked dominant disorder is incontinentia pigmenti, which is lethal in males. In young female carriers this disorder manifests as inflammatory whorls and blisters following Blaschko's lines, but later in life it causes hyperkeratotic pigmented lesions and then atrophic pale lesions. Linkage to Xp28 led to the identification of mutations in the *NEMO* gene (Aradhya *et al.*, 2001). Somatic mutations of an autosomal dominant single-gene disorder that result in localized lesions in Blaschko's lines can occur in epidermal keratins. Since these mutations may be carried by a proportion of germ cells, such individuals with congenital epidermolytic epidermal naevi may bear offspring at risk of having the full syndrome (Paller *et al.*, 1994). A number of other naevi, such as sebaceous and epidermal naevi, are only seen in a localized form, which suggests that a generalized form would be lethal. Inflammatory skin lesions do appear in adults, apparently distributed along Blaschko's lines, such as linear psoriasis, lichen planus, lichen striatus, morphoea and vitiligo. As these generalized disorders may have a polygenic inheritance with multiple susceptibility genes, it is hypothesized that linear lesions could reflect a somatic mutation in a susceptibility gene. As few of these genes have been identified to date, this is difficult to test.

Ectodermal dysplasias Ectodermal dysplasias are genetic disorders resulting in abnormalities of two or more ectodermal structures, including hair, teeth, nails and sweat glands. Hypohydrotic ectodermal dysplasia, usually X-linked recessive, results from mutations in the ectodysplasin gene on the X chromosome (Monreal *et al.*, 1999). Children have smooth skin without sweat pores, sparse fair hair, pointed teeth and a characteristic appearance due to frontal bossing and a saddle nose. The major functional problem is control of sweating, leading to heat intolerance. Hydrotic ectodermal dysplasia is also characterized by hair and nail changes, but no tooth deformities are present. Thick nails and keratoderma may be confused with PC-1, but are characteristically associated with hair loss (unlike PC-1). Autosomal dominantly inherited mutations in connexin 30 (*GBJ 6* gene) cause loss of gap junction function in the epidermis. Mutations in many connexins are also associated with hereditary sensorineural deafness (Table 14.2). Other even rarer ectodermal dysplasias result from mutations in the *p63* gene (AEC syndrome; EEC syndrome) and plakophilin 1, a desmosomal plaque protein involved in cell signalling.

Skin carcinogenesis: hereditary and acquired

Rare hereditary syndromes with an important predisposition to skin cancer have been critical in discovering DNA repair pathways, the importance of hedgehog signalling and cyclin-dependent kinases.

Xeroderma pigmentosum (XP, autosomal recessive) is characterized by defects in nucleotide excision repair. There are eight complementation groups (XPA, B, C, D, E, F, G, X) that result in a failure to repair UVR-induced DNA damage. DNA repair requires recognition of the damaged DNA, unwinding of DNA by helicase activity and then excision of the damaged DNA and its removal by endonucleases. Each of these steps of DNA repair can be affected by a defective *XP* gene. Photophobia and characteristic freckling of sun-exposed skin lead to both melanomas and non-melanoma skin cancers. Thorough sun protection may help to protect affected individuals.

In Muir–Torre syndrome (Lynch and Fusaro, 1999), in contrast, defects in mismatch repair genes, such as *MSH2* and *MLH1*, are due to autosomal dominant mutations. Sebaceous neoplasms develop in association with internal malignancies, particularly non-polyposis coli-associated colorectal and genitourinary cancers. The failure of mismatch repair produces microsatellite instability, which can be used in diagnosis. Acquired abnormalities of mismatch repair have been found in sporadic sebaceous carcinomas.

Naevoid basal cell carcinoma syndrome (NBCCS, Gorlin's syndrome) is an autosomal recessive disorder characterized by numerous naevoid basal cell carcinomas of the skin, with characteristic palmoplantar pits (Gorlin, 1995). Other associated neoplasms, such as medulloblastoma are present in these patients. Skeletal abnormalities commonly occur and include odontogenic keratocysts, bifid ribs and microcephaly. A classical genetic approach in affected kindreds identified linkage to chromosome 9q22.3-q31, and the candidate gene *PTCH* was found to be mutated. The PTCH protein inhibits the transmembrane protein Smoothened (SMO), which signals to the nucleus via the GLI1, GLI2 and GLI3 proteins. *PTCH* mutations cause failure of SMO repression and can be mimicked by activating mutations of *SMO*. Mutations in these components of the SHH signalling pathway have been found in basal cell carcinomas, emphasizing the importance of this signalling pathway in development, particularly of hair follicle epithelium (see earlier section). Acquired mutations of the SHH pathway have been found in sporadic basal cell carcinoma, which is thought to derive from follicular progenitors, as GLI1 and GLI3 are upregulated in this carcinoma (Owens and Watt, 2003).

Growth and differentiation disorders

Many inflammatory skin diseases perturb the normal epidermal homeostasis. In normal epidermis the basal cell compartment contains the epidermal stem cells and transit-amplifying cells (see earlier section). As they commence their complex terminal differentiation process, the cells leave the basal compartment and move suprabasally, becoming terminally differentiated corneocytes. Activation of the epidermis, for example due to the presence of activated T cells, results in an altered pattern of differentiation often associated with epidermal hyperproliferation. This occurs in the common skin disease psoriasis, a polygenic T cell-mediated disease.

Affected individuals have a thickened hyperproliferative epidermis which turns over very rapidly and fails to cornify fully, leading to loss of the stratum granulosum and to a parakeratotic stratum corneum retaining epidermal cell nuclei. The majority of progenitor cells (particularly transit-amplifying cells) are recruited into the cell cycle. The growth fraction approaches 100%, in contrast with the normally low percentage of cycling cells (<10%), and the cell cycle lasts around 50 hours. This expansion of the dividing cell population means that there are extensive suprabasal mitoses over two to three layers, resulting in a delay in the expression of suprabasal keratins. There is *de novo* expression of hyperproliferative keratins K6, K16 and K17 and premature expression of involucrin, the soluble precursor of the cornified envelope. These changes also occur in normal physiological wound healing, perhaps facilitating migration and proliferation of the epidermis to heal a defect, but they rapidly subside (10–14 days) when the epidermis stabilizes. It is likely that these changes in differentiation are secondary to the proliferative changes, although a few studies have suggested that expression of the high molecular weight keratins forming the more rigid keratin filaments suprabasally prevent cell division from taking place. Altered ceramide expression and barrier function have been reported in atopic eczema and in hyperproliferative genodermatoses, such as NBIE; however, a genetic basis for these changes has not been found. Much current research is aimed at identifying multiple susceptibility genes for diseases believed to be polygenic, particularly atopic eczema and psoriasis. Although several candidate loci have been described, few susceptibility genes have so far been identified.

Future perspectives

An interesting development in the last 15 years is the realization that the epidermis contains stem cells buried in the basal layer. This gives an entirely new perspective to developing treatments for skin diseases such as neoplasia and genodermatoses. It is now widely recognized that genetic defects must occur in stem cells, which survive long enough to accumulate mutations, whereas the transit-amplifying cells are lost during differentiation. The isolation and genetic manipulation of epidermal stem cells are among the challenges facing dermatologists for the treatment of skin diseases. Since the skin is the most accessible tissue, it provides the easiest source of adult stem cells. There is evidence to suggest that adult stem cells are plastic and can be reprogrammed to produce cells of other tissue lineages. Although the underlying mechanisms are poorly understood, the possibility of such a trans-differentiation opens new avenues for the use of epidermal stem cells for the treatment of other tissue degenerative disorders.

Technological advancements have allowed identification of genes involved in most genodermatoses. This knowledge has unfortunately not yet been translated into therapy, primarily because the technology required for correcting a diseased gene or replacing it with a healthy one either does not exist or it is at very early stages of development. There are two main technical problems for the development of gene therapy for genodermatosis patients: first, the delivery of specific genes into the skin is highly inefficient; and second, long-term maintenance of gene expression in the skin is currently impossible. The delivery of genes can be facilitated by genetically modifying the epidermal keratinocytes prior to grafting them back in the patient's skin. The drawback is that the keratinocytes must be enriched with epidermal stem cells, otherwise the grafted cells will be lost through epidermal differentiation. A grafting approach has been used to treat keratinocytes from recessive genodermatoses, including Herlitz junctional epidermolysis bullosa (Robbins *et al.*, 2001) and X-linked ichthyosis (Freiberg *et al.*, 1997). Other *in vivo* approaches being developed include the introduction of DNA constructs into the skin using a variety of techniques, such as direct injection (Ghazizadeh *et al.*, 1999), liposomes (Raghavachari and Fahl, 2002) or a gene gun (Lin *et al.*, 2000b).

A real challenge will be to develop therapeutic strategies for dominant negative disorders, such those caused by mutant keratin genes. Since the mutant gene is present in the germ line and therefore in epidermal stem cells, potential therapeutic approaches should either include silencing the gene or correcting the mutation. In this context, pharmacological substances such as 'designer' retinoids can be developed to suppress certain keratin genes. Specific keratin genes could also be inactivated by using designer ribozymes, small catalytic RNA molecules, or by siRNA. RNA trans-splicing, where a cellular splicing system can be exploited to correct a mutation, has been successful in the case of the collagen 17A1 gene (Dallinger *et al.*, 2000). The use of gene-specific small DNA–RNA hybrids to target a mutant gene for corrective purposes is another approach that has been successful in correcting albinism in experimental animals (Alexeev *et al.*, 2000). In spite of these advances, there are enormous hurdles to overcome before any of these approaches can reach the clinic for the treatment of genodermatoses.

References

Adams, J.C. and Watt, F.M. (1989) Fibronectin inhibits the terminal differentiation of human keratinocytes. *Nature* **340**: 307–309.

Agarwal, C., Efimova, T., Welter, J.F., Crish, J.F. and Eckert, R.L. (1999) CCAAT/enhancer-binding proteins. A role in regulation of human involucrin promoter response to phorbol ester. *J. Biol. Chem.* **274**: 6190–6194.

Akiyama, M., Smith, L.T. and Shimizu, H. (2000) Expression of transglutaminase activity in developing human epidermis. *Br. J. Dermatol.* **142**: 223–225.

Alexeev, V., Igoucheva, O., Domashenko, A., Cotsarelis, G. and Yoon, K. (2000) Localized *in vivo* genotypic and phenotypic correction of the albino mutation in skin by RNA–DNA oligonucleotide. *Nat. Biotechnol.* **18**: 43–47.

Allen, T.D. and Potten, C.S. (1974) Fine-structural identification and organization of the epidermal proliferative unit. *J. Cell. Sci.* **15**: 291–319.

Andersen, B., Schonemann, M.D., Flynn, S.E. *et al.* (1993) Skn-1a and Skn-1i: two functionally distinct Oct-2-related factors expressed in epidermis. *Science* **260**: 78–82.

Andersen, B., Weinberg, W.C., Rennekampff, O., McEvilly, R.J. *et al.* (1997) Functions of the POU domain genes *Skn-1a/i* and *Tst-1/Oct-6/SCIP* in epidermal differentiation. *Genes Dev.* **11**: 1873–1884.

Andriessen, M.P., Niessen, F.B., Van de Kerkhof, P.C. and Schalkwijk, J. (1998) Hypertrophic scarring is associated with epidermal abnormalities: an immunohistochemical study. *J. Pathol.* **186**: 192–200.

Aradhya, S., Woffendin, H., Jakins, T., Bardaro, T. *et al.* (2001) A recurrent deletion in the ubiquitously expressed *NEMO* (*IKK-γ*) gene accounts for the vast majority of incontinentia pigmenti mutations. *Hum. Mol. Genet.* **10**: 2171–2179.

Armstrong, D.K., McKenna, K.E., Purkis, P.E., Green, K.J. *et al.* (1999) Haplo-insufficiency of desmoplakin causes a striate subtype of palmoplantar keratoderma. *Hum. Mol. Genet.*: **8**: 143–148.

Baran, R. and Dawber, R.P.R. (1994) *Diseases of the Nails and Their Management*, 2nd edn. Blackwell Science: Oxford, 1994.

Barrandon, Y. and Green, H. (1987) Three clonal types of keratinocyte with different capacities for multiplication. *Proc. Natl Acad. Sci. USA* **84**: 2302–2306.

Bickenbach, J.R., McCutecheon, J. and Mackenzie, I.C. (1986) Rate of loss of tritiated thymidine label in basal cells in mouse epithelial tissues. *Cell Tissue Kinet.* **19**: 325–333.

Bloor, B.K., Tidman, N., Leigh, I.M., Odell, E. *et al.* (2003) Expression of keratin K2e in cutaneous and oral lesions: association with keratinocyte activation, proliferation, and keratinization. *Am. J. Pathol.* **162**: 963–975.

Bolognia, J.L., Orlow, S.J. and Glick, S.A. (1994) Lines of Blaschko. *J. Am. Acad. Dermatol.* **31**: 157–190.

Borradori, L. and Sonnenberg, A. (1999) Structure and function of hemidesmosomes: more than simple adhesion complexes. *J. Invest. Dermatol.* **112**: 411–418.

Botas, J. (1993) Control of morphogenesis and differentiation by *HOM/Hox* genes. *Curr. Opin. Cell Biol.* **5**: 1015–1022.

Botchkarev, V.A., Botchkareva, N.V., Roth, W., Nakamura, M. *et al.* (1999) Noggin is a mesenchymally derived stimulator of hair-follicle induction. *Nat. Cell Biol.* **1**: 158–164.

Brash, D.E., Ziegler, A., Jonason, A.S., Simon, J.A. *et al.* (1996) Sunlight and sunburn in human skin cancer: p53, apoptosis, and tumour promotion. *J. Invest. Dermatol. Symp. Proc.* **1**: 136–142.

Bronner-Fraser, M. (1993) Mechanisms of neural crest cell migration. *Bioessays* **15**: 221–230.

Chen, J.D., Lapiere, J.C., Sauder, D.N., Peavey, C. and Woodley, D.T. (1995) Interleukin-1 alpha stimulates keratinocyte migration through an epidermal growth factor/transforming growth factor-alpha-independent pathway. *J. Invest. Dermatol.* **104**: 729–733.

Chen, T.T., Wu, R.L., Castro-Munozledo, F. and Sun, T.T. (1997) Regulation of *K3* keratin gene transcription by Sp1 and AP-2 in differentiating rabbit corneal epithelial cells. *Mol. Cell Biol.* **17**: 3056–3064.

Chiang, C., Swan, R.Z., Grachtchouk, M., Bolinger, M. *et al.* (1999) Essential role for sonic hedgehog during hair follicle morphogenesis. *Dev. Biol.* **205**: 1–9.

Compton, C.C., Nadire, K.B., Regauer, S., Simon, M. *et al.* (1998) Cultured human sole-derived keratinocyte grafts re-express site-specific differentiation after transplantation. *Differentiation* **64**: 45–53.

Cotsarelis, G., Sun, T.T. and Lavker, R.M. (1990) Label-retaining cells reside in the bulge area of pilosebaceous unit: implications for follicular stem cells, hair cycle, and skin carcinogenesis. *Cell* **61**: 1329–1337.

Coulombe, P.A. (1997) Towards a molecular definition of keratinocyte activation after acute injury to stratified epithelia. *Biochem. Biophys. Res. Commun.* **236**: 231–238.

Dale, B.A., Holbrook, K.A., Kimball, J.R., Hoff, M. and Sun, T.T. (1985) Expression of epidermal keratins and filaggrin during human fetal skin development. *J. Cell Biol.* **101**: 1257–1269.

Dallinger, G., Puttaraju, M., Mitchell, L.G., Yancey, K.B. *et al.* (2000) Collagen 17A1 gene correction using spliceosome mediated RNA trans-splicing (SMaRT™) technology. *J. Invest. Dermatol.* **115**: 332.

DasGupta, R. and Fuchs, E. (1999) Multiple roles for activated LEF/TCF transcription complexes during hair follicle development and differentiation. *Development* **126**: 4557–4568.

DasGupta, R., Rhee, H. and Fuchs, E. (2002) A developmental conundrum: a stabilized form of β-catenin lacking the transcriptional activation domain triggers features of hair cell fate in epidermal cells and epidermal cell fate in hair follicle cells. *J. Cell Biol.* **158**: 331–344.

de Mare, S., van Erp, P.E., Ramaekers, F.C. and van de Kerkhof, P.C. (1990) Flow cytometric quantification of human epidermal cells expressing keratin 16 *in vivo* after standardized trauma. *Arch. Dermatol. Res.* **282**: 126–130.

Denning, M.F., Dlugosz, A.A., Williams, E.K., Szallasi, Z. *et al.* (1995) Specific protein kinase C isozymes mediate the induction of keratinocyte differentiation markers by calcium. *Cell Growth Differ.* **6**: 149–157.

Detmer, K., Lawrence, H.J. and Largman, C. (1993) Expression of class I homeobox genes in fetal and adult murine skin. *J. Invest. Dermatol.* **101**: 517–522.

Eckert, R.L. and Welter, J.F. (1996) Transcription factor regulation of epidermal keratinocyte gene expression. *Mol. Biol. Rep.* **23**: 59–70.

Eckes, B., Colucci-Guyon, E., Smola, H., Nodder, S. *et al.* (2000) Impaired wound healing in embryonic and adult mice lacking vimentin. *J. Cell Sci.* **113**(13): 2455–2462.

Elbashir, S.M., Martinez, J., Patkaniowska, A., Lendeckel, W. and Tuschl, T. (2001) Functional anatomy of siRNAs for mediating efficient RNAi in *Drosophila melanogaster* embryo lysate. *EMBO J.* **20**: 6877–6888.

English, R.S. and Shenefelt, P.D. (1999) Keloids and hypertrophic scars. *Dermatol. Surg.* **25**: 631–638.

Evan, G. and Littlewood, T. (1998) A matter of life and cell death. *Science* **281**: 1317–1322.

Faus, I., Hsu, H.J. and Fuchs, E. (1994) Oct-6: a regulator of keratinocyte gene expression in stratified squamous epithelia. *Mol. Cell Biol.* **14**: 3263–3275.

Feijen, A., Goumans, M.J. and van den Eijnden-van Raaij, A.J. (1994) Expression of activin subunits, activin receptors and follistatin in postimplantation mouse embryos suggests specific developmental functions for different activins. *Development* **120**: 3621–3637.

Fine, J.D.(1999) The classification of inherited epidermolysis bullosa: current approach, pitfalls, unanswered questions and future directions. In *Epidermolysis Bullosa: Clinical, Epidemiological and Laboratory Advances and the Findings of the National Epidermolysis Bullosa Registry*, Moshell, A. (ed.). Johns Hopkins University Press: Baltimore, MD; 114–146.

Fischer, D.F., Gibbs, S., van De Putte, P. and Backendorf, C. (1996) Interdependent transcription control elements regulate the expression of the *SPRR2A* gene during keratinocyte terminal differentiation. *Mol. Cell Biol.* **16**: 5365–5374.

Fistarol, S.K. and Itin, P.H. (2002) Nail changes in genodermatoses. *Eur. J. Dermatol.* **12**: 119–128.

Freiberg, R.A., Choate, K.A., Deng, H., Alperin, E.S. *et al.* (1997) A model of corrective gene transfer in X-linked ichthyosis. *Hum. Mol. Genet.* **6**: 927–933.

Fuchs, E. and Byrne, C. (1994) The epidermis: rising to the surface. *Curr. Opin. Genet. Dev.* **4**: 725–736.

Fuchs, E. and Raghavan, S. (2002) Getting under the skin of epidermal morphogenesis. *Nat. Rev. Genet.* **3**: 199–209.

Fuchs, E., Tumbar, T. and Guasch, G. (2004) Socializing with the neighbors: stem cells and their niche. *Cell* **116**: 769–778.

Gat, U., DasGupta, R., Degenstein, L. and Fuchs, E. (1998) *De novo* hair follicle morphogenesis and hair tumours in mice expressing a truncated β-catenin in skin. *Cell* **95**: 605–614.

Ghali, L., Wong, S.T., Green, J., Tidman, N. and Quinn, A.G. (1999) Gli1 protein is expressed in basal cell carcinomas, outer root sheath keratinocytes and a subpopulation of mesenchymal cells in normal human skin. *J. Invest. Dermatol.* **113**: 595–599.

Ghazizadeh, S., Harrington, R. and Taichman, L. (1999) *In vivo* transduction of mouse epidermis with recombinant retroviral vectors: implications for cutaneous gene therapy. *Gene Ther.* **6**: 1267–1275.

Ghazizadeh, S. and Taichman, L.B. (2001) Multiple classes of stem cells in cutaneous epithelium: a lineage analysis of adult mouse skin. *EMBO J.* **20**: 1215–1222.

Godwin, A.R. and Capecchi, M.R. (1999) Hair defects in *Hoxc13* mutant mice. *J. Invest. Dermatol. Symp. Proc.* **4**: 244–247.

Gorlin, R.J. (1995) Nevoid basal cell carcinoma syndrome. *Dermatol. Clin.* **13**: 113–125.

Griffiths, C.E., Rosenthal, D.S., Reddy, A.P., Elder, J.T. *et al.* (1992) Short-term retinoic acid treatment increases *in vivo*, but decreases *in vitro*, epidermal transglutaminase-K enzyme activity and immunoreactivity. *J. Invest. Dermatol.* **99**: 283–288.

Haake, A.R. and Scott, G.A. (1991) Physiologic distribution and differentiation of melanocytes in human fetal and neonatal skin equivalents. *J. Invest. Dermatol.* **96**: 71–77.

Happle, R. (1985) Lyonization and the lines of Blaschko. *Hum. Genet.* **70**: 200–206.

Hardy, M.H. (1992) The secret life of the hair follicle. *Trends Genet.* **8**: 55–61.

Headon, D.J. and Overbeek, P.A. (1999) Involvement of a novel Tnf receptor homologue in hair follicle induction. *Nat. Genet.* **22**: 370–374.

Hinchliffe, J.R. (1981) Cell death in embryogenesis. In *Cell Death in Biology and Pathology*, Lockshin, R.A. (ed.). Chapman & Hall: New York; 35–78.

Holbrook, K.A. and Minami, S.I. (1991) Hair follicle embryogenesis in the human. Characterization of events *in vivo* and *in vitro*. *Ann. N. Y. Acad. Sci.* **642**: 167–196.

Holbrook, K.A., Smith, L.T., Kaplan, E.D., Minami, S.A. *et al.* (1993) Expression of morphogens during human follicle development *in vivo* and a model for studying follicle morphogenesis *in vitro*. *J. Invest. Dermatol.* **101**: 39S–49S.

Horn, H.M. and Tidman, M.J. (2000) The clinical spectrum of epidermolysis bullosa simplex. *Br. J. Dermatol.* **142**: (2000) 468–472.

Huber, M., Floeth, M., Borradori, L., Schacke, H. *et al.* (2002) Deletion of the cytoplasmatic domain of BP180/collagen XVII causes a phenotype with predominant features of epidermolysis bullosa simplex. *J. Invest. Dermatol.* **118**: 185–192.

Huelsken, J., Vogel, R., Erdmann, B., Cotsarelis, G. and Birchmeier, W. (2001) β-Catenin controls hair follicle morphogenesis and stem cell differentiation in the skin. *Cell* **105**: 533–545.

Hunter, J.A.A., Savin, J.A. and Dahl, M.V. (1995) *Clinical Dermatology*. Blackwell Science: Oxford.

Irvine, A.D. and McLean, W.H. (1999) Human keratin diseases: the increasing spectrum of disease and subtlety of the phenotype–genotype correlation. *Br. J. Dermatol.* **140**: 815–828.

Jamora, C., DasGupta, R., Kocieniewski, P. and Fuchs, E. (2003) Links between signal transduction, transcription and adhesion in epithelial bud development. *Nature* **422**: 317–322.

Janes, S.M., Lowell, S. and Hutter, C. (2002) Epidermal stem cells. *J. Pathol.* **197**: 479–491.

Jang, S.I. and Steinert, P.M. (2002) Loricrin expression in cultured human keratinocytes is controlled by a complex interplay between transcription factors of the Sp1, CREB, AP1, and AP2 families. *J. Biol. Chem.* **277**: 42268–42279.

Jessen, B.A., Qin, Q. and Rice, R.H. (2000) Functional AP1 and CRE response elements in the human keratinocyte transglutaminase promoter mediating Whn suppression. *Gene* **254**: 77–85.

Jones, J.C., Hopkinson, S.B. and Goldfinger, L.E. (1998) Structure and assembly of hemidesmosomes. *Bioessays* **20**: 488–494.

Jones, P.H. and Watt, F.M. (1993) Separation of human epidermal stem cells from transit amplifying cells on the basis of differences in integrin function and expression. *Cell* **73**: 713–724.

Jonkman, M.F., Pas, H.H., Nijenhuis, M., Kloosterhuis, G. and Steege, G. (2002) Deletion of a cytoplasmic domain of integrin β4 causes epidermolysis bullosa simplex. *J. Invest. Dermatol.* **119**: 1275–1281.

Jung, H.S., Francis-West, P.H., Widelitz, R.B., Jiang, T.X. *et al.* (1998) Local inhibitory action of BMPs and their relationships with activators in feather formation: implications for periodic patterning. *Dev. Biol.* **196**: 11–23.

Kallioinen, M., Koivukangas, V., Jarvinen, M. and Oikarinen, A. (1995) Expression of cytokeratins in regenerating human epidermis. *Br. J. Dermatol.* **133**: 830–835.

Kanzler, B., Viallet, J.P., Le Mouellic, H., Boncinelli, E. *et al.* (1994) Differential expression of two different homeobox gene families during mouse tegument morphogenesis. *Int. J. Dev. Biol.* **38**: 633–640.

Karlsson, L., Bondjers, C. and Betsholtz, C. (1999) Roles for PDGF-A and sonic hedgehog in development of mesenchymal components of the hair follicle. *Development* **126**: 2611–2621.

Kobayashi, K. and Nishimura, E. (1989) Ectopic growth of mouse whiskers from implanted lengths of plucked vibrissa follicles. *J. Invest. Dermatol.* **92**: 278–282.

Kolodka, T.M., Garlick, J.A. and Taichman, L.B. (1998) Evidence for keratinocyte stem cells *in vitro*: long-term engraftment and persistence of transgene expression from retrovirus-transduced keratinocytes. *Proc. Natl Acad. Sci. USA* **95**: 4356–4361.

Komine, M., Freedberg, I.M. and Blumenberg, M. (1996) Regulation of epidermal expression of keratin K17 in inflammatory skin diseases. *J. Invest. Dermatol.* **107**: 569–575.

Komuves, L.G., Michael, E., Arbeit, J.M., Ma, X.K. *et al.* (2002) HOXB4 homeodomain protein is expressed in developing epidermis and skin disorders and modulates keratinocyte proliferation. *Dev. Dyn.* **224**: 58–68.

Kopan, R. and Fuchs, E. (1989) The use of retinoic acid to probe the relation between hyperproliferation-associated keratins and cell proliferation in normal and malignant epidermal cells. *J. Cell Biol.* **109**: 295–307.

Kupper, T.S. (1990) The activated keratinocyte: a model for inducible cytokine production by non-bone marrow-derived cells in cutaneous inflammatory and immune responses. *J. Invest. Dermatol.* **94**: 146S–150S.

La Celle, P.T. and Polakowska, R.R. (2001) Human homeobox *HOXA7* regulates keratinocyte transglutaminase type 1 and inhibits differentiation. *J. Biol. Chem.* **276**: 32844–32853.

Langbein, L., Rogers, M.A., Winter, H., Praetzel, S. and Schweizer, J. (2001) The catalog of human hair keratins. II. Expression of the six type II members in the hair follicle and the combined catalog of human type I and II keratins. *J. Biol. Chem.* **276**: 35123–35132.

Lavker, R.M. and Sun, T.T. (2000) Epidermal stem cells: properties, markers, and location. *Proc. Natl Acad. Sci. USA* **97**: 13473–13475.

Legg, J., Jensen, U.B., Broad, S., Leigh, I. and Watt, F.M. (2003) Role of melanoma chondroitin sulphate proteoglycan in patterning stem cells in human interfollicular epidermis. *Development* **130**: 6049–6063.

Leigh, I.M., Navsaria, H., Purkis, P.E., McKay, I.A. *et al.* (1995) Keratins (K16 and K17) as markers of keratinocyte hyperproliferation in psoriasis *in vivo* and *in vitro*. *Br. J. Dermatol.* **133**: 501–511.

Li, A., Simmons, P.J. and Kaur, P. (1998) Identification and isolation of candidate human keratinocyte stem cells based on cell surface phenotype. *Proc. Natl Acad. Sci. USA* **95**: 3902–3907.

Li, J., Tzu, J., Chen, Y., Zhang, Y.P. *et al.* (2003) Laminin-10 is crucial for hair morphogenesis. *EMBO J.* **22**: 2400–2410.

Lin, M.H., Leimeister, C., Gessler, M. and Kopan, R. (2000a) Activation of the Notch pathway in the hair cortex leads to aberrant differentiation of the adjacent hair-shaft layers. *Development* **127**: 2421–2432.

Lin, M.T., Pulkkinen, L., Uitto, J. and Yoon, K. (2000b) The gene gun: current applications in cutaneous gene therapy. *Int. J. Dermatol.* **39**: 161–170.

Lindner, G., Menrad, A., Gherardi, E., Merlino, G. *et al.* (2000) Involvement of hepatocyte growth factor/scatter factor and met receptor signalling in hair follicle morphogenesis and cycling. *FASEB J.* **14**: 319–332.

Lu, B., Rothnagel, J.A., Longley, M.A., Tsai, S.Y. and Roop, D.R. (1994) Differentiation-specific expression of human keratin 1 is mediated by a composite AP-1/steroid hormone element. *J. Biol. Chem.* **269**: 7443–7449.

Luetteke, N.C., Phillips, H.K., Qiu, T.H., Copeland, N.G. *et al.* (1994) The mouse waved-2 phenotype results from a point mutation in the EGF receptor tyrosine kinase. *Genes Dev.* **8**: 399–413.

Lyle, S., Christofidou-Solomidou, M., Liu, Y., Elder, D.E. *et al.* (1998) The C8/144B monoclonal antibody recognizes cytokeratin 15 and defines the location of human hair follicle stem cells. *J. Cell Sci.* **111**(21): 3179–3188.

Lynch, H.T. and Fusaro, R.M. (1999) The Muir–Torre syndrome in kindreds with hereditary nonpolyposis colorectal cancer (Lynch syndrome): a classic obligation in preventive medicine. *J. Am. Acad. Dermatol.* **41**: 797–799.

Ma, S., Rao, L., Freedberg, I.M. and Blumenberg, M. (1997) Transcriptional control of K5, K6, K14, and K17 keratin genes by AP-1 and NF-κB family members. *Gene Expr.* **6**: 361–370.

Machesney, M., Tidman, N., Waseem, A., Kirby, L. and Leigh, I. (1998) Activated keratinocytes in the epidermis of hypertrophic scars. *Am. J. Pathol.* **152**: 1133–1141.

Mackenzie, I.C. (1970) Relationship between mitosis and the ordered structure of the stratum corneum in mouse epidermis. *Nature* **226**: 653–655.

Mackenzie, I.C. (1997) Retroviral transduction of murine epidermal stem cells demonstrates clonal units of epidermal structure. *J. Invest. Dermatol.* **109**: 377–383.

Markey, A.C., Lane, E.B., Churchill, L.J., MacDonald, D.M. and Leigh, I.M. (1991) Expression of simple epithelial keratins 8 and 18 in epidermal neoplasia. *J. Invest. Dermatol.* **97**: 763–770.

Martin, P. (1997) Wound healing – aiming for perfect skin regeneration. *Science* **276**: 75–81.

Mathews, C.H., Detmer, K., Lawrence, H.J. and Largman, C. (1993) Expression of the *Hox 2.2* homeobox gene in murine embryonic epidermis. *Differentiation* **52**: 177–184.

Matzuk, M.M., Kumar, T.R., Vassalli, A., Bickenbach, J.R. *et al.* (1995) Functional analysis of activins during mammalian development. *Nature* **374**: 354–356.

Mauviel, A., Chen, Y.Q., Kahari, V.M., Ledo, I. *et al.* (1993) Human recombinant interleukin-1β upregulates elastin gene expression in dermal fibroblasts. Evidence for transcriptional regulation *in vitro* and *in vivo*. *J. Biol. Chem.* **268**: 6520–6524.

Maytin, E.V., Lin, J.C., Krishnamurthy, R., Batchvarova, N. *et al.* (1999) Keratin 10 gene expression during differentiation of mouse epidermis requires transcription factors C/EBP and AP-2. *Dev. Biol.* **216**: 164–181.

McKoy, G., Protonotarios, N., Crosby, A., Tsatsopoulou, A. *et al.* (2000) Identification of a deletion in plakoglobin in arrhythmogenic right ventricular cardiomyopathy with palmoplantar keratoderma and woolly hair (Naxos disease). *Lancet* **355**: 2119–2124.

McLean, W.H., Pulkkinen, L., Smith, F.J., Rugg, E.L. *et al.* (1996) Loss of plectin causes epidermolysis bullosa with muscular dystrophy: cDNA cloning and genomic organization. *Genes Dev.* **10**: 1724–1735.

Merrill, B.J., Gat, U., DasGupta, R. and Fuchs, E. (2001) Tcf3 and Lef1 regulate lineage differentiation of multipotent stem cells in skin. *Genes Dev.* **15**: 1688–1705.

Millar, S.E. (2002) Molecular mechanisms regulating hair follicle development. *J. Invest. Dermatol.* **118**: 216–225.

Moll, R., Franke, W.W., Schiller, D.L., Geiger, B. and Krepler, R. (1982) The catalog of human cytokeratins: patterns of expression in normal epithelia, tumours and cultured cells. *Cell* **31**: 11–24.

Monreal, A.W., Ferguson, B.M., Headon, D.J., Street, S.L. *et al.* (1999) Mutations in the human homologue of mouse *dl* cause autosomal recessive and dominant hypohydrotic ectodermal dysplasia. *Nat. Genet.* **22**: 366–369.

Morley, S.M., D'Alessandro, M., Sexton, C., Rugg, E.L. *et al.* (2003) Generation and characterization of epidermolysis bullosa simplex cell lines: scratch assays show faster migration with disruptive keratin mutations. *Br. J. Dermatol.* **149**: 46–58.

Morris, R.J. and Potten, C.S. (1994) Slowly cycling (label-retaining) epidermal cells behave like clonogenic stem cells *in vitro*. *Cell Prolif.* **27**: 279–289.

Nemes, Z. and Steinert, P.M. (1999) Bricks and mortar of the epidermal barrier. *Exp. Mol. Med.* **31**: 5–19.

Ng, D.C., Shafaee, S., Lee, D. and Bikle, D.D. (2000) Requirement of an AP-1 site in the calcium response region of the involucrin promoter. *J. Biol. Chem.* **275**: 24080–24088.

Niemann, C., Owens, D.M., Hulsken, J., Birchmeier, W. and Watt, F.M. (2002) Expression of DeltaNLef1 in mouse epidermis results in differentiation of hair follicles into squamous epidermal cysts and formation of skin tumours. *Development* **129**: 95–109.

Noramly, S., Freeman, A. and Morgan, B.A. (1999) β-Catenin signalling can initiate feather bud development. *Development* **126**: 3509–3521.

Norgett, E.E., Hatsell, S.J., Carvajal-Huerta, L., Cabezas, J.C. *et al.* (2000) Recessive mutation in desmoplakin disrupts desmoplakin-intermediate filament interactions and causes dilated cardiomyopathy, woolly hair and keratoderma. *Hum. Mol. Genet.* **9**: 2761–2766.

Okura, M., Maeda, H., Nishikawa, S. and Mizoguchi, M. (1995) Effects of monoclonal anti-c-kit antibody (ACK2) on melanocytes in newborn mice. *J. Invest. Dermatol.* **105**: 322–328.

Ortega, S., Ittmann, M., Tsang, S.H., Ehrlich, M. and Basilico, C. (1998) Neuronal defects and delayed wound healing in mice lacking fibroblast growth factor 2. *Proc. Natl Acad. Sci. USA* **95**: 5672–5677.

Oshima, H., Rochat, A., Kedzia, C., Kobayashi, K. and Barrandon, Y. (2001) Morphogenesis and renewal of hair follicles from adult multipotent stem cells. *Cell* **104**: 233–245.

Owens, D.M. and Watt, F.M. (2003) Contribution of stem cells and differentiated cells to epidermal tumours. *Nat. Rev. Cancer* **3**: 444–451.

Oyama, N., Takahashi, H., Tojo, M., Iwatsuki, K. *et al.* (2002) Different properties of three isoforms (alpha, beta, and gamma) of transcription factor AP-2 in the expression of human keratinocyte genes. *Arch. Dermatol. Res.* **294**: 273–280.

Paller, A.S. (2001) Expanding our concepts of mosaic disorders of skin. *Arch. Dermatol.* **137**: 1236–1238.

Paller, A.S., Syder, A.J., Chan, Y.M., Yu, Q.C. *et al.* (1994) Genetic and clinical mosaicism in a type of epidermal naevus *N. Engl. J. Med.* **331**: 1408–1415.

Park, G.T. and Morasso, M.I. (1999) Regulation of the *Dlx3* homeobox gene upon differentiation of mouse keratinocytes. *J. Biol. Chem.* **274**: 26599–26608.

Paus, R. and Cotsarelis, G. (1999) The biology of hair follicles. *N. Engl. J. Med.* **341**: 491–497.

Paus, R., Muller-Rover, S., Van Der Veen, C., Maurer, M. *et al.* (1999) A comprehensive guide for the recognition and classification of distinct stages of hair follicle morphogenesis. *J. Invest. Dermatol.* **113**: 523–532.

Paus, R. and Peker, S. (2003) *Biology of Hair and Nails.* Elsevier: London.

Pellegrini, G., Dellambra, E., Golisano, O., Martinelli, E. *et al.* (2001) p63 identifies keratinocyte stem cells. *Proc. Natl Acad. Sci. USA* **98**: 3156–3161.

Pennisi, D., Bowles, J., Nagy, A., Muscat, G. and Koopman, P. (2000) Mice null for *sox18* are viable and display a mild coat defect. *Mol. Cell Biol.* **20**: 9331–9336.

Potten, C.S. (1981) Cell replacement in epidermis (keratopoiesis) via discrete units of proliferation. *Int. Rev. Cytol.* **69**: 271–318.

Qin, J.Z., Chaturvedi, V., Denning, M.F., Bacon, P., Panella, J. *et al.* (2002) Regulation of apoptosis by *p53* in UV-irradiated human epidermis, psoriatic plaques and senescent keratinocytes. *Oncogene* **21**: 2991–3002.

Raghavachari, N. and Fahl, W.E. (2002) Targeted gene delivery to skin cells *in vivo*: a comparative study of liposomes and polymers as delivery vehicles. *J. Pharm. Sci.* **91**: 615–622.

Reddy, S., Andl, T., Bagasra, A., Lu, M.M. *et al.* (2001) Characterization of *Wnt* gene expression in developing and postnatal hair follicles and identification of Wnt5a as a target of Sonic hedgehog in hair follicle morphogenesis. *Mech. Dev.* **107**: 69–82.

Richard, G., Smith, L.E., Bailey, R.A., Itin, P. *et al.* (1998) Mutations in the human connexin gene *GJB3* cause erythrokeratodermia variabilis. *Nat. Genet.* **20**: 366–369.

Rieger, E., Bijl, J.J., van Oostveen, J.W., Soyer, H.P. *et al.* (1994) Expression of the homeobox gene *HOXC4* in keratinocytes of normal skin and epithelial skin tumours is correlated with differentiation. *J. Invest. Dermatol.* **103**: 341–346.

Robbins, P.B., Lin, Q., Goodnough, J.B., Tian, H. *et al.* (2001) *In vivo* restoration of laminin 5β3 expression and function in junctional epidermolysis bullosa. *Proc. Natl Acad. Sci. USA* **98**: 5193–5198.

Rogers, M.A., Langbein, L., Winter, H., Ehmann, C. *et al.* (2001) Characterization of a cluster of human high/ultrahigh sulfur keratin-associated protein genes embedded in the type I keratin gene domain on chromosome 17q12–21. *J. Biol. Chem.* **276**: 19440–19451.

Rosenthal, D.S., Griffiths, C.E., Yuspa, S.H., Roop, D.R. and Voorhees, J.J. (1992) Acute or chronic topical retinoic acid treatment of human skin *in vivo* alters the expression of epidermal transglutaminase, loricrin, involucrin, filaggrin, and keratins 6 and 13 but not keratins 1, 10, and 14. *J. Invest. Dermatol.* **98**: 343–350.

Saldanha, G., Fletcher, A. and Slater, D.N. (2003) Basal cell carcinoma: a dermatopathological and molecular biological update. *Br. J. Dermatol.* **148**: 195–202.

Scott, G.A. and Goldsmith, L.A. (1993) Homeobox genes and skin development: a review. *J. Invest. Dermatol.* **101**: 3–8.

Spradling, A., Drummond-Barbosa, D. and Kai, T. (2001) Stem cells find their niche. *Nature* **414**: 98–104.

Spritz, R.A. (1997) Piebaldism, Waardenburg syndrome, and related disorders of melanocyte development. *Semin. Cutan. Med. Surg.* **16**: 15–23.

Stasiak, P.C., Purkis, P.E., Leigh, I.M. and Lane, E.B. (1989) Keratin 19: predicted amino acid sequence and broad tissue distribution suggest it evolved from keratinocyte keratins. *J. Invest. Dermatol.* **92**: 707–716.

Steinert, P.M. (1993) Structure, function, and dynamics of keratin intermediate filaments. *J. Invest. Dermatol.* **100**: 729–734.

Stelnicki, E.J., Komuves, L.G., Kwong, A.O., Holmes, D. *et al.* (1998) *HOX* homeobox genes exhibit spatial and temporal changes in expression during human skin development. *J. Invest. Dermatol.* **110**: 110–115.

Stepp, M.A., Gibson, H.E., Gala, P.H., Iglesia, D.D. *et al.* (2002) Defects in keratinocyte activation during wound healing in the syndecan-1-deficient mouse. *J. Cell Sci.* **115**: 4517–4531.

Storey, A. (2002) Papillomaviruses: death-defying acts in skin cancer. *Trends Mol. Med.* **8**: 417–421.

Svingen, T. and Tonissen, K.F. (2003) Altered *HOX* gene expression in human skin and breast cancer cells. *Cancer Biol. Ther.* **2**: 518–523.

Takahashi, H., Oyama, N., Itoh, Y., Ishida-Yamamoto, A. *et al.* (2000) Transcriptional factor AP-2γ increases human cystatin A gene transcription of keratinocytes. *Biochem. Biophys. Res. Commun.* **278**: 719–723.

Talbot, D., Loring, J., Schorle, H. and Lorgin, J. (1999) Spatiotemporal expression pattern of keratins in skin of AP-2α-deficient mice. *J. Invest. Dermatol.* **113**: 816–820.

Tani, H., Morris, R.J. and Kaur, P. (2000) Enrichment for murine keratinocyte stem cells based on cell surface phenotype. *Proc. Natl Acad. Sci. USA* **97**: 10960–10965.

Taylor, G., Lehrer, M.S., Jensen, P.J., Sun, T.T. and Lavker, R.M. (2000) Involvement of follicular stem cells in forming not only the follicle but also the epidermis. *Cell* **102**: 451–461.

Telfer, N.R. (1991) Congenital and hereditary nail disorders. *Semin. Dermatol.* **10**: 2–6.

Tkatchenko, A.V., Visconti, R.P., Shang, L., Papenbrock, T. *et al.* (2001) Overexpression of Hoxc13 in differentiating keratinocytes results in downregulation of a novel hair keratin gene cluster and alopecia. *Development* **128**: 1547–1558.

Tomic-Canic, M., Day, D., Samuels, H.H., Freedberg, I.M. and Blumenberg, M. (1996) Novel regulation of keratin gene expression by thyroid hormone and retinoid receptors. *J. Biol. Chem.* **271**: 1416–1423.

Tomita, Y., Miyamura, Y., Kono, M., Nakamura, R. and Matsunaga, J. (2000) Molecular bases of congenital hypopigmentary disorders in humans and oculocutaneous albinism 1 in Japan. *Pigment Cell Res.* **13**(suppl 8): 130–134.

Toyofuku, K., Wada, I., Spritz, R.A. and Hearing, V.J. (2001) The molecular basis of oculocutaneous albinism type 1 (OCA1): sorting failure and degradation of mutant tyrosinases results in a lack of pigmentation. *Biochem J* **355**: 259–269.

Traupe, H. (1989) *The Ichthyoses. A Guide to Clinical Diagnosis, Genetic Counselling and Therapy*. Springer-Verlag: Berlin.

Tumbar, T., Guasch, G., Greco, V., Blanpain, C. *et al.* (2004) Defining the epithelial stem cell niche in skin. *Science* **303**: 359–363.

Vahlquist, A., Stenstrom, E. and Torma, H. (1987) Vitamin A and β-carotene concentrations at different depths of the epidermis: a preliminary study in the cow snout. *Ups. J. Med. Sci.* **92**: 253–257.

Vasioukhin, V., Bauer, C., Degenstein, L., Wise, B. and Fuchs, E. (2001) Hyperproliferation and defects in epithelial polarity upon conditional ablation of α-catenin in skin. *Cell* **104**: 605–617.

Vassar, R., Coulombe, P.A., Degenstein, L., Albers, K. and Fuchs, E. (1991) Mutant keratin expression in transgenic mice causes marked abnormalities resembling a human genetic skin disease. *Cell* **64**: 365–380.

Virtanen, M., Torma, H. and Vahlquist, A. (2000) Keratin 4 upregulation by retinoic acid *in vivo*: a sensitive marker for retinoid bioactivity in human epidermis. *J. Invest. Dermatol.* **114**: 487–493.

Webster, D., France, J.T., Shapiro, L.J. and Weiss, R. (1978) X-linked ichthyosis due to steroid-sulphatase deficiency. *Lancet* **1**: 70–72.

Weiss, L.W. and Zelickson, A.S. (1975) Embryology of the epidermis: ultrastructural aspects. 1. Formation and early development in the mouse with mammalian comparisons. *Acta Derm. Venereol.* **55**: 161–168.

Widelitz, R.B., Jiang, T.X., Chen, C.W., Stott, N.S. and Chuong, C.M. (1999) Wnt-7a in feather morphogenesis: involvement of anterior–posterior asymmetry and proximal–distal elongation demonstrated with an *in vitro* reconstitution model. *Development* **126**: 2577–2587.

Wojcik, S.M., Bundman, D.S. and Roop, D.R. (2000) Delayed wound healing in keratin 6a knockout mice. *Mol. Cell Biol.* **20**: 5248–5255.

Wyble, C.W., Hynes, K.L., Kuchibhotla, J., Marcus, B.C. *et al.* (1997) TNF-α and IL-1 upregulate membrane-bound and soluble E-selectin through a common pathway. *J. Surg. Res.* **73**: 107–112.

Yamamoto, H., Flannery, M.L., Kupriyanov, S., Pearce, J. *et al.* (1998) Defective trophoblast function in mice with a targeted mutation of *Ets2*. *Genes Dev.* **12**: 1315–1326.

Yang, L.C., Ng, D.C. and Bikle, D.D. (2003) Role of protein kinase C alpha in calcium induced keratinocyte differentiation: defective regulation in squamous cell carcinoma. *J. Cell Physiol.* **195**: 249–259.

Zhu, A.J. and Watt, F.M. (1999) β-Catenin signalling modulates proliferative potential of human epidermal keratinocytes independently of intercellular adhesion. *Development* **126**: 2285–2298.

Zhu, S., Oh, H.S., Shim, M., Sterneck, E. *et al.* (1999) C/EBPβ modulates the early events of keratinocyte differentiation involving growth arrest and keratin 1 and keratin 10 expression. *Mol. Cell Biol.* **19**: 7181–7190.

15

The Vertebral Column

David Rice and **Susanne Dietrich**

Introduction

The vertebral column is the central, and characteristic, skeletal structure of vertebrates, giving the name to the entire subphylum. The replacement of the notochord by the vertebral column can be considered a breakthrough during chordate evolution. The vertebrae confer optimal mechanical properties which facilitate numerous functions including locomotion and protection of the spinal cord. Given the functional importance of the vertebral column, it is astonishing how many vertebral column defects are compatible with life. Defects can range from mishappen intervertebral disks to chaotic vertebral patterns or to premature termination (agenesis) of the lumbar or sacral vertebral column (Table 15.1). As mild defects may only be detected by chance, the frequency of vertebral column defects is difficult to estimate. However, with the advances in prenatal diagnostics, vertebral column defects can be detected during early cartilage and bone formation. Vertebral column defects associated with *spina bifida* occur with a frequency of about 1:1000 newborns, the frequency of congenital scoliosis (side-to-side bending) and sacral agenesis are rarer (1:10,000 and 1:25,000, respectively; http://www.emedicine.com/orthoped/topic618.htm). Mild vertebral column defects may cause orthopaedic problems in the second half of life, and thus need medical attention only when problems arise. However, severe defects, in particular if they are associated with complex syndromes, or if they have serious knock-on effects, for instance when the spinal cord lies unprotected, require immediate attention. Thus, in order to diagnose and interpret vertebral column defects correctly, and to select the appropriate medial approach, we need to understand the normal and pathological development of the human vertebral column.

The characterization of events during ontogeny that lead to the establishment of the vertebral column has progressed steadily since the tissue the vertebral column

Embryos, Genes and Birth Defects, Second Edition Edited by Patrizia Ferretti, Andrew Copp, Cheryll Tickle and Gudrun Moore

Table 15.1 Examples of human congenital vertebral column malformations

Defective developmental process	Condition	Phenotype	Gene mutation	OMIM	Reference
Segmentation/somitogenesis	Klippel–Feil syndrome	Type I block fusion of many cervical and thoracic vertebrae		148900	Gunderson *et al.*, 1967;
	Autosomal dominant			118100	Raas-Rothschild *et al.*, 1988
	Autosomal recessive	Type II fusion of only 1 or 2 vertebrae (+/− hemivertebrae and occipito-atlantal fusion)		214300	
		Type III both cervical and lower thoracic or lumbar fusion			
		Type IV cervical fusion plus sacral agenesis			
	Alagille syndrome	Hemivertebrae	*Jagged 1/Serrate 1*	118450	Li *et al.*, 1997; Oda *et al.*, 1997
		Rib anomalies			
		Butterfly vertebral arch			
	Spondylocostal dysostosis, type 1 (autosomal recessive)	Segmentation defects throughout vertebral column	*DLL 3 2* MESP	277300	Turnpenny *et al.*, 2003; Whittock *et al.*, 2004
		Vertebral fusion, rib fusion, extra ribs, block/hemivertebrae			
		Abnormal odontoid process			
	Wolf–Hirschhorn syndrome (WHS)	Fused ribs	4p partial deletion, WHS candidate 1	194190	Gusella *et al.*, 1985; Naf *et al.*, 2001
		Fused vertebrae			
		Bifid vertebrae			

	Waardenburg syndrome, types I, III	Supernumerary vertebrae and ribs	*PAX3*	193500	Baldwin *et al.*, 1992; Tassabehji *et al.*, 1992
Segmentation and anteroposterior patterning defect	Simpson–Golabi–Behmel syndrome, type 1	Vertebral segmentation defects Fusion of C2–C3 posterior elements, cervical ribs, 13 pairs of thoracic ribs Extra thoracic and lumbar (6) vertebrae Thoracic hemivertebrae Sacral/coccygeal defects	*Glypican 3*	312870	Pilia *et al.*, 1996
Anteroposterior patterning defect	Cervical rib	Transformation of cervical into thoracic vertebrae (cervical ribs)		117900	Weston, 1956
	Cervical vertebrae agenesis	Seen as 'agenesis' of five cervical vertebrae with axis articulating with 1st thoracic vertebra. Phenotype suggests posterior homeotic transformation		214290	Nisan *et al.*, 1988
Specific sclerotome defects	Spinal dysplasia, Anhalt type	Mid-thoracic hemivertebrae Flat vertebrae, Narrow antero-posterior (A-P) vertebral body diameter Absent spinous processes of lower thoracic and lumbar vertebrael		601344	Anhalt *et al.*, 1995

OMIM, on-line Mendelian inheritance in man; http://www.ncbi.nlm.nih.gov

arises from, the segmented paraxial mesoderm or somites, was first noted by Malpighi in the 17th century. Significant advances have been made using developmental model organisms (Table 15.2), mainly amniote models (mouse and chick), but also non-amniote vertebrates such as frog and zebrafish. These studies established that the principles of vertebral column formation are the same in all vertebrates, although this is somewhat debated for the zebrafish (Fleming *et al.*, 2004). The combination of embryological and modern molecular techniques in recent years has led to the characterization of various genes that control vertebral column development, and progress has been made in deciphering molecular networks. Some of the molecular players control a single step in the development of the vertebral column, others have been recruited for numerous events. Thus, as we review the steps the embryo takes from the time the antero-posterior body axis is organized during gastrulation to the formation of mesenchymal pre-vertebrae, ready for ossification, it has to be taken into account that superficially similar phenotypes may have distinct, and also complex causes.

Developmental anatomy of the vertebral column

The vertebral column and ribs originate from the trunk paraxial mesoderm, the mesodermal tissue located between surface ectoderm and gut endoderm on either side of the spinal cord (trunk neuroectoderm) and notochord (axial mesoderm). Thus, vertebral column formation begins with gastrulation, the process that generates mesoderm (Figure 15.1a). Since the mesodermal lineage encompasses the axial (notochordal), paraxial, intermediate, lateral and extra-embryonic mesoderm, the next step that leads to the formation of the vertebral column is the specification of mesoderm as paraxial.

In all vertebrates, the trunk paraxial mesoderm is laid down as a strip of mesenchymal tissue (pre-somitic mesoderm or segmental plate). While gastrulation continuously adds new cells to the posterior end of the segmental plate, cells at its anterior end form re-iterated units termed somites, thus generating a segmental pattern for the first time (Figure 15.1a). This step, and the subsequent re-segmentation is essential for the formation of individual vertebrae (Figure 15.1e).

Somites along the axis have the same epithelial organisation, yet give rise to uniquely shaped vertebrae. Vertebral shape is controlled by intrinsic positional information. Somites generate more than vertebrae; they deliver the dorsal dermis and the entire musculature of the trunk. In order to do so, the somite receives instructions from the surrounding tissues that specify the ventral portion of the somite as sclerotome, the anlage of vertebral column and ribs, and the dorsal portion as dermomyotome, which yields the precursors for dermis and muscle (Figure 15.1b,c). Further patterning events are required to specify the individual components of each vertebra and to join them into a mesenchymal pre-vertebra (Figure 15.1d,f). Finally,

Table 15.2 Mutations in the mouse affecting early steps in vertebral column formation

Defective developmental process	Gene	Mutation	Phenotype	Reference
Gastrulation	*Nodal*	Insertional mutation, −/−	No endo-mesoderm	Conlon 1994, Brennan 2001
Gastrulation, mesoderm specification and targeted migration	*Fgfr1*	−/−	No endo-mesoderm, no emigration of cells	Ciruna 2001
Gastrulation, positional information	*Fgfr1*	Hypomorph	Homeotic transformations	Partanen 1998
Gastrulation, segmentation, positional information	*Wnt3a*	−/− Vestigial tail	Posterior truncation, irregular somites/vertebrae, homeotic transformation	Heston 1951; Grüneberg 1963; Takada 1994; Greco 1996; Aulehla 2003; Ikeya 2001
Gastrulation, segmentation	*Axin1*	Fused	Axis duplication, irregular and fused somites	Gluecksohn-Schoenheimer 1949; Grüneberg 1963; Zeng 1997
Gastrulation, positional information	*RARγ*	−/−	Resistance to RA induced posterior truncations, and block of RA signalling, hence anterior homeotic transformations	Iulianella 1997, 1999
Gastrulation	*Cyp26*	−/−	No clearance of RA, posterior truncation	Sakai 2001
Gastrulation, nototchord formation	*Brachyury/T*	Brachyury/T	Posterior truncation, notochord agenesis, in this region no sclerotome induction and hence no vertebral column	Dobrovolskaia-Zavadskaia 1927; Grüneberg 1963; Searle 1966; Rashbass 1991; Dietrich 1993; Rashbass 1994
Gastrulation, paraxial mesoderm specification	*Tbx6*	−/−	Posterior truncation, defective somites	Chapman 1998, 2003

(*continued*)

Table 15.2 (*continued*)

Defective developmental process	Gene	Mutation	Phenotype	Reference
Gastrulation, mesoderm specification	*Scleraxis*	–/–	No primitive streak, no mesoderm	Brown 1999
Gastrulation/convergence extension movements	*Wnt5a*	–/–	Posterior truncation	Yamaguchi 1999
Gastrulation/convergence extension movements, neurulation, specification of neural arch/spinous process	*Ltap/Lpp1*	Loop tail	No primitive streak retraction, widened floor plate, NTD, missing neural arches/spinous processes	Stein 1957; Kibar 2001; Murdoch 2001
Gastrulation/EMT	*Snail*	–/–	No EMT of mesoderm	Carver 2001
Axial mesoderm specification	*Foxa2 (HNF3β)*	–/–	No notochord	Ang 1994
Lateral and extraembryonic mesoderm specification	*Foxf1 (HFH-8, FREAC1)*	–/–	No lateral/extraembryonic mesoderm	Mahlapuu 2001
Extraembryonic mesoderm specification	*eed*	eed	Conversion of embryonic into extraembryonic mesoderm via deregulation of *Evx1* expression	Faust 1995; Schumacher 1996
Paraxial mesoderm specification or maturation, segmentation	*pMesogenin1*	–/–	Maintenance nascent mesoderm markers, no segmentation markers, defective somites	Yoon and Wold 2000
Segmentation, positional information	*Dll1*	–/–Overexpression of dominant negative *Dll1*	Segmentation defects plus homeotic transformations	Hrabe de Angelis 1997; Cordes 2004
Segmentation	*Dll3*	Pudgy	Irregular/fused somites and vertebrae	Kusumi 1998
Segmentation	*Notch1*	–/–	Irregular somites	Conlon 1995

Segmentation	*Hes7*	–/–	Irregular/fused somites and vertebrae	Bessho 2001
Segmentation	*Lunatic fringe*	–/–	Irregular/fused somites and vertebrae	Evrard 1998; Zhang 1998
Segmentation	*Presenilin1*	–/–	Irregular/fused somites and vertebrae	Shen 1997; Wong 1997
Segmentation	*Mesp2*	–/–	Irregular/fused somites and vertebrae	Saga 1997; Takahashi 2000; Nomura-Kitabayashi 2002
Segmentation	*Tbx18*	–/–	No maintenance of segmental boundaries	Bussen 2004
Epithelial somite formation	*Paraxis*	–/–	Irregular/fused somites and vertebrae	Burgess 1996
Epithelial somite formation, neural arch/spinous process formation	*Pax3*	Splotch	Irregular somites, NTD, reduction of neural arches, fusion of remainder	Tremblay 1998; Schubert 2001
Segmentation, epithelial somite formation	*Papc*	Treatment with dominant negative *Papc*	Irregular/fused somites	Rhee 2003
Segmentation, epithelial somite formation	*N Cadherin*	–/–	Small disrupted somites	Radice 1997
Positional information	*Hoxa7*	Gain of function	Posterior homeotic transformation	Kessel 1990
Positional information	*Hox10 paralogues*	–/–	Transformation of lumbar into thoracic vertebrae	Wellik 2003
Positional information	*Hox11 paralogues*	–/–	Transformation of sacral into lumbar vertebrae	Wellik 2003
Positional information	*Cdx1*	–/–	Delay *Hox* expression, anterior transformation	Subramanian 1995
Positional information	*Bmi1*	–/–	Precocious activation of *Hox*, posterior transformation	van der Lugt 1994
Positional information	*Mll*	–/–	bidirectional alterations of *Hox* expression and vertebral transformations	Yu 1995, 1998

(*continued*)

Table 15.2 (*continued*)

Defective developmental process	Gene	Mutation	Phenotype	Reference
Positional information, segmentation		Rachiterata	Posterior homeotic transformation beginning with C5/6 to T1 conversion, some irregular/fused somites and vertebrae	Schubert 2001
Notochord formation, sclerotome induction		Truncate	Failure or pausing of notochord formation, no sclerotome induction, no vertebral column	Theiler 1959; Dietrich 1993
Notochord maintenance, sclerotome induction/ maintenance		Danforth's short tail	Rapid degeneration of notochord, graded effect onto sclerotome depending on the development age of somite, loss to reduced vertebrae	Gluecksohn-Schoenheimer 1945; Paavola 1980; Dietrich 1993
Notochord maintenance, sclerotome maintenance, nucleus pulposus formation		Pintail	Slow degeneration of notochord, mild reduction of vertebrae, lack of nucleus pulposus	Berry 1960; Dietrich 1993
Notochord signalling	*Shh*	−/−	Induction but not maintenance of sclerotome, no vertebral column	Chiang 1996
Notochord/endoderm signalling	*Smo*	−/−	Block of all hedgehog (Hh) signalling, no sclerotome induction, no vertebral column	Zhang 2001
Notochord signalling	*Noggin*	−/−	Retarded sclerotome, loss of caudal vertebral column (note: additional malformations in line with multiple sites of expression)	McMahon 1998
Mediation of Hh signalling	*IFT172* *IFT88* *Kif3a*	Wimple polaris/flexo −/−	Like $Shh^{-/-}$	Huangfu 2003

Mediation of Hh signals, vertebral body, intervertebral disc formation	*Gli2*	–/–	Reduced ventromedial sclerotome, no vertebral bodies, intervertebral discs	Mo 1997; Buttitta 2003
Repression of Hh signalling	*Ptch*	–/–	Deregulated Hh signalling, expanded ventral phenotypes, NTD, (lethal at E9–10)	Goodrich 1997
Mediation of dorsal neural tube signals (?), repression of Hh signalling	*Rab23*	Open brain	Deregulated Hh signalling, reduction of dorsal neural tube signalling, reduction of neural arch	Eggenschwiler 2001
Mediation of dorsal neural tube signals, repression of Hh signalling	*Gli3*	Extra toes	Deregulated Hh signalling, reduction of dorsal neural tube signalling, reduction of neural arch	Mo 1997
Mediation of dorsal neural tube signals, repression of Hh signalling	*Zic1*	–/–	Deregulated Hh signalling, reduction of dorsal neural tube signalling, reduction of neural arch	Aruga 1999
Neural tube closure, neural arch/spinous process development	*Grhl3 (?)*	Curly tail	Possibly indirect effect on neural tube closure, NTD, absence of neural arches/spinous processes	Grüneberg 1963; Ting 2003
Formation of neural arch and spinous process	*Foxc2*	–/–	No lamina of the neural arches, no spinous processes	Iida 1997; Winnier 1997
Formation of ventromedial sclerotome components	*Pax1*	Undulated	No/reduced vertebral bodies and intervertebral discs	Wright 1947; Grüneberg 1963; Dietrich 1995; Wallin 1994
Formation of pedicles	*Uncx4.1*	–/–	No pedicles and proximal ribs	Mansouri 1997; Neidhardt 1997
Initiation of cartilage formation	*Bapx1 (Nkx3.2)*	–/–	No/reduced vertebral bodies and intervertebral discs, pedicles, proximal ribs	Lettice 1999; Tribioli 1999; Akazawa 2000

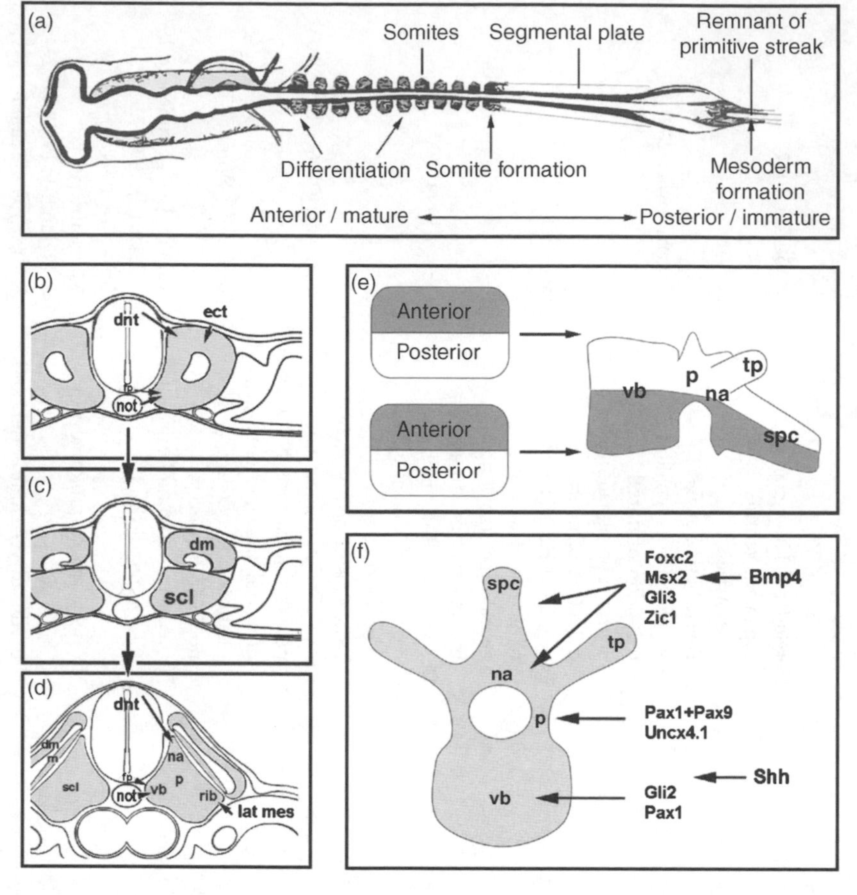

Figure 15.1 Developmental anatomy of vertebral column formation. (a) Schematic representation of a dorsal view of a 1 day old chick embryo, anterior to the left. The primitive streak, located at the posterior end of the embryo, has laid down the body down to mid-cervical levels. While the remnant of the primitive streak adds more paraxial mesoderm to the posterior end of the segmental plate, epithelial somites form at its anterior end. (b–d) Schematic cross-sections of differentiating amniote somites, dorsal to the top. (b) Upon epithelial somite formation, notochord and floor plate of the neural tube (Shh and Noggin signals) and dorsal neural tube and surface ectoderm (Wnt signals) specify dorsoventral cell fates (arrows). (c) Subsequently, the ventral area of the somite de-epithelializes and forms the sclerotome, the source of vertebral column and ribs. The dorsal area forms the dermomyotome, source of dorsal dermis and skeletal muscle. (d) While the dermomyotome generates the first embryonic muscle, the myotome, signals from the notochord/floor plate (Shh, Noggin), signals from the dorsal neural tube (Bmp4) and the lateral mesoderm (Bmp4) specify the individual components of the vertebrae (arrows). The shape of these components is controlled by the positional information intrinsic to the somite. (e) Scheme of two adjacent somites and a human lumbar vertebra, lateral view, anterior to the top. Note that the original somite boundaries are resolved, as two somites contribute to one vertebra. Also note that the previous somitic border has been plotted onto the vertebra, using data from the chick. (f) Anterior view of a schematic human lumbar vertebra. Genes implicated in the formation of individual vertebral components are indicated. dm, dermomyotome; dnt, dorsal neural tube; ect, surface ectoderm; fp, floor plate; lat mes, lateral mesoderm; m, myotome; na, neural arch; not, notochord; p, pedicle; scl, sclerotome; spc, spinous process; tp, transverse process; vb, vertebral body

once the future vertebra is laid down as a mesenchymal progenitor, endochondral ossification can begin.

Making the vertebral column

Gastrulation

The first morphological sign of gastrulation in amniotes is the formation of the primitive streak, an elongated structure characterised by the most active ingression of endo-mesodermal cells underneath the primitive ectoderm/epiblast (reviewed in Schoenwolf *et al.*, 1992; Tam *et al.*, 2000; Tam *et al.*, 2003), see also www.gastrulation.org. The activity of the primitive streak lays down the body in an anterior-posterior fashion, down to lumbosacral levels. From sacral levels to the tip of the tail, the body is generated by the activity of the tail bud (reviewed by Handrigan, 2003). Severe gastrulation defects are not compatible with life. However, hypomorph alleles of mouse gastrulation mutants, or heterozygotes of semidominant mutations often develop to term. They are affected by cessation rather than a total block of gastrulation, which results in posterior truncation, often associated with lumbosacral *spina bifida*, *atresia ani* or urogenital defects. Corresponding phenotypes have also been reported for human fetuses (http://www.emedicine.com/orthoped/topic618.htm), hence gastrulation defects have to be considered as causes of human posterior vertebral column agenesis (for examples in the mouse, see Figure 15.2A–C,E).

Gastrulation is a lengthy process, beginning with the specification of endo-mesodermal precursors within the epiblast. This follows a series of interactions between embryonic and extraembryonic tissues, which cannot be considered here (reviewed by Lu *et al.*, 2001; Tam *et al.*, 2003). However, it should be mentioned that endo-mesoderm specification crucially depends on the presence of the TGFβ signalling molecule Nodal (Brennan *et al.*, 2001; Conlon *et al.*, 1994), the Nodal co-receptor Cripto, the Nodal effector Smad2, the Forkhead transcription factor Foxh1/Fast, and the T-box transcription factor Eomes (Eomesodermin)/Tbr2 in the epiblast and primitive streak: in the absence of these factors, endo-mesodermal precursors are absent and hence gastrulation does not occur (reviewed in Lu *et al.*, 2001; Tam *et al.*, 2003). Nodal function is also required for the separation of endodermal and mesodermal lineages, with high Nodal signalling specifying cells as endodermal, and lower Nodal concentrations specifying cells as mesodermal (reviewed in Tam *et al.*, 2003). The different signalling strength of Nodal is achieved by the pro-protein convertases Spc1 and 4 which cleave Nodal into its active form, the activity of the Nodal repressor Drap1, and the reduction of the Nodal mediator and positive feedback regulator Foxh1. The endodermal lineage furthermore depends on the presence of the paired-like homeobox transcription factor Mixl1, the High mobility group (HMG) transcription factor Sox17, and the mediators of canonical Wnt signalling, Tcf2 and β-catenin, since embryonic stem (ES) cells lacking these genes cannot populate the gut. Expression of Mixl1 depends on continued expression of Eomes and Nodal; the main function of Mixl1 is to suppress factors that control the establishment of mesodermal fates.

Formation of mesodermal cells crucially depends on signalling by Fibroblast Growth Factors (Fgf) via the receptor Fgfr1, on canonical Wnt3a signalling, on Retinoic Acid (RA) signalling, on the members of the T-box transcription factors Brachyury/T, Tbx6, Tbx/Spadetail, and on the basic helix-loop helix transcription

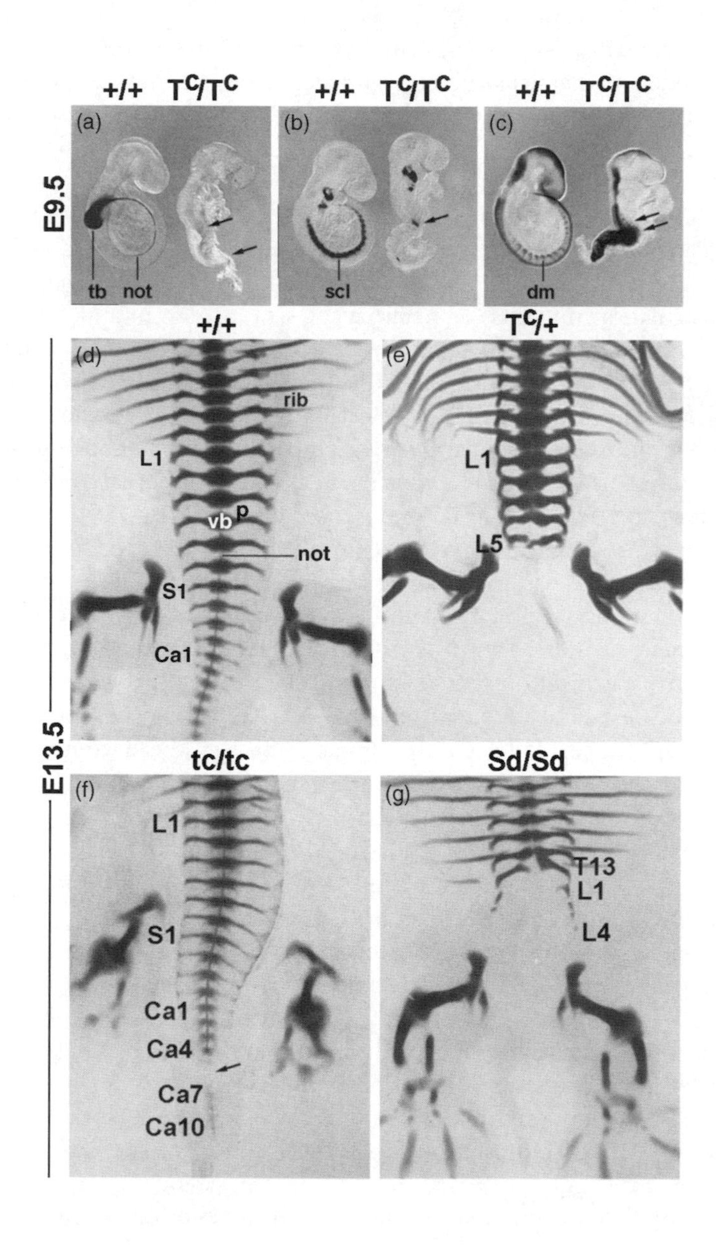

factor Scleraxis, since in the absence of any of these factors, cells do not contribute to mesoderm. They fail to leave the primitive streak and contribute to neural tissues instead, leading to multiple parallel neural tubes (reviewed in Tam *et al.*, 2003). Fgfs are extracellular signalling molecules that act via transmembrane tyrosine kinase receptors, triggering several pathways including the MAPK pathway (reviewed in Kannan and Givol, 2000; Powers *et al.*, 2000). Studies in Xenopus have established that Fgf molecules can induce mesoderm from cells otherwise developing as ectoderm (Amaya *et al.*, 1991; Kimelman and Kirschner, 1987; Kroll and Amaya, 1996; Paterno *et al.*, 1989; Slack *et al.*, 1989). Several *Fgfs* are expressed in the amniote primitive streak in an overlapping and dynamic fashion, with *Fgf8* in the chick labelling the whole streak and newly formed paraxial mesoderm, but becoming excluded from node and anterior streak over time, and *Fgf4* expression labelling the streak and the axial (prechordal and notochordal) mesoderm (Karabagli *et al.*, 2002; Lawson *et al.*, 2001; Shamim and Mason, 1999). While single *Fgf* mutants display relatively mild phenotypes, double mutants for *Fgf4/8* or mutants for the Fgf receptor I lack all

Figure 15.2 Conditions leading to vertebral agenesis. (a–c) *In situ* hybridization of wildtype mouse embryos (+/+, left) and homozygotes for the semidominant *T/Brachyury* mutation *Brachyury curtailed* (T^c/T^c, right) at day 9.5 of embryonic development, lateral views. (a) Wildtype embryos express the *T* gene in the tail bud, nascent mesoderm and notochord, (b) the *Pax1* gene in the developing sclerotomes and (c) the *Pax3* gene in the dermomyotomes and the dorsal neural tube. In T^c/T^c, gastrulation stops at fore limb levels, thus leading to posterior truncation of the body (a, bottom arrow). In anterior regions, a functional notochord is mostly lacking (A, absence of *T* staining, top arrow), *Pax1* expression is more or less absent (b, island of expression indicated by arrow) and *Pax3* expression is not dorsally restricted (c, arrows). Potentially, T^c/T^c would lack a vertebral column, but due to failure of the extraembryonic tissues connecting to the placenta, they do not develop to term. (d–g) Alcian Blue stained skeletal preparations of E13.5 mouse embryos, ventral views. (d) Wildtype. Note that the vertebral column, with the exception of the dorsal neural arches and spinous processes, is laid down in cartilage. (e) *Brachyury curtailed* heterozygote, $T^c/+$. The body is laid down to upper coccygeal levels. However, notochord formation has ceased already at lumbar levels. The vertebrae in reach of the notochordal signals are present though misshapen. Posterior to L5, the lack of the notochord prevented sclerotome induction, hence vertebrae are absent. (f) Homozygote for the recessive mutation *truncate* (*tc/tc*), affected by pausing of notochord formation at various axial levels. Note that the absence of the notochord between Ca4 and Ca7 prevented sclerotome induction, leading to local vertebral agenesis (arrow); where the notochord is present, vertebral column formation has resumed (Ca-7–Ca10). (g) Homozygote for the semidominant mutation *Danforth's short tail* (*Sd/Sd*), suffering from notochord degeneration at E10–11 of development. In posterior regions of the embryo, the notochordal breakdown meets undifferentiated somites, thus preventing the formation of any sclerotomes and hence vertebrae. At lower thoracic/upper lumbar levels, the degeneration process meets differentiating sclerotomes. Here, the development of ventromedial vertebral components requires prolonged notochordal signals in a dosage-dependent fashion. Thus, from posterior to anterior i.e. L4 to T13, the rudiments of the neural arches, pedicles, and finally vertebral bodies reappear. Ca1, 1st caudal (coccygeal) vertebra; L1, 1st lumbar vertebra; S1, 1st sacral vertebra; tb, tail bud; T13, 13th thoracic vertebra; others as in Fig. 15.1

endo-mesoderm, possibly due to a failure of cell migration (see below; Ciruna and Rossant, 2001; Ciruna *et al.*, 1997; Sun *et al.*, 1999). Nevertheless, recent studies supported a more direct role of Fgf signalling in amniote gastrulation. They showed that Fgf signals maintain cells in an immature state in the primitive streak, node, newly formed paraxial mesoderm and in the posterior neural plate, thereby granting the continuation of gastrulation (Dubrulle *et al.*, 2001; Mathieu *et al.*, 2004; Mathis *et al.*, 2001).

Wnt/wingless molecules are secreted glycoproteins, which, after binding to transmembrane Frizzled receptors and LRP co-receptors, trigger one or a combination of the canonical/β-Catenin pathway, the planar cell polarity/JNK pathway or the Calcium/Cam-kinase pathway (reviewed by Seto and Bellen, 2004). The canonical Wnt signalling molecule *Wnt3a* is expressed in the primitive streak and tail bud from mid-streak stages onwards; expression continues in the newly formed paraxial mesoderm (Greco *et al.*, 1996). In *Wnt3a*$^{-/-}$ mice the body is truncated at the level of the fore limbs. Interestingly, in the *Wnt3a* hypomorph *vestigial tail* (*vt/vt*) the truncation occurs at the level of the hind limb, indicating an increased requirement of this factor for the formation of more posterior body regions (Greco *et al.*, 1996; Grüneberg, 1963; Heston, 1951; Takada *et al.*, 1994). While canonical Wnt signalling is required for gastrulation/tail bud formation, the inhibitor of this signalling pathway, Axin1, has the opposite function. It restricts this process to a single site in the embryo as evidenced by the Axin mutant *Fused* whose phenotype ranges from bifurcated tails to almost complete twinning in homozygotes of the allele *Fused/kinky tail* (Gluecksohn-Schoenheimer, 1949; Grüneberg, 1963; Zeng *et al.*, 1997).

Retinoic acid is the principal biologically active form of Vitamin A and is required for antero-posterior patterning of the body axis (reviewed by Lazar, 1999). Its presence needs to be tightly regulated as excess of RA signalling via its nuclear receptor RARγ decreases, amongst others, Wnt3a levels, thus preventing the continuation of gastrulation (Iulianella *et al.*, 1999; Sakai *et al.*, 2001). After the cells, which due to elevated RA signalling accumulate in the primitive streak, have been cleared by apoptosis, epiblast cells are diverted towards a neural fate, leading to the same phenotype as displayed by *vt/vt* mutants (Shum *et al.*, 1999).

The T-box genes *Brachyury/T*, *Tbx6* and *Spadetail* all label the prospective mesoderm (reviewed by Showell *et al.*, 2004). *Brachyury/T* and *Tbx6* are expressed in nascent and newly formed mesoderm, with *Brachyury/T* expression continuing in the notochord and *Tbx6* expression in the paraxial mesoderm. In *Brachyury/T* homozygous mice, *Tbx6*, *Wnt3a*, *Wnt5a* and *Exv1* expression is not maintained in the primitive streak, cells fail to move away from the streak, thus gastrulation comes to an early halt, and the embryos are truncated at the level of the fore limb (Dobrovolskaia-Zavadskaia, 1927; Grüneberg, 1963; Rashbass *et al.*, 1991; Rashbass *et al.*, 1994; Searle, 1966; Figure 15.2a–c). Heterozygotes, depending on the allele, have shortened tails or are tailless, in the latter case often combined with lumbosacral *spina bifida* (Figure 15.1e). Similar phenotypes are displayed by mouse embryos mutant for *Tbx6* with misshapen cervical somites and posterior truncations (Chapman *et al.*, 2003; Chapman and Papaioannou, 1998), while in the zebrafish mutant, *Spadetail*, possibly due to defective gastrulation movements (see below), the centre of the embryo is affected (Yamamoto *et al.*, 1998).

The Twist-related class B basic helix-loop helix (bHLH) transcription factor Scleraxis, which later has a role in tendon formation (Brent *et al.*, 2003), is expressed in the early epiblast, primitive streak and newly formed paraxial mesoderm (Brown *et al.*, 1999). In *Scleraxis* deficient mice, the primitive streak is deficient and mesoderm is absent. However endodermal markers are expressed in the epiblast, suggesting that only mesoderm specification failed. *Scleraxis*$^{-/-}$ ES cells nevertheless can populate the primitive streak, suggesting that *Scleraxis* does not act cell-autonomously but rather via controlling the expression of FGF8, Cripto and Brachyury/T whose expression is drastically reduced (Brown *et al.*, 1999).

Gastrulation involves coordinated morphogenetic movements, which in amniotes encompass (a) the formation and anterior extension of the primitive streak by convergence extension movement of the epiblast, (b) when the streak has half-maximal extension, epithelial-to-mesenchymal transition (EMT) of cells that leave the epiblast to settle beneath, (c) targeted migration of cells leaving the streak and (d) convergence and extension movement of axial and paraxial mesoderm and the neural plate (Lawson and Schoenwolf, 2001). Impairment of these morphogenetic movements will bring gastrulation to a halt even if mesoderm is specified correctly. Recent studies in Xenopus and zebrafish established non-canonical Wnt signalling as the main regulator of convergence-extension movements (reviewed by Ip and Gridley, 2002; Tada *et al.*, 2002). In the fish, the *Wnt11* mutation *Silberblick* (planar cell polarity/JNK kinase pathway) shows impaired convergence-extension movements of cells at gastrulation stages, while in *Wnt5a/Pipetail* (Calcium/Cam-kinase pathway) mutants, these movements are affected at somitogenesis stages. In amniotes, both these Wnt factors are expressed in the primitive streak, with Wnt5a playing a conserved role in gastrulation movements (Yamaguchi *et al.*, 1999). In the zebrafish, further components of planar cell polarity pathway Wnt signalling have been shown to regulate convergence extension movements, including the four-pass transmembrane protein *Strabismus/van Gogh* (Ip and Gridley, 2002). Interestingly, the mouse homologue termed *Ltap/Lpp1* is affected in the mutant *Loop tail*, which suffers from craniorachischisis due to a significantly widened floor plate of the neural tube (Kibar *et al.*, 2001; Murdoch *et al.*, 2001; Stein and Mackensen, 1957). This phenotype possibly results from defective convergence extension movements in the neural plate and in the primitive streak that fails to retract; however underlying patterning defects in the ventral neural tube cannot be excluded. It should be noted that besides Wnt signalling, secreted Slit factors and possibly also their Robo receptors, better known for their role in axon guidance, regulate mesodermal cell movement, as overexpression of *Slit2* in zebrafish leads to impaired convergence-extension movements (Yeo *et al.*, 2001, reviewed in Piper and Little, 2003).

The EMT of gastrulating cells crucially depends on the Zinc-finger transcription factors Snail and Slug. They control the adhesive properties of cells in the primitive streak: excess of Snail turns off the expression of the calcium-dependent cell adhesion molecule E-Cadherin, leading to immediate EMT, while in the absence of Snail and

Slug, E-Cadherin prevails, preventing EMT (reviewed by Ip and Gridley, 2002). While conceptually separable events, EMT and mesoderm specification are linked in the embryo (Ciruna and Rossant, 2001): Fgf signalling via Fgfr1 triggers *Snail* expression, which blocks *E-Cadherin* expression. However, E-Cadherin negatively regulates Wnt signalling as it sequesters the mediator of canonical Wnt signalling, β-Catenin. Thus, in the presence of Snail, β-Catenin becomes available for Wnt3a signalling, which regulates *Fgf8* expression, and in conjunction with FGF, triggers expression of the mesodermal lineage genes *Brachyury/T* and *Tbx6*.

Several studies, for instance overexpressing the Fgf antagonist *Sprouty*, have established the dual role of Fgf in mesoderm specification and in the control of convergence-extension movements (Amaya *et al.*, 1991; Kroll and Amaya, 1996; Mathis *et al.*, 2001; Nutt *et al.*, 2001; reviewed by Ip and Gridley, 2002). However, a recent study also suggested the role of these signalling molecules in the targeted migration of primitive streak cells (Yang *et al.*, 2002): Fgf8 in the avian streak serves a repellent and Fgf4 in the notochord as an attractant for cells born at the anterior end of the streak and destined to give rise to paraxial mesoderm. This is in line with the finding that in the absence of both Fgf4/8 and Fgfr1, cells are unable to move away from the streak, thus not contributing to endo-mesoderm (Ciruna and Rossant, 2001; Sun *et al.*, 1999).

Specification of mesodermal cells as paraxial

During primitive streak elongation and retraction, different areas of the epiblast become engaged in the gastrulation process. Each of these areas is fated to give rise to distinct endodermal and mesodermal tissues. In the chick, the epiblast adjacent to the anterior end of the early primitive streak delivers the definitive endoderm. At later stages, the node predominantly provides the axial mesoderm, thus the fate of cells recruited for gastrulation changes over time such that all endodermal-mesodermal tissues are generated (reviewed by Tam *et al.*, 2003). Vital labelling of small groups of cells has shown that the primitive streak contains resident stem cells that may contribute to more than one tissue (Selleck and Stern, 1991). Moreover, when cells within the streak are heterotopically grafted, they fully integrate into the host site (i.e. they develop according to the new location). Thus, cells within the streak are not committed. Rather, they are plastic and their ultimate fate depends on extrinsic signals from the environment. Unfortunately, these signalling cascades are not well understood. Nevertheless, it is clear that mis-specification of mesodermal subtypes will deplete the embryo of all tissues derived hereof, in most cases leading to severe, non-viable phenotypes. As will be seen for the *eed* mouse mutant below, recessive mutations may run undetected, but may pose a problem if heterozygous carriers conceive.

The fate of the embryonic mesodermal precursor cells is controlled by Bmp4 from the lateral mesoderm and the Bmp antagonist Noggin from by the node: elevating BMP levels can turn the entire paraxial into lateral mesoderm, while elevated Noggin levels can cause conversion of prospective lateral and extra-embryonic mesoderm into paraxial mesoderm (Streit and Stern, 1999; Tonegawa *et al.*, 1997). Fgf signals have been suggested to control the balance between axial (notochordal) and paraxial mesoderm as in Fgfr1$^{-/-}$ mice, paraxial mesoderm is missing. It is conversely debated however, whether the cells destined to give rise to somites contribute to the axial mesoderm (Yamaguchi *et al.*, 1994) or to neural tissue instead (Ciruna *et al.*, 1997). Likewise, for a number of mouse and zebrafish mutants lacking a defined notochord (these mutants will be further discussed in the context of sclerotome induction) it is conversely debated whether the absence of the notochord is due to a defective node, to erroneous specification of cells normally contributing to notochord as paraxial mesoderm, or to defective convergence extension and differentiation of notochordal precursors.

Possibly acting downstream of the extrinsic cues, a number of transcription factors have been established as regulators of mesodermal fate. The zebrafish homeobox gene *Not1/floating head* for example, is essential for the specification of axial mesoderm, suppressing the paraxial fate (Yamamoto *et al.*, 1998). Similarly, the forkhead transcription factor Foxa2 (HNF3β) acts in the specification and maintenance of the notochord (Ang and Rossant, 1994). Other members of this gene family act in the specification of the paraxial mesoderm, which in the absence of the partially redundant *Foxc1* (*Mf1/congenital hydrocephalus* (Kume *et al.*, 1998) and *Foxc2* (*Mfh1*) genes, largely turns into intermediate mesoderm (Wilm *et al.*, 2004). *Foxf1* (*HFH-8, FREAC1*) on the other hand is crucial for lateral and extraembryonic mesoderm development (Mahlapuu *et al.*, 2001). A role for mesoderm specification has also been established for *T/Brachury*, crucial for notochord development, and for *Tbx6*, crucial for paraxial mesoderm formation (Chapman and Papaioannou, 1998; Dobrovolskaia-Zavadskaia, 1927; Grüneberg, 1963; Rashbass *et al.*, 1991; Rashbass *et al.*, 1994; Searle, 1966). Finally, the balance between embryonic and extra-embryonic mesoderm is controlled by the homeobox transcription factor *Evx1*, normally expressed in the primitive streak in an anterior (weak) to posterior (high) gradient (Bastian and Gruss, 1990). In the *eed* mouse mutant carrying a mutation in the mouse homologue of the Drosophila polycomb gene *extra sex combs*, *Evx1* expression is deregulated in the anterior streak, depleting the homozygous embryo of its embryonic mesoderm (Faust *et al.*, 1995; Schumacher *et al.*, 1996). *Evx1*$^{-/-}$ embryos however show early postimplantation lethality prior to gastrulation due to an earlier requirement of the gene for visceral endoderm development (Spyropoulos and Capecchi, 1994).

Paraxial mesoderm segmentation

Once the paraxial mesoderm is specified and laid down as segmental plate on either side of neural tube and notochord, it forms epithelially arranged somites from its anterior end at regular intervals. Additional mesenchymal cells are added to its posterior end from the primitive streak or tail bud (reviewed by Bessho and Kageyama, 2003; Pourquie, 2003). In the chick, new somites arise every 90 minutes, in the mouse every 2 hours, in humans every 8 hours. This process segments the body for the first time, and is a prerequisite for the formation of individual vertebrae. However, in the absence of coordinated segmentation and epithelial somite formation, the paraxial mesoderm is still capable of responding to environmental signals that trigger the formation of cartilage and bone. Thus, segmentation defects will cause chaotic vertebral patterns with irregular and frequently fused vertebrae and ribs as displayed by the mouse mutant for the signalling molecule Delta3, *Dll3/pudgy* (Kusumi *et al.*, 1998) (Figure 15.3b), or by patients suffering from congenital *spondylocostal dysostoses* caused by mutations of the human DLL3 gene (Turnpenny *et al.*, 2003) or the human MESP2 gene (Whittock *et al.*, 2004). Moreover, as the ribs superimpose a segmental pattern onto the sternum, sternal malformations may occur. In the same vein, segmentation defects can lead to extensive innervation defects, since the spinal nerves only project through the anterior half of the somite, which is set up during segmentation.

Classical embryological experiments demonstrated that the ability to segment is intrinsic to the paraxial mesoderm. Twenty years ago, based on theoretical evaluations, several models for the segmentation of the paraxial mesoderm were proposed (reviewed by Keynes and Stern, 1988). They all included a molecular oscillator, set up via molecular feedback mechanisms, for the repetition of the segmentation event, and a timing device, for instance through a maturation gradient, for the actual execution of the segmentation programme. Now, there is molecular evidence for both.

A number of genes show a remarkable cyclic expression in the segmental plate, switching on/off with the same frequency as the formation of somites (reviewed by Bessho and Kageyama, 2003; Pourquie, 2003). These include genes encoding members of the Hairy/Enhancer of split (Hes) family of bHLH transcription factors, and the gene for the cytoplasmic glycosyl transferase, Lunatic fringe. Notably, these factors are involved in signalling by the transmembrane receptor Notch upon binding to its (also membrane-based) ligands Delta (Dll) or Jagged/Serrate (reviewed by Schweisguth, 2004). Knock out mice for *Hes7* (Bessho *et al.*, 2001) or *Lunatic fringe* (Evrard *et al.*, 1998; Zhang and Gridley, 1998) develop severe segmentation defects. Importantly, they lack cyclic gene expression patterns, which are also absent when Notch-Delta signalling is perturbed directly (Jouve *et al.*, 2000), indicating that Notch-Delta signalling drives the molecular oscillator. Recent studies established that upon binding of Delta to Notch, the intracellular domain of Notch is released by the γ-secretases Presenilin1 and 2 and translocates to the nucleus. Here, the Notch intracellular domain initiates transcription of *Hes*, which in turn activates expression of *Lunatic Fringe*. Lunatic fringe travels to the cell surface and inhibits Notch activation. This leads to the downregulation of Notch targets including *Lunatic fringe* itself, such that Notch-Delta signalling becomes possible again, allowing a new cycle to start (Dale *et al.*, 2003; Lai *et al.*, 2003; Taniguchi *et al.*, 2002).

Unexpectedly, mutant mice deficient for *Notch1* still show some segmentation (Conlon *et al.*, 1995). Moreover, constitutive expression of *Lunatic fringe* in the mouse, while deregulating endogenous *Lunatic Fringe* and *Hes7* expression in the anterior segmental plate and hence causing segmentation defects, does not abolish

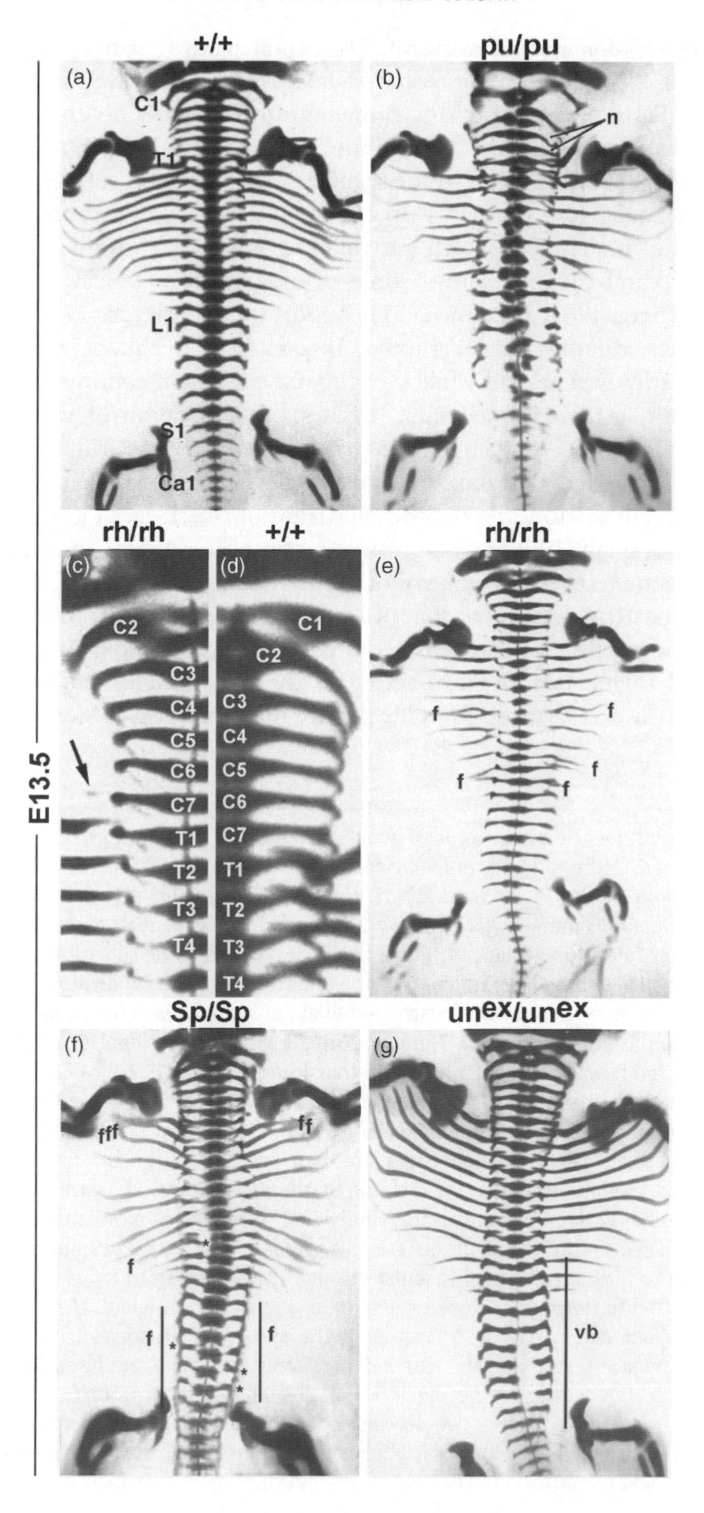

+/+
pu/pu
(a)
(b)
C1
T1
L1
S1
Ca1
n
rh/rh
+/+
rh/rh
(c)
(d)
(e)
C2
C3
C4
C5
C6
C7
T1
T2
T3
T4
C1
f
E13.5
Sp/Sp
un^ex/un^ex
(f)
(g)
fff
ff
vb

cyclic gene expression in the posterior segmental plate (Serth *et al.*, 2003). Thus signalling cascades other than the Notch-Delta system are involved in controlling the molecular oscillator and mesoderm segmentation. Indeed, a recent study (Aulehla *et al.*, 2003) demonstrated that the inhibitor of canonical Wnt signalling, Axin2, is also expressed in a cyclic fashion in the segmental plate, but out of phase with cycling Notch-Delta signalling components. In the hypomorph mouse mutants for *Wnt3a*, *vestigial tail* (i.e. mutants in which gastrulation proceeds to hind limb levels), the remaining vertebral column shows some disorganization typical of segmentation defects (Grüneberg, 1963; Heston, 1951). Importantly, the cyclic expression of both *Axin2* and *Lunatic Fringe* is interrupted. In contrast, in Notch pathway mutants, cyclic Notch pathway genes fail while cyclic *Axin2* expression continues (Aulehla *et al.*, 2003). Moreover, in *Axin1/Fused* mice, besides the bifurcation of the tail, profound segementation defects occur (Gluecksohn-Schoenheimer, 1949; Grüneberg, 1963; Zeng *et al.*, 1997). This suggests that Axin-mediated cyclic Wnt3a signalling may participate in the control of segmentation and may act upstream of Notch-Delta signalling.

Recent studies on the bHLH transcription factor pMesogenin1 indicated, that a yet further set of genes acts in the control of the molecular oscillator. This gene, though expressed in a pattern identical to that of *Tbx6*, seems not to act in the specification of the paraxial mesoderm. Rather, it is required for the expression of a large number of cycling and segmentation genes acting in the Notch-Delta pathway (Yoon and Wold, 2000). However, a more complex picture may emerge as pMesogenin is able to

Figure 15.3 Conditions leading to malformed vertebrae or homeotic transformations. (a–g) Alcian Blue stained mouse skeletal preparations at E13.5, ventral views. (a) Wildtype. Note the regular arrangement of the 7 cervical, (c) 13 thoracic (rib-bearing/jll), 6 lumbar and 4 sacral vertebrae. (b) *Pudgy/Dll3* homozygote (*pu/pu*). The disturbance of mesoderm segmentation and segment boundary formation results in highly irregular somites, which in turn leads to irregular, incomplete or fused vertebrae. Two normally shaped neck vertebrae are indicated (n). (c, d) Higher magnification of the neck and upper thoracic vertebral column of a *rachiterata* homozygote (*rh/rh*) (c) and a wildtype mouse embryo (d). Note that in the mutant, the identity of all vertebrae has shifted one position forward. Thus, C1 has been transformed into C2, C2 into C3, etc; C7 is further transformed as it carries the incomplete anlage of a rib (arrow). As a consequence of this posterior transformation, the neck vertebral column appears one vertebra short. However, as each somite has contributed to vertebra (even though the shape had been altered), this is not a sclerotome/vertebral agenesis phenotype. (e) Entire skeleton of the *rh/rh* animal shown in (d). Note that, in addition to the homeotic transformations, individual vertebrae are misshapen or fused (f), suggesting a mild segmentation/somitogenesis phenotype. (f) *Splotch/Pax3* homozygote, (*Sp/Sp*). These animals suffer from dorsal neural tube defects and lumbosacral *spina bifida* or *craniorachischisis*. The neural arches fail to surround the open neural tube, and instead fuse with their neighbours into a plate of cartilage (indicated by bars, f). In addition to the agenesis of the dorsal neural arches, a number of vertebrae are irregular (asterisks), the first 2–3 ribs, sometimes more, are fused (f), supporting the view of *Pax3* also plays a role in the regular formation of somites. (g) *Undulated extensive/Pax1* homozygote (*un^{ex}/un^{ex}*). Due to defective development of the ventromedial sclerotome, the vertebral bodies are strongly reduced. In the lumbosacral region, they are absent, with the pedicles of the neural arches barely reaching the notochord (bar, vb). f, fused, others as in Fig. 15.1, 15.2

convert non-mesodermal cells into paraxial mesoderm and to suppress notochordal fates in Xenopus *in vitro* assays (Yoon *et al.*, 2000) and in the mouse may promote the maturation of the paraxial mesoderm (see below, Yoon and Wold, 2000).

After a defined number of expression cycles, the cyclic gene expression stops and resolves into a stripe at the anterior or posterior border (depending on gene and vertebrate species) of a developing segment (reviewed by Pourquie, 2003). Thus, the clock is arrested and antero-posterior values within this segment are established. Moreover, the physical separation of this segment, indicated by the emergence of clefts, begins, in the presence of appropriate extrinsic signals (2.4.3) culminating in the formation of an epithelial somite. Notably, the point of clock arrest and morphological segmentation correlates with the shut down of markers for nascent mesoderm. This suggests the switch to a more mature state, in line with the model proposing a maturation gradient (reviewed by Keynes and Stern, 1988).

Recent studies confirmed this maturation gradient, which is set up by Fgf8. *Fgf8* is expressed by gastrulating cells in the primitive streak, keeping mesodermal cells and the adjacent neuroectoderm in an undifferentiated state. However, when cells leave the primitive streak to contribute to the segmental plate, they cease to transcribe *Fgf8*. Thus, only the previously made *Fgf8* mRNA can be translated, which is cleared from the cells over time, thereby creating an Fgf8 gradient, high posteriorly, low towards the anterior end of the segmental plate (Dubrulle and Pourquie, 2004). Notably, when Fgf8 levels are artificially elevated, cells in the segmental plate retain their immature state, delay the clock arrest, and form smaller or no somites. Conversely, when Fgf8 signalling is inhibited, cycling stops precociously and larger somites form (Dubrulle *et al.*, 2001; Sawada *et al.*, 2001).

Formation and maintenance of segmental boundaries

Upon arrest of the segmentation clock the expression of previously cycling genes becomes confined to distinct anterior or posterior domains within the developing somite. Moreover, the expression of further genes with antero-posteriorly restricted expression patterns begins, suggesting that distinct anterior-posterior values have been established. Indeed, antero-posterior rotation of the anterior segmental plate or newly formed somites does not change their original antero-posterior marker gene expression. Moreover, the vertebrae that develop from these inverted somites have an inverted anterior-posterior polarity, and the spinal nerves are redirected accordingly (Aoyoma and Asamoto, 1988; Dubrulle *et al.*, 2001; reviewed in Gossler and Hrabe de Angelis, 1998; Pourquie, 2003).

The establishment of anterior-posterior values within the somite has been seen as prerequisite for the formation and maintenance of somite boundaries, as anterior and posterior somite cells sort out, while anterior-anterior or posterior-posterior cells mix (Stern and Keynes, 1987). However anterior-posterior cells meet in the middle of each somite without forming a morphological boundary (until re-segmentation occurs, see below). Moreover, placing cells from anterior-posterior somitic borders into the middle of a somite triggers border formation. The same result can be obtained by misexpression of Lunatic fringe in small, discrete domains. In the embryo, the anterior border of the second somite to form instructs the first one in line to make a boundary (Sato *et al.*, 2002). Together this suggests that (a) it is not simply differential cell adhesion but rather active signalling across somite boundaries that is essential for boundary formation and (b) Notch-Delta signalling is involved in this process. Indeed, both in patients with *DLL3* mutations and in *Pudgy/Dll3* mouse mutants, somitic boundaries are severely disrupted and residual boundaries are not maintained, leading again to irregular and fused vertebrae and hence an immobile vertebral column (Kusumi *et al.*, 1998; Turnpenny *et al.*, 2003; reviewed in Gossler and Hrabe de Angelis, 1998; Pourquie, 2003).

The switch from cyclic to defined antero-posterior expression of Notch-Delta pathway components, essential for the establishment of segmental boundaries, is facilitated by bHLH transcription factors of the Mesp family (reviewed by Gossler and Hrabe de Angelis, 1998; Pourquie, 2003). Expression of these genes in segregating somites at the anterior end of the segmental plate depends on the earlier, cyclic Notch-Delta signalling (Barrantes *et al.*, 1999). In turn, *Mesp2* controls the establishment and maintenance of the antero-posteriorly restricted expression of Notch-Delta signalling components; and in the absence of *Mesp2*, chaotic vertebral patterns develop (Nomura-Kitabayashi *et al.*, 2002; Saga *et al.*, 1997; Takahashi *et al.*, 2000; Whittock *et al.*, 2004). It is noteworthy that combined activity of Mesp factors may be crucial already during gastrulation as *Mesp1,2* double mutants do not form mesoderm (Kitajima *et al.*, 2000). In a similar vein, mouse mutants lacking both forkhead transcription factors Foxc1 and Foxc2, or zebrafish embryos treated with morpholinos against *Foxc1a* lack the expression of factors associated with clock arrest and somite boundary formation, in line with the upregulated expression of the Fox genes in the anterior segmental plate (Kume *et al.*, 2001; Topczewska *et al.*, 2001). However, as discussed in the chapter on paraxial mesoderm specification, markers for the intermediate mesoderm are ectopically expressed in *Foxc1,2* double mutants, suggesting trans-fating of paraxial mesoderm and hence a mesoderm specification phenotype (Wilm *et al.*, 2004).

Downstream of Mesp2 and Notch-Delta signalling, the T-box transcription factor Tbx18 is involved in the maintenance of antero-posterior somite patterning as in *Tbx18* mutants; somite pattern is set correctly, but then the posterior somite half enlarges at the expense of the anterior, and larger posterior somite derivatives, i.e. pedicles of the neural arches and proximal ribs, form (Bussen *et al.*, 2004). This phenotype is opposite to the phenotype of *Uncx4.1* mutants, which so far have been seen as mutants with defects in the formation of vertebral pedicles only (see section on sclerotomal subdomains; Leitges *et al.*, 2000; Mansouri *et al.*, 2000).

Anatomical studies in particularly in the avian embryo have established that after the initial segmentation process, a new boundary forms inside the somite at the previous intra-somitic anterior-posterior interface (von Ebner's fissure; Huang *et al.*, 2000; Remak, 1850; Verbout, 1976; von Ebner, 1889). Moreover, the posterior half of one sclerotome will join the anterior half of the next sclerotome in line, such that each vertebra forms from two adjacent somites. This process is known as re-segmentation (Figure 15.1e). In segmentation mutants, re-segmentation is also disturbed, thus indicating the prolonged importance of antero-posterior somite patterning.

Epithelial somite formation

When the molecular oscillator stops and anterior-posterior values are established, three further events occur: (a) onset of expression of the transcription factors Paraxis (bHLH factor), Pax3 and Pax7 (paralogous paired and homeodomain containing transcription factors), Meox1 and Meox2 (paralogous homeodomain transcription factors), (b) increase of cell adhesion concomitant with the upregulation of the cell adhesion molecules NCAM, NCadherin, Cadherin11, Papc, (c) formation of an epithelially arranged somite with just a few mesenchymal cells in the somitocoele (reviewed by Gossler and Hrabe de Angelis, 1998). Microsurgical studies in the chick *in vivo*, and *in vitro* co-culture experiments have shown that these three events, although linked to the segmentation clock and clock arrest via Mesp2, require further, permissive signals from the environment. In the absence of the ectodermal covering of the segmental plate, *Paraxis* and *Pax3/7* expression and somite epithelialization fail (Correia and Conlon, 2000; Palmeirim *et al.*, 1998; Sosic *et al.*, 1997). The permissive signals from the surface ectoderm may be the pan-ectodermal signalling molecule Wnt6 (Schubert *et al.*, 2002), which, when expressed from certain tissue culture cell lines, can induce *Paraxis* and *Pax3/7* and the formation of somitic epithelia (Fan *et al.*, 1997; Schmidt *et al.*, 2004).

Functional studies on *Paraxis* and *Pax3/7* have demonstrated that in the absence of *Paraxis*, while the initial subdivision of the segmental plate into mesenchymal units occurs, somite epithelialization and the maintenance of boundaries fails, leading again to chaotic vertebral patterns (Barnes *et al.*, 1997; Burgess *et al.*, 1996; Johnson *et al.*, 2001; Sosic *et al.*, 1997). In the *Pax3* mouse mutant *Splotch*, irregular somites/vertebrae develop (Schubert *et al.*, 2001) (Figure 15.3F), in human PAX3/Waardenburg syndrome type I and III patients, supernumerary vertebrae and ribs are found; and in *Pax3-Pax7* double mutants, somite epithelialization fails altogether (Mansouri and Gruss, 1998; Tables 15.1, 15.2). For *Meox* genes, vertebral phenotypes occur in the absence of *Meox1*, and muscle and tendon phenotypes in the absence of *Meox2*, in line with a later role of the genes in sclerotome and dermomyotome/myotome differentiation (Mankoo *et al.*, 1999; Stamataki *et al.*, 2001). However, in *Meox1,2* double mutants, somites are ill-defined and boundary markers are lost (Mankoo *et al.*, 2003). These findings established *Paraxis* and *Pax3/7* and *Meox1/2* as regulators of somite epithelialization. Notably, in *Paraxis* mice, *Pax3* expression is not maintained (Burgess *et al.*, 1996), while in *Splotch*, *Paraxis* expression is reduced (Henderson *et al.*, 1999; Schubert *et al.*, 2001), suggesting that these transcription factors cross-talk.

A link between the segmentation machinery, clock arrest and increased cell adhesion has been established for the protocadherin Papc, which is expressed in a cyclic fashion in the actively forming somite and the next somite in line. *Papc* expression depends on the activity of Mesp2 and Lunatic Fringe, and when dominantly inhibited, disturbs somite boundary formation in the same fashion as loss of Mesp2 function (Rhee *et al.*, 2003). Similarly, loss of NCadherin function leads to small, disrupted somites (Linask *et al.*, 1998; Radice *et al.*, 1997); this phenotype is also drastically enhanced when Cadherin11 function is lost, too (Horikawa *et al.*, 1999). Notably, in these double mutants the somite is cleaved along the intrinsic anterior-posterior boundary, suggesting that the Cadherins are specifically required to hold the somite together until re-segmentation occurs.

Axial identity of somites and vertebral shape

The human vertebral column typically consists of seven cervical, twelve rib-bearing thoracic, five lumbar, five sacral and four-five caudal (coccygeal) vertebrae, each displaying a unique morphology (Gray, 1995). The axial formulae for other vertebrates differ, but are fairly constant for a given species (Goodrich, 1958). Classical embryological studies have established that the information as to which vertebra to make is intrinsic to the somite and linked to its position along the longitudinal body axis: when thoracic somites were transplanted to a different axial location, they still generated ribs. Thus, they developed according to their original position (Kieny *et al.*, 1972; reviewed by Burke, 2000). Therefore, although somites share the same developmental origin and morphology, they possess positional information that subsequently is translated into a distinct shape. How exactly this information is translated into shape is still not known. However, how positional values are established, is fairly well-characterised.

A wealth of studies, first performed in *Drosophila*, have established that the homeobox containing transcription factors of the *Hox/HOM* class confer positional information: in mutants for these genes, the affected body segments are turned into a phenocopy of another segment (homeotic transformation; reviewed by Burke, 2000; McGinnis and Krumlauf, 1992). In humans, the transformation of the 1st sacral into the last lumbar vertebra occurs quite frequently. Likewise, transformations at the neck-skull interface, which prevent the formation of the dens axis on the 2nd neck vertebra and hence the proper skull-vertebral column articulation, or which lead to fusion of the neck vertebral column to the base of the skull (occipitocervical synostosis) are frequent (1.4–2.5/1000; http://www.emedicine.com/orthoped/topic618.htm). This indicates that homeotic transformations are a widespread phenomenon throughout the animal kingdom (for a mouse example, see Figure 15.3c–e). Moreover, loss- and gain-of-function experiments of the mouse homologues of the Drosophila *Hox/HOM* genes frequently cause homeotic transformations, suggesting that *Hox/HOM* genes serve as universal homeotic selector genes.

Bilaterally symmetrical animals all harbour several *Hox/HOM* genes, which are organized in one or several gene clusters, with a linear arrangement of about 9 genes that belong to 13 paralogous groups. Amniote vertebrates, due to two rounds of genome duplication with a subsequent loss of a few duplicate genes established four clusters with 39 genes. In each cluster, 3′ genes have a more anterior border of expression than 5′ genes (spatial colinearity; reviewed by Burke, 2000; Kmita and Duboule, 2003; McGinnis and Krumlauf, 1992). Expression of Hox/HOM genes begins during gastrulation in the primitive streak (Deschamps *et al.*, 1999; Forlani *et al.*, 2003). The genes in each cluster are activated sequentially, with 3′ genes activated first (temporal colinearity; reviewed by Burke, 2000; Kmita and Duboule, 2003; McGinnis and Krumlauf, 1992). Due to the progress of gastrulation and sequential activation, the early expressed *Hox/HOM* genes have more anterior expression boundaries, the later expressed genes more posterior expression domains, nested within the expression domains of the earlier/anterior ones. Further refinement sets the final *Hox/HOM* expression domains, characterized by a sharp anterior, but ill-defined posterior boundary. Eventually, overlapping expression domains are established in neural and mesodermal tissues, with staggered anterior expression boundaries for a given Hox gene in the neural tube (most anterior), the somites (intermediate) and the lateral mesoderm (farthest posterior expression boundary).

In mammals, each pre-sacral somite/pre-vertebra expresses an assigned combination of Hox genes (reviewed by Burke, 2000; Kmita and Duboule, 2003; McGinnis and Krumlauf, 1992). If *Hox* gene expression is changed such that combinatorial expression occurs at a different site, vertebral identities change accordingly. This has led to the proposal of a Hox code that specifies axial identity (Kessel *et al.*, 1990; Kessel and Gruss, 1991). However, in non-mammalian vertebrates, individual somites/pre-vertebrae do not always have distinct, combinatorial *Hox* expression. Rather, conserved *Hox/HOM* expression boundaries are associated with anatomical landmarks (Burke *et al.*, 1995; Gaunt, 1994). Moreover, when posterior/late *Hox/HOM* genes are overexpressed, changes of vertebral identities predominantly occur at the anterior boundary of the expression domains of the paralogous genes (van den Akker *et al.*, 2001), while inactivation of all of the paralogous genes may lead to the transformation of vertebrae of one type into another, e.g. lumbar into thoracic (van den Akker *et al.*, 2001; Wellik and Capecchi, 2003). This suggests that Hox paralogues have related, redundant functions in the gross regionalization of the vertebral column. Nevertheless, posterior/late *Hox* genes can override programmes set by anterior *Hox* genes even in the absence of transcription, suggesting a *Hox* function of posterior prevalence (Schock *et al.*, 2000; reviewed by Kmita and Duboule, 2003).

Hox/HOM genes have relatively little specificity and affinity for their target sites in the genome. A number of studies have established, that the TALE homeoprotein Exd/Pbx interacts with Hox proteins, and cooperatively binds DNA, thereby improving the DNA binding of the Hox partner (Passner *et al.*, 1999; Piper *et al.*, 1999). A similar role is played by the paired-type homeodomain containing Meis/Perp1/HTH factors, which dimerize with Pbx (reviewed by Sagerstrom, 2004). Recent studies suggested that Pbx and Meis proteins bind to DNA to mark these sites for transcriptional activation. However, they associate with transcriptional repressors and keep the site silent, until binding of the sequence-specific activator, together with further co-activators takes place (reviewed by Sagerstrom, 2004). This system also acts in the regulation of *Hox/HOM* gene expression itself (Saleh *et al.*, 2000).

Hox/Hom genes are constitutively repressed until they escape this repression (Kmita *et al.*, 2000; van der Hoeven *et al.*, 1996; reviewed by Kmita and Duboule, 2003). Individual cis-acting elements have been identified in some *Hox* promoters that regulate expression in the hindbrain. However, the sequential activation of genes in a cluster suggests that individual regulation is the exception, and global regulation of the cluster is the rule. Indeed there is growing evidence that (a) the opening of the cluster creates a microenvironment that allows that next gene in the cluster to be expressed, (b) there is a graded sensitivity of genes in the cluster to signalling molecules (c) global enhancer elements outside the cluster facilitate the linkage of Hox expression to the segmental clock that is set in the primitive streak (Deschamps *et al.*, 1999; Forlani *et al.*, 2003) and that allows bursts of *Hox* expression in forming somites (Zakany *et al.*, 2001). Evidence is now accumulating that *Hox* gene expression is directly linked to mesoderm segmentation and Notch-Delta signalling, as overexpression of dominant-negative Dll1 in the paraxial mesoderm, besides causing segmentation defects also leads to changes in *Hox* gene expression and homeotic transformations (Cordes *et al.*, 2004).

Amongst the signalling molecules executing a global regulation, retinoic acid (RA) signalling, Fgf signalling and Wnt3a signalling are the best characterized. They all feature in the regulation of gastrulation, indicating that the process of axis formation and axial identity is intertwined. RA binds to heterodimers of the nuclear receptors RARα,β,γ and RXR, which then induce transcription from cis-acting RA response elements (RARE) in the promoters of target genes (Hansen *et al.*, 2000; Lazar, 1999). Excess of RA anteriorly shifts Hox expression boundaries, leading to posterior transformation of vertebrae (Kessel and Gruss, 1991), while block of RA signalling via mutant RAR/RXR anteriorizes vertebrae, suggesting that RA is a global regulator of *Hox/HOM* expression. Yet the relevance of RA/RARE in the direct control of Hox gene expression has only been established for the neural tube (Huang *et al.*, 1998; Marshall *et al.*, 1994; Morrison *et al.*, 1996; Packer *et al.*, 1998; Popperl and Featherstone, 1993; Zhang *et al.*, 1997), and in the somites, RA may act indirectly by (a) regulating *Cdx* genes, homeobox genes of the *Caudal* family, which in turn regulate *Hox* expression (Bel-Vialar *et al.*, 2002; Houle *et al.*, 2000; Lickert and Kemler, 2002; Subramanian *et al.*, 1995), and by (b) opening the *Hox* clusters (Bel-Vialar *et al.*, 2000; Kmita *et al.*, 2000), as RAR are part of HAT-HDAC protein complex that acts in chromatin remodelling (reviewed by Featherstone, 2002). This is supported by the observation that RA is found only in the early primitive streak of the mouse (Rossant *et al.*, 1991), that only the early Hox genes are sensitive to RA (e.g. Bel-Vialar *et al.*, 2002) and that Cdx function is attenuated in RAR mutants early, not later (Houle *et al.*, 2000). Thus RA's main function is to kickstart the Hox-system.

Evidence for Fgf molecules regulating axial identity stems from the observation that in *Xenopus*, Fgf2 activates *Xcad3/Cdx4* and late/posterior Hox genes, while dominant negative Fgf receptors delay and hence posteriorly shift the expression of *Cdx* and *Hox* (Cho and De Robertis, 1990; Isaacs *et al.*, 1998; Lamb and Harland, 1995; Pownall *et al.*, 1996). Likewise, in Fgfr1 hypomorph (i.e. gastrulation takes place) or gain-of-function mouse mutants, shifts of Hox expression boundaries in somites and vertebral transformations occur (Partanen *et al.*, 1998). Interestingly, in this study *Cdx1* expression was unaffected (other *Cdx* family members were not investigated), while a study on neural Hox expression established that the posterior *Hox/HOM* genes are sensitive to Fgf, which acts via *Cdx* (Bel-Vialar *et al.*, 2002). Since in contrast, only anterior Hox genes are sensitive to RA, it appears that Hox gene activation is under control of a combination of signalling molecules. As added level of complexity, these signalling molecules regulate each others' function: RA signalling for example regulates expression of Fgfr1 and 4, while Fgf signalling contributes to the regulation of RARα and further components of the RA signalling system (Shiotsugu *et al.*, 2004).

We previously saw that in the absence of Wnt3a, posterior structures fail to form (Greco *et al.*, 1996; Takada *et al.*, 1994; section on gastrulation). In hypomorph *vt/vt* mutants, those somites and vertebrae that do develop show vertebral transformations that correlate with altered Hox gene expression (Ikeya and Takada, 2001). Moreover, functional binding sites for the mediators of canonical Wnt signalling, Tcf/Lef, have been identified in the proximal *Cdx1* promoter (Lickert *et al.*, 2000; Lickert and Kemler, 2002; Prinos *et al.*, 2001), and Tcf4$^{-/-}$ and vt/vt mutants have reduced *Cdx4* expression (Ikeya and Takada, 2001; Lickert *et al.*, 2000; Prinos *et al.*, 2001). Notably, *in vitro* RA and Wnt3a act synergistically (Allan *et al.*, 2001; Prinos *et al.*, 2001).

Most of the signalling activity of RA, Fgf and Wnt3a seems to be mediated by *Caudal/Cdx* genes, homeobox transcription factors, which in the genome are located in the "para-Hox-clusters" together with *Pdx* and *Gsc* genes (Brooke *et al.*, 1998; reviewed by Lohnes, 2003). *Cdx* genes are expressed in the primitive streak, but then localize to somites and neural tube with defined anterior expression boundaries. In *Cdx1*$^{-/-}$ mice, *Hox* expression is delayed and the vertebrae show an anterior transformation (Subramanian *et al.*, 1995); this phenotype is enhanced in the allelic series of *Cdx1,2* double mutants (van den Akker *et al.*, 2002). Mis-expression of Cdx factors or Cdx-VP16 fusion proteins that are constitutively active leads to ectopic expression of *Hox* genes while dominant negative Cdx constructs prevent *Hox* expression (Charite *et al.*, 1998; Epstein *et al.*, 1997; Isaacs *et al.*, 1999; Isaacs *et al.*, 1998). Moreover, functional Cdx binding sites have been identified in *Hox* promoters (Charite *et al.*, 1998; Marom *et al.*, 1997; Subramanian *et al.*, 1995). Together this suggests that *Cdx* genes integrate RA, Wnt, and Fgf signalling and act as pivotal general posteriorizing factors.

In Drosophila, the long-term maintenance of spatially restricted *Hox/HOM* gene expression is controlled by the global transcriptional regulators of Polycomb group/PcG and Trithorax group/trxG, which form multiprotein complexes that modulate the structure of the chromatin by Histone methylation and either block or support the transcription machinery (reviewed by Orlando, 2003). In the mouse, loss of the PcG genes *Bmi1*, *M33*, *Mel18* or *Ring1B*, or partial loss of *Eed*, all lead to precautious/anterior activation of *Hox* genes and to posterior homeotic transformations with *Bmi1* acting synergistically with *M33* and *Mel18*, while *Bmi1* or *Ring1B* overexpression has the opposite effect. This indicates that PcG factors cooperate to keep *Hox/HOM* genes in a repressed state (Akasaka *et al.*, 1996; Akasaka *et al.*, 2001; Alkema *et al.*, 1995; Bel *et al.*, 1998; Schumacher *et al.*, 1996; Suzuki *et al.*, 2002; van der Lugt *et al.*, 1996; van der Lugt *et al.*, 1994). Loss-of-function of the murine trxG gene *Mll* leads to bidirectional homeotic transformations, with altered expression patterns of *Hox* genes in heterozygotes and loss of expression of certain *Hox* genes in homozygotes, suggesting that *Mll* maintains (some) *Hox* genes in an active state (Yu *et al.*, 1998; Yu *et al.*, 1995). This notion has been supported by the finding that posterior transformations associated with the loss of *Bmi1* are also compensated, when the function of *Mll* is lost (Hanson *et al.*, 1999).

During cell cycle, each daughter cell must receive one complete set of genetic information. This is achieved by a process termed replication licensing, and involves the sequential assembly of components of the replication complex at the origin of replication. This process is negatively controlled by the Geminin protein. Recent evidence suggests that Geminin, possibly via binding to PcG genes, negatively regulates *Hox*. Moreover, when Geminin is bound to Hox proteins it prevents DNA binding of Hox, thus exerting double-negative control over *Hox* genes (Tada *et al.*, 2001; Wohlschlegel *et al.*, 2000; reviewed by Li and Rosenfeld, 2004).

Somite differentiation and sclerotome formation

Upon the arrest of the segmentation clock, the downregulation of markers for nascent mesoderm and the formation of epithelial somites, the somite undergoes a remarkable morphological differentiation, with the ventral portion becoming mesenchymal again (reviewed by Brent and Tabin, 2002; Gossler and Hrabe de Angelis, 1998). Fate mapping experiments have established that this mesenchyme, the sclerotome, will provide the vertebral column and ribs, while the dorsal, epithelial part of the somite provides muscle and dermis. Prior to its morphological differentiation, the somite expresses markers associated with a particular fate: cells that will give rise to sclerotome express the paired box transcription factor *Pax1*, while dermomyotomal cells retain expression of *Pax3* and *Paraxis*.

Dorsoventral rotation of somites, ablation of tissues surrounding the somite, co-culture of somites plus/minus surrounding tissues or mouse mutants with defects in these surrounding tissues have established that the dorsoventral specification of somitic cells and the subsequent morphological differentiation of the somite are under the control of extrinsic (i.e. appositional) instructive cues (reviewed by Brent and Tabin, 2002; Gossler and Hrabe de Angelis, 1998). In particular, these studies established that the induction and maintenance of the sclerotome depends on the notochord: when the early notochord before floor plate induction, or alternatively the established notochord plus the functionally equivalent floor plate of the neural tube was removed, no sclerotome and hence no vertebral column formed (Dietrich *et al.*, 1997; Ebensperger *et al.*, 1995). Likewise, in mouse mutants with notochord agenesis such as *Brachyury curtailed* heterozygotes (T^c/+, Searle, 1966) or *truncate* homozygotes (*tc/tc*, Theiler, 1959), the sclerotome in the affected region is never induced (Dietrich *et al.*, 1993), while in mouse mutants displaying notochord degeneration, for example *Danforth's short tail* (*Sd/Sd*, Gluecksohn-Schoenheimer, 1945; Paavola *et al.*, 1980) or *Pintail* (*Pt/Pt*, Berry, 1960, all mutants reviewed in Grüneberg, 1963), the sclerotome is not or only partially maintained (Dietrich *et al.*, 1993). The resulting phenotypes often resemble the phenotypes of mild gastrulation mutations, as frequently, the vertebral columns are posteriorly truncated (Figure 15.2d–g). In fact, notochord agenesis seen in $T^c/+$ results from a combination of posterior gastrulation arrest plus failure of differentiation for notochord precursor cells (Figure 15.2e). However, it may be possible to distinguish gastrulation and notochord-dependent phenotypes since in the former, all somitic derivatives will be missing, while in the latter, somitic derivatives such as limb muscles will be unaffected. It also should be noted that notochord formation may be impaired locally, leading to a phenotype of "picked out" vertebrae (Figure 15.2f).

Much of the function of the notochord has been associated with the signalling molecule Sonic hedgehog (Shh), which is expressed by notochord and floor plate and signals to the somite via the transmembrane receptor Patched (Ptc). Shh-binding to Ptc releases the membrane-bound molecule Smoothened (Smo) from Ptc-mediated repression. This triggers a signal transduction cascade that culminates in the recruitment of the Gli Zinc finger transcription factors for the transcriptional control of Shh target genes, including the sclerotomal marker *Pax1* (reviewed by Brent and Tabin, 2002; Nybakken and Perrimon, 2002a). In $Shh^{-/-}$ mice somitic cells show a reduced rate of mitotis and enhanced rates of apoptosis, sclerotome development is arrested and no vertebral column develops (Chiang *et al.*, 1996). However, the initial activation of sclerotomal markers takes place, suggesting that further signals contribute to sclerotome induction. Indeed, another hedgehog family member, Indian hedgehog (Ihh), is expressed in the endoderm beneath the somite and may partially compensate for the loss of Shh, since in $Smo^{-/-}$ mice where all hedgehog signalling is eliminated, sclerotomal markers are never activated (Zhang *et al.*, 2001). Moreover, the BMP antagonist Noggin, expressed by node and notochord, is also able to trigger expression of *Pax1*, and in the absence of Noggin, sclerotome development is retarded (Capdevila and Johnson, 1998; McMahon *et al.*, 1998).

In the absence of notochordal signals, markers normally associated with the dermomyotome spread into the ventral somite. On the other hand, when the neural tube and surface ectoderm are removed, dermomyotome development fails and sclerotomal markers occupy dorsal areas of the somite instead (Dietrich *et al.*, 1993; Dietrich *et al.*, 1997; reviewed by Brent and Tabin, 2002). This suggests that the dorsal part of the somite is positively regulated by signals from neural and surface ectoderm, which at the same time antagonize notochord-induced sclerotome formation. This idea has been substantiated by the finding that neural tube and ectoderm-based Wnt molecules that signal via the canonical signalling pathway are crucial to induce dermomyotome formation (Capdevila *et al.*, 1998; Fan *et al.*, 1995; Fan and Tessier-Lavigne, 1994; reviewed by Brent and Tabin, 2002).

Though their range is limited due to interaction with the extracellular matrix, Hedgehog and Wnt molecules are secreted factors and most of them can diffuse from cells (Fan *et al.*, 1995; Fan and Tessier-Lavigne, 1994; reviewed by Nybakken and

Perrimon, 2002b). Thus, there are molecular mechanisms that protect the sclerotome from Wnt signalling and the dermomyotome from Shh signalling. Recent studies demonstrated that Shh induces in the sclerotome expression of *Sfrp2*, a secreted molecule that resembles Frizzled Wnt receptors but acts as dominant negative regulator of Wnt signalling (Lee *et al.*, 2000). On the other hand, Wnt signalling activates in the dermomyotome the glycosyl phosphate inositol (GPI) linked membrane glycoprotein Gas1, thought to sequester and inhibit Shh (Lee *et al.*, 2001). However, there is also a cross-talk between Shh and Wnt signalling, which is particularly important for the formation of muscle from the dermomyotome (Dietrich *et al.*, 1997; Münsterberg and Lassar, 1995; reviewed by Brent and Tabin, 2002): Shh and Wnt together upregulate the key component of the canonical Wnt signalling pathway, β-catenin (Schmidt *et al.*, 2000), and Wnt signals upregulate the expression of the mediators of Shh signalling, Gli2 and Gli3 (Borycki *et al.*, 2000). However, while Gli2 mostly activates Shh target genes, Gli3 predominantly serves in the dermomyotome to suppress sclerotomal markers (Buttitta *et al.*, 2003). In turn, the function of Gli3 is negatively regulated by IFT proteins better known for their function in cell cilia formation (Huangfu *et al.*, 2003). Similar to Gli3, Rab23, a member of the Rab family of vesicle transporters acts in the dorsal neural tube and dermomyotome to negatively regulate Shh signalling events. Mutation of this gene in the *open brain* mutant causes dorsal neural tube defects and severe somite patterning and vertebral column defects (Eggenschwiler *et al.*, 2001; Spörle and Schughart, 1998). Finally, Shh induces *Qsulf1* in the developing muscles, which desulphates heparin sulphate proteoglycans, thereby releasing locally bound Wnt molecules and hence locally boosting Wnt signalling (Dhoot *et al.*, 2001).

Specification of sclerotomal subdomains – neural arch, pedicle and proximal rib, vertebral body and intervertebral disc

Once the sclerotome has been induced, it becomes molecularly divided into subdomains, which are then sculpted into the shapes of the future vertebrae and ribs (i.e. mesenchymal pre-vertebrae; Figure 15.1). Thus, patterning events are required to ensure that the individual components of a future vertebra are made – the medial-most, peri-notochordal cells have to provide the vertebral body and the intervertebral disk, the laterally adjacent area of the sclerotome is destined to provide the pedicles of the neural arch and the proximal part of the ribs, the ventrolateral sclerotome gives rise to the distal ribs, and the dorsomedial portion of the sclerotome has to expand around the neural tube to give rise to the neural arch and the spinous process. Defects at this stage of vertebral column formation will lead to the absence of individual vertebral components (Figure 15.1f), for instance missing or reduced vertebral bodies as shown by the mouse *Pax1* mutant *undulated (un/un*; the loss-of-function allele *undulated extensive* is shown in Figure 15.3g, Dietrich and Gruss, 1995; Grüneberg, 1963; Wallin *et al.*, 1994; Wright, 1947), reduced or missing pedicles as displayed by mutants for the *Uncx4.1* gene (Leitges *et al.*, 2000; Mansouri *et al.*, 2000), or dorsally open neural arches without spinous processes and associated with either *spina bifida occulta* or *spina bifida aperta*, as displayed by mutants with neural tube defects such as the *Pax3* mutant *Splotch* (Figure 15.3f; Chapter 8 Auerbach, 1954; Grüneberg, 1963; Tremblay *et al.*, 1998).

The mediolateral subdivision of the sclerotome that discriminates between vertebral, body pedicle, neural arch, proximal and distal rib formation is initially morphologically concealed. However, marker gene expression patterns indicate an early subdivision on a molecular level: upon sclerotome induction, *Pax1* expression becomes restricted to the medial part of the sclerotome while expression of the basic helix-loop-helix transcription factor Sim1 and the Forkhead transcription factor Foxc2 (Mfh1) becomes confined to the lateral counterpart (Dietrich *et al.*, 1998; Pourquié *et al.*, 1996; Sudo *et al.*, 2001). Microsurgical manipulation of somites and surrounding tissues established that the medial-lateral patterning of the sclerotome, like the earlier dorsal ventral patterning of the somite, is under extrinsic control. Shh and Noggin emitted by the notochord cooperate to positively regulate *Pax1* expression and the formation of medial vertebral components, namely vertebral bodies and intervertebral discs (Capdevila and Johnson, 1998; Dietrich *et al.*, 1998; Johnson *et al.*, 1994; McMahon *et al.*, 1998). On the other hand, Bmp4 from the lateral mesoderm positively regulates *Foxc2* and *Sim1*, and the formation of distal ribs (Dietrich *et al.*, 1998; Pourquié *et al.*, 1995; Pourquié *et al.*, 1996; Sudo *et al.*, 2001). Rib formation may depend on further, yet unidentified factors since in mice with disturbed lateral outgrowth of the dermomyotome, rib defects occur (Kato and Aoyama, 1998; Tremblay *et al.*, 1998). Indeed, evidence is accumulating that the dermomyotome and myotome signal to the underlying sclerotome, although this signalling may be reserved to the specification of precursors for the connective tissue and tendons of skeletal muscle (Brent *et al.*, 2003; Henderson *et al.*, 1999).

When medial sclerotomal cells begin to surround the neural tube dorsally, they shut off *Pax1*, retain *Foxc2* and express the homeobox transcription factor Msx2, the Zinc-finger transcription factor Gli3 and the related Zinc-finger transcription factor Zic1 (Aruga *et al.*, 1999; Borycki *et al.*, 2000; Furumoto *et al.*, 1999; Monsoro-Burq *et al.*, 1994). Notably, Shh represses *Msx2* expression, while Bmp4, which besides the lateral mesoderm is also provided by the roof plate of the neural tube, stimulates *Msx2* and neural arch/spinous process formation (Monsoro-Burq *et al.*, 1994; Monsoro-Burq *et al.*, 1996; Watanabe *et al.*, 1998). This relationship between dorsal neural tube and neural arches provides an attractive explanation for the agenesis of neural arches upon defective dorsal neural tube closure as seen in the *Pax3* mutant *Splotch* (Figure 15.3f; Tremblay *et al.*, 1998), the *Gli3* mutant *extra toes* (Mo *et al.*, 1997) and the *Zic1* knock out mouse (Aruga *et al.*, 1999). It has to be noted, however, that absence of Bmp4 from the roof plate of the neural tube is not the only cause of neural arch/spinous process defects. For example, in the mouse mutant *curly tail (ct/ct)* and the knock out mouse for the transcription factor *Grhl3*, the candidate for *ct*, the primary defect seems to lie in insufficient growth of endoderm and notochord, mechanically preventing neural tube and vertebral closure (Grüneberg, 1963) or alternatively, in the surface ectoderm (Ting *et al.*, 2003). Likewise, in the mouse mutant carrying a mutation in the *Ltap/Lpp1* (*Strabismus/van Gogh*) gene, *Loop tail*, the primary defect may lie in the abnormally widened floor plate of the neural tube and/or in the elongated primitive streak (Kibar *et al.*, 2001; Murdoch *et al.*, 2001; Stein and Mackensen, 1957) (see Chapter 8). It also should be noted that dorsal neural tube patterning is little affected in the absence of *Zic1*, and both *Gli3* and *Zic1*, as they are expressed in the neural arch/spinous process anlage, may well act cell-autonomously in this tissue (Aruga *et al.*, 1999; Aruga *et al.*, 2002). Moreover, in mouse mutants in which the outgrowth of the sclerotome is impaired due to the lack of both *Pax1* and the *PDGF receptor* α, which on its own causes systemic skeletal defects (Grüneberg, 1963), a substantial *spina bifida* arises (Helwig *et al.*, 1995). This indicates that the sclerotome is not simply a recipient of signals from the neural tube, but actively promotes neural tube closure.

While the signals to dorsoventrally pattern the sclerotome have not been fully characterised, information is accumulating on the mediators of the extrinsic signals that act in the development of distinct vertebral components. As discussed in the section on somite differentiation, the Zinc finger transcription factor Gli2 acts as activator of Shh targets in the medial sclerotome, and when mutated, causes agenesis of the vertebral bodies and intervertebral discs (Buttitta *et al.*, 2003; Mo *et al.*, 1997). The paired box transcription factor Pax1 is expressed in the early somite throughout the prospective sclerotome (Dietrich and Gruss, 1995; Wallin *et al.*, 1994). Subsequently, expression becomes restricted to the ventromedial aspect of the sclerotome, with stronger expression within the zone of more active cell proliferation in the posterior sclerotome half, and two stripes at the borders of the anterior sclerotome half. In line with this expression, in *Pax1/undulated* mutants, the normal development and growth of ventromedial vertebral components is impaired, leading to strongly reduced or absent vertebral bodies and intervertebral discs (Figure 15.3g; Dietrich and Gruss, 1995; Grüneberg, 1963; Wallin *et al.*, 1994; Wright, 1947). The *Pax1* paralogue *Pax9* is expressed slightly more laterally in the prospective pedicles and proximal ribs (Neubüser *et al.*, 1995; Peters *et al.*, 1998). Loss of function mutants for *Pax9* do not show a vertebral column phenotype. However, in conjunction with loss of function for *Pax1*, the *undulated* phenotype is drastically enhanced, the medial sclerotome is affected by profound apoptosis, fails to compact, and vertebral bodies, intervertebral discs, pedicles and proximal ribs are all absent (Peters *et al.*, 1998; Peters *et al.*, 1999).

A gene that has emerged as essential for pedicle formation is *Uncx4.1*, which encodes a paired-type homeodomain transcription factor. It is expressed in the posterior half of newly formed somites, but then becomes confined to the posterior sclerotome (Mansouri *et al.*, 1997; Neidhardt *et al.*, 1997). In the absence of this gene, posterior sclerotome condensations are not maintained, *Pax1* and *Pax9* become downregulated in this area and pedicles and proximal ribs fail to develop (Leitges *et al.*, 2000; Mansouri *et al.*, 2000). Notably, the axonal repellent SemaD becomes downregulated to levels typical for anterior sclerotome halves, leading to disorganized dorsal root ganglia. This suggests that *Uncx4.*1 mutants may suffer from impaired maintenance of posterior somitic values and somitic boundaries, opposite to *Tbx18* mutants, in which anterior somitic values are not maintained (Bussen *et al.*, 2004). Thus, it may become necessary to reclassify *Uncx4.*1 mutants as segmentation mutants.

The possible role of the forkhead *Foxc1/Mf1/ congenital hydrocephalus* and *Foxc2/Mfh1* genes in paraxial mesoderm specification and segment boundary formation was discussed previously (sections on paraxial mesoderm specification and on segment boundary formation). Here it is noteworthy that continuing expression of both genes in the sclerotome, and the activation of a further forkhead gene, *Foxd1/Mf2*, is controlled by Shh (Furumoto *et al.*, 1999; Kume *et al.*, 2001; Winnier *et al.*, 1997; Wu *et al.*, 1998). Single loss-of function mutants show little vertebral column phenotypes with the exception of *Foxc2* mutants. Most striking is the loss of the lamina of the neural arches and the spinous processes (Iida *et al.*, 1997; Winnier *et al.*, 1997), associated with a *spina bifida* occulta. In *Foxc2-Pax1* double mutants, this phenotype is further enhanced, for the animals show extreme *spina bifida* plus subcutaneous myelomeningocoele and loss of neural arches, pedicles and vertebral bodies (Furumoto *et al.*, 1999). The proliferation rates in the sclerotome are drastically reduced, and the resulting small amount of progenitor tissue may be insufficient to allow subsequent cartilage and bone formation.

Similar to *Foxc* genes, *Meox* genes were mentioned previously for their function in epithelial somite and somite boundary formation (section on somite formation; Mankoo *et al.*, 2003). However, also the *Meox* genes play a prolonged role in somite differentiation, for *Meox1* mutants exhibit mild vertebral column defects, *Moex2* mutants have muscle and tendon defects, and in the double mutants, vertebral column formation is abolished altogether (Mankoo *et al.*, 1999, Mankoo *et al.*, 2003; Stamataki *et al.*, 2001). Interestingly, Meox factors interact with Pax proteins, with sclerotome-expressed Meox1 preferring Pax1 and dermomyotome-expressed Meox2 preferring Pax3, and evidence is accumulating that they cooperatively regulate the same downstream targets.

Onset of cartilage differentiation and osteogenesis

Once the mesenchymal pre-vertebrae are generated, they begin the process of bone formation by endochondral ossification. During this process, cartilage versions of the future vertebrae form, which are later replaced by bone. *In vitro* studies showed that this process is stimulated by BMP signalling molecules after exposure to Shh, suggesting that Shh, when it induces the sclerotome, makes cells susceptible to BMP and chondrogenesis (Murtaugh *et al.*, 1999). This also implies that mediating factors lie between Shh-Pax1-sclerotome induction and the onset of final differentiation. These mediators have been identified as members of the Nk family of homeobox transcription factors.

The homologues of the *Drosophila bagpipe* gene, *Nkx3.1* and *Bapx1/Nkx3.2*, are expressed in the early sclerotome, then in the posterior half and the developing cartilage condensations, in a Shh-dependent fashion (Kos *et al.*, 1998; Murtaugh *et al.*, 2001; Tanaka *et al.*, 1999; Tribioli *et al.*, 1997). While no vertebral column phenotype has been reported for *Nkx3.1*$^{-/-}$ mice (Bhatia-Gaur *et al.*, 1999; Schneider *et al.*, 2000; Tanaka *et al.*, 2000), the phenotype of *Bapx1*$^{-/-}$ resembles that of *Pax1,9* double mutants (Akazawa *et al.*, 2000; Lettice *et al.*, 1999; Tribioli and Lufkin, 1999). In mutants lacking *Bapx1*, expression of *Pax1*, *Pax9* and *Foxc2/Mfh1* is not changed, while markers for cartilage formation (Col2a1, FGFR3, Osf2, Ihh, Sox9) are lost, indicating that *Bapx1* acts downstream of the *Pax* and *forkhead* genes but upstream of chondrogenesis. *In vitro*, paraxial mesoderm cultures in the presence of either Shh or Pax1 initiate *Bapx1* expression and chondrogenesis, and Pax1/9 act synergistically on the *Bapx1* promoter (Murtaugh *et al.*, 2001; Rodrigo *et al.*, 2003). Nevertheless, the functions of *Bapx1* are certainly complex, as on the one hand side, *Bapx1* acts as a transcriptional repressor by forming a complex with HDAC1, Smad1 and Smad4 in a BMP-dependent manner (Kim and Lassar, 2003; Murtaugh *et al.*, 2001), and on the other hand it controls its own expression in an autoregulatory positive feedback loop (Zeng *et al.*, 2002).

Upon chondogenesis, the vertebral column then enters the final phase of its development, which comprises ossification of the bony elements, the formation of a marrow space and the final development of the intervertebral discs. The molecular basis of endochondral bone formation is starting to be understood, with Bmps, Fgfs, Hhs and Wnts all playing roles (reviewed by Kronenberg, 2003). The effects of these signals are mediated by Sox5/6/9, members of the Sry-related family of HMG box DNA-binding proteins, which cooperatively bind to the promoter of the collagen II gene *Col2a1* (Lefebvre *et al.*, 2001). In *Col2a1* null mice, no endochondral ossification takes place (Li *et al.*, 1995), similar to achondrogenesis type II observed in human patients lacking *Col2a1* function (Korkko *et al.*, 2000). This represents a systemic defect in cartilage and bone formation, and affects all bones made by the same final differentiation programme, which cannot be considered here. Nevertheless, it should be mentioned, that defects in cartilage and bone biology can also lead to fused vertebrae, with phenotypes reminiscent of segmentation defects. Possibly mutations in the human FILAMIN B gene belong to this category (Krakow *et al.*, 2004; Stern *et al.*, 1990).

As patterning events that shape the vertebral anlagen come to a close, it should be noted that the notochord has to play its final role. Squeezed out of the vertebral body, the notochord remains persist in the developing intervertebral disc as the highly hydrated nucleus pulposus and organize the annulus fibrosus, which gives the elastic strength to the intervertebral disc. The process is mediated by the transcription factors Sox5/6, which control the formation of the nucleus pulposus (Smits and Lefebvre, 2003).

Agenda for the future

Since this chapter was written for the first edition of "Embryos, genes and birth defects" 8 years ago, substantial progress has been made in deciphering molecular cascades that control the developmental steps towards vertebral column formation. The molecular players identified so far provide important diagnostic tools to identify mutations that cause vertebral column defects in humans. Moreover, these years of research have led to the understanding of combinatorial effects of mutations, which allow predictions regarding the susceptibility to vertebral column defects. In recent years, animal models have been developed for human neural tube defects, and it has been established that preventive treatment of females prior to conception with folic acid can lower the risk of the baby developing *spina bifida* (van Straaten and Copp, 2001). Now, various animal models for vertebral column defects are in place (Table 15.2), and we can start to develop pharmacological approaches for these.

This success must not conceal the fact that our understanding of vertebral column defects is still rather limited, and many more molecular players await discovery. In addition, research on morphogenetic movements shows that birth defects can arise as a knock-on effect from problems at a distant site in the embryo. Thus, continuation of the current line of embryological, molecular and cell biological research aimed at unravelling the regulation of vertebral column formation is of great importance.

We must not forget that substantial vertebral column defects are not limited to congenital diseases, but also occur after accidents or, for instance, in conjunction with certain types of cancer (e.g. multiple myeloma). Thus, strategies need to be developed to reconstitute a functional vertebral column in the adult. One strategy is to develop biomaterials that may substitute vertebral column components, and research is underway to develop synthetic intervertebral discs (e.g. Kotani *et al.*, 2004; Vernon *et al.*, 2003) or to perform tissue engineering to replace damaged discs (reviewed in Alini *et al.*, 2002). Furthermore, research into the use of stem cells may deliver new options. Interestingly, research on stem cells for skeletal muscle indicate that these cells employ much of their embryonic molecular tool kit (reviewed by Buckingham *et al.*, 2003). This serves as a reminder that regardless of the fashion waves sweeping through the biomedical sciences, sound knowledge of biological processes is imperative, and that the embryo is central to this.

Acknowledgements

It is a great honour for us to update this book chapter, based on the foundations laid by the original author, Michael Kessel. We are most grateful to Andrea Streit and Frank Schubert for critically reading the manuscript, for all their helpful comments and for their support. We also thank Lucy Di Silvio for helpful comments on biomaterials.

References

Akasaka, T., Kanno, M., Balling, R., Mieza, M. A., Taniguchi, M., and Koseki, H. (1996). A role for mel-18, a Polycomb group-related vertebrate gene, during the anteroposterior specification of the axial skeleton. *Development* **122**, 1513–1522.

Akasaka, T., van Lohuizen, M., van der Lugt, N., Mizutani-Koseki, Y., Kanno, M., Taniguchi, M., Vidal, M., Alkema, M., Berns, A., and Koseki, H. (2001). Mice doubly deficient for the

Polycomb Group genes Mel18 and Bmi1 reveal synergy and requirement for maintenance but not initiation of Hox gene expression. *Development* **128**, 1587–1597.

Akazawa, H., Komuro, I., Sugitani, Y., Yazaki, Y., Nagai, R., and Noda, T. (2000). Targeted disruption of the homeobox transcription factor Bapx1 results in lethal skeletal dysplasia with asplenia and gastroduodenal malformation. *Genes Cells* **5**, 499–513.

Alini, M., Roughley, P. J., Antoniou, J., Stoll, T., and Aebi, M. (2002). A biological approach to treating disc degeneration: not for today, but maybe for tomorrow. *Eur Spine J* **11 Suppl 2**, S215–220.

Alkema, M. J., van der Lugt, N. M., Bobeldijk, R. C., Berns, A., and van Lohuizen, M. (1995). Transformation of axial skeleton due to overexpression of bmi-1 in transgenic mice. *Nature* **374**, 724–727.

Allan, D., Houle, M., Bouchard, N., Meyer, B. I., Gruss, P., and Lohnes, D. (2001). RARgamma and Cdx1 interactions in vertebral patterning. *Dev Biol* **240**, 46–60.

Amaya, E., Musci, T. J., and Kirschner, M. W. (1991). Expression of a dominant negative mutant of the FGF receptor disrupts mesoderm formation in Xenopus embryos. *Cell* **66**, 257–270.

Ang, S. I., and Rossant, J. (1994). HNF-3 beta is essential for node and notochord formation in mouse development. *Cell* **78**, 561–574.

Anhalt, H., Parker, B., Paranjpe, D. V., Neely, E. K., Silverman, F. N., and Rosenfeld, R. G. (1995). Novel spinal dysplasia in two generations. *Am J Med Genet* **56**, 90–93.

Aoyoma, H., and Asamoto, K. (1988). Determination of somite cells: independence of cell differentiation and morphogenesis. *Development* **104**, 15–28.

Aruga, J., Mizugishi, K., Koseki, H., Imai, K., Balling, R., Noda, T., and Mikoshiba, K. (1999). Zic1 regulates the patterning of vertebral arches in cooperation with Gli3. *Mech Dev* **89**, 141–150.

Aruga, J., Tohmonda, T., Homma, S., and Mikoshiba, K. (2002). Zic1 promotes the expansion of dorsal neural progenitors in spinal cord by inhibiting neuronal differentiation. *Dev Biol* **244**, 329–341.

Auerbach, R. (1954). Analysis of the developmental effects of a lethal mutation in the house mouse. *Journal of Experimental Zoology* **127**, 305–329.

Aulehla, A., Wehrle, C., Brand-Saberi, B., Kemler, R., Gossler, A., Kanzler, B., and Herrmann, B. G. (2003). Wnt3a plays a major role in the segmentation clock controlling somitogenesis. *Dev Cell* **4**, 395–406.

Baldwin, C. T., Hoth, C. F., Amos, J. A., da-Silva, E. O., and Milunsky, A. (1992). An exonic mutation in the HuP2 paired domain gene causes Waardenburg's syndrome. *Nature* **355**, 637–638.

Barnes, G. L., Alexander, P. G., Hsu, C. W., Mariani, B. D., and Tuan, R. S. (1997). Cloning and characterization of chicken Paraxis: a regulator of paraxial mesoderm development and somite formation. *Dev Biol* **189**, 95–111.

Barrantes, I. B., Elia, A. J., Wunsch, K., De Angelis, M. H., Mak, T. W., Rossant, J., Conlon, R. A., Gossler, A., and de la Pompa, J. L. (1999). Interaction between Notch signalling and Lunatic fringe during somite boundary formation in the mouse. *Curr Biol* **9**, 470–480.

Bastian, H., and Gruss, P. (1990). A murine even-skipped homologue, Evx 1, is expressed during early embryogenesis and neurogenesis in a biphasic manner. *Embo J* **9**, 1839–1852.

Bel, S., Core, N., Djabali, M., Kieboom, K., Van der Lugt, N., Alkema, M. J., and Van Lohuizen, M. (1998). Genetic interactions and dosage effects of Polycomb group genes in mice. *Development* **125**, 3543–3551.

Bel-Vialar, S., Core, N., Terranova, R., Goudot, V., Boned, A., and Djabali, M. (2000). Altered retinoic acid sensitivity and temporal expression of Hox genes in polycomb-M33-deficient mice. *Dev Biol* **224**, 238–249.

Bel-Vialar, S., Itasaki, N., and Krumlauf, R. (2002). Initiating Hox gene expression: in the early chick neural tube differential sensitivity to FGF and RA signaling subdivides the HoxB genes in two distinct groups. *Development* **129**, 5103–5115.

Berry, R.J. (1960). Genetical studies on the skeleton of the mouse.XXVI. Pintail. Genet Res Camb **1**, 439–451.

Bessho, Y., and Kageyama, R. (2003). Oscillations, clocks and segmentation. *Curr Opin Genet Dev* **13**, 379–384.

Bessho, Y., Sakata, R., Komatsu, S., Shiota, K., Yamada, S., and Kageyama, R. (2001). Dynamic expression and essential functions of Hes7 in somite segmentation. *Genes Dev* **15**, 2642–2647.

Bhatia-Gaur, R., Donjacour, A. A., Sciavolino, P. J., Kim, M., Desai, N., Young, P., Norton, C. R., Gridley, T., Cardiff, R. D., Cunha, G. R., Abate-Shen, C., and Shen, M. M. (1999). Roles for Nkx3.1 in prostate development and cancer. *Genes Dev* **13**, 966–977.

Borycki, A., Brown, A. M., and Emerson, C. P., Jr. (2000). Shh and Wnt signaling pathways converge to control Gli gene activation in avian somites. *Development* **127**, 2075–2087.

Brennan, J., Lu, C. C., Norris, D. P., Rodriguez, T. A., Beddington, R. S., and Robertson, E. J. (2001). Nodal signalling in the epiblast patterns the early mouse embryo. *Nature* **411**, 965–969.

Brent, A. E., Schweitzer, R., and Tabin, C. J. (2003). A somitic compartment of tendon progenitors. *Cell* **113**, 235–248.

Brent, A. E., and Tabin, C. J. (2002). Developmental regulation of somite derivatives: muscle, cartilage and tendon. *Curr Opin Genet Dev* **12**, 548–557.

Brooke, N. M., Garcia-Fernandez, J., and Holland, P. W. (1998). The ParaHox gene cluster is an evolutionary sister of the Hox gene cluster. *Nature* **392**, 920–922.

Brown, D., Wagner, D., Li, X., Richardson, J. A., and Olson, E. N. (1999). Dual role of the basic helix–loop–helix transcription factor scleraxis in mesoderm formation and chondrogenesis during mouse embryogenesis. *Development* **126**, 4317–4329.

Buckingham, M., Bajard, L., Chang, T., Daubas, P., Hadchouel, J., Meilhac, S., Montarras, D., Rocancourt, D., and Relaix, F. (2003). The formation of skeletal muscle: from somite to limb. *J Anat* **202**, 59–68.

Burgess, R., Rawis, A., Brown, D., Bradley, A., and Olson, E. N. (1996). Requirement of the *paraxis* gene for somite formation and muscoloskeletal patterning. *Nature* **384**, 570–573.

Burke, A. C. (2000). Hox genes and the global patterning of the somitic mesoderm. *Curr Top Dev Biol* **47**, 155–181.

Burke, A. C., Nelson, C. E., Morgan, B. A., and Tabin, C. (1995). Hox genes and the evolution of vertebrate axial morphology. *Development* **121**, 333–346.

Bussen, M., Petry, M., Schuster-Gossler, K., Leitges, M., Gossler, A., and Kispert, A. (2004). The T-box transcription factor Tbx18 maintains the separation of anterior and posterior somite compartments. *Genes Dev* **18**, 1209–1221.

Buttitta, L., Mo, R., Hui, C. C., and Fan, C. M. (2003). Interplays of Gli2 and Gli3 and their requirement in mediating Shh-dependent sclerotome induction. *Development* **130**, 6233–6243.

Capdevila, J., and Johnson, R. L. (1998). Endogenous and ectopic expression of noggin suggests a conserved mechanism for regulation of BMP function during limb and somite patterning. *Dev Biol* **197**, 205–217.

Capdevila, J., Tabin, C., and Johnson, R. L. (1998). Control of dorsoventral somite patterning by Wnt-1 and beta-catenin. *Dev Biol* **193**, 182–194.

Carver, E. A., Jiang, R., Lan, Y., Oram, K. F., and Gridley, T. (2001). The mouse snail gene encodes a key regulator of the epithelial-mesenchymal transition. *Mol Cell Biol* **21**, 8184–8188.

Chapman, D. L., Cooper-Morgan, A., Harrelson, Z., and Papaioannou, V. E. (2003). Critical role for Tbx6 in mesoderm specification in the mouse embryo. *Mech Dev* **120**, 837–847.

Chapman, D. L., and Papaioannou, V. E. (1998). Three neural tubes in mouse embryos with mutations in the T-box gene Tbx6. *Nature* **391**, 695–697.

Charite, J., de Graaff, W., Consten, D., Reijnen, M. J., Korving, J., and Deschamps, J. (1998). Transducing positional information to the Hox genes: critical interaction of cdx gene products with position-sensitive regulatory elements. *Development* **125**, 4349–4358.

Chiang, C., Litingtung, Y., Lee, E., Young, K. E., Corden, J. L., Westphal, H., and Beachy, P. A. (1996). Cyclopia and defective axial patterning in mice lacking Sonic hedgehog gene function. *Nature* **383**, 407–413.

Cho, K. W., and De Robertis, E. M. (1990). Differential activation of Xenopus homeo box genes by mesoderm-inducing growth factors and retinoic acid. *Genes Dev* **4**, 1910–1916.

Ciruna, B., and Rossant, J. (2001). FGF signaling regulates mesoderm cell fate specification and morphogenetic movement at the primitive streak. *Dev Cell* **1**, 37–49.

Ciruna, B. G., Schwartz, L., Harpal, K., Yamaguchi, T. P., and Rossant, J. (1997). Chimeric analysis of fibroblast growth factor receptor-1 (Fgfr1) function: a role for FGFR1 in morphogenetic movement through the primitive streak. *Development* **124**, 2829–2841.

Conlon, F. L., Lyons, K. M., Takaesu, N., Barth, K. S., Kispert, A., Herrmann, B., and Robertson, E. J. (1994). A primary requirement for nodal in the formation and maintenance of the primitive streak in the mouse. *Development* **120**, 1919–1928.

Conlon, R. A., Reaume, A. G., and Rossant, J. (1995). *Notch1* is required for the coordinate segmentation of somites. *Development* **121**, 1533–1545.

Cordes, R., Schuster-Gossler, K., Serth, K., and Gossler, A. (2004). Specification of vertebral identity is coupled to Notch signalling and the segmentation clock. *Development* **131**, 1221–1233.

Correia, K. M., and Conlon, R. A. (2000). Surface ectoderm is necessary for the morphogenesis of somites. *Mech Dev* **91**, 19–30.

Dale, J. K., Maroto, M., Dequeant, M. L., Malapert, P., McGrew, M., and Pourquie, O. (2003). Periodic notch inhibition by lunatic fringe underlies the chick segmentation clock. *Nature* **421**, 275–278.

Deschamps, J., van den Akker, E., Forlani, S., De Graaff, W., Oosterveen, T., Roelen, B., and Roelfsema, J. (1999). Initiation, establishment and maintenance of Hox gene expression patterns in the mouse. *Int J Dev Biol* **43**, 635–650.

Dhoot, G. K., Gustafsson, M. K., Ai, X., Sun, W., Standiford, D. M., and Emerson, C. P., Jr. (2001). Regulation of Wnt signaling and embryo patterning by an extracellular sulfatase. *Science* **293**, 1663–1666.

Dietrich, S., and Gruss, P. (1995). undulated phenotypes suggest a role of Pax-1 for the development of vertebral and extravertebral structures. *Dev Biol* **167**, 529–548.

Dietrich, S., Schubert, F. R., and Gruss, P. (1993). Altered Pax gene expression in murine notochord mutants: the notochord is required to initiate and maintain ventral identity in the somite. *Mech Dev* **44**, 189–207.

Dietrich, S., Schubert, F. R., Healy, C., Sharpe, P. T., and Lumsden, A. (1998). Specification of the hypaxial musculature. *Development* **125**, 2235–2249.

Dietrich, S., Schubert, F. R., and Lumsden, A. (1997). Control of dorsoventral pattern in the chick paraxial mesoderm. *Development* **124**, 3895–3908.

Dobrovolskaia-Zavadskaia, N. (1927). Sur la mortification spontanee de la queue chez la souris nouveau-nee et sur l'existence d'un caractere hereditaire 'non-viable'. *C R Soc Biol* **97**, 114–116.

Dubrulle, J., and Pourquie, O. (2004). fgf8 mRNA decay establishes a gradient that couples axial elongation to patterning in the vertebrate embryo. *Nature* **427**, 419–422.

Dubrulle, J., McGrew, M. J., and Pourquie, O. (2001). FGF signaling controls somite boundary position and regulates segmentation clock control of spatiotemporal Hox gene activation. *Cell* **106**, 219–232.

Ebensperger, C., Wilting, J., Brand-Saberi, B., Mizutani, Y., Christ, B., Balling, R., and Koseki, H. (1995). Pax-1, a regulator of sclerotome development is induced by notochord and floor plate signals in avian embryos. *Anat Embryol (Berl)* **191**, 297–310.

Eggenschwiler, J. T., Espinoza, E., and Anderson, K. V. (2001). Rab23 is an essential negative regulator of the mouse Sonic hedgehog signalling pathway. *Nature* **412**, 194–198.

Epstein, M., Pillemer, G., Yelin, R., Yisraeli, J. K., and Fainsod, A. (1997). Patterning of the embryo along the anterior-posterior axis: the role of the caudal genes. *Development* **124**, 3805–3814.

Evrard, Y. A., Lun, Y., Aulehla, A., Gan, L., and Jonson, R. L. (1998). *lunatic fringe* is an essential mediator of somite segmentation and patterning. *Nature* **394**, 377–381.

Fan, C. M., Lee, C. S., and Tessier-Lavigne, M. (1997). A role for WNT proteins in induction of dermomyotome. *Dev Biol* **191**, 160–165.

Fan, C. M., Porter, J. A., Chiang, C., Chang, D. T., Beachy, P. A., and Tessier-Lavigne, M. (1995). Long-range sclerotome induction by sonic hedgehog: direct role of the amino-terminal cleavage product and modulation by the cyclic AMP signaling pathway. *Cell* **81**, 457–465.

Fan, C. M., and Tessier-Lavigne, M. (1994). Patterning of mammalian somites by surface ectoderm and notochord: evidence for sclerotome induction by a hedgehog homolog. *Cell* **79**, 1175–1186.

Faust, C., Schumacher, A., Holdener, B., and Magnuson, T. (1995). The eed mutation disrupts anterior mesoderm production in mice. *Development* **121**, 273–285.

Featherstone, M. (2002). Coactivators in transcription initiation: here are your orders. *Curr Opin Genet Dev* **12**, 149–155.

Fleming, A., Keynes, R., and Tannahill, D. (2004). A central role for the notochord in vertebral patterning. *Development* **131**, 873–880.

Forlani, S., Lawson, K. A., and Deschamps, J. (2003). Acquisition of Hox codes during gastrulation and axial elongation in the mouse embryo. *Development* **130**, 3807–3819.

Furumoto, T. A., Miura, N., Akasaka, T., Mizutani-Koseki, Y., Sudo, H., Fukuda, K., Maekawa, M., Yuasa, S., Fu, Y., Moriya, H., Taniguchi, M., Imai, K., Dahl, E., Balling, R., Pavlova, M., Gossler, A., and Koseki, H. (1999). Notochord-dependent expression of MFH1 and PAX1 cooperates to maintain the proliferation of sclerotome cells during the vertebral column development. *Dev Biol* **210**, 15–29.

Gaunt, S. J. (1994). Conservation in the Hox code during morphological evolution. *Int J Dev Biol* **38**, 549–552.

Gluecksohn-Schoenheimer, S. (1945). The embryonic development of mutants of the Sd-strain in mice. *Genetics* **30**, 29–38.

Gluecksohn-Schoenheimer, S. (1949). The effects of a lethal mutation responsible for duplications and twinning in mouse embryos. *J exp Zool* **110**, 47–76.

Goodrich, E. S. (1958). 'Studies on the structure and development of vertebrates.' Dover Publications Inc., New York.

Goodrich, L. V., Milenkovic, L., Higgins, K. M., and Scott, M. P. (1997). Altered neural cell fates and medulloblastoma in mouse patched mutants. *Science* **277**, 1109–1113.

Gossler, A., and Hrabe de Angelis, M. (1998). Somitogenesis. *Current Topics in Developmental Biology* **38**, 225–287.

Gray, H. (1995). 'Gray's Anatomy.' Churchill Livingstone, Edinburgh, London.

Greco, T. L., Takada, S., Newhouse, M. M., McMahon, J. A., McMahon, A. P., and Camper, S. A. (1996). Analysis of the vestigial tail mutation demonstrates that Wnt-3a gene dosage regulates mouse axial development. *Genes & Development* **10**, 313–324.

Grüneberg, H. (1963). 'The Pathology of Development.' Blackwell Scientific, Oxford.

Gunderson, C. H., Greenspan, R. H., Glaser, G. H., and Lubs, H. A. (1967). The Klippel-Feil syndrome: genetic and clinical reevaluation of cervical fusion. *Medicine (Baltimore)* **46**, 491–512.

Gusella, J. F., Tanzi, R. E., Bader, P. I., Phelan, M. C., Stevenson, R., Hayden, M. R., Hofman, K. J., Faryniarz, A. G., and Gibbons, K. (1985). Deletion of Huntington's disease-linked G8 (D4S10) locus in Wolf-Hirschhorn syndrome. *Nature* **318**, 75–78.

Handrigan, G. R. (2003). Concordia discors: duality in the origin of the vertebrate tail. *J Anat* **202**, 255–267.

Hansen, L. A., Sigman, C. C., Andreola, F., Ross, S. A., Kelloff, G. J., and De Luca, L. M. (2000). Retinoids in chemoprevention and differentiation therapy. *Carcinogenesis* **21**, 1271–1279.

Hanson, R. D., Hess, J. L., Yu, B. D., Ernst, P., van Lohuizen, M., Berns, A., van der Lugt, N. M., Shashikant, C. S., Ruddle, F. H., Seto, M., and Korsmeyer, S. J. (1999). Mammalian Trithorax and polycomb-group homologues are antagonistic regulators of homeotic development. *Proc Natl Acad Sci U S A* **96**, 14372-14377.

Helwig, U., Imai, K., Schmahl, W., Thomas, B. E., Varnum, D. S., Nadeau, J. H., and Balling, R. (1995). Interaction between undulated and Patch leads to an extreme form of spina bifida in double-mutant mice. *Nat Genet* **11**, 60–63.

Henderson, D. J., Conway, S. J., and Copp, A. J. (1999). Rib truncations and fusions in the Sp2H mouse reveal a role for Pax3 in specification of the ventro-lateral and posterior parts of the somite. *Dev Biol* **209**, 143–158.

Heston, W. E. (1951). The 'vestigial tail' mouse; a new recessive mutation. *J Hered* **42**, 71–74.

Horikawa, K., Radice, G., Takeichi, M., and Chisaka, O. (1999). Adhesive subdivisions intrinsic to the epithelial somites. *Dev Biol* **215**, 182–189.

Houle, M., Prinos, P., Iulianella, A., Bouchard, N., and Lohnes, D. (2000). Retinoic acid regulation of Cdx1: an indirect mechanism for retinoids and vertebral specification. *Mol Cell Biol* **20**, 6579–6586.

Hrabe de Angelis, M., McIntyre, J. I., and Gossler, A. (1997). Maintenance of somite borders in mice requires the *Delta* homologue *Dll1*. *Nature* **386**, 717–721.

Huang, D., Chen, S. W., Langston, A. W., and Gudas, L. J. (1998). A conserved retinoic acid responsive element in the murine Hoxb-1 gene is required for expression in the developing gut. *Development* **125**, 3235–3246.

Huang, R., Zhi, Q., Brand-Saberi, B., and Christ, B. (2000). New experimental evidence for somite resegmentation. *Anat Embryol (Berl)* **202**, 195–200.

Huangfu, D., Liu, A., Rakeman, A. S., Murcia, N. S., Niswander, L., and Anderson, K. V. (2003). Hedgehog signalling in the mouse requires intraflagellar transport proteins. *Nature* **426**, 83–87.

Iida, K., Koseki, H., Kakinuma, H., Kato, N., Mizutani-Koseki, Y., Ohuchi, H., Yoshioka, H., Noji, S., Kawamura, K., Kataoka, Y., Ueno, F., Taniguchi, M., Yoshida, N., Sugiyama, T., and Miura, N. (1997). Essential roles of the winged helix transcription factor MFH-1 in aortic arch patterning and skeletogenesis. *Development* **124**, 4627–4638.

Ikeya, M., and Takada, S. (2001). Wnt-3a is required for somite specification along the anteroposterior axis of the mouse embryo and for regulation of cdx-1 expression. *Mech Dev* **103**, 27–33.

Ip, Y. T., and Gridley, T. (2002). Cell movements during gastrulation: snail dependent and independent pathways. *Curr Opin Genet Dev* **12**, 423–429.

Isaacs, H. V., Andreazzoli, M., and Slack, J. M. (1999). Anteroposterior patterning by mutual repression of orthodenticle and caudal-type transcription factors. *Evol Dev* **1**, 143–152.

Isaacs, H. V., Pownall, M. E., and Slack, J. M. (1998). Regulation of Hox gene expression and posterior development by the Xenopus caudal homologue Xcad3. *Embo J* **17**, 3413–3427.

Iulianella, A., and Lohnes, D. (1997). Contribution of retinoic acid receptor gamma to retinoid-induced craniofacial and axial defects. *Dev Dyn* **209**, 92–104.

Iulianella, A., Beckett, B., Petkovich, M., and Lohnes, D. (1999). A molecular basis for retinoic acid-induced axial truncation. *Dev Biol* **205**, 33–48.

Johnson, J., Rhee, J., Parsons, S. M., Brown, D., Olson, E. N., and Rawls, A. (2001). The anterior/posterior polarity of somites is disrupted in paraxis-deficient mice. *Dev Biol* **229**, 176–187.

Johnson, R. L., Laufer, E., Riddle, R. D., and Tabin, C. (1994). Ectopic expression of Sonic hedgehog alters dorsal-ventral patterning of somites. *Cell* **79**, 1165–1173.

Jouve, C., Palmeirim, I., Henrique, D., Beckers, J., Gossler, A., Ish-Horowicz, D., and Pourquie, O. (2000). Notch signalling is required for cyclic expression of the hairy-like gene HES1 in the presomitic mesoderm. *Development* **127**, 1421–1429.

Kannan, K., and Givol, D. (2000). FGF receptor mutations: dimerization syndromes, cell growth suppression, and animal models. *IUBMB Life* **49**, 197–205.

Karabagli, H., Karabagli, P., Ladher, R. K., and Schoenwolf, G. C. (2002). Comparison of the expression patterns of several fibroblast growth factors during chick gastrulation and neurulation. *Anat Embryol (Berl)* **205**, 365–370.

Kato, N., and Aoyama, H. (1998). Dermomyotomal origin of the ribs as revealed by extirpation and transplantation experiments in chick and quail embryos. *Development* **125**, 3437–3443.

Kessel, M., Balling, R., and Gruss, P. (1990). Variations of cervical vertebrae after expression of a Hox-1.1 transgene in mice. *Cell* **61**, 301–308.

Kessel, M., and Gruss, P. (1991). Homeotic transformations of murine vertebrae and concomitant alteration of Hox codes induced by retinoic acid. *Cell* **67**, 89–104.

Keynes, R. J., and Stern, C. D. (1988). Mechanisms of vertebrate segmenation. *Development* **103**, 413–429.

Kibar, Z., Vogan, K. J., Groulx, N., Justice, M. J., Underhill, D. A., and Gros, P. (2001). Ltap, a mammalian homolog of Drosophila Strabismus/Van Gogh, is altered in the mouse neural tube mutant Loop-tail. *Nat Genet* **28**, 251–255.

Kieny, M., Mauger, A., and Sengel, P. (1972). Early regionalization of somitic mesoderm as studied by the development of axial skeleton of the chick embryo. *Dev Biol* **28**, 142–161.

Kim, D. W., and Lassar, A. B. (2003). Smad-dependent recruitment of a histone deacetylase/Sin3A complex modulates the bone morphogenetic protein-dependent transcriptional repressor activity of Nkx3.2. *Mol Cell Biol* **23**, 8704–8717.

Kimelman, D., and Kirschner, M. (1987). Synergistic induction of mesoderm by FGF and TGF-beta and the identification of an mRNA coding for FGF in the early Xenopus embryo. *Cell* **51**, 869–877.

Kitajima, S., Takagi, A., Inoue, T., and Saga, Y. (2000). MesP1 and MesP2 are essential for the development of cardiac mesoderm. *Development* **127**, 3215–3226.

Kmita, M., and Duboule, D. (2003). Organizing axes in time and space; 25 years of colinear tinkering. *Science* **301**, 331–333.

Kmita, M., van Der Hoeven, F., Zakany, J., Krumlauf, R., and Duboule, D. (2000). Mechanisms of Hox gene colinearity: transposition of the anterior Hoxb1 gene into the posterior HoxD complex. *Genes Dev* **14**, 198–211.

Korkko, J., Cohn, D. H., Ala-Kokko, L., Krakow, D., and Prockop, D. J. (2000). Widely distributed mutations in the COL2A1 gene produce achondrogenesis type II/hypochondrogenesis. *Am J Med Genet* **92**, 95–100.

Kos, L., Chiang, C., and Mahon, K. A. (1998). Mediolateral patterning of somites: multiple axial signals, including Sonic hedgehog, regulate Nkx-3.1 expression. *Mech Dev* **70**, 25–34.

Kotani, Y., Abumi, K., Shikinami, Y., Takahata, M., Kadoya, K., Kadosawa, T., Minami, A., and Kaneda, K. (2004). Two-year observation of artificial intervertebral disc replacement: results after supplemental ultra-high strength bioresorbable spinal stabilization. *J Neurosurg* **100**, 337–342.

Krakow, D., Robertson, S. P., King, L. M., Morgan, T., Sebald, E. T., Bertolotto, C., Wachsmann-Hogiu, S., Acuna, D., Shapiro, S. S., Takafuta, T., Aftimos, S., Kim, C. A., Firth, H., Steiner, C. E., Cormier-Daire, V., Superti-Furga, A., Bonafe, L., Graham, J. M., Jr., Grix, A., Bacino, C. A., Allanson, J., Bialer, M. G., Lachman, R. S., Rimoin, D. L., and Cohn, D. H. (2004). Mutations in the gene encoding filamin B disrupt vertebral segmentation, joint formation and skeletogenesis. *Nat Genet* **36**, 405–410.

Kroll, K. L., and Amaya, E. (1996). Transgenic Xenopus embryos from sperm nuclear transplantations reveal FGF signaling requirements during gastrulation. *Development* **122**, 3173–3183.

Kronenberg, H. M. (2003). Developmental regulation of the growth plate. *Nature* **423**, 332–336.

Kume, T., Deng, K. Y., Winfrey, V., Gould, D. B., Walter, M. A., and Hogan, B. L. (1998). The forkhead/winged helix gene Mf1 is disrupted in the pleiotropic mouse mutation congenital hydrocephalus. *Cell* **93**, 985–996.

Kume, T., Jiang, H., Topczewska, J. M., and Hogan, B. L. (2001). The murine winged helix transcription factors, Foxc1 and Foxc2, are both required for cardiovascular development and somitogenesis. *Genes Dev* **15**, 2470–2482.

Kusumi, K., Sun, E. S., Kerrebrock, A. W., Bronson, R. T., Chi, D.-C., Bulotsky, M. S., Spencer, J. B., Birren, B. W., Frankel, W. N., and Lander, E. S. (1998). The mouse pudgy mutation disrupts *Delta* homologue *Dll3* and initiation of early somite boundaries. *Nature Genetics* **19**, 274–277.

Lai, M. T., Chen, E., Crouthamel, M. C., DiMuzio-Mower, J., Xu, M., Huang, Q., Price, E., Register, R. B., Shi, X. P., Donoviel, D. B., Bernstein, A., Hazuda, D., Gardell, S. J., and Li, Y. M. (2003). Presenilin-1 and presenilin-2 exhibit distinct yet overlapping gamma-secretase activities. *J Biol Chem* **278**, 22475-22481.

Lamb, T. M., and Harland, R. M. (1995). Fibroblast growth factor is a direct neural inducer, which combined with noggin generates anterior-posterior neural pattern. *Development* **121**, 3627–3636.

Lawson, A., Colas, J. F., and Schoenwolf, G. C. (2001). Classification scheme for genes expressed during formation and progression of the avian primitive streak. *Anat Rec* **262**, 221–226.

Lawson, A., and Schoenwolf, G. C. (2001). New insights into critical events of avian gastrulation. *Anat Rec* **262**, 238–252.

Lazar, M. A. (1999). Nuclear hormone receptors: from molecules to diseases. *J Investig Med* **47**, 364–368.

Lee, C. S., Buttitta, L., and Fan, C. M. (2001). Evidence that the WNT-inducible growth arrest-specific gene 1 encodes an antagonist of sonic hedgehog signaling in the somite. *Proc Natl Acad Sci U S A* **98**, 11347-11352.

Lee, C. S., Buttitta, L. A., May, N. R., Kispert, A., and Fan, C. M. (2000). SHH-N upregulates Sfrp2 to mediate its competitive interaction with WNT1 and WNT4 in the somitic mesoderm. *Development* **127**, 109–118.

Lefebvre, V., Behringer, R. R., and de Crombrugghe, B. (2001). L-Sox5, Sox6 and Sox9 control essential steps of the chondrocyte differentiation pathway. *Osteoarthritis Cartilage* **9 Suppl A**, S69–75.

Leitges, M., Neidhardt, L., Haenig, B., Herrmann, B. G., and Kispert, A. (2000). The paired homeobox gene Uncx4.1 specifies pedicles, transverse processes and proximal ribs of the vertebral column. *Development* **127**, 2259–2267.

Lettice, L. A., Purdie, L. A., Carlson, G. J., Kilanowski, F., Dorin, J., and Hill, R. E. (1999). The mouse bagpipe gene controls development of axial skeleton, skull, and spleen. *Proc Natl Acad Sci U S A* **96**, 9695–9700.

Li, L., Krantz, I. D., Deng, Y., Genin, A., Banta, A. B., Collins, C. C., Qi, M., Trask, B. J., Kuo, W. L., Cochran, J., Costa, T., Pierpont, M. E., Rand, E. B., Piccoli, D. A., Hood, L., and Spinner, N. B. (1997). Alagille syndrome is caused by mutations in human Jagged1, which encodes a ligand for Notch1. *Nat Genet* **16**, 243–251.

Li, S. W., Prockop, D. J., Helminen, H., Fässler, R., Lapveteläinen, T., Kiraly, K., Peltarri, A., Arokoski, J., Lui, H., Arita, M., and Khillan, J. (1995). Transgenic mice with targeted inactivation of the Col2a1 gene for collagen II develop a skeleton with membranous and periosteal bone but no endochondral bone. *Genes Dev* **9**, 2821–2830.

Li, X., and Rosenfeld, M. G. (2004). Transcription: origins of licensing control. *Nature* **427**, 687–688.

Lickert, H., Domon, C., Huls, G., Wehrle, C., Duluc, I., Clevers, H., Meyer, B. I., Freund, J. N., and Kemler, R. (2000). Wnt/(beta)-catenin signaling regulates the expression of the homeobox gene Cdx1 in embryonic intestine. *Development* **127**, 3805–3813.

Lickert, H., and Kemler, R. (2002). Functional analysis of cis-regulatory elements controlling initiation and maintenance of early Cdx1 gene expression in the mouse. *Dev Dyn* **225**, 216–220.

Linask, K. K., Ludwig, C., Han, M. D., Liu, X., Radice, G. L., and Knudsen, K. A. (1998). N-cadherin/catenin-mediated morphoregulation of somite formation. *Dev Biol* **202**, 85–102.

Lohnes, D. (2003). The Cdx1 homeodomain protein: an integrator of posterior signaling in the mouse. *Bioessays* **25**, 971–980.

Lu, C. C., Brennan, J., and Robertson, E. J. (2001). From fertilization to gastrulation: axis formation in the mouse embryo. *Curr Opin Genet Dev* **11**, 384–392.

Mahlapuu, M., Ormestad, M., Enerback, S., and Carlsson, P. (2001). The forkhead transcription factor Foxf1 is required for differentiation of extra-embryonic and lateral plate mesoderm. *Development* **128**, 155–166.

Mankoo, B. S., Collins, N. S., Ashby, P., Grigorieva, E., Pevny, L. H., Candia, A., Wright, C. V., Rigby, P. W., and Pachnis, V. (1999). Mox2 is a component of the genetic hierarchy controlling limb muscle development. *Nature* **400**, 69–73.

Mankoo, B. S., Skuntz, S., Harrigan, I., Grigorieva, E., Candia, A., Wright, C. V., Arnheiter, H., and Pachnis, V. (2003). The concerted action of Meox homeobox genes is required upstream of genetic pathways essential for the formation, patterning and differentiation of somites. *Development* **130**, 4655–4664.

Mansouri, A., and Gruss, P. (1998). Pax3 and Pax7 are expressed in commissural neurons and restrict ventral neuronal identity in the spinal cord. *Mech Dev* **78**, 171–178.

Mansouri, A., Voss, A. K., Thomas, T., Yokota, Y., and Gruss, P. (2000). Uncx4.1 is required for the formation of the pedicles and proximal ribs and acts upstream of Pax9. *Development* **127**, 2251–2258.

Mansouri, A., Yokota, Y., Wehr, R., Copeland, N. G., Jenkins, N. A., and Gruss, P. (1997). Paired-related murine homeobox gene expressed in the developing sclerotome, kidney, and nervous system. *Dev Dyn* **210**, 53–65.

Marom, K., Shapira, E., and Fainsod, A. (1997). The chicken caudal genes establish an anterior-posterior gradient by partially overlapping temporal and spatial patterns of expression. *Mech Dev* **64**, 41–52.

Marshall, H., Studer, M., Popperl, H., Aparicio, S., Kuroiwa, A., Brenner, S., and Krumlauf, R. (1994). A conserved retinoic acid response element required for early expression of the homeobox gene Hoxb-1. *Nature* **370**, 567–571.

Mathieu, J., Griffin, K., Herbomel, P., Dickmeis, T., Strahle, U., Kimelman, D., Rosa, F. M., and Peyrieras, N. (2004). Nodal and Fgf pathways interact through a positive regulatory loop and synergize to maintain mesodermal cell populations. *Development* **131**, 629–641.

Mathis, L., Kulesa, P. M., and Fraser, S. E. (2001). FGF receptor signalling is required to maintain neural progenitors during Hensen's node progression. *Nat Cell Biol* **3**, 559–566.

McGinnis, W., and Krumlauf, R. (1992). Homeobox genes and axial patterning. *Cell* **68**, 283–302.

McMahon, J. A., Takada, S., Zimmerman, L. B., Fan, C. M., Harland, R. M., and McMahon, A. P. (1998). Noggin-mediated antagonism of BMP signaling is required for growth and patterning of the neural tube and somite. *Genes Dev* **12**, 1438–1452.

Mo, R., Freer, A. M., Zinyk, D. L., Crackower, M. A., Michaud, J., Heng, H. H., Chik, K. W., Shi, X. M., Tsui, L. C., Cheng, S. H., Joyner, A. L., and Hui, C. (1997). Specific and redundant functions of Gli2 and Gli3 zinc finger genes in skeletal patterning and development. *Development* **124**, 113–123.

Monsoro-Burq, A. H., Bontoux, M., Teillet, M. A., and Le Douarin, N. M. (1994). Heterogeneity in the development of the vertebra. *Proc. Natl. Acad. Sci. USA* **91**, 10435-10439.

Monsoro-Burq, A.-H., Duprez, D., Watanabe, Y., Bontoux, M., Vincent, C., Brickell, P., and Le Douarin, N. (1996). The role of bone morphogenetic proteins in vertebral development. *Development* **122**, 3607–3616.

Morrison, A., Moroni, M. C., Ariza-McNaughton, L., Krumlauf, R., and Mavilio, F. (1996). In vitro and transgenic analysis of a human HOXD4 retinoid-responsive enhancer. *Development* **122**, 1895–1907.

Münsterberg, A., and Lassar, A. (1995). Combinatorial signals from the neural tube, floor plate and notochord induce myogenic bHLH gene expression in the somite. *Development* **121**, 651–660.

Murdoch, J. N., Doudney, K., Paternotte, C., Copp, A. J., and Stanier, P. (2001). Severe neural tube defects in the loop-tail mouse result from mutation of Lpp1, a novel gene involved in floor plate specification. *Hum Mol Genet* **10**, 2593–2601.

Murtaugh, L. C., Chyung, J. H., and Lassar, A. B. (1999). Sonic hedgehog promotes somitic chondrogenesis by altering the cellular response to BMP signaling. *Genes Dev* **13**, 225–237.

Murtaugh, L. C., Zeng, L., Chyung, J. H., and Lassar, A. B. (2001). The chick transcriptional repressor Nkx3.2 acts downstream of Shh to promote BMP-dependent axial chondrogenesis. *Dev Cell* **1**, 411–422.

Naf, D., Wilson, L. A., Bergstrom, R. A., Smith, R. S., Goodwin, N. C., Verkerk, A., van Ommen, G. J., Ackerman, S. L., Frankel, W. N., and Schimenti, J. C. (2001). Mouse models for the Wolf-Hirschhorn deletion syndrome. *Hum Mol Genet* **10**, 91–8.

Neidhardt, L., Kispert, A., and Herrmann, B. G. (1997). A mouse gene of the paired-related homeobox class expressed in the caudal somite compartment and in the developing vertebral column, kidney and nervous system. *Dev Genes Evol* **207**, 330–339.

Neubuser, A., Koseki, H., and Balling, R. (1995). Characterization and developmental expression of Pax9, a paired-box-containing gene related to Pax1. *Dev Biol* **170**, 701–716.

Nisan, N., Hiz, M., and Saner, H. (1988). Total agenesis of five cervical vertebrae: brief report. *J Bone Joint Surg Br* **70**, 668–669.

Nomura-Kitabayashi, A., Takahashi, Y., Kitajima, S., Inoue, T., Takeda, H., and Saga, Y. (2002). Hypomorphic Mesp allele distinguishes establishment of rostrocaudal polarity and segment border formation in somitogenesis. *Development* **129**, 2473–2481.

Nutt, S. L., Dingwell, K. S., Holt, C. E., and Amaya, E. (2001). Xenopus Sprouty2 inhibits FGF-mediated gastrulation movements but does not affect mesoderm induction and patterning. *Genes Dev* **15**, 1152–1166.

Nybakken, K., and Perrimon, N. (2002a). Hedgehog signal transduction: recent findings. *Curr Opin Genet Dev* **12**, 503–511.

Nybakken, K., and Perrimon, N. (2002b). Heparan sulfate proteoglycan modulation of developmental signaling in Drosophila. *Biochim Biophys Acta* **1573**, 280–291.

Oda, T., Elkahloun, A. G., Pike, B. L., Okajima, K., Krantz, I. D., Genin, A., Piccoli, D. A., Meltzer, P. S., Spinner, N. B., Collins, F. S., and Chandrasekharappa, S. C. (1997). Mutations in the human Jagged1 gene are responsible for Alagille syndrome. *Nat Genet* **16**, 235–242.

Orlando, V. (2003). Polycomb, epigenomes, and control of cell identity. *Cell* **112**, 599–606.

Paavola, L. G., Wilson, D. B., and Center, E. M. (1980). Histochemistry of the developing notochord, perichordal sheath and vertebrae in Danforth's short-tail (sd) and normal C57BL/6 mice. *J Embryol Exp Morphol* **55**, 227–245.

Packer, A. I., Crotty, D. A., Elwell, V. A., and Wolgemuth, D. J. (1998). Expression of the murine Hoxa4 gene requires both autoregulation and a conserved retinoic acid response element. *Development* **125**, 1991–1998.

Palmeirim, I., Dubrulle, J., Henrique, D., Ish-Horowicz, D., and Pourquie, O. (1998). Uncoupling segmentation and somitogenesis in the chick presomitic mesoderm. *Dev Genet* **23**, 77–85.

Partanen, J., Schwartz, L., and Rossant, J. (1998). Opposite phenotypes of hypomorphic and Y766 phosphorylation site mutations reveal a function for Fgfr1 in anteroposterior patterning of mouse embryos. *Genes Dev* **12**, 2332–2344.

Passner, J. M., Ryoo, H. D., Shen, L., Mann, R. S., and Aggarwal, A. K. (1999). Structure of a DNA-bound Ultrabithorax-Extradenticle homeodomain complex. *Nature* **397**, 714–719.

Paterno, G. D., Gillespie, L. L., Dixon, M. S., Slack, J. M., and Heath, J. K. (1989). Mesoderm-inducing properties of INT-2 and kFGF: two oncogene-encoded growth factors related to FGF. *Development* **106**, 79–83.

Peters, H., Neubuser, A., Kratochwil, K., and Balling, R. (1998). Pax9-deficient mice lack pharyngeal pouch derivatives and teeth and exhibit craniofacial and limb abnormalities. *Genes Dev* **12**, 2735–2747.

Peters, H., Wilm, B., Sakai, N., Imai, K., Maas, R., and Balling, R. (1999). Pax1 and Pax9 synergistically regulate vertebral column development. *Development* **126**, 5399–5408.

Pilia, G., Hughes-Benzie, R. M., MacKenzie, A., Baybayan, P., Chen, E. Y., Huber, R., Neri, G., Cao, A., Forabosco, A., and Schlessinger, D. (1996). Mutations in GPC3, a glypican gene, cause the Simpson-Golabi-Behmel overgrowth syndrome. *Nat Genet* **12**, 241–247.

Piper, D. E., Batchelor, A. H., Chang, C. P., Cleary, M. L., and Wolberger, C. (1999). Structure of a HoxB1-Pbx1 heterodimer bound to DNA: role of the hexapeptide and a fourth homeodomain helix in complex formation. *Cell* **96**, 587–597.

Piper, M., and Little, M. (2003). Movement through Slits: cellular migration via the Slit family. *Bioessays* **25**, 32–38.

Popperl, H., and Featherstone, M. S. (1993). Identification of a retinoic acid response element upstream of the murine Hox-4.2 gene. *Mol Cell Biol* **13**, 257–265.

Pourquie, O. (2003). The segmentation clock: converting embryonic time into spatial pattern. *Science* **301**, 328–330.

Pourquié, O., Coltey, M., Bréant, C., and Le Douarin, N. M. (1995). Control of somite patterning by signals from the lateral plate. *Proc. Natl. Acad. Sci. USA* **92**, 3219–3223.

Pourquié, O., Fan, C. M., Coltey, M., Hirsinger, E., Watanabe, Y., Bréant, C., Francis-West, P., Brickell, P., Tessier-Lavigne, M., and Le Douarin, N. M. (1996). Lateral and axial signals involved in avian somite patterning: a role for BMP4. *Cell* **84**, 461–471.

Powers, C. J., McLeskey, S. W., and Wellstein, A. (2000). Fibroblast growth factors, their receptors and signaling. *Endocr Relat Cancer* **7**, 165–197.

Pownall, M. E., Tucker, A. S., Slack, J. M., and Isaacs, H. V. (1996). eFGF, Xcad3 and Hox genes form a molecular pathway that establishes the anteroposterior axis in Xenopus. *Development* **122**, 3881–3892.

Prinos, P., Joseph, S., Oh, K., Meyer, B. I., Gruss, P., and Lohnes, D. (2001). Multiple pathways governing Cdx1 expression during murine development. *Dev Biol* **239**, 257–269.

Raas-Rothschild, A., Goodman, R. M., Grunbaum, M., Berger, I., and Mimouni, M. (1988). Klippel-Feil anomaly with sacral agenesis: an additional subtype, type IV. *J Craniofac Genet Dev Biol* **8**, 297–301.

Radice, G. L., Rayburn, H., Matsunami, H., Knudsen, K. A., Takeichi, M., and Hynes, R. O. (1997). Developmental defects in mouse embryos lacking N-cadherin. *Dev Biol* **181**, 64–78.

Rashbass, P., Cooke, L. A., Herrmann, B. G., and Beddington, R. S. (1991). A cell autonomous function of Brachyury in T/T embryonic stem cell chimaeras. *Nature* **353**, 348–351.

Rashbass, P., Wilson, V., Rosen, B., and Beddington, R. S. (1994). Alterations in gene expression during mesoderm formation and axial patterning in Brachyury (T) embryos. *Int J Dev Biol* **38**, 35–44.

Remak, A. (1850). 'Untersuchungen über die Entwicklung der Wirbelthiere.' Reimer, Berlin.

Rhee, J., Takahashi, Y., Saga, Y., Wilson-Rawls, J., and Rawls, A. (2003). The protocadherin papc is involved in the organization of the epithelium along the segmental border during mouse somitogenesis. *Dev Biol* **254**, 248–261.

Rodrigo, I., Hill, R. E., Balling, R., Munsterberg, A., and Imai, K. (2003). Pax1 and Pax9 activate Bapx1 to induce chondrogenic differentiation in the sclerotome. *Development* **130**, 473–482.

Rossant, J., Zirngibl, R., Cado, D., Shago, M., and Giguere, V. (1991). Expression of a retinoic acid response element-hsplacZ transgene defines specific domains of transcriptional activity during mouse embryogenesis. *Genes Dev* **5**, 1333–1344.

Saga, Y., Hata, N., Koseki, H., and Taketo, M. M. (1997). *Mesp2*: a novel mouse gene expressed in the presegmented mesoderm and essential for segmentation initiation. *Genes & Development* **11**, 1827–1839.

Sagerstrom, C. G. (2004). PbX marks the spot. *Dev Cell* **6**, 737–738.

Sakai, Y., Meno, C., Fujii, H., Nishino, J., Shiratori, H., Saijoh, Y., Rossant, J., and Hamada, H. (2001). The retinoic acid-inactivating enzyme CYP26 is essential for establishing an uneven distribution of retinoic acid along the anterio-posterior axis within the mouse embryo. *Genes Dev* **15**, 213–225.

Saleh, M., Rambaldi, I., Yang, X. J., and Featherstone, M. S. (2000). Cell signaling switches HOX-PBX complexes from repressors to activators of transcription mediated by histone deacetylases and histone acetyltransferases. *Mol Cell Biol* **20**, 8623–8633.

Sato, Y., Yasuda, K., and Takahashi, Y. (2002). Morphological boundary forms by a novel inductive event mediated by Lunatic fringe and Notch during somitic segmentation. *Development* **129**, 3633–3644.

Sawada, A., Shinya, M., Jiang, Y. J., Kawakami, A., Kuroiwa, A., and Takeda, H. (2001). Fgf/MAPK signalling is a crucial positional cue in somite boundary formation. *Development* **128**, 4873–4880.

Schmidt, C., Stoeckelhuber, M., McKinnell, I., Putz, R., Christ, B., and Patel, K. (2004). Wnt 6 regulates the epithelialisation process of the segmental plate mesoderm leading to somite formation. *Dev Biol* **271**, 198–209.

Schmidt, M., Tanaka, M., and Munsterberg, A. (2000). Expression of (beta)-catenin in the developing chick myotome is regulated by myogenic signals. *Development* **127**, 4105–4113.

Schneider, A., Brand, T., Zweigerdt, R., and Arnold, H. (2000). Targeted disruption of the Nkx3.1 gene in mice results in morphogenetic defects of minor salivary glands: parallels to glandular duct morphogenesis in prostate. *Mech Dev* **95**, 163–174.

Schock, F., Reischl, J., Wimmer, E., Taubert, H., Purnell, B. A., and Jackle, H. (2000). Phenotypic suppression of empty spiracles is prevented by buttonhead. *Nature* **405**, 351–354.

Schoenwolf, G. C., Garcia-Martinez, V., and Dias, M. S. (1992). Mesoderm movement and fate during avian gastrulation and neurulation. *Developmental Dynamics* **193**, 235–248.

Schubert, F. R., Mootoosamy, R. C., Walters, E. H., Graham, A., Tumiotto, L., Munsterberg, A. E., Lumsden, A., and Dietrich, S. (2002). Wnt6 marks sites of epithelial transformations in the chick embryo. *Mech Dev* **114**, 143–148.

Schubert, F. R., Tremblay, P., Mansouri, A., Faisst, A. M., Kammandel, B., Lumsden, A., Gruss, P., and Dietrich, S. (2001). Early mesodermal phenotypes in splotch suggest a role for Pax3 in the formation of epithelial somites. *Dev Dyn* **222**, 506–521.

Schumacher, A., Faust, C., and Magnuson, T. (1996). Positional cloning of a global regulator of anterior-posterior patterning in mice. *Nature* **384**, 648.

Schweisguth, F. (2004). Notch signaling activity. *Curr Biol* **14**, R129–138.

Searle, A. G. (1966). Curtailed, a new dominant T-allele in the house mouse. *Genet Res* **7**, 86–95.

Selleck, M. A., and Stern, C. D. (1991). Fate mapping and cell lineage analysis of Hensen's node in the chick embryo. *Development* **112**, 615–626.

Serth, K., Schuster-Gossler, K., Cordes, R., and Gossler, A. (2003). Transcriptional oscillation of lunatic fringe is essential for somitogenesis. *Genes Dev* **17**, 912–925.

Seto, E. S., and Bellen, H. J. (2004). The ins and outs of Wingless signaling. *Trends Cell Biol* **14**, 45–53.

Shamim, H., and Mason, I. (1999). Expression of Fgf4 during early development of the chick embryo. *Mech Dev* **85**, 189–192.

Shen, J., Bronson, R. T., Chen, D. F., Xia, W., Selkoe, D. J., and Tonegawa, S. (1997). Skeletal and CNS Defects in *Presenilin-1*-Deficient Mice. *Cell* **89**, 629–639.

Shiotsugu, J., Katsuyama, Y., Arima, K., Baxter, A., Koide, T., Song, J., Chandraratna, R. A., and Blumberg, B. (2004). Multiple points of interaction between retinoic acid and FGF signaling during embryonic axis formation. *Development* **131**, 2653–2667.

Showell, C., Binder, O., and Conlon, F. L. (2004). T-box genes in early embryogenesis. *Dev Dyn* **229**, 201–218.

Shum, A. S., Poon, L. L., Tang, W. W., Koide, T., Chan, B. W., Leung, Y. C., Shiroishi, T., and Copp, A. J. (1999). Retinoic acid induces down-regulation of Wnt-3a, apoptosis and diversion of tail bud cells to a neural fate in the mouse embryo. *Mech Dev* **84**, 17–30.

Slack, J. M., Darlington, B. G., Gillespie, L. L., Godsave, S. F., Isaacs, H. V., and Paterno, G. D. (1989). The role of fibroblast growth factor in early Xenopus development. *Development* **107 Suppl**, 141–148.

Smits, P., and Lefebvre, V. (2003). Sox5 and Sox6 are required for notochord extracellular matrix sheath formation, notochord cell survival and development of the nucleus pulposus of intervertebral discs. *Development* **130**, 1135–1148.

Sosic, D., Brand-Saberi, B., Schmidt, C., Christ, B., and Olson, E. N. (1997). Regulation of paraxis expression and somite formation by ectoderm- and neural tube-derived signals. *Dev Biol* **185**, 229–243.

Spörle, R., and Schughart, K. (1998). Paradox segmentation along inter- and intrasomitic borderlines is followed by dysmorphology of the axial skeleton in the open brain (opb) mouse mutant. *Dev Genet* **22**, 359–373.

Spyropoulos, D. D., and Capecchi, M. R. (1994). Targeted disruption of the even-skipped gene, evx1, causes early postimplantation lethality of the mouse conceptus. *Genes Dev* **8**, 1949–1961.

Stamataki, D., Kastrinaki, M., Mankoo, B. S., Pachnis, V., and Karagogeos, D. (2001). Homeodomain proteins Mox1 and Mox2 associate with Pax1 and Pax3 transcription factors. *FEBS Lett* **499**, 274–278.

Stein, K. F., and Mackensen, J. A. (1957). Abnormal development of the thoracic skeleton in mice homozygous for the gene for looped-tail. *Am J Anat* **100**, 205–223.

Stern, C. D., and Keynes, R. J. (1987). Interactions between somite cells: the formation and maintenance of segment boundaries in the chick embryo. *Development* **99**, 261–272.

Stern, H. J., Graham, J. M., Jr., Lachman, R. S., Horton, W., Bernini, P. M., Spiegel, P. K., Bodurtha, J., Ives, E. J., Bocian, M., and Rimoin, D. L. (1990). Atelosteogenesis type III: a distinct skeletal dysplasia with features overlapping atelosteogenesis and oto-palato-digital syndrome type II. *Am J Med Genet* **36**, 183–95.

Streit, A., and Stern, C. D. (1999). Mesoderm patterning and somite formation during node regression: differential effects of chordin and noggin. *Mech Dev* **85**, 85–96.

Subramanian, V., Meyer, B. I., and Gruss, P. (1995). Disruption of the murine homeobox gene Cdx1 affects axial skeletal identities by altering the mesodermal expression domains of Hox genes. *Cell* **83**, 641–653.

Sudo, H., Takahashi, Y., Tonegawa, A., Arase, Y., Aoyama, H., Mizutani-Koseki, Y., Moriya, H., Wilting, J., Christ, B., and Koseki, H. (2001). Inductive signals from the somatopleure mediated by bone morphogenetic proteins are essential for the formation of the sternal component of avian ribs. *Dev Biol* **232**, 284–300.

Sun, X., Meyers, E. N., Lewandoski, M., and Martin, G. R. (1999). Targeted disruption of Fgf8 causes failure of cell migration in the gastrulating mouse embryo. *Genes Dev* **13**, 1834–1846.

Suzuki, M., Mizutani-Koseki, Y., Fujimura, Y., Miyagishima, H., Kaneko, T., Takada, Y., Akasaka, T., Tanzawa, H., Takihara, Y., Nakano, M., Masumoto, H., Vidal, M., Isono, K., and Koseki, H. (2002). Involvement of the Polycomb-group gene Ring1B in the specification of the anterior-posterior axis in mice. *Development* **129**, 4171–4183.

Tada, M., Concha, M. L., and Heisenberg, C. P. (2002). Non-canonical Wnt signalling and regulation of gastrulation movements. *Semin Cell Dev Biol* **13**, 251–260.

Tada, S., Li, A., Maiorano, D., Mechali, M., and Blow, J. J. (2001). Repression of origin assembly in metaphase depends on inhibition of RLF-B/Cdt1 by geminin. *Nat Cell Biol* **3**, 107–113.

Takada, S., Stark, K. L., Shea, M. J., Vassileva, G., McMahon, J. A., and McMahon, A. P. (1994). Wnt-3a regulates somite and tailbud formation in the mouse embryo. *Genes & Development* **8**, 174–189.

Takahashi, Y., Koizumi, K., Takagi, A., Kitajima, S., Inoue, T., Koseki, H., and Saga, Y. (2000). Mesp2 initiates somite segmentation through the Notch signalling pathway. *Nat Genet* **25**, 390–396.

Tam, P. P., Goldman, D., Camus, A., and Schoenwolf, G. C. (2000). Early events of somitogenesis in higher vertebrates: allocation of precursor cells during gastrulation and the organization of a meristic pattern in the paraxial mesoderm. *Curr Top Dev Biol* **47**, 1–32.

Tam, P. P., Kanai-Azuma, M., and Kanai, Y. (2003). Early endoderm development in vertebrates: lineage differentiation and morphogenetic function. *Curr Opin Genet Dev* **13**, 393–400.

Tanaka, M., Komuro, I., Inagaki, H., Jenkins, N. A., Copeland, N. G., and Izumo, S. (2000). Nkx3.1, a murine homolog of Ddrosophila bagpipe, regulates epithelial ductal branching and proliferation of the prostate and palatine glands. *Dev Dyn* **219**, 248–260.

Tanaka, M., Lyons, G. E., and Izumo, S. (1999). Expression of the Nkx3.1 homobox gene during pre and postnatal development. *Mech Dev* **85**, 179–182.

Taniguchi, Y., Karlstrom, H., Lundkvist, J., Mizutani, T., Otaka, A., Vestling, M., Bernstein, A., Donoviel, D., Lendahl, U., and Honjo, T. (2002). Notch receptor cleavage depends on but is not directly executed by presenilins. *Proc Natl Acad Sci U S A* **99**, 4014–4019.

Tassabehji, M., Read, A. P., Newton, V. E., Harris, R., Balling, R., Gruss, P., and Strachan, T. (1992). Waardenburg's syndrome patients have mutations in the human homologue of the Pax-3 paired box gene. *Nature* **355**, 635–636.

Theiler, K. (1959). Anatomy and development of the 'truncate' (boneless) mutation in the mouse. *Amer J Anat* **104**, 314–344.

Ting, S. B., Wilanowski, T., Auden, A., Hall, M., Voss, A. K., Thomas, T., Parekh, V., Cunningham, J. M., and Jane, S. M. (2003). Inositol- and folate-resistant neural tube defects in mice lacking the epithelial-specific factor Grhl-3. *Nat Med* **9**, 1513–1519.

Tonegawa, A., Funayama, N., Ueno, N., and Takahashi, Y. (1997). Mesodermal subdivision along the mediolateral axis in chicken controlled by different concentrations of BMP-4. *Development* **124**, 1975–1984.

Topczewska, J. M., Topczewski, J., Shostak, A., Kume, T., Solnica-Krezel, L., and Hogan, B. L. (2001). The winged helix transcription factor Foxc1a is essential for somitogenesis in zebrafish. *Genes Dev* **15**, 2483–2493.

Tremblay, P., Dietrich, S., Meriskay, M., Schubert, F. R., Li, Z., and Paulin, D. (1998). A Crucial Role for *Pax3* in the Development of the Hypaxial Musculature and the Long-Range Migration of Muscle Precursors. *Developmental Biology* **203**, 49–61.

Tribioli, C., Frasch, M., and Lufkin, T. (1997). Bapx1: an evolutionary conserved homologue of the Drosophila bagpipe homeobox gene is expressed in splanchnic mesoderm and the embryonic skeleton. *Mech Dev* **65**, 145–162.

Tribioli, C., and Lufkin, T. (1999). The murine Bapx1 homeobox gene plays a critical role in embryonic development of the axial skeleton and spleen. *Development* **126**, 5699–5711.

Turnpenny, P. D., Whittock, N., Duncan, J., Dunwoodie, S., Kusumi, K., and Ellard, S. (2003). Novel mutations in DLL3, a somitogenesis gene encoding a ligand for the Notch signalling pathway, cause a consistent pattern of abnormal vertebral segmentation in spondylocostal dysostosis. *J Med Genet* **40**, 333–339.

van den Akker, E., Forlani, S., Chawengsaksophak, K., de Graaff, W., Beck, F., Meyer, B. I., and Deschamps, J. (2002). Cdx1 and Cdx2 have overlapping functions in anteroposterior patterning and posterior axis elongation. *Development* **129**, 2181–2193.

van den Akker, E., Fromental-Ramain, C., de Graaff, W., Le Mouellic, H., Brulet, P., Chambon, P., and Deschamps, J. (2001). Axial skeletal patterning in mice lacking all paralogous group 8 Hox genes. *Development* **128**, 1911–1921.

van der Hoeven, F., Zakany, J., and Duboule, D. (1996). Gene transpositions in the HoxD complex reveal a hierarchy of regulatory controls. *Cell* **85**, 1025–1035.

van der Lugt, N. M., Alkema, M., Berns, A., and Deschamps, J. (1996). The Polycomb-group homolog Bmi-1 is a regulator of murine Hox gene expression. *Mech Dev* **58**, 153–164.

van der Lugt, N. M., Domen, J., Linders, K., van Roon, M., Robanus-Maandag, E., te Riele, H., van der Valk, M., Deschamps, J., Sofroniew, M., van Lohuizen, M., and et al. (1994). Posterior transformation, neurological abnormalities, and severe hematopoietic defects in mice with a targeted deletion of the bmi-1 proto-oncogene. *Genes Dev* **8**, 757–769.

van Straaten, H. W., and Copp, A. J. (2001). Curly tail, a 50-year history of the mouse spina bifida model. *Anat Embryol (Berl)* **203**, 225–237.

Verbout, A. J. (1976). A critical review of the 'Neugliederung' concept in relation to the development of the vertebral column. *Acta Biotheor* **25**, 219–258.

Vernon, B., Tirelli, N., Bachi, T., Haldimann, D., and Hubbell, J. A. (2003). Water-borne, in situ crosslinked biomaterials from phase-segregated precursors. *J Biomed Mater Res* **64A**, 447–456.

von Ebner, V. (1889). Urwirbel und die Neugliederung der Wirbelsäule. *Sitzungsber Akad Wiss Wien III* **97**, 194–206.

Wallin, J., Koseki, H., Wilting, J., Ebensperger, C., Christ, B., and Balling, R. (1994). The role of *Pax-1* in the development of the axial skeleton. *Development* **120**, 1109–1121.

Watanabe, Y., Duprez, D., Monsoro-Burq, A. H., Vincent, C., and Le Douarin, N. M. (1998). Two domains in vertebral development: antagonistic regulation by SHH and BMP4 proteins. *Development* **125**, 2631–2639.

Wellik, D. M., and Capecchi, M. R. (2003). Hox10 and Hox11 genes are required to globally pattern the mammalian skeleton. *Science* **301**, 363–367.

Weston, W. J. (1956). Genetically determined cervical ribs; a family study. *Br J Radiol* **29**, 455–456.

Whittock, N. V., Sparrow, D. B., Wouters, M. A., Sillence, D., Ellard, S., Dunwoodie, S. L., and Turnpenny, P. D. (2004). Mutated MESP2 causes spondylocostal dysostosis in humans. *Am J Hum Genet* **74**, 1249–1254.

Wilm, B., James, R. G., Schultheiss, T. M., and Hogan, B. L. (2004). The forkhead genes, Foxc1 and Foxc2, regulate paraxial versus intermediate mesoderm cell fate. *Dev Biol* **271**, 176–189.

Winnier, G. E., Hargett, L., and Hogan, B. L. (1997). The winged helix transcription factor MFH1 is required for proliferation and patterning of paraxial mesoderm in the mouse embryo. *Genes Dev* **11**, 926–940.

Wohlschlegel, J. A., Dwyer, B. T., Dhar, S. K., Cvetic, C., Walter, J. C., and Dutta, A. (2000). Inhibition of eukaryotic DNA replication by geminin binding to Cdt1. *Science* **290**, 2309–2312.

Wong, P. C., Zheng, H., Chen, H., Becher, M. W., Sirinathsinghji, D. J. S., Trumbauer, M. E., Chen, H. Y., Price, D. L., Van der Ploeg, L. H. T., and Sisodia, S. S. (1997). Presenilin1 is required for *Notch* and *Dll1* expression in the paraxial mesoderm. *Nature* **387**, 288–292.

Wright, M. E. (1947). Undulated: a new genetic factor in Mus musculus affecting the spine and tail. *Heredity* **1**, 137–141.

Wu, S. C., Grindley, J., Winnier, G. E., Hargett, L., and Hogan, B. L. (1998). Mouse Mesenchyme forkhead 2 (Mf2): expression, DNA binding and induction by sonic hedgehog during somitogenesis. *Mech Dev* **70**, 3–13.

Yamaguchi, T. P., Bradley, A., McMahon, A. P., and Jones, S. (1999). A Wnt5a pathway underlies outgrowth of multiple structures in the vertebrate embryo. *Development* **126**, 1211–1223.

Yamaguchi, T. P., Harpal, K., Henkemeyer, M., and Rossant, J. (1994). fgfr-1 is required for embryonic growth and mesodermal patterning during mouse gastrulation. *Genes Dev* **8**, 3032–3044.

Yamamoto, A., Amacher, S. L., Kim, S. H., Geissert, D., Kimmel, C. B., and De Robertis, E. M. (1998). Zebrafish paraxial protocadherin is a downstream target of spadetail involved in morphogenesis of gastrula mesoderm. *Development* **125**, 3389–3397.

Yang, X., Dormann, D., Munsterberg, A. E., and Weijer, C. J. (2002). Cell movement patterns during gastrulation in the chick are controlled by positive and negative chemotaxis mediated by FGF4 and FGF8. *Dev Cell* **3**, 425–437.

Yeo, S. Y., Little, M. H., Yamada, T., Miyashita, T., Halloran, M. C., Kuwada, J. Y., Huh, T. L., and Okamoto, H. (2001). Overexpression of a slit homologue impairs convergent extension of the mesoderm and causes cyclopia in embryonic zebrafish. *Dev Biol* **230**, 1–17.

Yoon, J. K., Moon, R. T., and Wold, B. (2000). The bHLH class protein pMesogenin1 can specify paraxial mesoderm phenotypes. *Dev Biol* **222**, 376–391.

Yoon, J. K., and Wold, B. (2000). The bHLH regulator pMesogenin1 is required for maturation and segmentation of paraxial mesoderm. *Genes Dev* **14**, 3204–3214.

Yu, B. D., Hanson, R. D., Hess, J. L., Horning, S. E., and Korsmeyer, S. J. (1998). MLL, a mammalian trithorax-group gene, functions as a transcriptional maintenance factor in morphogenesis. *Proc Natl Acad Sci U S A* **95**, 10632-10636.

Yu, B. D., Hess, J. L., Horning, S. E., Brown, G. A., and Korsmeyer, S. J. (1995). Altered Hox expression and segmental identity in Mll-mutant mice. *Nature* **378**, 505–508.

Zakany, J., Kmita, M., Alarcon, P., de la Pompa, J. L., and Duboule, D. (2001). Localized and transient transcription of Hox genes suggests a link between patterning and the segmentation clock. *Cell* **106**, 207–217.

Zeng, L., Fagotto, F., Zhang, T., Hsu, W., Vasicek, T. J., Perry, W. L., 3rd, Lee, J. J., Tilghman, S. M., Gumbiner, B. M., and Costantini, F. (1997). The mouse Fused locus encodes Axin, an inhibitor of the Wnt signaling pathway that regulates embryonic axis formation. *Cell* **90**, 181–192.

Zeng, L., Kempf, H., Murtaugh, L. C., Sato, M. E., and Lassar, A. B. (2002). Shh establishes an Nkx3.2/Sox9 autoregulatory loop that is maintained by BMP signals to induce somitic chondrogenesis. *Genes Dev* **16**, 1990–2005.

Zhang, F., Popperl, H., Morrison, A., Kovacs, E. N., Prideaux, V., Schwarz, L., Krumlauf, R., Rossant, J., and Featherstone, M. S. (1997). Elements both 5′ and 3′ to the murine Hoxd4 gene establish anterior borders of expression in mesoderm and neurectoderm. *Mech Dev* **67**, 49–58.

Zhang, N., and Gridley, T. (1998). Defects in somite formation in *lunatic fringe*-deficient mice. *Nature* **394**, 374–377.

Zhang, X. M., Ramalho-Santos, M., and McMahon, A. P. (2001). Smoothened mutants reveal redundant roles for Shh and Ihh signaling including regulation of L/R asymmetry by the mouse node. *Cell* **105**, 781–792.

16

The Kidney

Paul J. D. Winyard

Introduction

Mammalian kidneys perform a number of functions that are essential for normal post-natal life, including excretion of nitrogenous waste products, homeostasis of water, electrolytes and acid base balance, and the production of hormones. The main functioning units of mature mammalian kidneys are nephrons, each consisting of specialized segments including glomerulus, proximal tubule, loop of Henle and distal tubule, connected to the tree-like collecting duct system and intimately associated with the vascular supply. The precursor of the adult organ, the metanephros, arises in the 5th week of human development and consists of only two cell types: epithelia of the ureteric bud and mesenchyme of the metanephric blastema. Mutual interactions between these two tissues are essential for normal kidney development, or nephrogenesis, which involves precisely coordinated cell proliferation and death, morphogenesis and differentiation. A number of key nephrogenic molecules have been identified in the last few years in descriptive studies of normal and abnormal human development, whilst data from mice with targeted mutations has greatly increased our knowledge of underlying mechanisms (even though many rodent mutants do not appear to have human equivalents). These studies highlight three main ways in which kidney development can be perturbed: intrinsic genetic defects; extrinsic influences, such as teratogens and maternal diet; and physical insults, such as obstruction of the developing urinary tract, which may have a genetic component too. This chapter will outline gross kidney structure, consider fundamental processes in kidney development and then explore these three mechanisms of maldevelopment. For a more extensive account, readers are recommended to consult Vize *et al.* (2003).

Embryos, Genes and Birth Defects, Second Edition Edited by Patrizia Ferretti, Andrew Copp, Cheryll Tickle and Gudrun Moore

Structure and function

Mature human kidneys from adult males are around $11 \times 6 \times 2.5$ cm in size and weigh up to 170 g; females have slightly smaller organs (Ventatachalam *et al.*, 1998). The paired kidneys are located in the retroperitoneum lateral to and extending between the 12th thoracic and 3rd lumbar vertebrae. The renal pelvis lies medially and tapers into the ureter, which connects inferiorly to the bladder. The kidney parenchyma consists of nephrons, collecting ducts, blood vessels, lymphatics, nerves and interstitium. Each nephron consists of a glomerulus, proximal tubule, loop of Henle and a distal tubule, which is joined to a collecting duct via the connecting tubule.

Glomeruli consist of mesangial cells and matrix, supporting a tuft of capillaries directly surrounded by glomerular basement membrane and visceral epithelial cells (podocytes). The intimate relationship between the fenestrated capillary endothelium, basement membrane and podocyte foot processes facilitates glomerular filtration and restricts protein losses in healthy glomeruli; mutations of basement membrane and podocyte-associated genes compromise this function, resulting in protein and blood losses in the urine in conditions such as Alport's syndrome and some forms of nephrotic syndrome (Gubler, 2003). Proximal tubules have characteristic epithelia with a well-developed brush border, to increase the surface area for reabsorption, and numerous mitochondria and lysosomes, which reflect their high metabolic rate. Approximately two-thirds of the glomerular filtrate is reabsorbed in the proximal tubule, along with minerals, ions (Na^+, HCO_3^-, Cl^-, K^+, Ca^{++}, PO_4^{3-}), water, and organic solutes such as glucose and amino acids. The next segment of the nephron is the loop of Henle, which is lined by a flattened epithelium lacking a brush border. The main role of the loops is to generate an osmolar gradient to facilitate later reabsorption of water by the collecting ducts. Active Na^+,K^+-ATPase-driven ion transport, particularly sodium chloride via the Na^+,K^+,$2Cl^-$ co-transporter, occurs mainly in the thick ascending limbs, which lead into the distal convoluted tubule, where further fluid and electrolyte reabsorption and secretion take place. Between these segments is the macula densa, which consists of distinctive tall, columnar cells that detect sodium chloride delivery to the distal nephron. This forms part of the juxtaglomerular apparatus, along with segments of afferent and efferent glomerular arterioles from the same nephron, which modulates renin secretion via specialized granular myoepithelial cells in the afferent arteriole. Circulating renin activates the angiotensin–aldosterone system and has a major effect on control of systemic blood pressure.

The distal tubule is attached to the collecting duct by the connecting tubule, which is well defined in some species, such as the rabbit, but not so easily distinguishable in rodents and man. Collecting ducts arise in the cortex and consist of cortical, outer medullary and inner medullary segments. The initial collecting ducts each drain an average of 11 nephrons in humans and these join together to generate larger-calibre ducts as they pass into the medulla. The largest collecting ducts, the terminal papillary collecting ducts of Bellini, open into calyces at the papillary tips. Collecting duct epithelia consist of three specialized cell types: principal cells, which secrete potassium and regulate water transport via vasopressin binding to V2 receptors, which stimulate apical aquaporin-2 water channels; and α- and β-intercalated cells, which mediate hydrogen ion and bicarbonate secretion, respectively. There are few β-intercalated cells in meat-eaters, since the diet is so proton-rich that there is rarely a need to secret bicarbonate.

Most texts estimate that there are around 1 million nephrons in each human kidney, although definitive quantification is hampered by lack of material and differences in ascertainment techniques, which has led to formal estimates between 740 000 and 1 400 000 (Potter, 1972; Merlet-Benichou *et al.*, 1999; Pesce, 1998). Mice and rats have 10 000–20 000 nephrons (Welham *et al.*, 2002). All of the glomeruli are located in the cortex, a 1 cm thick strip which forms the outermost part of the kidney, whereas other nephron components extend into the medulla, towards the centre of the organ. In humans, the cortex is continuous, whereas the medulla consists of around 14 discrete pyramids. This is termed 'multipapillary' and contrasts with the 'unipapillary' kidneys found in rodents and rabbits.

Developmental anatomy of nephrogenesis

There are three pairs of 'kidneys' in the mammalian embryo: the pronephros, mesonephros and metanephros, which arise sequentially from intermediate mesoderm on the dorsal body wall (Vize *et al.*, 2003). The pronephros and mesonephros degenerate during fetal life in mammals, whereas the metanephros develops into the adult kidney. In contrast, the pronephros is the functioning kidney of adult hagfish and some amphibians, and the mesonephros in adult lampreys, some fishes and amphibians. Timing of nephrogenesis is outlined in Table 16.1.

Table 16.1 Timing of nephrogenic events: summary of time of appearance of renal structures during human and murine nephrogenesis. Timing in rat is generally 1 day later than in mice

Structure		Human	Mouse
Pronephros	Appears	22 days	9 days
	Regresses	25 days	10 days
Mesonephros	Appears	24 days	10 days
	Regresses	16 weeks	14 days
Metanephros		32 days	11.5 days
Renal pelvis		33 days	12.5 days
Collecting tubules/nephrons		44 days	13 days
Glomeruli		9 weeks	14 days
Nephrogenesis ceases		34–36 weeks	14 days after birth
Length of gestation		40 weeks	20 days

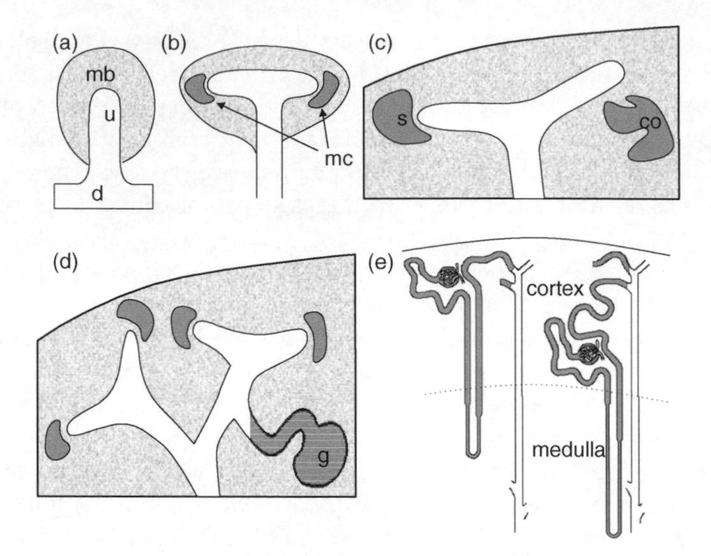

Figure 16.1 Schematic of nephrogenesis. Schematic view with mesonephric duct (d) and ureteric bud (u) shown in white, uninduced mesenchyme in the metanephric blastema (mb) in light grey, and mesenchymal condensates (mc) and nephron precursors, including comma shapes (co), S-shaped bodies (s) and immature glomeruli (g), in dark grey. (a) In the 5th week of human gestation the ureteric bud grows out from the mesonephric duct into metanephric mesenchyme. (b) Over the next week the bud branches once and mesenchyme condenses around the ampullae. (c) Comma- and S-shaped bodies are formed by the 8th week. (d) The first glomeruli are formed by the 9th week; further branching of the ureteric bud and mesenchymal condensation continues in the nephrogenic cortex until the 34th week of gestation. (e) Final nephron structure consists of glomeruli, proximal tubules, loops of Henle and distal tubules derived from the mesenchyme, and collecting ducts (grey) derived from the ureteric bud. Note the superficial and deep types of nephron

The pronephros

The pronephros develops from the 10-somite stage on day 22 after fertilization in humans (morphologically equivalent to E9 in mice). At this stage, it comprises a small group of nephrotomes with segmental condensations, grooves and vesicles between the second and sixth somites. The nephrotomes are non-functional vestiges of the pronephric kidney of lower vertebrates. The pronephric duct develops from the intermediate mesoderm lateral to the notochord (Gilbert, 1997) from around the level of the ninth somite. The duct elongates caudally and reaches the cloacal wall on day 26. It is renamed the mesonephric (or Wolffian) duct as mesonephric tubules develop. The nephrotomes and pronephric part of the duct involute and cannot be identified by day 24 or 25 after fertilization.

The mesonephros

In humans, the long sausage-shaped mesonephros develops around 24 days of gestation, and comprises the mesonephric duct and adjacent mesonephric tubules.

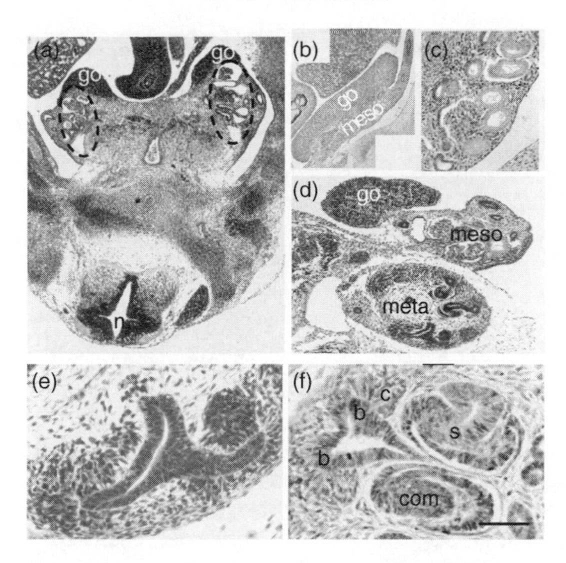

Figure 16.2 Anatomy of normal human renal development. Histological sections of developing human embryos at 38 (a), 42 (b, c), 56 (d, e) and 84 days of gestation (f). (a) Transverse section showing neural tube (n), mesonephroi (enclosed by dotted lines) and gonadal ridges (go). (b, c) Mesonephros showing elongated finger-like appearance and large glomeruli plus close relationship to developing gonad. (d) Gonad ventrally, mesonephros laterally and metanephros (meta) dorsally. Note that the ureteric bud has branched several times at this stage. (e) Higher-power view of peripheral ureteric bud branches with adjacent mesenchymal condensates. (f) Later stages of nephrogenesis showing peripheral bud branches (b) surrounded by condensing mesenchyme (c) with deeper (i.e. more mature) nephron precursors, including comma- (com) and S-shaped bodies (s). Bar corresponds to 500 μm in (a) (b--d), 150 μm in and 60 μm in the remainder

The mesonephric duct is initially a solid rod of cells, but this forms a lumen in a caudocranial direction after fusion with the cloaca. Mesonephric tubules develop from intermediate mesoderm medial to the duct by 'mesenchymal to epithelial' transformation, a process which is subsequently reiterated during nephron formation in metanephric development. In humans, a total of around 40 mesonephric tubules are produced (several per somite), but the cranial tubules regress at the same time as caudal ones are forming, hence there are never more than 30 pairs at any time.

Each human mesonephric tubule consists of a medial cup-shaped sac encasing a knot of capillaries, respectively analogous to the Bowman's capsule and glomerulus of the mature kidney, and a lateral portion in continuity with the mesonephric duct (Figure 16.2). Other segments of the tubule resemble mature proximal and distal tubules histologically but there is no loop of Henle. The human mesonephros is reported to produce small quantities of urine between weeks 6 and 10 that drains via the mesonephric duct, whereas the murine organ is much more rudimentary and does not contain well-differentiated glomeruli. Mesonephric structures involute during the 3rd month of gestation in humans, although caudal mesonephric tubules contribute to the efferent ducts of the epididymis and the mesonephric duct forms the duct of the epididymis, the seminal vesicle and ejaculatory duct.

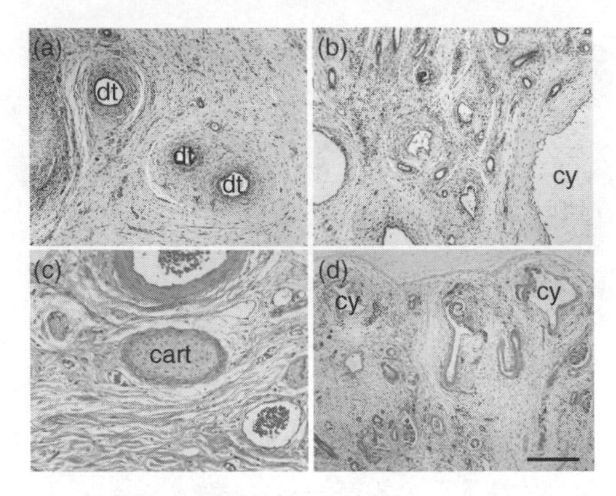

Figure 16.3 Histological sections of human dysplastic kidneys. (a–c) are postnatal and (d) is prenatal. (a, b) Complete lack of normal renal stuctures; instead replaced by dysplastic tubules (dt), surrounded by fibromuscular collarettes and cysts (cy) with expansion of loosely arrranged interstitium. (c) Metaplastic cartilage. (d) Disruption of nephrogenesis and subcortical cysts (cy) in an obstructed dysplastic kidney. Bar corresponds to 200 μm in (a), (b) and (d), and 60 μm in (c)

The metanephros

The adult human kidney develops from the metanephros, which consists of only two cell types at its inception: the epithelial cells of the ureteric bud, and the mesenchyme cells of the metanephric mesenchyme. A series of reciprocal interactions between these tissues cause the ureteric bud to branch sequentially to form the ureter, renal pelvis, calyces and collecting tubules, whilst the mesenchyme undergoes an epithelial conversion to form the nephrons from glomerulus to distal tubule. Other mesenchymal cells contribute to vascular development and give rise to interstitial cells in the mature kidney. This process is depicted graphically in Figure 16.1.

In humans, metanephric kidney development begins at day 28 after fertilization, when the ureteric bud sprouts from the distal part of the mesonephric duct. By day 32 the tip (ampulla) of the bud penetrates the metanephric blastema, a specialized area of sacral intermediate mesenchyme, and this condenses around the growing ampulla. The first glomeruli form by 8–9 weeks and nephrogenesis continues in the outer rim of the cortex until 34 weeks (Potter, 1972). Nephrons elongate and continue to differentiate postnatally but new nephrons are not formed. In mice, the ureteric bud enters the metanephric mesenchyme by embryonic day 10.5, the first glomeruli form by embryonic day 14 and nephrogenesis continues for 14 days after birth. Times for different stages of nephrogenesis are summarized in Table 16.1.

Differentiation of the ureteric bud

As the ureteric bud grows into the metanephric blastema, it becomes invested with condensed mesenchyme and mutual induction causes the ampullary tip to begin to divide. This process of growth and branching occurs repeatedly during nephrogenesis and leads to a tree-like collecting duct system, connected to nephrons which develop concurrently from condensing mesenchyme, as outlined below. Around 9–10 rounds of branching occur in mice and a further 10 generations in humans (Ekblom *et al.*, 1994). At their distal ends, collecting ducts drain into minor calyces, which connect to the major calyces of the renal pelvis and then the ureter. These intervening structures are formed by fusion or remodelling of early bud branches by apoptosis; Potter (1972) estimated that the first 3–5 generations form the pelvis and the next 3–5 give rise to the minor calyces and papillae.

Differentiation of the mesenchyme

Each nephron develops from mesenchyme adjacent to an ampullary tip of the ureteric bud. The mesenchyme is initially loosely arranged, but the cells destined to become nephrons condense around the bud tips and undergo phenotypic transformation into epithelial renal vesicles. Each vesicle elongates to form a comma shape, which folds back on itself to become an S-shaped body Figure 16.2. The proximal S-shape develops into the glomerulus, whilst the distal portion elongates and differentiates into all nephron segments from proximal convoluted tubule to distal convoluted tubule. Non-condensed mesenchymal cells give rise to renal interstitial cells and contribute to vessel development, although there is some debate over the relative importance of vasculogenesis (where vascular precursors develop *in situ*) and angiogenesis (where new vessels grow from outside) in kidney vessel development.

Vasculature development

Around 20% of blood flow in humans passes through the kidneys, via a complex system including glomerular capillaries adapted for filtration, the juxtaglomerular apparatus as above and vasa rectae, which pass alongside loops of Henle into the medulla. There are two potential sources of these vessels: vasculogenesis, in which mesenchyme differentiates *in situ* to form capillary endothelia, and angiogenesis, which involves ingrowth of existing capillaries (Woolf and Loughna, 1998). Renal capillaries were initially hypothesized to arise by angiogenesis, based on experiments showing that the glomeruli formed in organ culture are avascular (Bernstein *et al.*, 1981) and that capillary loops formed when mouse metanephroi were grafted onto avian chorioallantoic membranes are of host origin (Sariola *et al.*, 1983). This hypothesis has been challenged, however, by grafting experiments into the anterior eye chamber and under the capsule of neonatal mouse kidneys, where the glomerular endothelia are derived from the donor (Hyink *et al.*, 1996; Loughna *et al.*, 1997). Further support for vasculogenesis comes from recent reports that molecules characteristically expressed by endothelia are present in the metanephros from the inception of nephrogenesis (Woolf and Yuan, 2001). These results suggest that vasculogenesis definitely occurs during capillary formation in the metanephros, although they do not completely exclude formation of some vascular components by angiogenesis.

Renal malformations

Renal malformations are amongst the commonest congenital abnormalities and, reflecting the complexity of normal development, there are a number of different forms. These are discussed in detail elsewhere (Woolf *et al.*, 2004), but include:

- *Agenesis* – the kidney is absent. This often occurs in conjunction with absent/malformed ureters, which raises the question of whether early failure of ureteric bud branching caused the agenesis. Agenesis can be an isolated anomaly or form part of a multi-organ syndrome, such as Kallmann's syndrome (see below).

- *Dysplasia* – the kidney fails to undergo normal differentiation. These organs may be large and distended with cysts, as in multicystic dysplastic kidneys, or small with a few rudimentary tubules that resemble 'frustrated' ureteric bud branches (Potter, 1972) Figure 16.3. Occasional ectopic tissues may also be detected such as cartilage (Woolf and Price, 2004). Dysplasia can also occur as an isolated anomaly or in a multi-organ syndrome, such as the renal cysts and diabetes syndrome (see below).

- *Hypoplasia* – the kidney has significantly fewer nephrons than normal, but the formed nephrons appear normal; undifferentiated tissues are not present (otherwise these would be classified as dysplastic kidneys). Glomeruli and tubules are sometimes greatly enlarged, a condition called oligomeganephronia (Salomon *et al.*, 2001). Again, hypoplasia may be isolated or be a component of a wider malformation syndrome, such as the renal coloboma syndrome (see below).

- *Polycystic kidney disease* – early stages of renal development are normal but defects in terminal epithelial differentiation lead to cyst formation. The commonest is autosomal dominant polycystic kidney disease (ADPKD), where cysts arise from all nephron segments, but there is also autosomal recessive (AR) PKD, where cysts develop from collecting ducts, and glomerulocystic disease, where the histology is dominated by cystic dilatation of glomeruli. Since not a defect in primary development, further discussion of PKD is limited to the final 'New issues' section of this chapter, particularly focusing on recent reports linking cilia defects to cyst formation (Cantiello, 2004).

There are numerous human syndromes that include kidney and urinary tract malformations, as summarized in Table 16.2. This list is daunting at first glance, as there appear few logical connections between the different syndromes. Fortunately,

Table 16.2 Human genetic renal and urinary tract malformation syndromes

Known genes

- Apert syndrome (*FGFR2* mutation – growth factor receptor): hydronephrosis and duplicated renal pelvis with premature fusion of cranial sutures and digital anomalies
- Bardet–Biedl syndrome (several loci/genes implicated – includes a chaperonin and a centrosomal protein): renal dysplasia and calyceal malformations with retinopathy, digit anomalies, obesity, diabetes mellitus and male hypgonadism
- Beckwith–Wiedemann syndrome (in a minority of patients, *p57KIP2* mutation – cell cycle gene): widespread somatic overgrowth with large kidneys, cysts and dysplasia
- Branchio-oto-renal syndrome (*EYA1* mutation – transcription factor-like protein): renal agenesis and dysplasia with deafness and branchial arch defects, such as neck fistulae
- Campomelic dysplasia (*SOX9* mutation – transcription factor): diverse renal and skeletal malformations
- Carnitine palmitoyltransferase II deficiency (gene for this enzyme is mutated): renal dysplasia
- Congenital anomalies of the kidney and urinary tract (CAKUT) syndrome (*AT2* polymorphism – growth factor receptor): diverse, non-syndromic, renal and lower urinary tract malformations
- Denys–Drash syndrome (*WT1* mutation – transcription/splicing factor): mesangial cell sclerosis and calyceal defects
- Glutaric aciduria type II (*glutaryl-CoA dehydrogenase* mutation): cystic and dysplastic disease
- Hypoparathroidism, sensorineural deafness and renal anomalies (HDR) syndrome (*GATA3* mutation – transcription factor): renal agenesis, dysplasia and vesicoureteric reflux
- Fanconi anaemia (six mutant genes reported – involved in DNA repair): renal agenesis, ectopic/horseshoe kidney, anaemia and limb malformations
- Fraser syndrome (*FRAS1* mutation – putative cell adhesion molecule): renal agenesis and dysplasia, digit and ocular malformations
- Kallmann's syndrome (X-linked form – *KAL1* mutation – cell adhesion molecule; autosomal form – *FGFR1* mutation – growth factor receptor): renal agenesis and dysplasia in X-linked form
- Mayer–Rokitansky–Kuster–Hauser syndrome (*WNT4* mutation – growth factor signalling pathway): renal agenesis and absence of Mullerian-derived structures in females
- Nail–patella syndrome (*LMX1B* mutation – transcription factor): malformation of the glomerulus and renal agenesis
- Oral–facial–digital syndrome type 1 (*OFD1* mutation – centrosomal protein): glomerular cysts with facial and digital anomalies
- Renal–coloboma syndrome (*PAX2* mutation – transcription factor): renal hypoplasia and vesicoureteric reflux
- Renal cysts and diabetes syndrome (*HNF1β* mutation – transcription factor): renal dysplasia, cysts and hypoplasia
- Simpson–Golabi–Behmel syndrome (*GPC3* mutation – proteoglycan): renal overgrowth, cysts and dysplasia
- Smith–Lemli–Opitz syndrome (*7-dehydrosterol-delta(7)-reductase* mutation – cholesterol biosynthesis): renal cysts and dysplasia
- Townes–Brockes syndrome (*SALL1* mutation – transcription factor): renal dysplasia and lower urinary tract malformations
- Urogenital adysplasia syndrome (some cases have *HNF1β* mutation): renal dysplasia and uterine anomalies
- von Hippel Lindau disease (*VHL* mutation – tumour suppressor gene): renal and pancreatic cysts, renal tumours
- WAGR syndrome (*WT1* and *PAX6* contiguous gene defect – transcription factors): Wilms' tumour, aniridia and genital and renal malformations
- Zellweger syndrome (peroxisomal protein mutation): cystic dysplastic kidneys

(*continued*)

Table 16.2 (*continued*)

Genes unknown

- CHARGE association (genetic basis unknown): coloboma, heart malformation, choanal atresia, retardation, genital and ear anomalies; diverse urinary tract malformations can occur
- DiGeorge syndrome (microdeletion at 22q11 – probably several genes involved): renal agenesis, dysplasia, vesicoureteric reflux, with heart and branchial arch defects
- Duplex kidney and ureter (loci and genes unknown): non-syndromic familial cases are recognized
- Meckel syndrome (loci at 11q and 17q – genes unknown): cystic renal dysplasia, central nervous system and digital malformations
- Posterior urethral valves (PUV; loci and genes unknown): non-syndromic cases in siblings and male relatives (full phenotype not seen in females)
- Urofacial (Ochoa) syndrome (locus on 10q – gene undefined): congenital obstructive bladder and kidney malformation with abnormal facial expression
- VACTERL association (basis unknown apart from one report of mitochondial gene mutation): vertebral, cardiac, tracheoesophageal, renal, radial and other limb anomalies
- Vesicoureteric reflux (genetically heterogeneous – one locus on chromosome 1 but gene undefined): non-syndromic familial cases with no secondary cause (e.g. urinary flow impariment) are recognized

however, there are many shared pathways of maldevelopment and there are only a limited number of causes, such as genetic defects, obstruction of the developing urinary tract, and extrarenal factors such as teratogens and altered maternal diet. These will be considered individually below, although it must be remembered that different perturbing influences can occur in the same fetus, which might have a synergistic effect on renal development. This is illustrated graphically in Figure 16.4, and examples include mutations in angiotensin 2 type 2 receptors, which not only have primary genetic effects on kidney/urinary tract development but can also cause lower urinary tract obstruction (Pope *et al.*, 1999).

Basic processes during nephrogenesis

Nephrogenesis involves a balance between basic cellular and tissue processes, such as proliferation, death, differentiation and morphogenesis, all controlled by regulated gene expression. There is extensive cell proliferation as the adult mammalian kidney develops from less than 1000 cells at its inception to many millions in the mature organ, but this is mainly confined to the narrow rim of cortex containing actively branching ureteric bud tips and adjacent condensing mesenchyme (Winyard *et al.*, 1996b). Fine tuning of cell numbers occurs by apoptosis, with as many as 50% of the cells produced in the developing kidney deleted via this process (Coles *et al.*, 1993; Winyard *et al.*, 1996a). The major sites of apoptosis are early nephron precursors, such as comma- and S-shaped bodies and the medulla, locations in which cell death may be important for morphogenesis and collecting duct remodelling. Several levels of differentiation occur during normal nephrogenesis, ranging from early mesenchymal–

epithelial differentiation to form renal vesicles, through to terminal differentiation, where different cells in the same nephron segments acquire different functions (e.g. the α- and β-intercalated and principal cells in collecting ducts). Morphogenesis is the process whereby groups of cells acquire complex three-dimensional shapes. This is clearly important in the kidney, where there is such an intimate relationship between different nephron and collecting duct segments and the renal vasculature, but little is known about controlling factors (Woolf, Price etc., 2004).

Molecular control of nephrogenesis

Several classes of molecules work in concert to ensure normal nephrogenesis, including transcription factors, growth factors, survival factors and adhesion molecules. These categories are not mutually exclusive, since certain molecules fall into two categories; examples being PAX2, a transcription factor, and epidermal growth factor (EGF), both of which also promote cell survival. Potential roles of a few of these factors were identified from human genetic syndromes, but a large proportion have come to light in recent years as mutant mice have been selectively generated.

Transcription factors

Transcription factors are the conductors of the 'nephrogenic chorus', regulating expression of other genes to set up embryonic patterning. Most contain DNA sequence-specific binding domains which modulate target gene mRNA transcription, although precise targets are often surprisingly unknown. PAX2, WT1, EYA1, HOX, BF2 and HNF1β are reviewed below, whilst other important factors include EMX2 (Miyamoto *et al.*, 1997), LIM1 (Shawlot and Behringer, 1995), LMX1B (Chen *et al.*, 1998), POD1 (Quaggin *et al.*, 1998), retinoic acid receptors (Mendelsohn *et al.*, 1999) and SOX9 (Bell *et al.*, 1997).

PAX2 (and PAX8)

The PAX family of transcription factors control diverse aspects of embryonic patterning and cell specification in a number of organisms, including *Drosophila*, zebrafish, frog, chick, mouse and human (Halder *et al.*, 1995), but only PAX2 and PAX8 are expressed in the developing kidney. PAX2 has a critical role in normal and abnormal renal development, particularly during mesenchymal condensation and epithelial transformation (Rothenpieler and Dressler, 1993). PAX2 is expressed in intermediate mesoderm, the nephric duct, then the mesonephric duct, and finally in the tips of the ureteric bud and the condensing mesenchyme in the metanephros (Dressler *et al.*, 1990). This pattern is repeated in the outer cortex throughout nephrogenesis, and expression persists in nephron precursors, such as the comma- and S-shaped bodies, but then decreases as these epithelia mature in deeper parts of the kidney (Winyard *et al.*, 1996a; Dressler and Douglas, 1992).

There is now compelling evidence that controlled PAX2 expression is essential for kidney development, with abnormal kidney phenotypes resulting from either too little or too much expression. Human *PAX2* mutations involving a single nucleotide deletion within the conserved octapeptide sequence in exon five have been described in the human 'renal–coloboma' syndrome, which consists of optic nerve colobomas, renal anomalies and vesicoureteral reflux (Sanyanusin *et al.*, 1995). Mice with decreased levels of Pax2 have aberrant kidney development: heterozygous mutations cause hypoplastic kidneys with reduced branching of the ureteric bud, reduced numbers of nephrons and cortical thinning, whilst homozygous null mutants lack mesonephric tubules and the metanephroi fail to form because the ureteric buds are absent (Torres *et al.*, 1995). Interestingly, some of these defects can be abrogated by overexpression of Pax5, a highly related Pax factor not normally expressed in the kidney (Bouchard *et al.*, 2000). Overexpression of Pax2 also causes murine kidney abnormalities, including cystic tubular changes, proteinuria and renal failure (Dressler *et al.*, 1993).

Pax2 may have two functions in nephrogenesis. First, it appears critical in mesenchymal–epithelial transformation in both the mesonephros and metanephros; this process can be specifically blocked in metanephric organ culture using antisense pax2 oligonucleotides (Dressler *et al.*, 1993). Second, it may alter the proliferation/apoptosis balance in favour of the former. In the normal developing kidney, for example, PAX2 expression is confined to areas with the highest rates of cell division, which also have low apoptotic rates; it is overexpressed in highly proliferative human cystic renal epithelia (Winyard *et al.*, 1996a) and in diverse tumours (Dressler and Douglas, 1992; Gnarra and Dressler 1995), whilst reduced expression decreases cyst expansion in cpk ARPKD mice (Ostrom *et al.*, 2000) and overexpression causes cell transformation *in vitro* (Maulbecker and Gruss, 1993). Moreover, apoptosis inhibitors partially rescue defects in nephrogenesis in a mouse model of the renal–coloboma syndrome (Clark *et al.*, 2004).

Mechanisms that initiate and control PAX2 expression are uncertain, although there is recent evidence that Yin Yang 1, another transcription factor, binds to part of the Pax2 promoter, leading to increased expression (Patel and Dressler, 2004). WT1, on the other hand, appears to downregulate PAX2 (Ryan *et al.*, 1995); this function may complete a negative feedback loop, since PAX2 binds to two sites in the WT1 promoter sequence and causes up to a 35-fold increase in expression, as assessed using reporter genes (McConnel *et al.*, 1997). A further downstream target of PAX2 may be glial cell line-derived neurotrophic factor (GDNF), a growth factor also essential for nephrogenesis (see below): Pax2 binds to upstream regulatory elements within the GDNF promoter and, as again assessed using reporter genes, transactivates expression (Brophy *et al.*, 2001).

Pax-8 is expressed in the developing mesonephric tubules and then in the condensing mesenchyme of the murine metanephros (Plachov *et al.*, 1990), and is downregulated in maturing nephron epithelia (Poleev *et al.*, 1992). Its roles in nephrogenesis are uncertain, since null mutants have normal kidneys, instead having thyroid maldevelopment (Mansouri *et al.*, 1998). PAX8 may be important as a co-factor in very early kidney development, however: intermediate mesoderm does not undergo the mesenchymal–epithelial transformation required for nephric duct formation in double *PAX8/PAX2* mutants (Bouchard *et al.*, 2000), and combined overexpression of PAX8 and LIM1 cause enlargement and induction of an ectopic pronephros in *Xenopus*. Interestingly, the latter effect can be abrogated by additional overexpression of HNF1β (Wu *et al.*, 2004), a transcription factor implicated in the human renal cysts and diabetes syndrome (Kolatsi-Joannou *et al.*, 2001).

WT1

The Wilms' tumour 1 (*WT1*) gene was discovered in these tumours, but is paradoxically only mutated in a small percentage. *WT1* encodes a transcription factor protein containing four zinc-finger DNA-binding motifs (Pritchard-Jones and Hawkins, 1997). Several studies have documented expression of both *WT1* mRNA and protein during mammalian development (Armstron *et al.*, 1993) and specifically in the urogenital system and Wilms' tumours (Winyard *et al.*, 1996a; Pelletier *et al.*, 1991). During early human nephrogenesis, *WT1* mRNA is expressed in the mesonephric glomeruli and at low levels in condensing metanephric mesenchyme. As the comma- and S-shaped bodies develop, WT1 levels increase but become restricted to the visceral glomerular epithelia, with podocytes remaining strongly positive in the mature kidney (Winyard *et al.*, 1996a).

Complete lack of WT1 causes death *in utero* in null-mutant mice secondary to defects in mesothelial-derived components where WT1 is normally expressed, including the heart and lungs (Kreidberg *et al.*, 1993). Renal development is also severely disrupted: small numbers of normal-appearing mesonephric tubules form but the ureteric bud fails to branch from the Wolffian duct and the intermediate mesoderm, which should form the metanephric blastema, dies by apoptosis. Several human syndromes are associated with *WT1* mutations. Denys–Drash syndrome consists of genitourinary abnormalities, including ambiguous genitalia in 46 XY males, nephrotic syndrome with mesangial sclerosis leading to renal failure, and a predisposition to Wilms' tumour (Little and Wells, 1997). This is caused by point mutations of *WT1*, predominantly affecting the zinc finger DNA-binding domains. WAGR syndrome consists of Wilms' tumour, aniridia, genitourinary abnormalities including gonadoblastoma and mental retardation. Frasier syndrome is characterized by focal glomerular sclerosis with progressive renal failure and gonadal dysgenesis. This is caused by intronic point mutations of *WT1*, which affect the balance between different WT1 splice isoforms (Klamt *et al.*, 1998).

These data suggests that *WT1* has at least two functions in the kidney, namely control of ureteric bud outgrowth and mesenchyme survival during early development, and a later role in glomerular mauration. Recent studies using small interfering RNA (siRNA) suggest an additional potential role in nephron differentiation: early

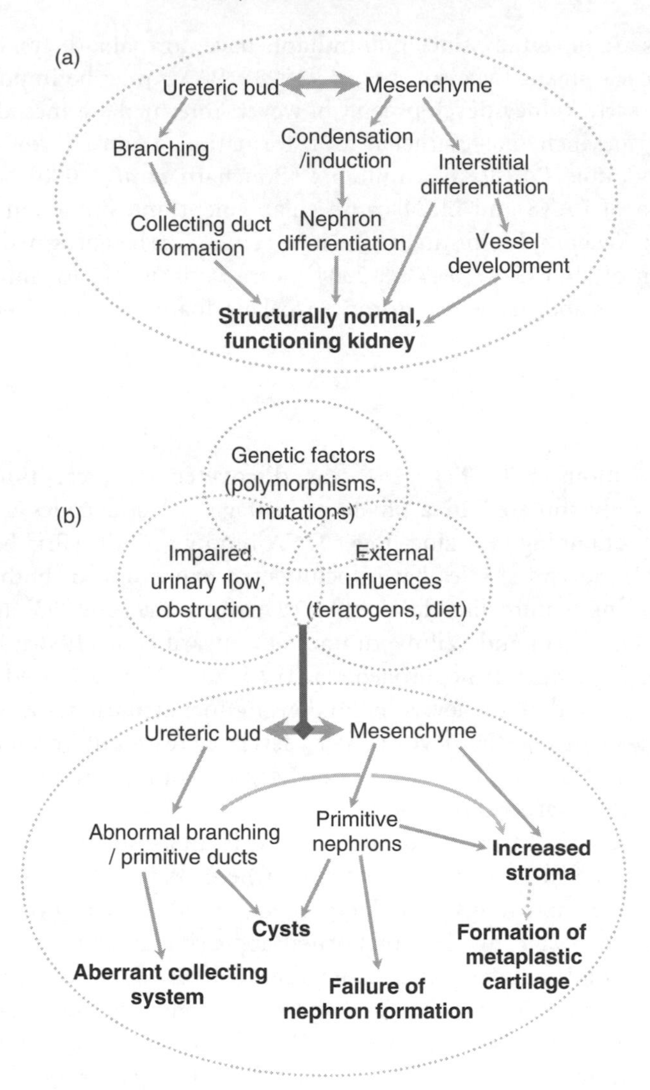

Figure 16.4 Schematic of general processes in normal and dysplastic renal development. (a) Mutual interaction between the ureteric bud and mesenchymal lineages leads to normal renal development. (b) Perturbing influences, which may be genetic, physical (as in urinary tract obstruction) or chemical (such as teratogens and diet leading to changes in surrounding milieu) or any combination of these, lead to perturbed interactions between the bud and mesenchymal lineages, which result in dysplasia

siRNA blockade phenocopies null mutants, but later *WT1* knock-down causes abnormal nephron proliferation, perhaps mimicking aspects of Wilms' tumours (Davies *et al.*, 2004). *WT1* is capable of such manifold actions because multiple isoforms are generated by alternative splicing, RNA editing and alternative translation initiation sites (Wagner *et al.*, 2003). Functions as a transcription factor are mainly mediated via the isoform lacking the amino acids lysine, threonine, and serine between zinc fingers 3 and 4, which is known as the KTS form. Downstream repression targets include the insulin-like growth factor axis (Drummond *et al.*, 1997), early growth response gene 1 (*EGR1*) (Rackley *et al.*, 1995), plus *PAX2* (Ryan *et al.*, 1995) and *WT1* itself (Rupprecht *et al.*, 1994). One further target is *Spry1*, part of the sprouty family identified in *Drosophila* as antagonists to fibroblast and epidermal growth factor signalling (Gross *et al.*, 2003). Spry1 expression is mainly in the ureteric bud, whilst Spry2 and Spry4 are more widespread, encompassing ureteric bud, mesenchyme and glomeruli (Zhang *et al.*, 2001). Recent experiments overexpressing Spry2 in the ureteric bud suggest that these factors may regulate ureteric branching via interactions with several critical nephrogenic growth factors (Chi *et al.*, 2004).

EYA1

EYA1 is the mammalian homologue of the transcriptional co-activator 'eyes absent' gene, which is required for normal eye specification in *Drosophila.* Mutations of the human *EYA1* gene occur in 20–25% of patients with branchio-oto-renal syndrome (Abdelhak *et al.*, 1997). This is characterized by a combination of hearing loss, preauricular pits, branchial fistulae and variable renal anomalies including agenesis, hypoplasia and dysplasia.

In mice, eya1 is widely expressed in the ear, branchial arches and metanephric mesenchyme during development. Homozygous *Deya1* null mutant mice die at birth with multiple abnormalities, including craniofacial and skeletal defects and absent ears (Xu *et al.*, 1999). They also lack kidneys because of defective ureteric bud outgrowth, leading to failure of metanephric induction and increased mesenchymal apoptosis. Rare, strain-dependant, renal defects occur in $eya1^{+/-}$ heterozygotes, including hypoplasia and unilateral agenesis. Parallels in the development of other organs suggest that there is an evolutionarily conserved regulatory cascade involving *Pax* family genes (i.e *Pax6* in the eye, *Pax2* in the kidney), *EYA1* and homologues of the *Drosophila* sine oculis (*so*) gene in the '*Six*' family (Xu *et al.*, 1999). *Six1* null mutants also have multiple developmental defects including absent kidneys (Laclef *et al.*, 2003). Recent data has added the *Drosophila* dachshund (*dach*)-related family to this regulatory network Li *et al.*, 2003), and it is believed that EYA1 functions to convert the *six/dach* complex from a transcriptional repressor to an activator.

HOX

Vertebrate *hox* genes encode homeodomain transcription factors which specify positional information along the anterior–posterior axis. Expression of 37 *Hox* genes within the developing kidney has been recently reported (Patterson and Potter, 2004), revealing expression throughout ureteric bud and nephron segments. Four conclusions were drawn from this survey: (a) *Hox* genes from the more 3′ positions in clusters were more often expressed in the ureteric bud; (b) there was no segment specificity throughout the ureteric bud; (c) overlapping domains of Hox expression were not observed in nephron segments; and (d) paralogous *Hox* genes often showed surprisingly diverse distribution patterns. These findings may explain why double knock-outs generated by interbreeding of *Hoxa11* and *Hoxd11* mutants have renal agenesis or hypoplasia whilst single mutants do not (Davis *et al.*, 1995).

Forkhead/winged helix transcription factors

The conserved forkhead/winged helix transcription factor gene family has several members implicated in kidney and urinary tract development. Mutations in the *Foxc1* gene are responsible for the classical congenital hydrocephalus mouse, and homozygous mutants have markedly abnormal early nephrogenesis with ectopic mesonephric tubules and anterior ureteric buds, often leading to duplex kidneys and ureters (Kume *et al.*, 2000). *Foxc1* also appears to interact with Foxc2 during renal (and heart) development, since most *Foxc1/Foxc2* compound heterozygotes have hypoplastic kidneys and a single hydroureter, while all heterozygotes are normal.

Foxd1, previously known as *BF2*, is one of the only genes known to be implicated in interstitial differentiation. Foxd1 is expressed in the cells immediately surrounding condensed mesenchyme cells which express Pax2, and mice with null mutations have rudimentary, fused kidneys and die soon after birth (Hatini *et al.*, 1996). Interestingly, the mesenchyme condenses in the null-mutant mice but does not develop any further, and neither comma- nor S-shaped bodies are formed. The ureteric bud also fails to branch normally and ret (see below) is widely distributed in the bud epithelium, rather than confined to the bud tips. Hence, Foxd1 may modulate expression of a yet unknown factor, or factors, from the ‘uninduced’ cells which is essential for both ureteric bud growth and maturation of pretubular aggregates.

Hepatocyte nuclear factor (HNF) 1β

Mutations of the gene encoding the transcription factor hepatocyte nuclear factor 1β (HNF1β) cause the renal cysts and diabetes (RCAD) syndrome in humans (Kolatsi-Joannou *et al.*, 2001). HNF1β is expressed widely during embryogenesis, including the mesonephric duct, ureteric bud lineage and early nephron epithelia, and adjacent paramesonephric ducts which should differentiate into the uterus and Fallopian tubes (Kolatsi-Joannou *et al.*, 2001; Coffinier *et al.*, 1999). Renal malformations in RCAD are highly variable, ranging from grossly cystic dysplastic kidneys, through hypoplasia with oligomeganephronia to apparent unilateral agenesis and, in females, are accompanied by similarly diverse uterine abnormalities. Absence of HNF1β expression at the very tips of the branching ureteric tree has led to speculation that it is a 'maturation factor' rather than a 'branching factor', but investigation of this potential role has been difficult, since null mutants die in early embryogenesis (Barbacci *et al.*, 1999). HNF1β is, however, also expressed in the *Xenopus* developing excretory system and overexpression of specific mutants has proved informative: using mutants that retained DNA binding, dimerization and transactivation activities, the pronephros was smaller than normal, whereas mutants lacking these properties generated larger kidneys (Bohn *et al.*, 2003; Wild *et al.*, 2000). These data suggest that mutated proteins which lack DNA binding are not inactive but must interact with some, unknown, regulatory components.

Growth factors and their receptors

Growth factors have important functions in nephrogenesis via three potential roles: as paracrine factors secreted by one cell and acting on neighbouring cells; as autocrine factors acting on the producing cell; and as juxtacrine factors that become inserted into the plasma membrane of the producing cell to interact with receptors on adjoining cells. The growth factors bind to specific cell surface receptors, mainly receptor tyrosine kinases, which dimerize and become autophosphorylated and transduce signals into the cell. These signals may stimulate many different processes, including cell division, cell survival, apoptosis, differentiation and morphogenesis. Many of the signalling systems are stucturally related; several, for example, belong to the extended transforming growth factor (TGF) β family, including glial cell line-derived neurotrophic factor (GDNF), bone morphogenetic proteins (BMP) and TGFβ1 itself. There is also frequent promiscuity between ligands and receptors, with several ligands potentially activating multiple receptors. Reviewed here are GDNF, BMPs, Wnts, hepatocyte growth factor (HGF) and epidermal growth factor (EGF). Other growth factors described elsewhere include: fibroblast growth factors (FGF; Dudley *et al.*, 1999), insulin-like growth factors (IGF; Rogers *et al.*, 1991), platelet-derived growth factor (PDGF; Leveen *et al.*, 1994), TGFα (Rogers and Ryan, 1992) and vascular endothelial growth factor (Tufro *et al.*, 1999).

GDNF

The GDNF signalling pathway is one of the most important in nephrogenesis, and involves binding of GDNF to the Ret receptor tyrosine kinase (Vega *et al.*, 1996) in association with an adapter molecule, GDNF receptor α (GFRα). Genetic ablation of any one of these factors in mice causes either complete failure of metanephric development or severe dysplasia (Schuchardt *et al.*, 1994; Pichel *et al.*, 1996b). This pathway is also critical in neural development, and studies of neuronal survival/differentiation have identified several additional ligands, including persephin, neurturin and artemin, that signal via Ret with specificity determined by different members of the GFRα family (GFRα1–4; Sariola and Saarma, 2003). Moreover,

Table 16.3 Mutant and transgenic mice with aberrant kidney development

Gene/molecule	Function	Renal phenotype	Additional affected tissues	References
Transcription factors				
Emx2	Transcription factor	Absent kidneys, ureters, gonads and genital tracts	Brain, genital tract	Miyamoto *et al.* (1997)
Eya1	Transcription factor	Bilateral agenesis (−/−); hypoplasia (+/−)	Craniofacial structures, ears, spine	Xu *et al.* (1999)
Foxc1 (*Mf1*; also spontaneous mutation in congenital hydrocephalus), *Foxc2* (*Mfh1*)	Forkhead/winged helix transcription factor	Ureteral anomalies ($Foxc1^{-/-}$ and $Foxc1^{+/-}$; $\text{Foxc2}^{+/-}$)	Eye, skeleton, heart	Kume *et al.* (2000)
Foxd1 (*Bf2*)	Forkhead/winged helix transcription factor	Hypoplasia, midline fusion	Eye, brain, adrenals (subtle)	Hatini *et al.* (1996)
Hoxa11,Hoxd11 double mutation	Homeobox-containing transcription factor	Renal dysplasia or hypoplasia in double null mutants	Absence of radius/ulna, spine	Patterson and Potter (2004); Davis *et al.* (1995)
Lim1	Homeodomain-containing transcription factor	Bilateral agenesis	Absent gonads, head structures	Shawlot and Behringer (1995)
Pax2	Paired box transcription factor	Bilateral agenesis (−/−); hypoplasia (+/−)	Brain, eyes, genital tract	Torres *et al.* (1995)
*Pax21Neu**	Paired box transcription factor	Hypoplasia (+/−)	Eye	Favor *et al.* (1996)
Pod1	Basic helix–loop–helix transcription factor	Hypoplasia/ dysplasia	Lung	Quaggin *et al.* (1998)
Rarα, *Rarβ2*	Retinoic acid receptor transcription factors	Renal agenesis, hypoplasia ($Rara^{-/-}$; $Rarb2^{-/-}$ only)	Heart, lung, thymus, diaphragm, limb	Mendelsohn *et al.* (1994, 1999)
Sall1	Zinc finger transcription factor	Renal agenesis, hypoplasia	None	Nishinakamura *et al.* (2001)
Six1	Homeobox-containing transcription factor	Renal agenesis	Thymus, craniofacial	Laclef *et al.* (2003)

Table 16.3 (*continued*)

Gene/molecule	Function	Renal phenotype	Additional affected tissues	References
Tbx1 (Di George region on 22q11)	T-box transcription factor	Multicystic dysplasia, agenesis, hydronephrosis	Heart/vessels, thymus, immune system	Jerome and Papaioannou (2001); Lindsay *et al.* (2001)
Wilms' tumour-1 (*Wt1*)	Zinc finger transcription factor	Bilateral agenesis	Respiratory system, heart, gonads	Kreidberg *et al.* (1993)
Growth factor systems				
ActRIIB	Activin type II receptor B	Agenesis, hypoplasia	Heart, spleen, vertebrae	Oh and Li (1997)
Bmp4	Peptide growth factor	Hypoplasia, ureteral abnormalities	Craniofacial structures, eye	Miyazaki *et al.* (2000); Dunn *et al.* (1997)
Short ear (*se*)*	BMP5, peptide growth factor	Hydronephrosis, hydroureter	Ear, spine, ribs	Green (1968)
Bmp7	Peptide growth factor	Dysplasia, hydroureter	Eye, skeleton	Dudley *et al.* (1995); Luo *et al.* (1995)
Gremlin	BMP antagonist	Renal agenesis	Lung and limb defects	Michos *et al.* (2004); (see also *ld* mice)
EGFR	Growth factor receptor	Cystic dilatation of collecting ducts	Hair, skin, gastrointestinal tract, brain	Threadgill *et al.* (1995)
Fgf10	Growth factor	Hypoplasia, dysplasia	Ear, limb, lung, gastrointestinal tract	Ohuchi *et al.* (2000)
Fgf7	Growth factor	Hypoplasia, hypoplastic papilla	Hair coat	Qiao *et al.* (1999)
Fgfr2(IIIb) (dominant negative transgenic)	Receptor tyrosine kinase	Unilateral or bilateral agenesis	Ear, limb, lung, gastrointestinal tract	Chen *et al.* (1998)
Gdf11	Growth/ differentiation factor			Esquela and Lee (2003)
Gdnf	Peptide growth factor	Bilateral agenesis (−/−); severe dysgenesis (+/−)	Enteric nervous system	Moore *et al.* (1996); Pichel *et al.* (1996a)

(*continued*)

Table 16.3 (*continued*)

Gene/molecule	Function	Renal phenotype	Additional affected tissues	References
Gfra1	GDNF co-receptor (with ret)	Renal agenesis, dysplasia	Enteric nervous system	Cacalano *et al.* (1998)
Pdgfr-β	Growth factor receptor	Absent mesangial cells	Heart	Soriano (1994)
Pdgf-β	Growth factor	Absent mesangial cells	Heart	Leveen *et al.* (1994)
Ret	Receptor tyrosine kinase	Renal agenesis or dysplasia	Enteric nervous system	Schuchardt *et al.* (1994)
Shh	Signalling molecule	Dysplasia	Parts of VATER syndrome	Chiang *et al.* (1996)
Wnt4	Secreted signalling molecule	Renal agenesis or dysplasia		Stark *et al.* (1994)
Survival molecules				
Bcl2	Anti-apoptotic factor	Hypoplasia and cystic kidneys	Hair, immune system	Veis *et al.* (1993)
p57KIP2	Cyclin-dependent kinase inhibitor	Medullary dysplasia	Eye, skeleton, muscle, adrenal	Zhang *et al.* (1997)
Matrix/adhesion molecules				
Fras1	Extracellular matrix protein	Agenesis, cystic kidneys		McGregor *et al.* (2003)
glypican-3 (*Gpc3*)	Cell surface heparan sulphate proteoglycan	Nephromegaly, medullary dysplasia, hydroureter	Lung	Cano-Gauci *et al.* (1999)
Heparan sulphate 2-sulphotransferase (*Hs2st*)	Proteoglycan synthesis enzyme	Bilateral agenesis	Eye, skeleton	Bullock *et al.* (1998)
Itga3	α-3 Integrin	Hypoplasia	Lung	Kreidberg *et al.* (1996)
Itga8	α-8 Integrin	Bilateral agenesis	Not described	Muller *et al.* (1997)
L1	Cell adhesion molecule	Duplex kidneys, overgrowth (note, on X chromosome, hence in –/Y males, and +/– females)	Nervous system	Debiec *et al.* (2002)

Table 16.3 *(continued)*

Gene/molecule	Function	Renal phenotype	Additional affected tissues	References
Miscellaneous				
Agtr2	Angiotensin-2 receptor	Hypoplasia/ dysplasia, ureteral anomalies (CAKUT)	Not described	Nishimura *et al.* (1999)
Angiotensinogen (*Agt-1*)	Secreted peptide	Hypoplastic papillae, widening of renal pelvis/calyces	Not described	Niimura *et al.* (1995)
Cox2	Prostaglandin synthesis enzyme	Dysplasia	Heart, gonads	Dinchuk *et al.* (1995)
Limb deformity (*ld*)*$^{\Delta}$	Formin protein, function unknown	Aplasia, hypoplasia, hydroureter	Limb	Maas *et al.* (1994)

*Spontaneous, rather than genetically engineered.
$^{\Delta}$Now thought to be caused by perturbation of the BMP antagonist gremlin (Zuniga *et al.*, 2004).

GDNF can also signal via GFRα proteins without Ret by activating Met, another receptor tyrosine kinase (Popsueva *et al.*, 2003).

Ret and GFRα are expressed along the entire nephric/mesonephric duct, in the ureteric bud which arises from it and also in branching bud tips in the metanephros (Towers *et al.*, 1998). GDNF, on the other hand, is initially restricted to the mesenchyme in the vicinity of the nascent ureteric bud, and is then expressed in condensing renal mesenchyme adjacent to actively branching bud tips in the metanephros (Pachnis *et al.*, 1993). There is strong evidence for at least two functions for this signalling system in nephrogenesis: (a) initiation of ureteric outgrowth from the mesonephric duct; and (b) promoting branching/arborealization of the ureteric tree in the metanephros. At least three groups have demonstrated that excess GDNF stimulates ectopic ureteric bud formation *in vitro* (Brophy *et al.*, 2001; Sainio *et al.*, 1997). Moreover, *Foxc1* mutants (see above) have anterior expansion of the GDNF expression domain in the mesenchyme surrounding the origin of the ureteric bud, and these mice have an increased incidence of duplex ureters (Kume *et al.*, 2000). Recent evidence suggests that Slit2/Robo signalling, another pathway identified in neural development, normally restricts GDNF expression to more posterior nephrogenic mesenchyme (Griesshammer *et al.*, 2004). GDNF has equally potent effects within the metanephros: excess growth factor increases ureteric bud branching in organ culture (Towers *et al.*, 1998) and stimulates branching morphogenesis of a ureteric bud-derived cell line (Qiao *et al.*, 1999) whilst, conversely, GDNF-neutralizing antibodies inhibit branching morphogenesis and heterozygous *GDNF* mutant mice have decreased ureteric bud branches (Vega *et al.*, 1996). A further possible role for

GDNF may be modulation of cell survival, since GDNF prevents apoptosis of ureteric bud cultured as a monolayer (Towers *et al.*, 1998).

HGF/Met

Hepatocyte growth factor (HGF), or 'scatter factor' as it is also known, is the ligand for the Met receptor tyrosine kinase (Bottaro *et al.*, 1991). Expression patterns of HGF and Met are similar in human and murine nephrogenesis: HGF is expressed in the renal mesenchyme, particularly in the cortex, as the kidney matures, whilst Met mainly localizes to developing epithelia (Sonnenberg *et al.*, 1993; Kolatsi-Joannou *et al.*, 1997). This system is intimately related to the GDNF–Ret pathway: Met can be activated directly by GDNF without Ret as an intermediary (Popsueva *et al.*, 2003) and both pathways are potent inducers of ureteric bud tubulogenesis (Montesano *et al.*, 1991a).

Mice with homozygous *HGF* or *Met* null mutations die around embryonic day 13–14 with placental, liver and muscle abnormalities (Schmidt *et al.*, 1995). Early nephrogenesis appears grossly normal in these animals, which is surprising, since blockade of this signalling system perturbs metanephric development in organ culture (Santos *et al.*, 1994; Woolf *et al.*, 1994). Exogenous HGF, moreover, causes branching morphogenesis of the Madin Darby canine kidney (MDCK) collecting duct-derived cell line in culture (Montesano *et al.*, 1991b) and overexpression leads to prominent tubular cystic disease and progressive glomerulosclerosis (Takayama *et al.*, 1997). This system is therefore likely to be important, but not critical, for stimulating cell proliferation and organ growth during nephrogenesis.

EGF

Epidermal growth factor (EGF) and its embryonic homologue transforming growth factor alpha (TGFα) both bind to the epidermal growth factor receptor (EGFR). Rogers *et al.* (1992) showed that E13 rat metanephroi produce TGFα and that nephrogenesis is perturbed by blocking antibodies against TGFα. EGF is a potent inhibitor of cell death within the developing kidney *in vivo* (Coles *et al.*, 1993) and rescues isolated renal mesenchyme from apoptosis *in vitro* (Koseki *et al.*, 1992). EGF has also been reported to increase Pax2 levels in rabbit proximal tubule cells (Liu *et al.*, 1997). Interestingly, mice with null mutations of EGF receptors have different renal phenotypes, depending on their genetic background (Threadgill *et al.*, 1995).

TGFβ

Transforming growth factor beta 1 (TGFβ1) is the prototypic molecule of a large growth factor family that includes both GDNF and the BMPs. TGFβ signalling is transduced via cell surface type I and type II receptors (TGFβR1 and TGFβR2; Wrana,1998). In early mouse kidney development, *TGFβ1* mRNA is expressed in the metanephric mesenchyme (Lehnert and Akhurst, 1998; Schmid *et al.*, 1991) and the protein is distributed in both the mesenchyme and ureteric bud (Rogers *et al.*, 1993). At similar stages, receptor transcripts are expressed in both renal mesenchyme and immature epithelia (Mariano *et al.*, 1998). Both the ligand and receptors are also expressed in the developing vasculature.

Homozygous null mutants of *TGFβ1* have normal kidneys, although these are not true functional nulls because fetuses are exposed to TGFβ1 via transplacental transfer of maternal circulating growth factor (Letterio *et al.*, 1994). Knock-out of the closely related TGFβ2 ligand cause urogenital abnormalities, although only a limited number of animals were described (Sanford *et al.*, 1997). Excess TGFβ1 inhibits tubulogenesis of MDCK cells *in vitro* (Sakurai *et al.*, 1997) and markedly disrupts nephrogenesis in organ culture, where blocking antibodies promote new nephron formation (Rogers *et al.*, 1993). This suggests a modest role for TGFβ1 in controlling nephron number in normal development (Bush *et al.*, 2004), but this system may become much more important in pathological conditions such as renal dysplasia: Yang *et al.* (2000) described upregulation of TGFβ1 in human dysplastic kidneys and demonstrated that exogenous factor caused dysplastic epithelia to differentiate towards a mesenchymal lineage *in vitro*. Excess TGFβ1 may therefore be part of the underling mechanism that causes reduced nephron numbers and expansion of the mesenchymal/stromal compartment in dysplastic kidneys. Moreover, renal TGFβ1 is also upregulated in urinary tract obstruction, as described later.

BMPs

Bone morphogenetic proteins (BMPs) comprise the largest subfamily of the TGFβ group, and at least three have been widely studied in nephrogenesis, namely BMP2, BMP4 and BMP7 (Martinez and Bertram, 2003). BMP2 is expressed in condensing mesenchyme and subsequent structures derived from it, but null mutants die early in development; conditional knock-outs with which one could examine effects on nephrogenesis have not yet been reported. Excess BMP2 has similar effects to TGFβ1 *in vitro*, however, causing inhibition of metanephric growth and ureteric branching (Piscione *et al.*, 1997). BMP4 is initially expressed in stromal cells surrounding the mesonephric duct, including at the site of ureteric bud outgrowth, and then in mesenchyme around the stalk of the ureteric bud. BMP4 knock-out mice also die early, but heterozygous mutants survive with hypoplastic or dysplastic kidneys. Based on these studies and *in vitro* up- or downregulation of the BMP4 axis, there appear to be at least three potential functions in nephrogenesis: (a) control of ureteric bud outgrowth and elongation; (b) prevention of apoptosis in metanephric mesenchyme; and (c) promotion of smooth muscle development in the ureters (Miyazaki *et al.*, 2000, 2003).

BMP7 expression is more widespread in the kidney, encompassing the ureteric bud, loose and condensed mesenchyme and primitive nephrons (Dudley and Robertson, 1997). *BMP7* null mutants survive until birth but then die from renal failure as a result of defective mesenchymal differentiation and increased apoptosis, leading to a severe nephron deficit (Dudley *et al.*, 1995; Luo *et al.*, 1995). Unsurprisingly in view of these findings, BMP7 is also thought to act as a mesenchymal survival factor. In addition, *in vitro* data suggests that a low concentration of BMP7 increases metanephric growth and ureteric bud branching, whereas higher doses inhibit arborealization (Piscione *et al.*, 2001). BMP5 is also implicated in kidney development, since this gene is mutated in mouse short ear mutants which have hydronephrosis and hydroureters (Green, 1968).

BMPs signal via BMP type 1 (BMPRIA, BMPRIB) and type II (BMPRII) receptors. These are expressed in broadly similar locations in the kidney: in the ureteric tree, particularly in bud tips, and mesenchymal condensates, developing vesicles and comma-shaped bodies. Such overlapping distributions is unsurprising, since the receptors form heterotetrameric complexes. *In vitro* studies have identified roles for BMP2 and BMPR1A (also known as activin-like kinase 3, ALK3) in inhibiting branching morphogenesis, with downstream signalling via cytoplasmic SMAD1 (Piscione *et al.*, 2001). Moreover, transgenic mice expressing constitutively active BMPR1A in the ureteric bud developed medullary cystic dysplasia, which was associated with increased β-catenin–SMAD1 complexes. This led to an investigation of these factors in human dysplastic kidneys, and both SMAD1 and β-catenin were found to be overexpressed (Hu *et al.*, 2003a).

BMP signalling is refined by antagonists such as the cystein knot protein gremlin (Grem1), which preferentially antagonizes BMP2 and BMP4. Grem1 is upregulated by sonic hedgehog (Shh) in the limb, where it sets up the FGF4 feedback loop and is implicated in *Xenopus* kidney development (Hensy *et al.*, 2002). *Grem1* null mutants have renal agenesis caused by defective ureteric bud outgrowth/branching, a failure to establish GDNF/Ret signalling and apoptotic death of the metanephric mesenchyme (Michos *et al.*, 2004). Moreover, it has recently been reported that mouse *limb deformity* mutants, which are associated with kidney malformations and were thought originally to be due to mutations in the formin gene, have disruption of gremlin regulatory regions (Zuniga *et al.*, 2004).

WNT

The *WNT* gene family consists of over 20 members, and five of these secreted signalling molecules (Wnt2b, Wnt4, Wnt6, Wnt7b and Wnt11) have been implicated in normal nephrogenesis (Vainio, 2003). In addition, Wnt1-transfected fibroblasts induce nephron formation when co-cultured with isolated renal mesenchyme, even though Wnt1 is not an endogenous metanephric molecule (Herzlinger *et al.*, 1994).

Unlike the broadly similar BMPs, the Wnts have very different expression patterns during nephrogenesis with Wnt2b and Wnt4 in the mesenchyme and Wnt6, Wnt7b and Wnt11 in the ureteric bud (for full review, see Vainio, 2003). Wnt4 is upregulated in renal mesenchymal cells as they differentiate and nephrogenesis is 'frozen' at the condensate stage in null mutant mice (Stark *et al.*, 1994). Wnt4 can also trigger tubulogenesis in isolated metanephric mesenchyme, in a process which is dependent on cell contact and sulphated glycosaminoglycans (Kispert *et al.*, 1998). A loss-of-function *WNT4* mutation has also been recently described in an 18 year-old woman with Mayer–Rokitansky–Kuster–Hauser syndrome, which comprises defects in Mullerian-derived structures and renal agenesis (Biason-Lauber *et al.*, 2004). Wnt11 is expressed at the tips of the ureteric bud (Lako *et al.*, 1998) but is not sufficient to induce tubulogenesis (Kispert *et al.*, 1998). Targeted mutations of this locus disrupt ureteric branching morphogenesis, which leads to kidney hypoplasia, and it has recently been suggested that Wnt11 and GDNF/Ret cooperate in a positive autoregulatory feedback loop (Majumdar *et al.*, 2003).

Wnt signalling is transduced in at least two systems: activation of β-catenin-mediated transcription via frizzled receptors, which is termed the canonical pathway, and Rho/Rac and Ca^{2+} signalling in the non-canonical pathway. β-Catenin-mediated signalling ties in with the BMPs as above and this pathway is also implicated in Townes–Brocks syndrome (ear, limb, heart and renal anomalies), as the zinc-finger containing transcriptional repressor *SALL1* mutated in this syndrome synergistically activates canonical Wnt signalling (Sato *et al.*, 2004).

Survival/proliferation factors

Several survival and proliferation-related genes have been identified in renal development.

BCL2

BCL2 was originally discovered in human follicular B cell lymphomas and is the prototypic member of an evolutionarily conserved gene family involved in the control of apoptotic cell death (Knudson and Korsmeyer, 1997): BCL2 and homologues prevent death, whereas 'BH3-only' family members heterodimerize with them and block antiapoptotic activity. BCL2 is upregulated in the mesenchyme as it condenses around the ureteric bud tips during nephrogenesis, but is then rapidly downregulated in the comma- and S-shaped bodies and is barely detectable in the adult organ (Winyard *et al.*, 1996a; LeBrun *et al.*, 1993). Homozygous *Bcl2* null mutants have a number of defects in haematopoiesis and hair development, and were initially reported to have polycystic kidneys with epithelial hyperproliferation and dilation of proximal and distal tubular segments (Veis *et al.*, 1993). Further analysis, however, suggests that the abnormalities are more complex: prior to birth, 'fulminant' apoptosis leads to small hypoplastic kidneys, with fewer nephrons and smaller nephrogenic zones (Sorenson *et al.*, 1995), whilst cysts develop postnatally accompanied by increased proliferation in both cortex and medulla and apoptosis of interstitial cells (Sorenson *et al.*, 1996). Hence, primary deregulation of cell survival in the *Bcl2* null mutant mice appears to cause a secondary upregulation of proliferation in the kidney. These changes can be abrogated by concurrently knocking out either one or both alleles of the pro-apoptotic partner *BIM* (Bouillet *et al.*, 2001). We have observed upregulation of BCL2 in cystic epithelia in dysplastic kidneys (Winyard *et al.*, 1996b), which also express PAX2; hence these epithelia are exposed to concurrent stimuli to survive and proliferate, providing at least one mechanism for cyst expansion. Intriguingly, we also observed upregulation of WT1 in mesenchyme surrounding dysplastic epithelia, and this transcription factor upregulates BCL2 in both Wilms's tumours and cell culture studies (Mayo *et al.*, 1999).

P57-KIP2

Cyclin-dependent kinases (CDKs) are essential for regulation of the cell cycle and proliferation. Cyclin kinase inhibitors such as p57-KIP2 (also termed CDKN1C) block proliferation by binding to the CDKs in G_1–S phase. P57-kip2 is expressed in podocytes in glomeruli and stromal cells between the renal tubules during nephrogenesis; null mutants have fewer renal tubules and small inner medullary pyramids (Zhang *et al.*, 1997). Poorly formed medullary pyramids are also seen in human Beckwith–Wiedemann syndrome, which is caused by loss of the imprinted, expressed maternal allele on chromosome 11p15.5. This site is close to the *p57-KIP2* gene and heterozygous mutations have been reported in some Beckwith–Wiedemann (B-W) patients. It was therefore suggested that dysregulated cell proliferation and apoptosis secondary to *p57-KIP2* mutations might be implicated in aberrant medulla development. Definitive proof of this hypothesis is lacking, however, and a recent review of 159 B-W patients linked uniparental disomy to renal abnormalities, whereas mutations in *p57-KIP2* were not associated (Goldman *et al.*, 2002); this suggests that other imprinted genes on 11p15.5 are more critical for renal development (Goldman *et al.*, 2002).

Cell adhesion molecules

Adhesion molecules mediate cell–cell and cell–matrix adhesion. Examples of the former are the calcium independent neural cell adhesion molecule (NCAM; Bellais *et al.*, 1995; Klein *et al.*, 1988) and calcium-dependent E-cadherin (also known as uvomorulin) (Vestweber and Kemler, 1985), whilst molecules involved in cell–matrix adhesion include the collagens, fibronectin (Bellairs *et al.*, 1995), galectin 3 (Winyard *et al.*, 1997; Bullock *et al.*, 2001), KAL (Duke *et al.*, 1995; Soussi-Yanicostas *et al.*, 1996), laminins (Klein *et al.*, 1988), nidogen (Ekblom *et al.*, 1994), tenascin (Aufderheide *et al.*, 1987) and integrin cell surface receptors (Kreidberg *et al.*, 1996; Muller *et al.*, 1997).

Many of these molecules also have additional non-adhesion-related functions, particularly proteoglycans, which can modulate binding of growth factors to their receptors. Heparan sulphate, for example, binds fibroblast growth factors, which prevents degradation and facilitates receptor binding (Kiefer *et al.*, 1990), whilst syndecan binds FGF2 (Elenius *et al.*, 1992) and β-glycan binds TGFβ (Ruoslahti and Yamaguchi, 1991). A recently described example involves GDNF and heparan sulphate (Davies *et al.*, 2003): GDNF signalling requires cell surface heparan sulphate glycosaminoglycans, and exogenous modified heparins demonstrated that 2-O-sulphate groups impart high activity to this system. These findings may explain the common finding of renal agenesis in *GDNF* null mutants and mice with a gene trap mutation in the enzyme heparan sulphate 2-sulphotransferase, which is essential for 2-O-sulphate biosynthesis (Bullock *et al.*, 1998). A brief overview of adhesion molecules is given here, followed by specific discussion of integrins, galectin 3 and KAL. For a more detailed account, see recent reviews by Kanwar *et al.* (2004).

Uninduced metanephrogenic mesenchyme expresses numerous matrix molecules, including collagens I and III, fibronectin, neural cell adhesion molecule (NCAM) and the proteoglycan syndecan (Klein *et al.*, 1988; Ekblom, 1981; Vainio *et al.*, 1989). Laminin, the large multidomain cruciform glycoprotein, is also expressed, particularly triplet combinations of B1 and B2 chains. This profile changes as the mesenchyme condenses during the transition to polarized epithelia: syndecan levels increase transiently but then decrease, NCAM decreases and cells stop expressing interstitial collagens and fibronectin and begin to express uvomorulin, collagen type IV and laminin A chain (Klein *et al.*, 1988; Vestweber and Kemler, 1985; Ekblom, 1981; Ekblom *et al.*, 1981; Vainio *et al.*, 1989). Mesenchymal cells that do not undergo epithelial transformation continue to express interstitial collagens and fibronectin and upregulate tenascin as the cells around them are condensing (Aufderheide *et al.*, 1987).

Tubule formation is perturbed in mouse metanephric organ culture by antibodies to fragments E3 and E8 of the laminin A chain, because the mesenchymal cells are unable to convert to polarized epithelial cells (Klein *et al.*, 1988). Another basement membrane glycoprotein, known as either nidogen or entactin, is produced by mesenchymal cells and binds to domain III on the laminin B2 chain. This laminin B2–nidogen binding is critical for the production of epithelial basement membranes of epithelial structures formed during early nephrogenesis (Ekblom *et al.*, 1994).

Integrins

Integrins are transmembrane heterodimeric glycoprotein complexes consisting of α and β chains with diverse roles in cell–cell and cell–matrix adhesion, polarity, migration and angiogenesis (Wallner *et al.*, 1998). Specific ligand binding is determined by the particular combination of α and β chains. Integrin subunits expressed during nephrogenesis include $\alpha 1$ in uninduced mesenchyme (Korhonen *et al.*, 1990b), $\alpha 2$ in glomerular endothelium and distal tubules (Korhonen *et al.*,

1990b), $\alpha3$ in maturing podocytes (Ekblom, 1996), $\alpha6$ and $\alpha8$ in the mesenchyme undergoing condensation and epithelial transformation adjacent to the ureteric bud (Muller *et al.*, 1997; Falk *et al.*, 1996), $\beta1$ in undifferentiated mesenchyme and glomerular endothelial cells, $\beta3$ in Bowman's capsule and $\beta4$ in fetal collecting ducts (Korhonen *et al.*, 1990a). Co-expression of different subunits includes $\alpha1\beta1$ and $\alpha4\beta1$ in uninduced mesenchyme, $\alpha2\beta1$ in endothelia and $\alpha6\beta1$ in epithelia.

Homozygous $\alpha3$ and $\alpha8$ integrin mutant mice have marked disruption of kidney development. Mice with mutations in the $\alpha3$ subunit have very abnormal glomeruli containing wide capillaries, disorganized basement membrane and aberrant podocyte foot processes, plus microcystic proximal tubules and decreased branching of the medullary collecting ducts, hence suggesting roles for $\alpha3$ in basement membrane organization and branching morphogenesis (Kreidberg *et al.*, 1996). Mice lacking the $\alpha8$ subunit exhibit profound kidney defects, with aberrant branching of the ureteric bud and formation of nephrons, which suggests, in conjunction with the expression data (Muller *et al.*, 1997), that this molecule is essential for mesenchymal–epithelial transformation (Kreidberg *et al.*, 1996). Other integrins, such as $\alpha2$ and $\alpha6$, have also been functionally implicated in tubulogenesis using diverse strategies including organ culture, isolated ureteric bud and cell line culture (Sorokin *et al.*, 1990; Zent *et al.*, 2001). In accord with these findings, aberrant expression of $\alpha1$, $\alpha2$ and $\alpha6$ subunits occurs in human dysplastic kidneys (Daikha-Dahmane *et al.*, 1997).

Galectin 3

Galectin 3, is a calcium-independent water-soluble β-galactoside binding lectin, which interacts with embryonic glycoforms of laminin and fibronectin that have polylactosamine side chains and modulates laminin–integrin interactions (Sato *et al.*, 1992). It also has diverse additional postulated roles in pre-mRNA splicing, oncogenic transformation and prevention of apoptosis (Perillo *et al.*, 1998).

Galectin 3 expression is highly lineage-specific during nephrogenesis, being almost exclusively confined to the ureteric bud and collecting ducts derived from it, in both humans and mice (Winyard *et al.*, 1997; Bullock *et al.*, 2001). Based on *in vitro* experiments, galectin-3 has at least two functions at different developmental stages in this lineage. First, it is involved in control of ureteric bud branching – anti-galectin-3 antibodies or exogenous galectin-3 protein, for example, but not control galectins, perturb ureteric branching in organ culture (Bullock *et al.*, 2001). Second, it is involved in terminal differentiation of collecting ducts – the lectin is essential for assembly of multimeric hensin, which is required for such differentiation steps (Hikita *et al.*, 2000). In view of this evidence, it is surprising that galectin-3 null mutant mice have grossly normal kidney development (Colnot *et al.*, 1998a). Although galectin-3 does not appear to be required for normal development *in vivo*, it does become important in pathological situations where it is upregulated. An extrarenal example is in the inflammatory response: one of the original names for galectin-3 was Mac-2, because of expression in activated macrophages, and the function of these is clearly perturbed in null mutants (Colnot *et al.*, 1998b). We previously reported galectin-3 upregulation in cystic/dysplastic human kidneys and have recently investigated its potential functions in congenital polycystic (cpk) mice: intriguingly, exogenous galectin-3 retards cyst development *in vitro*, whilst lack of the lectin accelerates cyst formation *in vivo* (Chiu *et al.*, 2004). This finding is consistent with previous experiments in three-dimensional cell culture, where MDCK cyst growth is influenced by galectin 3 expression: blocking antibodies increase cyst growth and exogenous galectin-3 slows cyst growth (Bao and Hughes, 1995). This raises the possibility that galectin-3 may be a potential therapy for polycystic kidney disaese, and we are currently testing this strategy in cpk mice.

KAL

KAL-1 is the gene mutated in the X-linked form of Kallmann's syndrome (KS), which comprises hypogonadotrophic hypogonadism and anosmia. It affects 1/8000 males and 1/40 000 females, with other forms of inheritance also described (Hu *et al.*, 2003b). Renal aplasia, generally unilateral, occurs in 40% of patients (Kirk *et al.*, 1994), but cystic dysplastic kidneys are also reported (Deeb *et al.*, 2001). *KAL-1* encodes the extracellular matrix protein, anosmin-1. KAL-1 transcripts occur in the human metanephros and olfactory bulb from 45 days of gestation (Duke *et al.*, 1995), sites consistent with organs affected in KS, whilst the protein anosmin-1 immunolocalizes to the basement membrane of human UB branches (Hardelin *et al.*, 1999). Anosmin-1 is modular, consisting of an N-terminal cysteine-rich region, a whey acidic protein-like 4-disulphide core motif (WAP), four contiguous fibronectin-like type III (FnIII) domains and a histidine-rich C-terminus. Similar WAP- and FnIII-encoding domains occur in predicted KAL proteins in birds, fish, flies and worms. There are no rodent homologues of *KAL-1*, hence its function has been investigated in *Caenorhabditis elegans* (Bulow *et al.*, 2002; Rugarli *et al.*, 2002): worm *Kal-1* mutants have defects in ventral closure and male tail formation, partially rescued by the human gene, suggesting conservation of function across species, and neuronal targeting studies implicated *FnIII* domains in the control of axon branching and both *FnIII* and *WAP* domains in axon misrouting. The *FnIII* domains are predicted to be involved in anosmin-1/heparan sulphate interactions and heparan-6-O-sulphotransferase, an enzyme required for the formation of cell membrane-associated HSPG, was identified as a modifier of *KAL-1*-induced axonal defects in *C. elegans*. Intriguingly, loss-of-function mutations in FGF receptor 1 have recently been reported in dominantly inherited KS, and binding of heparan sulphate to FGF and its receptors is also required for FGF signalling (Dode *et al.*, 2003).

Glypicans

The glypican family consists of heparan sulphate proteoglycans linked to the cell surface through a glycosyl-phosphatidylinositol anchor. The first member linked to renal development was glypican-3: the gene encoding this protein is mutated in patients with the Simpson–Golabi–Behmel syndrome, which consists of pre- and postnatal overgrowth, and organ abnormalities including renal malformations; whilst glypican-3-deficient mice replicate many of these features, including cystic and dysplastic kidneys (Cano-Gauci *et al.*, 1999), with underlying dysregulation of proliferation and apoptosis during ureteric bud branching and medullary differentiation (Grisaru *et al.*, 2001). More recently, glypican 4 has been shown to have potential roles in modulating HGF-mediated extracellular signal-regulated kinase (ERK) activation in collecting duct cells *in vitro* (Karihaloo *et al.*, 2004).

Other molecules

There are a number of functionally important metanephric molecules which do not fit comfortably into any of the categories outlined above, including cyclo-oxygenase 2 (cox2), an enzyme involved in the synthesis of prostaglandins (Morham *et al.*, 1995), and the renin–angiotensin system, which has diverse effects on growth and apoptosis, as described below.

Renin–angiotensin system

The renin–angiotensin system has an important role in renal development. Renin, which converts angiotensinogen to angiotensin I, is widely expressed in perivascular cells in the arterial system during early nephrogenesis, but is restricted to the juxtaglomerular apparatus in the mature kidney (Gomez *et al.*, 1988). Angiotensin II, generated from angiotensin I by angiotensin converting enzyme, binds to two types of G protein-coupled receptors, AT1 and AT2. AT1 mediates the majority of the traditionally recognized functions of angiotensin II, such as vasoconstriction and stimulation of cell growth, while AT2 antagonizes some of these actions and has postulated roles in the control of apoptosis. Both receptors are expressed in the developing kidney: AT1 is expressed in S-shaped bodies, developing tubules and mature glomeruli, whereas AT2 is restricted to mesenchymal cells, initially around the stalk of the ureteric bud but then extending to just outside the nephrogenic cortex, and cells between collecting ducts (Kakuchi *et al.*, 1995). This distribution of AT2 receptors corresponds to areas with high levels of apoptosis. The wide spectrum of renal/urological abnormalities that result from targeted mutation of different components of the renin–angiotensin system have recently been reviewed (Sequeira Lopez and Gomez, 2004) and mutations have now been described in humans with autosomal recessive renal tubular dysgenesis (Gribouval *et al.*, 2005).

Non-genetic causes of renal malformations

Genetic defects are not the sole explanation for renal malformations, and two other potential causes will be considered here: urinary tract obstruction, and teratogens/

maternal diet. In both cases, however, it must be remembered that there may be a combination of several perturbing influences. In the case of urinary flow impairment, for example, there may be genetic factors that either generate the obstruction initially, with secondary effects in the kidney, or there may be altered gene activity in both upper and lower urinary tract. The latter gives rise to a 'field effect', and has been postulated for genes such as *BMP4* and *AT2*, which have widespread expression throughout the kidney and urinary tract (Miyazaki *et al.*, 2003; Nishimura *et al.*, 1999).

Urinary tract obstruction

The broad category 'renal dysplasia' is the most important cause of childhood renal failure, requiring long-term dialysis and kidney transplantation, but the second largest group is dysplasia associated with urinary tract obstruction, with posterior urethral valves being the commonest specific diagnosis (Lewis, 1999; Woolf and Thiruchevlam, 2001; Jenkins *et al.*, 2005). Indeed, almost every structural abnormality that impairs urinary flow has been linked to dysplasia, although most are unilateral or not severe enough to cause severe renal failure. Associations include urethral atresia, and obstructive lesions at the pelvi-ureteric and uretero-vesical junctions, the latter category including obstructive mega-ureters and ureterocoeles (Woolf *et al.*, 2004). In addition, multicystic dysplastic kidneys have frequently been described in conjunction with 'atretic' non-patent ureters, and obstruction has been invoked in the early aetiopathogenesis of the prune belly syndrome, a condition in boys with urinary tract dilatation and dysplasia, plus cryptorchidism and incomplete development of the abdominal wall muscles.

These associations are intriguing, but is there any proof that obstruction can cause kidney malformations? Yes, is the unequivocal answer to this question, based on diverse experiments in several animal models. Over 30 years ago, Beck demonstrated that early surgically-generated obstruction in fetal sheep perturbed kidney development and replicated many features of dysplasia (Beck, 1971). Attar *et al.* (1998) and Peters *et al.* (1992) reiterated these experiments to investigate molecular aspects of this maldevelopment, and we report a remarkable concordance with our findings in non-obstructed human dysplastic kidneys. Common patterns of dysregulation of proliferation in cyst epithelia and apoptosis in the surrounding tissues were observed (Attar *et al.*, 1998), for example, plus upregulation of key nephrogenic molecules, including PAX2 and TGFβ1 (Yang *et al.*, 2000; 2001). This series of experiments led us to propose that fetal urinary tract obstruction initiates a series of events, summarized in Figure 16.2B, which causes gene dysregulation that leads to renal malformations. In this obstruction-induced scheme, increased stretch of developing kidney tissues causes upregulation of PAX2 expression in epithelial tubules which, if unchecked, causes enhanced growth and cyst formation. At the same time, however, TGFβ1 upregulation limits epithelial overgrowth but only at a cost, with loss of potential precursor cells by apoptosis or by diversion into an abnormal metaplastic lineage (e.g. smooth muscle and cartilage). Although applied to obstruction here, it is likely that there are common pathways involving other molecules, as well as PAX2 and TGFβ1, irrespective of underlying causation. A recent example reflecting this commonality concerns the secreted matrix metalloproteinase matrilysin, a target gene of Wnt signalling, which is upregulated in cystic kidneys, toxin-induced nephropathy and obstruction (Surendran *et al.*, 2004).

Several other obstruction models have also been used. Opossums are marsupials and metanephroi have been experimentally obstructed neonatally in the pouch, which corresponds to a very early stage of organogenesis. Broadly similar changes are observed to the ovine studies cited above (Liapis *et al.*, 2000). This model also gives scope for therapeutic intervention: administration of IGF, for example, ameliorates the renal fibrosis, tubular cystic changes and calyceal dilatation which follow obstruction (Steinhardt *et al.*, 1995). Nephrogenesis continues for 2 weeks after birth in rodents, hence neonatal obstruction still perturbs the developing kidney: Chevalier *et al.* (1998) have extensively investigated ureteric obstruction in neonatal rats, and reported that obstructive nephropathy is attenuated by EGF or IGF. Animal models have also been used to assess whether decompression of the developing urinary tract can improve renal outcome. In the sheep, Glick *et al.* (1984) found that *in utero* decompression prevents renal dysplasia, but this conflicts with Chevalier *et al.*'s (2002) reports of neonatal rats, where relief of obstruction attenuated but did not reverse renal injury resulting from 5 days of ureteric obstruction. These data are consistent with the generally poor results for *in utero* intervention to relieve obstruction in humans (Holmes *et al.*, 2001), since the kidneys may well be too far advanced along the path of maldevelopment before the abnormalities are detectable.

Teratogens/maternal diet

Teratogens have been implicated in the pathogenesis of diverse kidney and lower urinary tract malformations (Solhaug *et al.*, 2004), and can be divided into two broad categories: exogenous factors, such as drugs, and endogenous factors, which become teratogenic when present in excess. A classic drug example is angiotensin-converting enzyme inhibitors (ACE-I) which, when used to treat hypertension during pregnancy, can cause skull malformations termed hypocalvaria plus neonatal renal failure from a combination of haemodynamic compromise and renal tubular dysgenesis (Barr and Cohen, 1991). These effects are unsurprising when one considers the importance of the renin–angiotensin system in renal development, described above, but such logical teratogenicity does not always occur, as evidenced with non-steroidal anti-inflammatory drugs: these agents inhibit cyclo-oxygenases (COX) such as COX2, and mice lacking this factor have severe renal dysplasia (Dinchuk *et al.*, 1995), but it is relatively rare to find renal malformations after exposure during pregnancy in humans (Cuzzolin *et al.*, 2001).

High glucose levels in diabetic mothers are associated with an increased incidence of kidney and lower urinary tract malformations, plus abnormalities in the nervous, cardiovascular and skeletal systems (Chugh *et al.*, 2003). Experimental exposure of embryonic kidneys to high glucose alters the expression of laminin $\beta 2$ in the basement membrane (Abrass *et al.*, 1997) and expression of IGF receptors (Duong *et al.*, 2003), but there are at least two other factors that may contribute to congenital diabetic nephropathy. First, there is often an association with caudal regression syndrome, which causes reduction of distal structures such as the sacrum and hindlimbs and might clearly affect the lower urinary tract. Second, it is likely that many of the cases ascribed to maternal diabetes had HNF1β mutations, as reported in the RCAD syndrome above. High doses of vitamin A and its derivative retinoids have also been linked to human and rodent renal malformations such as renal agenesis (Rothman *et al.*, 1995), although these are also sometimes associated with caudal regression (Padmanbhan, 1998). Too little vitamin A also perturbs renal development, however, via multiple effects, including modulation of GDNF/Ret, Wnt11 and WT1 signalling in the metanephros (Vilar *et al.*, 2002), and interaction with Ret, which patterns distal ureter and bladder trigone development (Batourina *et al.*, 2002).

Maternal diet may have a subtle effect on nephogenesis, affecting nephron number rather than inducing gross changes. This hypothesis arose from epidemiological data suggesting that individuals born to mothers with poor diets are prone to develop hypertension and cardiovascular disease in adulthood (Barker *et al.*, 1989; Barker, 2004) and the proposed link between congenital 'nephron deficits' and later hypertension (Brenner *et al.*, 1988). Using animal experiments, it is now well established that moderate to severe dietary protein restriction during pregnancy impairs somatic growth and reduces the numbers of glomeruli per kidney (Langley-Evans *et al.*, 1999) and this has been linked to early deletion of renal mesenchymal cells, which may reduce the pool of renal precursor cells (Welham *et al.*, 2002). Vitamin A may also have a regulatory effect on nephron number (Gilbert and Merlet-Benichou, 2000), whilst other factors such as dietary iron may be important: maternal iron restriction, for example, causes hypertension, which has been linked in part to a deficit in nephron number (Lisle *et al.*, 2003), consistent with proposed roles for iron in the development of renal epithelial (Yang *et al.*, 2003).

Agenda for the future

Primary cilium–basal body complex

Primary cilia are small membrane-enclosed tubular structures that project from cells (Wheatley *et al.*, 1996). They comprise nine microtubule doublets emanating from one of the basal bodies, which is a modified form of the centriole. The ciliary axoneme is referred to as a 9 + 0 arrangement, in contrast to motile cilia, which have an additional two central doublets, giving them a 9 + 2 structure. Primary cilia are predominantly associated with epithelial cells, but have been described in virtually all mammalian cells at some point during their development (Wheatley, 1995). Few researchers were particularly interested in primary cilia until recent convergent data implicated them in polycystic kidney disease. One of the first intimations of their potential importance came in *C. elegans*, when Barr and Sternberg reported that the PKD1 homologue is required for male mating behaviour, which involves modified cilia (Barr and Sternberg, 1999). Subsequently, expression of virtually every PKD-associated protein has been described in the primary cilium, including polycystin 1, polycystin 2, polaris and inversin (Zhang *et al.*, 2004). Inversin is particularly interesting because, in addition to renal malformations, mice with these mutations have perturbed left–right asymmetry, which has been attributed to defects in motile 9 + 0 cilia found around the embryonic node (Nonaka *et al.*, 1998; Otto *et al.*, 2003).

The cilium is now believed to function as a flow-sensitive mechanoreceptor which stimulates calcium influx via several channels (including polycystin-2), and activation of diverse intracellular signalling cascades to transduce differentiation signals into renal epithelia (Ong and Wheatley, 2003). Various functions have been ascribed to the individual PKD-associated proteins in the cilium, including control of calcium influx, intraflagellar transport and assembly of the cilium (Cantiello, 2004; Snell *et al.*, 2004). The ciliary basal body may also be important, particularly where multi-organ abnormalities are associated with the renal malformations, because recent reports have described expression of the oral-facial-digital (OFD) 1 and Bardet–Biedl syndrome proteins in this location (Romio *et al.*, 2004; Li *et al.*, 2004). Our understanding of the roles of the cilium and basal body in normal renal development and non-PKD malformations are likely to increase significantly over the next few years.

Therapy for malformations – a role for 'progenitor' cells?

There has been a dramatic increase in the amount of research into the use of 'stem' cells in treatment for diverse diseases, as reviewed recently by Rookmaaker *et al.* (2004) for renal conditions. Much of this has focused on diseases of the mature kidney, however, with particular emphasis on bone marrow-derived cells in renal repair (Poulsom *et al.*, 2003). In these conditions, the problem is how to prevent destruction of pre-existing structures or at least effect a rapid, non-destructive repair. This is quite distinct from the malformation situation, where functioning renal structures never develop. The traditional approach to severe malformations is dialysis and then the introduction of 'new' renal function by kidney transplant. Results from this procedure are continually improving with better immunosuppression and management, but most transplanted kidneys last for a finite life (usually 10–15 years) and it is likely that a patient receiving transplants during childhood will need several more during his/her lifetime. This raises the question of whether there is any other way to generate functional renal tissue and, if so, whether there any way to make it less immunogenic so that chances of rejection are reduced. This research is in its infancy but there are two general, non-mutually exclusive approaches: 'rescue' of normal development, and tissue engineering to generate new functional renal tissue.

At first glance, the most appealing approach is to intervene *in utero* to restore normal differentiation. This is certainly theoretically possible, utilizing our increased knowledge of control of normal (and also abnormal) nephrogenesis to target specific growth factors or other nephrogenic pathways, but there are problems of timing and access: most human renal abnormalities are detected by antenatal ultrasound scan, often around mid-gestation, which may be too late for intervention, as evidenced by the poor returns for relief of urinary tract obstruction *in utero* (Holmes *et al.*, 2001), whilst it is currently not possible to target therapies to the kidney specifically without potentially affecting the development of other organs. Hence, several groups have focused on the steps required to generate functional renal tissue *de novo*. Hammerman and colleagues, for example, have demonstrated that early metanephroi will undergo organogenesis *in situ* and develop a measurable filtration rate when transplanted into adult animals (Rogers *et al.*, 1998). Although this technique is effective across both concordant (rat-to-mouse) and highly disparate (pig-to-rodent) xenogeneic barriers, successful development of the transplants only occurred in these experiments, when host renal function was compromised, by preceding nephrectomy; this finding may make transplantation of metanephroi particularly suitable for congenital malformations (Hammerman, 2004). Parallel studies by Dekel *et al.* (2003) have demonstrated a gestational 'window of opportunity' for transplanting human (and pig) kidney precursors into mice: transplants from 7–8 weeks of human gestation survive, grow and form a functional organ able to produce urine without generating a significant immunological response, whereas earlier samples differentiate poorly and older ones are more immunogenic. Unless xenotrasnplantation is used, which has its own technical and moral issues, these transplantation experiments are unlikely to translate into effective therapies, since there is never going to be a large enough supply of early human metanephroi.

As an alternative strategy, we hypothesized that it might be feasible to generate human renal precursor cell lines *in vitro*, as a reproducible supply for later transplantation *in vivo*. Thus far, four lines have been generated from normal developing human kidneys at 10–12 weeks of gestation, supplied by the Human Developmental Biology Resource (HDBR) at the Institute of Child Health, London, and characterized in monolayer culture at passages 4–6. All four lines have now been propagated for 20 passages without immortalization (Romio *et al.*, 2003). Representative images of two of the lines and their expression of key nephrogenic molecules are shown in Figure 16.5: all lines appeared mesenchymal in shape, with an irregular, elongated outline, and western blots demonstrated differential expression of key nephrogenic markers. In the examples shown, the line from the 73 day gestation kidneys (N73) expressed high levels of the WT1 and PAX2 transcription factors and secreted high levels of glial cell line-derived neurotrophic factor (GDNF) and hepatocyte growth factor (HGF). Production of these classic renal mesenchyme-derived morphogens is consistent with N73 having an identity of 'induced' mesenchyme that is just about to undergo epithelial transformation. In contrast, the 84 day gestation line (N84) had barely detectable levels of WT1, PAX2, GDNF or HGF, consistent with 'uninduced' mesenchyme. Two mesenchymal cell lines were also

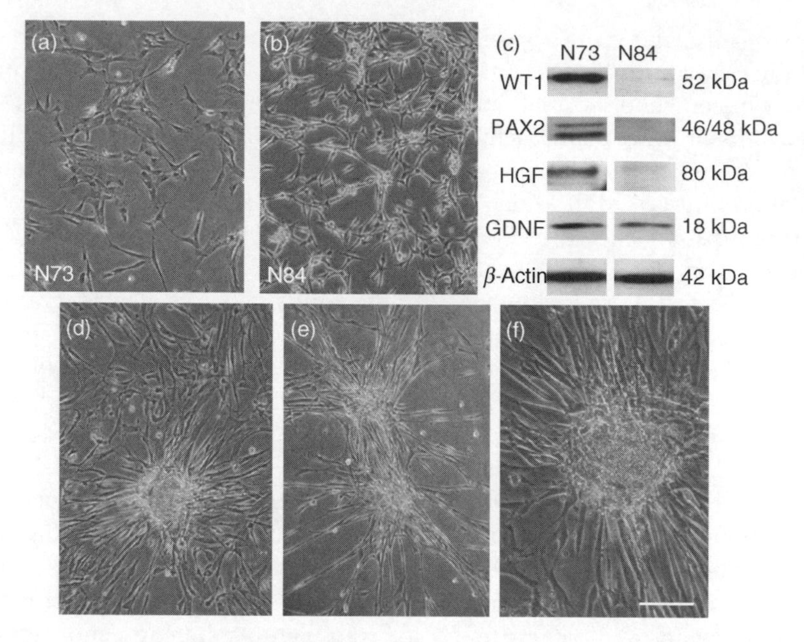

Figure 16.5 Human renal precursor cells. (a, b) Cell lines derived from 73 and 84 day gestation kidneys (N73 and N84, respectively). These have an irregular, elongated shape characteristic of mesenchymal cells and have been propagated for 20 passages without immortalization (Romio *et al.*, 2003). (c) Western blots demonstrating that N73 expresses high levels of WT1 and PAX2, and secretes high levels of GDNF and HGF, consistent with an identity of 'induced' mesenchyme that is just about to undergo epithelial transformation, whilst N84 has barely detectable levels of WT1, PAX2, GDNF or HGF, consistent with 'uninduced' mesenchyme. (d–f) N73 cells treated with FGF2, LIF and TGFα, which caused aggregation and clumping as though the cells were about to condense. Increased expression of the early epithelial marker E-cadherin was detected, but the cells did not progress further towards epithelial differentiation. Bar corresponds to 20 μm in (a), (b), (d) and (e) and 10 μm in (f)

generated from postnatal dysplastic kidneys: both expressed low levels of WT1, but not PAX2, results consistent with an 'uninduced' state and similar to dysplastic stroma *in vivo*.

These non-transduced human renal cell lines are an unparalleled resource with which to investigate normal and abnormal nephrogenesis in future studies. A first step will be to determine whether these precursor cells have the capacity to undergo partial or full epithelial transformation. This is demonstrable with rat renal mesenchymal cells *in vitro*, using a combination of FGF2, LIF and TGFα or $\beta 2$ (Barasch *et al.*, 1999; Plisov *et al.*, 2001), but our early attempts have met with only partial success. Although cell morphology changed and the cells clumped together, as though undergoing condensation (Figure 16.5d–f), and the early epithelial marker E-cadherin was upregulated (not shown), epithelial transformation did not proceed further. Hence, we are now exploring different cocktails of growth factors, with the aim of inducing complete epithelialization as a first step to tubule formation in three-dimensional culture. Furthermore, we are investigating whether any of our growth factor combinations can promote conversion of the dysplastic kidney-derived lines towards a 'normal' phenotype. These studies should generate at least two important pieces of information. First, they may confirm that the precursor cells have the potential for normal differentiation. In the long term, this raises the tantalizing possibility of using them in renal replacement therapy, either by transplanting them directly into 'failing' kidneys, perhaps after *in vitro* manipulation to ensure prolonged expression of the correct sequence of pro-nephrogenic factors, or as part of a bioartificial kidney, as envisaged by Atala and colleagues for several organ systems (Koh and Atala, 2004). Second, if further differences are observed between the normal and dysplastic lines, then this may allow us to develop rational treatment strategies to rescue normal cellular growth and differentiation in these important congenital malformations.

Conclusion

Renal development is ostensibly a simple process involving mutual inductive interactions between only two major cell types. Nevertheless, there are many ways in which nephrogenesis can go wrong, with an increasing catalogue of molecules implicated from human syndromes and murine mutants. Future advances in therapies for renal malformations will be dependent on a better understanding of the basic mechanisms involved in kidney formation.

References

Abdelhak, S., Kalatzis, V., Heilig, R. *et al.* (1997) A human homologue of the *Drosophila* eyes absent gene underlies branchio-oto-renal (BOR) syndrome and identifies a novel gene family. *Nat. Genet.* **15**: 157–164.

Abrass, C.K., Spicer, D., Berfield, A.K., St John, P.L. and Abrahamson, D.R. (1997) Diabetes induces changes in glomerular development and laminin-$\beta 2$ (s-laminin) expression. *Am. J. Pathol.* **151**: 1131–1140.

Armstrong, J.F., Pritchard-Jones, K., Bickmore, W.A., Hastie, N.D. and Bard, J.B. (1993) The expression of the Wilms' tumour gene, *WT1*, in the developing mammalian embryo. *Mech. Dev.* **40**: 85–97.

Attar, R., Quinn, F., Winyard, P.J., Mouriquand, P.D *et al.* (1998) Short-term urinary flow impairment deregulates PAX2 and PCNA expression and cell survival in fetal sheep kidneys. *Am. J. Pathol.* **152**: 1225–1235.

Aufderheide, E., Chiquet-Ehrismann, R. and Ekblom P. (1987) Epithelial–mesenchymal interactions in the developing kidney lead to expression of tenascin in the mesenchyme. *J. Cell Biol.* **105**: 599–608.

Bao, Q. and Hughes, C. (1995) Galectin-3 expression and effects on cyst enlargement and tubulogenesis in kidney epithelial MDCK cells cultured in three-dimensional matrices *in vitro. J. Cell Sci.* **108**: 2791–2800.

Barasch, J., Yang, J., Ware, C.B., Taga, T. *et al.* (1999) Mesenchymal to epithelial conversion in rat metanephros is induced by LIF. *Glycobiology* **99**: 377–386.

Barbacci, E., Reber, M., Ott, M.O., Breillat, C. *et al.* (1999) Variant hepatocyte nuclear factor 1 is required for visceral endoderm specification. *Development* **126**: 4795–4805.

Barker, D.J., Osmond, C., Golding, J., Kuh, D. and Wadsworth, M.E. (1989) Growth *in utero*, blood pressure in childhood and adult life, and mortality from cardiovascular disease. *Br. Med. J.* **298**: 564–567.

Barker, D.J. (2004) The developmental origins of well-being. *Phil. Trans. R. Soc. Lond. B Biol. Sci.* **359**: 1359–1366.

Barr, M.J. and Cohen, M.M.J. (1991) ACE inhibitor fetopathy and hypocalvaria: the kidney–skull connection. *Teratology* **44**: 485–495.

Barr, M.M., Sternberg, P.W. (1999) A polycystic kidney-disease gene homologue required for male mating behaviour in *C. elegans*. *Nature* **401**: 386–389.

Batourina, E., Choi, C., Paragas, N., Bello, N. *et al.* (2002) Distal ureter morphogenesis depends on epithelial cell remodeling mediated by vitamin A and Ret. *Nat. Genet.* **32**: 109–115.

Beck, A.D. (1971) The effect of intra-uterine urinary obstruction upon the development of the fetal kidney. *J. Urol.* **105**: 784–789.

Bell, D.M., Leung, K.K., Wheatley, S.C., Ng, L.J. *et al.* (1997) SOX9 directly regulates the type-II collagen gene. *Nat. Genet.* **16**: 174–178.

Bellairs, R., Lear, P., Yamada, K.M., Rutushauser, U. and Lash, J.W. (1995) Posterior extension of the chick nephric (Wolffian) duct: the role of fibronectin and NCAM polysialic acid. *Dev. Dyn.* **202**: 333–342.

Bernstein, J., Cheng, F. and Roszka, J. (1981) Glomerular differentiation in metanephric culture. *Lab. Invest* **45**: 183–190.

Biason-Lauber, A., Konrad, D., Navratil, F. and Schoenle, E.J. (2004) A *WNT4* mutation associated with Mullerian duct regression and virilization in a 46,XX woman. *N. Engl. J. Med.* **351**: 792–798.

Bohn, S., Thomas, H., Turan, G., Ellard, S. *et al.* (2003) Distinct molecular and morphogenetic properties of mutations in the human *HNF1β* gene that lead to defective kidney development. *J. Am. Soc. Nephrol.* **14**: 2033–2041.

Bottaro, D.P., Rubin, J.S., Faletto, D.L., Chan, A.M.L. *et al.* (1991) Identification of the hepatocyte growth factor as the *c-met* proto-oncogene product. *Science* **251**: 802–804.

Bouchard, M., Pfeffer, P. and Busslinger, M. (2000) Functional equivalence of the transcription factors Pax2 and Pax5 in mouse development. *Development* **127**: 3703–3713.

Bouillet, P., Cory, S., Zhang, L.C., Strasser, A. and Adams, J.M. (2001) Degenerative disorders caused by Bcl-2 deficiency prevented by loss of its BH3-only antagonist Bim. *Dev. Cell* **1**: 645–653.

Brenner, B.M., Garcia, D.L. and Anderson, S. (1988) Glomeruli and blood pressure. Less of one, more the other? *Am. J. Hypertens.* **1**: 335–347.

Brophy, P.D., Ostrom, L., Lang, K.M. and Dressler, G.R. (2001) Regulation of ureteric bud outgrowth by Pax2-dependent activation of the glial-derived neurotrophic factor gene. *Development* **128**: 4747–4756.

Bullock, S.L., Fletcher, J.M., Beddington, R.S. and Wilson, V.A. (1998) Renal agenesis in mice homozygous for a gene trap mutation in the gene encoding heparan sulfate 2-sulfotransferase. *Genes Dev.* **12**: 1894–1906.

Bullock, S.L., Johnson, T.M., Bao, Q., Hughes, R.C. *et al.* (2001) Galectin-3 modulates ureteric bud branching in organ culture of the developing mouse kidney. *J. Am. Soc. Nephrol.* **12**: 515–523.

Bulow, H.E., Berry, K.L., Topper, L.H., Peles, E. and Hobert, O. (2002) Heparan sulfate proteoglycan-dependent induction of axon branching and axon misrouting by the Kallmann syndrome gene *kal-1*. *Proc. Natl Acad. Sci. USA* **99**: 6346–6351.

Bush, K.T., Sakurai, H., Steer, D.L., Leonard, M.O. *et al.* (2004) TGF-β superfamily members modulate growth, branching, shaping, and patterning of the ureteric bud. *Dev. Biol.* **266**: 285–298.

Cacalano, G., Farinas, I., Wang, L.C., Hagler, K. *et al.* (1998) GFRα1 is an essential receptor component for GDNF in the developing nervous system and kidney. *Neuron* **21**: 53–62.

Cano-Gauci, D.F., Song, H.H., Yang, H., McKerlie, C. *et al.* (1999) Glypican-3-deficient mice exhibit developmental overgrowth and some of the abnormalities typical of Simpson–Golabi–Behmel syndrome. *J. Cell Biol.* **146**: 255–264.

Cantiello, H.F. (2004) Regulation of calcium signaling by polycystin-2. *Am. J. Physiol. Renal Physiol.* **286**: F1012–F1029.

Chen, H., Lun, Y., Ovchinnikov, D., Kokubo, H. *et al.* (1998) Limb and kidney defects in *Lmx1b* mutant mice suggest an involvement of LMX1B in human nail patella syndrome. *Nat. Genet.* **19**: 51–55.

Chevalier, R.L., Goyal, S., Wolstenholme, J.T. and Thornhill, B.A. (1998) Obstructive nephropathy in the neonatal rat is attenuated by epidermal growth factor. *Kidney Int.* **54**: 38–47.

Chevalier, R.L., Thornhill, B.A., Chang, A.Y., Cachat, F. and Lackey, A. (2002) Recovery from release of ureteral obstruction in the rat: relationship to nephrogenesis. *Kidney Int.* **61**: 2033–2043.

Chi, L., Zhang, S., Lin, Y., Prunskaite-Hyyrylainen, R. *et al.* (2004) Sprouty proteins regulate ureteric branching by coordinating reciprocal epithelial Wnt11, mesenchymal Gdnf and stromal Fgf7 signalling during kidney development. *Development* **62**: 3345–3356.

Chiang, C., Litingtung, Y., Lee, E., Young, K.E. *et al.* (1996) Cyclopia and defective axial patterning in mice lacking Sonic hedgehog gene function. *Nature* **383**: 407–413.

Chiu, M., Woolf, A.S., Johnson, T.M., Hillman, K.A *et al.* (2004) Galectin-3 retards the development of autosomal recessive polycystic kidney disease and localizes to the primary cilium. *J. Am. Soc. Nephrol.* **15**: 12A.

Chugh, S.S., Wallner, E.I. and Kanwar, Y.S. (2003) Renal development in high-glucose ambience and diabetic embryopathy. *Semin. Nephrol* **23**: 583–592.

Clark, P., Dziarmaga, A., Eccles, M. and Goodyer, P. (2004) Rescue of defective branching nephrogenesis in renal–coloboma syndrome by the caspase inhibitor, Z-VAD-fmk. *J. Am. Soc. Nephrol.* **15**: 299–305.

Coffinier, C., Barra, J., Babinet, C. and Yaniv, M. (1999) Expression of the vHNF1/HNF1β homeoprotein gene during mouse organogenesis. *Mech. Dev.* **89**: 211–213.

Coles, H.S.R., Burne, J.F. and Raff, M.C. (1993) Large-scale normal death in the developing rat kidney and its reduction by epidermal growth factor. *Development* 1993; **118**: 777–784.

Colnot, C., Fowlis, D., Ripoche, M.A., Bouchaert, I. and Poirier, F. (1998a) Embryonic implantation in galectin 1/galectin 3 double mutant mice. *Dev. Dyn.* **211**: 306–313.

Colnot, C., Ripoche, M.A., Milon, G., Montagutelli, X. *et al.* (1998b) Maintenance of granulocyte numbers during acute peritonitis is defective in galectin-3-null mutant mice. *Immunology* **94**: 290–296.

Cuzzolin, L., Dal Cere, M. and Fanos, V. (2001) NSAID-induced nephrotoxicity from the fetus to the child. *Drug Safety* **24**: 9–18.

Daikha-Dahmane, F., Narcy, F., Dommergues, M., Lacoste, M. *et al.* (1997) Distribution of α-integrin subunits in fetal polycystic kidney diseases. *Pediatr. Nephrol.* **11**: 267–273.

Davies, J.A., Ladomery, M., Hohenstein, P., Michael, L. *et al.* (2004) Development of an siRNA-based method for repressing specific genes in renal organ culture and its use to show that the Wt1 tumour suppressor is required for nephron differentiation. *Hum. Mol. Genet.* **13**: 235–246.

Davies, J.A., Yates, E.A. and Turnbull, J.E. (2003) Structural determinants of heparan sulphate modulation of GDNF signalling. *Growth Factors* **21**: 109–119.

Davis, A.P., Witte, D.P., Hsieh-Li, H.M., Potter, S.S. and Capecchi, M.R. (1995) Absence of radius and ulna in mice lacking hoxa-11 and hoxd-11. *Nature* **375**: 791–795.

Debiec, H., Kutsche, M., Schachner, M. and Ronco, P. (2002) Abnormal renal phenotype in L1 knock-out mice: a novel cause of CAKUT. *Nephrol. Dial. Transpl.* **17**(suppl 9): 42–44.

Deeb, A., Robertson, A., McColl, G., Bouloux, P.M.G. *et al.* (2001) X-linked Kallmann syndrome and multicystic dysplastic kidney – a new association. *Nephrol. Dial. Transpl.* **16**: 1170–1175.

Dekel, B., Burakova, T., Arditti, F.D., Reich-Zeliger, S. *et al.* (2003) Human and porcine early kidney precursors as a new source for transplantation. *Nat. Med.* **9**: 53–60.

Dinchuk, J.E., Car, B.D., Focht, R.J., Johnston, J.J. *et al.* (1995) Renal abnormalities and an altered inflammatory response in mice lacking cyclooxygenase II. *Nature* **378**: 406–409.

Dode, C., Levilliers, J., Dupont, J.M., De Paepe, A. *et al.* (2003) Loss-of-function mutations in *FGFR1* cause autosomal dominant Kallmann syndrome. *Nat. Genet.* **33**: 463–465.

Dressler, G.R., Deutsch, U., Chowdhury, K., Nornes, H.O. and Gruss, P. (1990) Pax2, a new murine paired-box-containing gene and its expression in the developing excretory system. *Development* **109**: 787–795.

Dressler, G.R., Douglas, E.C. (1992) Pax-2 is a DNA-binding protein expressed in embryonic kidney and Wilms tumor. *Proc. Natl Acad. Sci. USA* **89**: 1179–1183.

Dressler, G.R., Wilkinson, J.E., Rothenpieler, U.W., Patterson, L.T. *et al.* (1993) Deregulation of Pax-2 expression in transgenic mice generates severe kidney abnormalities. *Nature* **362**: 65–67.

Drummond, I.A., Madden, S.L., Rohwer-Nutter, P., Bell, G.I. *et al.* (1997) Repression of the insulin-like growth factor II gene by Wilms tumor suppressor WT1. *Science* **257**: 674–677.

Dudley, A.T., Godin, R.E. and Robertson, E.J. (1999) Interaction between FGF and BMP signaling pathways regulates development of metanephric mesenchyme. *Genes Dev.* **13**: 1601–1613.

Dudley, A.T., Lyons, K.M. and Robertson, E.J. (1995) A requirement for bone morphogenetic protein-7 during development of the mammalian kidney and eye. *Genes Dev.* **9**: 2795–2807.

Dudley, A.T. and Robertson, E.J. (1997) Overlapping expression domains of bone morphogenetic protein family members potentially account for limited tissue defects in BMP7-deficient embryos. *Dev. Dyn.* **208**: 349–362.

Duke, V.M., Winyard, P.J.D., Thorogood, P., Soothill, P. *et al.* (1995) *KAL*, a gene mutated in Kallmann's syndrome, is expressed in the first trimester of human development. *Mol. Cell Endocrinol.* **110**: 73–79.

Dunn, N.R., Winnier, G.E., Hargett, L.K., Schrick, J.J. *et al.* (1997) Haploinsufficient phenotypes in *Bmp4* heterozygous null mice and modification by mutations in *Gli3* and *Alx4*. *Dev. Biol.* **188**: 235–247.

Duong Van Huyen, J.P., Amri, K., Belair, M.F., Vilar, J. *et al.* (2003) Spatiotemporal distribution of insulin-like growth factor receptors during nephrogenesis in fetuses from normal and diabetic rats. *Cell Tissue Res.* **314**: 367–379.

Ekblom, P., Ekblom, M., Fecker, L., Klein, G. *et al.* (1994) Role of mesenchymal nidogen for epithelial morphogenesis *in vitro. Development* 1994; **120**: 2003–2014.

Ekblom, P., Lehtonen, E., Saxen, L. and Timpl, R. (1981) Shift in collagen type as an early response to induction of the metanephric mesenchyme. *J. Cell Biol.* **89**: 276–283.

Ekblom, P. (1996) Extracellular matrix and cell adhesion molecules in nephrogenesis. *Exp. Nephrol.* **4**: 92–96.

Ekblom, P. (1981) Formation of basement membranes in the embryonic kidney: an immunohistological study. *J. Cell Biol.* **91**: 1–10.

Elenius, K., Maatta, A., Salmivirta, M. and Jalkanen, M. (1992) Growth factors induce 3T3 cells to express bFGF-binding syndecan. *J. Biol. Chem.* **267**: 6435–6441.

Esquela, A.F. and Lee, S.J. (2003) Regulation of metanephric kidney development by growth/differentiation factor 11. *Dev. Biol.* **257**: 356–370.

Falk, M., Salmivirta, K., Durbeej, M., Larsson, E. *et al.* (1996) Integrin α6Bβ 1 is involved in kidney tubulogenesis *in vitro. J. Cell Sci.* **109**(12): 2801–2810.

Favor, J., Sandulache, R., Neuhauser-Klaus, A., Pretsch, W. *et al.* (1996) The mouse *Pax2(1Neu)* mutation is identical to a human *PAX2* mutation in a family with renal–coloboma syndrome and results in developmental defects of the brain, ear, eye, and kidney. *Proc. Natl Acad. Sci. USA* **93**: 13870–13875.

Gilbert, S.F. (1997) *Developmental Biology*. Sinauer Associates: Sunderland, MA.

Gilbert, T. and Merlet-Benichou, C. (2000) Retinoids and nephron mass control. *Pediatr. Nephrol.* **14**: 1137–1144.

Glick, P.L., Harrison, M.R., Adzick, N.S., Noall, R.A. and Villa, R.L. (1984) Correction of congenital hydronephrosis *in utero* IV: *in utero* decompression prevents renal dysplasia. *J. Pediatr. Surg.* **19**: 649–657.

Gnarra, J.R. and Dressler, G.R. (1995) Expression of Pax-2 in human renal cell carcinoma and growth inhibition by antisense oligonucleotides. *Cancer Res.* **55**: 4092–4098.

Gribouval, O., Gonzales, M., Neuhaus, T., Aziza, J. *et al.* (2005) Mutations in genes in the renin–angiotensin system are associated with autosomal recessive renal tubular dysgenesis. *Nat. Genet.* **37**: 964–968.

Goldman, M., Smith, A., Shuman, C., Caluseriu, O. *et al.* (2002) Renal abnormalities in Beckwith–Wiedemann syndrome are associated with 11p15.5 uniparental disomy. *J. Am. Soc. Nephrol.* **13**: 2077–2084.

Gomez, R.A., Lynch, K.R., Chevalier, R.L., Wilfong, N. *et al.* (1988) Renin and angiotensinogen gene expression in maturing rat kidney. *Am. J. Physiol.* **254**: F582–F587.

Green, M.C. (1968) Mechanism of the pleiotropic effects of the short-ear mutant gene in the mouse. *J. Exp. Zool.* **167**: 129–150.

Grieshammer, U., Le, M., Plump, A.S., Wang, F. *et al.* (2004) SLIT2-mediated ROBO2 signaling restricts kidney induction to a single site. *Dev. Cell* **6**: 709–717.

Grisaru, S., Cano-Gauci, D., Tee, J., Filmus, J. and Rosenblum, N.D. (2001) Glypican-3 modulates BMP- and FGF-mediated effects during renal branching morphogenesis. *Dev. Biol.* **231**: 31–46.

Gross, I., Morrison, D.J., Hyink, D.P., Georgas, K. *et al.* (2003) The receptor tyrosine kinase regulator Sprouty1 is a target of the tumor suppressor WT1 and important for kidney development. *J. Biol. Chem.* **278**: 41420–41430.

Gubler, M.C. (2003) Podocyte differentiation and hereditary proteinuria/nephrotic syndromes. *J. Am. Soc. Nephrol.* **14**(suppl 1): S22–S26.

Halder, G., Callaerts, P. and Gehring, W.J. (1995) Induction of ectopic eyes by targeted expression of the eyeless gene in *Drosophila* [see comments]. *Science* **267**: 1788–1792.

Hammerman, M.R. (2004) Renal organogenesis from transplanted metanephric primordia. *J. Am. Soc. Nephrol.* **15**: 1126–1132.

Hardelin, J.P., Julliard, A.K., Moniot, B. *et al.* (1999) Anosmin-1 is a regionally restricted component of basement membranes and interstitial matrices during organogenesis: implications for the developmental anomalies of X chromosome-linked Kallmann syndrome. *Dev. Dyn.* **215**: 26–44.

Hatini, V., Huh, S.O., Herzlinger, D., Soares, V.C. and Lai, E. (1996) Essential role of stromal mesenchyme in kidney morphogenesis revealed by targeted disruption of Winged Helix transcription factor BF-2. *Genes Dev.* **10**: 1467–1478.

Hensey, C., Dolan, V., Brady, H.R. (2002) The *Xenopus* pronephros as a model system for the study of kidney development and pathophysiology. *Nephrol. Dial. Transpl.* **17**(suppl 9): 73–74.

Herzlinger, D., Qiao, J., Cohen, D., Ramakrishna, N. and Brown, A.M. (1994) Induction of kidney epithelial morphogenesis by cells expressing Wnt-1. *Dev. Biol.* **166**: 815–818.

Hikita, C., Vijayakumar, S., Takito, J., Erdjument-Bromage, H. *et al.* (2000) Induction of terminal differentiation in epithelial cells requires polymerization of hensin by galectin 3. *J. Cell Biol.* **151**: 1235–1246.

Holmes, N., Harrison, M.R., Baskin, L.S. (2001) Fetal surgery for posterior urethral valves: long-term postnatal outcomes. *Pediatrics* **108**: E7.

Hu, M.C., Piscione, T.D. and Rosenblum, N.D. (2003a) Elevated SMAD1/β-catenin molecular complexes and renal medullary cystic dysplasia in *ALK3* transgenic mice. *Development* **130**: 2753–2766.

Hu Y, Tanriverdi F, MacColl GS, Bouloux PM. (2003b) Kallmann's syndrome: molecular pathogenesis. *Int. J. Biochem. Cell Biol.* **35**: 1157–1162.

Hyink, D.P., Tucker, D.C., St John, P.L., Leardkamolkarn, V. *et al.* (1996) Endogenous origin of glomerular endothelial and mesangial cells in grafts of embryonic kidneys. *Am. J. Physiol.* **270**: F886–F899.

Jenkins, D., Bitner-Glindzicz, M., Malcolm, S., Hu, C.C. *et al.* (2005) *De novo* Uroplakin IIIa heterozygous mutations cause human renal adysplasia leading to severe kidney failure. *J. Am. Soc. Nephrol.* **16**: 2141–2149.

Jerome, L.A. and Papaioannou, V.E. (2001) DiGeorge syndrome phenotype in mice mutant for the T-box gene, *Tbx1*. *Nat. Genet.* **27**: 286–291.

Kakuchi, J., Ichiki, T., Kiyama, S., Hogan, B.L. *et al.* (1995) Developmental expression of renal angiotensin II receptor genes in the mouse. *Kidney Int.* **47**: 140–147.

Kanwar, Y.S., Wada, J., Lin, S., Danesh, F.R. *et al.* (2004) Update of extracellular matrix, its receptors, and cell adhesion molecules in mammalian nephrogenesis. *Am. J. Physiol. Renal Physiol.* **286**: F202–F215.

Karihaloo, A., Kale, S., Rosenblum, N.D. and Cantley, L.G. (2004) Hepatocyte growth factor-mediated renal epithelial branching morphogenesis is regulated by glypican-4 expression. *Mol. Cell Biol.* **24**: 8745–8752.

Kiefer, M.C., Stephans, J.C., Crawford, K., Okino, K. and Barr, P.J. (1990) Ligand-affinity cloning and structure of a cell surface heparan sulfate proteoglycan that binds basic fibroblast growth factor. *Proc. Natl Acad. Sci. USA* **87**: 6985–6989.

Kirk, J.M.W., Grant, D.B., Besser, G.M., Shalet, S. *et al.* (1994) Unilateral renal aplasia in X-linked Kallmann's syndrome. *Clin. Genet.* **46**: 260–262.

Kispert, A., Vainio, S., McMahon, A.P. (1998) Wnt-4 is a mesenchymal signal for epithelial transformation of metanephric mesenchyme in the developing kidney. *Development* **125**: 4225–4234.

Klamt, B., Koziell, A., Poulat, F., Wieacker, P. *et al.* (1998) Frasier syndrome is caused by defective alternative splicing of WT1 leading to an altered ratio of WT1 $\pm$ KTS splice isoforms. *Hum. Mol. Genet.* **7**: 709–714.

Klein, G., Langegger, M., Garidis, C. and Ekblom, P. (1988) Neural cell adhesion molecules during embryonic induction and development of the kidney. *Development* **102**: 749–761.

Klein, G., Langegger, M., Timpl, R. and Ekblom, P. (1997) Role of laminin A chain in the development of epithelial cell polarity. *Glycobiology* **55**: 331–341.

Knudson, C.M., Korsmeyer, S.J. (1997) Bcl-2 and Bax function independently to regulate cell death. *Nat. Genet.* **16**: 358–363.

Koh, C.J., Atala, A. (2004) Tissue engineering, stem cells, and cloning: opportunities for regenerative medicine. *J. Am. Soc. Nephrol.* **15**: 1113–1125.

Kolatsi-Joannou, M., Bingham, C., Ellard, S., Bulman, M.P. *et al.* (2001) Hepatocyte nuclear factor-1β: a new kindred with renal cysts and diabetes and gene expression in normal human development. *J. Am. Soc. Nephrol.* **12**: 2175–2180.

Kolatsi-Joannou, M., Moore, R., Winyard, P.J. and Woolf, A.S. (1997) Expression of hepatocyte growth factor/scatter factor and its receptor, MET, suggests roles in human embryonic organogenesis. *Pediatr. Res.* **41**: 657–665.

Korhonen, M., Ylanne, J., Laitinen, L. and Virtanen, I. (1990a) Distribution of β1 and β3 integrins in human fetal and adult kidney. *Lab. Invest.* **62**: 616–625.

Korhonen, M., Ylanne, J., Laitinen, L. and Virtanen, I. (1990b) The α1–α6 subunits of integrins are characteristically expressed in distinct segments of developing and adult human nephron. *J. Cell Biol.* **111**: 1245–1254.

Koseki, C., Herzlinger, D. and Al-Awqati, Q. (1992) Apoptosis in metanephric development. *J. Cell Biol.* **119**: 1327–1333.

Kreidberg, J.A., Donovan, M.J., Goldstein, S.L., Rennke, H. *et al.* (1996) $\alpha3\beta1$ integrin has a crucial role in kidney and lung organogenesis. *Development* 1996; **122**: 3537–3547.

Kreidberg, J.A., Sariola, H., Loring, J.M., Maeda, M. *et al.* (1993) WT-1 is required for early kidney development. *Glycobiology* **74**: 679–691.

Kume, T., Deng, K. and Hogan BL. (2000) Murine forkhead/winged helix genes *Foxc1* (*Mf1*) and *Foxc2* (*Mfh1*) are required for the early organogenesis of the kidney and urinary tract. *Development* **127**: 1387–1395.

Laclef, C., Souil, E., Demignon, J. and Maire, P. (2003) Thymus, kidney and craniofacial abnormalities in Six 1-deficient mice. *Mech. Dev.* **120**: 669–679.

Lako, M., Strachan, T., Bullen, P., Wilson, D.I. *et al.* (1998) Isolation, characterisation and embryonic expression of *WNT11*, a gene which maps to 11q13.5 and has possible roles in the development of skeleton, kidney and lung. *Gene* **219**: 101–110.

Langley-Evans, S.C., Welham, S.J. and Jackson, A.A. (1999) Fetal exposure to a maternal low protein diet impairs nephrogenesis and promotes hypertension in the rat. *Life Sci.* **64**: 965–974.

LeBrun, D.P., Warnke, R.A. and Cleary, M.L. (1993) Expression of BCL-2 in fetal tissues suggests a role in morphogenesis. *Am. J. Pathol.* **142**: 743–753.

Lehnert, S.A. and Akhurst, R.J. (1988) Embryonic expression pattern of TGFβ type-1 RNA suggests both paracrine and autocrine mechanisms of action. *Development* **104**: 263–273.

Letterio, J.J., Geiser, A.G., Kulkarni, A.B., Roche, N.S. *et al.* (1994) Maternal rescue of transforming growth factor-β1 null mice. *Science* **264**: 1936–1938.

Leveen, P., Pekny, M., Gebre-Medhin, S., Swolin, B. *et al.* (1994) Mice deficient for PDGF-β show renal, cardiovascular, and hematological abnormalities. *Genes Dev.* **8**: 1875–1887.

Lewis, M. (1999) Report of the Paediatric Renal Registry 2003. In *UK Renal Registry Report 2003*, Ansell, D., Feest, T. (eds). UK Renal registry: Bristol, UK; 175–188.

Li, J.B., Gerdes, J.M., Haycraft, C.J., Fan, Y. *et al.* (2004) Comparative genomics identifies a flagellar and basal body proteome that includes the BBS5 human disease gene. *Glycobiology* **117**: 541–552.

Li, X., Oghi, K.A., Zhang, J., Krones, A. *et al.* (2003) Eya protein phosphatase activity regulates Six1–Dach–Eya transcriptional effects in mammalian organogenesis. *Nature* **426**: 247–254.

Liapis, H., Yu, H., Steinhardt, G.F. (2000) Cell proliferation, apoptosis, Bcl-2 and Bax expression in obstructed opossum early metanephroi. *J. Urol.* **164**: 511–517.

Lindsay, E.A., Vitelli, F., Su, H., Morishima, M. *et al.* (2001) Tbx1 haploinsufficiency in the DiGeorge syndrome region causes aortic arch defects in mice. *Nature* **410**: 97–101.

Lisle, S.J., Lewis, R.M., Petry, C.J., Ozanne, S.E. *et al.* (2003) Effect of maternal iron restriction during pregnancy on renal morphology in the adult rat offspring. *Br. J. Nutr.* **90**: 33–39.

Little, M. and Wells, C. (1997) A clinical overview of *WT1* gene mutations. *Hum. Mutat.* **9**: 209–225.

Liu, S., Cieslinski, D.A., Funke, A.J. and Humes, H.D. (1997) Transforming growth factor-β1 regulates the expression of *Pax-2*, a developmental control gene, in renal tubule cells. *Exp. Nephrol.* **5**: 295–300.

Loughna, S., Hardman, P., Landels, E., Jussila, L. *et al.* (1997) A molecular and genetic analysis of renal glomerular kidney development. *Angiogenesis* **1**: 84–101.

Luo, G., Hofmann, C., Bronckers, A.L., Sohocki, M. *et al.* (1995) BMP-7 is an inducer of nephrogenesis, and is also required for eye development and skeletal patterning. *Genes Dev.* **9**: 2808–2820.

Maas, R., Elfering, S., Glaser, T. and Jepeal, L. (1994) Deficient outgrowth of the ureteric bud underlies the renal agenesis phenotype in mice manifesting the limb deformity (ld) mutation. *Dev. Dyn.* **199**: 214–228.

Majumdar, A., Vainio, S., Kispert, A., McMahon, J. and McMahon, A.P. (2003) Wnt11 and Ret/Gdnf pathways cooperate in regulating ureteric branching during metanephric kidney development. *Development* **130**: 3175–3185.

Mansouri, A., Chowdhury, K. and Gruss, P. (1998) Follicular cells of the thyroid gland require *Pax8* gene function. *Nat. Genet.* **19**: 87–90.

Mariano, J.M., Montuenga, L.M., Prentice, M.A., Cuttitta, F. and Jakowlew, S.B. (1998) Concurrent and distinct transcription and translation of transforming growth factor-β type I and type II receptors in rodent embryogenesis. *Int. J. Dev. Biol.* **42**: 1125–1136.

Martinez, G. and Bertram, J.F. (2003) Organization of bone morphogenetic proteins in renal development. *Nephron Exp. Nephrol.* **93**: e18–e22.

Maulbecker, C.C. and Gruss, P. (1993) The oncogenic potential of *Pax* genes. *EMBO J.* **12**: 2361–2367.

Mayo, M.W., Wang, C.Y., Drouin, S.S., Madrid, L.V. *et al.* (1999) WT1 modulates apoptosis by transcriptionally upregulating the *bcl-2* proto-oncogene. *EMBO J.* **18**: 3990–4003.

McConnell, M.J., Cunliffe, H.E., Chua, L.J., Ward, T.A. and Eccles, M.R. (1997) Differential regulation of the human Wilms tumour suppressor gene (*WT1*) promoter by two isoforms of PAX2. *Oncogene* **14**: 2689–2700.

McGregor, L., Makela, V., Darling, S.M., Vrontou, S. *et al.* (2003) Fraser syndrome and mouse blebbed phenotype caused by mutations in *FRAS1/Fras1* encoding a putative extracellular matrix protein. *Nat. Genet.* **34**: 203–208.

Mendelsohn, C., Batourina, E., Fung, S., Gilbert, T. and Dodd, J. (1999) Stromal cells mediate retinoid-dependent functions essential for renal development. *Development* **126**: 1139–1148.

Mendelsohn, C., Lohnes, D., Decimo, D., Lufkin, T. *et al.* (1994) Function of the retinoic acid receptors (RAR) during development. *Development* **120**: 2749–2771.

Merlet-Benichou, C., Gilbert, T., Vilar, J., Moreau, E. *et al.* (1999) Nephron number: variability is the rule. Causes and consequences. *Lab. Invest.* **79**: 515–527.

Michos, O., Panman, L., Vintersten, K., Beier, K. *et al.* (2004) Gremlin-mediated BMP antagonism induces the epithelial–mesenchymal feedback signaling controlling metanephric kidney and limb organogenesis. *Development* **123**: 3401–3410.

Miyamoto, N., Yoshida, M., Kuratani, S., Matsuo, I., Aizawa, S. (1997) Defects of urogenital development in mice lacking Emx2. *Development* **124**: 1653–1664.

Miyazaki, Y., Oshima, K., Fogo, A., Hogan, B.L. and Ichikawa, I. (2000) Bone morphogenetic protein 4 regulates the budding site and elongation of the mouse ureter. *J. Clin. Invest.* **105**: 863–873.

Miyazaki, Y., Oshima, K., Fogo, A. and Ichikawa, I. (2003) Evidence that bone morphogenetic protein 4 has multiple biological functions during kidney and urinary tract development. *Kidney Int.* **63**: 835–844.

Montesano, R., Matsumoto, K., Nakamura, T. and Orci, L. (1991a) Identification of a fibroblast-derived epithelial morphogen as hepatocyte growth factor. *Glycobiology* **67**: 901–908.

Montesano, R., Schaller, G. and Orci, L. (1991b) Induction of epithelial tubular morphogenesis *in vitro* by fibroblast-derived soluble factors. *Glycobiology* **66**: 697–711.

Moore, M.W., Klein, R.D., Farinas, I., Sauer, H. *et al.* (1996) Renal and neuronal abnormalities in mice lacking GDNF. *Nature* **382**: 76–99.

Morham, S.G., Lanhenbach, R., Loftin, C.D., Tiano, H.F. *et al.* (1995) Prostaglandin synthase 2 gene disruption cuases severe renal pathology in the mouse. *Glycobiology* **83**: 473–482.

Muller, U., Wang, D., Denda, S., Meneses, J.J. *et al.* (1997) Integrin $\alpha 8 \beta 1$ is critically important for epithelial–mesenchymal interactions during kidney morphogenesis. *Glycobiology* **88**: 603–613.

Niimura, F., Labosky, P.A., Kakuchi, J., Okubo, S. *et al.* (1995) Gene targeting in mice reveals a requirement for angiotensin in the development and maintenance of kidney morphology and growth factor regulation. *J. Clin. Invest.* **96**: 2947–2954.

Nishimura, H., Yerkes, E., Hohenfellner, K., Miyazaki, Y. *et al.* (1999) Role of the angiotensin type 2 receptor gene in congenital anomalies of the kidney and urinary tract, CAKUT, of mice and men. *Mol. Cell* **3**: 1–10.

Nishinakamura, R., Matsumoto, Y., Nakao, K., Nakamura, K. *et al.* (2001) Murine homolog of *SALL1* is essential for ureteric bud invasion in kidney development. *Development* **128**: 3105–3115.

Nonaka, S., Tanaka, Y., Okada, Y., Takeda, S. *et al.* (1998) Randomization of left–right asymmetry due to loss of nodal cilia generating leftward flow of extraembryonic fluid in mice lacking KIF3B motor protein. *Glycobiology* **95**: 829–837.

Oh, S.P. and Li, E. (1997) The signaling pathway mediated by the type IIB activin receptor controls axial patterning and lateral asymmetry in the mouse. *Genes Dev.* **11**: 1812–1826.

Ohuchi, H., Hori, Y., Yamasaki, M., Harada, H. *et al.* (2000) FGF10 acts as a major ligand for FGF receptor 2 IIIb in mouse multi-organ development. *Biochem. Biophys. Res. Commun.* **277**: 643–649.

Ong, A.C. and Wheatley, D.N. (2003) Polycystic kidney disease – the ciliary connection. *Lancet* **361**: 774–776.

Ostrom, L., Tang, M.J., Gruss, P. and Dressler, G.R. (2000) Reduced *pax2* gene dosage increases apoptosis and slows the progression of renal cystic disease. *Dev. Biol.* **219**: 250–258.

Otto, E.A., Schermer, B., Obara, T., O'Toole, J.F. *et al.* (2003) Mutations in INVS encoding inversin cause nephronophthisis type 2, linking renal cystic disease to the function of primary cilia and left–right axis determination. *Nat. Genet.* **34**: 413–420.

Pachnis, V., Mankoo, B. and Costantini, F. (1993) Expression of the *c-ret* proto-oncogene during mouse embryogenesis. *Development* **119**: 1005–1017.

Padmanabhan, R. (1998) Retinoic acid-induced caudal regression syndrome in the mouse fetus. *Reprod. Toxicol.* **12**: 139–151.

Patel, S.R. and Dressler, G.R. (2004) Expression of Pax2 in the intermediate mesoderm is regulated by YY1. *Dev. Biol.* **267**: 505–516.

Patterson, L.T. and Potter, S.S. (2004) Atlas of *Hox* gene expression in the developing kidney. *Dev. Dyn.* **229**: 771–779.

Pelletier, J., Bruening, W., Kashtan, C.E., Mauer, S.M. *et al.* (1991) Germline mutations in the Wilms' tumor suppressor gene are associated with abnormal urogenital development in Denys–Drash syndrome. *Glycobiology* **67**: 437–447.

Perillo, N.L., Marcus, M.E., Baum, L.G. (1998) Galectins: versatile modulators of cell adhesion, cell proliferation, and cell death. *J. Mol. Med.* **76**: 402–412.

Pesce, C. (1998) Glomerular number and size: facts and artefacts. *Anat. Rec.* **251**: 66–71.

Peters, C.A., Carr, M.C., Lais, A., Retik, A.B. and Mandell, J. (1992) The response of the fetal kidney to obstruction. *J. Urol.* **148**: 503–509.

Pichel, J.G., Shen, L., Sheng, H.Z., Granholm, A.C. *et al.* (1996a) Defects in enteric innervation and kidney development in mice lacking GDNF. *Nature* **382**: 73–76.

Pichel, J.G., Shen, L., Sheng, H.Z., Granholm, A.C. *et al.* (1996b) GDNF is required for kidney development and enteric innervation. *Cold Spring Harb. Symp. Quant. Biol.* **61**: 445–457.

Piscione, T.D., Phan, T. and Rosenblum, N.D. (2001) BMP7 controls collecting tubule cell proliferation and apoptosis via Smad1-dependent and -independent pathways. *Am. J. Physiol. Renal Physiol.* **280**: F19–F33.

Piscione, T.D., Yager, T.D., Gupta, I.R., Grinfeld, B. *et al.* (1997) BMP-2 and OP-1 exert direct and opposite effects on renal branching morphogenesis. *Am. J. Physiol.* **273**: F961–F975.

Plachov, D., Chowdhury, K., Walther, C., Simon, D. *et al.* (1990) *Pax8*, a murine paired box gene expressed in the developing excretory system and thyroid gland. *Development* **110**: 643–651.

Plisov, S.Y., Yoshino, K., Dove, L.F., Higinbotham, K.G. *et al.* (2001) TGFβ2, LIF and FGF2 cooperate to induce nephrogenesis. *Development* **128**: 1045–1057.

Poleev, A., Fickenscher, H., Mundlos, S., Winterpacht, A. *et al.* (1992) *PAX8*, a human paired box gene: isolation and expression in developing thyroid, kidney and Wilms' tumors. *Development* **116**: 611–623.

Pope, J.C., Brock, J.W. III, Adams, M.C., Stephens, F.D. and Ichikawa I. (1999) How they begin and how they end: classic and new theories for the development and deterioration of congenital anomalies of the kidney and urinary tract, CAKUT. *J. Am. Soc. Nephrol.* **10**: 2018–2028.

Popsueva, A., Poteryaev, D., Arighi, E., Meng, X. *et al.* (2003) GDNF promotes tubulogenesis of GFRα1-expressing MDCK cells by Src-mediated phosphorylation of Met receptor tyrosine kinase. *J. Cell Biol.* **161**: 119–129.

Potter, E.L. (1972) *Normal and Abnormal Development of the Kidney.* Year Book Medical Publishers: Chicago, IL.

Poulsom, R., Alison, M.R., Cook, T., Jeffery, R. *et al.* (2003) Bone marrow stem cells contribute to healing of the kidney. *J. Am. Soc. Nephrol.* **14**(suppl 1): S48–S54.

Pritchard-Jones, K. and Hawkins, M.M. (1997) Biology of Wilms' tumour. *Lancet* **349**: 663–664.

Qiao, J., Sakurai, H., Nigam, S.K. (1999) Branching morphogenesis independent of mesenchymal–epithelial contact in the developing kidney. *Proc. Natl Acad. Sci. USA* **96**: 7330–7335.

Quaggin, S.E., Vanden Heuvel, G.B. and Igarash, P. (1998) Pod-1, a mesoderm-specific basic helix–loop–helix protein expressed in mesenchymal and glomerular epithelial cells in the developing kidney. *Mech. Dev.* **71**: 37–48.

Rackley, R.R., Kessler, P.M., Campbell, C. and Williams, B.R. (1995) *In situ* expression of the early growth response gene-1 during murine nephrogenesis. *J. Urol.* **154**: 700–705.

Rogers, S.A., Lowell, J.A., Hammerman, N.A. and Hammerman, M.R. (1998) Transplantation of developing metanephroi into adult rats. *Kidney Int.* **54**: 27–37.

Rogers, S.A., Ryan, G. and Hammerman, M.R. (1991) Insulin-like growth factors I and II are produced in the metanephros and are required for growth and development *in vitro*. *J. Cell Biol.* **113**: 1447–1453.

Rogers, S.A., Ryan, G. and Hammerman, M.R. (1992) Metanephric transforming growth factor-α is required for renal organogenesis *in vitro*. *Am. J. Physiol.* **262**: F533–F539.

Rogers, S.A., Ryan, G., Purchio, A.F. and Hammerman, M.R. (1993) Metanephric transforming growth factor-β1 regulates nephrogenesis *in vitro*. *Am. J. Physiol.* **264**: F996–F1002.

Romio, L., Fry, A.M., Winyard, P.J.D., Malcolm, S. *et al.* (2004) OFD1 is a centrosomal/basal body protein expressed during mesenchymal–epithelial transition in human nephrogenesis. *J. Am. Soc. Nephrol.* **15**: 2556–2568.

Romio, L., Wright, V., Price, K., Winyard, P.J. *et al.* (2003) *OFD1*, the gene mutated in oral–facial–digital syndrome type 1, is expressed in the metanephros and in human embryonic renal mesenchymal cells. *J. Am. Soc. Nephrol.* **14**: 680–689.

Rookmaaker, M.B., Verhaar, M.C., van Zonneveld, A.J. and Rabelink, T.J. (2004) Progenitor cells in the kidney: biology and therapeutic perspectives. *Kidney Int.* **66**: 518–522.

Rothenpieler, U.W. and Dressler, G.R. (1993) *Pax-2* is required for mesenchymal-to-epithelium conversion during kidney development. *Development* **119**: 711–720.

Rothman, K.J., Moore, L.L., Singer, M.R., Nguyen, U.S. *et al.* (1995) Teratogenicity of high vitamin A intake. *N. Engl. J. Med.* **333**: 1369–1373.

Rugarli, E.I., Di Schiavi, E., Hilliard, M.A., Arbucci, S. *et al.* (2002) The Kallmann syndrome gene homolog in *C. elegans* is involved in epidermal morphogenesis and neurite branching. *Development* **129**: 1283–1294.

Ruoslahti, E. and Yamaguchi, Y. (1991) Proteoglycans as modulators of growth factor activities. *Glycobiology* **64**: 867–869.

Rupprecht, H.D., Drummond, I.A., Madden, S.L., Rauscher, F.J. and Sukhatme, V.P. (1994) The Wilms' tumor suppressor gene *WT1* is negatively autoregulated. *J. Biol. Chem.* **269**: 6198–6206.

Ryan, G., Steele-Perkins, S.V., Morris, J.F., Rausher, F.J. III and Dressler, G.R. (1995) Repression of Pax-2 by WT1 during normal kidney development. *Development* **121**: 867–875.

Sainio, K., Suvanto, P., Davies, J., Wartiovaara, J. *et al.* (1997) Glial cell line-derived neurotrophic factor is required for bud initiation from ureteric epithelium. *Development* **124**: 4077–4087.

Sakurai, H., Barros, E.J., Tsukamoto, T., Barasch, J. and Nigam, S.K. (1997) An *in vitro* tubulogenesis system using cell lines derived from the embryonic kidney shows dependence on multiple soluble growth factors. *Proc. Natl Acad. Sci. USA* **94**: 6279–6284.

Salomon, R., Tellier, A.L., Attie-Bitach, T., Amiel, J. *et al.* (2001) *PAX2* mutations in oligomeganephronia. *Kidney Int.* **59**: 457–462.

Sanford, L.P., Ormsby, I., Gittenberger-de Groot, A.C., Sariola, H. *et al.* (1997) TGFβ2 knock-out mice have multiple developmental defects that are non-overlapping with other TGFβ knock-out phenotypes. *Development* **124**: 2659–2670.

Santos, O.F.P., Barros, E.J.G., Yang, X.M., Matsumoto, K. *et al.* (1994) Involvement of hepatocyte growth factor in kidney development. *Dev. Biol.* **163**: 525–529.

Sanyanusin, P., Schimmentl, L.A., McNoe, L.A., Ward, T.A. *et al.* (1995) Mutations of the *PAX2* gene in a family with optic nerve colobomas, renal anomalies and vesicoureteral reflux. *Nat. Genet.* **9**: 358–364.

Sariola, H., Ekblom, P., Lehtonen, E. and Saxen, L. (1983) Differentiation and vascularization of the metanephric kidney grafted on the chorioallantoic membrane. *Dev. Biol.* **96**: 427–435.

Sariola, H. and Saarma, M. (2003) Novel functions and signalling pathways for GDNF. *J. Cell Sci.* **116**: 3855–3862.

Sato, A., Kishida, S., Tanaka, T., Kikuchi, A. *et al.* (2004) Sall1, a causative gene for Townes–Brocks syndrome, enhances the canonical Wnt signaling by localizing to heterochromatin. *Biochem. Biophys. Res. Commun.* **319**: 103–113.

Sato, S. and Hughes, R.C. (1992) Binding specificity of a baby hamster lectin for H type I and II chains of polylactosamine glycans and appropriately glycosylated forms of laminin and fibronectin. *J. Biol. Chem.* **267**: 6983–6990.

Schmid, P., Cox, D., Bilbe, G., Maier, R. and McMaster, G.K. (1991) Differential expression of *TGFβ1*, *-β2* and *-β3* genes during mouse embryogenesis. *Development* **111**: 117–130.

Schmidt, C., Bladt, F., Goedecke, S., Brinkmann, V. *et al.* (1995) Scatter factor/hepatocyte growth factor is essential for liver development. *Nature* **373**: 699–702.

Schuchardt, A., D'Agati, V., Larsson-Blomberg, L., Costantini, F. *et al.* (1994) Defects in the kidney and enteric nervous system of mice lacking the tyrosine kinase receptor Ret. *Nature* **367**: 380–383.

Sequeira Lopez, M.L. and Gomez, R.A. (2004) The role of angiotensin II in kidney embryogenesis and kidney abnormalities. *Curr. Opin. Nephrol. Hypertens.* **13**: 117–122.

Shawlot, W. and Behringer, R.R. (1995) Requirement for Lim1 in head-organizer function. *Nature* **374**: 425–430.

Snell, W.J., Pan, J. and Wang, Q. (2004) Cilia and flagella revealed: from flagellar assembly in *Chlamydomonas* to human obesity disorders. *Glycobiology* **117**: 693–697.

Solhaug, M.J., Bolger, P.M. and Jose, P.A. (2004) The developing kidney and environmental toxins. *Pediatrics* **113**: 1084–1091.

Sonnenberg, E., Meyer, D., Weidner, K.M. and Birchmeier, C. (1993) Scatter factor/hepatocyte growth factor and its receptor, the c-met tyrosine kinase, can mediate a signal exchange between mesenchyme and epithelia during mouse development. *J. Cell Biol.* **123**: 223–235.

Sorenson, C.M., Padanilam, B.J. and Hammerman, M.R. (1996) Abnormal postpartum renal development and cystogenesis in the bcl-$2^{-/-}$ mouse. *Am. J. Physiol.* **271**: F184–F193.

Sorenson, C.M., Rogers, S.A., Korsmeyer, S.J. and Hammerman, M.R. (1995) Fulminant metanephric apoptosis and abnormal kidney development in bcl-2-deficient mice. *Am. J. Physiol.* **268**: F73–F81.

Soriano, P. (1994) Abnormal kidney development and hematological disorders in platelet derived growth factor b-receptor mutant mice. *Genes Dev.* **8**: 1888–1896.

Sorokin, L., Sonnenberg, A., Aumailley, M., Timpl, R. and Ekblom, P. (1990) Recognition of the laminin E8 cell-binding site by an integrin possessing the $\alpha 6$ subunit is essential for epithelial polarization in developing kidney tubules. *J. Cell Biol.* **111**: 1265–1273.

Soussi-Yanicostas, N., Hardelin, J.P., Arroyo-Jimenez, M.M., Ardouin, O. *et al.* (1996) Initial characterization of anosmin-1, a putative extracellular matrix protein synthesized by definite neuronal cell populations in the central nervous system. *J. Cell Sci.* **109**: 1749–1757.

Stark, K., Vainio, S., Vassileva, G. and McMahon, A.P. (1994) Epithelial transformation of metanephric mesenchyme in the developing kidney regulated by Wnt-4. *Nature* **372**: 679–683.

Steinhardt, G.F., Liapis, H., Phillips, B., Vogler, G. *et al.* (1995) Insulin-like growth factor improves renal architecture of fetal kidneys with complete ureteral obstruction. *J. Urol.* **154**: 690–693.

Surendran, K., Simon, T.C., Liapis, H. and McGuire, J.K. (2004) Matrilysin (MMP-7) expression in renal tubular damage: association with *Wnt4*. *Kidney Int.* **65**: 2212–2222.

Takayama, H., LaRochelle, W.J., Sabnis, S.G., Otsuka, T. and Merlino, G. (1997) Renal tubular hyperplasia, polycystic disease, and glomerulosclerosis in transgenic mice overexpressing hepatocyte growth factor/scatter factor. *Lab. Invest.* **77**: 131–138.

Threadgill, D.W., Dlugosz, A.A., Hansen, L.A., Tennenbaum, T. *et al.* (1995) Targeted disruption of mouse EGF receptor. Effect of genetic background on mutant phenotype. *Science* **269**: 230–234.

Torres, M., Gomex-Pardo, E., Dressler, G.R. and Gruss, P. (1995) *Pax-2* controls multiple steps of urogenital development. *Development* **121**: 4057–4065.

Towers, P.R., Woolf, A.S. and Hardman, P. (1998) Glial cell line-derived neurotrophic factor stimulates ureteric bud outgrowth and enhances survival of ureteric bud cells *in vitro*. *Exp. Nephrol.* **6**: 337–351.

Tufro, A., Norwood, V.F., Carey, R.M. and Gomez, R.A. (1999) Vascular endothelial growth factor induces nephrogenesis and vasculogenesis. *J. Am. Soc. Nephrol.* **10**: 2125–2134.

Vainio, S., Lehtonen, E., Jalkanen, M., Bernfield, M. and Saxen, L. (1989) Epithelial–mesenchymal interactions regulate the stage-specific expression of a cell surface proteoglycan, syndecan, in the developing kidney. *Dev. Biol.* **134**: 382–391.

Vainio, S.J. (2003) Nephrogenesis regulated by Wnt signaling. *J. Nephrol.* **16**: 279–285.

Vega, Q.C., Worby, C.A., Lechner, M.S., Dixon, J.E. and Dressler, G.R. (1996) Glial cell line-derived neurotrophic factor activates the receptor tyrosine kinase RET and promotes kidney morphogenesis. *Proc. Natl Acad. Sci. USA* **93**: 10657–10661.

Veis, D.J., Sorenson, C.M., Shutter, J.R. and Korsmeyer, S.J. (1993) Bcl-2-deficient mice demonstrate fulminant lymphoid apoptosis, polycystic kidneys and hypopigmented hair. *Glycobiology* **75**: 229–240.

Ventatachalam, M.A. and Kriz, W. (1998) Anatomy. In *Heptinstall's Pathology of the Kidney*, Jennette, J.C., Olson, J.L., Schwartz, M.M. and Silva, F.G. (eds). Lippincott-Raven: Philadelphia, PA; 3–66.

Vestweber, D. and Kemler, R. (1985) Identification of a putative cell adhesion domain of uvomorulin. *EMBO J.* **4**: 3393–3398.

Vilar, J., Lalou, C., Duong, V.H., Charrin, S. *et al.* (2002) Midkine is involved in kidney development and in its regulation by retinoids. *J. Am. Soc. Nephrol.* **13**: 668–676.

Vize, P.D., Woolf, A.S. and Bard, J.B.L. (eds) (2003) *The Kidney: From Normal Development to Congenital Disease*. Academic Press: London.

Wagner, K.D., Wagner, N. and Schedl, A. (2003) The complex life of WT1. *J. Cell Sci.* **116**: 1653–1658.

Wallner, E.I., Yang, Q., Peterson, D.R., Wada, J. and Kanwar, Y.S. (1998) Relevance of extracellular matrix, its receptors, and cell adhesion molecules in mammalian nephrogenesis. *Am. J. Physiol.* **275**: F467–F477.

Welham, S.J., Wade, A. and Woolf, A.S. (2002) Protein restriction in pregnancy is associated with increased apoptosis of mesenchymal cells at the start of rat metanephrogenesis. *Kidney Int.* **61**: 1231–1242.

Wheatley, D.N., Wang, A.M. and Strugnell, G.E. (1996) Expression of primary cilia in mammalian cells. *Cell Biol. Int.* **20**: 73–81.

Wheatley, D.N. (1995) Primary cilia in normal and pathological tissues. *Pathobiology* **63**: 222–238.

Wild, W., Pogge, V.S., Nastos, A., Senkel, S. *et al.* (2000) The mutated human gene encoding hepatocyte nuclear factor 1β inhibits kidney formation in developing *Xenopus* embryos. *Proc. Natl Acad. Sci. USA* **97**: 4695–4700.

Winyard, P.J.D., Bao, Q., Hughes, R.C. and Woolf, A.S. (1997) Epithelial galectin-3 during human nephrogenesis and childhood cystic diseases. *J. Am. Soc. Nephrol.* **8**: 1647–1657.

Winyard, P.J.D., Nauta, J., Lirenman, D.S., Hardman, P. *et al.* (1996a) Deregulation of cell survival in cystic and dysplastic renal development. *Kidney Int.* **49**: 135–146.

Winyard, P.J.D., Risdon, R.A., Sams, V.R., Dressler, G. and Woolf, A.S. (1996b) The PAX2 transcription factor is expressed in cystic and hyperproliferative dysplastic epithelia in human kidney malformations. *J. Clin. Invest.* **98**: 451–459.

Woolf, A.S., Kolatsi-Joannou, M., Hardman, P., Andermarcher, E. *et al.* (1995) Roles of hepatocyte growth factor/scatter factor and the met receptor in the early development of the metanephros. *J. Cell Biol.* **128**: 171–184.

Woolf, A.S. and Loughna, S. (1998) Origin of glomerular capillaries: is the verdict in? *Exp. Nephrol.* **6**: 17–21.

Woolf, A.S., Price, K.L., Scambler, P.J. and Winyard, P.J. (2004) Evolving concepts in human renal dysplasia. *J. Am. Soc. Nephrol.* **15**: 998–1007.

Woolf, A.S. and Thiruchelvam, N. (2001) Congenital obstructive uropathy: its origin and contribution to end-stage renal disease in children. *Adv. Renal Replacem. Ther.* **8**: 157–163.

Woolf, A.S. and Yuan, H.T. (2001) Angiopoietin growth factors and Tie receptor tyrosine kinases in renal vascular development. *Pediatr. Nephrol.* **16**: 177–184.

Wrana, J.L. (1998) TGF-β receptors and signalling mechanisms. *Miner. Electrolyte Metab.* **24**: 120–130.

Wu, G., Bohn, S. and Ryffel, G.U. (2004) The HNF1β transcription factor has several domains involved in nephrogenesis and partially rescues Pax8/lim1-induced kidney malformations. *Eur. J. Biochem.* **271**: 3715–3728.

Xu, P.X., Adams, J., Peters, H., Brown, M.C. *et al.* (1999) Eya1-deficient mice lack ears and kidneys and show abnormal apoptosis of organ primordia. *Nat. Genet.* **23**: 113–117.

Yang, J., Mori, K., Li, J.Y. and Barasch, J. (2003) Iron, lipocalin, and kidney epithelia. *Am. J. Physiol. Renal Physiol.* **285**: F9–18.

Yang, S.P., Woolf, A.S., Quinn, F. and Winyard, P.J.D. (2001) Deregulation of renal transforming growth factor-β1 after experimental short-term ureteric obstruction in fetal sheep. *Am. J. Pathol.* **159**: 109–117.

Yang, S.P., Woolf, A.S., Yuan, H.T., Scott, R.J. *et al.* (2000) Biological role of transforming growth factor-β1 in human congenital kidney malformations. *Am. J. Pathol.* **157**: 1633–1647.

Zent, R., Bush, K.T., Pohl, M.L., Quaranta, V. *et al.* (2001) Involvement of laminin binding integrins and laminin-5 in branching morphogenesis of the ureteric bud during kidney development. *Dev. Biol.* **238**: 289–302.

Zhang, P., Liegeois, N.J., Wong, C., Finegold, M. *et al.* (1997) Altered cell differentiation and proliferation in mice lacking p57KIP2 indicates a role in Beckwith–Wiedemann syndrome. *Nature* **387**: 151–158.

Zhang, Q., Taulman, P.D. and Yoder, B.K. (2004) Cystic kidney diseases: all roads lead to the cilium. *Physiology (Bethesda)* **19**: 225–230.

Zhang, S., Lin, Y., Itaranta, P., Yagi, A. and Vainio, S. (2001) Expression of Sprouty genes 1, 2 and 4 during mouse organogenesis. *Mech. Dev.* **109**: 367–370.

Zuniga, A., Michos, O., Spitz, F., Haramis, A.P. *et al.* (2004) Mouse limb deformity mutations disrupt a global control region within the large regulatory landscape required for Gremlin expression. *Genes Dev.* **18**: 1553–1564.

17

The Teeth

Irma Thesleff

Developmental anatomy

Teeth develop as appendages of embryonic ectoderm and their early morphogenesis shares similar anatomical features with other ectodermal appendages, such as hair follicles and various glands. Initiation of individual teeth is preceded by the formation of the dental lamina, a stripe of ectoderm located at the sites of future dental arches in the maxilla and mandible. The first morphological signs of tooth initiation are thickenings of the dental lamina epithelium at the sites of tooth development. Subsequently the *dental placodes* form within the thickened epithelia. They are multilayered epithelial condensations resembling, both morphologically and functionally, the placodes of other ectodermal organs. The placodal epithelium then forms a *bud* and this is accompanied by condensation of neural-crest derived mesenchymal cells around the bud (Figure 17.1).

The transition of the bud to the *cap stage* starts when the epithelial bud invaginates at its tip. The *enamel knot* forms at this location as an aggregation of epithelial cells. The flanking epithelium grows down, forming the cervical loops. The mesenchymal cells that become surrounded by the epithelium form the *dental papilla.* These events determine the extent of the tooth crown. The epithelium differentiates to distinct cell layers and forms the *enamel organ.* The peripheral part of condensed dental mesenchyme generates the *dental follicle* that surrounds the enamel organ epithelium and gives rise to periodontal tissues.

During the following *bell stage*, the tooth germ grows rapidly and the shape of the tooth crown becomes evident. The location of the cusps is determined by the *secondary enamel knots.* They form as epithelial thickenings and specify the points of epithelial folding. The terminal differentiation of the tooth-specific secretory cells also starts during this stage. The mesenchymal cells of the dental papilla directly

Embryos, Genes and Birth Defects, Second Edition Edited by Patrizia Ferretti, Andrew Copp, Cheryll Tickle and Gudrun Moore

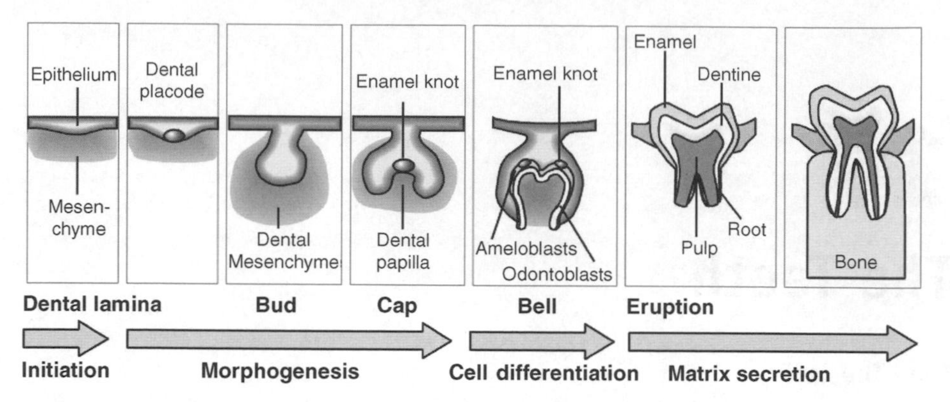

Figure 17.1 Main stages of tooth development. Interactions between the oral ectoderm and neural crest-derived mesenchyme direct morphogenesis and cell differentiation. The tooth shape results from growth and folding of the epithelial sheet. The epithelial signalling centres in the dental placode and enamel knots are key regulators of morphogenesis

underlying the dental epithelium differentiate into *odontoblasts*, laying down the organic matrix of dentine, and the juxtaposed epithelial cells differentiate into *ameloblasts*, depositing the enamel matrix. Cell differentiation and matrix deposition always start at the tips of future cusps, i.e. at the sites of enamel knots. During the entire morphogenesis of the tooth crown, a gradient of differentiation is seen in which the stage of differentiation decreases in the cuspal tip-to-cervical direction.

The root forms after completion of crown development in all human teeth. Root morphogenesis is guided by the growth of the cervical part of the dental epithelium. These cells do not differentiate into ameloblasts and they remain on the root surface as a network of so-called Malassez epithelial cells. The mesenchymal dental follicle cells contacting the root surface differentiate into *cementoblasts*. They secrete a thin layer of bone-like *cementum*, which covers the entire root. The dental follicle cells also form the periodontal ligament, linking the tooth to alveolar bone. The tooth subsequently erupts to the oral cavity (Figure 17.1).

Dentine resembles bone in its biochemical composition, although its histological appearance is different. Unlike the bone-forming osteoblasts, the odontoblasts do not become incorporated into the dentine matrix. Instead, each odontoblast leaves behind a cytoplasmic process, which becomes embedded in dentine and thereby contributes to the formation of a dentine tubule. Odontoblast cell bodies remain as a confluent layer between the dentine and the cells of the dental pulp. The *enamel* matrix is composed of unique enamel proteins, including amelogenin, enamelin and ameloblastin, which direct the formation and mineralization of enamel into the hardest tissue in the body. After the end of the secretory phase, the ameloblasts regulate the maturation of enamel, and they degenerate with the other layers of enamel epithelium during tooth eruption.

The period during which the human teeth develop is extremely long, starting from the 2nd month of embryonic development until completion during adolescence. The first *deciduous teeth* are initiated during the 5th week of gestation, and their mineralization starts during the 14th week. At this time the first *permanent teeth* have reached the bud stage, and they start to mineralize prior to birth. The first deciduous teeth normally erupt in children at 6 months of age, and the first permanent teeth at the age of 5–6 years. The last teeth to be formed, the third molars, are initiated postnatally and their crown development is completed between 12 and 16 years of age. Hence, tooth germs representing many different stages of development are present in fetuses and children.

The deciduous dentition of humans comprises 20 teeth and the permanent dentition 32 teeth. It is important to note that the human teeth, like mammalian teeth in general, are heterodont, i.e. teeth in different tooth groups differ in their shapes (as compared to the homodont dentitions in lower vertebrates, e.g. fish and reptiles, in which all teeth are conical). The mammalian teeth fall into three groups: incisiform, caniniform and molariform. The shapes of individual teeth are remarkably constant and the variation in tooth shape has been an important tool in studies on the evolution of man as well as other mammals. Detailed descriptions of the histology and timing of tooth development can be found in several textbooks (e.g. Nanci, 2002; Koch and Poulsen, 2001)

Main classes of defects

For most mammals, a complete and well-functioning dentition is essential for survival. Although dental anomalies do not threaten human lives in modern society, they may be aesthetically disturbing and affect the quality of life. Importantly, they are sometimes valuable diagnostic signs of congenital syndromes and various diseases, such as metabolic disorders and cancer as well as environmental insults. Dental anomalies include variations in number, shape and size as well as structural defects of dental hard tissues. The most common dental defects are described in detail in textbooks of clinical dentistry (e.g. Koch and Poulsen, 2001). There are also recent comprehensive descriptions of the genetics of dental anomalies and their associations with other congenital defects (OMIM database; Gorlin *et al.*, 1990; Thesleff and Pirinen, 2005).

Aberrations in number, shape and size of teeth

Hypodontia or missing teeth is a common anomaly that occurs in many different forms and is also seen as a trait in malformation syndromes. Third molars (wisdom teeth) are lacking in 20 % of people and one or more other permanent teeth are missing in 7–9 %. In the majority of cases, fewer than six teeth (excluding third molars) are missing. This, so-called incisor–premolar hypodontia is perhaps the most common congenital anomaly in humans. It is inherited in autosomal dominant

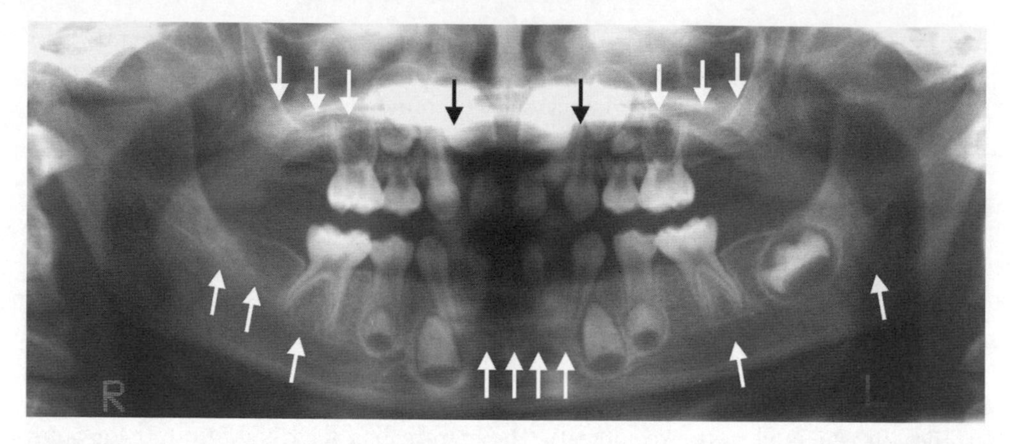

Figure 17.2 Severe dental agenesis (oligodontia). In this case mutations in the *AXIN2* gene have caused the lack of most permanent teeth (indicated by arrows). The same mutation predisposes to colorectal cancer (Lammi *et al.*, 2004)

manner but the genes involved have not yet been identified. Hypodontia is very rare in the deciduous dentition.

Oligodontia refers to hypodontia of more than six teeth (besides the missing wisdom teeth) and is much less common than the mild incisor–premolar hypodontia. Oligodontia may be seen as an isolated trait, although it is usually associated with other congenital defects. It is common in ectodermal dysplasia syndromes, affecting a variety of ectodermal organs, including hair, nails and glands such as the sweat, salivary and mammary glands (see Chapter 14). Oligodontia and milder hypodontia are also often seen in patients with cleft lip and palate and in syndromes such as Down's syndrome and Rieger syndrome.

The genetic basis of oligodontia is known in many cases (Thesleff and Pirinen, 2003). Mutations in *PAX9* cause non-syndromic oligodontia (Stockton *et al.*, 2005). *MSX1* mutations cause oligodontia, which may be associated with cleft lip and palate (Vastardis *et al.*, 1996; van den Boogaard *et al.*, 2000). Mutations in *AXIN2* were recently identified in two families and they were associated with severe oligodontia, affecting only the permanent teeth. In addition they caused colorectal cancer (Figure 17.2, Lammi *et al.*, 2004). The most common form of ectodermal dysplasia, i.e. hypohidrotic ectodermal dysplasia (HED), is associated with mutations in genes of the ectodysplasin signalling pathway (Mikkola and Thesleff, 2003). Mutations in *p63* and *PVRL1* have been found in other forms of ectodermal dysplasias with oligodontia. Rieger syndrome is caused by mutations in the *PITX2* gene.

Supernumerary teeth, or *hyperdontia*, is much less common than hypodontia. Hyperdontia of a single tooth is usually seen in the form of 'mesiodens' developing between the upper central incisors or as paramolars. The best-known syndrome with hyperdontia is cleidocranial dysplasia, affecting mainly bone development. The number of supernumerary teeth varies and they appear to constitute a partial third dentition (Jensen and Kreiborg, 1990). The gene mutated in this autosomal dominant syndrome is *RUNX2*, encoding a transcription factor necessary for osteoblast differentiation and bone formation.

Abnormal shapes and sizes of teeth are mostly linked with deviations in tooth number. A reduction of tooth size (microdontia) as well as abnormal shapes, such as conical teeth and missing cusps, are always seen in association with oligodontia and frequently also with less severe hypodontia (Arte *et al.*, 2001). For instance, small and peg-shaped lateral incisors are typical features of the common incisor–premolar hypodontia. Macrodontia and various types of fusions of teeth are less common and they are also mostly associated with numerical variations of teeth.

Although the anomalies in tooth number, shape and size are mainly caused by gene mutations; also, environmental factors may play a role as aetiological factors. This concerns only the permanent teeth, which develop postnatally. They are therefore more sensitive to harmful external influences during morphogenesis than deciduous teeth, which develop prenatally and are well protected in the uterus. For example, treatment of childhood cancer by chemotherapy or radiation can cause hypodontia, microdontia and defective root formation in the permanent dentition (Hölttä *et al.*, 2002).

Defects in structure of dentine and enamel

Mild variations in the colour and structure of enamel are common. They are generally caused by environmental factors disturbing ameloblast function and the deposition of enamel. Permanent teeth are usually affected because their enamel formation continues several years postnatally. Environmental influences such as chemicals, e.g. dioxins and fluorides, and medicines, including tetracyclines, affect enamel formation (Alaluusua *et al.*, 1999). Diseases such as fever and metabolic diseases may also cause enamel hypoplasias.

Inherited defects in dentine and enamel structure are rare. They are usually more severe than the environmentally caused defects and always affect both the deciduous and the permanent dentition. Amelogenesis imperfecta refers to hereditary defects in enamel formation. It appears in several clinically different forms, including hypoplastic and hypomineralized enamel, pits and colour changes. Mutations have been identified in genes encoding enamel proteins such as amelogenin and enamelin. In addition, enamel defects occur as traits in several syndromes, mostly in association with skin diseases and metabolic diseases (Thesleff and Pirinen, 2005).

Heritable dentine defects may affect both the structure of teeth and their colour, which is due to the transparency of enamel. Dentinogenesis imperfecta and dentine dysplasia are severe dentine defects affecting both the crowns and roots of teeth. Mutations in the dentine matrix component dentine sialophosphoprotein (DSPP) have been identified as causes (Xiao *et al.*, 2002). Coloured or opalescent teeth are seen in association with osteogenesis imperfecta, a syndrome affecting bones. This is caused by mutations in type I collagen, the main component of both bone and dentine matrix.

Cellular and molecular mechanisms affecting development

Communication between dental cells

The odontogenic fate is programmed very early in the cells forming the teeth. This was demonstrated decades ago in classical studies on experimental embryology. When the tissue in the molar area was dissected from mouse embryos before the buds of the first molar had formed and transplanted to the anterior chamber of the eye, all three molars developed (Lumsden, 1988). Teeth also undergo morphogenesis when cultured *in vitro* if they are explanted at bud stage or later in development. Glasstone (1963) was the first to show that molars develop a normal cusp pattern when grown in isolation in organ culture. Tooth morphogenesis, therefore, does not depend on interactions with bone or other surrounding tissues. Hence, the teeth are determined at an early stage and thereafter the tissue which has been committed to tooth development has inherent developmental potential.

The morphogenesis of all organs is regulated by communication between the cells of the organ rudiment. *Interactions* between the epithelial and mesenchymal tissue components have particularly important functions in developing teeth, as well as in all other organs forming as ectodermal appendages. The epithelial component of teeth has its origin in oral ectoderm and the mesenchymal cells derive from the neural crest. As shown in many experimental studies in which the epithelial and mesenchymal tissues have been recombined and cultured in different heterotypic and heterochronic combinations, the interactions are sequential and reciprocal and there is a chain of interactive events between the two tissues driving advancing tooth morphogenesis (Kollar and Baird, 1969; Ruch, 1987; Mina and Kollar, 1987; Lumsden, 1988). In addition to the interactions between the two tissue types, there is also important signalling taking place within cells in each tissue. For example, ectodysplasin, a signal molecule in the tumour necrosis factor family, mediates communication between ectodermal cell compartments (Laurikkala *et al.*, 2001; see below).

Cell and tissue interactions regulate various of cellular functions. In particular, cell division and cell adhesion are affected resulting in condensations of cells, cell shape changes and growth, and folding of the epithelial sheets. Apoptosis is also regulated, as shown during the removal of the enamel knots and the degeneration of dental epithelium after morphogenesis, as well as in maturation stage ameloblasts (Vaahtokari *et al.*, 1996b; Joseph *et al.*, 1999). The differentiation of odontoblasts and ameloblasts is regulated by epithelial–mesenchymal signalling as well, and this results in exit from the cell cycle and marked changes in cellular morphology as the cells polarize, become columnar and start the secretion of extracellular matrix.

Molecules participating in tooth development and their developmental roles

Most data on the molecules regulating tooth development have derived from gene expression studies using *in situ* hybridization or immunohistology. Expression patterns of over 350 genes can be viewed in the graphical database (http://bite-it.helsinki.fi). Although the descriptive expression studies obviously do not indicate functions for the genes, they have pinpointed numerous potentially important regulatory molecules of tooth development and they have also been instrumental in enhancing the understanding of the principles of tooth morphogenesis. In particular, expression studies have underlined the sequential and reciprocal nature of the epithelial–mesenchymal interactions by showing that several regulatory genes are repeatedly expressed at different stages of morphogenesis. They have also revealed developmentally significant co-expression patterns of genes in various dental cell populations suggestive of genetic pathways. It is also noteworthy that the majority of the genes in the database are associated with signalling networks mediating cell communication.

Signals in most of the currently known signal families are expressed in developing teeth, and in many cases downstream targets of the signals have been identified and networks between different pathways have been elucidated. Specific functions in tooth development have been demonstrated for an increasing number of genes encoding molecules of signalling pathways. Hence, it has become obvious that the molecular basis of cell communication regulating tooth morphogenesis is based on the same tool-kit of molecules as in all other embryonic organs. Of particular importance are signal molecules in four families, including bone morphogenetic proteins (BMPs), fibroblast growth factors (FGFs), hedgehogs (Hhs) and Wnts (Thesleff, 2003). Most information has come from studies on mouse tooth development but, as the developmental regulatory molecules are conserved between mammals and even across the animal kingdom, it can be safely assumed that the acquired knowledge can be largely applied to human tooth development as well.

Dental placodes and enamel knots as signalling centres

The observation that several signal molecules are co-expressed in a small, non-dividing epithelial cell population in the cap stage tooth germ led to the discovery of the enamel knot as a signalling centre (Jernvall *et al.*, 1994; Vaahtokari *et al.*, 1996a). The first signal detected was FGF4, followed by Shh (sonic hedgehog), BMP-2, BMP-4 and BMP-7, and later other FGFs and several Wnts were localized in the enamel knots (Thesleff and Mikkola, 2002) (Figure 17.3). In addition, the cells of the enamel knot express a number of targets of molecules involved in signal mediation, pinpointing signalling pathways that regulate enamel knot formation and function. Such molecules include the Wnt pathway target lef1, BMP targets p21 and msx2 and the ectodysplasin signal mediators edar and edaradd.

There are also transient signalling centres in dental epithelium during the initiation of tooth development which were initially revealed by co-expression of signal molecules (Keränen *et al.*, 1998). They appear in the dental placodes, which in this respect resemble the placodes of other ectodermal organs (Pispa and Thesleff, 2003). In addition, a third set of signalling centres, the secondary enamel knots, appear in the epithelium of molar tooth germs during the bell stage and are associated with cusp morphogenesis (Jernvall *et al.*, 1994). Mostly the same signals are expressed in

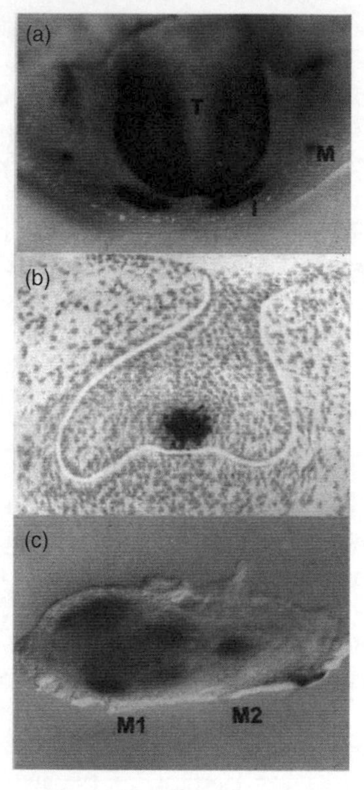

Figure 17.3 Signalling centres in the dental epithelium of embryonic mouse teeth. (a) The dental placodes regulate initiation of budding. *Shh* expression visualized by whole mount *in situ* hybridization in a 12 day mouse embryo mandible. (b) The primary enamel knot regulates the transition from bud to cap stage. *Fgf4* expression shown by *in situ* hybidization in a tissue section of a 14 day embryo molar. (c) Secondary enamel knots regulate cusp formation. *Shh* expression in a whole mount of dental epithelium of 17 day embryo molars. Four secondary enamel knots are seen in the occlusal view of a bell stage 1st molar. The primary enamel knot expresses *Shh* in the cap stage 2nd molar. I, incisor placode; M, molar placode; T, tongue; M1, 1st molar; M2, 2nd molar

all three centres. The reiterative appearance of epithelial signalling centres marks critical stages of tooth development. The dental placodes in the incisor and molar regions of embryonic day 12 mouse embryos precede epithelial budding and initiate epithelial morphogenesis. The (primary) enamel knots appear during the transition from the bud to the cap stage and initiate the morphogenesis of the tooth crowns. They regulate proliferation in the underlying mesenchyme and in the flanking epithelium forming the cervical loops. The secondary enamel knots precede the folding of the inner enamel epithelium during the bell stage and initiate cusp formation (Figure 17.3; Jernvall *et al.*, 2000).

Careful three-dimensional comparisons of gene expression patterns and tooth morphologies have revealed close associations between the primary and secondary enamel knots and tooth shapes and have indicated central roles for the enamel knots in patterning the tooth cusps (Jernvall *et al.*, 2000). Interestingly, a wide range of tooth shapes in different mammalian species and cusp patterns can be reproduced by mathematical modelling of enamel knot signalling (Salazar-Ciudad and Jernvall, 2002).

The ectodysplasin–edar signal pathway plays a key role in the regulation of dental signalling centres. The edar receptors are locally expressed in the dental placodes and enamel knots, as well as in the placodes of other ectodermal organs (Figure 17.3; Laurikkala *et al.*, 2001, 2002). The tooth phenotypes of eda, edar and edaradd mutants (mouse models for human hypohidrotic ectodermal dysplasia) were analysed in detail decades ago. They often lack the third molars and they have a cusp defect in the first molars, characterized by fused and lacking cusps. This phenotype appears to be due to defective enamel knot signalling, since their primary enamel knots are small and the secondary enamel knots are fused, and this cusp phenotype was partially rescued by FGF4 (Pispa *et al.*, 1999). Interestingly, mice overexpressing ectodysplasin in the ectoderm have large molar placodes and extra teeth in front of the first molars, and their cusp patterns are abnormal (Mustonen *et al.*, 2003, 2004; Kangas *et al.*, 2004). Addition of ectodysplasin protein on cultured skin explants also increased the size of placodes. Hence, the main function of eda–edar signalling is the stimulation of placode growth, which occurs by change of cell fate rather than increased cell proliferation (Mustonen *et al.*, 2004).

Conserved signalling pathways and their networks regulate morphogenesis

The genetic modification of mouse development has generated important data concerning the functions of numerous genes in tooth development. In most cases where dental defects have been observed in transgenic mice, or in spontaneous mouse mutants, the genes have been associated with cell–cell signalling. Such genes include those for the signals FGF8, activinA, Shh and Eda. More often, however, the dental defects have resulted from the lack of function of modulators or targets of signalling molecules. Such molecules include gli2 and gli3 in the Shh pathway, pax9, pitx2, dlx1, dlx2, msx1, runx2, and FgfR2b in the FGF pathway, lef1 in the Wnt pathway, msx1, msx2, and dlx2 in the BMP pathway, edar and edaradd in the Eda pathway, and the activin target follistatin (Figure 17.4). It is also noteworthy that the arrested development in most knock-outs occurs either at E11, prior to dental placode development and budding, or at E13, prior to primary enamel knot formation and transition to cap stage. It is conceivable that the formation of epithelial signalling centres integrating several signalling pathways requires the coordinated functions of numerous genes, and that their appearance therefore represents sensitive stages of tooth morphogenesis.

The importance of signal inhibitors for morphogenetic regulation has become increasingly apparent. Follistatin, an inhibitor of activin and BMP signalling, is required for normal tooth crown development. The patterning of secondary enamel knots is abnormal in follistatin knock-outs, resulting in irregular folding of the dental epithelium (Wang *et al.*, 2004a). Hence, the fine-tuning of BMP and activin signalling

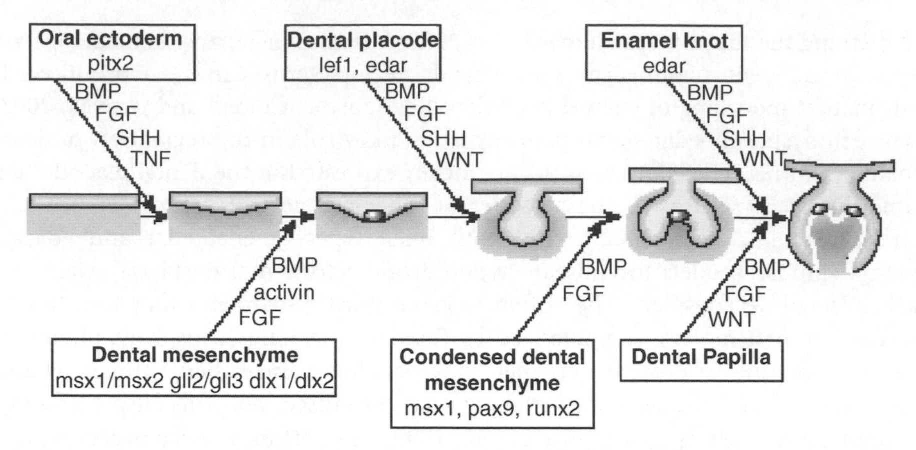

Figure 17.4 Schematic presentation of signalling networks mediating sequential and reciprocal interactions between the dental epithelium and mesenchyme. The same signalling molecules regulate development at many stages. The genes indicated in the boxes have been shown to be necessary for the advancement of tooth morphogenesis in knock-out mice. Mutations in several of the same genes cause oligodontia in humans. BMP, bone morphogenetic protein; FGF, fibroblast growth factor; SHH, sonic hedgehog; TNF, tumour necrosis factor

is important for normal cusp patterning. Recently another BMP inhibitor, ectodin, was shown to be necessary for normal cusp patterning (Laurikkala *et al.*, 2003; Kassai *et al.*, 2005). The Wnt signal inhibitor AXIN2 is also required for tooth development, since its mutations cause severe tooth agenesis in humans (Lammi *et al.*, 2004). However, Wnt signalling is necessary for tooth development, since the loss-of-function mutant of *lef1*, a target and mediator of Wnt signalling, causes missing teeth in mice (van Genderen *et al.*, 1994). These results indicate that fine tuning of Wnt signalling is important for normal tooth development.

The function of signalling molecules and their downstream targets has been elucidated by *in vitro* studies in which recombinant signalling molecules are introduced into dissected dental tissues, using agarose or heparin acrylic beads (Sahlberg *et al.*, 2002). For example, msx1 and msx2 were shown to be regulated in the early dental mesenchyme by epithelial BMP4 signals (Vainio *et al.*, 1993). The examination of signal effects in dental tissues of mouse mutants has allowed further dissection of signalling pathways and the functions of individual genes in the networks. Bei *et al.* (2000) showed that msx1 is required for the induction of *Bmp4* expression by BMP4 in the mesenchyme at bud stage and that the arrest in tooth development in *msx1* mutants could be rescued by BMP4. This indicated roles for msx1 both upstream and downstream of BMP4. The role of lef1 in the enamel knot was demonstrated by bead experiments and tissue recombination between wild-type and mutant tissues by Kratochwil *et al.* (1996, 2002). They showed that lef1 regulates *Fgf4* expression in the enamel knot and that this is required for *Fgf3* expression in the mesenchyme. Gene constructs have recently been introduced into the tissues by viral vectors and by electroporation (Angeli *et al.*, 2002). By using various experimental approaches, the targets of many signalling molecules have been identified at different stages of tooth development in mesenchyme and epithelium. A general picture of the molecular regulation of tooth morphogenesis has emerged, based on studies from many laboratories during the last 10 years (Figure 17.4; Thesleff, 2003).

Regulation of ameloblast and odontoblast differentiation

The differentiation of ameloblasts and odontoblasts is regulated by epithelial–mesenchymal interactions, like tooth morphogenesis, and the same signalling molecules have been implicated. TGFβ superfamily signals regulate both enamel and dentine formation (Ruch *et al.*, 1995; Coin *et al.*, 1999). BMPs in particular have been used successfully for dentine regeneration *in vivo* (Rutherford *et al.*, 1993). Recent evidence from transgenic mice indicates that BMP4 is the major signalling molecule regulating ameloblast differentiation and enamel formation (Wang *et al.*, 2004b). This study also revealed an inhibitory function for the dental follicle in amelogenesis. It was shown that activin from the dental follicle induces follistatin expression in pre-ameloblasts, and that follistatin in turn antagonizes the function of odontoblast-derived BMP4 as an ameloblast inducer.

The possibility of using stem cells for dentine and enamel regeneration has gained much attention during recent years. Stem cells have been isolated from the dental pulp that can differentiate into odontoblasts when transplanted to ectopic locations (Gronthos *et al.*, 2000). Epithelial stem cells have been identified in continuously growing mouse incisors (Harada *et al.*, 1999). It was shown that FGF and Notch signalling participate in the maintenance and early differentiation of these epithelial stem cells. FGF-10, in particular, is a mesenchymal signal that is necessary for the continued renewal of epithelial stem cells in the mouse incisor (Harada *et al.*, 2002).

How cellular and molecular developmental mechanisms assist in elucidating the causes of abnormal development

Dental defects are commonly associated with abnormal development of other organs

It has become clear that the cellular and molecular mechanisms regulating tooth development are not unique to teeth. On the contrary, they are remarkably similar to the mechanisms regulating other organs, in particular those developing as ectodermal appendages. It is noteworthy that so far no regulatory gene unique to teeth has been identified. Therefore, abnormal tooth development can in most cases be expected to be associated with defects in other tissues or organs. Exceptions are enamel and dentine defects caused by mutations in tooth-specific genes encoding unique components of the dental hard tissues.

Dental defects may also be linked to cancer, as indicated by colorectal cancer in patients with oligodontia caused by *AXIN2* mutations (Figure 17.2; Lammi *et al.*, 2004). Axin2 is a modulator of the Wnt pathway and, as mutations in Wnt and other signalling pathways have been associated with malignant development, it is possible that more links between dental defects and tumours will be discovered. Associations between hypodontia and cancer are probably uncommon, but it is important to note that oligodontia may sometimes be used as an indicator of cancer susceptibility.

Causative genes cannot be predicted from the phenotype of a dental defect

Hundreds or even thousands of genes are likely to participate in the regulation of tooth morphogenesis. Complex integrated signalling networks constitute the main driving force of morphogenesis and, in principle, mutations in any of the genes of the signalling pathways may be harmful to tooth development. As shown by mouse mutants, deletion of the function of many different genes can result in an arrest of tooth morphogenesis at the same stage. It is also evident that normal development requires fine-tuning of signalling and that both inhibition and stimulation of the same signalling pathways may result in abnormal development. Therefore, the mutant phenotype is seldom an indicator of the causative gene.

Mouse mutants have shown that genes often have redundant functions in developing teeth. This is exemplified by *dlx1/dlx2*, *msx1/msx2*, as well as *gli2/gli3* double mutant mice. Tooth development is arrested prior to dental placode development in the double mutants but not in single mutants (Thomas *et al.*, 1997; Bei and Maas, 1998; Hardcastle *et al.*, 1998). All these genes encode transcription factors; they are likely to regulate the same target genes and can therefore compensate for each others' functions. It is likely that there is also redundancy between different signalling pathways. Therefore, although a gene may be an important regulator of tooth development, the loss of its function may not result in a dental phenotype at all, or it may cause only mild defects.

Furthermore, as the same genes are used reiteratively during morphogenesis (Figure 17.4), their deletion may have harmful effects only at an advanced developmental stage if their function is compensated by other genes at earlier stages. An example is the mouse *msx1* gene. Its function is compensated by *msx2* during initiation of morphogenesis but, since *msx2* is not expressed at bud stage and therefore cannot compensate for the lack of *msx1* function, *msx1* mutant tooth germs are arrested during the transition from bud to cap stage.

The mutations that have so far been identified in human cases of isolated tooth agenesis are loss-of-function mutations in *MSX1*, *PAX1* and *AXIN2* and they are thought to result in haplo-insufficiency. Hence, the function of the respective genes is reduced and, typically, not all teeth are missing in any of the patients. However, there are quite extensive differences in the number of missing teeth between patients with the same mutation. Also, *PAX9* mutations seem to affect the posterior teeth more severely than do *MSX1* mutations. Since *msx1* and *pax9* apparently interact genetically (both regulate *Bmp4* expression; Bei *et al.*, 2000; Peters *et al.*, 1998), the different phenotypes may be due to differences in expression patterns of *MSX1*and *PAX1* during tooth development. *AXIN2* mutations, on the other hand, do not affect the deciduous teeth at all but the development of the secondary dentition is severely impaired. The mechanism remains unclear, as the process of tooth renewal is poorly understood. This is mainly because mice, which are commonly used as model animals, do not have a secondary dentition.

Deficient function of dental placodes and enamel knots can explain why tooth agenesis and morphological defects are linked

Clinical studies have shown that the defects in tooth number, size and shape are linked in human conditions. In particular, smaller teeth, often peg-shaped incisors, and molars with fewer cusps are seen in association with hypodontia. Similar associations are apparent also in mouse mutants, for example in the mouse model of ectodermal dysplasia syndromes, the *Tabby* mouse (Figure 17.5c). The most obvious explanation for these associations lies in the role of these genes in regulating the formation of signalling centres at the dental placodes and enamel knots (Figure 17.3). The total absence of a dental placode would result in lack of organogenesis, whereas a hypoplastic dental placode or enamel knot would give rise to a smaller tooth. The size of the dental placode may determine a graded threshold specifying the number of teeth developing from that placode. Therefore, the last teeth that develop from the dental placode, e.g. the third molars or lateral incisors, are most vulnerable and will be more severely affected than the earlier developing teeth. Impaired enamel knot signalling, on the other hand, can lead to the reduction of crown size and a fewer number of cusps. Since the placodes and enamel knots express many molecules in different signalling pathways, it is conceivable that mutations in many different genes may affect both the number and shape of teeth. In addition to the Eda pathway, BMP and activin signalling has also been linked to both tooth number and shape (Figure 17.6; Ferguson *et al.*, 1998; Wang *et al.*, 2004a).

Defects in dental hard tissues may be associated with defects in number and shape

Enamel formation can be prevented in mice by affecting many signalling pathways. Overexpression of Wnt3 (Millar *et al.*, 2003), ectodysplasin (Mustonen *et al.*, 2003), edar (Pispa *et al.*, 2004; Tucker *et al.*, 2004) and follistatin (Figure 17.6; Wang *et al.*, 2004b) in the dental epithelium prevents the differentiation of ameloblasts. Defective enamel formation has been reported also in *Shh* and *msx2* knock-out mice (Dassule

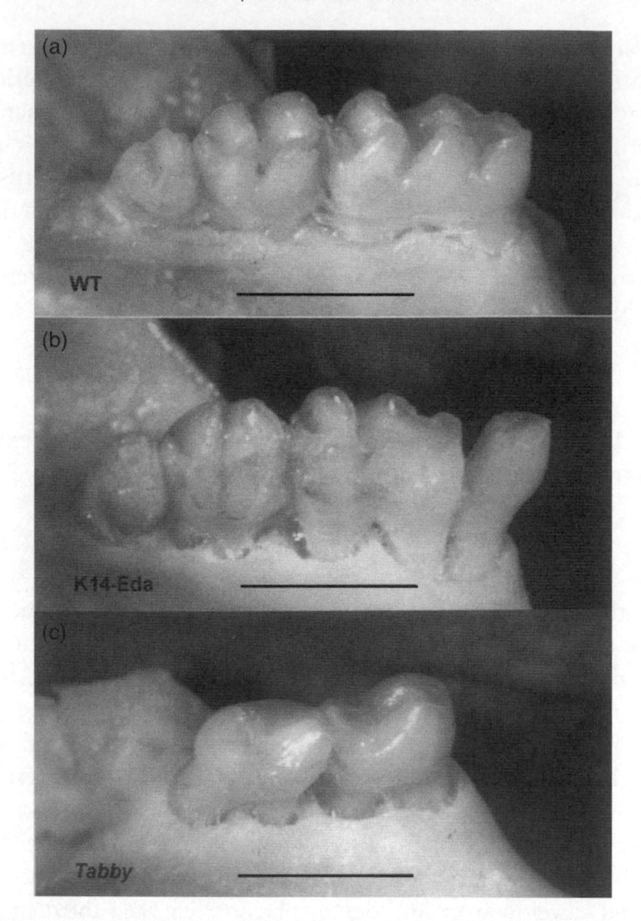

Figure 17.5 Effects of ectodysplasin, the TNF signal missing in ectodermal dysplasia syndrome, in mouse tooth formation. (a) Wild-type mouse has three molars. (b) When ectodysplasin is overexpressed in the dental epithelium of transgenic mice (K14-Eda), an extra tooth forms in front of the first molar (star) and there are major changes in cusp patterning of the first molar (Kangas *et al.*, 2004). In addition, enamel formation is inhibited in the incisors of this mouse (not shown; Mustonen *et al.*, 2003). (c) The *Tabby* mutant, which has no functional ectodysplasin, lacks the 3rd molar and the 1st molar is small with missing or fused cusps. M1, 1st molar; M2, 2nd molar; M3, 3rd molar

et al., 2000; Bei *et al.*, 2004). In most of these mouse mutants, the number and shape of teeth is also affected (Figures 17.5 and 17.6; Grüneberg, 1965; Dassule *et al.*, 2000; Mustonen *et al.*, 2003; Wang *et al.*, 2004a, 2004b). Hence, although the genes that have so far been associated with human enamel and dentine defects encode mostly structural molecules, it can be expected that in the future more genes regulating the differentiation and function of ameloblasts and odontoblasts will be discovered, and also that their mutations may cause defects in dental hard tissues, together with variations in tooth number and shape.

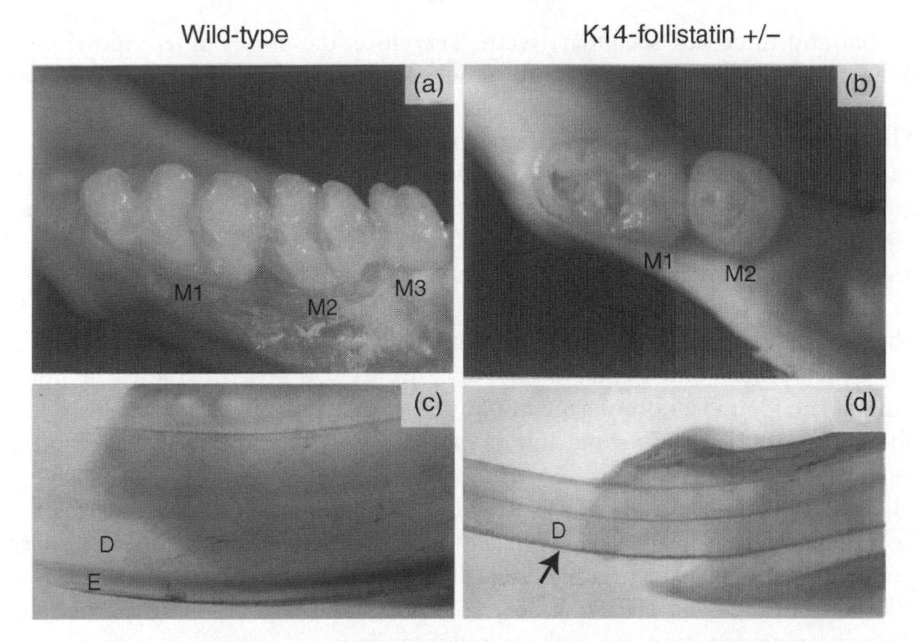

Figure 17.6 Inhibition of TGFβ signalling in transgenic mice results in defects in tooth number, shape and structure. (a) Wild-type molars. (b) Transgenic mice overexpressing the TGFβ inhibitor follistatin (K14-follistatin $+/-$) lack third molars and the cusp pattern of their first and second molars is severely disturbed. (c) Dentine and enamel in wild-type mouse. (d) Enamel formation has been inhibited in the incisors of the K14-follistatin $+/-$ mouse (arrow) (Wang *et al.*, 2004a, 2004b). M1, M2, M3, 1st, 2nd and 3rd molars; D, dentine; E, enamel

Agenda for the future

The ultimate goal of research on the mechanisms and genetic basis of tooth development and the pathogenesis of dental defects is to generate knowledge that could be applied in clinical practice for the diagnosis, prevention and treatment of defects. The regeneration of whole teeth seems a very distant goal and may not be feasible; even if achievable, it may be too complicated and expensive to replace the available prosthetic tooth replacement therapies. However, regeneration of parts of teeth may be a more realistic vision for the future. Work on dental stem cells should therefore be continued. Mesenchymal stem cells from the adult dental pulp and follicle can regenerate dentine and periodontal tissues upon transplantation (Gronthos *et al.* 2000; Seo *et al.*, 2004) and multipotential mouse embryonic stem cells can form the mesenchymal components of teeth when transplanted with embryonic mandibular epithelium (Ohazama *et al.*, 2004). The epithelial stem cells of continuously growing mouse incisors provide a model for studies on epithelial stem cell maintenance and their differentiation into ameloblasts and root sheath cells (Harada *et al.*, 1999; Tummers and Thesleff, 2003). However, mouse teeth may not be optimal models for regeneration studies, since they have a quite limited regenerative potential and mice lack secondary teeth. Therefore, the use of other animal models (e.g. zebrafish) should be explored for studies on tooth regeneration and continuous tooth replacement (Huysseune and Thesleff, 2004).

Although prevention of dental defects will probably not be possible in the majority of cases, it may be feasible in certain specific defects. The injection of Eda protein into pregnant *Tabby* mice (i.e. *Eda* mutants representing the mouse model for X-linked HED) corrected many symptoms in the offspring (Gaide and Schneider, 2003). This intriguing experiment is a wonderful example of research starting from identification of a human disease gene (Kere *et al.*, 1996) and leading to potential clinical applications. It also advocates the continuation of both experimental analysis of tooth development and dental defects and molecular genetic studies in humans aiming at the identification of gene mutations underlying dental defects. In addition, analysis of tooth defects in some animal models, in particular zebrafish, will provide tools for gene discovery.

The identification of *AXIN2* as a cause of dental agenesis led to uncovering a link between oligodontia and cancer. This allowed early diagnosis of colorectal cancer in oligodontia patients and surgical removal of the tumours at an early stage. Future studies should focus on clarifying how common the links between hypodontia and malignant transformation are at the population level.

Although the molecular basis of tooth development is understood in great detail, the picture is far from complete. The genomic approaches now available should be used more widely for the identification of new genes involved in tooth development. Microarray analysis is under way in many laboratories to search for novel genes that regulate specific developmental events. In many cases, gene expression is compared between wild-type mice and mutants with tooth phenotypes to discover genes downstream of the mutated genes.

So far, the emphasis in molecular studies has been on genes involved in signalling networks. However, besides cell proliferation, the responses that different signals elicit on cell behaviour are poorly understood. Microarray analysis will probably identify genes associated with cellular functions, such as adhesion, polarization and migration. Gene expression studies have already revealed developmentally regulated patterns for many genes involved in cell adhesion and cell–matrix interactions (http://bite-it.helsinki.fi). The study of the functions of these genes and proteins will also require the development of cell biology methodology, in particular imaging and labelling techniques, and of cell culture techniques that can be applied to dental cells.

The modulation of gene function in transgenic mice will continue to be a powerful method for the elucidation of gene function. Confirmation of the function of many signalling molecules has not been possible by the traditional knock-out approach because of early embryonic lethality. Conditional mutagenesis has not been widely used for the analysis of tooth development because of the lack of suitable promoters. Conditional deletion of Shh function in the ectoderm, using the K14 promoter targeting expression to oral ectoderm, demonstrated the necessary role of Shh in tooth morphogenesis from bud stage onwards (Dassule *et al.*, 2000). However, recombination did not occur early enough to address the role of Shh in the dental lamina and dental placode. It is obvious that other promoters are needed for targeting transgene expression to early dental epithelium and mesenchyme. A nestin1–Cre promoter construct was unexpectedly found to target expression to the early mandibular arch epithelium (Trumpp *et al.*, 1999). Deletion of *Fgf8* function using this promoter resulted in the inhibition of molar tooth initiation but the incisors formed. This was presumed to be due to a redundant function of *Fgf9* in the anterior region. The analysis of genes that are used reiteratively during tooth development will require the use of inducible promoters allowing silencing of gene function at desired stages of development.

In vitro methods will continue to be valuable for elucidating the biological function of molecules involved in tooth development. The morphogenesis of dissected tooth germs continues in organ culture from bud to late bell stage of development, and it can be manipulated in different ways. Gene function can be inhibited by antibodies and antisense technologies, and RNAi methodology may allow functional deletion of several genes simultaneously. Gene constructs can be introduced into developing teeth by electroporation (Angeli *et al.*, 2002). In addition to loss-of-function analysis, this approach can also be used for gain-of-function experiments, which will be useful for elucidating normal gene function.

Gene function redundancy poses difficult problems in the interpretation of experimental studies. Several genes may have synergistic, additive and antagonistic effects in one signalling pathway, and also genes in different signalling pathways may compensate for each others' functions. It is obvious that functional ablation of single genes, or even of two or three genes at a time, will not be enough to understand complex processes. Systemic approaches and computer modelling of the biological processes are needed. The signalling networks regulating cusp patterning are a perfect example of a complex system in which multiple pathways are integrated and the concerted action of inhibitory and stimulatory signals determines the final crown shape. Computer modelling has already been applied to recapitulate the networks in this process (Salazar-Ciudad and Jernvall, 2002) and such studies can be expected to increase our understanding of the complex signalling processes and their roles in tooth morphogenesis.

References

Alaluusua, S., Lukinmaa, P.-L., Torppa, J., Tuomisto, J. and Vartiainen, T. (1999) Developing teeth as biomarkers of dioxin exposure. *Lancet* **353**: 206.

Angeli, I., James, C.T., Morgan, P.J. and Sharpe,P.T. (2002) Misexpression of genes in mouse tooth germs using *in vitro* electroporation. *Connect. Tissue Res.* **43**: 180–185.

Arte, S., Nieminen, P., Apajalahti, S., Haavikko, K. *et al.* (2001) Characteristics of incisor–premolar hypodontia in families. *J. Dent. Res.* **80**: 1445–1450.

Bei, M. and Maas, R. (1998) FGFs and BMP4 induce both Msx1-independent and Msx1-dependent signalling pathways in early tooth development. *Development* **125**: 4325–4333.

Bei, M., Kratochwil, K. and Maas, L. (2000) BMP4 rescues a non-cell-autonomous function of Msx1 in tooth development. *Development* **127**: 4711–4718.

Bei, M., Stowell, S. and Maas R. (2004) Msx2 controls ameloblast terminal differentiation. *Dev. Dyn.* **231**: 758–765.

Coin, R., Haikel, Y. and Ruch, J.V. (1999) Effects of apatite, transforming growth factor β-1, bone morphogenetic protein-2 and interleukin-7 on ameloblast differentiation *in vitro*. *Eur. J. Oral Sci.* **6**: 487–495.

Dassule, H.R., Lewis, P., Bei, M., Maas, R. and McMahon, A.P. (2000) Sonic hedgehog regulates growth and morphogenesis of the tooth. *Development* **127**: 4775–4785.

Ferguson, C.A., Tucker, A.S., Christensen, L., Lau, A.L. *et al.* (1998) Activin is an essential early mesenchymal signal in tooth development that is required for patterning of the murine dentition. *Genes Dev.* **12**: 2636–2649.

Gaide, O. and Schneider, P. (2003) Permanent correction of an inherited ectodermal dysplasia with recombinant EDA. *Nat. Med.* **9**: 614–618.

Glasstone, S. (1963) Regulative changes in tooth germs grown in tissue culture. *J. Dent. Res.* **42**: 1364–1368.

Gorlin, R.J., Cohen, M.M. and Levin, L.S. (1990) *Syndromes of the Head and Neck*. Oxford University Press: Oxford.

Gronthos, S., Mankani, M., Brahim, J., Robey, P.G. and Shi, S. (2000) Postnatal human dental pulp stem cells (DPSCs) *in vitro* and *in vivo*. *Proc. Natl Acad. Sci. USA*. **97**: 13625–13630.

Grüneberg, H. (1965) Genes and genotypes affecting the teeth of the mouse. *J. Embryol. Exp. Morph.* **14**: 137–159.

Harada, H., Kettunen, P., Jung, H.S., Mustonen, T. *et al.* (1999) Localization of putative stem cells in dental epithelium and their association with Notch and FGF signalling. *J. Cell Biol.* **147**: 105–120.

Harada, H., Toyono, T., Toyoshima, K., Yamasaki, M. *et al.* (2002) FGF10 maintains stem cell compartment in developing mouse incisors. *Development* **129**: 1533.

Hardcastle, Z., Mo, R., Hui, C.C. and Sharpe, P.T. (1998) The shh signalling pathway in tooth development – defects in *gli2* and *gli3* mutants. *Development* **125**: 2803–2811.

Huysseune, A. and Thesleff, I. (2004) Continuous tooth replacement: the possible involvement of epithelial stem cells. *BioEssays* **26**: 665–671.

Hölttä, P., Alaluusua, S., Saarinen-Pihkala, U.M., Wolf, J. *et al.* (2002) Long-term adverse effects on dentition in children with poor-risk neuroblastoma treated with high dose chemotherapy and autologous stem cell transplantation with or without total body irradiation. *Bone Marrow Transpl.* **29**: 121–127.

Jensen, B.L. and Kreiborg, S. (1990) Development of the dentition in cleidocranial dysplasia. *J. Oral Pathol. Med.* **19**: 89–93.

Jernvall, J., Kettunen, P., Karavanova, I., Martin, L.B. and Thesleff, I. (1994) Evidence for the role of the enamel knot as a control centre in mammalian tooth cusp formation: non-dividing cells express growth stimulating *Fgf-4* gene. *Int. J. Dev. Biol.* **38**: 463–469.

Jernvall, J., Keränen, S.V. and Thesleff, I. (2000) Evolutionary modification of development in mammalian teeth: quantifying gene expression patterns and topography. *Proc. Natl. Acad. Sci. USA* **97**: 14444–14448.

Joseph, B.K., Harbrow, D.J., Sugerman, P.B., Smid, J.R. *et al.* (1999). Ameloblast apoptosis and IGF-1 receptor expression in the continuously erupting rat incisor model. *Apoptosis* **6**: 441–447.

Kangas, A., Mustonen, T., Mikkola, M., Thesleff, I. and Jernvall, J. (2004) Non-independence of mammalian dental characters. *Nature* **432**: 211–214.

Kassai, Y., Munne, P., Hotta, Y., Penttilä, E., Kavanagh, K., Ohbayashi, N., Takada, S., Thesleff, I., Jernvall, J., Itoh, N. (2005): Regulation of mammalian cusp patterning by ectodin. Science **309**: 2067–2070.

Kere, J., Srivastava, A.K., Montonen, O., Zonana, J. *et al.* (1996) X-linked anhydrotic (hypohydrotic) ectodermal dysplasia is caused by mutation in a novel transmembrane protein. *Nat. Genet.* **13**: 409–416.

Keränen, S.V.E., Åberg, T., Kettunen, P., Thesleff, I. and Jernvall, J. (1998) Association of developmental regulatory genes with the development of different molar tooth shapes in two species of rodents. *Dev. Gen. Evol.* **208**: 477–486.

Koch G. and Poulsen S. (2001) *Pediatric Dentistry.* Munksgaard: Copenhagen.

Kollar, E.J. and Baird, G.R. (1969) The influence of the dental papilla on the development of tooth shape in embryonic mouse tooth germs. *J. Embryol. Exp. Morph.* **21**: 131–148.

Kratochwil, K., Dull, M., Farinas, I., Galceran, J. and Grosshedl, R. (1996) Lef1 expression is activated by BMP-4 and regulates inductive tissue interactions in tooth and hair development. *Genes Dev.* **10**: 1382–1394.

Kratochwil, K., Galceran, J., Tontsch, S., Roth, W. and Grosschedl, R. (2002) FGF4, a direct target of LEF1 and Wnt signalling, can rescue the arrest of tooth organogenesis in Lef1$^{-/-}$ mice. *Genes Dev.* **16**: 3173–3185.

Lammi, L., Arte, S., Somer, M., Järvinen, H. *et al.* (2004) Mutations in *AXIN2* cause familial tooth agenesis and predispose to colorectal cancer. *Am. J. Hum.Genet.* **74**: 1043–1050.

Laurikkala, J., Mikkola, M., Mustonen, T., Åberg, T. *et al.* (2001) TNF signalling via the ligand–receptor pair ectodysplasin and edar controls the function of epithelial signalling centres and is regulated by Wnt and activin during tooth organogenesis. *Dev. Biol.* **229**: 443–455.

Laurikkala, J., Pispa, J., Jung, H-S., Nieminen, P. *et al.* (2002) Regulation of hair follicle development by the TNF signal ectodysplasin and its receptor Edar. *Development* **129**: 2541–2553.

Laurikkala, J., Kassai, Y., Pakkasjärvi, L., Thesleff, I. and Itoh, N. (2003) Identification of a secreted BMP antagonist, ectodin, integrating BMP, FGF, and SHH signals from the tooth enamel knot. *Dev. Biol.* **264**: 91–105.

Lumsden, A.G. (1988) Spatial organization of the epithelium and the role of neural crest cells in the initiation of the mammalian tooth germ. *Development* **103**: 155–169.

Mikkola, M.L. and Thesleff, I. (2003) Ectodysplasin signalling in development. *Cytok. Growth Fact. Rev.* **14**: 211–224.

Millar, S.E., Koyama, E., Reddy, S.T., Andl, T. *et al.* (2003) Over- and ectopic expression of Wnt3 causes progressive loss of ameloblasts in postnatal mouse incisor teeth. *Connect. Tissue Res.* **44**(suppl. 1): 124–129.

Mina, M. and Kollar, E.J. (1987) The induction of odontogenesis in non-dental mesenchyme combined with early murine mandibular arch epithelium. *Arch.Oral. Biol.* **32**: 123–127.

Mustonen, T., Pispa, J., Mikkola, M.L., Pummila, M. *et al.* (2003) Stimulation of ectodermal organ development by ectodysplasin-A1. *Dev. Biol.* **259**: 123–136.

Mustonen, T., Ilmonen, M., Pummila, M., Kangas, A. *et al.* (2004) Ectodysplasin-A1 promotes placodal cell fate during early morphogenesis of ectodermal appendages. *Development* **131**: 4907–4919.

Nanci, A. (2002) *Ten Cate's Oral Histology*, 6th edn. Mosby: St. Louis, MO.

Ohazama, A., Modino, S.A., Miletich, I., Sharpe, P.T. (2004) Stem-cell-based tissue engineering of murine teeth. *J Dent Res* **83**: 518–522.

Peters, H., Neubuser, A., Kratochwil, K. and Balling, R. (1998) Pax9-deficient mice lack pharyngeal pouch derivatives and teeth and exhibit craniofacial and limb abnormalities. *Genes Dev.* **12**: 2735–2747.

Pispa, J., Jung, H-S., Jernvall, J., Kettunen, P. *et al.* (1999) Cusp patterning defect in Tabby mouse teeth and its partial rescue by FGF. *Dev. Biol.* **216**: 521–534.

Pispa, J. and Thesleff, I. (2003) Mechanisms of ectodermal organogenesis. *Dev. Biol.* **262**: 195–205.

Pispa, J., Mustonen, T., Mikkola, M.L., Kangas, A.T. *et al.* (2004) Tooth patterning and enamel formation can be manipulated by misexpression of TNF receptor Edar. *Dev. Dyn.* **231**: 432–440.

Ruch, J.V. (1987) Determinisms of odontogenesis. *Cell Biol. Rev.* **14**: 1–112.

Ruch, J.V., Lesot, H. and Begue.Kirn, C. (1995) Odontoblast differentiation. *Int. J. Dev. Biol.* **39**: 51–68.

Rutherford, R.B., Wahle, J., Tucker, M., Rueger, D. and Charette, M. (1993) Induction of reparative dentine formation in monkeys by recombinant human osteogenic protein-1. *Arch. Oral Biol.* **38**: 571.

Sahlberg, C., Mustonen, T. and Thesleff, I. (2002) Explant cultures of embryonic epithelium: analysis of mesenchymal signals. In *Epithelial Cell Culture Protocols. Methods of Molecular Biology*, Wise, C. (ed.). Humana: Totowa, NJ; 373–382.

Salazar-Ciudad, I. and Jernvall, J. (2002) A gene network model accounting for development and evolution of mammalian teeth. *Proc. Natl Acad. Sci. USA* **99**: 8116–8120.

Seo, B., M. Miura, M., Gronthos, S., Bartold, P.M., Batouli S., Brahim, J., Young, M., Robey, P.G., Wang, C.Y., Shi, S. (2004) Investigation of multipotent postnatal stem cells from human periodontal ligament. Lancet **364**: 149–155.

Stockton, D.W., Das, P., Goldenberg, M., D'Souza, R.N. and Patel, P.I. (2000) Mutation of *PAX9* is associated with oligodontia. *Nat. Genet.* **24**: 18–19.

Thesleff, I. (2003) Developmental biology and building a tooth. *Quintessence Int.* **34**: 613–620.

Thesleff, I. and Mikkola, M. (2002) Death receptor signalling giving life to ectodermal organs. Science's STKE: http://www.stke.org, sigtrans; 2002/131/pe22

Thesleff, I. and Pirinen, S. (2005) Genetics of dental defects. In *Encyclopedia of Human Genome*, Nature Publishing Group, www.ehgonline.net

Thomas, B.L., Tucker, A.S., Qui, M., Ferguson, C.A. *et al.* (1997) Role of *Dlx-1* and *Dlx-2* genes in patterning of the murine dentition. *Development* **124**: 4811–4818.

Trumpp, A., Depew, M.J., Rubenstein, J.L., Bishop, J.M. and Martin, G.R. (1999) Cre-mediated gene inactivation demonstrates that FGF8 is required for cell survival and patterning of the first branchial arch. *Genes Dev.* **13**: 3136–3148.

Tucker, A.S., Headon, D.J., Courtney, J.M., Overbeek P. and Sharpe, P.T. (2004) The activation level of the TNF family receptor, Edar, determines cusp number and tooth number during tooth development. *Dev. Biol.* **268**: 185–194.

Tummers, M. and Thesleff, I. (2003) Root or crown: a developmental choice orhestrated by the differential regulation of the epithelial stem cell niche in the tooth of two rodent species. *Development* **130**: 1049–1057.

Vaahtokari, A., Åberg, T., Jernvall, J., Keränen, S. and Thesleff, I. (1996a) The enamel knot as a signalling centre in the developing mouse tooth. *Mech. Dev.* **54**: 39–43.

Vaahtokari, A., Åberg, T. and Thesleff, I. (1996b) Apoptosis in the developing tooth: association with an embryonic signalling centre and suppression by EGF and FGF-4. *Development* **122**: 121–129.

Vainio, S., Karavanova, I., Jowett, A. and Thesleff, I. (1993) Identification of BMP-4 as signal mediating secondary induction between epithelial and mesenchymal tissues during early tooth development. *Cell* **75**: 45–58.

van den Boogaard, M.J., Dorland, M., Beemer, F.A. and van Amstel, H.K. (2000) *MSX1* mutation is associated with orofacial clefting and tooth agenesis in humans. *Nat. Genet.* **24**: 342–343.

van Genderen, C., Okamura, R.M., Farinas, I., Quo, R.G. *et al.* (1994) Development of several organs that require inductive epithelial–mesenchymal interactions is impaired in LEF-1-deficient mice. *Genes Dev.* **8**: 2691–2703.

Vastardis, H., Karimbux, N., Guthua, S.W., Seidman, J.G. and Seidman, C.E. (1996) A human *MSX1* homeodomain missense mutation causes selective tooth agenesis. *Nat. Genet.* **13**: 417–421.

Wang, X-P., Suomalainen, M., Jorgez, C.J., Matzuk, M.M. *et al.* (2004a) Modulation of activin/bone morphogenetic protein signalling by follistatin is required for the morphogenesis of mouse molar teeth. *Dev. Dyn.* **231**: 98–108.

Wang, X-P., Suomalainen, M., Jorgez, C.J., Matzuk, M.M. *et al.* (2004b). Follistatin regulates enamel patterning in mouse incisors by asymmetrically inhibiting BMP signaling and ameloblast differentiation. *Dev. Cell* **7**: 719–730.

Xiao, S., Yu, C., Chou, X., Yuan, W. *et al.* (2002) Dentinogenesis imperfecta 1 with or without progressive hearing loss is associated with distinct mutations in DSPP. *Nat. Genet.* **27**: 201–204.

Index

Page numbers in italic, e.g. *3*, refer to figures. Page numbers in bold, e.g. **101**, signify entries in tables.

Embryos, Genes and Birth Defects, Second Edition Edited by Patrizia Ferretti, Andrew Copp, Cheryll Tickle and Gudrun Moore

Index compiled by John Holmes